Mechanical Engineering Series

Series Editor

Francis A. Kulacki, University of Minnesota

The Mechanical Engineering Series presents advanced level treatment of topics on the cutting edge of mechanical engineering. Designed for use by students, researchers and practicing engineers, the series presents modern developments in mechanical engineering and its innovative applications in applied mechanics, bioengineering, dynamic systems and control, energy, energy conversion and energy systems, fluid mechanics and fluid machinery, heat and mass transfer, manufacturing science and technology, mechanical design, mechanics of materials, micro- and nano-science technology, thermal physics, tribology, and vibration and acoustics. The series features graduate-level texts, professional books, and research monographs in key engineering science concentrations.

More information about this series at http://www.springer.com/series/1161

Rakesh Kumar Maurya

Characteristics and Control of Low Temperature Combustion Engines

Employing Gasoline, Ethanol and Methanol

Springer

Rakesh Kumar Maurya
Department of Mechanical Engineering
Indian Institute of Technology Ropar
Rupnagar, Punjab, India

ISSN 0941-5122 ISSN 2192-063X (electronic)
Mechanical Engineering Series
ISBN 978-3-319-68507-6 ISBN 978-3-319-68508-3 (eBook)
https://doi.org/10.1007/978-3-319-68508-3

Library of Congress Control Number: 2017953552

Printed on acid-free paper

This Springer imprint is published by Springer Nature
The registered company is Springer International Publishing AG
The registered company address is: Gewerbestrasse 11, 6330 Cham, Switzerland

Preface

World demand for energy, in general, and for transportation, in particular, is increasing rapidly. The internal combustion engines, fuelled mostly by petroleum-derived liquid fuels, have been the main source of transportion power over the past century and are likely to remain so in the foreseeable future. Automotive engines and fuels are facing challenge to reduce emissions for improving local city air quality as well as improving the fuel conversion efficiency, which decreases the CO_2 emission to reduce global warming risk. The grand challenges for engine researchers are to develop technologies for maximizing engine efficiency, minimizing the exhaust emissions and optimize the tolerance to a wide variety of fuels in combustion engines.

The most frequently used types of internal combustion engines are compression ignition (CI) engines and spark ignition (SI) engines. Conventional CI diesel engines have higher fuel conversion efficiency than SI engines. Diesel engines typically emit comparatively higher NO_x and particulate matter (PM). Costly after-treatment devices are used to meet the stringent emission legislations of different countries. It is difficult to meet future norms with currently available technologies, and if possible, the cost will also be very high. Therefore, it is important to reduce NO_x and PM emissions inside the engine cylinder. Low temperature combustion (LTC) mode is known to be a promising approach for simultaneous reduction of NO_x and PM emissions along with high fuel conversion efficiency. This technology drastically reduces the cost of after-treatment and saves the fuel due to higher fuel conversion efficiency. Depending on combustion control strategies, degree of premixing and stratification of charge, several technologies such as homogeneous charge compression ignition (HCCI), partially premixed combustion (PPC) and reactivity controlled compression ignition (RCCI) are demonstrated as low temperature engine combustion strategies. This book presents comprehensive discussion on fundamental aspects of LTC engines and recent advances in this novel technology. This book has been designed for senior undergraduate and postgraduate students, researchers, practising engineers and professionals, who have the basic knowledge of internal combustion engines.

Fundamentals of conventional engines and fuels are briefly described in Chap. 1. This chapter also introduces the alternative engines and alternative combustion modes. Chapter 2 explains the LTC principles, basic autoignition reaction kinetics involved in combustion and different strategies to enable premixed LTC engines. Advantage and limitations of different LTC strategies are also discussed in Chap. 2. The LTC process involves various physical processes (atomization, evaporation and mixing) and complex chemical reactions occurring in the cylinder. Fuel properties and its composition play an important role in all the physical and chemical processes involved in LTC process. Chapter 3 discusses the autoignition characteristics (autoignition chemistry, impact of fuel molecular structure on autoignition, fuel autoignition quality), fuels effect on autoignition and several fuel index developed for LTC engines. Fuel design and fuel properties/quality required for LTC engines are also discussed in Chap. 3. Premixed charge preparation strategies for LTC engines are discussed in Chap. 4. Quality of air–fuel mixture governs the combustion process and its rate in LTC engines. Premixed LTC engines need different enabling technologies depending on the fuel and strategy used to achieve combustion in the cylinder. To achieve higher fuel conversion efficiency, desired phasing of combustion timings is essential even at moderate combustion rates. Chapter 5 describes the combustion control variables and strategies, demonstrated for LTC engines.

Chapter 6 presents the combustion characteristic of LTC engines using conventional and alternative fuels. Ignition and heat release characteristics of LTC combustion process are described by analysis of ignition delay, heat release rate, ringing intensity, combustion phasing, combustion duration and combustion efficiency. Combustion stability and cyclic variations analysis of combustion parameters using statistical and nonlinear dynamics methods is also discussed in this chapter. Chapter 7 presents the performance characteristics such as operating range, thermal efficiency and exhaust gas temperature in LTC engine employing gasoline-like fuels. Chapter 8 presents the regulated and unregulated emission characteristics of LTC engines using different fuels. Formation and emission trend of NOx, unburned hydrocarbons (HC), carbon monoxide (CO) and PM from LTC engines are discussed in detail.

Typically, LTC engines use premixed air–fuel mixture and combustion is mainly governed by chemical kinetics. Small variations in engine operating conditions may lead to a large effect on the combustion phasing. Therefore, HCCI engine cannot be operated reliably based on engine map. The combustion instability conditions in HCCI combustion make a compulsory closed-loop combustion control for the very operation of HCCI engine. Chapter 9 presents the closed-loop control of LTC engines. For closed-loop combustion control, sensors, actuators and control strategies for LTC engines are also discussed in Chap. 9. Summary of main findings regarding performance, combustion and emissions characteristics of various LTC strategies is presented in Chap. 10, and recommendations for further work are also outlined. Large-scale implementation of LTC combustion modes could generate some significant change in fuel technology as well as exhaust after treatment technologies. Implementation of this technology will save fuel and have significant environmental impact by reducing emissions.

This book represents the material that has been collected for a period of several years during my research and teaching work on this subject. The presentation of the book is influenced by technical literature published by Society of Automotive Engineers (SAE), Elsevier, SAGE, American Society of Mechanical Engineers (ASME) and Taylor & Francis Group. I wish to acknowledge with thanks the permission given by SAE International, Elsevier, ASME, SAGE and Taylor & Francis Group to reprint some figures from their publications. I also express my thanks to Prof. Andreas Cronhjort, KTH Royal Institute of Technology, Sweden, for giving permission to reproduce a picture from his research.

I wish to express my thanks to my students Mr. Mohit Raj Saxena and Mr. Yogendra Vishwakarma for helping in the preparation of figures for the book. I extend my deepest gratitude to my parents for their invaluable love, encouragement and support. I offer greatest thanks to my wife Suneeta for her contributions of patience, love, faith and constant encouragement. Thanks for being there and bringing happiness into my life. I wish to express a special appreciation to my wife for taking care of our lovely son Shashwat and managing family affairs while I am busy with writing the book. Thanks to little Shashwat for making life joyful and sorry for spending less time with you during the book writing.

Last but not the least, I would like to express my thanks to all those helped me directly or indirectly for successful completion of this book. Finally, I thank the readers for choosing this book. I welcome any feedback or questions on the topics discussed in this book, or discussions of emerging engine or vehicle powertrain technologies.

Rupnagar, Punjab, India
August 2017

Rakesh Kumar Maurya

Contents

About the Author

Dr. Rakesh Kumar Maurya has been a faculty member in the Department of Mechanical Engineering, Indian Institute of Technology Ropar, since August 2013. Before joining IIT Ropar, he was working as senior research associate (Pool Scientist, CSIR) at IIT Kanpur. He received his bachelor's, master's and PhD degrees in mechanical engineering from Indian Institute of Technology Kanpur, India. He received Early Career Research Award from Science and Engineering Research Board (SERB), Government of India, New Delhi. He is also a recipient of Young Scientist Award (2016) from International Society for Energy, Environment and Sustainability. He has served as a journal referee and committee member/ session co-chair of international conferences on multiple occasions. He teaches and conducts research in the area of internal combustion engines. His areas of interest are low temperature engine combustion, alternative fuels, engine combustion diagnostics, engine instrumentation, combustion and emission control, particulate matter characterization, engine management systems and philosophy of science.

Chapter 1
Introduction

Abstract Energy is a fundamental prime mover for the economic growth of any country and essential for sustainability of modern economy as well as society. Long-term availability of environment-friendly, affordable and accessible energy sources is desirable for economic growth in the future. Presently, humanity is at crossroads and requires the radical and novel approach for the utilization of energy. The goal of this book is to present a novel approach for internal combustion (IC) engines, which are one of the most important machines for transforming the energy of hydrocarbon fuels into useful mechanical work. Two different viewpoints exist on energy production using IC engines. A number of people think in terms of mobility advantages, while others associate with emissions of harmful exhaust gases and large-scale consumption of limited fossil fuel reserves. Irrespective of one's viewpoint, the number of vehicles running on IC engine will be increasing in the future due to the rapid economic growth. Furthermore, on-board power requirement on vehicle will increase due to the growing number of accessories and electronic devices. These factors lead to the increase in worldwide fuel consumption and gaseous emissions. Therefore, an IC engine with alternative combustion mode having superior characteristics than conventional engines needs to be developed. Ideally, such an alternative combustion mode should be operated on renewable fuels and have a better fuel conversion efficiency and no harmful emissions. Present book deals with the detailed analysis of performance, combustion and emissions characteristics of novel low temperature combustion (LTC) concept using conventional as well as alternative fuels. This chapter first discusses the motivation for engine research in general and LTC mode in particular. The LTC mode is an alternative to conventional, well-known and frequently used combustion concepts, i.e. compression ignition (CI) and spark ignition (SI) combustion modes. Brief description of conventional combustion mode is also provided in this chapter. An overview of alternative engine concepts and alternative fuels is also presented in this chapter for setting the stage for the discussion of various LTC engines.

Keywords Combustion • Autoignition • HCCI • Alternative fuels • Powertrain • SI • CI • PPC • RCCI • Engine

R.K. Maurya, *Characteristics and Control of Low Temperature Combustion Engines*, Mechanical Engineering Series, https://doi.org/10.1007/978-3-319-68508-3_1

1.1 Motivation

Global economy and modern society are dependent on the availability of reliable transportation systems. Modern civilization would not have reached existing living standards (in terms of physical facilities) without the transportation by millions of automotive vehicles. Combustion engines are the prime mover for the automotive vehicles, ships, construction equipment, agricultural machines and gensets. Currently, a vast majority of combustion engines used in automotive vehicles are reciprocating piston engines powered by combustion of petroleum-based fossil fuels. Reciprocating IC engines are well accepted and the most significant source of energy since the last century because of their superior performance, controllability, robustness, durability and absence of other viable alternatives.

Assured supply of transportation energy is required from sources with lower carbon footprint to ensure sustainable development. Increasing mechanization of the world has led to a steep rise in demand for fossil fuels and increase in number of automotive vehicles [1, 2]. The International Energy Outlook 2016 shows that the fuel demand is expected to rise over the next three decades and fossil fuels share 78% of energy use in 2040 [2]. As a result of stringent emission legislations and higher oil prices, reciprocating IC engine vehicles are expected to continue to become more efficient. In addition, several new technologies (hybrid cars, electric vehicles, fuel cells, etc.) are being developed for fuel economy improvement and reduction in exhaust emissions locally. However, presently shares of these new technologies are very small. According to future projections, these new vehicle technologies collectively would account for only 6% of new passenger vehicle sales by 2020 and 19% by 2035 (hybrids would have major share) [1]. Currently, there are no realistic alternatives that could completely replace the reciprocating IC engines. Electric and hybrid electric vehicles are potential technologies, which can be alternative to IC engines. Electric and hybrid electric vehicles will be however suitable for short-range journeys and more suitable in light-duty vehicle category. However, the volumetric and gravimetric density of modern batteries is still inferior to that of the fuel used in any IC engines [3]. Therefore, combustion engines are expected to be around for several decades or maybe even centuries to come, as long as more fuel-efficient and cleaner alternative is made available. Hence, research focusing on improving the fuel conversion efficiency and reducing harmful emissions from the IC engine is justified and required in the current scenario.

Over the years, improving the performance in terms of fuel conversion efficiency and power density of IC engine has been the major driving force for research and development (R&D). Exhaust emissions from automobiles were recognized as major contributor to urban air pollution for the first time in California during 1950s [4]. Now, the users of IC engines are aware of the air pollution from vehicles, and consequently they demand compliance to environmental consideration and regulatory legislations prevailing around the world. The main governing factors for engine research are described in the next subsections.

1.1.1 Environmental Concerns

The world is presently facing crisis of depletion in fossil fuel resources and degradation of environmental conditions. Environmental pollution is a key public health issue in most of the cities. Epidemiological studies show that air pollution causes large number of deaths, huge medical costs and lost productivity every year. These losses and the accompanying degradation in quality of life enforce a substantial burden on humanity [5]. There are several issues related to environmental pollution from combustion engines, which are summarized in Fig. 1.1. Major concerns that appear due to heavy use of combustion engines are global warming, photochemical smog, carcinogenic particles, acid rains and ozone depletion.

Currently global warming is an important environmental challenge. The phenomenon of global warming occurs due to thermal energy imbalance because of heat trapped in the earth's environment by greenhouse gases. The principal greenhouse gases generated due to human activities are carbon dioxide (CO_2), methane (CH_4), nitrous oxide (N_2O) and fluorinated gases (hydrofluorocarbons, perfluorocarbons and sulphur hexafluoride) [6]. IC engines burn fossil fuels in the combustion chamber and produce CO_2 along with small amounts of methane and nitrous oxide also. From an IC engine point of view, reduction in the CO_2 emissions can be achieved by developing more efficient engines.

Photochemical smog is a brownish-grey haze caused by reactions between unburned hydrocarbons and nitrogen oxides in the presence of solar radiation. It comprises of various organic compounds, ozone and nitrogen oxides (NO_x) confined above the ground level due to temperature inversion conditions [4]. The vehicles contribute to smog by emitting nitrogen oxides and unburned

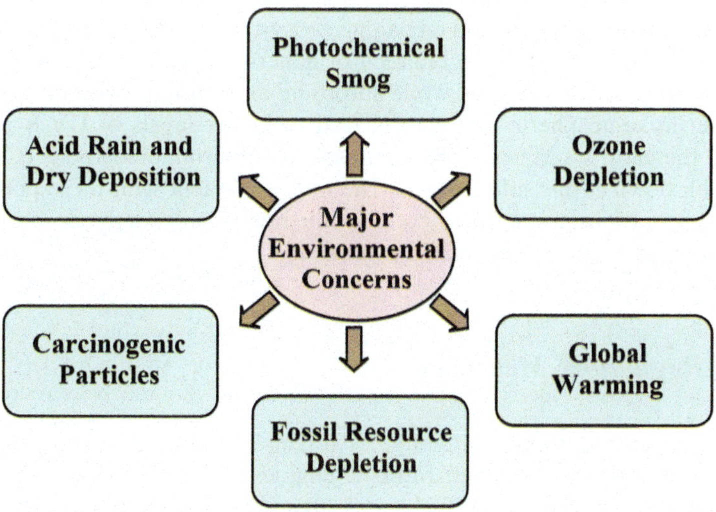

Fig. 1.1 Major environmental concerns arise due to extensive use of internal combustion engines for automotive and stationary applications

hydrocarbons (HC). Therefore, it is required to develop technologies which reduce the emissions of NO_x and unburned hydrocarbons. Acid rain has a negative impact on vegetation by accelerating acidification of the soil and directly attacking plant leaves. Acid rain broadly refers to a mixture of wet and dry deposition of nitric and sulphuric acids from the environment. Emission of sulphur dioxide and NO_x from combustion of fossil fuel contributes to acid rain [7].

Solid particles emitted from automotive vehicles mainly consist of carbonaceous matter (soot) comprising a small fraction of inorganic matters. Different types of liquid-phase substances and other hydrocarbon species are either adsorbed or absorbed on solid soot particles [4]. One of the main sources of soot particles introduced into the atmospheric air is diesel engine. Soot particles are produced in the diesel engine due to diffusion-controlled heterogeneous combustion of the locally rich fuel–air mixture. For human beings, larger particles are not a serious health threat because they are taken care of by the body's defence system. Smaller particles less than 2.5 microns (μm) are the main concern because they take long time to settle and remain airborne for days altogether. Therefore, smaller soot particles could reach the respiratory system of human beings. Particles particularly less than 1 μm are too small to be trapped in the upper portion of lungs, and they penetrate deep into the lungs [4]. They pose a serious threat to the human health since carcinogenic compounds such as polycyclic aromatic hydrocarbons (PAHs) adsorbed on the surface of these soot particles are carried to lungs and can potentially cause cancer. An improvement based on the measured mass of particulate matter may actually lead to an increased number of smaller particles, and result could be misleading. Therefore, it is required to develop engine technologies, which use premixed flames and emit lower number of particles.

Ozone depletion is another key environmental concern as ozone layer guards against adverse effects on humans (e.g. skin cancer and cataracts), on biosphere (e.g. inhibiting plant growth and damaging ecosystems) and on physical infrastructure of the modern era (e.g. degradation of materials) [8]. Ozone is broken down continuously in the stratosphere while absorbing the harmful UV-B solar radiation. Reduction in stratospheric ozone level lead to higher levels of UV-B radiations reaching the earth's surface. In the stratosphere, chlorofluorocarbons (CFCs) and NO_x break down ozone into oxygen. The major problem with these pollutants is that they do not form a stable compound while breaking down the ozone; therefore new compounds continue to break down ozone.

1.1.2 Regulatory Measures

Utilization of fossil fuels in IC engines affects the regional (local) and global environment. Pollutants emitted from IC engines have adverse effect on human health and ambient air quality. Health and other hazards of pollutants depend on its concentration as well as time of exposure to human body. Harmful health effects of various pollutants emitted from combustion engines (adapted from references [4, 9,

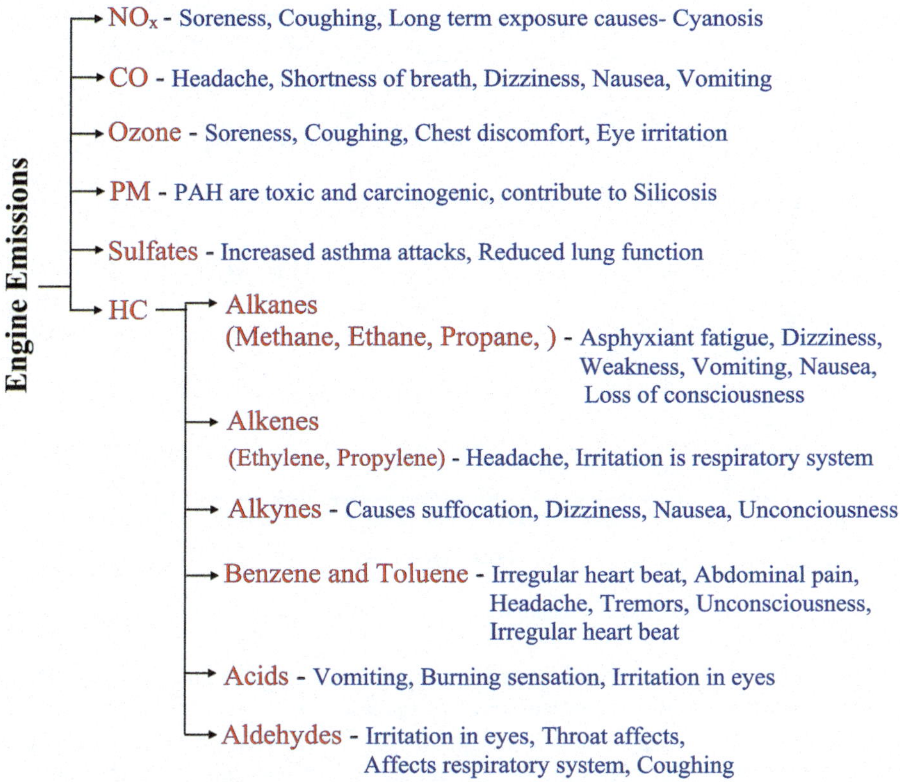

Fig. 1.2 Effect of principal pollutants emitted from IC engines on human health

10]) are summarized in Fig. 1.2. Governments worldwide have gradually imposed increasingly stringent restrictions on emission levels and tougher quality norms for fuel composition in order to reduce the influence of pollutants on the environment and human health. Furthermore, there are demands for a required emission durability and in-use inspection and obligatory maintenance [11].

Vehicular exhaust emission standards are specified in terms of pollutant mass per unit of distance travelled (g/km). For light- and medium-duty vehicles, emission standards are based on driving cycles that represent typical driving pattern in any specific country. The test driving cycles are composed of a cold start period, idling, moderate acceleration and deceleration and cruise modes. For heavy-duty vehicles, engines are tested on different combinations of speed and load condition for steady test cycle. To make the test more representative of actual road driving, conditions on transient tests are also conducted. One major drawback is that the test methods and emission standard often differ from one country to another and direct comparison is generally not possible [4]. Emission legislations include pollutant species such as NO_x, unburned hydrocarbons, CO and particulate matter (PM), and these pollutants are frequently referred to as “regulated pollutants”. However, NO_x and

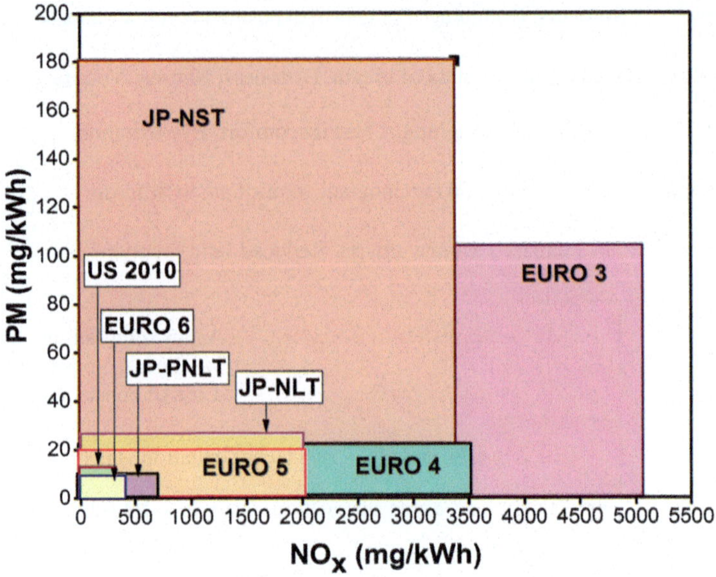

Fig. 1.3 NO_x and PM emission standard for heavy-duty vehicles in different regions of the world

PM emissions have been a challenge for conventional diesel engines because of the heterogeneous combustion process in the cylinder [12]. NO_x and PM emission standards for heavy-duty vehicles (adapted from Worldwide Emissions Standards [13]) are presented in Fig. 1.3. The significant reductions that are made continuously in every country are worth noting. Due to concerns regarding the environmental effects and stringent emission legislations, the research for next-generation combustion mode for IC engines has gained increasing attention worldwide. Detailed discussion on next-generation alternative combustion modes is provided in Chap. 2.

1.1.3 Engine Fuel Challenge

Automotive engines and fuels are facing challenges to reduce emissions for improving local city air quality as well as reduce CO_2 to reduce global warming risk. Two-thirds of the oil consumption in the world is presently used in the transportation sector, and half of that goes to passenger cars and light trucks [14]. This heavy consumption of the fossil fuels results in the emission of large amount of CO_2 which is identified as greenhouse gas (GHG). To address this issue, the IEA's (International Energy Agency) roadmap is to reduce the fuel use per kilometre by 30–50% in new road vehicles worldwide by 2030 [15]. All these goals stimulate new developments in both conventional and alternative engines as well as fuels as illustrated in Fig. 1.4 (adapted from [16]).

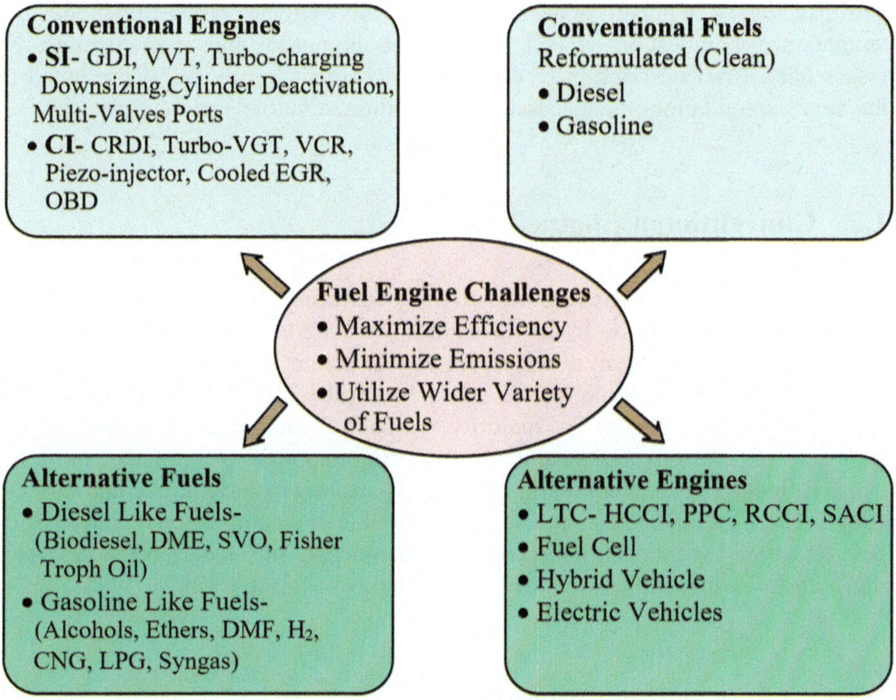

Fig. 1.4 Overview of engine and fuel technologies to tackle technical challenges for future automotive engines

Grand challenges in IC engine research are summarized as to develop technologies for maximizing engine efficiency, minimize exhaust emissions and optimize the tolerance for utilization of wide variety of fuels [17]. To address the challenges, there are four possible approaches, i.e. (i) improvement of conventional engines, (ii) improvement of conventional fuels, (iii) development of alternative engine and (iv) utilization of alternative fuels and their combinations. Over a century, there are significant developments in conventional engine and fuel technology for improving their performance to present level (Fig. 1.4). Another possible approach is to explore new engine concepts that use alternative fuels (such as biodiesel, alcohols, natural gas, etc.) for meeting future emission legislation. The utilization of alternative fuel results in reducing the monopoly of fossil fuels as well as increasing the engine efficiency. Improvement in IC engine efficiency is still continued, which represents the richness of the engine combustion process. In recent study, Reitz [18] summarized engine combustion as "…. a low Mach number, compressible, multiphase, high-Reynolds number turbulent flow with chemical reactions and heat transfer, confined in a time-varying geometry. The combustion process spans multiple regimes that include turbulent flame propagation, mixing-controlled burning, and chemical kinetics-controlled processes, and their combinations". There is still space for understanding the engine combustion process and opportunities for new discoveries. In pursuit of developing alternative engine combustion mode, low temperature combustion (LTC) modes are proposed and demonstrated by researchers. Brief description of conventional combustion modes is presented in the next section before the discussion on the alternative fuels and engines.

1.2 Conventional Engine Concepts

Internal combustion engines are heat engine converting chemical energy bound in fuel into mechanical work. In the IC engines, working fluid is burned and combustion products directly apply force on the engine piston. The most frequently used types of IC engine in automotive vehicles are compression ignition (CI) engines and spark ignition (SI) engines. The majority of CI and SI engines are four-stroke cycle engines, i.e. there are four distinct strokes in a complete cycle of CI/SI engine, namely, intake stroke, compression stroke, expansion or power stroke and exhaust stroke. There are differences in charge preparation and combustion characteristics between these two conventional combustion concepts (discussed in the next section); however the engine cycle principle remains the same for both.

1.2.1 Spark Ignition Engines

In conventional SI engine, fuel is injected into the intake manifold by port fuel injector/carburettor where fuel atomizes, vaporizes and mixes with intake air present in the manifold. Modern SI engines are often port injected, and the fuel is injected by a low-pressure (2-5 bar) fuel injection system. The fuel-air mixture is then inducted into the combustion chamber during intake stroke, where fuel–air mixture mixes with residual gases of the previous combustion cycle. During this period, mixing continues and a close to homogeneous mixture is created in the engine cylinder. Homogeneous mixture of vaporized fuel, air and residual gases is then compressed, and by the end of the compression stroke well before top dead centre (TDC), the mixture is ignited by a single intense, high temperature spark which initiates the flame kernel. The charge mixture composition and motion around the spark plug at the time of spark discharge is decisive for the flame development and subsequent flame propagation. This makes early flame development and subsequent propagation vary from cycle to cycle. Flame kernel generated by spark grows, and a turbulent flame propagates throughout the mixture until it reaches the combustion chamber walls, where it extinguishes (Fig. 1.5). Figure 1.5 shows the flame propagation at different crankshaft positions for different fuel injection techniques used for charge preparation (reproduced from [19]). In advanced modern SI engine, engine can be operated on heterogeneous and homogeneous combustion modes using direct injection (DI) of fuel to meet the emission legislation and achieve higher fuel conversion efficiency. At higher engine loads,

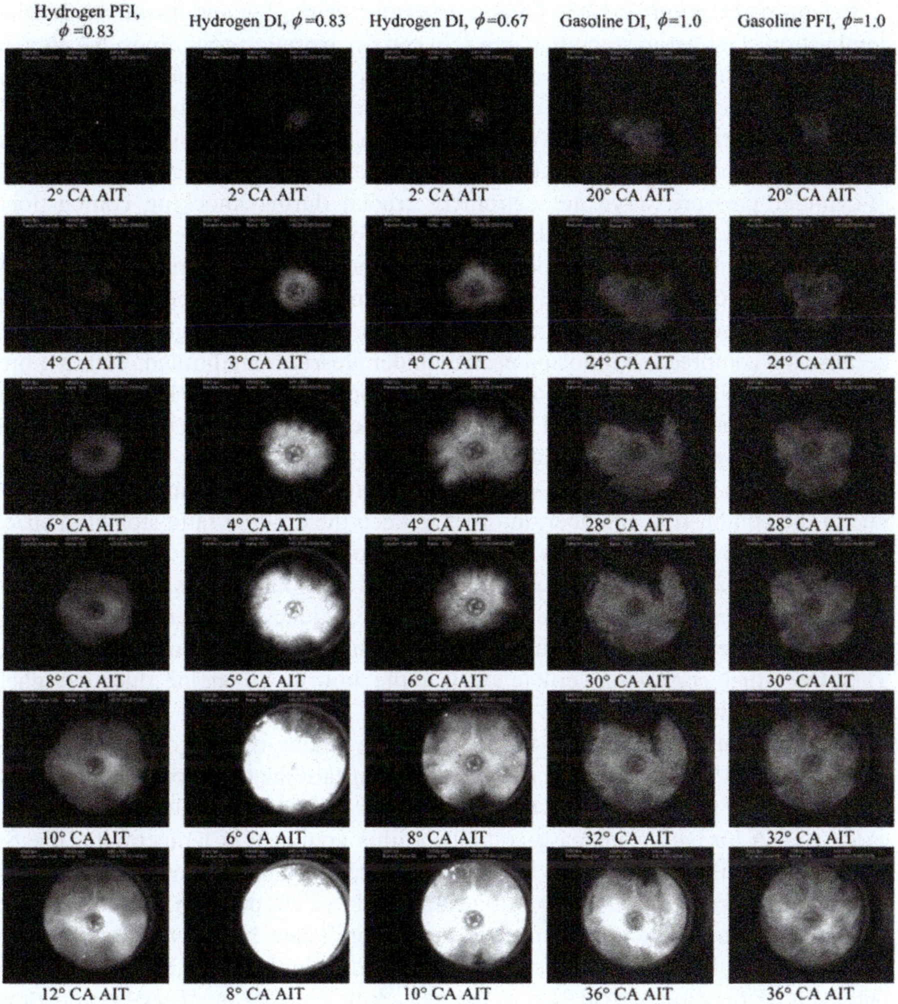

Fig. 1.5 Flame propagation (chemiluminescence image) in SI engine for different charge preparation techniques [19]

fuel is injected during the intake stroke of cycle which provides sufficient time to evaporate and mix with air leading to create a homogeneous mixture before ignition. This operation mode is similar to conventional port fuel injection (PFI) system. During lower engine loads, stratified charge mode is used to take the advantages of wide open throttle operation without pumping loss. In stratified charge operation mode, injection is staged to ensure that a combustible charge must be present close to the spark plug at the time of ignition through appropriate fuel–air mixture preparation processes [20].

In a normal combustion, the flame starts from spark plug and travels across the combustion chamber in smooth manner. In certain operating conditions, the abnormal combustion or knocking can occur in the engine cylinder. During knocking in SI engine, the end charge auto-ignites before the flame front consumes it, which may result in structural damage mainly to piston due to very high pressure rise in the cylinder. The interactions of flame front propagation, end-gas auto-ignition and in-cylinder pressure wave are extremely crucial during knocking combustion, which affect the features of local pressure mutation, combustion regime transitions and knocking intensity [21]. Knocking phenomenon limits the compression ratio of SI engines, which in turn limits the achieved thermal efficiency. The SI engine has a possibility to use higher engine rotational speed in order to get high specific power because the compression ratio and peak cylinder pressures are limited. Thus a more lightweight design can be used for engine construction that allows higher rotational speed. Additionally, flame speed in SI engine scales very well with the engine speed which also allows the higher rotational speed.

In SI engine, load control is achieved by throttling, which changes the air flow rate in the combustion chamber, in order to keep the air–fuel ratio stoichiometric. The air–fuel mixture needs to be close to stoichiometric for complete flame propagation [22]. Throttling leads to increased pumping losses during the gas exchange process, which reduces the part load efficiency of SI engines. In a car, majority of the engine operating points are in low to medium engine loads. Therefore, fuel conversion efficiency is quite low in SI engine due to higher pumping loses by throttling.

The mechanism of emission formation in the engine is governed by the combustion process and combustion chemistry. To sustain the flame propagation in SI engine, burned gas temperature needs to be over 1900 K [23]. Since the nitrogen oxide (NO_x) formation increases rapidly at this combustion temperature, the SI engine emits higher amount of NO_x. The burned gases in the cylinder are compressed during compression stroke till piston reaches TDC position and temperature attained in combustion chamber leads to the significant NO_x formation. Carbon monoxide (CO) is primarily a result of oxygen deficiency in the air–fuel mixture that leads to incomplete oxidation of the fuel. With decrease in air–fuel ratio below stoichiometric value ($\lambda<1$), CO formation sharply increases in the cylinder [24]. Combustion flame quenches near cold cylinder walls and it leaves a very thin quench layer of unburned fuel–air mixture. Flame is also unable to burn the fuel–air mixture present in the crevices (between piston top land and cylinder wall above top ring, around spark plug threads, cylinder head gasket) of combustion chamber. Adsorption of fuel vapours in the lubricating oil film on cylinder walls and combustion chamber deposits are another source of unburned hydrocarbon (HC) emissions in conventional SI engines [22]. Engine design and operating parameters also affect the NO_x and HC emissions. CO emissions are mainly affected by fuel–air equivalence ratio, and other parameters influence CO formation indirectly [24]. The effect of some of the important design and operating parameters on NO_x and HC emissions is qualitatively summarized and presented in Table 1.1 (adapted from [24], and for more details see original reference).

Table 1.1 Effect of operating and design parameters on NO_x and HC emissions in SI engines

Increase in parameter	NO_x	HC
Operating parameters		
Engine load	Increase	Decrease
Engine speed	Uncertain	Increase
Coolant temperature	Increase	Decrease
EGR	Decrease	Increase
Intake swirl and turbulence	Increase	Decrease
Advanced ignition timings	Increase	Increase
Design parameters		
Compression ratio	Increase	Increase
Surface-to-volume ratio	Decrease	Increase
Bore/stroke ratio	Decrease	Increase
Valve overlap	Decrease	Increase

To improve the performance of conventional SI engines and meet the emission legislation limits, various technologies are developed and implemented over the several decades. The development of SI engines over the last five decades with view on their control is presented in Fig. 1.6 [25]. The SI engines are mechanically controlled with electromechanical coil ignition till around 1965. Subsequently, replacement of carburettors with manifold fuel injection systems with electronic analog control started. Emission legislations significantly governed the developments of various technologies (electronic control, direct injection, etc.). Conventional SI ignition engines use three-way catalytic converter to meet emission legislation. Three-way catalytic converters simultaneously oxidize CO and HC and reduce NO_x emission from engine exhaust. The essential condition to use three-way catalytic converter is that the engine operates at or very close to stoichiometric air–fuel ratio. This condition is required to ensure that the enough reducing CO and HC species are present to reduce NO_x to N_2 and enough oxygen is available to oxidize CO and HC emissions [4]. A closed-loop feedback management system with an oxygen (λ) sensor in the exhaust is used for precise control of air–fuel ratio in SI engine.

1.2.2 Compression Ignition Engines

Combustion in compression ignition (CI) engine is fundamentally different from SI engines. Unlike SI engines, in the diesel engine, only air is drawn into the cylinder during intake stroke, and no throttle is required for engine operation. The inducted air is then compressed, and towards the end of the compression stroke, shortly before TDC, the fuel is injected at high pressure into the hot compressed air in the combustion chamber. The highly pressurized fuel is introduced into the combustion chamber via five to eight fuel sprays, depending on the size of the cylinder. The injected fuel atomizes, evaporates and mixes with the hot compressed air and auto-

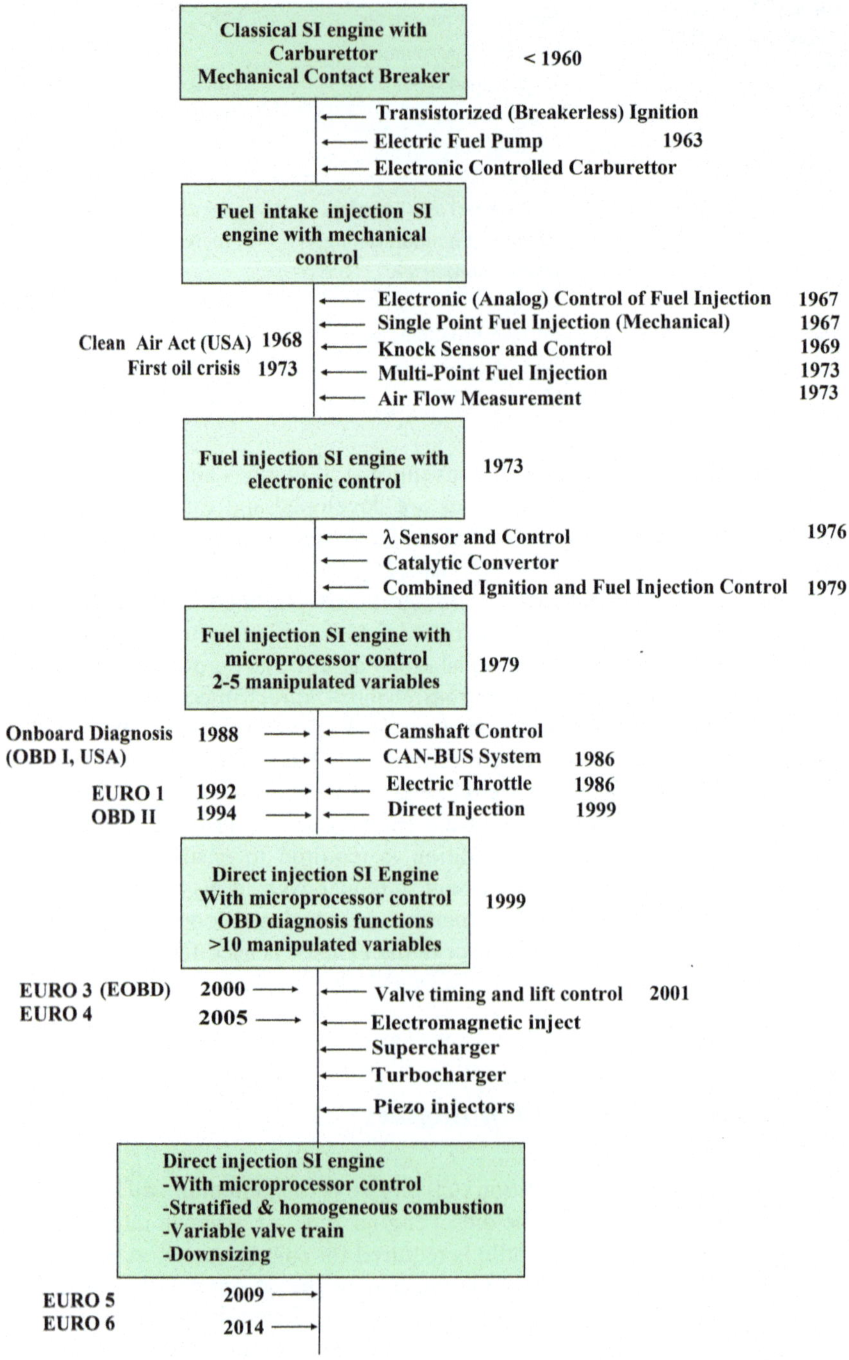

Fig. 1.6 Historical development of spark ignition engines [25]

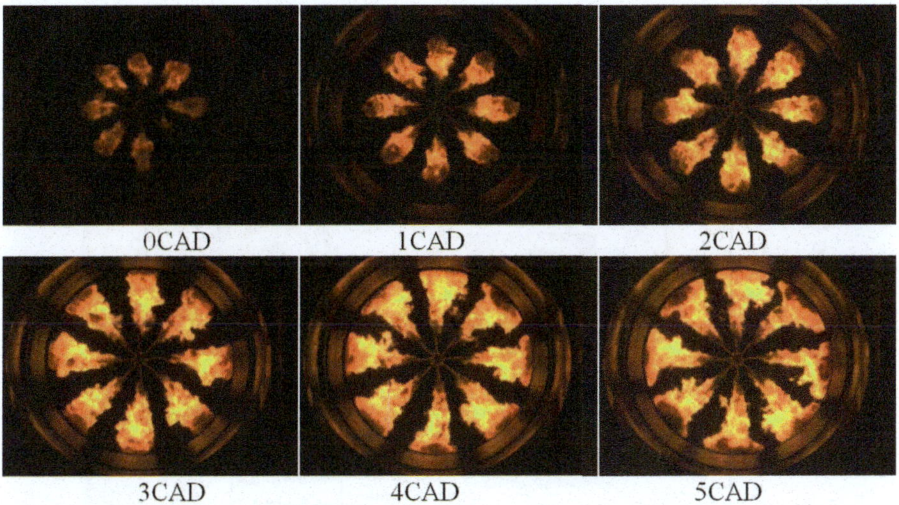

Fig. 1.7 Image sequence of diesel combustion at 1200 rpm and 160 MPa injection pressure (Courtesy of Andreas Cronhjort). *Yellow* luminosity of flames originates from hot radiating soot particles

ignites in the cylinder. In contrast to SI engines, time available for fuel–air mixture formation is very short in diesel engines. Therefore, a fast injection and the best possible atomization are required for intensive mixture formation in a diesel engine [25]. When fuel injection starts, some fuel atomizes, vaporizes and mixes partially with the air before auto-ignition occurs. Partially premixed mixture is then rapidly consumed when auto-ignition starts and the remaining combustion takes place under non-premixed conditions. The flames can be characterized as a "turbulent unsteady diffusion flames" [22]. After premixed phase combustion, diesel combustion is controlled by turbulent mixing of fuel and air at the outskirts of the diesel spray. Mixing-controlled combustion phase continues until the fuel injection is ended. After the end of fuel injection, remaining fuel or combustion intermediates burn in a diffusion flame called late mixing-controlled combustion phase [26]. In this combustion phase, the fuel spray no longer governs the fuel–air mixing process. Heat release rate (HRR) decreases fast and then gradually reduces to a zero because of dissipation of spray turbulence. Cylinder volume increases and charge cools down due to movement of piston, which may result into poor combustion efficiency in case of very low late-cycle mixing [26]. Figure 1.7 and 1.8 illustrates the typical combustion process in diesel engines.

Figure 1.7 depicts the sequence of combustion images from a diesel engine at 1200 rpm. Fuel is injected into the cylinder at 160 MPa fuel injection pressure using eight-hole injector having 190 μm orifice diameter. Heterogeneous nature of diesel combustion and diffusion flame development along with its progress is clearly illustrated (Fig. 1.7). At high combustion temperature (2000–2500 °C), carbon

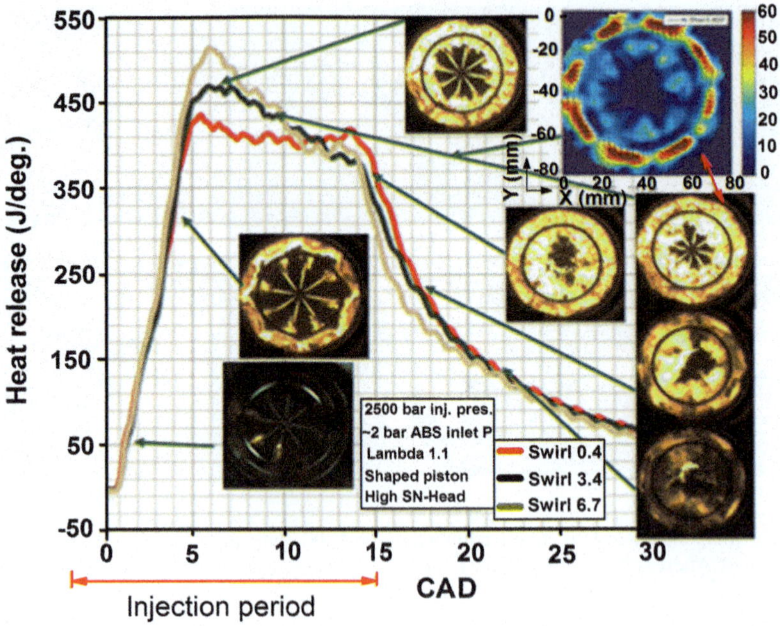

Fig. 1.8 HRR and flame luminescence images at load of 20 bar IMEP and 2500 bar injection pressure in diesel engine [27]

particles in the diffusion flame have sufficient luminosity and appear as yellow region. As flame cools, the radiation from the particles changes colour through orange to red [22]. The appearance of brown region indicates excessively rich mixture region, where substantial soot production has occurred. Initially (0 and 1 CAD; Fig. 1.7), images are mostly brownish colour due to richer mixture and relatively lower mixing with air. As combustion proceeds, more air entrainment takes place, the mixing occurs and their combustion leads to higher temperature, which changes the image colour in subsequent crank angle position.

Figure 1.8 shows heat release rate (HRR) and the progress of diesel combustion with respect to crankshaft position from the start of combustion to the end of combustion at 1000 rpm. In the first image, combustion has just started, and in the second image, all sprays have developed diffusion flames. The next two combustion images have the highest intensity, based on the HRR. It can also be observed from Fig. 1.8 that HRR is affected by the different swirl numbers (SN) [27]. The HRR increases with SN during the diffusion combustion duration, and the opposite is found at the post-oxidation duration. Premixed combustion duration is not seen prominently in modern engines (Fig. 1.8). The premixed burn appears more prominently in textbook HRR curves that is typically shown for low-pressure mechanical fuel injection (e.g. Heywood [22]). Modern diesel engines show only a very small portion (depending on engine load) of premixed heat release because of accelerated mixing process and reduced ignition delay period due to very high fuel injection pressures [26]. Diesel fuel injection system and its control are developed for very precise fuel injection at very high injection pressure (~2000 bar). The development of diesel engines over the last five decades with respect to fuel injection system and their control is presented in Fig. 1.9 [25]. Around 1989 diesel engines started using microprocessor controlled direct fuel injection along with wastegate turbochargers. In later years, exhaust gas recirculation (EGR), oxidation catalyst and turbochargers with variable geometry developed. Current diesel engines are equipped with high fuel injection pressure (~2000 bar), piezo-injectors, common rail direct injection, high EGR rates, twin turbochargers or VGT and recent after-treatment technologies (DeNOx catalyst, DPF, SCR).

In the diesel engines, combustion process is a very complex phenomenon. Fuel injection, atomization, vaporization, mixing and combustion occur at the same time in the combustion chamber. Mixing and vaporization are the slowest processes in diesel engine combustion and hence it determines the combustion rate. Therefore combustion timing and rate can be easily controlled by diesel injection timings [22]. The speed of diesel combustion is limited by the mixing between the injected fuel and the air in the combustion chamber, which also limits the maximum engine speed. Engine load in CI engine is controlled by varying the amount of fuel injected, and hence, airflow rate remains constant for all engine loads in naturally aspirated diesel engines. Since CI engine combustion is a mixing-controlled process, engine knock can be avoided, and knocking does not limit the compression ratio of the diesel engine. Due to lower pumping losses and higher compression ratio, diesel engines offer higher part load efficiency as compared to SI engines. Lower cylinder temperatures due to leaner engine operation lead to lower heat loss to coolant as well as exhaust, which also contributes to higher CI engine efficiency. As an additional benefit, the ratio of specific heats (γ) is higher for lean mixture operated engines, and therefore lower thermal energy is lost in the excitation states of larger triatomic species (CO_2 and H_2O) [28]. This leads to the availability of higher amount of thermal energy in working fluid, and more work can be extracted, which contributes to higher fuel conversion efficiency of the engine.

Major emission concerns from diesel engines are HC, NO_x and particulates (PM). However, HC emission can be easily mitigated by oxidation catalyst. Based on laser sheet imaging studies of diesel spray combustion, Dec [12] proposed a conceptual model for soot formation. According to this model, liquid fuel jet penetration is rather short in length, and all the fuel in combustion zone is in vapour phase. Soot first appears just downstream of liquid jet in rich premixed combustion region. Figure 1.10 shows the temperature distribution and combustion chemistry of diesel fuel spray in the conceptual DI diesel combustion model [29]. The concentration of soot increases and particle size grows as soot flows downstream towards the spray boundary. The model proposed that the formation of soot precursors and consequent generation of soot particles happen in the rich premixed flame ($\varphi = 2$–4) and the soot particle size increases as they pass through spray towards the head vortex [24]. Some of the soot particles reach the diffusion flame at the periphery of the jet, where they can be oxidized by OH radicals. Most of the

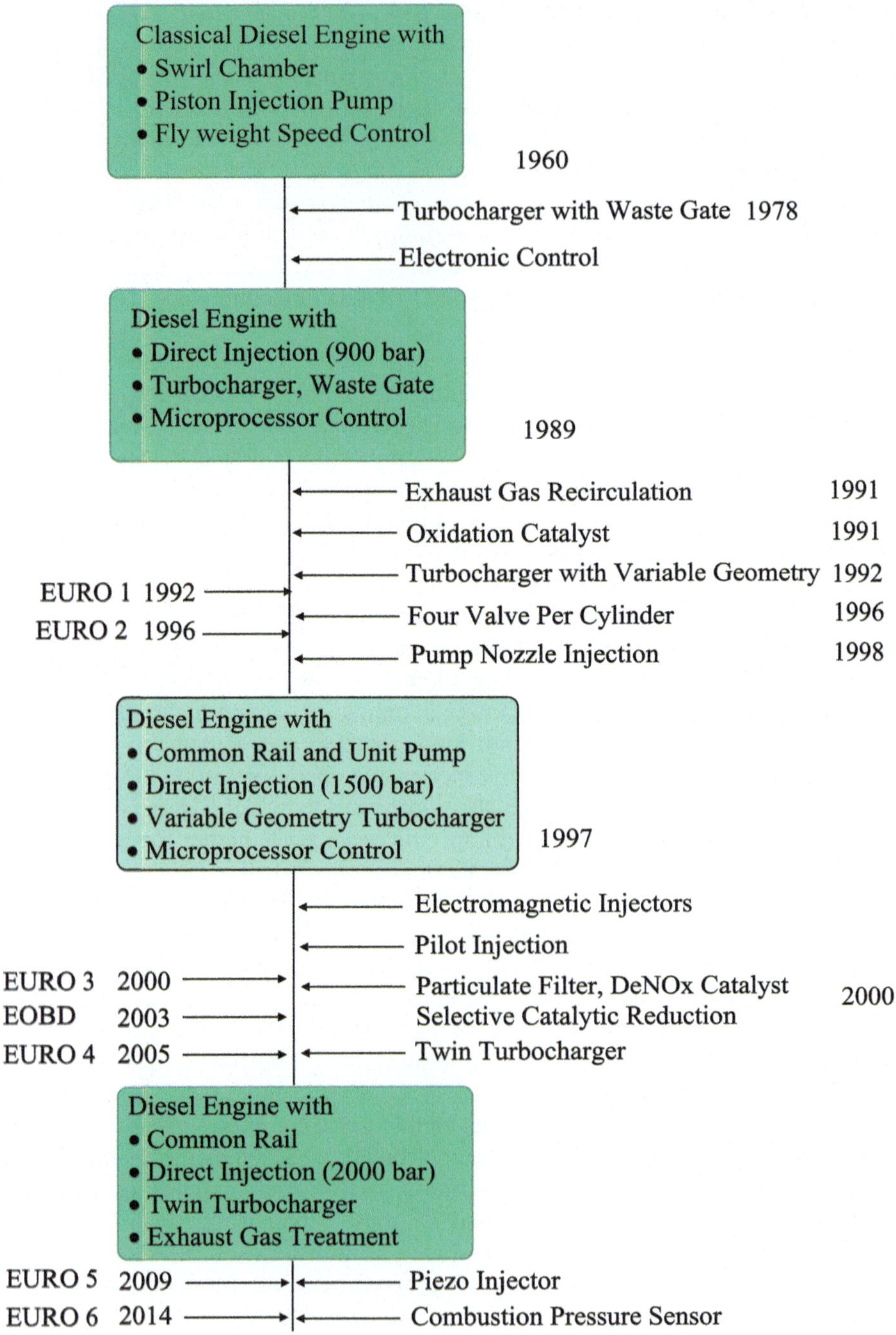

Fig. 1.9 Historical development of diesel engines [25]

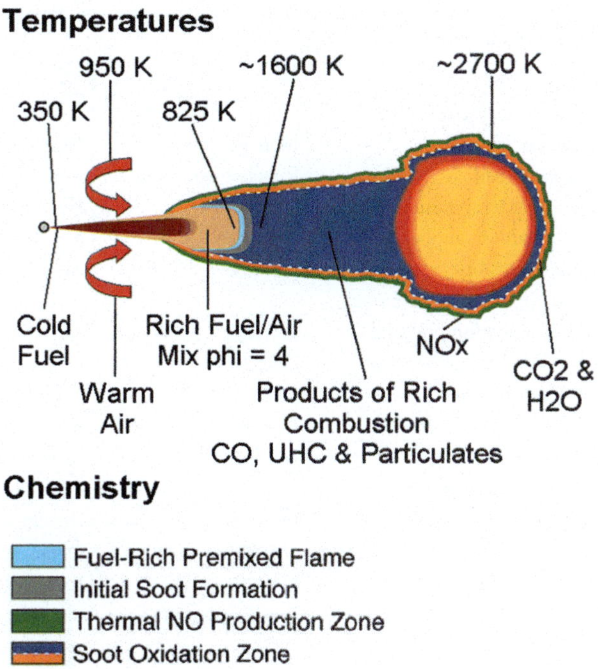

Fig. 1.10 Conceptual model of spray combustion in DI diesel engine. Reprinted with permission Copyright © 1999 SAE International [29]

NO_x formation happens during the time of high-energy release rates associated with the diffusion burning process, where temperature reaches around 2700 K [29].

Diesel engine emission formation is affected by engine design, operating and fuel injection parameters. The effect of some of the important engine operating, fuel injection and design parameters on NO_x, PM and HC emissions is qualitatively summarized and presented in Table 1.2 (adapted from [24, 30]).

Simultaneous reduction of NO_x and PM presents the biggest challenge as most of the engine design strategies reduce either NO_x emissions or PM. The reduction of one (NO_x or PM) causes an increase in the other, which is typically known as soot-NO_x trade-off in a diesel engine [4]. Several design changes for NO_x reduction result in higher fuel consumptions. To meet the emission legislations, various exhaust after-treatment technologies are used in diesel engines. Advanced exhaust after-treatments such as NO_x storage reduction (NSR), selective catalytic reduction (SCR), diesel particulate filter (DPF), continuously regenerating trap (CRT), etc. are developed and still continuously improved to meet the emission legislations for diesel engine.

Table 1.2 Effect of engine operating, fuel injection and design parameters on NO_x, PM and HC emissions in diesel engines [24, 30]

Parameter	NO_x	PM	HC
Operating parameters			
Engine speed	Increase	Reaches optimum	Increase
Engine load	Reaches optimum	Reaches optimum	Decrease
Coolant temperature	Increase	Decrease	Decrease
Fuel cetane number	Decrease	No effect	Decrease
Fuel volatility	Increase	No effect	No effect
Fuel injection parameters			
High injection pressure	Increase	Decrease	Decrease
Retarded injection timing	Decrease	Increase	Increase
Pilot injection	Decrease	No effect	No effect
Spray angle	Reaches optimum	Reaches optimum	No effect
Sac volume	No effect	Reaches optimum	Increase
Design parameters			
Compression ratio	Reaches optimum	Reaches optimum	Decrease
Stroke/bore ratio	No effect	Decrease	Decrease
Crevice volume	No effect	Increase	No effect
Swirl	Reaches optimum	Reaches optimum	Reaches optimum

1.3 Automotive Fuels

Chemical bond energy of a combustible fuel is converted to useful mechanical work by IC engines. As long as fuel is convenient to handle, transport and store, relatively safe, its utilization in the efficient IC engines is the best choice for automotive applications. Fuel quality affects the engine design and exhaust emissions. Apart from the environmental considerations, several other fuel quality requirements such as better combustion quality, high heat of combustion, high volumetric energy, low and high temperature performance, good oxidation stability, material capability, good flow characteristics, etc. are to be met by potential automotive engine fuels. Relationship between fuel quality and engine vehicle performance is discussed in detail by Pundir [4].

Four basic requirements for a future automotive fuel are (i) high power density, (ii) certain supply, (iii) overall economic feasibility and (iv) the incorporation of environmental and climate protection requirements [31]. Engine fuels as well as fuel constituents can be obtained from various feedstocks such as crude oil, natural gas, biomass, coal, oil shale, methane hydrates, oil sands and even carbon oxides (CO and CO_2) reacted with hydrogen [32]. Most of liquid engine fuel components are produced from distillation of petroleum-derived crude oil. Fischer–Tropsch process is another important method of producing engine fuels using syngas or natural gas. Natural gas (predominantly methane) is the only fuel that needs almost no processing for automotive engine applications. It requires only drying and removal of hydrogen sulphide (H_2S) from "sour" gas. Biomass is another key feedstock for the production of biofuels for combustion engines using thermochemical and biochemical routes. Thermochemical routes include pyrolysis and gasification process which convert biomass to liquid or gases, and they can be utilized in combustion engines by significant post-processing [32]. Biochemical routes involve using microorganisms to produce fuels. Biodiesel and ethanol are the other major biofuels derived from renewable materials.

Physical and chemical properties of a fuel are governed by its chemical compositions. Modern automotive fuel is a combination of different kinds of components including hydrocarbons, oxygenates, additives and sulphur. Present automotive gasoline and diesel have a higher compositional complexity and typically consist of over 100 and 1000 different components, respectively [32]. Major components of automotive fuels summarized from Mueller et al. [31] are shown in Fig. 1.11, and detailed description of each component can be found in original source. Petroleum crude oil mainly contains hydrocarbons, categorized as paraffins, naphthenes and aromatics. Gasoline and diesel fuels also consist of olefins that are formed during refining process, which are not originally present in crude oil [24]. Each hydrocarbon group has its own characteristic chemical structure and properties. Organic compounds consist of oxygen mainly alcohols, ethers, and esters (which are termed as oxygenates) are added to automotive fuels to enhance their fuel properties and reduce the emission of harmful combustion species. To further improve the fuel performance, some additives such as cetane improvers, lubricity improvers, oxidation inhibitors, detergents and metal deactivators are added to fuels at ppm levels [32].

Fuel properties affect the performance, combustion and emission characteristics of the engine. Mueller et al. categorized and discussed the key fuel properties into five categories, namely, combustion properties, physical properties, material

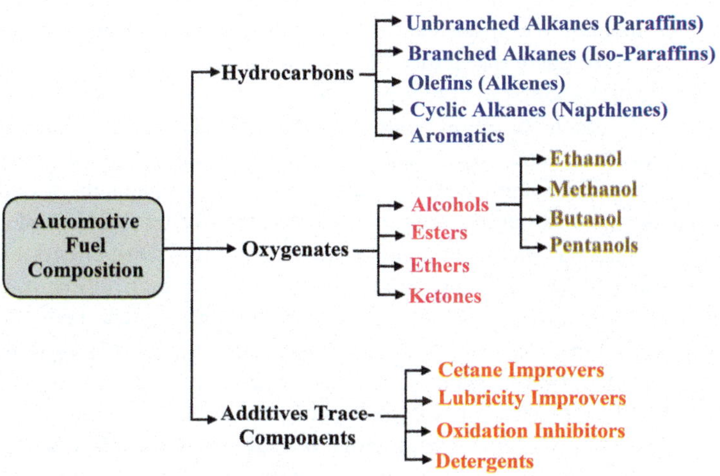

Fig. 1.11 Summary of automotive fuel composition

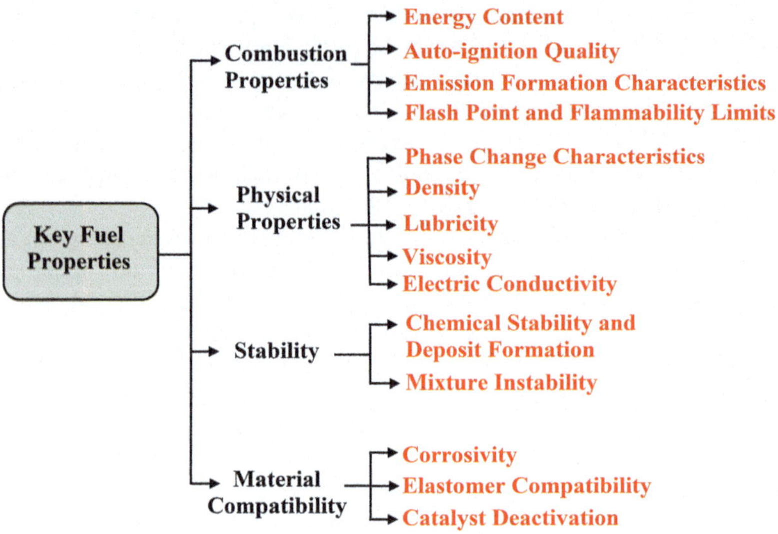

Fig. 1.12 Summary of important properties of automotive fuels

compatibility, stability and environmental considerations [32]. Important automotive fuel properties are listed in Fig. 1.12. Main combustion properties of an automotive fuel are its autoignition quality, energy content and emission formation characteristics. Better ignition quality leads to improved fuel economy and reduction in exhaust emissions. High octane number (ON) fuels for SI engine and high cetane number (CN) fuels for CI engines are required for good combustion characteristics. High energy content of automotive fuels leads to requirement of smaller amount of onboard fuel storage on vehicle for the same operating range. Emission formation is affected by fuel composition which affects the physical process happening in the engine combustion such as fuel injection, atomization, vaporization, mixing and combustion kinetics. Physical properties such as density, viscosity, lubricity and phase change characteristics are important for fuel injection system and charge preparation for engine combustion. Fuel resistance to variations in its chemical composition and/or mixture characteristics with time is known as fuel stability [32]. Good low temperature oxidation stability reduces the fuel deterioration during storage and deposit formation in the fuel system [24]. Material compatibility is necessary to prevent corrosion of metallic components as well as deterioration of rubber and elastomeric components which come into contact from point of production to engine combustion (e.g. storage tanks, pipelines, vehicle fuel tank, injection system, engine components). Fuel quality requirements specific to low temperature combustion (LTC) engines are discussed in Chap. 3.

Depletion of fossil fuels, forex expenditure, energy security and environmental pollution are the key factors to commence critical steps for substituting conventional gasoline and diesel fuels by alternative and renewable fuels at this juncture of time. Several alternative fuels are being demonstrated and proposed to substitute

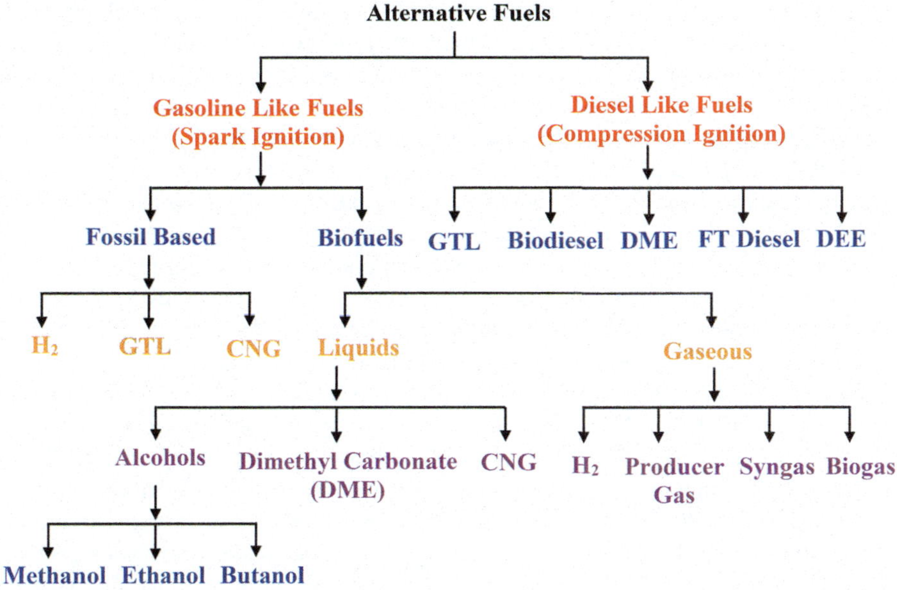

Fig. 1.13 Summary of alternative fuels presently used/proposed for automotive engines

conventional automotive fuels (gasoline and diesel). Alternative fuels for automotive transportation summarized from Ref. [33] are presented in Fig. 1.13, and detailed description of each fuel property and production process is provided in the original source.

Alternative fuels can be categorized as gasoline-like fuels (for spark ignition) and diesel-like fuels (for compression ignition) based on their utilization in two broad categories of IC engine currently used for automotive transportation. Gasoline-like fuels typically have higher octane numbers, and auto-ignition temperature is higher, which requires an external energy source to ignite the fuel-air mixture in combustion chamber (spark ignition and laser ignition). Alternative gasoline-like fuels can be fossil base (GTL, CNG, LPG, coal-based H_2) and renewable biofuels (alcohols, 2-5-dimethylfuran (DMF), syngas, biogas, H_2). Diesel-like fuels have higher cetane number, and auto-ignition temperature is comparatively lower, which can be auto-ignited by itself in combustion chamber due to compression of charge. Typical diesel-like alternative fuels are biodiesel (ethyl/methyl esters), dimethyl ether (DME), diethyl ether (DEE) and Fischer–Tropsch (FT) diesel. Biodiesel is a renewable fuel obtained from a variety of edible and nonedible vegetable oils as well as animal fat. Biodiesel is a clean burning mono-alkyl ester-based oxygenated fuel produced by transesterification process. Transesterification is the reaction of triglycerides present in the vegetable oils with primary alcohols in presence of a catalyst, which produces primary esters (biodiesel) and glycerol [34]. Physical and combustion properties of biodiesel are very similar to diesel fuel. DME is a colourless gas (easy to liquefy and transport) at standard ambient temperature with a typical odour and has emerged only recently as an automotive fuel. DME is one of the best alternative fuels for CI engines, and its clean combustion also decreases local air pollution. It can be easily auto-ignited due to its high cetane number and results in practically soot-free combustion due to easy vaporization and absence of carbon-to-carbon bonds [35].

Hydrogen emerged as promising gasoline-like alternative fuels. Hydrogen can be produced mainly by two routes involving electrolysis of water and steam reforming or gasification of a hydrogen-containing raw material. Hydrogen requires very low ignition energy and has very wide flammability limits. The main benefit of hydrogen as fuel over other alternative fuels is its carbon-free exhaust species from engines. In hydrogen combustion engine, the CO_2 emission is not there, which is the major species emitted after combustion of other alternative fuels. Hydrogen utilization in automotive vehicles is proposed mainly in two ways through (1) hydrogen fuel cell (FC) vehicles and (2) hydrogen-based IC engine vehicles. Presently, cost and fuel conversion efficiency of hydrogen fuel cell-based powertrains is higher than corresponding hydrogen IC engine powertrains [36].

Primary alcohols and their gasoline blends are typically used as alternative fuels for SI engines [37]. Methanol is a high oxygen content fuel among all primary alcohols. Methanol-fueled engines have clean combustion because of high oxygen content and simple chemical structure. Methanol is a highly toxic compound and careful utilization is required as fuel. Ethanol properties are similar to methanol, but it is less toxic and less corrosive [34]. In view of the attractive attributes of biofuels especially alcohols, efforts have been made to use them into the energy conversion systems; this book presents their utilization in alternative engine combustion concept, namely, LTC engines. Detailed discussion on combustion, performance and emissions characteristics of ethanol and methanol in LTC engines is presented in Chaps. 6, 7 and 8, respectively.

1.4 Alternative Engines

As discussed in Sect. 1.1.3, one of the possible ways to meet the engine fuel challenge is to develop an alternative engine. Factors such as global warming, air quality improvement in urban areas, energy efficiency and energy security govern the development of newer and alternative vehicle propulsion system. This challenge can be addressed by developing alternative powertrains (fuel cells, hybrid, electric vehicles) or by improving the performance of conventional IC engines by developing alternative engine combustion modes. This section explains alternative powertrains as well as alternative engine concepts.

1.4.1 Alternative Powertrains

The design of vehicle powertrain is governed by the type of energy to be used. The complete energy supply chain (source, processing, transportation and distribution) and its impact on the environment (well-to-wheel analysis) as well as cost are to be evaluated for the development of vehicle powertrain. The vehicle powertrains such as fuel cell, hybrid electric propulsion, batteries for electric vehicles and sterling engines are mainly investigated and evaluated over the years [24].

Hybrid electric vehicle (HEV) is proposed as a solution to achieve higher fuel economy and lower exhaust emissions in comparison to conventional internal combustion engines (ICE). The HEV is an intermediate step between ICE and full electric vehicle. The HEVs use internal combustion engines in combination with one or more electric motors connected to battery pack providing power to vehicle wheels either separately or jointly. Two basic configurations of HEV are series type and parallel type. In the series hybrid, only electric motor powers the wheels, and ICE runs a generator which powers the motor or charges the batteries. In this case, engine is not subjected to the transient operation of vehicles. In the parallel hybrid, engine and motor both are connected to wheels. The ICE powers one set of wheels and motor powers the other set of wheels [24]. In this mode, ICE is subjected to vehicle transient operation and leading to penalty on fuel conversion efficiency and exhaust emissions. Most of the ICE-operated cars have engines of higher rating (for, e.g. in India 25–70 kW), and average power usage in city driving is very low (Indian cities driving around 5 kW or lower). The fuel conversion efficiency of ICE is very poor at lower power output. The HEV uses a smaller rating engine, which can be operated at higher fuel conversion efficiency, and transients are managed by electric motor, which leads to higher fuel economy. Lower exhaust emission is achieved with HEV due to the operation of ICE at fixed conditions of engine operating load and speed point, and exhaust after-treatment is also more efficient at steady-state conditions [24].

Fuel cell (FC) is another potential option for powering vehicles. Fuel cell is an electrochemical device that produces DC electrical energy through a chemical reaction of hydrogen and oxygen [38]. Electrical energy can be used to drive the motor to power the vehicle wheels similar to HEV. Hydrogen FC have higher power density therefore preferred for fuel cell vehicles (FCV). The FCVs are comparatively less noisy. The FCV is promising if hydrogen is produced from renewable sources. Hydrogen can be produced from reforming of fossil fuels, oil and natural gas, renewable biomass, etc. in stand-alone stationery units and supplied to vehicles. There are challenges for large-scale production of hydrogen, and well-to-wheel efficiency of FCV is not significantly better than their counterparts [39]. On-board hydrogen storage is also an important factor for commercial success of FCV. There are many parameters still needs to be addressed before full swing market of HEVs [40] such as (i) high power density renewable energy sources; (ii) higher cost; (iii) detailed analysis of hydrogen production for FC, including delivery and storage tank systems; (iv) rapid recharging systems for plug-in BEVs; and (v) limited life cycle of batteries.

Implementation of most of the new ideas of efficient powertrain to replace ICE is restricted due to cost, complexity and overlooked real-world design shortcomings. The HEVs or BEVs have advantages in some light-duty applications depending on duty cycle [18]. Additionally, electric power station efficiency is less than 50% along with losses in power distribution, mining and transportation of fuel, if electricity is not coming through renewable sources. There is no apparent alternative to the ICE for the medium- and heavy-duty commercial vehicle markets, which are mostly high-efficiency compression ignition diesel engines. However, fossil fuel utilization in combustion engines is also not a sustainable option. These issues govern the requirement of the development of more efficient alternative combustion mode engines, which can be operated on renewable non-fossil fuels. This is the main subject of the present book.

1.4.2 Alternative Combustion Concepts

Recent developments in conventional SI and CI engines has shown significant improvements in fuel economy and exhaust emissions. Conventional SI engines are required to operate on homogeneous stoichiometric mixture and throttled operation, which results into lower fuel conversion efficiency (Sect. 1.2.1). In SI engine, compression ratio is limited by combustion knocking, which further causes lower thermal efficiency. To overcome some of the disadvantages of conventional SI engines, lean SI combustion engines are developed. Lean combustion has thermodynamically higher efficiency due to higher ratio of specific heats (γ), and lean combustion also has lower combustion temperature, thus lower heat transfer loss [22]. Unfortunately, there exists a lean combustion limit and beyond this dilution combustion flame is extinguished. Cyclic variations in indicated mean effective pressure (IMEP) also increase for lean mixture engine operation [22]. Another major demerit of leaner SI engine operation is that standard three-way catalytic converter cannot be used for after-treatment of exhaust gases due to lesser availability of CO and HC for reduction of NO_x.

Stratified charge spark ignition (SCSI) combustion process is developed to mitigate limits of lean homogeneous spark-ignited combustion by gasoline direct injection (GDI) technology. In this combustion strategy, locally near-stoichiometric mixture is created in the vicinity of spark plug, and globally (overall) fuel-air mixture is lean. This combustion strategy has the advantages of lean combustion such as lower pumping loss, lower heat transfer losses and higher ratio of specific heats. Additionally, direct fuel injection in cylinder leads to charge cooling due to fuel evaporation, and thus, chances of knocking are reduced. Due to this reason, engine can also be operated at higher compression ratios [41]. Higher compression ratio thermodynamically leads to higher thermal efficiency, but still, the compression ratio of GDI is much lower than the CI engines. In SCSI engines, fuel-rich zone also formed, which leads to higher soot formation. The SCSI engines also require lean NO_x after-treatment technology for reduction of NO_x.

The conventional CI engines are operated on lean fuel-air mixture (globally) and higher compression ratios, which leads to higher thermal efficiency in comparison to SI engines. In conventional CI engine, fuel is directly injected into combustion chamber, and turbulent heterogeneous combustion occurs in the cylinder. Heterogeneous (combustion of locally rich fuel-air mixture) combustion leads to the formation of NO_x and soot in the diesel engine.

Ideally, an engine should be operated on homogeneous leaner mixture at higher compression ratios for higher thermal efficiency and simultaneously lower NO_x and soot emissions. The homogeneous charge compression ignition (HCCI) strategy is precisely using the same concept of burning leaner homogeneous mixture at higher compression ratios. In HCCI combustion, homogeneous charge preparation is similar to SI engines, and ignition is similar to CI engines (auto-ignition of charge in the cylinder). In HCCI engines fuel–air mixture is always diluted by excess air, residuals or a combination of air and residuals. In HCCI engine, auto-ignition process is indirectly controlled by pressure, temperature, composition and the time history of the fuel–air mixture, which can be controlled by regulating the air–fuel ratio, inlet manifold temperature and pressure, EGR or amount of residuals, coolant temperature, fuel properties or fuel blend and direct fuel injection timings [42]. Due to controlled auto-ignition of the fuel–air mixture in HCCI combustion, it is also referred by another name called controlled auto-ignition (CAI). Detailed discussion on HCCI combustion and its control is presented in Chaps. 2 and 5 of the present book.

Onishi et al. performed the first systematic investigation on HCCI combustion using a two-stroke engine in 1979 and named the combustion process as active thermo-atmospheric combustion (ATAC) [43]. In this combustion process, whole fuel–air mixture in the cylinder auto-ignited simultaneously, and no flame propagation was observed. During the same time frame, another study conducted by Noguchi et al. presented the same combustion event in a two-stroke opposed piston engine by measurements of radical concentration in the cylinder [44]. Najt and Foster investigated HCCI combustion process in a four-stroke variable compression ratio (VCR) engine using mixture of iso-octane and n-heptane in 1983 [45]. Over the last three decades, the HCCI combustion process has been investigated with significant achievement in two- and four-stroke engines with liquid and gaseous fuels [46]. Considering the common features of auto-ignition and premixed fuel–air mixture, several technologies with different names such as PCCI (premixed charge compression ignition), ATAC (active thermo-atmospheric combustion), HCCI (homogeneous charge compression ignition), OKP (optimized kinetic process), TS (Toyota-Soken), CAI (controlled auto-ignition), PREDIC (premixed lean diesel combustion), ARC (active radical combustion), MK (modulated kinetics), UNIBUS (uniform bulky combustion system), CIHC (compression-ignited homogeneous charge), etc. are investigated over the year, and detailed reference (for further study) of each technology can be found in Ref. [47].

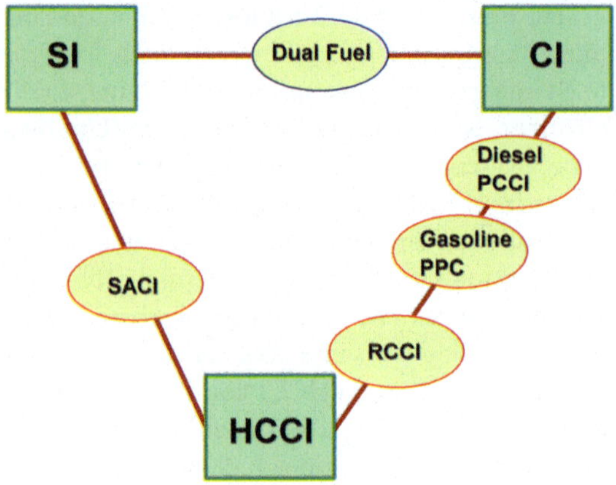

Fig. 1.14 Schematic representation of three major combustion modes and intermediate processes in IC engines (Adapted from [47])

Major drawback in HCCI mode is lack of combustion control and limited power output. To overcome the limitations of lower power density and combustion control, partially stratified charge compression ignition (SCCI) mode is proposed. In this development process, first diesel partially premixed compression ignition (PPCI, sometimes also referred as PCCI) and then gasoline partially premixed combustion (PPC) are proposed to achieve higher engine load while keeping the benefits of HCCI mode. To achieve higher power output along with better combustion phasing and duration control, dual fuel reactivity-controlled compression ignition (RCCI) strategy is most recently developed. Comprehensive discussion on these combustion strategies is presented in Chap. 2. Figure 1.14 schematically shows the three main combustion modes (SI, CI and HCCI) and five intermediate combustion processes in IC engines. The spark-assisted compression ignition (SACI) is an intermediate combustion process between SI and HCCI modes. However, RCCI, PPC and PCCI are intermediate combustion process between CI and HCCI modes. The difference between dual fuel RCCI and conventional dual fuel combustion process is the fuel injection strategy and premixed charge engine operation. In RCCI combustion process, diesel is premixed (longer ignition delay) to increase the reactivity of charge and reactivity distribution of charge in the engine cylinder, whereas in conventional dual fuel combustion process, ignition delays are very short (in comparison to RCCI) and diesel burns as conventional diffusion flame. Therefore, conventional dual fuel combustion process is an intermediate process between CI and SI modes. However, RCCI combustion is intermediate combustion process between CI and HCCI mode.

A common classification to all the mentioned premixed combustion technologies is low temperature combustion (LTC) due to significantly lower combustion temperature in comparison to conventional CI and SI engines. All the premixed LTC strategies have common characteristic of comparatively higher thermal efficiency and simultaneously reducing NO_x and soot emissions to an ultralow level. Figure 1.15 depicts the different engine combustion strategies based on fuel reactivity of charge used. It is observed that conventional combustion strategies are

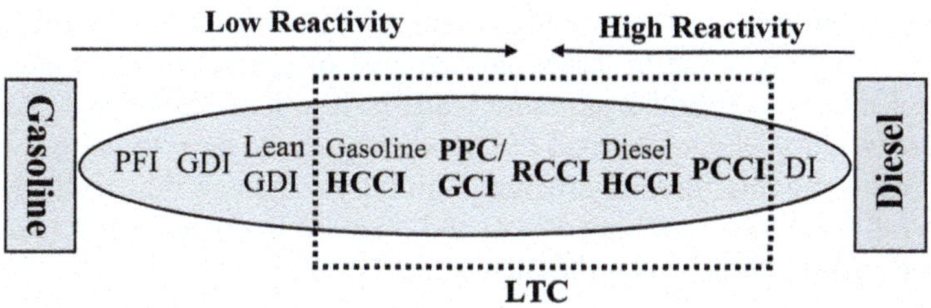

Fig. 1.15 Proposed combustion strategies based on fuel reactivity (Adapted from [49])

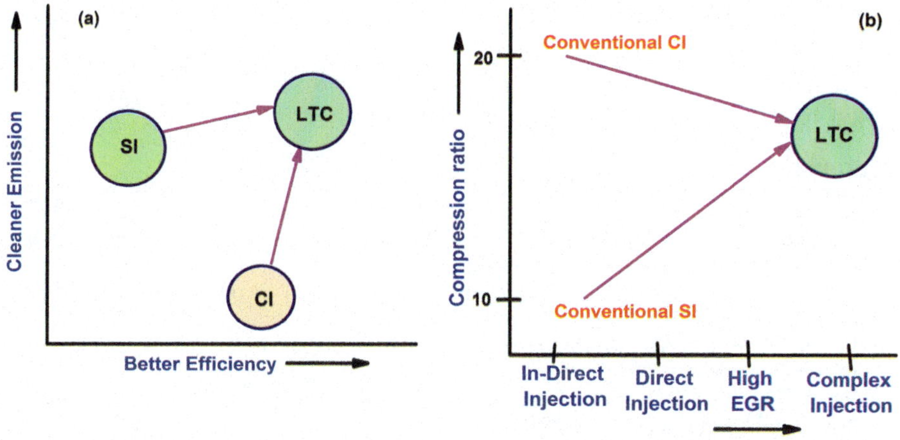

Fig. 1.16 Merging trend of conventional diesel and gasoline engine technology into LTC engine (Adapted from [48, 50])

converging towards the fuel reactivity between gasoline and diesel by the development of newer LTC engines. This provides the fuel flexibility to use both low reactivity gasoline-like fuels and high reactivity diesel-like fuels in LTC engines. Additionally, the future engine management system can use best control strategies for the fuel utilization based on vehicle operating conditions using the advanced multi-fuel engine technology [49].

Another important observation for LTC strategy is that it combines the best features of conventional CI and SI strategies, i.e. better efficiency of CI engine and cleaner emissions of SI engine (Fig. 1.16a). Figure 1.16 shows the merging trend of conventional diesel (CI) and gasoline (SI) engine technologies. Future advanced LTC engines required complex fuel injection strategies using both port and direct fuel injection, and engine is operated at moderate compression ratio (Fig. 1.16b). In CI engines, compression ratio has a key role in auto-ignition of fuel especially under cold start conditions by providing higher gas temperature at higher compression ratio. This is one of the factors; the older indirect diesel injection (IDI) engines (prechamber injection) have very high compression ratios (20–24) to compensate for higher heat losses due to larger surface area of prechamber. Development of direct fuel injection at higher pressure leads to lower heat loss, and reduction in compression ratio. Engine thermal efficiency has nonlinear correlation with compression ratio. Engine thermal efficiency initially increases drastically with compression ratio (up to approximately 14) and increase in compression ratio has less significant effect at higher compression ratios [22].

In premixed LTC engines, longer ignition delay is required to provide sufficient time for mixing fuel with air. Engine operation at comparatively lower compression ratio leads to lower cylinder gas temperature, which is helpful to have longer ignition delay period. Therefore LTC engines are moving towards complex fuel injection strategies along with moderate compression ratios. Detailed discussion on characteristics and control of LTC engines is the subject of the following chapters of this book.

References

1. International Energy Agency (2010) World energy outlook. International Energy Agency, France. isbn 978-92-64-08624-1. www.iea.org
2. U.S. Energy Information Administration (2016) International energy outlook, . http://www.eia.gov/outlooks/ieo/pdf/0484(2016).pdf
3. Fischer M, Werber M, Schwartz PV (2009) Batteries: higher energy density than gasoline? Energy Policy 37(7):2639–2641
4. Pundir BP (2007) Engine emissions- pollutant formation and advances in control technology. Narosa Publishing House, New Delhi
5. Ramadhas AS (2011) Fuels and trends. In: Ramadhas AS (ed) Alternative fuels for transportation. CRC Press, Boca Raton
6. Environmental Protection Agency, USA. https://www.epa.gov/ghgemissions/overview-greenhouse-gases. Accessed 29 Jan 2017
7. Environmental Protection Agency, USA. https://www.epa.gov/acidrain/what-acid-rain. Accessed 30 Jan 2017
8. United Nations Environment Programme (2010) Scientific assessment of ozone depletion. http://ozone.unep.org/Assessment_Panels/SAP/Scientific_Assessment_2010/00-SAP-2010-Assement-report.pdf
9. Agarwal AK, Shukla PC, Gupta JG, Patel C, Prasad RK, Sharma N (2015) Unregulated emissions from a gasohol (E5, E15, M5, and M15) fuelled spark ignition engine. Appl Energy 154:732–741
10. Agarwal AK, Shukla PC, Patel C, Gupta JG, Sharma N, Prasad RK, Agarwal RA (2016) Unregulated emissions and health risk potential from biodiesel (KB5, KB20) and methanol blend (M5) fuelled transportation diesel engines. Renew Energy 98:283–291
11. Bansal G, Bandivadekar A (2013) Overview of India's vehicle emissions control program. ICCT, Beijing/Berlin/Brussels/San Francisco/Washington, DC
12. Dec JE (1997) A conceptual model of DI diesel combustion based on laser-sheet imaging (No. 970873). SAE technical paper
13. Delhpi, Worldwide Emissions Standards: Heavy Duty and Off-Highway Vehicle (2015–16). http://www.delphi.com/manufacturers/auto/powertrain/emissions_standards
14. Malikopoulos AA (2014) Supervisory power management control algorithms for hybrid electric vehicles: a survey. IEEE Trans Intell Transp Syst 15(5):1869–1885

15. IEA (2012) Energy technology perspectives. International Energy Agency, Paris. http://www.iea.org/publications/freepublications/publication/name,31269,en.html
16. Duret P (2002) 18 gasoline CAI and diesel HCCI: the way towards zero emission with major engine and fuel technology challenges (No. 2002-32-1787). SAE technical paper
17. Reitz RD (2015) Grand challenges in engine and automotive engineering. Front Mech Eng 1:1
18. Reitz RD (2013) Directions in internal combustion engine research. Combust Flame 1 (160):1–8
19. Aleiferis PG, Rosati MF (2012) Flame chemiluminescence and OH LIF imaging in a hydrogen-fuelled spark-ignition engine. Int J Hydrog Energy 37(2):1797–1812
20. Spicher U (2014) Spark ignition combustion. Enc Automot Eng 1:209–230
21. Pan J, Shu G, Zhao P, Wei H, Chen Z (2016) Interactions of flame propagation, auto-ignition and pressure wave during knocking combustion. Combust Flame 164:319–328
22. Heywood JB (1988) Internal combustion engine fundamentals. McGraw-Hill, New York
23. Flynn PF, Hunter GL, Durrett RP, Farrell LA, Akinyemi WC (2000) Minimum engine flame temperature impacts on diesel and spark-ignition engine NOx production (No. 2000-01-1177). SAE technical paper
24. Pundir BP (2010) IC engines combustion and emissions. Narosa Publishing House, New Delhi
25. Isermann R (2014) Engine modeling and control, modeling and electronic management of internal combustion engines. Springer-Verlag, Berlin/Heidelberg. ISBN:978-3-642-39934-3
26. Andersson Ö, Miles PC (2014) Diesel and diesel LTC combustion. Enc Automot Eng 1:231–266
27. Dembinski HW (2014) The effects of injection pressure and swirl on in-cylinder flow pattern and combustion in a compression–ignition engine. Int J Engine Res 15(4):444–459
28. Ciatti SA (2015) Compression ignition engines–revolutionary technology that has civilized frontiers all over the globe from the industrial revolution into the twenty-first century. Front Mech Eng 1:5
29. Flynn PF, Durrett RP, Hunter GL, zur Loye AO, Akinyemi OC, Dec JE, Westbrook CK (1999) Diesel combustion: an integrated view combining laser diagnostics, chemical kinetics, and empirical validation (No. 1999-01-0509). SAE technical paper
30. Borman GL, Ragland KW (1998) Combustion engineering. McGraw Hill, Newyork. ISBN:0-07-006567-5
31. Hagenow G, Reders K, Heinze HE, Steiger W, Zigan D, Mooser D (2010) Fuels. In: Mollenhauer K, Tschoeke H (eds) Handbook of diesel engines. Springer, Heidelberg/Berlin
32. Mueller CJ, Cannella WJ, Kalghatgi GT (2014) Fuels for engines and the impact of fuel composition on engine performance. Enc Automot Eng 1:359–386
33. Babu MG, Subramanian KA (2013) Alternative transportation fuels: utilisation in combustion engines. CRC Press, Boca Raton
34. Agarwal AK (2007) Biofuels (alcohols and biodiesel) applications as fuels for internal combustion engines. Prog Energy Combust Sci 33(3):233–271
35. Park SH, Lee CS (2013) Combustion performance and emission reduction characteristics of automotive DME engine system. Prog Energy Combust Sci 39(1):147–168
36. Verhelst S, Wallner T (2009) Hydrogen-fueled internal combustion engines. Prog Energy Combust Sci 35(6):490–527
37. Balki MK, Sayin C, Canakci M (2014) The effect of different alcohol fuels on the performance, emission and combustion characteristics of a gasoline engine. Fuel 115:901–906
38. Basu S (2007) Fuel cell science and technology. Anamaya Publishers, New Delhi
39. Weiss M, Heywood JB, Schafer A, Natarajan VK (2003) Comparative assessment of fuel cell cars. Report MIT LFEE 2003–001 RP, Laboratory for Energy and the Environment, MIT
40. Hannan MA, Azidin FA, Mohamed A (2014) Hybrid electric vehicles and their challenges: a review. Renew Sust Energ Rev 29:135–150
41. Spicher U, Reissing J, Kech JM, Gindele J (1999) Gasoline direct injection (GDI) engines-development potentialities (No. 1999-01-2938). SAE technical paper

42. Zhao H (2007) Overview of CAI/HCCI engines. In: Zhao H (ed) HCCI and CAI engines for the automotive industry. Woodhead Publishing Limited, Cambridge
43. Onishi S, Jo SH, Shoda K, Do Jo P, Kato S (1979) Active thermo-atmosphere combustion (ATAC)-a new combustion process for internal combustion engines (No. 790501). SAE technical paper
44. Noguchi M, Tanaka Y, Tanaka T, Takeuchi Y (1979) A study on gasoline engine combustion by observation of intermediate reactive products during combustion (No. 790840). SAE technical paper
45. Najt PM, Foster DE (1983) Compression-ignited homogeneous charge combustion (No. 830264). SAE technical paper
46. Maurya RK, Agarwal AK (2011) Experimental study of combustion and emission characteristics of ethanol fuelled port injected homogeneous charge compression ignition (HCCI) combustion engine. Appl Energy 88(4):1169–1180
47. Zhao H (2007) Motivation, definition and history of HCCI/CAI engines. In: Zhao H (ed) HCCI and CAI engines for the automotive industry. Woodhead Publishing Limited, Cambridge
48. Johansson B (2016) Fuels and combustion. In: Boot M (ed) Biofuels from lignocellulosic biomass: innovations beyond bioethanol. Wiley-VCH Verlag GmbH & Co, Weinheim
49. Hongming X (2012) Present and future of premixed compression ignition engines. J Automot Saf Energ 3(3):185–199
50. Curran S, Prikhodko V, Gao Z, Parks J, Wagner R (2013) High efficiency clean combustion in multi-cylinder light-duty engines. DOE hydrogen program and vehicle technologies annual merit review, May 14
51. http://www.nissan-global.com/EN/TECHNOLOGY/OVERVIEW/hcci.html. Accessed 2 Feb 2017

Chapter 2
Low Temperature Combustion Engines

Abstract Current stringent emission legislation, market and environmental concern governs the automotive research for developing high-efficiency and low emission engines. Vehicle manufacturers meet the present emission norms using a combination of in-cylinder emission reduction methods and exhaust after-treatment devices. Proposed future emission norms are even more stringent; and thus, newer technologies are required to satisfy emission standards worldwide. Low temperature combustion (LTC) engines have potential to deliver higher fuel conversion efficiency and simultaneous reduction of NO_x and soot emissions to an ultralow level. The LTC engines can also reduce the heavy dependence on NO_x and soot after-treatment devices for meeting the emission norms. In this chapter, LTC principles along with different proposed LTC strategies are discussed in detail. The LTC strategies are broadly categorized into homogeneous charge compression ignition (HCCI) and partially stratified charge compression ignition (SCCI). The SCCI strategy is further classified as thermal stratification- and fuel concentration stratification-based combustion process. The partially premixed compression ignition (PPCI) and reactivity-controlled compression ignition (RCCI) are the two main engine combustion modes with fuel concentration stratification are discussed in this chapter. Thermally stratified compression ignition mode is also described in the present chapter.

Keywords LTC • HCCI • Stratification • Combustion • Heat release • LTHR • HTHR • PCCI • EGR • SACI • Ignition • RCCI • Dual fuel • Direct injection

2.1 Low Temperature Combustion Principle

Presently, internal combustion (IC) engines are the most preferred option for automotive powertrain. Two commonly used engine types are spark ignition (SI) and compression ignition (CI) engines, running on gasoline and diesel, respectively. The CI engines have comparatively higher thermal efficiency due to higher compression ratio and lean (equivalence ratio$<$1) and unthrottled engine operation. Therefore, CI engines are preferred choice for medium- and heavy-duty applications. The in-cylinder NO_x, CO and unburned hydrocarbon (HC) formation is higher in SI engines, and these emissions from tailpipe can be reduced to legislation

R.K. Maurya, *Characteristics and Control of Low Temperature Combustion Engines*, Mechanical Engineering Series, https://doi.org/10.1007/978-3-319-68508-3_2

limits level by effective operation of three-way catalytic converter (TWC). The conventional CI engines have mainly high emissions of NO_x and particulate matter (PM), and also suffer from the NO_x-PM trade-off [1]. The regulated limits of NO_x and PM emissions are drastically reduced all over the world in the last 20 years (see Fig. 1.3). The early emission norms were complied by in-cylinder emission reduction strategies such as higher fuel injection pressure, exhaust gas recirculation (EGR), optimal fuel injection timings, improved design of combustion chamber, boosting intake air, etc., without any exhaust after-treatment technologies [2]. The present and future emission norms are fulfilled or expected to be accomplished by a combination of in-cylinder strategies as well as exhaust after-treatment devices. Adaptation of presently available after-treatment technologies faces challenges of higher cost, fuel economy penalties, durability issues and larger space requirements in comparison to the engines without exhaust after-treatment technologies [3]. Additionally, there is a requirement of higher fuel conversion efficiency at minimal cost to compete in the market. Consequently, in-cylinder combustion strategies need to be significantly improved for higher thermal efficiency as well as further reduction in engine-out emissions so that dependency on the after-treatment technologies can also be reduced.

In-depth understanding of in-cylinder processes is required for the development of appropriate technology to meet the present and future emission legislation limits while maintaining higher thermal efficiency [4]. Various studies are conducted using advanced laser-based combustion diagnostic techniques to understand the in-cylinder processes during diesel combustion in CI engines. A conceptual model for quasi-stationary diesel spray combustion was proposed based on laser sheet imaging (Fig. 2.1a) [5]. This model presented a liquid fuel penetration length much shorter than penetration length estimated by earlier studies. Furthermore, no droplets are present in the combustion zone; only fully vaporized fuel is present. Just downstream of the liquid part of the diesel spray, a rich premixed combustion zone appeared. Soot particles are formed in regions of locally rich mixture, where fuel and air first reacted, and then this rich mixture burns out in a high temperature diffusion flame at the jet periphery, leading to NO_x formation [10]. Soot formation starts in the fuel-rich premixed zone (Fig. 2.1a). Rate of soot formation is dependent on equivalence ratio in this zone, which is determined by flame lift-off length.

Flame lift-off characteristics play an important role in combustion and emission processes in diesel engines [11]. Flame lift-off length is the distance between the nozzle hole and the most upstream part of the turbulent diffusion flame. Most of the surrounding air entrainments into the jet are through this distance (lift-off length). Air entrained downstream of the lift-off length reacts (oxygen is consumed) in the diffusion flame, and therefore entrainment into the jet interior is very limited in the burning portion of the jet (Fig. 2.1a). Equivalence ratio (φ) inside the jet is lower in case the more air entrained into the spray up to the lift-off length, which leads to lesser soot formation [12]. Thus, equivalence ratio at lift-off plays an important role in soot formation during spray combustion. Flame lift-off characteristics depend on ambient (cylinder gas) temperature and density, injector orifice diameter, injection pressure and oxygen concentration [13, 14]. Pickett et al. [13] derived an empirical relation to estimate lift-off length (H) given by Eq. (2.1):

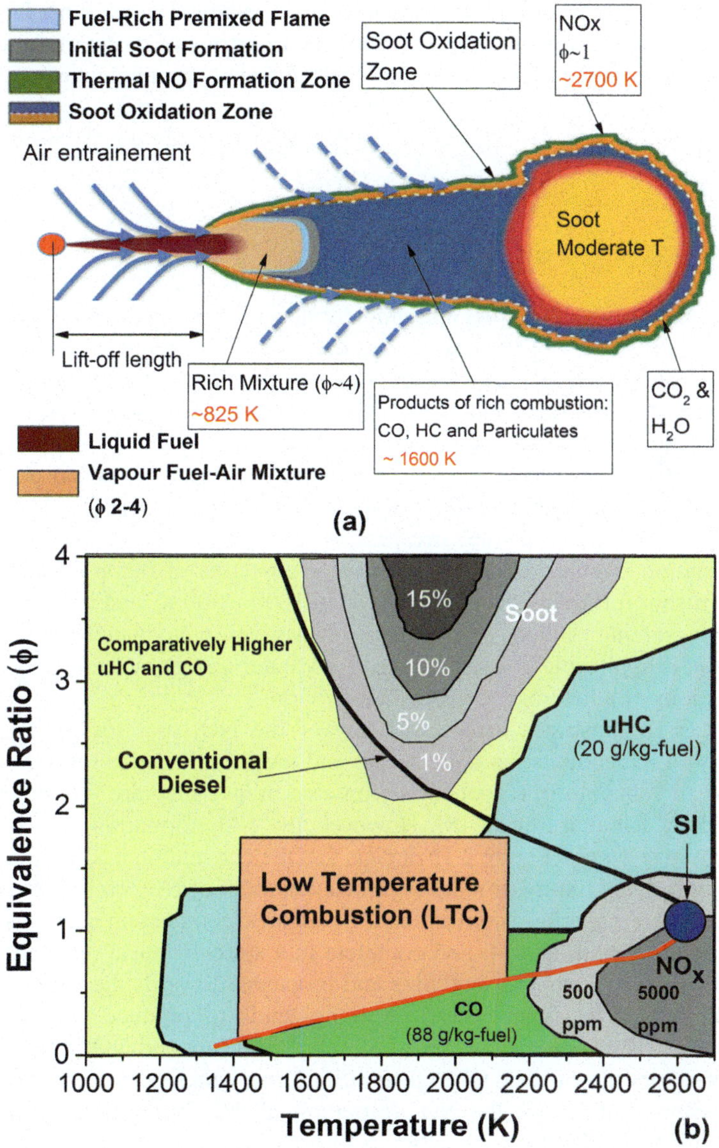

Fig. 2.1 (**a**) Illustration of NO_x and soot formation processes in diesel combustion in CI engines (Adapted from [5–7]). (**b**) LTC operating region on φ-T map (Adapted from [4, 6, 8, 9])

$$H = CT_a^{-3.74}\rho a^{-0.85}d^{0.34}U^1Z_{st}^{-1} \tag{2.1}$$

where C is a proportionality constant, T_a [K] is ambient gas temperature, ρ_a [kg/m3] is ambient gas density, d [μm] is injector tip orifice diameter, U[m/s] is injection velocity and Z_{st} is stoichiometric mixture fraction, which accounts for effects of ambient gas oxygen concentration. Mean cross-sectional φ of a model jet as a function of distance from nozzle of injector can be estimated by Eq. (2.2) (proposed by Naber and Siebers [15]) for quantitative estimate air entrainment effect in spray:

$$\bar{\Phi} = \frac{2(A/F)_s}{\sqrt{1 + 16(x/x^{+})^2} - 1} \tag{2.2}$$

where $x^{+} = \sqrt{\frac{\rho_f}{\rho_a}} \frac{\sqrt{C_a} d}{\tan\left(\frac{\alpha}{2}\right)}$ and $\tan\left(\frac{\alpha}{2}\right) = ac_1\left(\left(\frac{\rho_a}{\rho_f}\right)^{0.19} - C_2\sqrt{\frac{\rho_a}{\rho_f}}\right)$

where ρ_f and ρ_a are the fuel and ambient densities, respectively. C_a is the area contraction coefficient of the nozzle, d is the nozzle hole diameter and α is the jet spreading angle. Actual equivalence ratio (φ) at flame lift-off length can be estimated using Eqs. (2.1) and (2.2). This estimated φ can give an idea about the soot formation during spray combustion.

Computational studies analysed the variations of soot formation on the local equivalence ratio–temperature (φ-T) map as depicted in Fig. 2.1b [8, 16]. It is found that for $\varphi < 2$, no soot is formed irrespective of the combustion temperature and no soot formation is observed below 1500 K irrespective of the φ (Fig. 2.1b). The maximum soot formation tendency is around 1800–2000 K, and at higher temperatures, the sooting tendency is reduced for a given equivalence ratio. The φ-T correlation is applicable to a homogeneous cylinder charge or the locally homogeneous regions of a heterogeneous cylinder charge.

Figure 2.1b illustrates the dependency of the NO_x-soot formation and soot oxidation on the combustion temperature and local equivalence ratio in the cylinder. Size and shape of the soot formation area in φ-T map are fuel dependent as presented by Kitamura et al. [8]. However, the NO_x formation area is not fuel dependent in φ-T map. Figure 2.1b depicts that the conventional SI engine operated on stoichiometric mixture produce significant amount of NO_x emissions (shown in NO_x region in φ-T map). In conventional CI engines, combustion reactions start in a richer mixture (about $\varphi = 4$) and complete in a stoichiometric ($\varphi = 1$) diffusion flame [4]. Flame is prone to initialize and propagate towards the locally stoichiometric regions. The stoichiometric burning tends to produce very high flame temperatures. The combustion zones fall in the high soot and NO_x formation areas in φ-T map for conventional CI engines (Fig. 2.1b), which leads to higher NO_x and PM emissions. To reduce the NO_x formation, if the combustion temperature is lowered, then soot formation kinetics is also slowed down [17, 18]. However, at lower combustion temperature, in-cylinder soot oxidation rate also decreases more than the formation rates [19], and hence the exhaust gas contains more PM in the diesel engines. Studies confirm that lowering combustion temperature by using EGR in CI engines initially increases the emission of PM [20]. For very high EGR rate operating conditions, soot formation rate can be too low so that PM emission can decrease even with lower oxidation rates [16, 17, 19, 20]. Another potential challenge arises by low engine combustion temperature is higher levels of CO and HC emissions. The CO and HC formations are dependent on temperature and can form in lean as well as rich fuel–air mixtures. Higher combustion temperature leads to near-complete oxidation of CO and HC in leaner conditions, but richer conditions result in partial oxidation of HC to CO resulting into higher CO emissions. At lower combustion temperature, richer mixture leads to higher CO and HC emissions (Fig. 2.1b). There exists a region in φ-T map with comparatively lower CO and HC emissions apart from soot–NO_x region called low temperature combustion (LTC) region (Fig. 2.1b). Thus, both NO_x and soot formations can be simultaneously avoided in LTC, which is inevitable in conventional high temperature combustion (HTC) processes in compression ignition engines.

In the pursuit of developing an engine having ability to yield ultralow NO_x and soot emission along with higher thermal efficiency, several new CI combustion strategies have been demonstrated in the last two decades. A common name for these combustion strategies is LTC mode. Figure 2.2 presents the evolution of

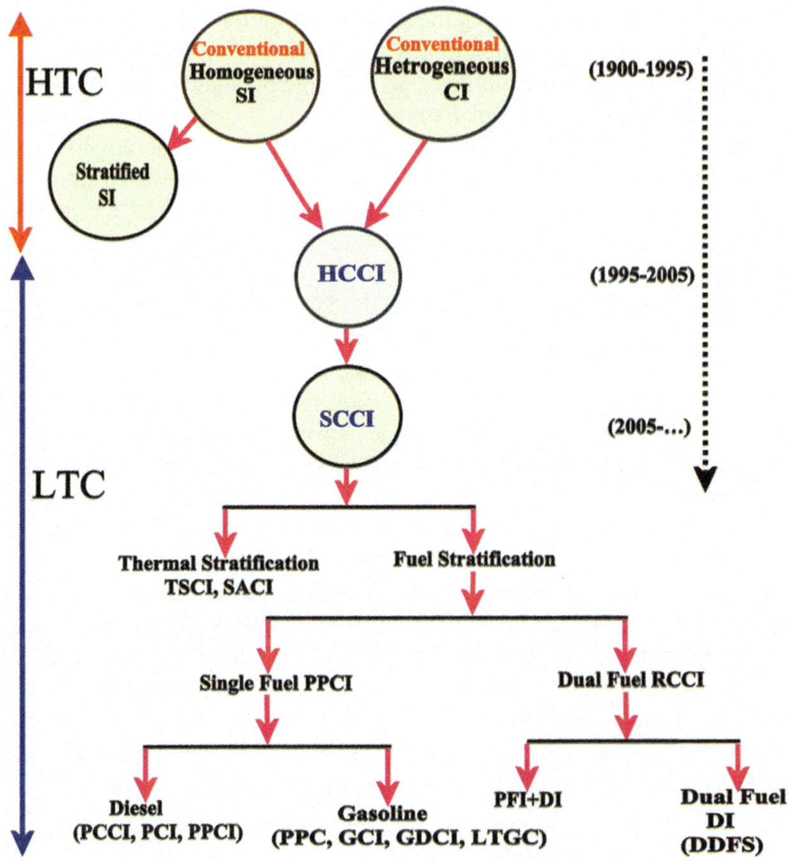

Fig. 2.2 Evolution of different combustion strategies in IC engines

different combustion strategies for IC engines. In LTC strategy, adequate premixing between the fuel and air is created to avoid diffusion combustion in a globally dilute environment. The combustion of lean fuel–air mixture leads to lower combustion temperatures in the cylinder. In LTC engines, charge dilution (with either air or EGR) is typically used to avoid high NO_x formation by lowering combustion temperature in the cylinder. In LTC engines, providing more time (3–20 ms) for premixing fuel and air evades high soot formation rates due to lower local φ achieved by mixing. However, overmixing of fuel and air results into very low φ, which leads to very high HC and CO emission due to very low combustion temperature [6]. In-cylinder control of soot emissions is a very complex process in LTC engines. Soot emission depends on quality of fuel–air mixing, combustion temperature and availability of oxygen in the charge. Therefore, these parameters can be utilized for controlling the in-cylinder soot formation. Lowering combustion temperature by dilution and increasing premixing by prolonged ignition delay are two typical methods for soot control. Another strategy to control the soot emission is by controlling the reactivity of fuel–air mixture by reformulating the fuel composition (oxygen content, C/H ratio, octane/cetane) because soot region in φ-T map is dependent on fuel also (Fig. 2.1b). Most of the LTC concepts have utilized all the three parameters with a special focus on any one parameter in a particular strategy.

Various LTC concepts have been investigated with several acronyms and names (Fig. 2.2) in published literature. The LTC strategies can be divided into two main categories, namely, homogeneous charge compression ignition (HCCI) and partially stratified charge compression ignition (SCCI) based on degree of premixing of fuel–air mixture. In the HCCI strategy, charge is well mixed before compression stroke and engine is operated on lean mixture. In the SCCI strategy, possibly two types of stratification occur, namely, thermal stratification and fuel concentration stratification in the cylinder. Thermal stratification is used in thermally stratified compression ignition (TSCI) and spark-assisted compression ignition (SACI) strategy. Fuel concentration stratification is possible by single as well as dual fuel. In the fuel stratification category, gasoline partially premixed combustion (PPC) and dual fuel reactivity-controlled compression ignition (RCCI) are the two main combustion concepts, which are widely investigated due to their potential of better combustion phasing control and lower emission along with higher thermal efficiency. In the last few years, PPC and RCCI strategies gained significant advancement such as multiple injection, dual direct injection fuel stratification (DDFS), etc. Other combustion strategies investigated with gasoline-like fuels and different injection strategies are published with different acronyms such as gasoline compression ignition (GCI), low temperature gasoline combustion (LTGC), gasoline direct-injection compression-ignition (GDCI) (see Fig. 2.2). Detailed discussion of all the mentioned LTC strategies is presented in the next sections of the present chapter.

2.2 Homogeneous Charge Compression Ignition

2.2.1 HCCI Fundamentals

Homogeneous charge compression ignition (HCCI) is one of the engine operating modes in LTC region (Fig. 2.1b). The HCCI mode means the compression ignition (auto-ignition/self-ignition) of homogeneous (premixed) fuel and air mixture in the engine cylinder. Hence, HCCI operating mode exhibits characteristics from both CI and SI engines. The HCCI operating mode is similar to conventional SI engine in charge preparation strategy (both modes use premixed charge) and similar to CI engines in combustion initiation process (in both modes combustion starts with autoignition). Even though HCCI operation mode called "homogeneous" uses premixed charge, in real engines, there are always thermal inhomogeneities (due to wall heat transfer and convection) and fuel concentration inhomogeneities (difficulties in preparing fully homogeneous charge) [21]. Figure 2.3b shows the HCCI combustion process vis-à-vis CI and SI combustion processes. In HCCI combustion, nearly homogeneous charge (fuel–air mixture) enters in the engine cylinder during intake stroke (Fig. 2.3a). During compression stroke pressure and temperature of charge increases, and auto-ignition reactions start when temperature reaches towards auto-ignition temperature (close to TDC). When auto-ignition temperature reaches, entire charge in cylinder reacts simultaneously. However, combustion process takes finite amount of time, even though entire cylinder charge is active. In-cylinder imaging using laser-induced fluorescence (LIF) of OH radical (for reaction activity) and planar laser-induced fluorescence (PLIF) of fuel showed that onset of HCCI combustion has very large local spatial and temporal variations. These variations are present even with premixed charge created using either external mixing tank or port fuel injection (PFI) system [22]. Generally, premixed charge created using PFI has higher inhomogeneity in comparison to premixed charge created by external mixing tank. Nonoccurrence of simultaneous combustion throughout the combustion chamber is also a result of thermal stratification caused by wall heat transfer and turbulent mixing during compression stroke, which is experimentally confirmed by chemiluminescence imaging [23, 24]. HCCI combustion occurs sequentially starting with hottest zone (also referred to as "hot spots" [25], see Fig. 2.3a) where most favourable conditions for auto-ignition exist, followed by next hottest zone [21]. Due to the start of exothermic reactions, charge temperature increases, and hence, oxidation reactions become even faster providing local positive feedback to the charge temperature [25]. Sufficient time is not available to distribute all the generated heats (in the hot spot) to surrounding cold bulk gases (due to fast temperature feedback), which leads to gradual amplification of small inhomogeneities in the cylinder [24, 25]. The amplification of small inhomogeneities is confirmed and observed as a large structure in experimental chemiluminescence images of HCCI combustion. Hot spot size in HCCI combustion is found to be of the same order as the turbulence integral length scale in combustion chamber [25]. The HCCI combustion changes the combustion

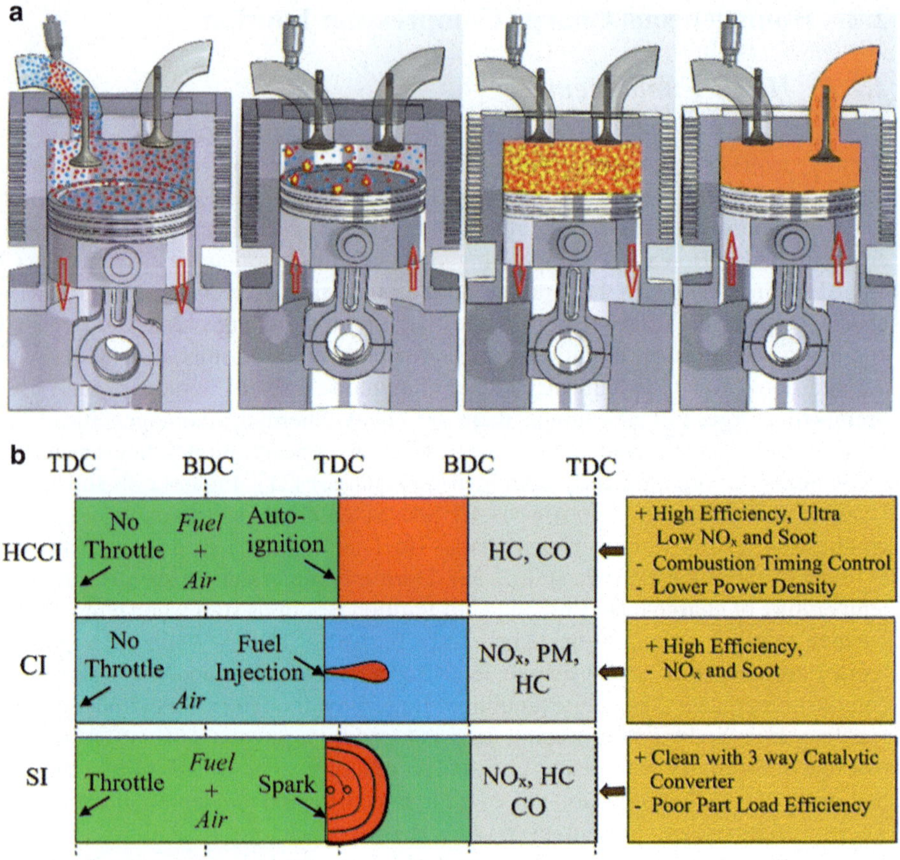

Fig. 2.3 (**a**) The HCCI combustion process. (**b**) Combustion process comparison of SI, CI and HCCI engines in four-stroke cycle

behaviour during the process, and Hultqvist et al. [24] observed three distinct combustion modes in the HCCI combustion process. First, the initial combustion process starts with evenly distributed number of regions in the combustion chamber, where small inhomogeneities in fuel or temperature are favourable to the auto-ignition reactions. Fuel is gradually consumed by local reactions in these regions without flame propagation because of no sharp borders are observed between burned and unburned gases, and the fuel concentration gradually decreases with little or no expansion of the reaction zone. Second, increased chemical activity is due to the start of exothermic reactions lead to positive feedback to the charge temperature. The initial ignition kernels grow larger as combustion progresses. Study also suggested that the number and locations of kernels are reliant on global parameters like distribution of air, fuel, residual gas and charge temperature. Third, the formation of new ignition kernels at locations that have become favourable because of the global temperature and pressure rise is due to combustion progress.

These new ignition kernels grow until they merge with other kernels in the combustion chamber.

Due to combustion progress in entire combustion chamber, combustion duration is comparatively shorter in HCCI engine, and combustion occurs close to TDC position. Work is obtained in expansion stroke, and cycle is completed by exhaust stroke similar to conventional engine by expelling the exhaust gases out of the cylinder (Fig. 2.3a). Comparison of HCCI, CI and SI combustion in a four-stroke cycle is presented in Fig. 2.3b. In HCCI operation mode, power output is controlled by varying the fuel flow rate, and engine is operated unthrottled similar to CI engines. The HCCI engine is operated on globally leaner mixture in order to control the combustion rate, which leads to lower combustion temperature. Premixed charge and lower combustion temperature in HCCI combustion process leads to ultralow NO_x and soot emissions. However, HCCI engines emit comparatively higher HC and CO emissions, which can be easily mitigated by oxidation catalyst. There is no direct control on ignition timing and combustion rate in HCCI combustion process similar to direct control of spark timing and injection timing in the conventional engine combustion processes. In HCCI engines, combustion timing and combustion rate are governed by chemical kinetics [26, 27]. The HCCI combustion process does not have either visible flame propagation (like SI engines) or highly stratified diffusion combustion (like diesel engines). Idea of flame propagation in HCCI combustion process is rejected by analysing the PLIF and chemiluminescence images as new "hot spots" generated randomly and the structures appeared in the images are not moving from image to image [23, 24]. A study summarized the two main combustion chamber condition requirements for the HCCI combustion process as follows: (i) auto-ignition temperature of premixed charge should be reached towards the end of compression stroke, and (ii) the fuel–air mixture should be diluted enough either by air or residuals to achieve reasonable HCCI combustion rate [25].

Historically, combustion process similar to HCCI combustion is known from last more than 100 years [28]. A hot-bulb two-stroke oil engine was designed where premixed vaporized fuel–air mixture was created in a heated prechamber by injecting fuel onto the surface of hot bulb. Fuel–air mixture created in prechamber was autoignited in the main chamber (cylinder) [28]. Another study achieved controlled autoignition in cylinder by injecting active species created from partially burned mixture in a prechamber [29]. In the early investigations of HCCI combustion process, the most recognized and first systematic studies were conducted on two-stroke engines [30, 31]. To overcome the problems of two-stroke engines (high residuals at low loads and tendency for run-on combustion), Onishi et al. developed a combustion process named "active thermo-atmospheric combustion (ATAC)" which depends on high residual levels and high initial charge temperature [30]. They found significant improvement in fuel economy and significantly lower emissions in ATAC mode. Concurrent with this study, Nouguchi et al. conducted spectroscopic analysis of combustion process in opposed piston two-stroke engine named Toyota-Soken (TS) [31]. They noted the overall fast combustion rate, excellent fuel economy and lower engine emissions. Study detected the high levels of intermediate species (CH_2O, HO_2 and O radicals) well before auto-ignition in the combustion chamber, which was the result of low temperature oxidation chemistry of larger paraffinic hydrocarbon fuels. After the start of combustion, high concentrations of CH, H and OH radicals are found, which indicates high temperature oxidation chemistry during bulk combustion. After pioneering the work of Onishi and Noguchi, several follow-up studies were conducted [32–34], and first production of two-stroke HCCI engine was developed with combustion name "active radical combustion (ARC)" [35, 36]. Based on previous investigations on two-stroke engine, Najt and Foster conducted first study on four-stroke engine to advance the understanding of underlying physics of HCCI combustion [37]. Another study further extended the HCCI combustion investigation on four-stroke engine, and first introduced the terminology "homogeneous charge compression ignition (HCCI)" [38]. Since the 1990s, the HCCI combustion had developed into a great research topic worldwide [39–41]. The HCCI combustion process is also known with another terminology called "controlled auto-ignition (CAI)" process [42]. The CAI process used variable valve timing (VVT) to trap large amount of residual gases in chamber to achieve auto-ignition of premixed gasoline in the cylinder. In CAI process, ignition timing is controlled by amount of residual gases trapped in the cylinder.

The HCCI engine operation has been investigated with significant success in two-stroke and four-stroke engines using liquid and gaseous fuels [40–43]. Both gasoline-like fuels and diesel-like fuels can achieve the HCCI combustion using appropriate strategies for charge preparation (see Chap. 4) and combustion phasing control strategy (see Chap. 5). Volatile fuels make premixed charge preparation strategy simple and easier (PFI, fumigation). Less volatile fuels require complex fuel injection strategies (direct injection) for charge preparation. To achieve auto-ignition in the combustion chamber, several actuation strategies such as variable compression ratio (VCR), intake heating, VVT, EGR, etc. are used, though each technique has its own merits and demerits [44]. Auto-ignition characteristics in HCCI combustion process also depend on fuel, and HCCI combustion characteristics are governed by fuel oxidation chemistry [41]. Auto-ignition of different fuels is achieved through few categories of reaction pathways characterized by temperature range over which certain reactions occur.

2.2.2 *HCCI Auto-Ignition and Heat Release*

Autoignition quality of the fuel is an important factor that influences all the current HCCI engine designs. Auto-ignition can be defined as the ignition of a fuel–air mixture because of heat produced from exothermic oxidation reactions without the involvement of external energy sources such as a spark or a flame [45, 46]. The HCCI engine operation has been achieved with fuels having a wide range of auto-ignition properties ranging from higher octane gasoline-like fuels to higher cetane diesel-like fuels. Several studies used primary reference fuels (PRFs) containing a

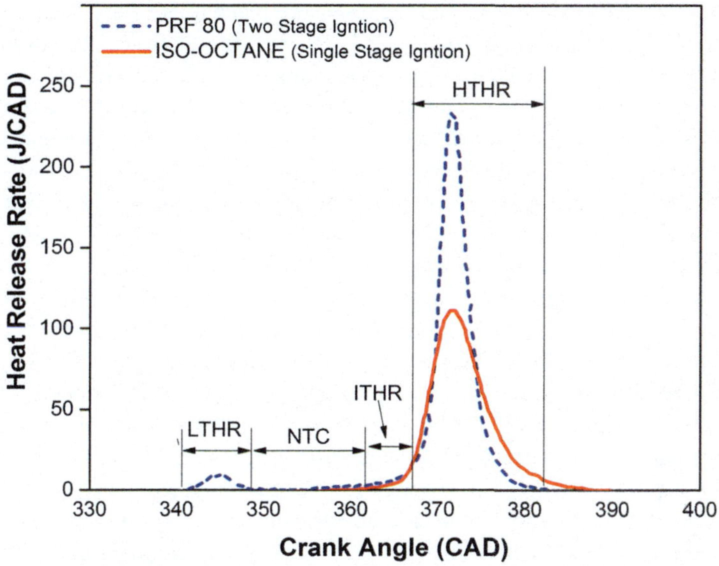

Fig. 2.4 Illustration of single- and two-stage heat release in HCCI combustion for two different fuels (Adapted from [50])

blend of n-heptane and iso-octane for investigation of fundamental HCCI combustion characteristics [47, 48]. Generally, iso-octane is used as a surrogate for gasoline and n-heptane is used as surrogate fuel for diesel. Auto-ignition characteristics of fuel or fuel constituent in HCCI engine operation can be typically categorized into single-stage ignition and two-stage ignition process [49]. Figure 2.4 demonstrates the two different types of ignition in HCCI combustion for two different auto-ignition characteristics of fuels. Two-stage ignition fuel has separate region with a small heat release in low temperature regime around 10–20 CAD before the main heat release in high temperature regime. The time delay between two stages of heat release is known as "negative temperature coefficient (NTC)" regime. However, single-stage ignition fuels do not have early heat release region. Typically, lower octane fuels (less auto-ignition resistant) such as diesel, n-heptane, PRF80 (80% iso-octane and 20% n-heptane) and DME exhibit two-stage ignition, while higher octane fuels (high auto-ignition resistant) such as ethanol, methanol, butanol, iso-octane and natural gas typically exhibit single-stage ignition [27, 39, 51]. It is important to note that the main autoignition process does not take place instantaneously for both types of fuels. In single-stage auto-ignition fuels, oxidation reactions start progressively once charge temperature reaches around 950–1050 K due to compression of the charge by piston [49, 50]. The charge temperature initially increases slowly because of intermediate temperature heat release (ITHR) reactions [49, 52], and after attaining the thermal runaway point, charge temperature increases quickly because of the high temperature heat release (HTHR) reactions [49]. In two-stage auto-ignition fuels, low temperature heat release

(LTHR) or "cool-flame" oxidation reactions take place at the start of ignition when temperature is in the range of 760–880 K before attaining the ITHR phase [53]. After LTHR, similar to single-stage heat release, ITHR takes place in two-stage ignition fuels also. The HCCI engine operation by two-stage ignition fuels needs lower charge preheating or lower compression ratios because LTHR raises the charge temperature in the cylinder [49].

Combustion/oxidation of hydrocarbon fuels occurs through a classical chain reaction mechanism consisting of several individual reactions with many intermediate species created during oxidation reaction, which converts fuel into combustion products. Individual reactions during combustion are also known as elementary reactions. Typically, four fundamental types of elementary reactions, namely, initiation, propagation, branching and termination reactions, occur during combustion. Elementary initiation reactions produce unstable radical species from the breakdown of stable species. The step in which free radical produces another radical by reaction is called chain propagation reaction. There is no change in the radical concentration in chain propagation reactions. Reactions producing more free radicals than consumption of radicals are known as chain branching reactions, and this type of reactions results in very high combustion rates. When entire reactants are consumed, free radicals combine with one another to produce stable species by terminating the reactions known as chain termination reactions [54].

During the compression stroke, temperature inside the cylinder increases gradually, and no significant reaction occurs until the charge temperatures reach about 550 K [51]. As piston further moves towards TDC (charge temperature further increases), first initiation reactions create small amount of radical species, which abstract hydrogen atom from fuel molecule at temperature below 850 K leading to production of alkyl radicals (R·) by reaction R1 [27]:

$$RH + OH\cdot \rightarrow R\cdot + H_2O \quad (R1)$$

where radical species are denoted as the "·" symbol next to character. Different molecular structures of paraffin lead to the different types of alky radicals by hydrogen abstraction. The alky radicals are consumed in two parallel paths shown by reactions R2 and R3 [55]. One pathway is to produce a conjugate olefin and a hydroperoxy radical (from reaction R2), and an alternative other pathway is to form alkylperoxy radical ($RO_2\cdot$) by reaction R3.

$$R\cdot + O_2 \rightarrow \text{olefin} + HO_2\cdot \quad (R2)$$
$$R\cdot + O_2 \leftrightarrow RO_2\cdot \quad (R3)$$

The reaction R3 is a reversible reaction, and very important reaction for the HCCI auto-ignition process. Reaction R3 has different reaction rates constant in reverse and forward directions. Reverse reaction requires high activation energy because the oxygen–alkyl radical bond must be broken to proceed the reaction. However, forward reaction needs lower activation energy to proceed the reaction, and hence its rate is affected only by the concentration of alkyl radical and oxygen [51]. Equilibrium of the reversible reaction R3 depends on both the temperature and pressure of cylinder. At higher temperature, equilibrium shifts towards reactants side because reverse reaction is faster than forward reaction. In this condition, concentration of alkylperoxy radical decreases drastically, and low temperature chain branching reactions are quenched leading to very low combustion rate [51]. This phenomenon is responsible for NTC regime in HCCI combustion using two stage heat release fuel.

The alkylperoxy radical undergoes internal isomerization by the oxygen abstracting a hydrogen atom from a C–H bond elsewhere within the oxygen molecule to form a hydroperoxyalkyl radical (·QOOH) [55]:

$$RO_2\cdot \rightarrow \cdot QOOH \quad (R4)$$

The succeeding reactions of the ·QOOH depend on the structure of the ·QOOH, and several possibilities exit for the next reaction. The ·QOOH radical is consumed through two basic routes [55]. In the first route, an OH· radical is produced when ·QOOH radical either spontaneously decomposes into a lower-molecular-weight alkene or forms a cyclic ether and OH· radical (R5):

$$\cdot QOOH \rightarrow CARBONYL + R'\cdot + OH\cdot \quad (R5)$$

In the other route, the ·QOOH radical undergoes a second oxygen addition to form a hydroperoxyalkylperoxy radical (·OOQOOH), followed by an internal hydrogen atom abstraction to produce an alkyl hydroperoxide which decomposes into an aldehyde and two OH· radicals (R6–R7):

$$\cdot QOOH + O_2 \leftrightarrow \cdot O_2QOOH \quad (R6)$$
$$\cdot O_2QOOH \rightarrow HOOROOH \quad (R7a)$$
$$HOOROOH \rightarrow RCHO + R'O + OH\cdot + OH\cdot \quad (R7b)$$

It can be noticed that reaction R6 is also a reversible reaction and has similar characteristics of reaction R3. Equilibrium of these reactions depends on pressure and temperature of the combustion chamber, and when charge temperature reaches about 800–850 K, the net addition reactions of oxygen molecule to radicals completely stop. Silke et al. also confirmed that reactions R3 and R6 had the largest effect on the required BDC temperature for constant combustion phasing at different boost pressure conditions because LTHR affects the requirement of inlet temperature for the auto-ignition of fuels in HCCI engines [56]. Reactions R1–R7 define the low temperature oxidation mechanism of a large paraffinic fuel. Figure 2.5 summarized the reaction pathway at different temperature ranges for oxidation of n-heptane [57, 58].

Due to further increase in compression temperature (above 850 K) intermediate temperature regime reaction R2 produces more olefinic hydrocarbons and hydroperoxy radicals ($HO_2\cdot$). Hydrogen peroxide (H_2O_2) formation becomes significant by $HO_2\cdot$ radicals by abstracting hydrogen atom from fuel (by reaction R8), and concentration of H_2O_2 increases gradually until charge temperature reaches to 1000 K [51]:

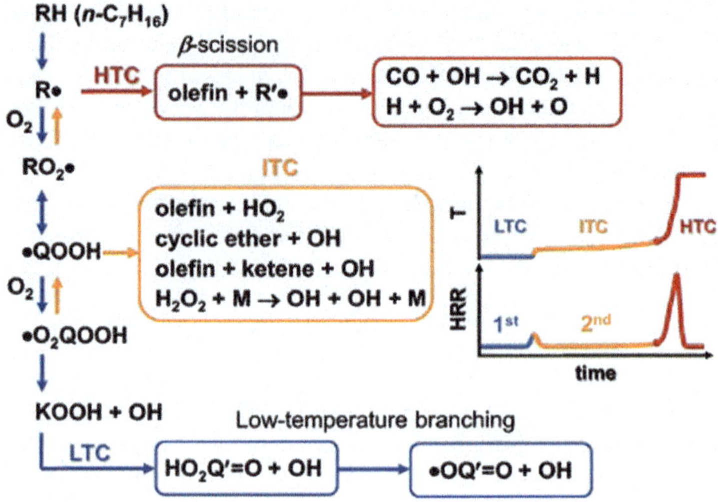

Fig. 2.5 Reaction pathways of n-heptane oxidation at various temperature ranges [57, 58]

$$RH + HO_2\cdot \leftrightarrow R\cdot + H_2O_2 \quad (R8)$$

Further increase in the charge temperature (above 1000 K) leads to the decomposition of H_2O_2 into hydroxyl radical by reaction R9, and this process quickly releases amount of hydroxyl radical resulting into higher overall reaction rate and heat release:

$$H_2O_2 + M \rightarrow OH\cdot + OH\cdot + M \quad (R9)$$

The decomposition of H_2O_2 into OH radicals and their consequent reactions leads to the increase in charge temperature, which further accelerates the decomposition of H_2O_2. This phenomenon leads to an auto-ignition process, which increases the temperature above 1200 K in the high temperature reaction regime [27]. Figure 2.6 presents the experimental and simulated distribution of H_2O_2 and HO_2 in the cylinder with respect to crank angle (CA) position in HCCI engines [59]. It is observed that concentration of H_2O_2 is higher during intermediate heat release stage. Another consequence of the hydrogen peroxide branching reaction is that the intermediate and hot ignition temperatures are comparatively independent of the fuel characteristics [27].

The high temperature oxidation of hydrocarbons can be described in sequential three steps [55] shown in reaction R10. First parent fuel is converted into lower hydrocarbon and then, intermediate species further converted to produce CO and water. Lastly, CO is oxidized to CO_2 releasing large fraction of heat energy:

$$RH + OH\cdot \rightarrow R'\cdot R'\cdot + H_2O \rightarrow CO + H_2O \rightarrow CO_2 + \text{heat} \quad (R10)$$

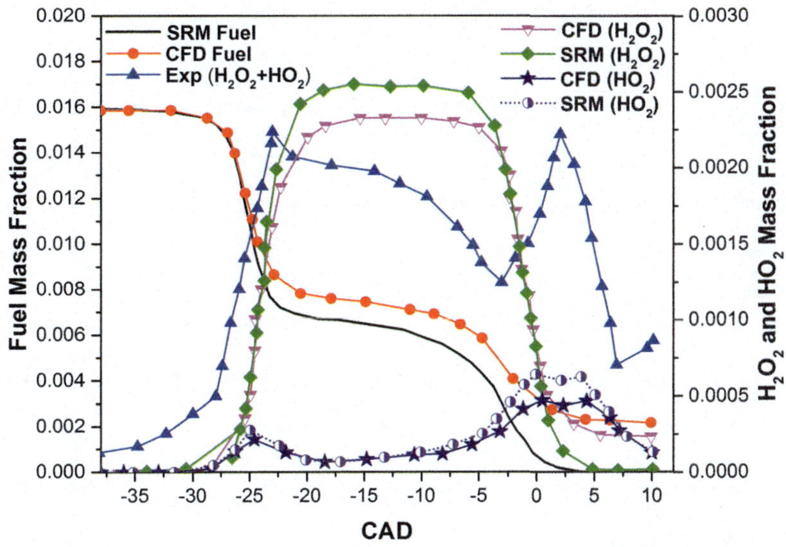

Fig. 2.6 Variation of measured and simulated H_2O_2 and HO_2 with crank shaft position in HCCI engine [59]

The important chain carrying radicals in high temperature regime are OH, H and O [55]. These radicals are very reactive, and hence, hydrocarbon oxidation rates are extremely fast. The important chain branching reactions in the high temperature regime are R11, R12 and R13:

$$H\cdot + O_2 \rightarrow OH \cdot O\cdot \quad (R11)$$
$$O\cdot + H_2 \rightarrow OH\cdot + H\cdot \quad (R12)$$
$$H_2O + O\cdot \rightarrow OH\cdot + OH\cdot \quad (R13)$$

The final stage of the energy release at the high temperature is the oxidation of CO, and principal reactions involved in CO oxidation are R14, R15 and R16 [55]:

$$CO + OH\cdot \rightarrow CO_2 + H\cdot \quad (R14)$$
$$CO + HO_2\cdot \rightarrow CO_2 + OH\cdot \quad (R15)$$
$$CO + O\cdot + M \rightarrow CO_2 + M \quad (R16)$$

At high temperature, OH· (R14) mainly consumes the entire CO and other reactions have only small contribution in CO oxidation.

Different chemical kinetic mechanisms are used to analyse combustion characteristics and auto-ignition timings for various fuels. Chemical kinetic mechanisms can be grouped into four general classes: (i) single-step mechanism (no elementary reactions), (ii) generalized mechanisms (no track of individual reactive species), (iii) reduced mechanisms and (iv) detailed mechanisms. First two types of mechanisms are empirical approaches. Reduced mechanisms are derived from detailed mechanisms by conducting the sensitivity analysis. Sensitivity analysis determines

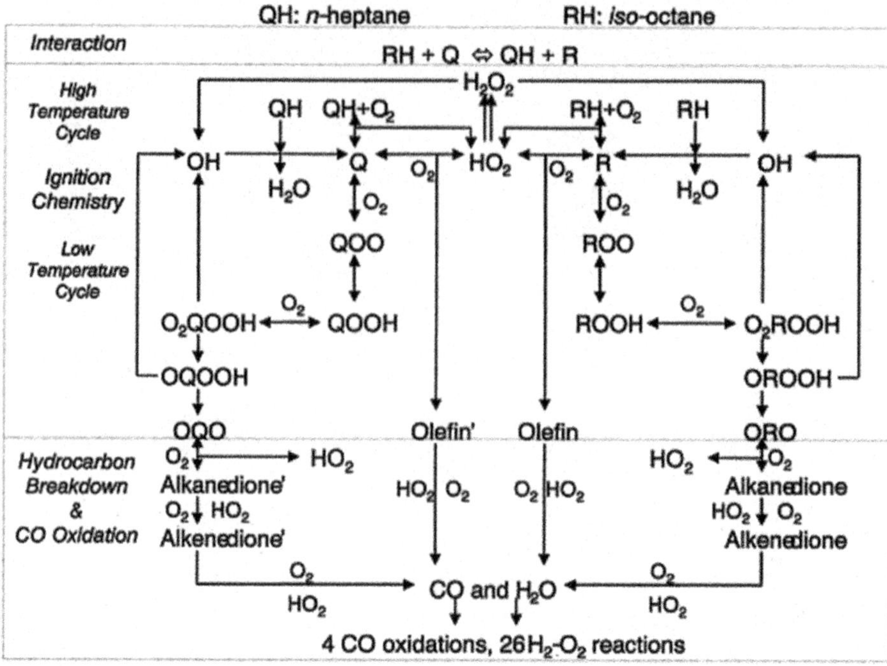

Fig. 2.7 Reduced oxidation mechanism of n-heptane and iso-octane mixtures in HCCI combustion [48]

the most important reactions for a given process. Detailed mechanisms are the most complex reaction mechanism. Detailed oxidation mechanism attempt to incorporate all the chemical details consisting hundreds of species, well over a thousands of elementary reactions [55]. To understand the heat release process or predict the emissions from HCCI combustion, detailed mechanism or appropriately reduced mechanism is essential. Mostly surrogate fuels are used for chemical kinetic analysis consisting of n-heptane and/or iso-octane. Figure 2.7 presents a proposed reduced oxidation reaction mechanism for mixtures of n-heptane, and iso-octane in HCCI combustion consists of 32 species and 55 reactions [48]. The reaction mechanisms for both paraffins are quite similar. The figure shows that the HCCI combustion occurs, when H_2O_2 starts to decompose at a significant rate (which triggers ignition) and produce OH radicals.

The high octane industrial gasoline exhibits different characteristics than PRFs in HCCI ignition process. An industry average gasoline consists of roughly 60% paraffins, 10% olefins and 30% aromatics [60]. Aromatic reactions play an important role in gasoline auto-ignition process. Figure 2.8a depicts the heat release curve for gasoline at three different equivalence ratios in HCCI engine operated at an inlet temperature of 70 °C and compression ratio of 13.5 [61]. Figure depicts three different stages of heat release: (i) weak cool flame heat release (denoted as C),

(ii) in-between pre-ignition (denoted as P) and (iii) final ignition (denoted as F). To explain this phenomenon, chemical kinetic simulation is conducted using gasoline 95 (gasoline octane number 95), PRF95 (contains 95% iso-octane and 5% n-heptane) and a surrogate fuel containing 11 vol% n-heptane, 59 vol% iso-octane and 30 vol% toluene. It is found that gasoline 95 and surrogate fuel show similar three-stage heat release which is not there in PRF95 [61]. A phenomenon called "obstructed pre-ignition (OPI)" is defined, which is responsible for delay in the final ignition of gasoline and surrogate. Typically, HTHR occurs due to decomposition of H_2O_2 leading to formation of large amount of OH radicals. This process is obstructed by consuming the OH radicals in formation of relatively stable benzyl radical by toluene present in the fuel. The benzyl radical formation competes with the reaction of OH radicals with the other alkanes and olefins. Therefore, benzyl radical acts as a sink for the OH radicals formed in the cylinder, which leads to reduction in the combustion rate that delayed the final heat release [61]. Figure 2.8b summarized the autoignition process of surrogate fuel consisting of n-heptane, iso-octane and toluene. The figure depicts the influence of the obstructed pre-ignition phenomenon with the most important reactions and species (shown within a dashed square). The obstructed pre-ignition heat release occurs between the cool flame and final ignition stages in gasoline as well as three-component surrogate fuel of gasoline (containing toluene). This type of heat release is not observed in the fuel PRF95 (containing n-heptane and iso-octane) having same

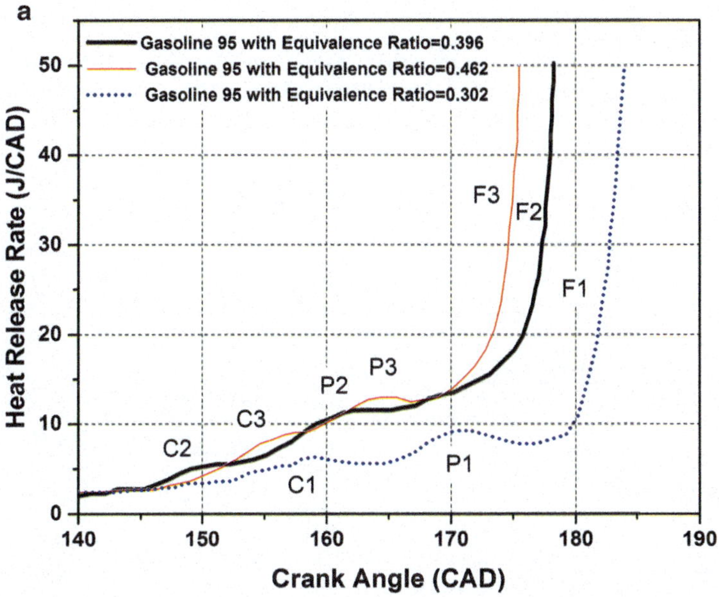

Fig. 2.8 (**a**) Comparison of the heat release rates at three equivalence ratios for gasoline 95 in HCCI engine [61]. (**b**) Summary of the reaction pathway showing the interaction of the "obstructed pre-ignition" phenomenon in HCCI combustion [61]

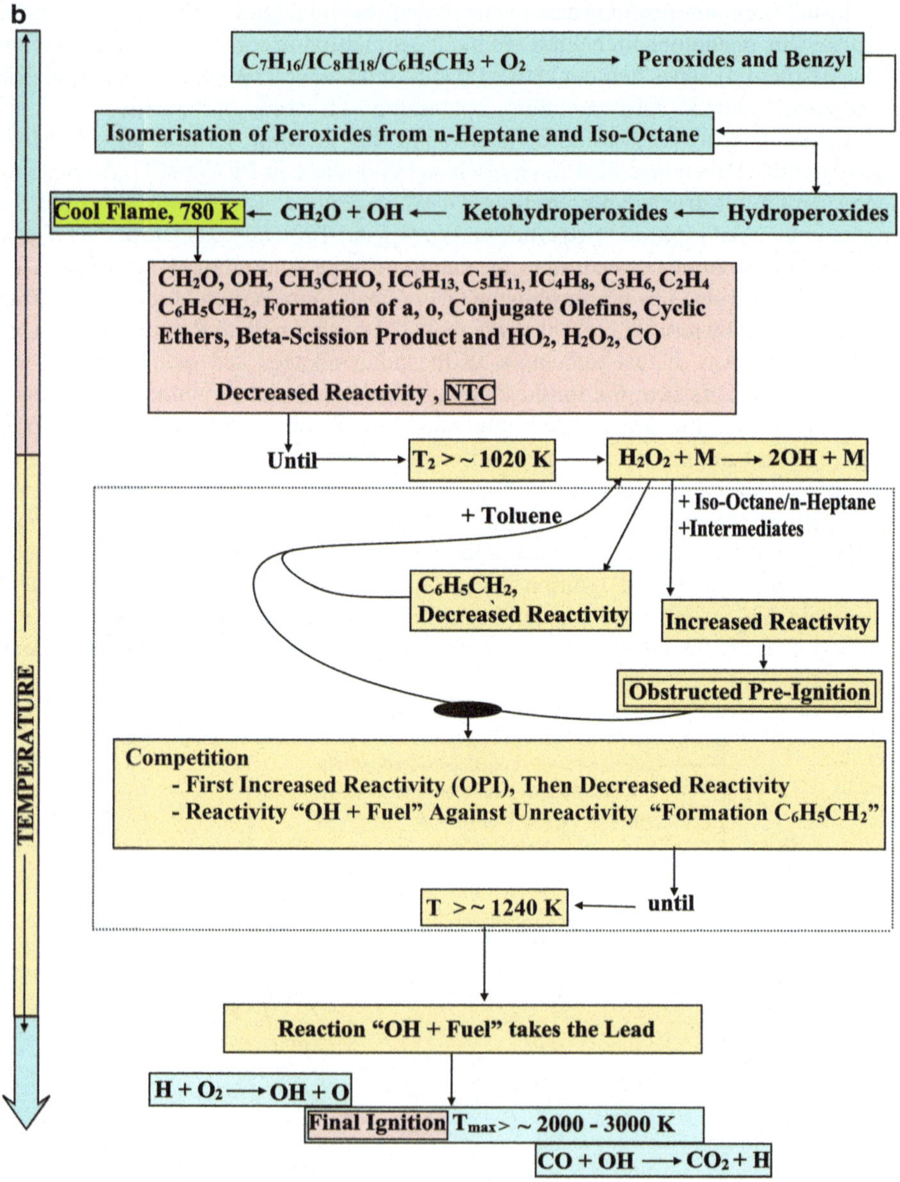

Fig. 2.8 (continued)

octane number. This observation indicates that toluene has an important role in the chemical kinetics of the gasoline autoignition.

Understanding of ignition characteristics and heat release process is important for effective control and operation of HCCI engine. The ignition characteristics can also be altered by changing the fuel composition. For example, the ITHR is induced by adding small amount of n-heptane in ethanol mixtures [62]. The ITHR can play a key role in sustaining high load HCCI engine operation at retarded combustion phasing [62, 63]. To achieve higher engine load in HCCI engine, a large amount of fuel needs to be burned in cylinder, which results into very high pressure rise rate and excessive noise. To avoid the very high combustion rate, combustion phasing needs to be delayed after TDC. The cylinder volume expansion after TDC tries to cool down the charge before HTHR. This charge cooling can be avoided by ITHR, which can allow charge temperature to keep increasing shortly after the TDC. Hence, HCCI combustion can be achieved at delayed combustion phasing for fuels with ITHR. A study investigated the ITHR using ethanol and n-heptane mixture in HCCI combustion, depicts that ITHR can be in the range of 2–5% of total heat release. Figure 2.9 shows the variations of ITHR with the n-heptane percentage in fuel–air mixture for different boost pressures in HCCI engine [62]. The ITHR is found minimal at pure ethanol case, and moderate (2–3%) up to 35% n-heptane in the mixture. When n-heptane percentage increases more than 35, ITHR rapidly increases to 5% of total heat release.

Reactions responsible for ITHR are investigated by chemical kinetic analysis. Figure 2.10 exhibits the contribution of different reaction groups in heat release rate with respect to crank shaft position for different compositions of n-heptane at intake pressure of 2.2 bar and $\varphi = 0.4$ [62]. The reactions are grouped according to the number of carbon atoms in the oxidized species, namely, C1–C2, c1–c2, C3–C4 and C7. Again, "C1–C2" denotes species with carbon derived from n-heptane, and "c1–c2" is for species with carbon derived from ethanol. Another study also suggested that

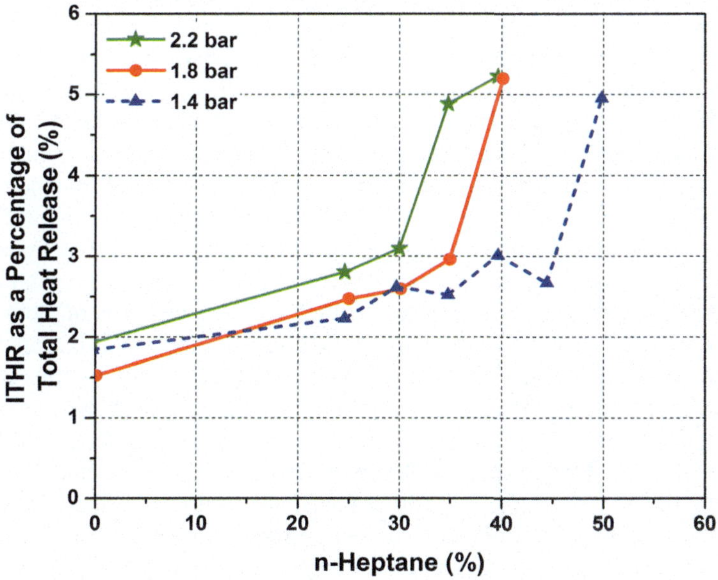

Fig. 2.9 ITHR as a function of n-heptane percentage in fuel and intake pressures ($\varphi = 0.4$ and $CA_{50} = 10$ °aTDC) in HCCI combustion [62]

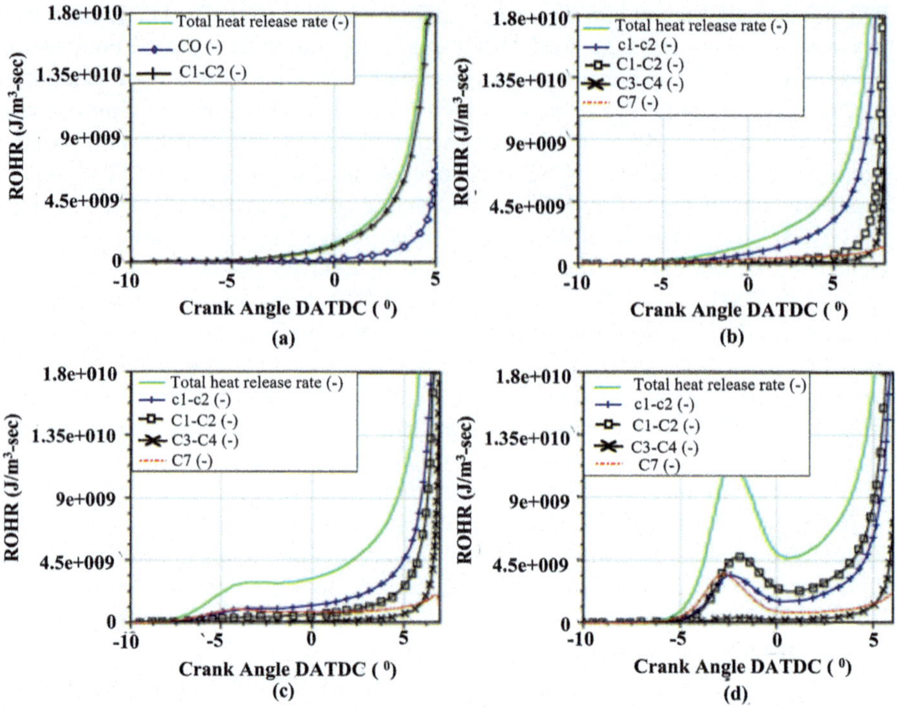

Fig. 2.10 Variation of heat release rates in HCCI combustion for (**a**) 100% ethanol, (**b**) 25% n-heptane, (**c**) 45% n-heptane and (**d**) 55% n-heptane from various reaction groups [62]

the number of carbons in the hydrocarbon species undergoing partial oxidation is correlated with primary heat-releasing reactions. Reactions consisting of small molecules (zero, one and two carbon atoms) produce heat at different crank angle positions and magnitudes, than species with three and four carbon atoms, and they also show different trends than larger hydrocarbons (five or more carbon atoms) [49].

In case of pure ethanol, very low ITHR observed (Fig. 2.10a) and ethanol reactions c1–c2 are the largest single groups of heat-producing reactions in 25% n-heptane case (Fig. 2.10b). At the 45% n-heptane, a more prominent ITHR characteristic is noticed, which is correlated with an increase of both heptane oxidation reactions and mostly ethanol oxidation reactions. A shift from ITHR to LTHR characteristics is noticed at 55% the n-heptane fraction, and this characteristic corresponds to an increase in importance of C1–C2 reactions (derived from n-heptane). However, heat release in ethanol reaction groups also increases. It is concluded that the blending of the highly reactive fuel increases the ITHR and at very high blend percentage, LTHR can also be triggered. The ITHR can be generated by adding small amount of reactive fuel, in the parent fuel, which does not exhibit ITHR characteristics in HCCI combustion [62].

2.2.3 HCCI Advantages and Challenges

The HCCI combustion process attracted the attention of researchers from the last few decades due to their benefits of higher thermal efficiency along with simultaneous reduction of NO_x and PM to a ultralow levels. The level of benefits gained from HCCI combustion process depends on the fuel and strategy used to achieve the auto-ignition in the engine cylinder. Figure 2.11 presents the main advantages and challenges of HCCI combustion process in IC engines. One of the main strengths and benefits of HCCI engine is the higher thermal efficiency. The main factors affecting the thermal efficiency of engine are compression ratio, pumping loss, specific heat ratio, heat transfer from combustion chamber, gas molecular dissociation, combustion phasing and mechanical friction [64]. The first three factors (higher compression ratio, lower pumping loss, higher specific heat ratio) leading to higher thermal efficiency are common to HCCI and conventional CI engines. The conventional CI engines have higher thermal efficiency because of engine operation at higher compression ratio, lower pumping loss (due to unthrottled operation) and higher ratio of specific heat (due to lean engine operation). The HCCI engines are operated at lower combustion temperature leading to lower heat transfer loss and less dissociation of gas molecules. Additionally, premixed charge ignition leads to

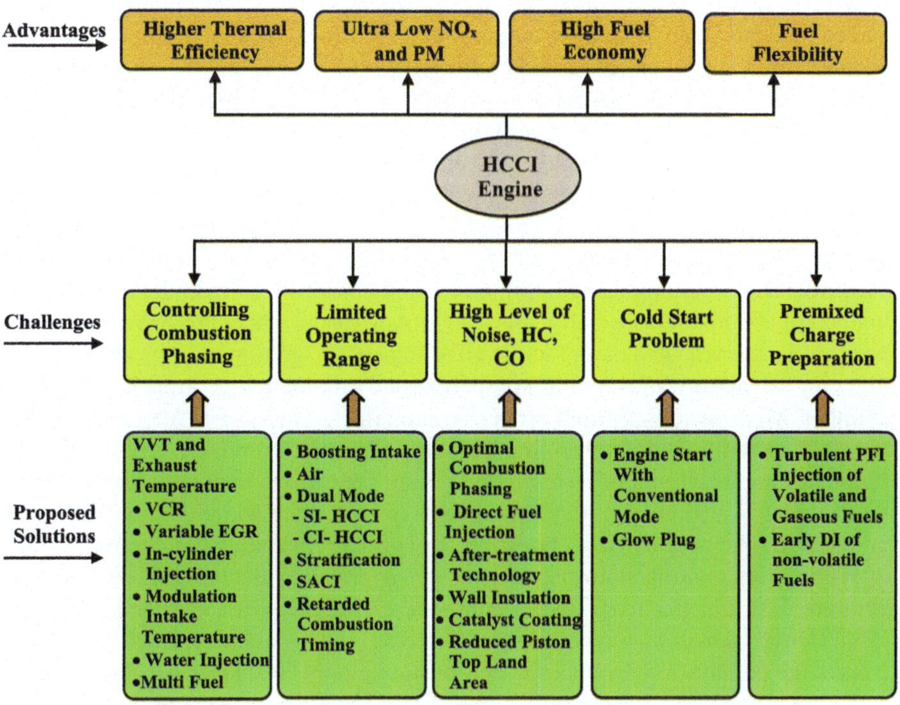

Fig. 2.11 The HCCI engine advantages, major challenges and their proposed solutions

sootless (smokeless) combustion, which also reduces the radiation heat loss in HCCI engines. Combustion phasing dictates the thermal efficiency by its effect on combustion timing loss and heat transfer [64]. The HCCI combustion has smaller combustion duration because of simultaneous ignition of entire charge in the cylinder, which leads to near constant volume combustion cycle. Constant volume combustion in HCCI engine also contributes to higher thermal efficiency. Mechanical friction leads to reduction in brake thermal efficiency of engine. Mechanical friction mainly depends on the peak cylinder pressure and engine speeds. Generally, advantage of lower mechanical friction depends on compression ratio of engine used for HCCI combustion process.

Another major strength of HCCI combustion engine is simultaneous reduction of NO_x and PM emission to ultralow levels due to lower combustion temperature (lower NO_x formation reactions) and premixed charge (lower soot formation due to absence of local rich zone in charge) operation in LTC region (Fig. 2.1). The HCCI combustion process has fuel flexibility to operate on both gasoline-like and diesel-like fuels. To operate on particular fuel, suitable auto-ignition control strategy needs to be used. In advanced compression ignition engines such as HCCI and extended HCCI, required fuel reactivity is converging to in-between reactivity of gasoline and diesel (see Fig. 1.13). Higher fuel economy is achieved in HCCI engines due to higher thermal efficiency. The HCCI engines have higher part load efficiency and limited operating range. This characteristic makes HCCI engine well suited for hybrid electric vehicles (HEVs), where lower-power engine is required and engine can avoid the transient operation. Therefore, HCCI can offer higher fuel economy and is a potential candidate for HEVs [65].

Although HCCI combustion has advantages over conventional engines, it also has several technical challenges (Fig. 2.11) that need to be addressed before utilizing it for production engines. One of the main challenges in HCCI combustion is to control the combustion phasing, which governs the combustion rate and peak pressure in the cylinder. As discussed (Sect. 2.2.2), HCCI combustion timing is governed by chemical kinetics, and there is no direct control over the combustion phasing. Hence, simultaneous ignition of entire charge in the cylinder leads to very high combustion rate especially at higher engine loads, where large amount of fuel needs to be burned. The main factors affecting HCCI ignition timings and combustion rate are mixture auto-ignition properties and mixture temperature history in the cylinder. Mixture auto-ignition properties are affected by type of fuel, fuel concentration, fuel additives and reforming, blending of two or more fuels, residual (and/or EGR) rate and reactivity of residuals [55]. Similarly, mixture temperature history in the cylinder is affected by intake conditions (temperature and pressure), EGR and/or residual, thermal barriers, boosting, in-cylinder fuel injection timings, water injection, compression ratio, latent heat of vaporization of the fuel, coolant temperature, heat transfer from engine components, engine speed and load conditions [55]. Hence, actuation methods that affect the stated parameters can be used to control the combustion timings and combustion rate in HCCI engine. The proposed actuation strategies for controlling the HCCI combustion phasing are VVT and residual/exhaust gas trapping, VCR, variable EGR, modulation of intake temperature, water injection, in-cylinder injection timings, variable coolant temperature and modulating two or more fuels [66]. Although there exists the sufficient number of proposed strategies to control the combustion phasing in HCCI combustion, every strategy has its own operating limits due to thier response time, on-board control on engine, etc.

The other major challenge is to extend the limited HCCI operating range, while maintaining the full advantages of HCCI combustion. To achieve higher engine load operation, comparatively larger amount of fuel needs to be burned at appropriate combustion timings. Due to oxidation reactions in entire cylinder simultaneously, very high pressure rise rate is observed at higher engine load, which also leads to higher pressure oscillations, combustion noise and higher NO_x emissions. High pressure rise rate limits the power output and defines high load boundary (upper operating limit) of HCCI engine operation. Additionally, lower load operation is also limited, which defines lower operating limit. At low engine load operation, insufficient temperature is achieved to activate auto-ignition of the mixture, which results into emissions of higher CO and HC [40]. Parameters affecting high load limit are systematically investigated. Several strategies are proposed to extend the upper limit of HCCI engine operation. Fundamental parameters affecting high load HCCI operation are fuel properties and chemical kinetics (LTHR, ITHR and HTHR), in-cylinder charge conditions (residuals, charge and thermal stratification), combustion timing and heat transfer [27]. Solutions proposed to extend HCCI operating range are boosting intake air (turbocharging/supercharging), engine operation in dual mode (SI–HCCI or CI–HCCI), spark-assisted HCCI, charge stratification and late combustion phasing engine operation.

Another major issue with HCCI engines is the higher CO and unburned HC emissions especially at lower and moderate engine loads. Higher unburned HC and CO emissions mainly occur because of lower combustion temperature due to lean–burn or highly diluted charge combustion. The fuel conversion efficiency is also hampered by higher HC and CO emissions at low engine load conditions. The CO-to-CO_2 conversion reactions are sensitive to temperature and oxygen availability in the cylinder (Fig. 2.1), and high temperature is required to oxidize CO in to CO_2 (Sect. 2.2.2). Exhaust after-treatment by oxidation catalyst is one of the best proposed methods for reduction of HC and CO from HCCI engines. However, oxidation catalyst with low catalyst light-off temperature is required for HCCI engines due to lower exhaust gas temperatures particularly at lower engine loads. There exist methods to reduce catalyst light-off time, which can be used to achieve the required operating temperature of catalytic converter [67–69]. Other strategies proposed for reduction of HC emissions are wall insulation, catalyst coating and reduced piston topland crevices [55].

In HCCI engine, the problem of weak cold start capability exists because auto-ignition process is very sensitive to intake charge temperature. Initial charge temperature needed for auto-ignition is dependent on fuel properties and engine operation conditions. At cold start conditions, charge temperature is very low, and heat transfer loss from compressed charge is also higher due to colder cylinder walls [40]. Without any compensating mechanism for low lower charge temperature, initiation of combustion is very difficult, and it can lead to misfire. Several strategies are proposed to address the cold starting in HCCI engines such as using glow plugs, increasing the compression ratio using VCR or VVT and using a different fuel or fuel additive. The most practical approach to address the cold start in HCCI engine is to start the engine in conventional SI or CI mode, and switch to HCCI mode after warm-up [40, 66]. Effective premixed charge preparation and avoiding fuel–wall interactions are essential in HCCI engine for higher fuel conversion efficiency and lower emissions. This issue typically comes up for poor volatility diesel-like fuels in HCCI engine because charge inhomogeneity affects the auto-ignition process and governs the ignition timing in HCCI engine. To prepare sufficiently premixed charge in HCCI engine, port injection of volatile and gaseous fuel, and early in-cylinder injection with sophisticated fuel injector for less volatile diesel-like fuels are proposed [66].

2.2.4 Parameters Influencing HCCI Combustion

2.2.4.1 Equivalence Ratio

The amount of fuel burned in the cylinder mainly controls the power obtained from HCCI engine. In HCCI engine, throttle is not used to control the inducted air quantity; therefore changing the quantity of fuel at different engine loads leads to the change in equivalence ratio. To represent the quality of fuel–air mixture with respect to stoichiometric air–fuel ratio, two metrics, namely, equivalence ratio (φ) and relative air–fuel ratio (λ), are typically used. These two parameters are inversely correlated ($\varphi = 1/\lambda$) with each other. Figure 2.12 presents the effect of λ on in-cylinder pressure and heat release rate (HRR) in HCCI combustion at constant intake air temperature. Figure shows that maximum cylinder pressure (P_{max}) and maximum HRR (HRR_{max}) increase as mixture becomes richer (lower λ values). Maximum pressure rise rate (PRR_{max}) is also very high for richer mixture operation of HCCI combustion because higher quantity of premixed fuel is burned simultaneously in the cylinder. Combustion duration also decreases with richer fuel–air mixture, and lowest combustion duration (<10 CAD) observed for richest mixture ($\lambda = 2.0$). Combustion phasing becomes very late for leaner mixture, which leads to lower HRR. Chances of misfire increases for very lean conditions in HCCI combustion. The combustion phasing advances with richer fuel–air mixture, and increasing λ from 2.0 to 2.6 (towards leaner mixture) leads to the decrease in HRR_{max} by an order of magnitude (Fig. 2.12).

The main factors affected by a change in equivalence ratio (quantity of fuel per cycle) are the temperature and composition of the residuals, wall temperatures and related heat transfer rates, heating/cooling during induction, burn duration and combination of chemical–kinetic and thermodynamic properties of the charge mixture [71]. During HCCI combustion operation, increase in fuelling rate (or φ) also leads to increase in the residual and cylinder wall temperatures. To understand

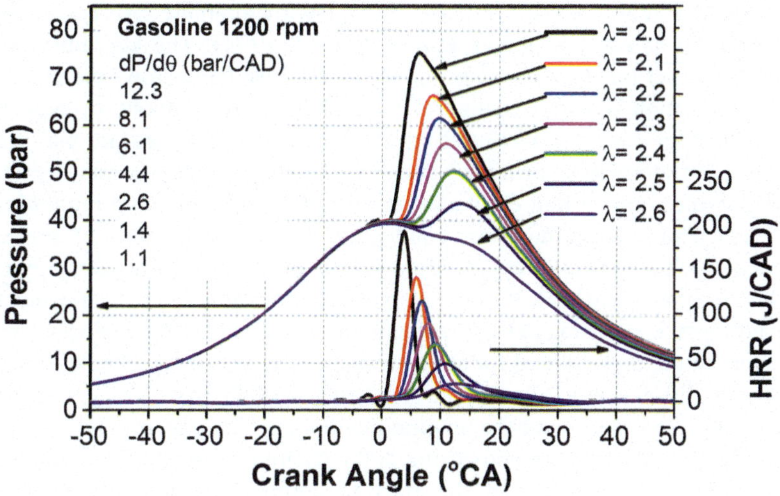

Fig. 2.12 Variations of in-cylinder pressure and HRR at constant intake air temperature using gasoline HCCI combustion (Adapted from [70])

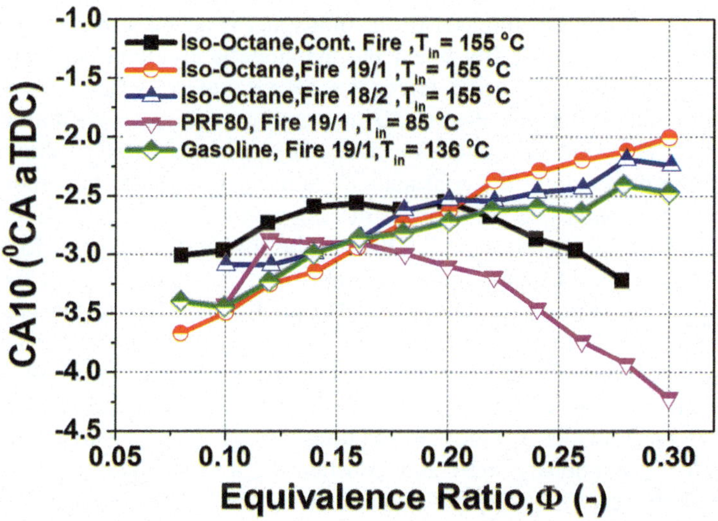

Fig. 2.13 Variations in CA_{10} for iso-octane with fuelling for the continuously fired, fire 19/1 and fire 18/2 schemes with a constant $T_{in} = 155$ °C, and for gasoline and PRF80 with fuelling fire 19/1 scheme (Adapted from [71])

the individual effect of these parameters, studies used (i) alternate firing operation to remove changes in wall temperature and residuals [71–73] and (ii) a method for determining the effective intake temperature to remove the effect of heating/cooling during induction of charge [71]. Figure 2.13 presents variations in crank angle position for 10% heat release (CA_{10}) with alternate firing schemes for different fuels at constant intake temperature (T_{in}) [71]. The key advantages of conducting experiments at constant T_{in} are as follows: (i) it shows the response of HCCI combustion to rapid change in equivalence ratio without T_{in} adjustment; (ii) heat transfer during induction does not vary with φ for constant wall temperature case. In alternative firing schemes (19/1 or 18/2), engine was operated on 19 or 18 cycles on constant equivalence ratio ($\varphi = 0.2$), and then remaining 1 or 2 cycles (where measurements are conducted) at equivalence ratio of interest and sequence repeats after 20th cycle. More details of alternate firing strategy can be found in original study [71]. Variations in wall temperature (T_{wall}) and residuals are removed together by using the 19/1 scheme. In this scheme, 20th cycle has the same wall temperature and residual for all the tested φ. In 18/2 firing scheme (two cycles are fired at the desired φ and data is taken for 20th cycle), T_{wall} can be kept approximately constant, and the acquired cycle has residuals of 19th cycle which is typically present in continuous firing at the desired φ for which measurement is taken. Figure 2.13 shows that the start of combustion (CA_{10}) for iso-octane becomes more retarded almost linearly as φ is increased at constant T_{in}, when the effect of variables T_{wall} and residuals are removed using the 19/1 firing scheme. It means the higher requirement of T_{in} with increase in φ for constant CA_{10} phasing using iso-octane fuel because of lower combustion temperature due to the reduction in ratio of specific heat (γ) as more fuel is added at higher φ. Gasoline data (19/1 firing) is also fairly linear but the slope is less than iso-octane as γ is similar. For PRF80 in 19/1 firing, effect of γ is similar to iso-octane, however strong cool flame kinetics advances the CA_{10} phasing with φ, particularly $\varphi>0.16$ (Fig. 2.13). The 18/2 firing scheme data (for iso-octane) trend follows closely 19/1 scheme data than to the continuously fired data. It means that under normal continuous fire conditions, the changes in T_{wall} with φ have a larger effect on CA_{10} than the effects of residuals [71]. This can be justified for typical low residual engines, as generally residual mass fraction is low (5.5% for the test case).

The CA_{10} phasing is a good measure of start of combustion and hence useful in isolating the effect of fuel chemistry (auto-ignition chemistry and thermodynamic properties) on ignition. However, the effect of changing equivalence ratio on CA_{50} (crank angle position corresponding to 50% heat release) phasing is important from an engine performance perspective. Figure 2.14 shows the variations in CA_{50} phasing for continuously fired and constant T_{wall} and residual (19/1 scheme) using three different fuels in HCCI combustion. The 19/1 scheme data at constant T_{in} presents the immediate response of combustion phasing to a rapid change in φ. This observation is useful for interpretation of fast transient operation where T_{in} and T_{wall} not necessarily vary at that pace. The shift in CA_{50} phasing with φ can be taken as a measure of the amount of compensation immediately required for constant phasing operation of engine with a change in load/fuelling. For each fuel, combustion phasing advances with increase in φ with different sensitivities for each fuel.

Figure 2.14 presents the magnitude of the effects of fuel type on CA_{50} (or the required compensation) during a fuelling transient. It was found that iso-octane needs essentially no immediate compensation to maintain the CA_{50} at TDC, and

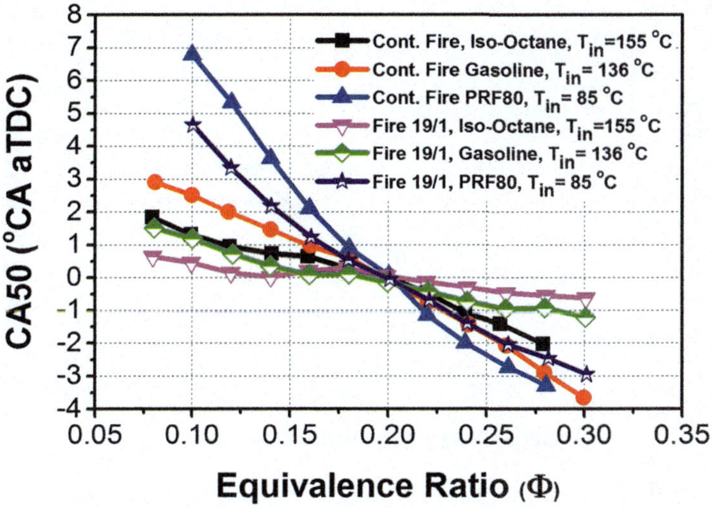

Fig. 2.14 Variations in CA_{50} phasing for continuous and 19/1 firing scheme for iso-octane, gasoline and PRF80 (Adapted from [71])

only a relatively small amount of compensation after T_{wall} equilibrates. Gasoline also shows similar trend with little more variation in CA_{50} with φ than iso-octane as both fuels do not show cool flame chemistry [71]. This observation indicates that controlling combustion phasing during fuelling transients should be relatively easy and straightforward for iso-octane or gasoline. The PRF80 data exhibits substantial changes in the CA_{50} with φ (Fig. 2.14) due to the significant cool flame chemistry. This means substantially more temperature compensation would be required to maintain constant CA_{50} with this fuel, and controlling fuelling transients could be more difficult with these kinds of fuels. Since iso-octane exhibits less than a 1.2° CA variation in CA_{50} from $0.1 \leq \varphi \leq 0.3$ (Fig. 2.14), fuel stratification offers little potential for adjusting the CA_{50} phasing. However, fuel stratification in the charge has the potential to shift combustion phasing by several degrees in PRF80 because it is more reactive and φ-sensitive fuel.

The φ-sensitivity of fuels also depends on the intake air pressure. Figure 2.15 presents the φ-sensitivity of three different fuels (gasoline, iso-octane and PRF73) for naturally aspirated (NA) and boosted (high intake pressure) conditions. High negative slope in the curves indicate higher fuel reactivity (and earlier ignition) and more sensitive fuel. In naturally aspirated case, PRF73 shows significant advancement in CA_{10} with increased φ. Therefore, PRF73 is a highly φ-sensitive fuel because of the strong dependence of its robust pre-ignition reactions (reactions responsible for the LTHR and ITHR) on the fuel concentration [72]. However, iso-octane and gasoline show comparatively less sensitivity at naturally aspirated case due to weak pre-ignition reactions of both fuels. In contrast to PRF73, gasoline and iso-octane show delayed CA_{10} timing with increasing φ because of lower compressed gas temperature achieved with higher fuel quantity due to decrease in

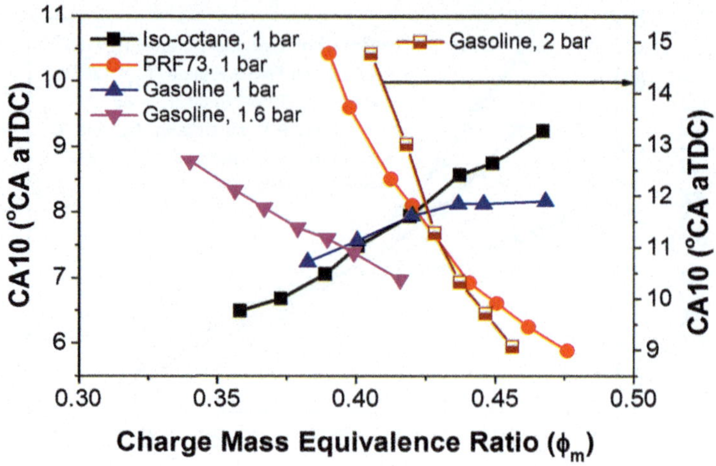

Fig. 2.15 φ-Sensitivity of different fuels for naturally aspirated engine operation and at a range of boost pressures (Adapted from [72, 73])

γ (c_p/c_v) which dominates over pre-ignition heat release [63, 72]. Intake pressure boosting significantly increases the auto-ignition reactivity of gasoline [63]. The φ-sensitivity of gasoline increases (even stronger than PRF73) at $P_{\text{in}} = 2$ bar (Fig. 2.15) due to large increase in pre-ignition reactions in terms of increased ITHR, while gasoline remains as a single-stage heat release fuel. High φ-sensitivity of gasoline suggests that partial fuel stratification (PFS) should be effective for reducing the HRR of boosted HCCI operation. The PFS can allow higher φ without excessive PRR (knock) and/or more advanced combustion phasing for higher efficiency. Additionally, these results also suggest that the ITHR pre-ignition reactions are as important (sometimes more important as case of PRF73) for determining φ-sensitivity as the LTHR reactions [72–74].

2.2.4.2 Fuel Properties

In general, several physiochemical and combustion properties are important for automotive fuels (see Fig. 1.11). Volatility and auto-ignition reactivity are the two essential fuel properties required for HCCI combustion. To prepare well-premixed homogeneous charge, volatile fuel is required. Sufficient fuel reactivity is required for obtaining auto-ignition at appropriate combustion phasing in the cylinder within control parameters of the HCCI engine. In case of very reactive fuels, limited operating range is obtained, and fuel reactivity also limits the engine compression ratio. For advance LTC engines, fuel reactivity requirement is merging in between diesel and gasoline (see Fig. 1.13). Gasoline-like fuels (high volatility fuels) are more suitable for well-premixed HCCI combustion engines. Fuel volatility is controlled by intermolecular forces depending on molecular size and structure.

There is no significant electrostatic attraction in hydrocarbon molecules as they are not positively or negatively charged molecules. The dispersion forces govern the molecular attraction in hydrocarbons generated by the change in electron density in the molecular periphery [27]. The strength of attraction forces depends on the molecular surface area and intermolecular forces. Attraction forces increase with higher surface area, which requires comparatively large energy to reach boiling point. Therefore, volatility of fuel decreases with increasing number of carbon atom in straight-chained hydrocarbons (such as diesel consists of straight-chained and lightly branched 10–19 C atom hydrocarbons), and these kinds of fuels create difficulty in premixed charge preparation in HCCI engines. The chain branching increases the volatility by reducing molecular surface area leading to lower intermolecular forces [27].

Auto-ignition reactivity of HCCI fuels depends on the intensity of oxidation reactions taking place before the start of HTHR (hot ignition). Two main categories of single- and two-stage heat release fuels in HCCI combustion are illustrated in Fig. 2.4. Two-stage ignition fuels exhibit cool flame reaction (LTHR) followed by ITHR, which leads to HTHR reactions (detailed discussion in Sect. 2.2.2). However typical gasoline-like fuels exhibit single-stage heat release, where no LTHR is observed, and HTHR is governed by heat release during ITHR regime in naturally aspirated conditions. At higher intake pressure, fuel such as gasoline changes the auto-ignition reactivity due to higher ITHR (Fig. 2.15). Typically, fuels having larger amount of heat release during LTHR (if present) and ITHR before onset of HTHR auto-ignite easily. This type of fuels requires less temperature and pressure of intake charge at intake valve closing position to achieve auto-ignition in the cylinder. As discussed in Sect. 2.2.2, hydrogen abstraction and isomerization reactions are the important reactions during LTHR for exhibiting two-stage heat release for a particular fuel. The molecular structure of the fuel and type of "C–H" bond (primary, secondary or tertiary bond) control the occurrence of the hydrogen abstraction and isomerization reactions [27, 75, 76]. Primary "C–H" bond is the strongest bond (among all C–H bonds), and for hydrogen abstraction requires highest activation energy [75]. Therefore, fuel molecule having more number of secondary C–H bond (long straight-chain molecules) is more prone to hydrogen abstraction during LTHR reactions. During isomerization reaction also, secondary hydrogen removal is easy (due to comparatively weaker bond) from long straight-chain molecule as isomerization reaction takes place when the O_2 chain reaches around to remove a hydrogen atom from another site [27]. Additionally, due to long chain length and flexibility of the fuel molecule, hydrogen abstraction by O_2 is easy and more probable in long-chain molecules [77]. Hence, long straight-chain alkanes (like n-heptane) are more prone to LTHR in comparison to branched alkane (like iso-octane).

The temperatures characterizing the LTHR and ITHR zone depend on particular fuel and engine operating conditions like boost pressure and engine speed [21, 49]. Figure 2.16a shows the comparative in-cylinder pressure and mean cylinder charge temperature for two different kinds of fuels (iso-octane, single-stage heat release, and PRF80, two-stage heat release) [50]. The LTHR is observed at 340 CAD in PRF80 fuel while no significant heat release in iso-octane until

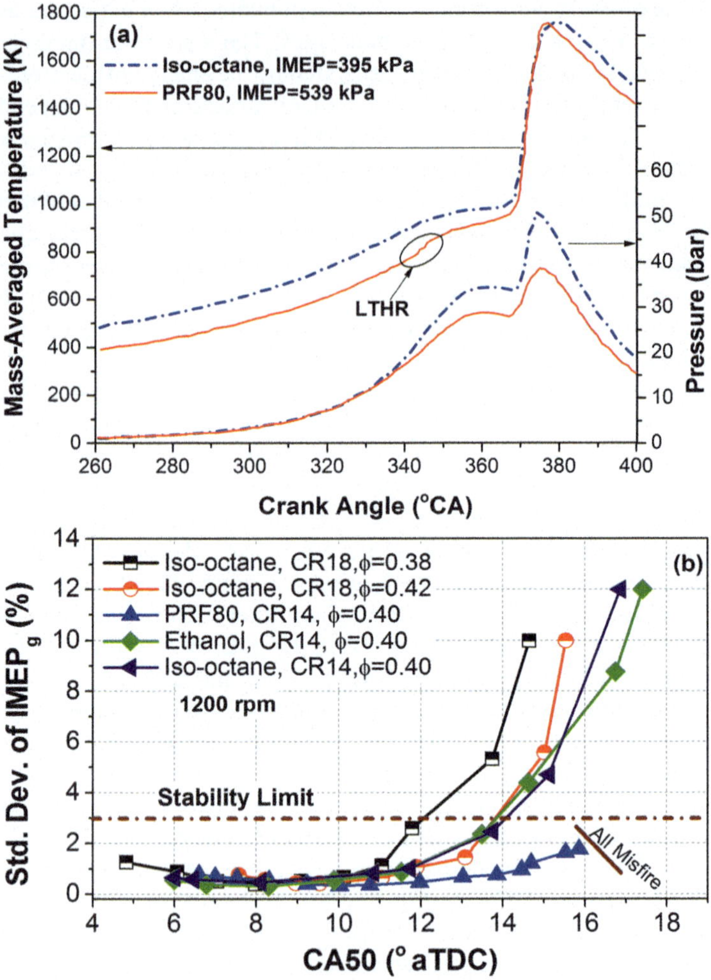

Fig. 2.16 (**a**) Variation of in-cylinder pressure and temperature with crank angle for PRF80 and iso-octane in HCCI engine [50]. (**b**) Standard deviation of $IMEP_g$ as a function of CA_{50} for iso-octane, PRF80 and ethanol (Adapted from [78, 79])

HTHR at 367 CAD. The cylinder charge temperature is comparatively lower for PRF80 before LTHR. The LTHR increases the temperature as well as pressure before HTHR of PRF80 at about same 367 CAD (Fig. 2.16a). The cylinder charge temperature is lower in PRF80 for same combustion phasing because of lower intake temperature requirement due to LTHR and strong ITHR. In case of PRF80, strong ITHR is observed because of increased intermediate temperature kinetics due to heat released during LTHR. The lower charge temperature in case of PRF80 can take the advantage of having higher charge density by inducting more charge in the combustion chamber, which leads to higher work output for same equivalence ratio.

The cycle-to-cycle variations in start of combustion (CA_{10}) are reduced in PRF80 due to higher rate of temperature rise before HTHR [50]. The cyclic variations are associated with small random charge temperature variations, and these temperature variations are dampened out because of heat release during LTHR in the cylinder. Figure 2.16b shows the variation in indicated mean effective pressure (IMEP) with CA_{50} for iso-octane, PRF80 and ethanol. The figure shows that PRF80 can tolerate higher CA_{50} retard with lower cyclic variations (acceptable standard deviation) in IMEP because of lower cyclic temperature fluctuations and higher temperature rise rate before HTHR [4, 21, 50]. Increased combustion phasing retards (which allows higher φ without knock), and higher inducted mass leads to achieve higher engine load operation in HCCI combustion using two-stage fuels like PRF80. In richer mixture condition (φ=0.38 to 0.42, Fig. 2.16b), engine operation with more retarded CA_{50} is possible for the same limit of IMEP variations.

Fuel auto-ignition reactivity also affects the high load limit in HCCI combustion engines. Figure 2.17 demonstrates the effect of auto-ignition reactivity on maximum achievable high load in naturally aspirated HCCI engine using iso-octane, PRF80 and PRF60 fuels. Figure shows that maximum IMEP (6.5 bar) is obtained for higher auto-ignition reactivity fuels (PRF60 and PRF80). Comparatively lower IMEP is obtained using lower reactivity fuel (iso-octane). The different fuels have different mechanisms of limiting high load boundary in HCCI operation. Iso-octane has two limiting factors (excessive NO_x and wandering CA_{50}) for high load boundary at compression ratio (CR) 14 and 3bar/CAD pressure rise rate (PRR) (Fig. 2.17). With increase in fuelling rate, IMEP and combustion temperature increases, which leads to excessive NO_x emission beyond certain limit of IMEP.

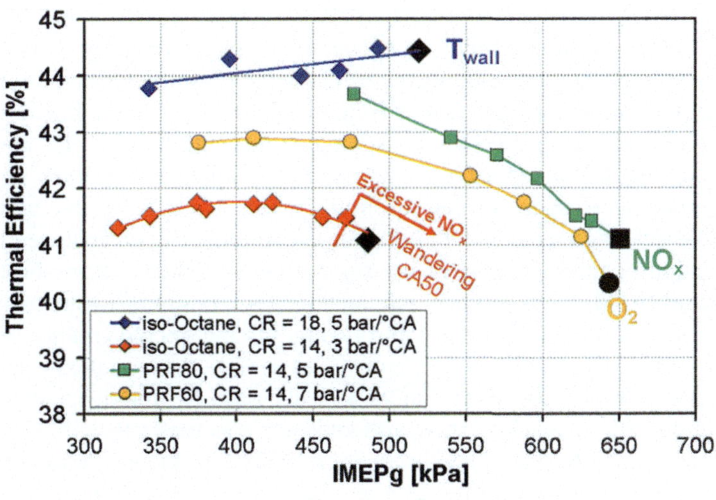

Fig. 2.17 Illustration of high load limits for different fuels for naturally aspirated conditions in HCCI combustion [80]

On further increase of fuelling rate, steady-state condition cannot be established and constant combustion phasing cannot be maintained (i.e., wandering CA_{50}). Unsteadiness is caused by high sensitivity of the ignition timings to random temperature fluctuations rather than the large changes in wall temperature with CA_{50} [80]. At higher compression ratio (CR = 18), wall heating runaway occurs beyond 5.2 bar IMEP. Cylinder walls heat up at advanced combustion phasing than average, due to higher heat transfer to the wall. The higher wall temperature (T_{wall}) leads to even more advanced combustion causing even higher heat transfer and so on. Small random variations in the charge temperature (at retarded CA_{50} and low ITHR) cause relatively large variation in CA_{50} which leads to higher T_{wall} variations in comparison to normal non-limiting cycles [21]. This higher thermal sensitivity leads to runaway at this condition. At lower PRR (3 bar/CAD) and CR (14), variations in T_{wall} are not adequate to initiate runaway to knock but cause wandering of CA_{50}. PRFs are less sensitive to variations in temperature due to LTHR followed by increased ITHR. PRF60 is a higher reactivity fuel and requires large amount of EGR to maintain acceptable PRR limit by retarding CA_{50}. In this case, high load boundary is limited by availability of oxygen, which decreases with increase in EGR fraction. At higher loads, faster runaway to knock occurs in PRF80 in comparison to iso-octane. Due to two-stage heat release in PRF80, CA_{50} is less sensitive to the increase in inlet charge temperature. Therefore, in this case, wall heating is not mainly responsible for fast runaway to knock. Study proposed NO_x-induced runaway caused by chemically enhanced auto-ignition due to NO_x present in the residuals (EGR) [80]. For detailed explanation and justification of data, refer to original study [80].

2.2.4.3 Intake Temperature and Pressure

The intake charge conditions (equivalence ratio, temperature and pressure) and fuel composition in charge have significant effect on combustion phasing, heat release and other performance parameters in HCCI engine. The intake temperature modulation is one of the most common strategies to control the HCCI combustion phasing. In the beginning of the compression stroke, temperature of fuel–air mixture increases to speed up the reaction kinetics and achieve the auto-ignition at desired crank angle position. This higher temperature can be achieved by intake air preheating or retaining part of hot residuals inside the cylinder. Fast thermal management (FTM) is an another technique for fast control of intake temperature. In FTM, intake air temperature is controlled by mixing of cold and hot air coming from different sources [27, 81]. Figure 2.18 shows the variations of cylinder pressure and HRR at constant relative air–fuel ratio (λ) for different intake air temperatures [70]. Increasing intake air temperature increases the cylinder pressure and HRR very rapidly (Fig. 2.18). Pressure rise rate also increases drastically with increase in intake temperature. Figure also depicts that start of combustion advances with the increase in intake air temperature. Higher intake air temperature increases the charge temperature in the cylinder, which increases the reaction rate

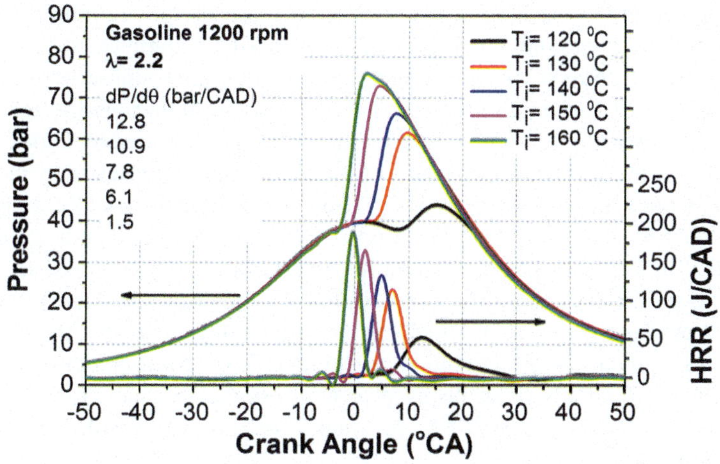

Fig. 2.18 Variation of in-cylinder pressure and HRR at 1200 rpm and constant λ for gasoline HCCI combustion (Adapted from [70])

kinetics resulting into advance combustion and faster HRR. Thus, intake air preheating significantly affects the combustion phasing in HCCI combustion [70].

The requirement of particular intake air temperature depends on the engine operating parameters (λ, speed, coolant temperature) and fuel and engine design parameters such as compression ratio. By selecting appropriate parameters, optimal combustion phasing can be obtained. At particular compression ratio and engine speed, thermal conditions of charge in the cylinder are determined by intake charge temperature and wall temperature. The intake charge temperature has direct impact on the thermal state of the bulk gas in the core of cylinder, and the coolant (wall) temperature creates more substantial effect on the charge in the thermal boundary layer. Thus, the wall temperature affects the thermal stratification in the combustion chamber near-wall regions [82]. The study also found that variations in combustion phasing and HRR created by change in wall temperature can be compensated by varying intake temperature in opposite direction with factor of 1.11 [82]. The decrease of coolant temperature has strong effect on combustion instability in comparison to the decrease in intake charge temperature, and hence combustion stability limits are more dependent on wall temperature in HCCI combustion engine. The requirement of intake air temperature is affected by compression ratio of engine and octane number of fuel. Compression ratio has strong influence on intake air temperature. Study shows a reduction in compression ratio from 21.4 to 17 results into increased requirement of intake air temperature from 30 °C to 130 ° C for iso-octane fuel [39]. Increase in octane number from 0 to 100 requires increment compression ratio by a factor of two in HCCI engine [39]. An increased compression ratio of engine increases HRR and advances the combustion phasing [83]. Increase in boost pressure resulted into strong decrease in inlet temperature requirement [84].

Intake pressure boost is typically used to increase the power output of engine, and it significantly extends the high load limit of HCCI engines. At higher intake pressures, same equivalence ratio and dilution level are achieved by inducting higher amount of fuel in every cycle, which leads to higher engine loads in the engine. The application of high intake pressure (boost) is limited by knocking (very high PRR). The knocking propensity of HCCI engine increases with increase in intake pressure mainly by two reasons: (i) higher amount of fuel inducted generates high PRR in combustion for the same combustion duration and phasing, and (ii) higher intake pressure increases the auto-ignition reactivity by an amount depending on the fuel [21]. The φ-sensitivity of fuels can also increase at higher intake air pressure (Fig. 2.15), which can be utilized for increasing the higher load range using fuel stratification. Higher intake pressure in combination with external exhaust gas recirculation can be used to attain higher engine load, and a study achieved $IMEP_g$ of ~16.3 bar at 3.25 bar intake pressure with gasoline fuel in HCCI combustion [63].

2.2.4.4 Engine Speed

The engine speed and fuel type affect the start of bulk gas reactions in HCCI combustion. There exists a strong coupling between fuel auto-ignition quality and required intake charge temperature to have an optimal combustion phasing. The intake charge temperature directly affects the completeness of combustion in the cycle because CO-to-CO_2 conversion reactions are highly sensitive to the maximum combustion temperatures [85]. The onset of bulk quenching of CO-to-CO_2 conversion reactions can be used as fuelling rate limits at particular operating condition, and below this fuelling rate, combustion is incomplete. Figure 2.19 shows the variation of required intake charge temperature with engine speed for different fuelling rates for PRF80 (two-stage heat release) and gasoline (single-stage heat release) fuel. Typically, higher intake charge is required at higher engine speed for maintaining the combustion phasing because at higher engine speed, there is less time available for chemical reactions to set off the auto-ignition process [70, 85, 86]. Additionally, the small inlet valve diameter, the short cam duration and the pressure drop over the intake air heater (if present) lead to the decrease in volumetric efficiency (η_v) of engine as engine speed increases. Lower volumetric efficiency results into lower pressure during intake stroke. The higher charge temperature is required to compensate for pressure drop (due to lower η_v) and to accelerate the chemical kinetics of reactions due to less availability of time at higher engine speed [70, 86]. Gasoline requires 20° C higher intake temperature as engine speed increases 600–1200 rpm. However, between engine speeds 1200–2400 rpm, intake charge requirement is almost constant (Fig. 2.19). Several possible reasons for this inconsistency are as follows [85]: (i) lower heat transfer from gas to cylinder due to less time available at higher engine speed provides higher compression temperature at constant T_{in}; (ii) cylinder surfaces are significantly hotter at higher engine speed which contributes to higher charge

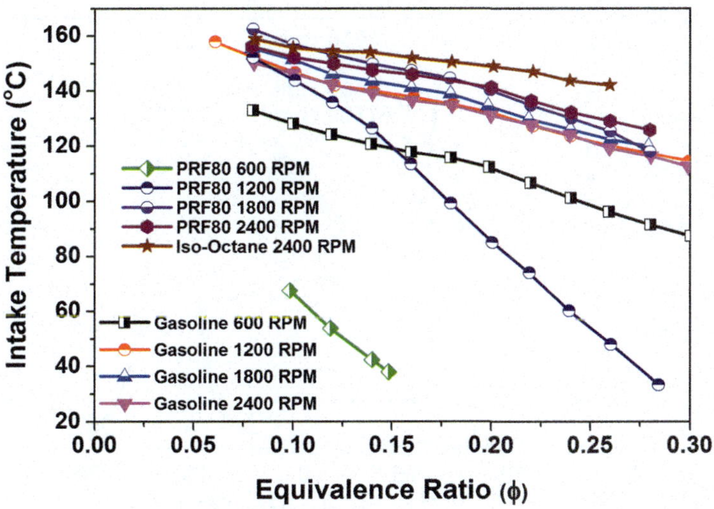

Fig. 2.19 Variations of required intake charge temperature with engine speed for PRF80 and gasoline in HCCI combustion (Adapted from [85])

temperatures; (iii) the "ram" effect is significant above 1800 rpm engine speed, which effectively compresses and heats the intake charge towards the end of the intake stroke; and (iv) charge heating due to turbulent dissipation of kinetic energy during the charge induction increases with engine speed.

In case of PRF80, intake charge requirement is highly dependent on engine speed and equivalence ratio (Fig. 2.19). The intake charge temperature for PRF80 has a much stronger dependence on φ at 1200 rpm and lower engine speeds. The required intake charge temperature drops quickly with increased fuelling (φ). There is a vast difference in necessary intake temperature for maintaining the combustion phasing at high equivalence ratios. As the φ decreased, the two curves appear to be merging [85]. This trend can be understood in terms of LTHR (cool flame activity) variations of PRF80 as it is a two-stage heat release fuel. The cool flame activity increases with lower engine speed, higher equivalence ratio and higher n-heptane content [85]. The heat release during LTHR duration leads to the reduction in the intake charge temperature requirement.

Fuels with two-stage heat release have some significant advantage to overcome the major challenge of extending high load boundary and improving combustion phasing control. Fast combustion phasing control can be accomplished with LTHR fuels because of the φ-dependence of the LTHR. The combustion phasing can be controlled by varying the amount of fuel stratification for a fixed φ because the amount of LTHR depends on the actual mixture undergoing combustion [53, 71, 87]. The φ-dependence of the LTHR in two-stage fuels also offers a means for lowering the maximum HRR by partial stratification using dual injection [78]. The cycle-to-cycle variations are also lower at extensively retarded combustion phasing operation of engine using LTHR fuels [50]. The LTHR also have higher power

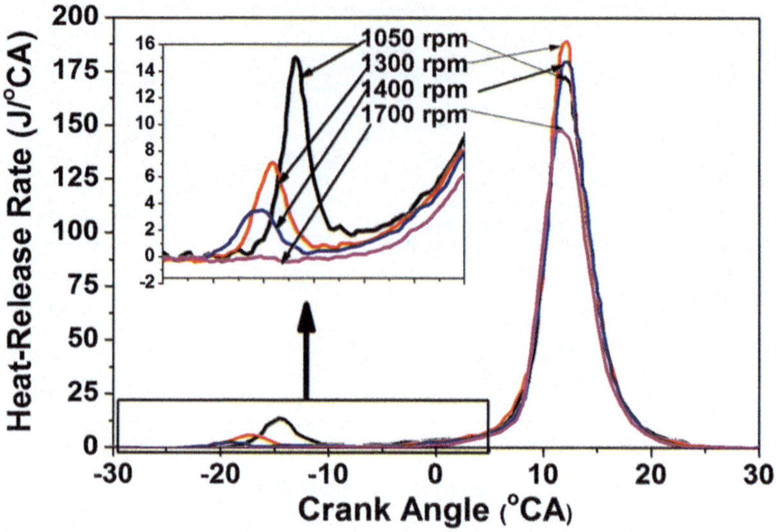

Fig. 2.20 Variations of HRR for PRF80 at φ = 0.38 for different engine speeds at constant combustion phasing in HCCI combustion (Adapted from [53])

density (Fig. 2.17). The LTHR significantly reduces the required T_{in}, which increases the charge density for a fixed intake pressure. The amount of LTHR is highly dependent on engine speed. Figure 2.20 shows the variations of HRR with engine speed for PRF80 at constant combustion phasing. The amount of LTHR decreases drastically with increasing engine speed (Fig. 2.20). There are two main reasons for this observation. First, the LTHR is fairly slow, oxidation reactions occur actively in 760–880 K temperature range, and there is less time for the LTHR to take place before 880 K as the engine speed is increased [53]. The faster progression of charge temperature from lower to higher temperatures at higher engine speed is due to the combination of LTHR and compression heating. Second, engine is operated at higher charge temperature and at higher engine speeds to maintain the combustion phasing constant, which leads to reduction in cylinder pressure, while the charge passes through the 760–880 K range. The low temperature heat release is sensitive to pressure (Sect. 2.2.2), which further decreases the amount of LTHR. The combination of these two effects results into the rapid reduction of the amount of LTHR with increasing engine speed. Figure 2.20 also shows that the LTHR disappears completely in engine speed range of 1400–1500-rpm, where the T_{in} requirement increases most rapidly [53].

Fuel composition has strong effect on the engine speed beyond which LTHR disappears. Figure 2.21 shows the variations of required intake temperature to maintain the constant combustion phasing for different fuels at constant equivalence ratio in HCCI combustion. It can be noticed that the change of n-heptane/iso-octane blending ratio influences the autoignition reactivity of the fuel, which affects the T_{in} required to maintain combustion phasing. Figure shows that the LTHR is

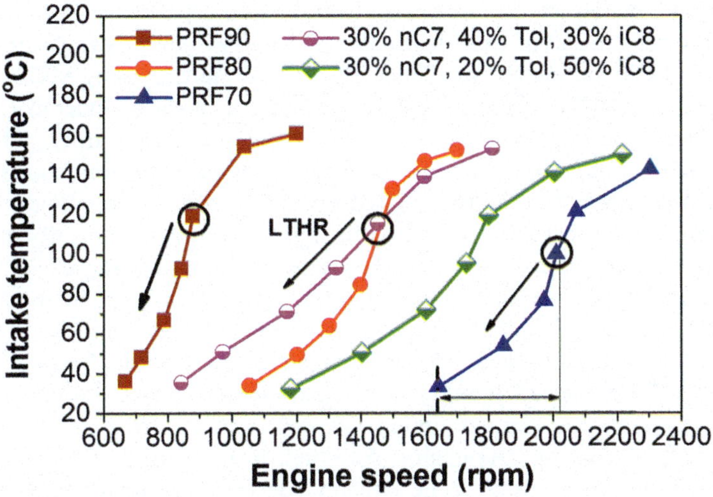

Fig. 2.21 Variations of intake temperature requirement with engine speed for different fuels at $\Phi = 0.38$ (Adapted from [53])

present in the ranges of 660–870 rpm, 1050–1440 rpm and 1640–2000 rpm for PRF90, PRF80 and PRF70, respectively. Figure 2.21 also illustrates that the toluene addition reduces the engine speed sensitivity of the PRF blends. This happens because the LTHR-suppression effect of toluene is roughly proportional to the amount of LTHR that the n-heptane would have produced and had toluene not been present [53].

Figure 2.21 also demonstrates the several issues related to LTHR fuels: (i) the very reactive fuel (PRF70) generates too much LTHR at lower engine speed, and the required T_{in} drops below the ambient temperature, which is impractical; (ii) the disappearance of LTHR at higher engine speeds renders the partial fuel stratification ineffective, which is used to lower the PRR to achieve higher engine loads, and it also limits the potential for combustion phasing control; and (iii) the engine speed range where LTHR is present imposes high demands on the thermal management as T_{in} varies with engine speed [53].

2.2.4.5 Exhaust Gas Recirculation

Exhaust gas recirculation (EGR) is a practical means to control the charge temperature in HCCI engine by adding high levels of exhaust gases into the intake manifold of the engine. The EGR also provides the dilution of charge to control the excessive HRR of HCCI combustion. Typical composition of the EGR is CO_2, H_2O, N_2, O_2, CO, unburned HC, particulate matter, NO_x and some combustion reaction intermediates. Two different EGR strategies are used in HCCI engine, namely, internal EGR and external EGR. In the internal EGR strategy, hot combustion products are captured in the cylinder by early closing the exhaust valve (EVC) towards the end of the exhaust stroke. The EVC timing decides the amount of retained EGR in the cylinder. The early EVC is realized using VVT system of engine. Earlier EVC timing leads to more hot residuals in the cylinder that raises the charge temperature at BDC (bottom dead center) of intake stroke. The higher residual mass fraction in the charge has higher heat capacity, which tries to delay the ignition timing. The higher residual mass captured in the cylinder also has higher temperature that tries to advance the ignition timings. Generally, the higher temperature effect of retained residuals in the cylinder dominates in the auto-ignition process [27]. A study investigated the charge heating, dilution (oxygen reduction), heat capacity, chemical (only CO_2 and H_2O) and stratification effects of internal EGR in CAI engines [87]. It is concluded that charge heating is mainly responsible for advanced ignition timings, dilution and heat capacity have no significant effects on ignition timing but decrease the HRR and increase the combustion duration of CAI engine. Stratified residual gases facilitate the CAI combustion. In the external EGR strategy, desired amount of exhaust gas is recirculated back in the intake manifold from the exhaust manifold through appropriate cooling system. The amount of EGR can be varied using exhaust valve in the exhaust line and changing the exhaust back pressure. The effects of combined internal and external EGR on gasoline CAI have been used to increase the attainable load in a multi-cylinder engine [88]. Study found that highest output achieved using internal EGR is limited by excessive PRR, and additional external EGR retarded the ignition, which lowers the HRR and PRR.

The use of substantial EGR/residuals is central to the most gasoline HCCI applications for maintaining the combustion phasing at desired position. The EGR is also desirable for diluting the charge for operating the engine on stoichiometric conditions, so that three-way catalytic converter can be used to mitigate the undesired NO_x at higher engine load conditions [89]. The diesel-like fuels easily auto-ignite by the rise in charge temperature during compression stroke. In diesel HCCI, substantial quantities of EGR are required for reducing the auto-ignition propensity of the fuel. Typically, substantial EGR addition is used to increase the ignition delay for getting sufficient time for fuel/charge mixing in the cylinder. The use of EGR also sufficiently reduces the oxygen concentration in the cylinder. Therefore, the peak combustion temperatures in the fuel-rich regions (due to inhomogeneity) do not reach to a level required for significant NO_x formation [90]. Hence, the use of EGR is also central to diesel HCCI combustion. The EGR can also be used to manage LTHR at higher engine speeds, where LTHR vanishes (Fig. 2.21). Therefore, a potential exists to use the benefits of LTHR for combustion phasing and HRR control over a wide range of speeds and boost conditions using EGR.

The addition of EGR typically retards the ignition timings for all the fuels. The amount of retard in ignition timings by EGR depends on the type of fuel and quantity of EGR used. The different mechanisms exist to produce thermodynamic and chemical effect of EGR during combustion. A study identified the various mechanisms of EGR as [89] (i) thermodynamic cooling effect due to increased

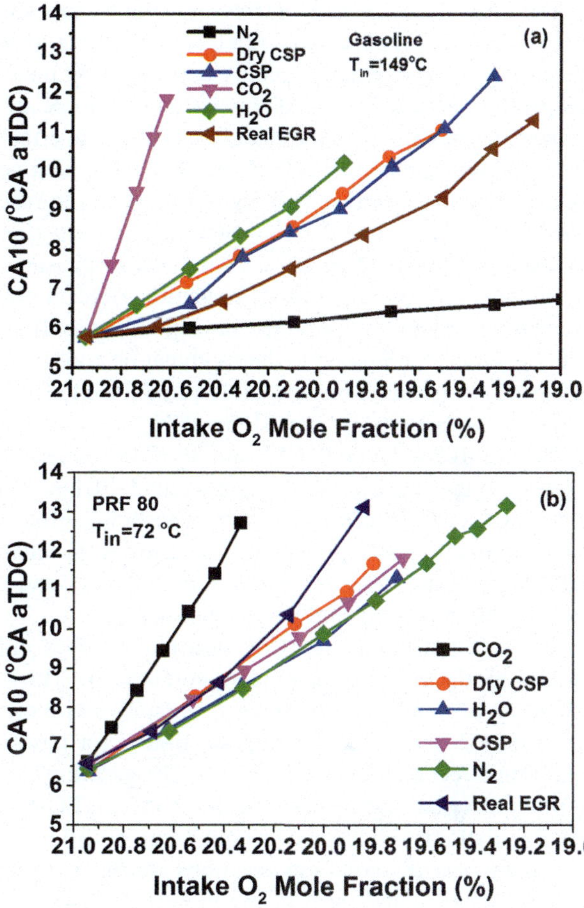

Fig. 2.22 Effect of EGR and its constituents on ignition timing for (**a**) gasoline and (**b**) PRF80 in HCCI combustion (Adapted from [89])

specific heat capacity, (ii) O_2 concentration reduction effect, (iii) enhancement of auto-ignition due to the presence of H_2O and (iv) enhancement or suppression of auto-ignition due to the presence of trace species (CO and NO). Figure 2.22 shows the effect of EGR and its constituents on ignition timing retard for gasoline and PRF80 fuels. To separate the effect of trace species present in real EGR due to incomplete combustion, the complete stoichiometric products (CSP) containing CO_2, H_2O and N_2 are used for the study. Addition of CSP) in intake has strong effect for both gasoline and PRF80 (Fig. 2.22). A detailed study conducted using gasoline, iso-octane, PRF80 and PRF60 shows that CSP has strongest effect on pure iso-octane and lowest sensitivity to gasoline, while PRF80 and PRF60 have in-between sensitivity [89]. Retarding effect of real EGR is higher for PRF80 than gasoline and iso-octane. In practical EGR system, water content in EGR may condense on cold surfaces (mainly at the time of cold starting and/or cold weather), which will change the composition of real EGR. This effect is illustrated by the trend between dry CSP and CSP. Loss of water makes less effective in iso-octane while more effective in gasoline and PRF80 (Fig. 2.22).

The strongest cooling effect is found for CO_2 for both gasoline and PRF80 (Fig. 2.22). The cooling effect of H_2O is in between CO_2 and CSP. This observation is also experimentally confirmed by measuring motored compression temperature data at 350° CAD by varying the EGR and its constituents [89]. In the same experiment, it is found that N_2 has opposite effect and it slightly increases the compression temperature. However, using N_2 as diluent retarded the combustion phasing due to oxygen sensitivity of auto-ignition reactions not by thermal effect. Therefore, N2 is used to separate the O_2 reduction effect of EGR in engine combustion. In all other cases, combustion phasing retard is a combination of thermodynamic cooling and O_2 reduction effect. Detailed description with experimental observation is provided in original study [89]. The O_2 reduction effect is relatively weak compared to thermodynamic cooling (seen by comparing the slope of curves) in gasoline but has more pronounced effects in PRF80 (Fig. 2.22). The higher O_2 sensitivity (higher slope of N_2 curve in Fig. 2.22b) is attributed to the LTHR in PRF80. Trend in CSP and dry CSP indicates that the dry CSP has slightly stronger retarding effect in gasoline and PRF80 (Fig. 2.22). This observation suggests that H_2O has a chemical enhancing effect on hot ignition more than thermodynamic cooling particularly at lower concentration. However, in case of gasoline, adding water increases both chemical and thermodynamic effects, and CA_{10} becomes significantly more retarded.

The difference between CSP and real EGR is only by the presence of trace species produced from incomplete combustion. The concentrations of these trace species in the intake manifold increase rapidly with increasing EGR rate. Real EGR has less retarding effect on ignition timing for gasoline and comparatively more retarding effect on PRF80 (Fig. 2.22). The gasoline has an enhancement of the auto-ignition reactions due to trace species from real EGR. Gasoline is a single-stage less reactive fuel, and addition of whole range of different hydrocarbons might lead to new pathways of oxidation reactions during auto-ignition process. PRF80 is a reactive species, and most of the molecules produced by incomplete combustion might be comparatively less reactive [89]. This explains the behaviour of gasoline and PRF80 with real EGR. The CO is a major specie present in the EGR. Study conducted on pure n-heptane and two surrogate fuels (80% n-heptane–20% toluene, 75% n-heptane–25% iso-octane) showed that CO addition in the intake manifold has no significant effect on auto-ignition process for up to 2000 ppm concentrations [91]. Other studies also experimentally as well as numerically confirmed that CO addition up to 170 ppm does not have significant effect on auto-ignition process [47, 92]. However, a numerical study showed that the CO additions of 1000 ppm or more leads to delayed the auto-ignition timings. The same study also showed that adding more than 10,000 ppm CO in the manifold, the chemical energy supplied by CO starts to increase the peak pressure, which is less likely a situation in HCCI engine [92].

Formaldehyde is also emitted in significant quantity in HCCI engines [93, 94]. Adding formaldehyde in intake manifold delays the ignition timings because formaldehyde consumes the hydroxyl radicals and decreases overall reactivity [92]. Although HCCI engine is known for ultralow NO_x emission but at certain engine operating conditions, NO emission can be in significant amount. The effect of NO in the intake during EGR is systematically investigated [95, 96]. Study conducted using PRF, toluene reference fuel (TRF) and gasoline showed that sharp advance of LTHR as well as the main heat release at low concentrations (~15 ppm) of NO is added to the intake air. The rate of this change decreases with increasing NO concentration. At higher concentrations of NO, heat release is retarded with higher NO concentration) mainly for PRFs [95]. Study also concluded that combustion phasing in the HCCI engine is more sensitive to variations of charge NO concentration compared to the SI engine [95]. Another study also concluded that NO addition in intake up to 45 ppm advances the auto-ignition process, and beyond this value, the auto-ignition is retarded for PRF [96]. This observation indicates that the NO can be used to control HCCI combustion phasing by varying the amount of EGR (containing adequate NO concentration). Table 2.1 presents the summary of effect of EGR constituents on ignition timings for different autoignition characteristics of fuels as discussed in the current section.

In naturally aspirated operating conditions without EGR, the amount of LTHR is dependent on fuel type. Reactive fuels show LTHR in a very narrow range of engine speed and intake temperature (T_{in}) (Fig. 2.21). The EGR can be used very effectively to operate with a constant amount of LTHR at fixed T_{in}, across wide ranges of engine speeds because EGR counteracts autoignition, especially the amount of LTHR [53]. The EGR can also be used to extend the high load boundary in combination with intake boost. Figure 2.23 shows the maximum achieved IMEP (with ringing intensity limit < 5 MW/m^2) with and without EGR at CR $= 14$ and 1200 rpm. Figure shows that the engine load is significantly increased with the use of cooled EGR and high boost pressure. The maximum load line with EGR (Fig. 2.23) shows that the IMEP increases almost linearly with boost up to $P_{in} = 200$ kPa. Above 200 kPa boost pressure, the slope of the maximum load line becomes less steep. Main mechanism suggested by authors for this reduction in the slope is as follows [63]. Initially, 200 kPa $\leq P_{in} \geq 240$ kPa, reduction in slope occurs because CA_{50} could not be retarded beyond the amount used for $P_{in} = 200$ kPa with acceptable combustion stability. Hence, allowable fuelling

Table 2.1 Effect of EGR constituents on auto-ignition for single- and two-stage heat release fuels [97]

Effect items	Single-stage fuels (gasoline and iso-octane)		Two-stage fuels (PRF60 and PRF80)	
	Effect on auto-ignition	Sensitivity	Effect on auto-ignition	Sensitivity
Thermodynamic cooling	Retarding	High	Retarding	Low
$[O_2]$ reduction	Retarding	Low	Retarding	High
Presence of H_2O	Enhancement	Low	Enhancement	High
Presence of trace species	Enhancement	–	Suppression	–

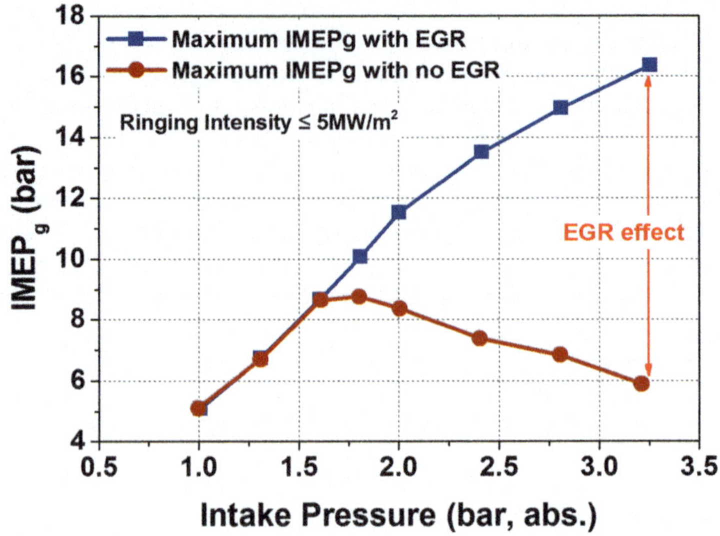

Fig. 2.23 Variation of maximum IMEP obtained with EGR and without EGR using gasoline in HCCI combustion at 1200 rpm (Adapted from [63])

rate cannot be increased to achieve higher IMEP. For higher intake pressures $P_{in} \geq$ 260 kPa, the maximum load is no longer limited by the ability to retard CA_{50} because fuelling is limited by the oxygen content of the charge due to higher amount of EGR used to retard the combustion phasing. Therefore, maximum achievable load (IMEP) has different limiting mechanisms at different operating conditions.

The maximum achievable steady IMEP depends on the fuel type, selected PRR (or knock metric), EGR quality and EGR/fresh gas unmixedness [98]. As the fuelling is increased beyond the maximum achievable steady condition, it is increasingly difficult to maintain sufficiently retarded combustion) to prevent runaway knock without having poor combustion stability that can also drift into misfire. The five load limiting factors/mechanisms which are identified leading to runaway knock are [98] (i) residual NO_x-induced runaway advancement of the combustion phasing, (ii) EGR-NO_x/wall heating-induced runaway, (iii) EGR-NO_x-induced runaway, (iv) EGR-induced oxygen deprivation and (v) excessive partial-burn occurrence due to EGR unmixedness.

Figure 2.24a summarized the main runaway mechanism for PRF80 and PRF60 at different PRR in combination of the type of EGR (well mixed or hetrogeneous, simulated or real EGR). In PRF80 fuel, main limiting mechanisms are residual NO_x-induced runaway and EGR-NO_x-induced runaway. Using simulated EGR (complete stoichiometric products), very fast knock runaway is found in comparison to wall heating-induced runaway. Authors suggested that the runaway is so rapid because of auto-ignition enhancement) due to residual NO_x [98]. When

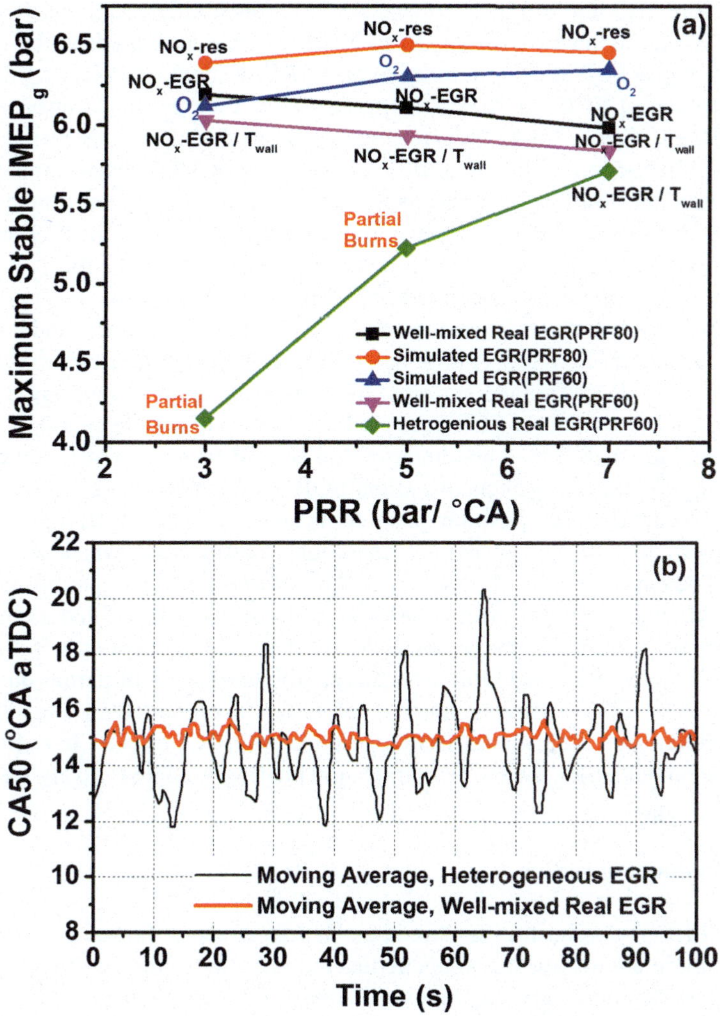

Fig. 2.24 (**a**) Main knock runaway mechanism for PRF80 and PRF60 at different PRR in HCCI combustion (Adapted from [98]). (**b**) Illustration of stable and unstable CA_{50} with well-mixed real EGR and heterogeneous EGR at PRR 3 bar/CAD (Adapted from [98])

some part of NO_x generated in one cycle is retained within the cylinder as a part of the residual gases, the autoignition enhancement comes into effect immediately during the following cycle. With real EGR in PRF80, runaway is found 21 times slower than for the residual NO_x-induced runaway observed for the PRF80, and 4 times faster than the wall heating-induced runaway observed for the iso-octane. It is concluded that largest contributor to this moderately fast runaway with real EGR is the enhancement of the autoignition by NO_x recirculated to the intake with the EGR gases. Wall heating-induced runaways also have small contribution to this process. The load limiting mechanisms are different for the two fuels (PRF80 and PRF60). With simulated EGR (CSP) in PRF60, IMEP is limited by the available oxygen because PRF60 is more reactive than PRF80, therefore requiring the use of higher amount of EGR. In PRF60, comparatively slow runaway is observed using real EGR. In PRF60, lower NO_x formation due to the high EGR rate leads to a substantial reduction of the contribution from EGR-NO_x enhancement of the autoignition, and wall heating mechanism also almost equally contributes to runaway process with real EGR. A more complete discussion with suitable justification can be found in original study for all the mentioned mechanisms [98].

Effect of well-mixed and heterogeneous EGR needs to be investigated (Fig. 2.24b). The stringent requirements for uniform cylinder-to-cylinder distribution of EGR are challenging. Additionally, due to the pulsating nature of the flow in the exhaust, EGR and intake systems, there is an inherent risk of cycle-to-cycle variations of the amount of EGR gases inducted into each of the cylinders. A study investigated the well-mixed and heterogeneous EGR using two different routes of EGR supplied [98]. The effects of a slight EGR unmixedness are clearly noticeable at more retarded CA_{50} position used to achieve a lower PRR of 3 bar/°CA (Fig. 2.24a). In this type of engine operations, partial-burn cycles are likely to appear. The data in Fig. 2.24a used strict limit of maximum one partial-burn cycle which is allowed in 1000 cycles [98]. Partial-burn cycle is defined as having an IMEP lower than 95% of the average IMEP. If engine is operated with heterogeneous EGR above this load limit, an unacceptable frequency of partial-burn cycles appears (Fig. 2.24b). Moving average of CA_{50} is presented in Fig. 2.24b for heterogeneous and well-mixed EGR operation of HCCI engine. It is found that with well-mixed EGR), the HCCI combustion phasing is comparatively well stable.

2.2.4.6 Inhomogeneities

Although HCCI combustion is termed as "homogeneous," however in real HCCI engines, there are always inhomogeneities mainly due to fuel stratification and/or thermal stratifications. Investigations are conducted to prove the presence of inhomogeneity by comparing the simulated homogeneous combustion (single zone) and real experimental data at same operating conditions. Figure 2.25 demonstrates the effect of inhomogeneities in real engine by comparing PRR at different equivalence ratios in fully homogeneous and real engines. Single-zone homogeneous charge model is used for simulation of pure "homogeneous" case because it is not possible to create a charge in a real engine, which is fully homogeneous in both mixture and temperature. Figure 2.25a shows that PRR increases very fast with increase of equivalence ratio in case of homogeneous model than experimental HCCI engine. For the same knocking limit, it is possible to operate real engines at higher equivalence ratio at both combustion phasing (Fig. 2.25a). Higher equivalence ratio operation leads to higher engine loads. It is believed that the lower PRR in real engine is because of naturally occurring charge stratification due to either mixture or thermal inhomogeneities [23]. Inhomogeneities in the charge can appear

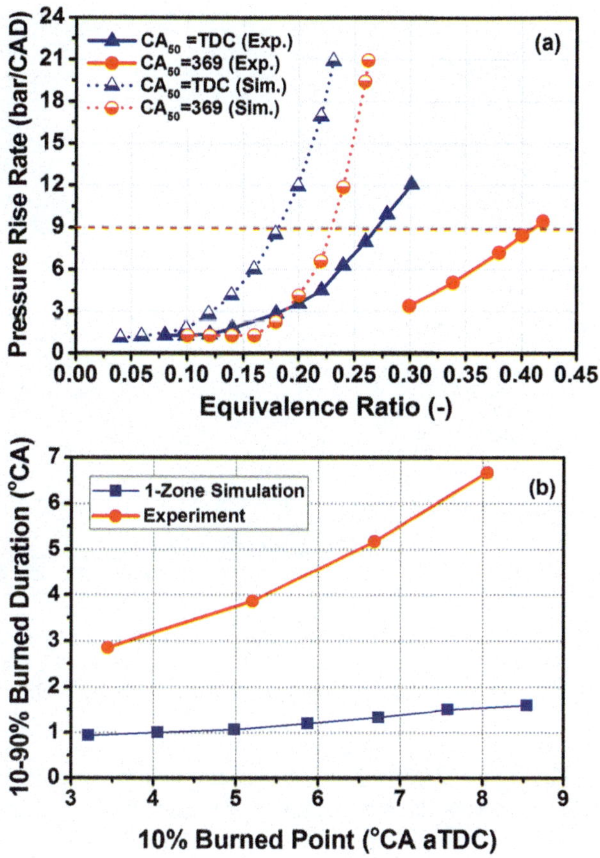

Fig. 2.25 (**a**) Variation of PRR with φ at different CA_{50} for simulated fully homogeneous and real HCCI engine (Adapted from [23]). (**b**) Combustion duration ($CA_{90\text{-}10}$) vs. CA_{10} for homogeneous and real experimental data at φ = 0.367 (Adapted from [99])

due to various reasons such as incomplete mixing of fresh charge and residuals, incomplete fuel/air mixing or heat transfer, non-isothermal intake conditions and turbulent mixing during the compression stroke [23]. Thermal and/or mixture stratification in HCCI engine leads to spatially and temporally distributed combustion in the cylinder rather than simultaneous combustion in purely homogeneous charge. Therefore, the natural stratification in the cylinder plays an important role in controlling the high load limit of an HCCI engine. This benefit of natural stratification is increased at retarded combustion phasing because with same knock limit, it is possible to operate the engine at even higher φ (Fig. 2.25a). Due to this reason, it is concluded that even without enhancement of naturally occurring thermal stratification, combustion phasing retard has potential to operate the engine at higher fuelling rate and higher engine speeds [100]. At retarded combustion phasing, charge temperature is lower, and combustion duration increases rapidly with delayed ignition timings (CA_{10}). The change in combustion duration is mainly affected by rapidly changing thermal stratification of the charge around TDC and less affected by slowing chemical kinetics at particular ignition timing

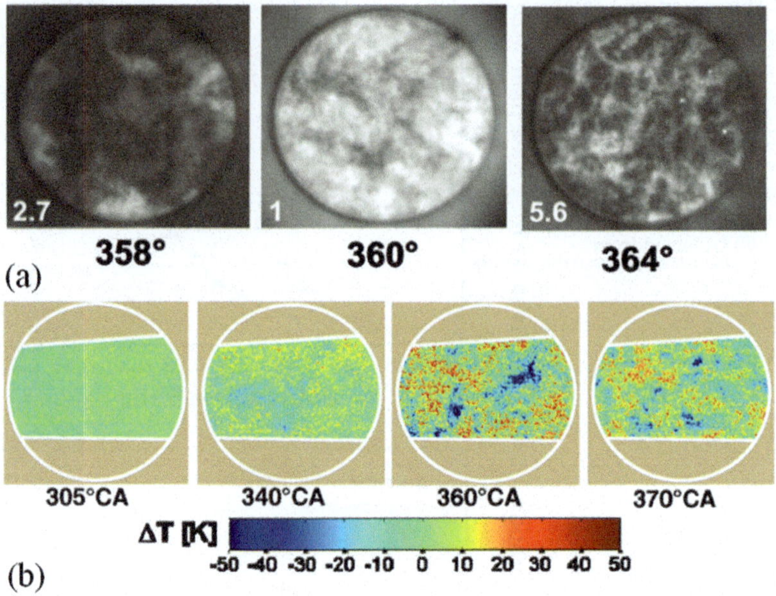

Fig. 2.26 (**a**) Illustration of inhomogeneities in HCCI combustion using chemiluminescence image sequences (Adapted from [23]). (**b**) Illustration of evolution of thermal stratification in HCCI combustion using T map images (Adapted from [101])

[99]. Figure 2.25b shows that combustion duration for experimental engine is changing rapidly with ignition timings (CA_{10}) than single-zone homogeneous model. The single-zone model represents the effect of combustion phasing on the chemical kinetics alone. A more complete discussion on the effect of combustion retard and thermal stratification on PRR is present in the reference [99].

Various imaging studies have indicated the significant inhomogeneities in the HCCI combustion [23, 24, 101, 102]. Figure 2.26 illustrates the inhomogeneities in the HCCI combustion, which uses premixed charge. Typically, heat transfer through cylinder wall, which affects the thermal boundary layers, is considered responsible for thermal stratification. Chemiluminescence imaging and PLIF imaging studies showed that thermal stratification extends in entire bulk charge in the cylinder [23, 101, 102]. Figure 2.26a shows the chemiluminescence images of HCCI combustion using iso-octane at $\varphi = 0.24$ for CA_{50} at TDC (360 CAD). Strong turbulent structures are noticed from figures, which indicate that the combustion is not homogeneous. Initially, the chemiluminescence images suggest that combustion starts at multiple locations throughout the cylinder with no specific pattern. The images clearly show that several regions have ignited (bright spots) while other regions have not started to ignite (Fig. 2.26a). As combustion progresses (during 358°–360° CAD), more regions start to react resulting into much brighter chemiluminescence images. At 360° CAD position, most regions are producing strong chemiluminescence, indicating intense reactions (Fig. 2.26a).

After this position, intensity starts decreasing due to burn-out of first ignited regions, while lately ignited regions are burning. Images shown in Fig. 2.26a are of not the same engine combustion cycle (aquired at same setting in different cycles). In the same study, authors also acquired the sequential images of one particular combustion event, and images show that autoignition begins with the hottest zone and progresses towards comparatively cooler zones in a sequential manner. It is also speculated that turbulence is important in mixing the cooler gas near the cylinder wall with the hotter gas in the bulk of the cylinder [23].

Study conducted to investigate the primary cause of inhomogeneities concluded that the inhomogeneities are caused primarily by thermal stratification due to heat transfer during compression, combined with turbulent transport [23]. The incomplete mixing of premixed charge and incomplete mixing during direct fuel injection have insignificant role because images for three different kinds of injection strategies are giving essentially the same appearance and the same development through the engine cycle [23]. Another study investigated the role of the thermal stratification produced during the intake stroke on the temperature distribution at TDC of compression stroke [103]. It is concluded that the variations in the initial temperature distributions have almost no effect on the temperature distribution and the cylinder wall heat transfer development during compression stroke. The convective transport of cold boundary-layer gases towards the cylinder centre is found as the dominant mechanism. Therefore, thermal distribution at BDC has no effect on evolution of temperature field during compression stroke. However, very small variations in the initial flow field amplified in the compression stroke, and the resulting differences close to TDC have significant effect on temperature and the wall heat flux distribution [103]. The thermal stratification created by different intake temperatures for each port in HCCI combustion has limited effect in reducing PRR [104–106]. This observation is clearly explained by the fact that thermal distribution at BDC has minor effect on the temperature distribution at TDC.

Majority of combustion takes place in the central part of cylinder (between the cylinder head and piston crown surfaces) from the start of combustion until peak PRR. Separate boundary-layer combustion along the cylinder head and piston crown surfaces takes place during the later part of combustion (after peak PRR). Therefore, thermal stratification between the bulk gases and the boundary layer has less contribution in controlling the peak PRR. The peak PRR is mainly controlled by thermal stratification within bulk gases [23]. To get the information about the magnitude of the thermal stratification and its spatial distribution in the entire chamber and boundary layer, PLIF imaging study is conducted. Chemiluminescence imaging provides valuable information on the effect of thermal stratification, but do not show the magnitude of the thermal stratification and its spatial distribution. [101]. Figure 2.26b shows the evolution of thermal stratification in the cycle using T maps created by PLIF image in motored cycle. The colour mapping in the figure shows the stratification between the coolest and hottest regions can be in the order of 100 K. Natural thermal stratification occurs mostly from wall heat transfer and turbulent convection [23]; therefore thermal stratification is not expected to be significant until the charge temperature has increased well above the wall temperature. Additionally, thermal stratification of the bulk gas needs time for convection to transport the cooler near-wall gases into the bulk of charge. There is no noticeable stratification at 305° CA (Fig. 2.26b) as expected, and some distinct thermal stratification appears at 340° CA. The thermal stratification increases until piston reaches TDC and then starts decreasing.

Researchers explored various strategies to create the thermal stratification artificially in the cylinder to increase engine load because natural stratification controls the maximum PRR. Few studies tried to increase the wall heat transfer by lowering coolant temperature or increasing the turbulence or swirl (by changing geometrical design) in the cylinder to create thermal stratification [27, 107]. However, increased heat transfer can decrease the thermal efficiency also. The combustion phasing delay provides more time for the evolution of naturally occurring thermal stratifications, and hence delayed combustion phasing is the most effective strategy to control the maximum PRR in HCCI combustion [27].

Strictly pure HCCI combustion means that it uses completely homogeneous charge, and there are no inhomogeneities. As discussed, inhomogeneities are present even with externally fully premixed charge (using mixing tank) or PFI in intake manifold [22, 23]. Naturally occurring fuel stratification in premixed charge does not have significant effect on combustion development [23]. Artificially created fuel stratification using direct injection has the potential to increase the high load limit and control the peak PRR [108]. It is very difficult to control the bulk thermal stratification in real engine to control the PRR; therefore intentional fuel stratification strategy is explored to increase the high engine load limit. This strategy is commonly referred to as partial fuel stratification (PFS) strategy, and their extension is typically known as partially premixed compression ignition (PPCI). To control the maximum HRR and PRR, the PFS strategy needs to create sequential auto-ignition (similar to thermal stratification), which requires local φ-sensitive fuel (Fig. 2.15). Detailed discussion on the determination of φ-sensitivity is provided in Sect. 2.2.4.1. In PFS strategy, majority of fuel (80-90%) is injected in the manifold, and small amount of fuel is injected directly in the cylinder to create fuel stratification in the cylinder. Figure 2.27 shows the effect of fuel stratification on cylinder pressure and heat release curve. Figure 2.27a shows that HRR decreases significantly for split injection (stratified) case in comparison to single PFI injection at same overall φ. The sequential auto-ignition (because of φ-sensitivity of fuel) due to fuel stratification leads to the increase in the combustion duration and lowers the HRR in case of split injection strategy. Figure 2.27b shows the effect of direct injection (DI) quantity on pressure and HRR using gasoline at intake pressure (P_{in}) 2 bar as gasoline is φ-sensitive at higher pressure (Fig. 2.15). Figure 2.27b shows that increasing DI fraction has significant effect on HRR and cylinder pressure. Zones with higher φ increases with higher DI quantity and ignite early. Early autoignition of rich zone charge advances auto-ignition for same CA_{50}, which effectively increases combustion duration [72]. Higher combustion duration lowers the peak pressure as well as peak HRR (Fig. 2.27b). Fuel stratification in the cylinder is also used for improving the low

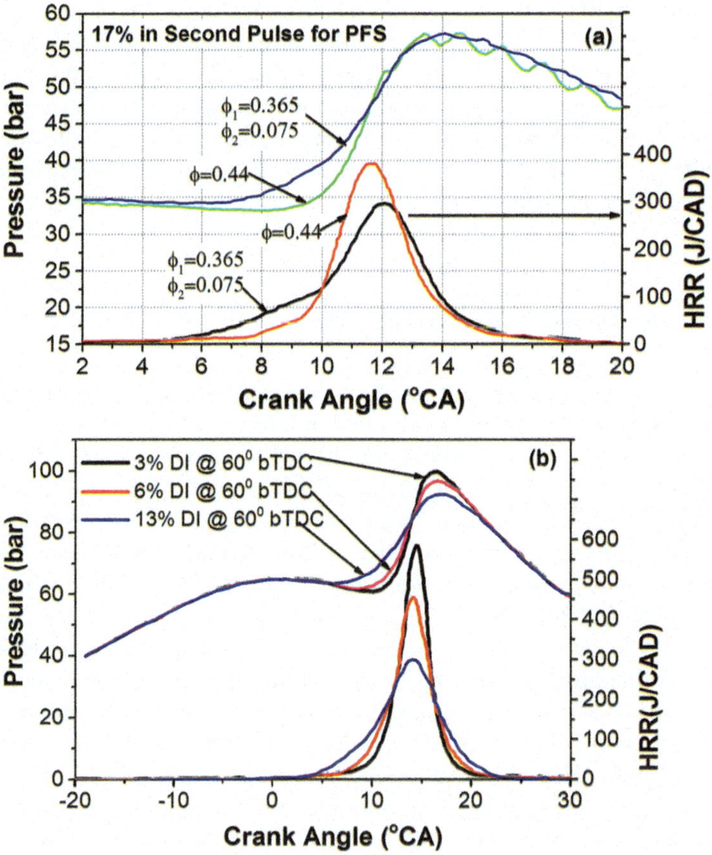

Fig. 2.27 (**a**) Variation of pressure and HRR for homogeneous (premixed) and fuel-stratified operation (Adapted from [78]). (**b**) Effect of DI quantity on pressure and HRR using gasoline at $P_{\text{in}} = 2$ bar (Adapted from [72])

load combustion efficiency (by advancing the combustion phasing) [109], and therefore this method can be used to increase the lower load limit of HCCI engine.

In PFS strategy, the partial mixture stratification also forms an associated thermal stratification (in addition to the typical thermal stratification created by heat transfer and convection) that tends to partially counteract the ϕ-sensitivity effect of the fuel on autoignition timing. This thermal stratification is produced by two reasons. First, rich regions (containing more directly injected fuels) have relatively greater cooling due to fuel vaporization of the DI fuel. Second, the richer regions have a lower γ, therefore their temperature increase with compression is relatively lower. Both effects act in the same direction, causing temperatures at the time of autoignition to be lowest in the richest (highest ϕ_m) regions and to be progressively hotter as the local ϕ_m decreases. Thus, this temperature distribution acts opposite of the chemical effect for ϕ-sensitive fuels that causes the richest regions to autoignite first, followed by sequentially leaner regions [110]. Fortunately, the ϕ-sensitivity effect is sufficiently strong for many fuel-type/operating condition combinations such that it is more than to overcome the temperature effects associated with mixture stratification. For these conditions, the richest regions autoignite first followed by sequentially leaner regions, allowing PFS to reduce the peak HRR and provide significant improvements in thermal efficiency as well as higher engine load [110].

2.3 Spark-Assisted HCCI Engine

Spark-assisted compression ignition (SACI) is a technique to set off HCCI combustion in existing engine using an external ignition source [111]. This technique is a key enabler for SI–HCCI hybrid system [21]. The engine is started in SI mode to tackle the cold start problem of HCCI and then shift to HCCI mode for lower engine load operations. At higher engine load requirements, engine can be shifted back to SI operation mode. Frequent SI–HCCI transition is required due to load variations during real engine operation. This transition is achieved with SACI by beginning a flame by spark and progressively varying the quantity of residuals using valve timing, which leads to combustion of fraction of charge in SI or HCCI mode [21, 112]. Although SACI is considered mainly as combustion mode transition method in the beginning [112], further in-depth investigation demonstrated that SACI strategy can be used to increase high and low load limits of pure HCCI combustion mode [21].

The SACI process can be divided into four different combustion phases (shown in Fig. 2.28a), namely, the spark discharge (region A), the early kernel growth (EKG) (region B), the flame propagation (region C) and the compression ignition phase (region D) [113]. In SACI, dilute homogeneous mixture is prepared similar to HCCI engine in the manifold or early intake stroke. However, a small amount of fuel is injected late in compression stroke to achieve a fuel-rich mixture near the spark plug for promoting SI and initiate flame [113]. Another study achieved SACI by direct injection of small amount of fuel (additional to PFI charge) near the spark plug, facilitating a propagating flame which increases charge pressure and temperature in the cylinder, leading to auto-ignition in the surrounding homogeneous diluted charge [114]. After the spark discharge, during EKG period (during CA position shown as B), a small reaction kernel formed starts growing slowly with very little heat release. During EKG period, flame is too weak for strong propagation. The end of EKR period is defined as point where second derivative of the flame area takes a maximum value and eventually rapid heat release takes place [113]. From this point flame propagation starts in the engine cylinder. Flame propagation releases additional heat by fuel burning, leading to further compression heating which drives the compression ignition in the remaining charge. The end of the flame propagation period is defined as a point, where the combustion transition takes place to CI combustion. The compression ignition HRR is faster than

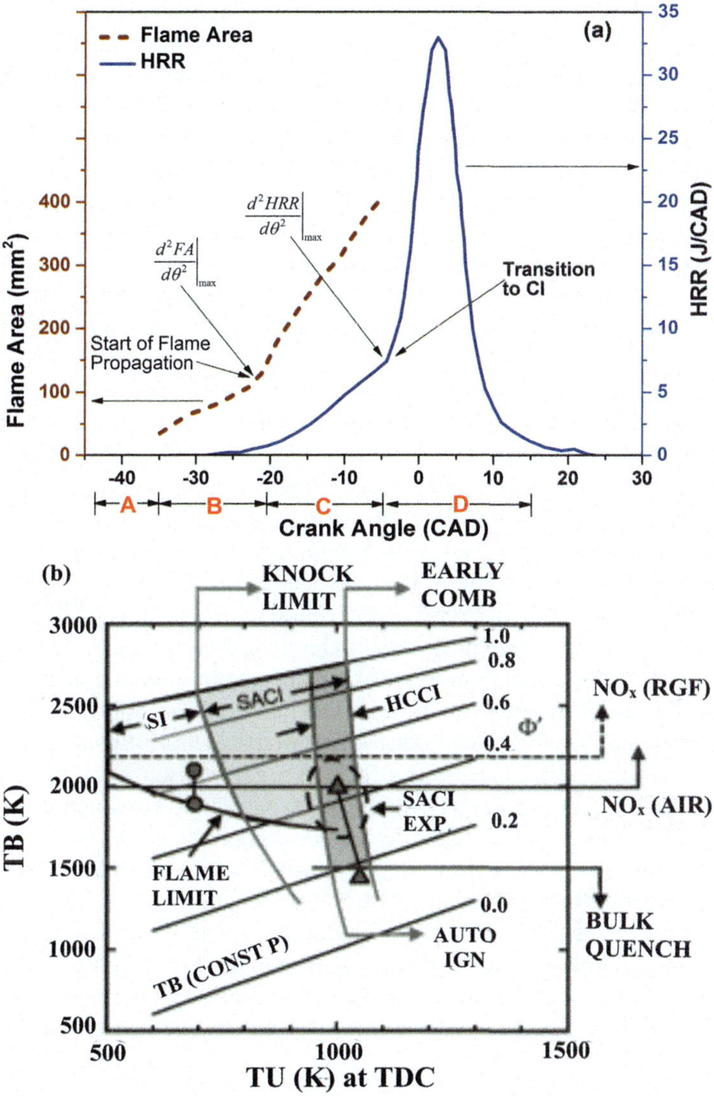

Fig. 2.28 (**a**) Illustration of different combustion phases in SACI process (Adapted from [113]). (**b**) Multimode combustion diagram presenting SI, SACI and HCCI combustion zones in terms of burned and unburned gas temperature for iso-octant [111]

turbulent flame propagation (due to combustion occurrence over reaction time scale). Therefore, maxima of second derivative of HRR curve can be used as transition point to CI combustion phase (Fig. 2.28a) [113, 115]. The CI combustion phase is comparatively shorter, and the EKG phase is a primary source of combustion phasing cyclic variability in SACI engine [113].

Figure 2.28b shows a conceptual diagram defining the HCCI, SACI and SI combustion zones in terms of unburned gas temperature (TU) near TDC and burned gas temperature (TB), which is the corresponding constant pressure adiabatic temperature. The SACI region is enclosed by operating constraint lines of limits for excessive knock, flame propagation, auto-ignition, bulk quenching and excessive NO_x emissions [111]. The shape of region or temperature range depends on fuel and engine operating conditions. The parameter φ' (fuel to charge equivalence ratio [111]) is used as a primary load metric for air and/or EGR diluted mixtures. The fuel to charge equivalence ratio (φ') is approximately related to fuel–air equivalence ratio (φ) by relation $\varphi' \approx \varphi$ (1 residual gas fraction). The most successful demonstrations of SACI take place at higher equivalence ratios ($\varphi' \geqslant 0.4$) and above the proposed turbulent SI flame propagation limit line. For $\varphi' \geqslant 0.4$, spark can be used to advance and modify the HRR significantly [111, 116, 117]. The spark near TDC is used to improve the combustion stability for lower equivalence ratios near $\varphi' = 0.3$, but it is not effective in advancing HRR. Analysis of SACI data in terms of the multimode combustion diagram suggests that spark assist is mainly useful at moderate and high engine loads for ignition and HRR control [111]. The two basic requirements for SACI are as follows: (i) the charge equivalence ratio and temperature must be adequately high to sustain flame propagation, which is fast enough to consume fraction of charge, and (ii) the compressed gas pressure and temperature must be sufficiently low such that the charge does not auto-ignite until it is further compressed by flame combustion and heat release [21].

The SACI combustion has the potential of partially decoupling the combustion phasing from peak HRR, which is one of the major technical challenges in HCCI engines. Maximum HRR can be controlled by varying the fraction of fuel consumed by flame propagation/auto-ignition at particular combustion phasing of SACI [118]. The simulations of SACI combustion predicted 43% reduction in the maximum HRR with increasing spark advance while only 23% reduction in unburned charge at the time of autoignition. The variation in spark advance changes fraction of charge consumed by flame leading to variation in the end-gas thermal and compositional distributions prior to auto-ignition [118]. Figure 2.29 shows the flame front location and spatial distribution of end-gas temperature in the cylinder at 5° aTDC in SACI. Temperature is shown by colour and flame front location by

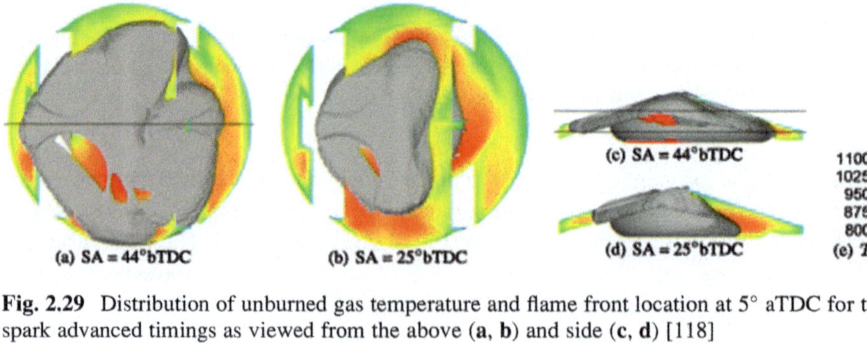

Fig. 2.29 Distribution of unburned gas temperature and flame front location at 5° aTDC for two spark advanced timings as viewed from the above (**a**, **b**) and side (**c**, **d**) [118]

grey surface (Fig. 2.29). More charge is consumed by flame for advanced spark timings. Therefore, the reduction in maximum HRR during CI regime is dependent on both the mass of the end-gas charge and the end-gas charge reactivity [118].

An experimental study conducted at constant equivalence ratio by independently varying spark timing and unburned temperature (by internal and external EGR) in SACI regime concluded that at constant CA_{50}, a ~40% reduction in maximum HRR and ~75% reduction in ringing intensity can be achieved while no penalty on thermal efficiency [119]. Longer combustion duration and a larger portion of flame-based heat release are found with decrease in unburned gas temperature and spark advance. The study also found that bulk end-gas reactions start at a similar unburned zone temperature (~1040 K), regardless of the temperature at spark timing [119]. Another experimental study concluded that flames could be used to consume a substantial portion (20–30%) of the charge mass to modulate HRR in SACI combustion [120]. The effect of spark assist on HCCI combustion varies with equivalence ratio for mixtures diluted with preheated intake air. At higher equivalence ratio ($\varphi > 0.62$), spark assist has much larger effect on combustion phasing, and lower equivalence ratio ($\varphi < 0.38$) spark assist does not have significant effect on HCCI combustion behaviour [121]. Flame propagation cannot occur, or the flame is very weak to have significant effect on the combustion phasing at excessively dilute (70%) low load engine operating conditions [111]. In the SACI process, the deflagrative reaction fronts are sustained in ultra-dilute mixtures (30–60% dilution [122]) when unburned mixture temperature is increased using charge preheating (e.g. with intake heaters and/or hot residual gases) [116, 123]. Typically, unheated conventional SI combustion cannot use such highly dilute mixtures, and there exists a lean limit (~30% air dilution) for flame initiation/propagation [124]. The elevated unburned gas temperatures are purposefully used to achieve flame propagation within a highly dilute mixture and to ensure end-gas auto-ignition in SACI.

The dilution method affects the combustion process in LTC engine, and a variety of dilution methods (air, EGR, etc.) are used. In HCCI combustion, EGR is more effective to reduce peak HRR and increases the combustion duration for HCCI than air at constant combustion phasing. A computation study showed that laminar burning velocities decrease with increase in EGR dilution (vs. air dilution) due to thermal and chemical effects of EGR [125]. Study found that the flame speed is most sensitive to the high temperature chain branching and chain-terminating reaction rates, where O_2 concentration plays an important role. Thus, EGR diluted charge is less affected by spark assist than air dilute charge for same energy content. An experimental study reported that early HRR consistent with flame propagation are more rapid for air dilute SACI condition than EGR dilute engine operating conditions at constant equivalence ratio, spark timing and combustion phasing [120].

One of the main advantages of SACI is comparatively higher engine load operation without excessive peak PRR/knock. In SACI engine, higher load operation is possible due to mainly three reasons [21]. First, SACI operates at higher charge density than HCCI by inducting more charge in the cylinder due to comparatively lower intake heating/induction of hot residuals. Higher intake charge

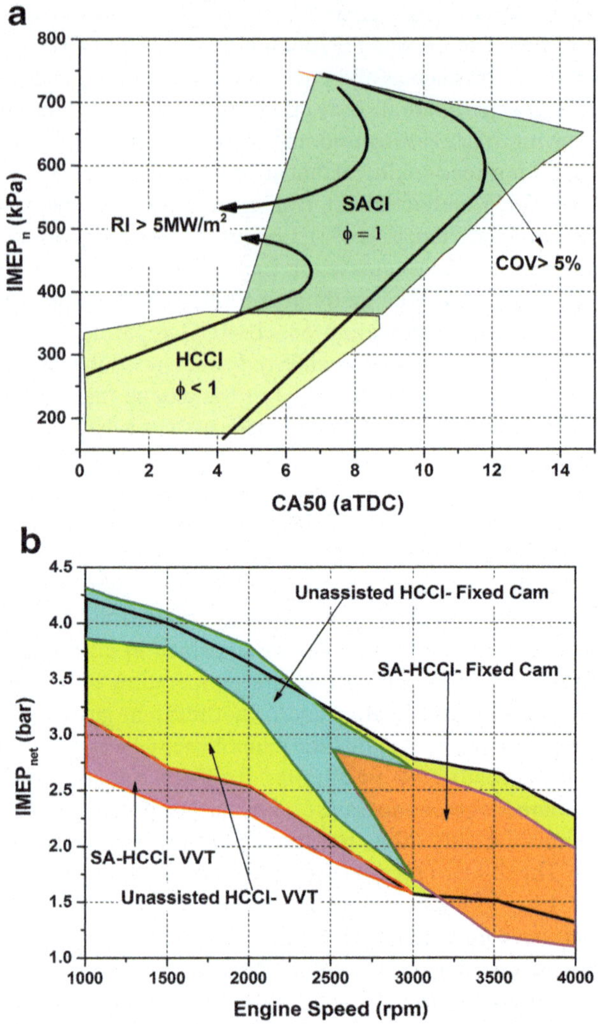

Fig. 2.30 (**a**) The HCCI operating range extension by SACI using NVO in naturally aspirated engine (Adapted from [122]) and (**b**) extension of operating load regime by SACI using fixed camshafts and VVT with asymmetric NVO (Adapted from [127])

requirement is compensated by heat release during flame propagation. Second, overall lower HRR leads to higher fuelling rate and/or comparatively more advanced combustion phasing without excessive PRR. The overall HRR is lower because significant amount of charge is burned by flame propagation, which is generally slower than bulk auto-ignition; overall burn duration increases. Typical flame combustion portions are 30–40% for negative valve overlap (NVO) and 40–55% for positive valve overlap (PVO) [126]. Third, the SACI can be operated on more retarded combustion phasing, which reduces the HRR during HCCI burn. Retarded HCCI phasing can allow higher fuelling rates leading to higher engine load. Figure 2.30 demonstrates the operating range extension of HCCI combustion by SACI combustion.

Figure 2.30a shows that the engine load can be well extended above the usable range of HCCI using SACI flame-based heat release by controlling the overall HRR. This is achieved in naturally aspirated automotive engine with CR 12.5 using NVO for gasoline. It is reported that ringing intensity increases with CA_{50} advance and combustion stability decreases with retarded combustion phasing. The engine operating load constraints also depend on control strategy used to achieve SACI. Figure 2.30b shows the operating regime for HCCI and SACI with fixed cam engine and with VVT along with asymmetric NVO. Fixed cam operating range is found comparatively lower in SACI combustion [127]. A study achieved nearly 10 bar IMEP by continuously retarding spark timing and combustion phasing using large PVO to control residual concentration [126]. Additionally, SACI offers a more robust level of cycle-to-cycle control on ignition timings and increases the applicability of advanced LTC in transient multimode engine operation. The spark assistance can improve the combustion stability near misfire limit [128]. The spark assistance has been also demonstrated as valuable misfire prevention mechanism during rapid combustion mode transition between HCCI and conventional SI [129].

2.4 Thermally Stratified Compression Ignition

As discussed in Sect. 2.2.4.6, thermal stratification in HCCI combustion significantly affects the HRR and ringing intensity. It is also discussed that cylinder wall condition has a small influence on the thermal stratification in combustion chamber. Instead, the TDC temperature, combustion phasing and swirl all have a significant effect on the level of thermal stratification before auto-ignition [130]. Due to ease of control, several LTC strategies (Sects. 2.5 and 2.6) use intentional fuel stratification created by direct fuel injection to control the peak HRR and PRR in premixed compression ignition. However, fuel stratification leads to a risk of higher PM and NO_x emissions due to the locally rich zones present in the cylinder. A new strategy called thermally stratified compression ignition (TSCI) is proposed to artificially create the thermal stratification in the cylinder [131]. In this combustion mode, water is directly injected in the cylinder to control the mean charge temperature along with temperature distribution in the cylinder, which offers cycle-to-cycle control of start of combustion and HRR in premixed compression ignition. Water injection in the port is also investigated in HCCI engine, and study indicated that evaporative cooling in the intake manifold offers the control of ignition timings by controlling the intake valve closing temperature [132]. Another study also used direct water injection using diesel fuel injector to extend the load range premixed compression ignition [133]. The study showed that the combustion suppression effect increases with increase in water injection quantity and excessive water injection leads to higher THC emission along with lower thermal efficiency.

Direct water injection significantly retards the combustion phasing in HCCI engines [131]. Thermal stratification analysis is used to determine the unburned temperature distribution before the ignition process. The results indicate that the temperature distribution is widened by direct water injection because evaporation of water in the spray locally cools down the areas targeted by the spray. The areas in the chamber not targeted by water spray remain unaffected. The effectiveness of cooling is affected by injection timing of water [131]. The study showed that early injection timing (at or earlier than 80° CA bTDC) is not effective in spreading the temperature distribution because local cooling achieved by water spray got mix-out with charge due to larger available time before ignition. For late injection timing (at 10° CA bTDC or later), water spray not gets enough timing for evaporation, and effects on temperature distribution are minimal. It is found that approximately 20 CAD at 2000 rpm are required to water spray for breakup and evaporation. Therefore direct water injection can be used for cycle-to-cycle control of HRR in LTC engines. The study showed TSCI has a 370% larger operating range than pure HCCI combustion at particular operating conditions [131]. The TSCI strategy is less explored in published literature in comparison to fuel stratification strategies and needs further investigations to be practically implemented. Additionally, this strategy might have also corrosive effect on contained components, which further needs to be investigated.

2.5 Partially Premixed Compression Ignition

The LTC strategies can be divided into two broad categories, namely, HCCI and SCCI, based on degree of charge premixing. In the HCCI engines, fuel vapour is normally well mixed with the air and residual gases before compression, and charge is over all lean (Sect. 2.2.1). Laser-based combustion diagnostics and modelling investigations indicate that HCCI combustion is far from homogeneous even though charge is well-mixed (Sect. 2.2.4.6). The brake thermal efficiency of combustion engine is a product of combustion, thermodynamic, gas exchange and mechanical efficiencies. Higher brake thermal efficiency in HCCI engines is achieved due to lower pumping losses and higher thermodynamic efficiency because of higher compression ratio and higher γ of working fluid in the cylinder due to lean mixture engine operation. A study showed that the thermodynamic efficiency of HCCI can reach up to 50% or even slightly higher, but there are some issues with combustion, gas exchange and mechanical losses [134]. The combustion efficiency is around 90% for most of the operating conditions. The HCCI engine is mainly operated in low load range, and hence parameters such as friction losses and cooling losses have a large impact on the achieved brake efficiency. The overall brake thermal efficiency is found in the range of 32–42% [134]. The other novel strategies investigated to overcome the limitation of HCCI mode are SACI and SCCI modes. Several studies are conducted to investigate various aspects of SACI (see Sect. 2.3), but the applicability is limited because it is very hard to master, and it has higher cyclic variations, higher pumping loss and relatively higher NO_x emissions.

In SCCI mode, intentional thermal or fuel stratification is used to overcome the limitations of HCCI mode. The HCCI load control by creating bulk thermal stratification is still a challenge, and research efforts continue (Sect. 2.4). Fuel stratification creation in the combustion chamber is easy using direct fuel injection, and presently all required technologies to achieve fuel stratification are available. This category of combustion mode is generally termed as partially premixed compression ignition (PPCI) or sometimes abbreviated as partially premixed combustion (PPC), which uses direct fuel injection to modulate mixing time. The PPCI can be considered as an intermediate process, which is a mix of conventional diesel combustion and HCCI combustion [135]. In PPCI fuel is injected during the compression stroke to allow sufficient premixing before ignition. Ideally, all the fuels should be injected prior to start of ignition (positive ignition dwell), but at the same time, fuel injection should not be too early to produce excessive HRR once ignition starts. The ignition dwell is the time duration from end of injection to start of ignition. Typically, conventional diesel engine has negative ignition dwell. The positive ignition dwell in PPCI mode provides time for some degree of premixing to all the injected fuels. Generally, positive ignition dwell is obtained by increasing ignition delay using EGR [17]. Shorter fuel injection durations also help to obtain positive ignition dwell in PPCI mode. The PPCI strategy is investigated for diesel-like (high cetane fuels) and gasoline-like (low cetane fuels) fuels. First diesel PPCI and then gasoline PPCI mode are discussed in the following sections.

2.5.1 Diesel PPCI

Conventional diesel engine combustion has a soot–NO_x trade-off, and generally, methods of reducing soot lead to increases in NO_x emissions and vice versa. To obtain simultaneous reduction in soot and NO_x emissions, charge must be well mixed (to avoid regions with unfavourable carbon–oxygen ratios, charge must also be rich enough to prevent misfire), and combustion results into lower combustion temperatures (below NO_x formation temperature). To achieve lower combustion temperature, most of the strategies used cooled EGR. To achieve premixed charge combustion, high degree of fuel–air mixing is required before ignition. The formation of premixed mixture is mainly determined by location and timing of fuel injection. Since diesel is a non-volatile fuel, direct injection is preferred for premixed combustion. In diesel PPCI mode, two different strategies of direct injection are used, namely, early direct injection and late direct injection with respect to conventional diesel combustion (CDC) injection timings. In early direct injection strategy, fuel is injected directly into the cylinder during the compression stroke, well before TDC to allow time for sufficient fuel-air mixing. The generic combustion with early direct injection is typically known as premixed charge compression ignition (PCCI) [136]. In late direct injections, fuel is injected just before TDC or towards the start of expansion stroke. This injection concept is the basis of the modulated kinetics (MK) and highly premixed late injection (HPLI) combustion strategies in LTC regime [137, 138]. In MK combustion mode, high swirl ratio is used to improve the mixture formation; and to increase the ignition delay, a reduction in compression ratio and cooled EGR is used. High injection pressure in combination with higher nozzle orifice diameter is used to reduce the fuel injection duration. The combination of lower oxygen concentration (by EGR) and retarded fuel injection timings leads to simultaneous reduction in NO_x and soot emissions in MK concept.

In early direct injection PCCI combustion, various strategies exist such as premixed lean diesel combustion (PREDIC), multiple stage diesel combustion (MULDIC), homogeneous charge intelligent multiple injection combustion (HiMICS), premixed compression ignition (PCI), uniform bulky combustion system (UNIBUS) and narrow angle direct injection (NADI). These strategies use one or more injectors, and single or multiple injections in an engine combustion cycle. The start of fuel injection is limited by misfiring due to overmixing (in very early injection) in most of the early injection systems. In the first study of PREDIC strategy, different injector configurations and positions are investigated in combination with variation of injection timings for mixing the fuel [139]. Fuel is injected simultaneously with two side injectors in order to avoid collision of the fuel spray with the cylinder liner. The mixing of air and fuel is promoted by using long ignition delay. Low NO_x and soot emission is achieved because of the good fuel-air mixing and lower combustion temperature. However, results are mainly affected by the fuel injection timings, geometrical arrangement of injectors and charge dilution. Another study presented several improvements in PREDIC strategy and shows that maximum achievable load in PREDIC mode is 50% of naturally aspirated conventional diesel combustion [140]. This limit comes from the limit in excess air ratio to maintain low NO_x and from maximum cylinder pressure limitations (knocking). To extend the PREDIC operation range, a second injection around TDC is introduced in the MULDIC system, which combines a lean premixed combustion (PREDIC) with a more conventional diffusion combustion separated in time [141]. Similarly, the PCI combustion concept uses a conventional direct injection diesel injector, an early start of injection and a reduced compression ratio to achieve low NO_x emission [142]. The UNIBUS system used a fuel injector with a pintle-type nozzle featuring a large hole and a bulbous protrusion to reduce penetration and keep the fuel mixture in the centre of the cylinder away from the cylinder wall [143]. In NADI combustion strategy, the combustion chamber design and spray formation are optimized for early injection by using nozzles with a narrow spray include angle (around 80°) such that fuel can be injected in early compression stroke at relatively lower charge density and temperature environment [144]. In this system, engine can operate in conventional mode at full load with a careful design of the piston bowl geometry. Most of the early direct injection systems can achieve lower NO_x and soot emissions only at lower engine load conditions.

The ignition characteristics and heat release in PPCI combustion mode are significantly different from conventional diffusion-controlled diesel combustion. In LTC mode, initial ignition reactions occur at slower rate than conventional diesel

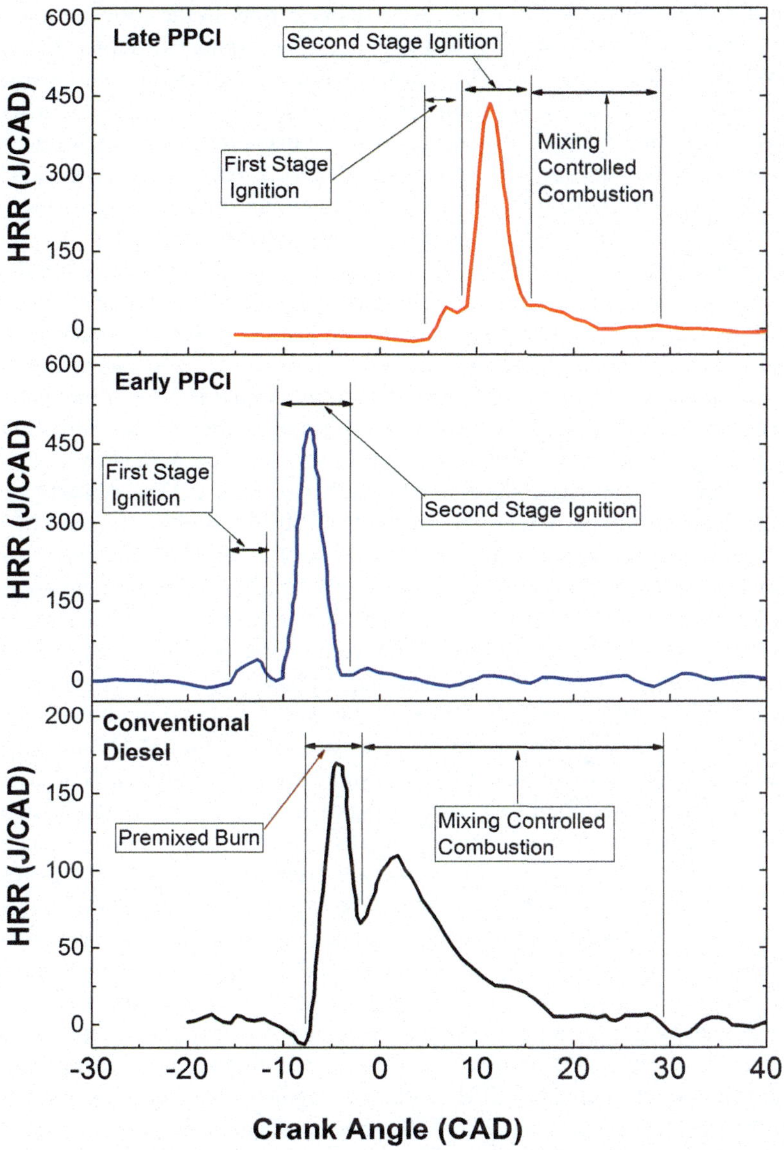

Fig. 2.31 Variations of HRR for early and late direct injection PPCI and conventional diesel engine at similar engine load conditions (Adapted from [5, 145, 146])

combustion due to higher dilution and premixing along with relatively cooled cylinder gases (due to EGR). Figure 2.31 shows the heat release characteristics in early and late direct injection PPCI vis-à-vis conventional diesel engine at similar engine load conditions (~4 bar IMEP). In conventional diesel combustion, two distinct combustion phases of premixed combustion and mixing-controlled combustion are observed after the ignition delay period. In the conventional diesel engine, fuel injection starts near TDC (~11° bTDC) in compressed hot (~990 K) and high density (16 kg/m^3) air [5]. The fuel spray mixes with hot air, and vaporization starts by taking thermal energy from surrounding hot air. The decrease in sensible thermal energy due to vaporization manifests as decrease in HRR during ignition delay period. The fuel–air mixture prepared during ignition delay period auto-ignites simultaneously during premixed phase combustion leading to high peak HRR. The HRR during this period is mainly controlled by chemical kinetics of the mixture formed during the ignition delay period. The HRR decreases significantly after premixed combustion phase due to limited charge preparation by diffusion-controlled process. In mixing-controlled combustion phase, combustion rate is mainly controlled by evaporation and mixing process in the spray instead of chemical kinetics. In typical PPCI combustion mode, the HRR exhibits additional characteristics (two-stage heat release), which is not present in conventional diesel HRR. In the early injection PPCI, fuel injection starts comparatively early (~22° bTDC), where cylinder charge is relatively cooler (~770 K) and less dense (11 kg/m^3) [17]. The first-stage heat release begins near 15° bTDC and ends near 11° bTDC (Fig. 2.31). The second-stage heat release rate is very high (nearly three times of conventional HRR) and is observed in early PPCI mode. In this case, no significant mixing-controlled heat release is observed. The late injection PPCI strategy has similar characteristics to early PPCI with a comparatively lower peak HRR and more significant mixing-controlled combustion phase.

Large fraction of diesel fuel typically consists of long straight-chain alkanes, which mainly governs the two-stage heat release in LTC regime. Although diesel is a mixture of hundreds of chemical species, significant insight into kinetics of diesel combustion can be obtained by investigation of single straight-chain alkane surrogate. The n-heptane is typically considered as surrogate fuel for diesel as its ignition delay and HRR characteristics in engine are similar to diesel [17]. Figure 2.32 shows the ignition behaviour and temperature evolution in early and late injection PPCI mode using n-heptane as fuel. Important insight into heat release process in PPCI mode can be obtained by analysing the evolution of few important species during combustion. Additionally, the behaviour of species like formaldehyde (HCHO) and hydroxyl radical assists in the analysis of laser diagnostic data as these species are accessible in optical diagnostic techniques [17]. Figure shows that a pool of ketohydroperoxide (KPH) species is formed in pre-ignition reaction just before first-stage heat release. The KPH species plays an important role in first-stage ignition in PPCI combustion. The sequence of reaction starts with O_2, and subsequently, hydrogen atom abstraction from parent fuel molecule occurs by hydroxyl and hydroperoxy radicals that results into alkyl radical formation. The alkyl radical reacts with oxygen molecule to form alkylperoxy radicals (RO_2). After several isomerization reactions, a relatively stable KPH is formed. The decomposition of the RO_2 and KPH molecules generates OH and other radicals, which further reacts with the parent fuel molecules. This process accelerates the ignition reactions. At higher temperatures (above 850), formation of RO_2 is curtailed (reaction R3, Sect. 2.2.2) due to higher reverse reaction rate, and this slows down

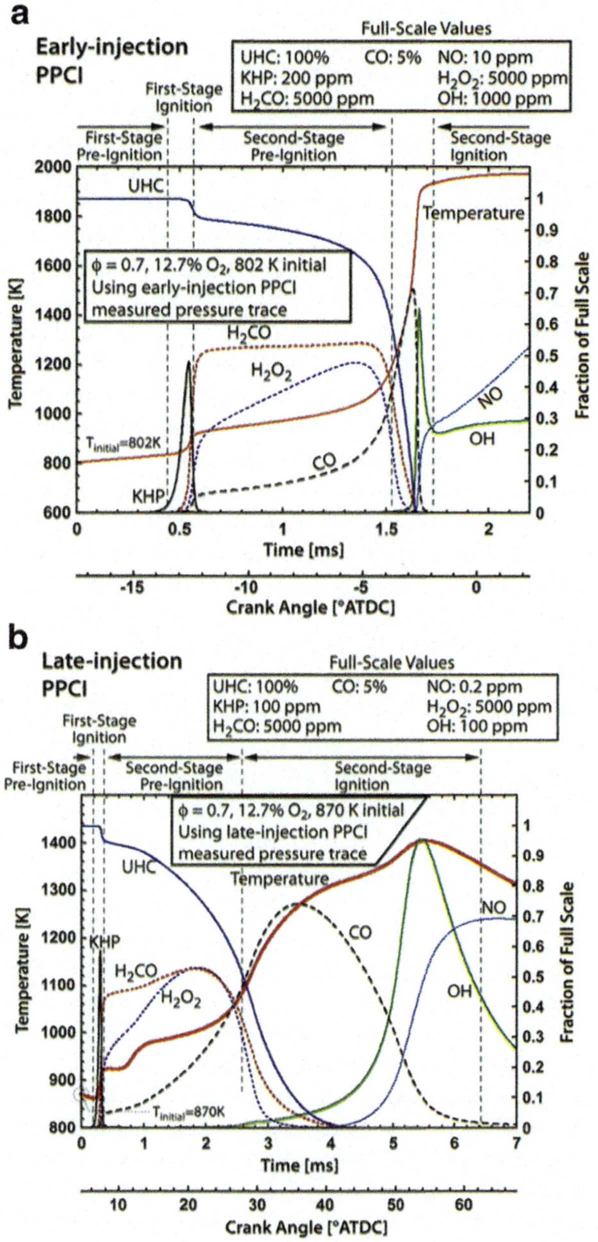

Fig. 2.32 Ignition behaviour of vapour-phase n-heptane for (**a**) early injection (initial temperature 802 K) and (**b**) late injection (initial temperature 870 K) PPCI conditions [17]

the reactions building KHP pool (Fig. 2.32). Three important species (formaldehyde, hydrogen peroxide and CO) are formed in first-stage heat release [17]. The KHP decomposition leads to the formation of oxygenated hydrocarbons, which forms formaldehyde after decomposition. A pool of H_2O_2 is formed and it has an important role in HTHR (Sect. 2.2.2). The H_2O_2 decomposition exceeds production, as charge temperature rises above 1000 K leading to formation of pool of OH radicals, which accelerates the overall oxidation rate. The CO is also formed during the first-stage heat release. However, CO is mostly formed in the early part of the second-stage heat release (Fig. 2.32). Figure shows that second-stage heat release in late injection PPCI has comparatively longer span than early injection PPCI, which is also shown in Fig. 2.31. However first-stage heat release is comparatively smaller. Early injection PPCI mode does not suffer from incomplete combustion in leaner zones to the same degree as late injection PPCI mode. In the late injection PPCI conditions, the charge cools during the expansion stroke that prevents the progress of second-stage ignition. The more complete discussion on combustion reactions in PPCI can be found in original study [17].

Progress of combustion process in PPCI mode is significantly different from conventional diesel combustion mode as illustrated by conceptual model presented in Fig. 2.33. In the early injection conditions, fuel is injected at low temperature and less dense environment leading to comparatively higher liquid length than conventional combustion mode because of fast jet penetration. The instantaneous liquid length decreases by vaporization of downstream portion of fuel jet due to mixing by propagation of entrainment wave (indicated by vertical lines) around 7° ASI after peak injection rate. The fuel is fully vaporized within 1° after the end of injection (EOI) in the PPCI condition investigated in the study [17]. Ignition processes for the PPCI jet diverge from the conventional diesel jet after the initial jet penetration. In conventional diesel combustion (CDC), first fuel-rich portions of the spray jet ignite, and soot precursors formed during the premixed combustion, followed by the soot formation. In PPCI mode, fuel injection ends before second-stage heat release, and OH radical formed near the peak of the second-stage heat release. The OH and formaldehyde distributions do not overlap spatially in the chamber. Some fuel-rich zones are apparent in the downstream jet, where formaldehyde is consumed, but OH does not appear due to the too rich mixtures for accumulating significant OH radical. After few CAD, soot precursors and soot appear within the rich pockets, typically surrounded by OH radical. The soot pockets in PPCI jets are formed in far downstream, near the head vortex, while in conventional diesel jets, soot fills the jet cross-section. Refer to the original study for detailed description of the PPCI model in LTC regime [17].

The main disadvantage of diesel PPCI mode is requirement of very high EGR (upto 80%) for achieving simultaneously low soot and NO_x at higher engine load conditions [147]. The combustion efficiency is very low (~85%) at this very high EGR operating conditions. The compression ratio also needs to be reduced to have lower soot concentration, which limits the expansion ratio. Figure 2.34 shows the variations of NO_x and soot with EGR at different compression ratios. The figure shows that at higher compression ratio, soot formation starts increasing for EGR greater than 50% (Fig. 2.34b) and below 50% EGR, NO_x emission is higher. For simultaneous reduction of the NO_x and soot emissions, lower compression ratio and higher EGR are required in diesel partially premixed combustion. The use of thermodynamically better compression ratio similar to present diesel engines

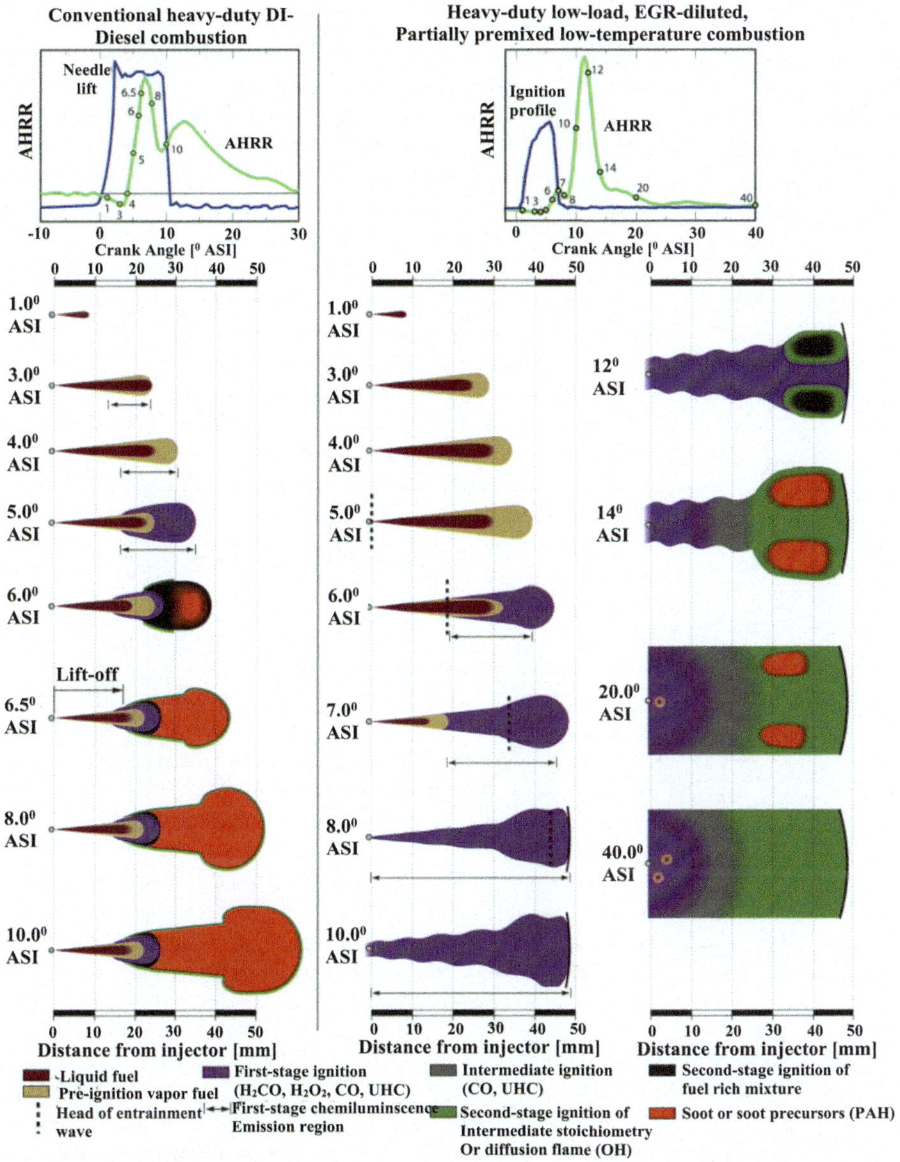

Fig. 2.33 Conceptual model for conventional and LTC (single injection EGR diluted) diesel combustion for heavy-duty engine [17]

(~CR = 18) significantly lowers the ignition delay due to the increase in temperature and pressure in the cylinder. Lower ignition delay limits the premixing of fuel, and this will limit the achievable high load with PPCI using diesel.

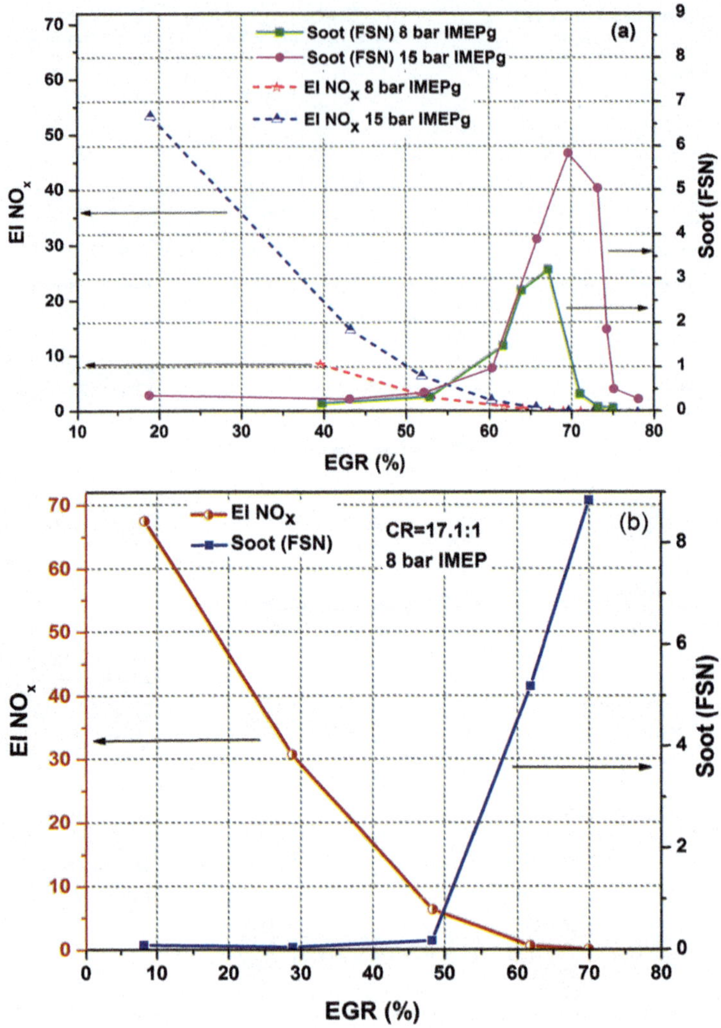

Fig. 2.34 Variation of NO_x and soot emissions with EGR at 8 bar IMEP for (**a**) CR = 12.4 and (**b**) CR = 17.1 (Adapted from [147])

To overcome the problem of shorter ignition delay, low cetane (high octane) fuel is used in PPCI mode, and very low NO_x and soot are obtained at moderate EGR (50%) at a 15 bar net IMEP [148]. The combustion efficiency approaches to 98% even at very close to stoichiometric engine operation with 50% of EGR. This result leads to the further research in low cetane (high octane, gasoline-like) fuels in PPCI engines. The gasoline PPCI is discussed in the next section.

2.5.2 Gasoline PPCI

Inhomogeneities (thermal and fuel stratifications) play a key role in controlling PRR in premixed compression ignition engines (Sect. 2.2.4.6). Thermal stratification has a potential to control HRR in premixed compression ignition engines. Natural thermal stratification exists in cylinder, but it is difficult to create or control thermal stratification in the bulk charge of cylinder. The very unusual methods along with sophisticated control are required to achieve variations in the bulk temperature distribution that utilizes the thermal stratification strategy well (Sect. 2.4). Fuel stratification is typically achieved by varying the fuel injection timings or by using multiple injection strategies. This strategy creates zones of different fuel concentrations in the combustion chamber, which leads to sequential auto-ignition. The zones with higher fuel concentration auto-ignite first and comparatively lower concentration zone next. Sequential auto-ignition in the distribution of equivalence ratio leads to the reduction in peak HRR and extends the combustion duration. The advantage of fuel stratification depends on the φ-sensitivity of fuel (Fig. 2.15). The more φ-sensitive fuel has higher potential to utilize the fuel stratification strategy. Fuels like gasoline show higher φ-sensitivity at higher intake pressure in comparison to naturally aspirated conditions. Therefore, gasoline can be used in PPCI combustion at higher intake pressure conditions more effectively depending on the stratification levels in the cylinder. As discussed in Sect. 2.5.1, diesel PPCI has limitations on higher load limits. To overcome the limitations of diesel PPCI combustion, gasoline PPCI combustion is proposed. This strategy has various names by different groups of researchers such as PPCI [149], gasoline compression ignition (GCI) [150, 151], partially premixed combustion (PPC) [135], low temperature gasoline combustion (LTGC) [110] and gasoline direct injection compression ignition (GDCI) [152]. All these strategies used gasoline-like single fuel in the cylinder. Dual fuel-stratified PPCI is known as reactivity-controlled compression ignition (RCCI), which uses two different reactivity fuels for combustion (discussed in Sect. 2.6).

Figure 2.35 demonstrates the LTC strategies based on the level of fuel stratification in the cylinder from fully premixed to heterogeneous conventional diesel combustion. Using diesel, two different LTC strategies exist, namely, RCCI and PCCI. In RCCI combustion, low reactivity (gasoline-like) fuel is premixed, and high reactivity (diesel-like) fuel is directly injected to create reactivity stratification in the cylinder. Partially premixed combustion using diesel is known as PCCI (see Sect. 2.5.1). The gasoline PPCI strategies can be divided into three main categories: (i) partial fuel stratification (PFS), (ii) moderate fuel stratification (MFS) and (iii) heavy fuel stratification (HFS) (Fig. 2.35) based on degree of fuel stratification in the combustion chamber [6]. The gasoline PPCI is first introduced and experimentally demonstrated in 2005 [148, 153]. Consequently several studies are conducted in PFS [21, 71–73, 78, 154–157], MFS [152, 158–163] and HFS [164–172] regime. In PFS strategy, homogeneous part of charge is prepared by either port fuel injection or very early direct fuel injection, and stratification is created using subsequent direct injection of fuel in small quantity. Stratification is created in such a way that sequential auto-ignition takes place while keeping ultralow NO_x and soot emissions.

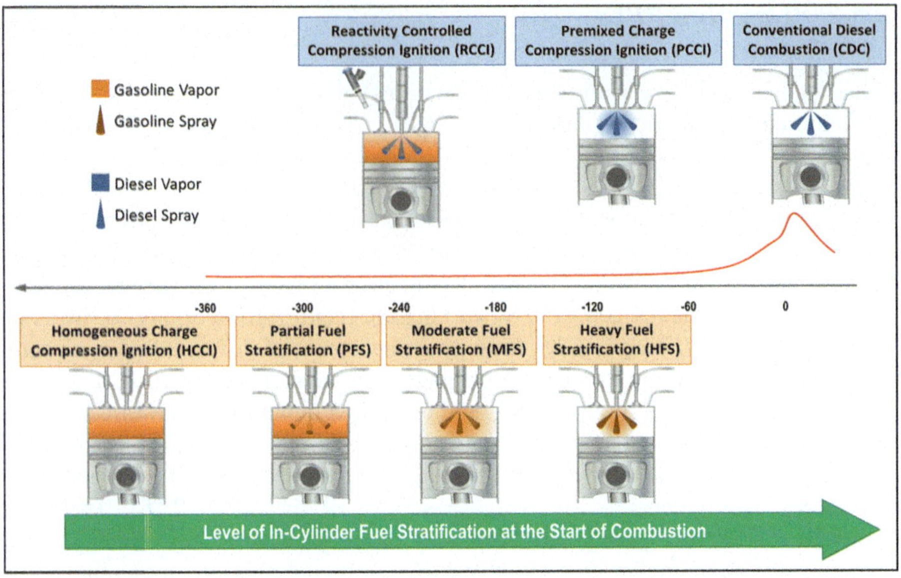

Fig. 2.35 Illustration of LTC strategies in advanced CI engines using gasoline and diesel based on level of stratification in the cylinder [6]

In the MFS strategy, stratification level is higher in comparison to PFS, and it is increased by reducing the amount of premixed fuel. Generally in MFS strategy, all the fuels are injected directly in the cylinder during compression stroke and a fuel injection event near TDC (around 15% of total fuel) to trigger combustion [6]. The HFS strategy has highest level of fuel stratification and it typically has very less premixed fuel. Multiple direct injection events relatively close to TDC is used for charge preparation in HFS strategy. The HFS strategy uses higher fuel injection pressure in comparison to PFS and MFS for completing the injection events before the start of ignition.

The PFS and MFS strategies have comparatively lower combustion efficiency especially at lower engine loads due to lower combustion temperature achieved using leaner mixture. However, reduced heat loss and higher ratio of specific heat (due to leaner mixture) of charge lead to higher thermal efficiency. The NO*x* emissions are strongly connected with the level of fuel stratification and decrease as mixture becomes more homogeneous [6]. Therefore, higher amount of EGR is required in HFS strategy for LTC operation and lower NO_x emissions. A study conducted for a variety of gasoline fuels (in PPC combustion mode with HFS-like strategy) shows that around 50% EGR is required to keep NO_x emission at low level [166]. This EGR percentage is still below than the EGR required for diesel PPCI. This requirement puts pressure to have efficient air handling system to supply high level of EGR and overall lean mixture especially at higher engine load operations. In HFS strategy, ignition timing is governed by injection timings; therefore combustion phasing can be controlled on a cycle-to-cycle basis. The MFS strategy has comparatively lower PRR and longer combustion duration than HFS strategy. In MFS strategy, the combustion phasing is typically controlled by the last fuel injection near TDC.

In the gasoline PPCI mode, fuel injection strategy is central to overall gasoline compression ignition concept. A study suggested the multiple late injections of gasoline (RON91) in GDCI strategy (in MFS regime). In GDCI mode, directly injected gasoline vaporizes and mixes rapidly at low injection pressure typical of gasoline direct injection (GDI) engines [152]. Injection strategy used in GDCI mode is shown in Fig. 2.36a on φ-T map. The colour contours in φ-T map show simulated CO emissions. The fuel injection process can have one, two or three injections per cycle in cylinder during compression stroke (indicated as quantities

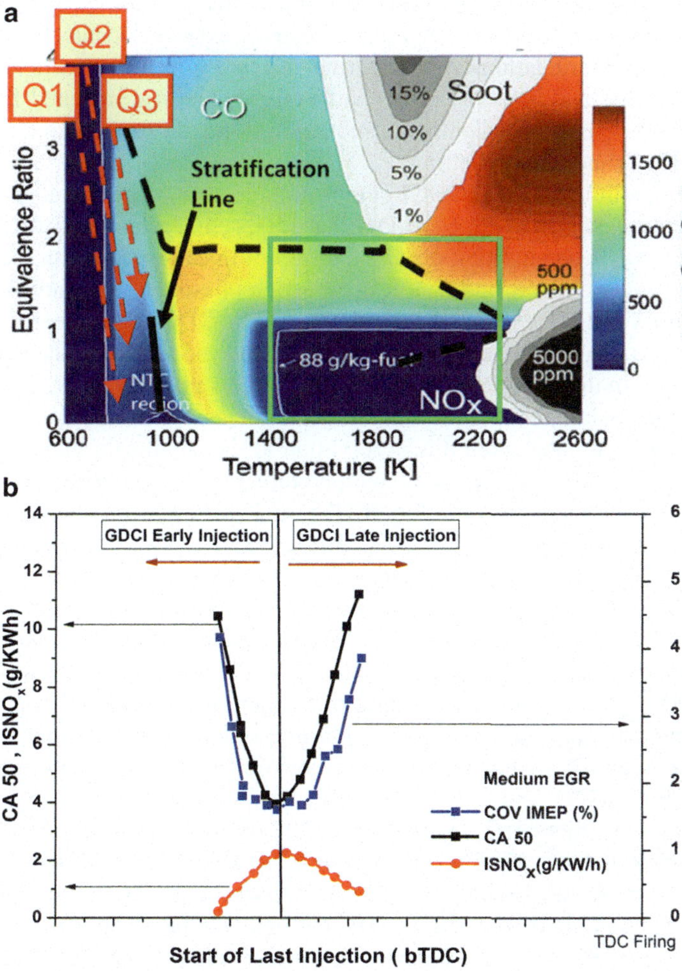

Fig. 2.36 (**a**) Illustration of GDCI fuel injection strategy. Reprinted with permission Copyright © 2014 SAE International [161]. (**b**) Variation of combustion phasing, COV of IMEP and $ISNO_x$ as function of last injection timing at moderate EGR (Adapted from [152])

Q1, Q2 and Q3 in φ-T map). Every injection event starts in the upper left of the φ-T map (liquid), which rapidly vaporizes and mixes to equivalence ratio less than two prior to the start of ignition. At the beginning of ignition, fuel–air mixture is stratified to obtain controlled heat release and stable ignition [161]. The ideal fuel injection process is represented by ideal "stratification line" in Fig. 2.36a. The combustion process must occur in the rectangular region (Fig. 2.36) to achieve low NO_x and PM emissions simultaneously. To minimize CO emissions also, combustion must occur in the region $0 < \varphi < 1.2$ and temperature $1300° < T < 2200°$ K. Therefore, at the beginning of combustion (around TDC), all parcels in the cylinder should be no richer than φ of approximately 1.2 [152]. To achieve the GDCI combustion mode, study used centrally mounted injector, a shallow pent roof combustion chamber, and 15:1 compression ratio engine. To support fuel injection-controlled mixture stratification, a quiescent open chamber design is used. Swirl, tumble and squish motions are minimized because excessive charge motion may destroy the fuel stratification created during the fuel injection events [161]. The piston bowl shape and injector spray characteristics are also matched for typical GDCI injection timings for significant reduction in piston surface area and lower propensity for cylinder wall wetting.

Figure 2.36b illustrates the GDCI combustion process using a double injection strategy and moderate EGR for 6 bar IMEP at 1500 rpm. A characteristic U-shaped response of CA_{50} is obtained by varying the start of injection (SOI) for the last fuel injection even in the cycle, and hence, combustion phasing to be controlled in one of two regimes (GDCI early and late injection). In GDCI late injection regime, combustion phasing retards as injection timing is delayed similar to the response of a CDC system. In this regime, ignition dwell (IDW) is relatively low. In the GDCI early injection regime, combustion phasing retards as injection timing advances. Authors of the study attributed this observation to the significant increase in IDW due to lower cylinder pressure and temperature [152]. Combustion stability is measured by coefficient of variation (COV) of IMEP, and combustion stability decreases as combustion phasing retards (Fig. 2.36b).

Another study investigated the nature of heat release in gasoline PPCI engines using PFI/DI split injection (60/40) strategy at 1200 rpm, 15.1 compression ratio and 1.8 bar intake pressure [149]. The PFI is used for preparing the premixed charge, and DI is used to create stratification which controls the combustion phasing. Figure 2.37 shows the variation of maximum pressure rise rate (MPRR), ignition timing (CA_{10}) and combustion phasing (CA_{50}) with DI timings at 15% EGR. Three regimes (A, B and C) are found in behaviour of maximum PRR (marked in Fig. 2.37).

In regime A, as injection is retarded from 30° aBDC (intake) to 105° aBDC, the maximum PRR reduces significantly (from 9 MPa/ms to 1.8 MPa/ms), and the corresponding retard is also observed in the combustion phasing (CA_{10} and CA_{50}). However, the combustion becomes erratic and ignition fails with further injection retard (beyond 105°aBDC) but retarded injection timings beyond 140°aBDC; the combustion stabilizes again [149]. In the regime A, ignition delay is longer due to injection in low temperature and pressure charge. With SOI retard, ignition delay

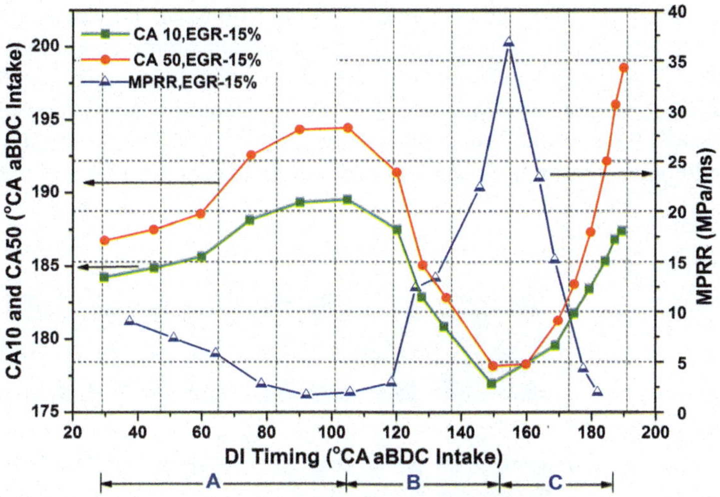

Fig. 2.37 Maximum PRR, CA_{10} and CA_{50} versus DI timings in PPCI engine (Adapted from [149])

decreases, but it is not enough to compensate for the retard of SOI and hence CA_{10} delays with retarded SOI. In regime B, further injection retard leads to the increase in maximum PRR, and it goes up to very high peak of 37 MPa/ms. In this regime, fuel stratification is also higher, and ignition delay is shorter due to injection at relatively higher cylinder pressure and temperature. The fuel-rich zones in the high temperature environment have a shorter ignition delay, which compensates for the retarded SOI. Therefore, retarded injection timing results in advancement of ignition (CA_{10}). In regime C, maximum PRR decreases with further retard of DI injection timing, and with late DI timings, the heat release becomes more a two-part process, in which the mass fraction burned curve first shows a slower burn followed by a faster burn [149]. In this regime, ignition delay is very short due to fuel injection in higher temperature and pressure conditions. The time needed to physically evaporate and mix fuel overpowers the gain in shortening the delay at this stratification level. Therefore, trend reverses again and the combustion retards with the retard of SOI. This study further concluded that maximum PRR correlates mostly with the combustion phasing that changes the mixture temperature at point of ignition. It is also suggested that the main effect of SOI is to change the combustion phasing, and the MPRR responds to the combusting phasing change [149].

A recent study developed a reduced chemical mechanism for five-component gasoline surrogate and investigated the heat release characteristics in GCI engine [173]. This study concluded that with comparatively early direct injection timings (−52° aTDC), the GCI combustion process can be understood as a partially sequential auto-ignition by the competition between the charge cooling effect and the ϕ-sensitivity effect for the stratified mixture. The partial sequential autoignition lowers the peak PRR. In case of late DI (−5° aTDC) timing, the GCI combustion duration is increased due to the mixing process between the late injected fuel and the well-mixed mixture. Thus, the GCI combustion can be decoupled into a partially sequential auto-ignition and a subsequent non-premixed combustion.

A study investigated the degree of stratification at low load engine conditions for different injection strategies using OH chemiluminescence imaging at constant NVO, inlet pressure (1.3) and λ (3.5) and inlet temperature (80) [162]. Figure 2.38 shows the sequential combustion images and HRR for single injection. In this case, there is a

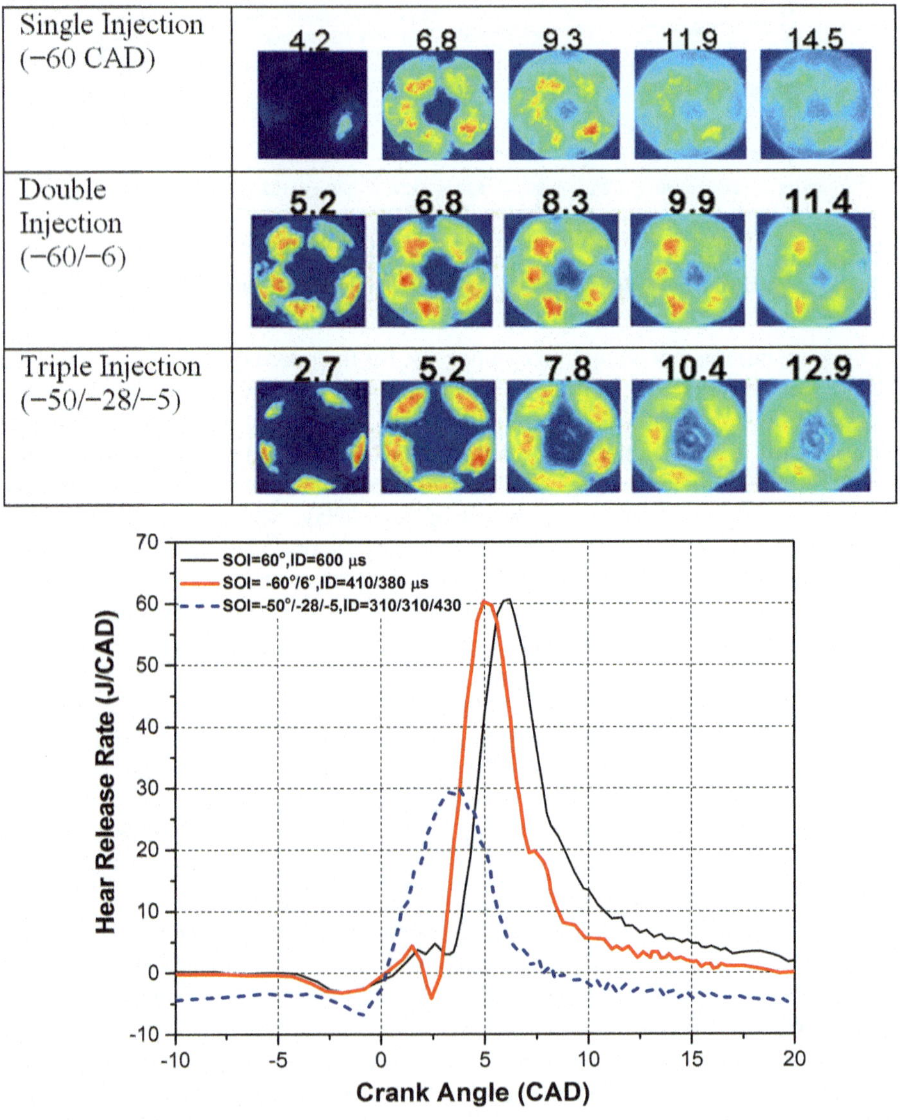

Fig. 2.38 OH – chemiluminescence and heat release rate for single, double and triple injection in PPC combustion (Adapted from [162])

zone with rich mixture starting auto-ignition at 4° aTDC, and consequently five different combustion zones are appearing at around 7° aTDC. There is no significant difference observed in HRR from single and double injection. The early first injection in the compression stroke gives sufficient time for the stratified charge to mix with the surroundings before start of combustion [162]. In case of three injection, peak HRR is much lower, and combustion duration is comparatively longer. In this case, auto-ignition initiates at the outer rim of the combustion chamber. The first fuel injection pulse creates a rather homogeneous charge, and second injection pulse produces a stratification in the combustion chamber and generating zones with higher reactivity. The fuel injection in third pulse burns high reactivity zones first, and then the reaction extends to the relatively leaner zones [162].

Fuel effects on ignition delay and low temperature reactions (LTR) during PPC are analysed using different research octane number (RON) surrogate fuels [174]. The study found that the increase in ignition delay as RON value increases. Ignition delay in PPC engine is more influenced by alcohol than aromatics due to increases in RON value. A proportional correlation between LTR and ignition delay is observed. Another interesting finding of the study is that the ethanol and toluene amplified the LTR phase, while n-heptane suppressed it in PPC mode which is in contradiction to what was reported in HCCI mode [174]. Another study investigated nine fuels in the boiling point range of gasoline but with an octane number spanning from 69 to 99 in PPC mode in the range of 5 and 25 bar gross IMEP at 1250 rpm [166]. The study concluded that a fuel in the boiling point range of gasoline with an octane number in the range of 70 is the best fuel for PPC engine from maximum load to idle condition. The study conducted in HFS regime used high boost level combined with ~50% of EGR and ~1.4 of λ, which allowed around 53.5% gross indicated efficiency throughout the entire test load sweep. The boost pressure, λ, injection pressure and main injection timings used for different loads in PPC are illustrated in Fig. 2.39 for RON70 fuel. Figure shows that very high boost pressure along with higher EGR rate is required for PPC mode combustion at higher engine loads. Fuel injection pressure used is also very high (~2400 bar). Therefore, to run in PPC mode at higher engine load, very efficient air handling and fuel injection system is required.

2.6 Reactivity-Controlled Compression Ignition

2.6.1 RCCI Fundamentals

Reactivity-controlled compression ignition (RCCI) strategy is also in the category of partially premixed compression ignition, and it uses two fuels of different reactivities. Thus, RCCI can be named as dual fuel partially premixed combustion. The premixed compression ignition modes have two major challenges of combustion phasing and combustion rate control. To have better control on combustion

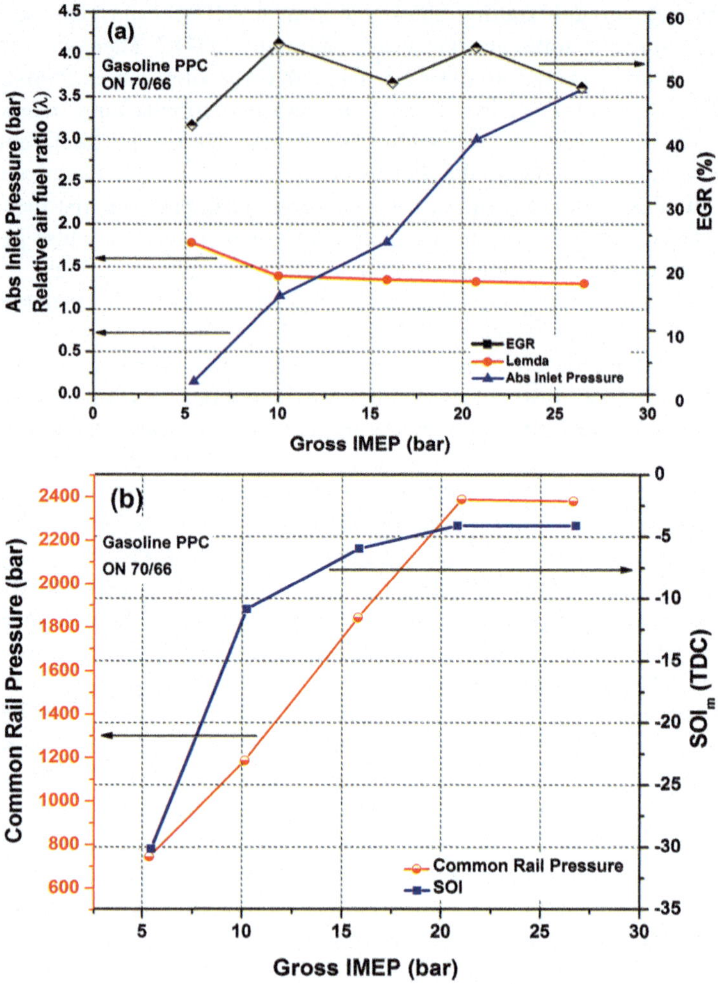

Fig. 2.39 EGR, λ, absolute inlet pressure, common rail pressure and start of main injection as a function of load in PPC engine (Adapted from [162])

phasing, direct fuel injection strategy is typically used, but sufficiently long ignition delay is provided for mixing. Due to positive ignition dwell in these strategies, injection and ignition process is decoupled. Thus, chemistry time scales should be controlled to achieve appropriate combustion phasing. The GCI strategies (Sect. 2.5.2) have shown significant potential to achieve higher engine load at higher efficiency as well as lower NO_x and PM emissions. In GCI strategies, achieving combustion at lower engine load is difficult due to lower auto-ignition reactivity of gasoline. Diesel has higher reactivity but it has difficulty of controlling combustion phasing at higher engine loads in PPCI mode (Sect. 2.5.1). Therefore, studies explored the premixed compression ignition using blends of different fuel reactivities [175–177]. A study concluded that the best fuel for HCCI combustion has autoignition reactivity between diesel and gasoline fuels [176]. Studies suggested that different fuel blends are required at different engine operating conditions such as a high reactivity fuel at lower engine loads and a low reactivity fuel at higher engine loads. Thus, in-cylinder blending of two different reactivities can be beneficial in PPCI combustion.

Thermal and fuel stratification in the cylinder is a promising strategy to control the HRR in premixed charge CI engines (Sect. 2.2.4.6). Partial fuel stratification and thermal stratification introduce the gradients into the auto-ignition characteristics (i.e. the ignition delay) of the charge [175]. The HRR depends on speed of propagation of multiple ignition fronts in the chamber. Speed of a propagating subsonic spontaneous ignition front (S_{ig}) is related to spatial gradient in ignition delay ($\nabla\tau$) by inverse relation ($S_{ig} = 1/|\nabla\tau|$) [175, 178, 179]. Inverse relation of ignition delay gradient to ignition front propagation shows that controlling the gradients in ignition delay is critical to control the HRR of premixed combustion. Ignition delay gradient depends on state of mixture and the state of surrounding mixtures that interact through transport and compression heating. Ignition delay is a function of temperature, pressure, composition, equivalence ratio and fuel type. The dependency of ignition delay and associated HRR on fuel type suggests that in-cylinder fuel blending can be beneficial to control the HRR also. Thus, in-cylinder fuel blending of two fuels with different reactivities has the potential for simultaneous control of the HRR and combustion phasing, and hence, RCCI combustion is investigated in detail [175].

The RCCI combustion is a dual fuel LTC strategy. In this strategy, two fuels of different auto-ignition reactivities (high and low reactivity) are used for premixed charge preparation by in-cylinder fuel blending. This strategy adopts multiple injection strategies and appropriate EGR level for controlling the in-cylinder reactivity to optimize the combustion phasing and combustion duration, which leads to higher thermal efficiency along with simultaneous reduction of NO_x and soot emissions to ultralow level [180, 181]. Figure 2.40 schematically illustrates RCCI combustion concept. The low reactivity fuels (gasoline-like fuels) are injected in intake manifold using port fuel injection (PFI) system, which premixes with air during intake stroke, due to their volatile nature. The high reactivity fuel (diesel-like fuels) is directly injected in the cylinder using DI system with a single, double or triple injection strategy during compression stroke. The early injected high reactivity fuel targets the squish region of the combustion chamber, whereas the relatively late injected high reactivity fuel acts as an ignition source [181, 182]. The dilute gasoline–air mixture does not auto-ignite without diesel injection in cylinder. The ratio of low and high reactivity fuel and injection timings of high reactivity fuel can be used to control the RCCI combustion process. Injection of two different auto-ignition reactivity fuels creates both the reactivity and equivalence ratio (φ) stratifications in the entire combustion chamber. Reactivity stratification adds an additional fast-response control parameter in the form of fuel chemistry, enabling the global fuel properties to change across the operating map.

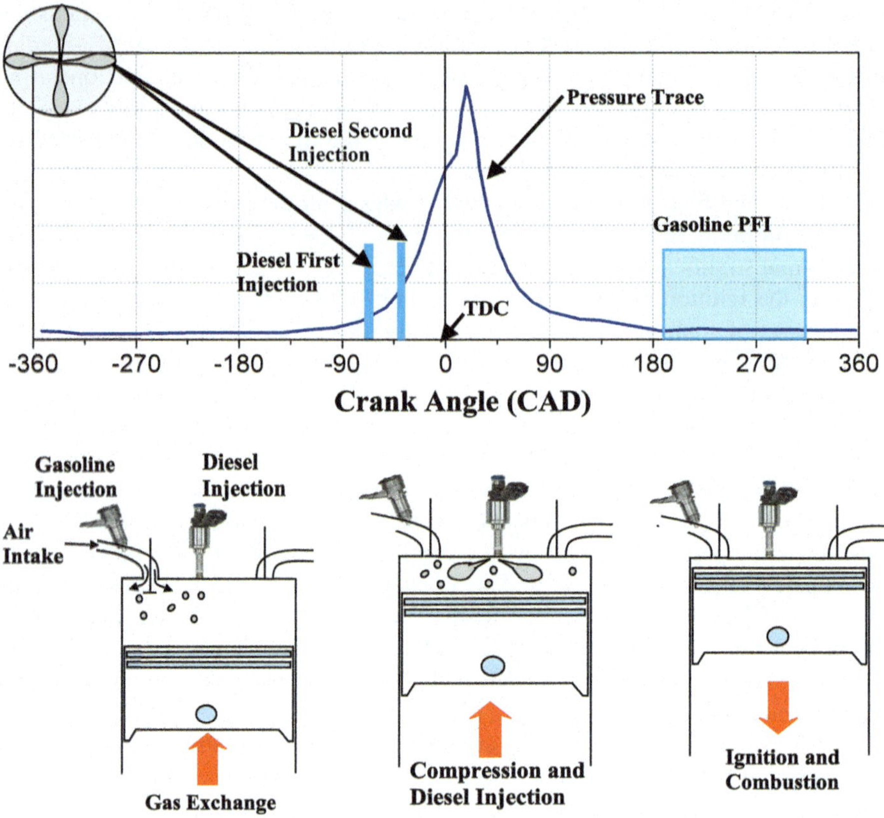

Fig. 2.40 Schematic diagram of RCCI combustion mode (Adapted from [182])

Effect of fuel distribution due to start of injection (SOI) timings of high reactivity fuel in RCCI combustion (with iso-octane as the premixed low reactivity fuel and n-heptane as the direct injected high reactivity fuel) is demonstrated by optical imaging study [183]. The high-speed combustion luminosity imaging in Fig. 2.41 shows distinctly different combustion events for the early, mid and late SOI timings. Figure 2.41 shows the effect of injection timing on the combustion process as the combustion event progresses in the cylinder due to φ- and reactivity stratification. The differences in combustion characteristics are due to differences in the mixing times of the three cases. The early SOI timing results in rapid energy release due to an under-stratified (overmixed) charge, and the late SOI timing results in rapid energy release due to an over-stratified charge (under-mixed) [183]. The mid SOI timing (−50° ATDC) is a representative of typical RCCI injection timings and demonstrated a long heat release duration due to the stratified nature of the moderately premixed charge.

Figure 2.42 depicts the PRF (primary reference fuel), equivalence ratio and ignition delay distributions near the cylinder wall from the optical study

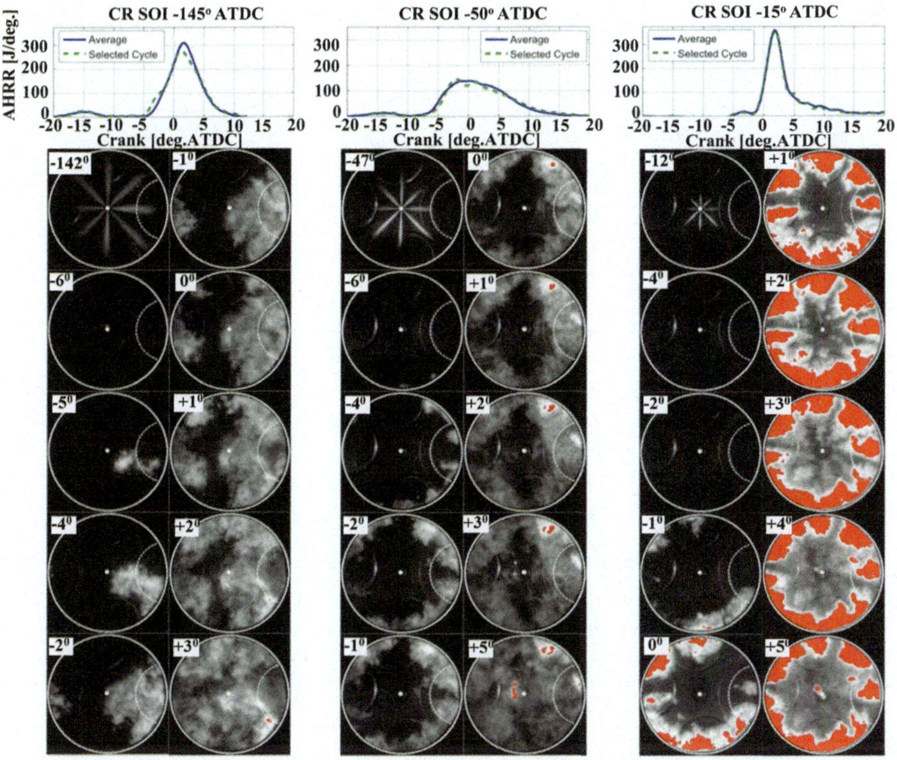

Fig. 2.41 High-speed movie sequence showing liquid fuel sprays and natural luminosity for the cases with common rail SOI timings of −155°, −50° and −15° ATDC. The number in the upper *left*-hand corner of each image shows the crank angle at which the image is acquired. Reprinted with permission Copyright © 2012 SAE International [183]. Regions coloured *red* indicate saturation

[183]. The numbers shown agaist PRF are octane number of PRF fuel (Fig. 2.42). Large variations in PRF distribution are observed for −15° SOI, and very rich regions exist for this injection timing. Study concluded that the high equivalence ratio regions ($\varphi > 0.6$) are accountable for the high HRR (for SOI −15°) because of the almost same ignition delay in these regions (Fig. 2.42b, rightmost image). The ignition delay sensitivity to φ decreases with increased φ beyond about $\varphi = 0.6$. Due to lack of ignition delay sensitivity, almost entire mixture ignited in a very small combustion duration [183, 184]. Combustion timing and duration are dependent on the ignition delay gradients in RCCI combustion. It is well known that short ignition delays lead to advanced combustion phasing, while long ignition delays result in retarded combustion phasing. The smaller ignition delay gradients result in short combustion durations and large ignition delay gradients result in long combustion durations in RCCI engine [184]. The ignition delay gradient is very small

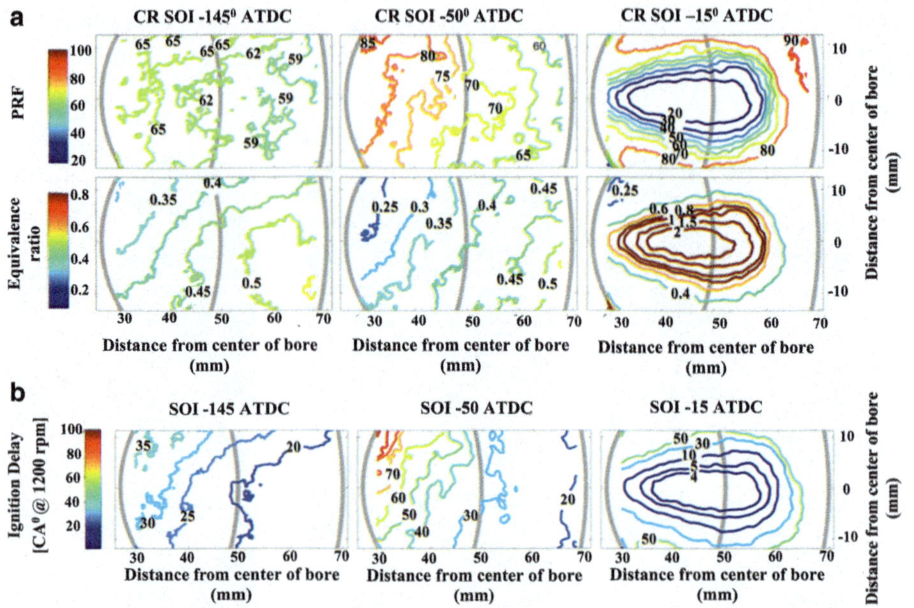

Fig. 2.42 (**a**) Ensemble-averaged PRF and equivalence ratio distributions and (**b**) corresponding ignition delay (calculated from constant volume adiabatic simulations) at −5° CA ATDC for different SOI timings. Reprinted with permission Copyright © 2012 SAE International [183]

for overmixed case (−145° SOI case) leading to short combustion duration and high HRR. The relatively larger ignition delay gradient (produced by both the reactivity and equivalence ratio distributions (Fig. 2.42)) is responsible for the longer combustion duration and lower HRR for the −50° SOI condition.

The evolution of PRF distribution with respect to crank angle position is shown in Fig. 2.43 for RCCI combustion with two direct injections. The figure shows that at −50°ATDC (shortly after the first injection), vapour fuel has penetrated near to the cylinder walls. As piston moves by −45°ATDC position, the fuel vapour starts spreading along the cylinder liner. The PRF number in the upstream zone rapidly starts decreasing due to the further penetration of vapour fuel and entraining ambient gas (air and iso-octane) [185]. At −45°ATDC position, second injection occurs and mixes with cylinder charge with slower penetration because of the higher ambient densities at time of second injection. The ensemble-averaged PRF number in the head of the jet is approximately 40 (i.e. a 40–60 blend of iso-octane and n-heptane) at −25°ATDC position. It can be noticed that much of the fuel from the first injection has mixed to produce a relatively uniform mixture in the squish region of the chamber with a PRF number ranging from 55 to 60, outside region is influenced by the second fuel injection [185]. Figure 2.43 also shows that the charge continues to mix in the duration between −21° and −5° ATDC. Highest reactivity region remains positioned in the downstream portion of the jet, near the cylinder

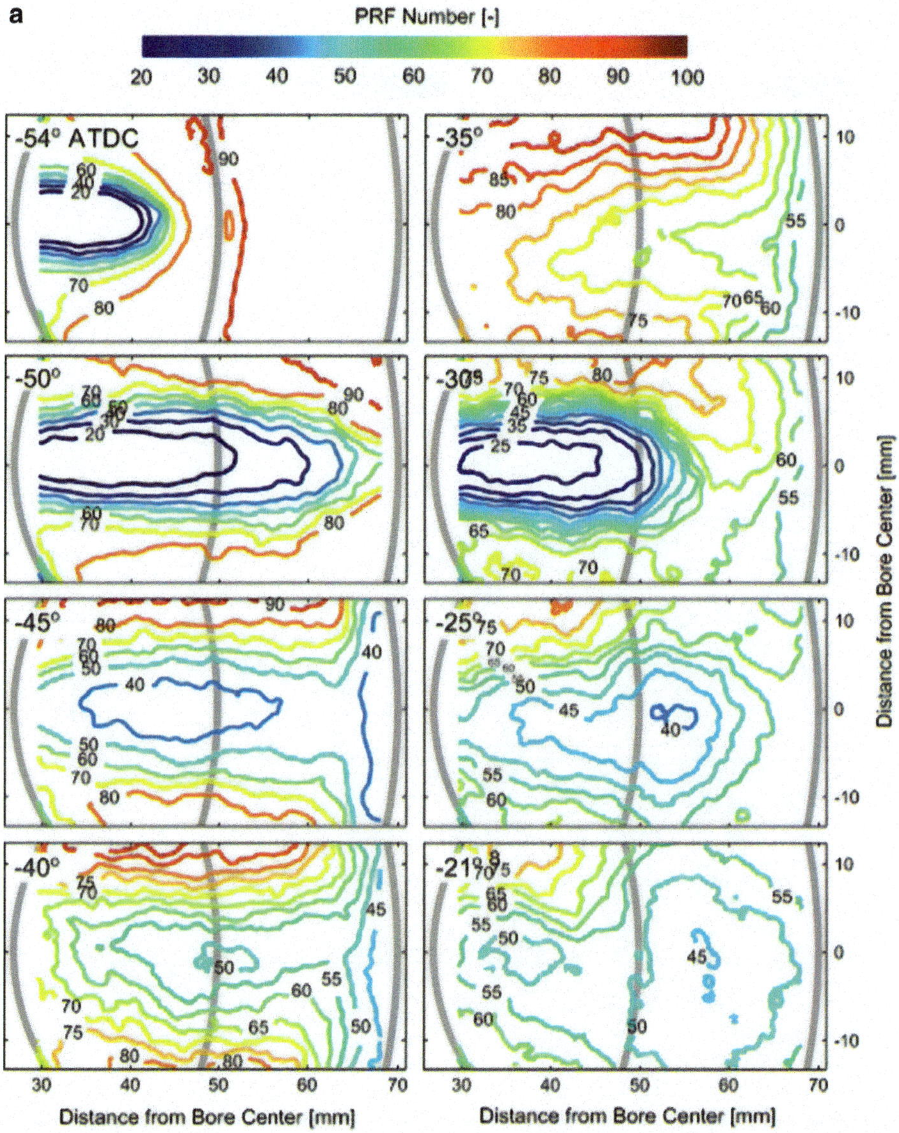

Fig. 2.43 Sequence of ensemble-averaged PRF maps at different crank angle positions during the common rail injection event and prior to ignition event in RCCI engine [185]

wall. The fuel reactivity reduces with decreasing distance from the cylinder centre. A more complete discussion can be found on original study [185].

The reactivity in RCCI combustion chamber can be characterized into global reactivity and reactivity gradient/distribution [181]. Global reactivity of charge is purely estimated by the amount of each fuel and the auto-ignition reactivity indices of fuels. The reactivity gradient is associated with the spray penetration of directly

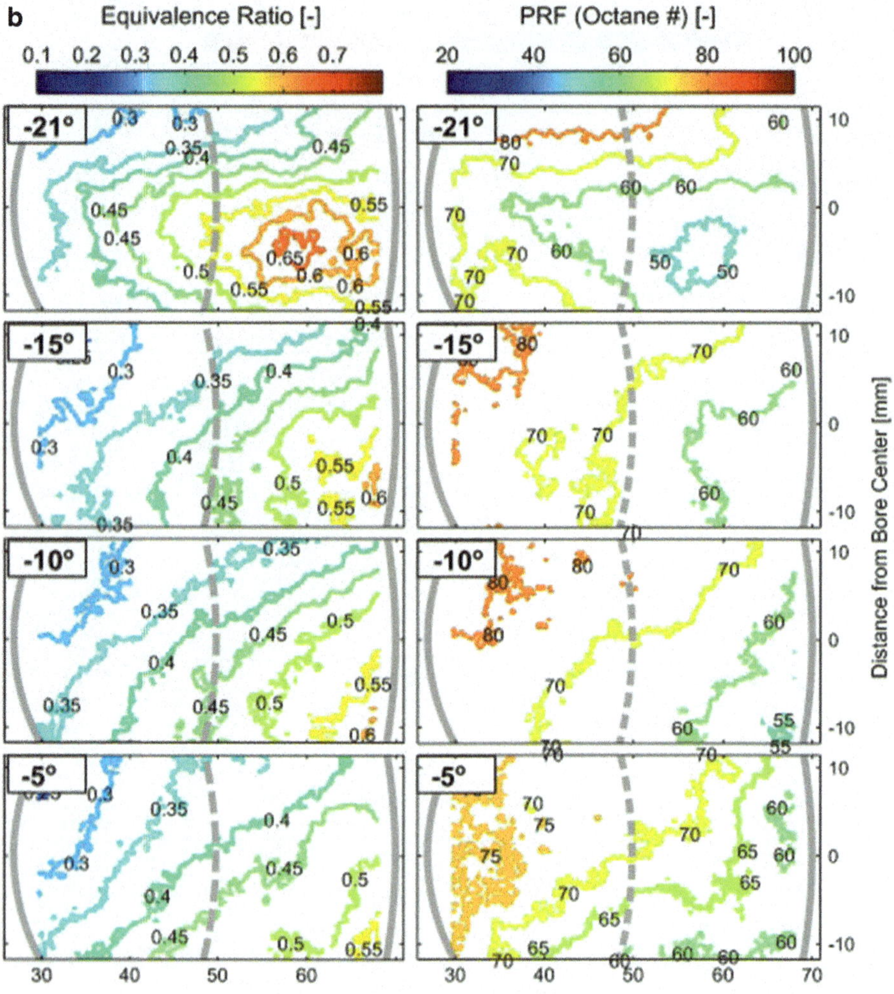

Fig. 2.43 (continued)

injected high reactivity fuel and the entrainment of surrounding charge. A study investigated the individual effect of reactivity, equivalence ratio stratification and temperature stratifications using combination of optical diagnostics and chemical kinetic modelling [185]. Figure 2.44a depicts the predicted ignition delays (estimated using chemical kinetic simulations in constant volume chamber) as a function of distance from the cylinder centre to various combinations of φ-, PRF and temperature stratification. Figure 2.44 depicts that temperature stratification competes with the φ- and PRF stratification (i.e. zones having high φ and low PRF numbers achieve lower temperatures due to evaporative cooling). Thus, the absence of the temperature stratification increases the spread between the least reactive and most reactive regions (Fig. 2.44a). Temperature stratification alone has a small

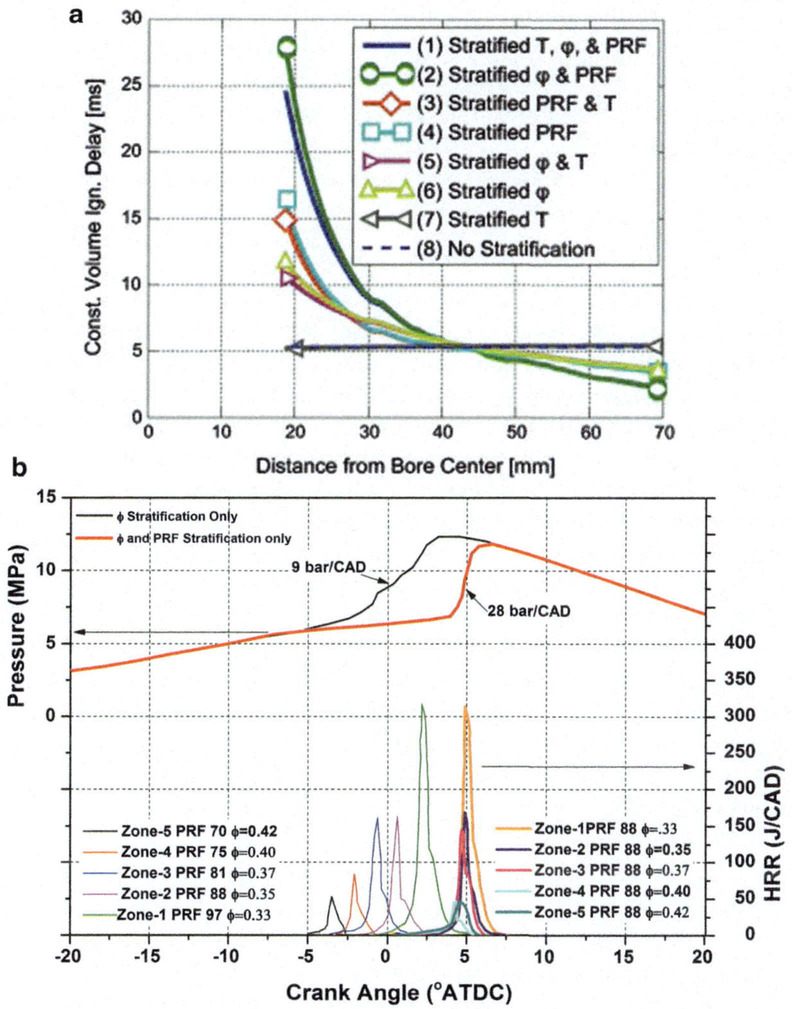

Fig. 2.44 (**a**) Effects of equivalence ratio (φ), PRF and temperature stratification on radial variation of the predicted ignition delay [185] and (**b**) calculated cylinder and HRR with and without fuel reactivity gradient (Adapted from [186])

effect on the ignition delay (Fig. 2.44a, case7), and PRF stratification has most significant effects (Fig. 2.44a, case4). The study concluded that PRF number stratification is the dominant factor for controlling the ignition location and growth rate of the reaction region [185]. The equivalence ratio (φ) has comparatively smaller effect but it is significant.

Figure 2.44b demonstrates the effect of fuel reactivity gradient on HRR and PRR. In case of only φ-stratification, every zone ignites almost simultaneously, which leads to shorter combustion duration and higher PPR. In case of φ- and PRF stratification, zones that are more reactive ignite significantly early in the cycle, and energy release between zones is staged such that the combustion duration is considerably increased leading to lower PRR (Fig. 2.44b). Therefore, effects of φ- and reactivity stratifications must be considered simultaneously in the RCCI combustion mode [187].

2.6.2 RCCI Fuel Management

Fuel reactivity gradient has a crucial role in RCCI combustion characteristics. Researchers have extensively investigated different types of low and high reactivity fuels with their different combinations in RCCI combustion. The RCCI combustion fuelling strategy is of mainly two types: dual fuel strategy and single fuel strategy (along with cetane improver). In dual fuel strategy, two different reactivity fuels are used to create the reactivity gradient in the combustion chamber, and therefore two fuel tanks are required in this particular strategy [181]. This requirement to refuel two fuel systems can result in decreased market acceptance. An alternative to dual fuel strategy is suggested using a single fuel stock with the same fuel delivery system but with the addition of a highly reactive cetane improver in small concentration to the DI fuelling system [187]. High reactivity cetane improver creates reactivity gradient in the cylinder, and φ-gradients are created by direct injection similar to dual fuel strategy. Two common cetane number (CN) improvers are 2-ethylhexyl nitrate (2-EHN) and ditertiary butyl peroxide (DTBP). Significant increase in CN is achieved by using only a small percentage of either of these chemicals [188, 189].

In the dual fuel strategy, gasoline and natural gas are most widely used as low reactivity fuel in RCCI combustion with diesel as high reactivity fuel [180, 181, 190]. Renewable fuels such as methanol, ethanol and iso-butanol are also considered as substitute of gasoline for low reactivity fuel in RCCI combustion [191–194]. A study compared the combinations of methanol/diesel and gasoline/diesel and indicated that the use of methanol could extend the RCCI engine operating load with relatively lower requirement of EGR [193]. Another study investigated different types of low reactivity fuels (various blends of ethanol/gasoline) in a RCCI engine and found that lower premixed ratio of E85 (85% of ethanol in volume) is required to obtain a stable combustion [195]. Typically, diesel and biodiesel are used as high reactivity fuels in RCCI engine [181, 196]. The RCCI engine fuelled by gasoline/ULSD (ultralow sulphur diesel) and gasoline/biodiesel showed that gasoline/biodiesel has comparatively lower NO_x and HC emission in comparison to gasoline/ULSD at the selected operating conditions [196]. Another study showed that the more stability to the cycle-to-cycle transitions could be achieved in the tested NG/biodiesel fuelled RCCI engine due to higher cetane and oxygen containing biodiesel fuel [197].

Similarly, several studies investigated the effect of cetane improvers in single fuel strategy of RCCI combustion. The effect of CN improvers in RCCI engine is investigated by port fuel injection (PFI) of gasoline and direct injection (DI) of gasoline mixed with small amounts of DTBP [198]. Study showed that the RCCI combustion could be realized by adding as little as 2% by volume DTBP into the direct injected gasoline. Another study showed similar trend with EHN [199]. Although there is availability of DTBP, 2-EHN is a more common additive, and it is used in refineries to increase CN of commercial ULSD blends [187]. Although EHN is a more effective CN improver than DTBP, the fuel-bound nitrogen in EHN can increase engine-out NO_x emissions [200]. Detailed discussion on fuel management strategies (dual and single fuel) can be found in review studies [180, 181, 187].

2.6.3 RCCI Engine Management

Fuel injection strategy plays an important role in combustion and performance characteristics of RCCI combustion engine. The main controllable parameter is the injection parameters of high reactivity fuel such as number of injection (single, double and triple injection events), injection timings, injection pressure and quantity of injection in each event. The effect of SOI timing of high reactivity fuel on combustion characteristics and charge quality before ignition is discussed in Sect. 2.6.1. The number of injection events affects the reactivity gradient distribution in the combustion chamber (Fig. 2.43). The number of pulses in the fuel injection strategy can also affect the engine load. A study investigated three different injection strategies, early single injection, early double injection and late double injection, in RCCI engine at constant combustion phasing ($CA_{50} = 6°$ ATDC) [201]. This study found that late double injection could be used to extend operating load range while maintaining acceptable emission levels and peak pressure rise rate (PPRR) values. A numerical study showed that the split injection fraction of high reactivity fuel could affect the originating region (bowl or squish) of combustion and consequently the PPRR and ringing intensity [190]. However, study found there is less effect of split fuel fraction on engine performance and emissions.

A study used three different fuel injection strategies to achieve the dual fuel operation in the entire engine operating map with constraint on maximum PRR and peak cylinder pressure below 15 bar/CAD and 190 bar, respectively [202]. Figure 2.45 demonstrates the proposed fuel injection strategies and typical HRR from each strategy at 1500 rpm. Based on the level of charge stratification, three regimes are characterized as fully premixed RCCI, highly premixed RCCI and dual fuel diffusion combustion. At higher engine load, diffusion-controlled HRR can be noticed (Fig. 2.45). In fully premixed RCCI (around 8 bar IMEP), a double diesel injection pulse with highly advanced injection timings is used. Ultralow NO_x and soot emissions obtained in fully premixed RCCI regime due to the high mixing time available before start of combustion (SOC). In highly premixed strategy (40% up to 75% load), a double diesel injection pulse is used with the second injection event occur close to TDC. The first injection is used to improve the reactivity in the

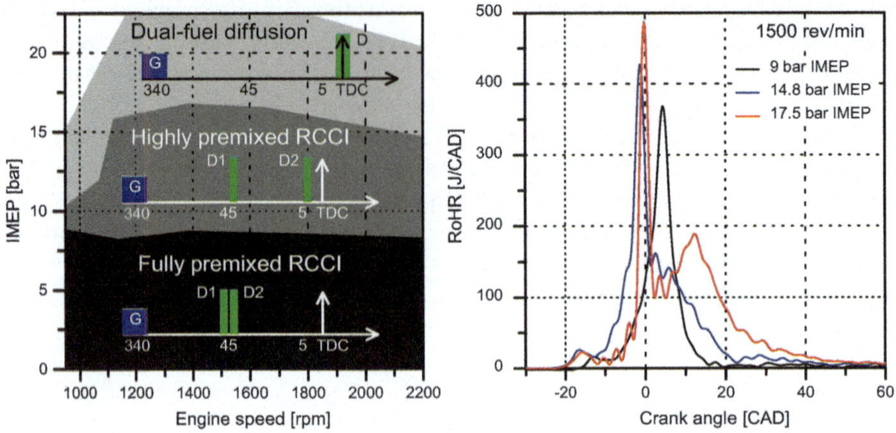

Fig. 2.45 Fuel injection strategies on the engine map (*left*) and typical HRR curves (*right*) from each injection strategy at 1500 rpm [202]

crevice zone. The second diesel injection acts as an ignition source as injection occurs at the higher pressure and temperature in the cylinder [202]. Diffusive fuel strategy (from 75% up to full load) used a single diesel injection with a retarded fuel injection timing that permits injecting higher amount of diesel to achieve the required load.

Most of the RCCI studies used higher fuel injection pressure (greater than 500 bar) using common rail injection (CRI) hardware for injection of high reactivity fuels [180]. The feasibility of using low-pressure fuelling strategy by commercial gasoline direct injector (GDI) hardware is investigated [203]. Low-pressure fuelling strategy has capability to provide cost savings. Study showed that with conventional diesel combustion (CDC) piston, low-pressure direct injection of high reactivity fuels lowers the combustion efficiency by 1% while maintaining benefits of NO_x and soot emission at 6 bar IMEP in RCCI engine. Study indicated that low-pressure direct injection of high reactivity fuel is not able to provide sufficient mixing to offer equivalent engine performance as with the use of high-pressure direct injection with CDC piston [203]. Same study optimized the RCCI piston geometry and found comparable engine performance for both low-pressure direct injection and high-pressure direct injection RCCI engine operations. Thermal efficiencies are increased 5% absolute in comparison to CDC piston with 500 bar CRI operation by using optimized piston with 150 bar GDI operation. This gain in thermal efficiency is attributed to significant reductions in heat transfer via lower piston surface area coupled with reduced and elongated heat release rates for the peak efficiency cases [203].

Appropriate combustion chamber design can be effective in reduction of emission formation, without affecting the RCCI engine performance. The piston bowl geometry affects the charge mixing process, which can influence the reactivity, and φ-stratification in RCCI engine that results into different combustion characteristics. Piston surface area also affects the heat transfer process, which has an impact on combustion temperature and engine efficiency. A study optimized the piston shape for premixed fuel by reducing surface to volume ratio for lower heat transfer [204]. Decreased surface to volume ratio of RCCI piston bowl can increase the throat diameter of the piston bowl which is subsequently beneficial to the early SOI timing of RCCI injection strategy [205]. A study investigated RCCI combustion in a light duty multi-cylinder engine at three operating points according to the US-FTP cycle by comparing OEM (original equipment manufacturer) piston and custom-machined pistons designed for RCCI operation [206]. The piston bowl profile was optimized to reduce unburned fuel emissions and piston bowl surface area using genetic algorithm optimization. The custom RCCI piston was designed for lower HC emissions by reducing the squish area and crevice volume. Heat transfer losses are reduced by reducing the surface area and by increasing the piston bowl radius and decreasing the bowl depth (for constant compression ratio (CR)). Reduction in heat transfer loss leads to higher thermal efficiency. Additionally, the high levels of piston-induced mixing produced by narrow or deep piston bowl increases the heat transfer coefficient and further add the heat transfer losses. Therefore, the more quiescent combustion chamber created by the wide or shallow bowl is beneficial for premixed charge compression ignition (PCI) strategies [187].

The RCCI mode was proposed to reduce the requirement of EGR level in premixed compression ignition engines [177]. However, the assistance of EGR is required mainly at higher engine load operation, to mitigate the high PRR in RCCI engines [181]. Studies demonstrated that PPRR, NO_x and soot emissions are reduced by increasing EGR rate [207, 208]. The unburned hydrocarbons (UHC) and CO emissions are increased with the use of EGR in most of the studies due to the formation of local fuel-rich region resulted from the reduced the oxygen concentration in the cylinder by increasing EGR rate [181]. A study numerically evaluated the necessity of EGR utilization in RCCI engine fuelled with methanol and diesel at medium loads [209]. Study suggested that the EGR requirement depends on the initial cylinder temperature and initial temperature below a critical value (380 K in this case); EGR can be unemployed and the methanol fraction can be varied to maintain the optimal performance.

Another important issue in RCCI engine management is cylinder balancing. In a multi-cylinder engine, there exist significant cylinder-to-cylinder variations in the initial conditions such as EGR, cylinder temperature, fuel rail pressure, trapped mass, etc. [180]. These imbalances between cylinders lead to comparatively poor performance due to excessively delayed or advanced combustion timings in every cylinder because RCCI combustion is a kinetically controlled process. The variations in the combustion phasing results in significant variation of exhaust emissions in each cylinder. To regulate these variations in RCCI engine, the total amount of fuel and ratio of low to high reactivity fuel for each cylinder can be varied to match the combustion phasing and IMEP over all the cylinders [180]. A study used an "adjustment factor" multiplier to the PFI and DI duration commands for each cylinder to achieve the cylinder balancing in RCCI engine [206].

2.6.4 Direct Injection Dual Fuel Stratification

A variety of PPCI strategies are developed to increase the level of control on ignition timing and HRR while maintaining the benefits of HCCI combustion. Typically, PPCI strategies produce fuel stratification in cylinder by direct injection of either a portion or all of the fuels, in order to increase the operating range and combustion stability. Fuel stratification leads to φ- and temperature stratification in the cylinder resulting into variation in local ignition delay in entire cylinder. The variation in local ignition delay increases the combustion duration and retard the combustion phasing, which leads to the reduction in HRR and combustion noise. There exists a continuum of stratification levels between PFS (mostly premixed) and PPC (partially premixed), and each strategy has their relative merits (Sects. 2.2.4.6 and 2.5.2). Dual fuel RCCI combustion has shown high thermal efficiency and superior controllability with low NO_x and soot emissions. The RCCI strategy addresses the combustion control by extending the concept of control by additional fuel reactivity stratification. The RCCI combustion has demonstrated gross indicated thermal efficiencies over 56% and levels of soot and NO_x below the EPA 2010 heavy-duty emission regulation limits [204]. The RCCI combustion emits relatively high levels of HC and CO, due to low wall temperatures and the inability for complete oxidation of fuel in crevice regions. In RCCI strategy, shape and phasing of heat release can be adjusted with much greater fidelity (than is possible in single fuel strategies), by adjusting the ratio of high and low reactivity fuel. At particular engine operating condition, the combustion phasing can be controlled by the fuel ratio, and to a lesser degree, by the timing of the high reactivity fuel injections [210]. However, the reactivity range of the fuels being used limits the operating range of RCCI combustion. The amount of high reactivity fuel required tends to decrease towards zero as engine load increases (well before full load), and there is no longer any direct control over the combustion process [211]. To extend the high load operation of RCCI combustion, high octane fuels such as E85 (an 85% ethanol and 15% gasoline mixture) [212] and methane [190] are used as low reactivity fuels to have longer ignition delay in comparison to gasoline. Engine operation at lower compression ratio is another strategy to increase the load of RCCI engine at the expense of fuel conversion efficiency [204, 213]. However, stratification of both (high and low reactivity) fuels is one possible path to achieve additional control on RCCI combustion for engine operation with conventional fuels and compression ratios of typical diesel engines. This strategy is known as direct injection dual fuel stratification (DDFS) in published literature [210, 211].

The DDFS strategy combines the benefits of RCCI and PPC by injecting both gasoline and diesel directly, enabling control over the in-cylinder distribution of both fuels [211]. In this strategy, gasoline is injected during the intake stroke and it is effectively premixed. Similar to RCCI, diesel is injected 40–60° before TDC to create a reactivity gradient that allows for precise control over the LTHR. After LTHR is complete, and just as the main heat release begins (0–10° BTDC), gasoline

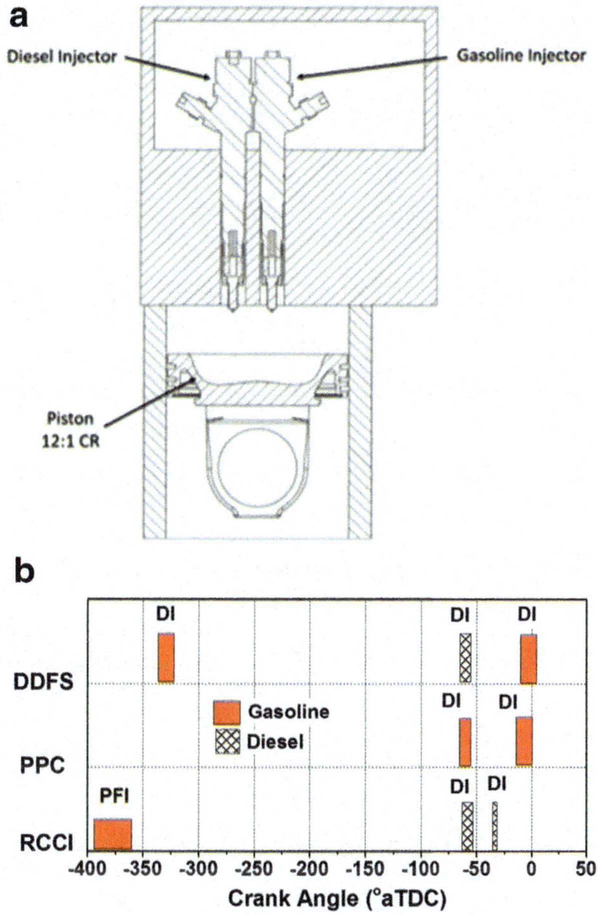

Fig. 2.46 (**a**) Drawing of cylinder showing the injector locations and orientations in typical DDFS [214]. (**b**) Conceptual illustration of injection strategies in DDFS (Adapted from [210])

is injected a second time and burns in a diffusion-controlled manner, allowing for considerable increase in load without increasing peak HRR and combustion noise [211]. Figure 2.46 illustrates the injector locations and orientations along with fuel injection strategy for implementation of DDFS strategy.

In DDFS strategy, the timing and quantity of the diesel injection afford a large degree of control over the early heat release. The reactivity and equivalence ratio stratification created by direct diesel injection extends the heat release and gives control over when combustion begins. The relative impact of reactivity and equivalence ratio on second injection is evaluated by injecting gasoline and diesel as both have direct injection arrangement [210]. Figure 2.47a shows the effect of injecting gasoline instead of diesel on cylinder pressure and HRR. Retarded combustion phasing and high instability of the gasoline operation are observed when second

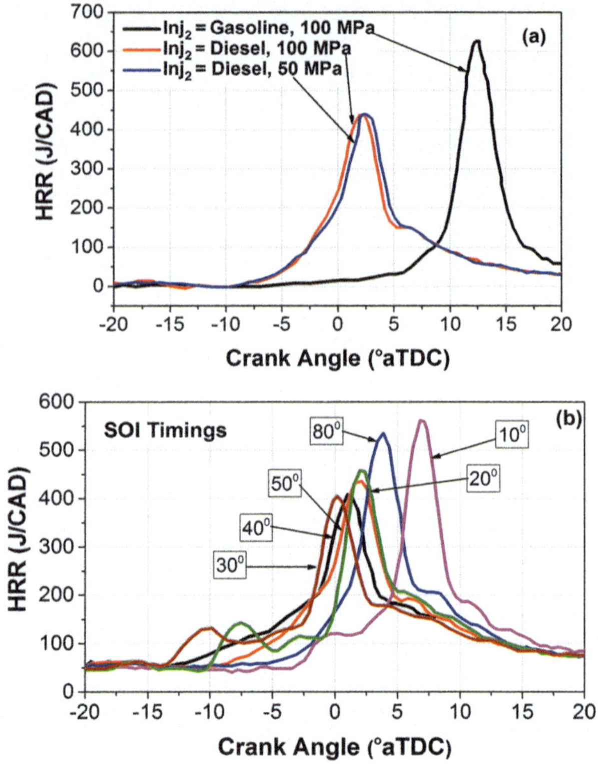

Fig. 2.47 (**a**) Cylinder pressure and HRR for DDFS with either diesel or gasoline as the second injection fuel. (**b**) Cylinder pressure and HRR for DDFS with a sweep of SOI-2 from −200 to −30° ATDC (Adapted from [210])

injection fuel was switched from diesel to gasoline. Higher HRR and shorter combustion duration is also observed for gasoline injection case. Since fuel injection pressure and injection duration are the same for both the fuels, therefore this difference in combustion characteristics is purely a fuel effect. The main fuel differences are volatility and reactivity, which influence mixing and ignition delay, respectively. The figure also illustrates the limited control availability with equivalence ratio stratification alone and the flexibility gained by direct injection of highly reactivity fuel [210]. Figure 2.47a also depicts that diesel injection pressure had little effect on HRR. Relatively lower thermal efficiency and higher unburned HC and CO emissions are also observed for the case of gasoline injection. This observation suggests that φ-stratification alone is insufficient for full potential of this strategy.

The same study also conducted the effect of diesel fuel ratio and start of injection (SOI) sweep on combustion characteristics of DDFS. Figure 2.47b shows the effect of diesel SOI timings on cylinder pressure and HRR in DDFS strategy. Few points are omitted for clarity and all data points can be seen in original study [210]. Figure depicts that before −40° ATDC SOI timings, combustion phasing advances with increase in diesel stratification. At SOI of −30°ATDC, the large bump at the beginning of the heat release suggests that the diesel may be transitioning to a diffusion-limited combustion. After this SOI timing, combustion phasing retards with increase in diesel stratification in the cylinder. Detailed analysis of heat release curve shows that LTHR is advanced and increased slightly in magnitude for SOI up to −50°ATDC, and a slight shift is observed in the beginning of the LTHR at −40°ATDC. Major change in LTHR occurs at −30°ATDC, and LTHR was significantly diminished after this SOI, which is interpreted by authors as the onset of diffusion-limited combustion of the diesel injection [210].

This study identified three possible combustion regimes for the second injection of high reactivity fuel: premixed, RCCI and diffusion-limited. Study found the RCCI regime is superior in several ways, including improved control over CA_{50} and noise; substantial reduction of HC, CO and soot emissions; improved indicated efficiency; and NO_x emission comparable to the premixed regime. DDFS has a unique ability of independent control on the combustion duration (with a near TDC gasoline injection) and the combustion phasing (with high reactivity diesel injection). Advantage of this ability can be availed by operating engine at optimum combustion phasing across the engine map, and combustion duration can be controlled without retarding the combustion phasing. Engine can be operated at higher compression ratios, and higher thermal efficiency can be achieved using this strategy. The combination of the diesel and late gasoline direct injections offers robust fast-response control mechanisms with a wide range of authority [210].

2.6.5 RCCI vis-à-vis Other LTC Strategies

Several studies have demonstrated the superior performance of RCCI strategy over conventional diesel combustion (CDC) strategy. A study comprehensively investigated the RCCI combustion on a HD engine over a wide range of engine operating loads by varying the gasoline-to-diesel ratio at fixed diesel injection strategy [215]. The study found that RCCI operation achieved NO_x reduction of nearly three orders of magnitude, six times lower soot and 16.4% higher gross indicated efficiency in comparison to CDC without EGR. A detailed comparison between CDC and RCCI combustion at matched conditions of load, speed, boost pressure and combustion phasing is conducted [216]. The study found 24% lower integrated piston heat transfer and 25 °C lower mean surface temperature for RCCI engine in comparison to CDC. The SOI timing strongly influences CDC heat flux but has a negligible effect on RCCI heat flux, even in the limit of near TDC high reactivity fuel injection timings. These observations indicate that the high reactivity fuel injection does not have significant effect on the thermal environment.

Figure 2.48 depicts the distributions of the in-cylinder temperature for the CDC, HCCI and RCCI regimes for constant combustion phasing at 3°CA ATDC. In CDC engine, the fuel vapour is mainly concentrated around the piston lip and bowl

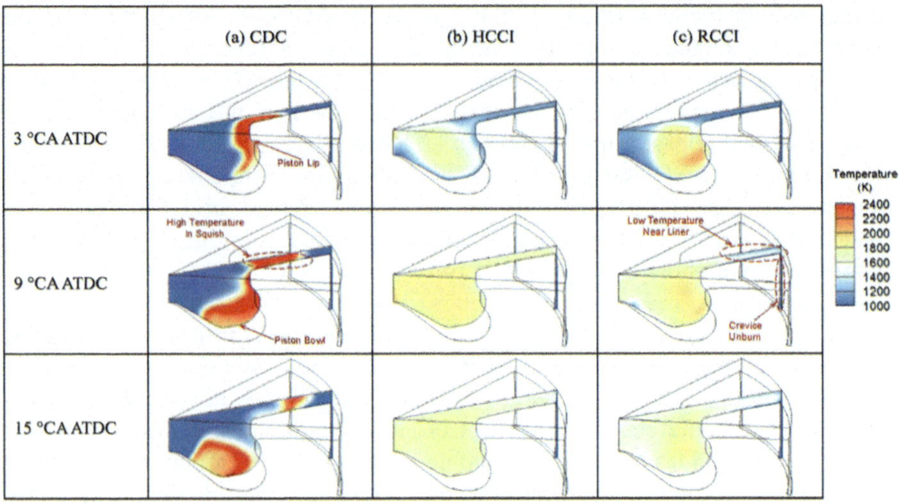

Fig. 2.48 Distributions of the in-cylinder temperature for the CDC, HCCI and RCCI regimes for constant combustion phasing at 3°CA ATDC [217]

during the combustion process due to close to TDC direct injection of fuel. This results into the high temperature region in the cylinder (Fig. 2.48a), which leads to higher NO_x emissions. Sufficiently premixed fuel–air mixture and the avoidance of local high temperature regions in HCCI and RCCI strategy lead to substantially lower combustion temperature (Fig. 2.48b, c) and lower NO_x emissions. Some premixed and low temperature fuel vapour that exists near the cylinder liner and in the crevice region cannot be fully oxidized, resulting in high HC and CO emissions [217].

A comprehensive comparative investigation of RCCI, PPC and HCCI mode is conducted using single-cylinder engine for comparison of performance and emissions, combustion sensitivity to intake conditions and the control ability using fuel injection strategy [218]. This study used PFI for low reactivity fuel and direct injection of high reactivity fuel in RCCI and only PFI for HCCI combustion. In PPC strategy, single fuel is used in both PFI and DI systems. All the three strategies have shown higher indicated thermal efficiency and acceptable coefficient of variation (COV) of IMEP. The RCCI combustion has longer combustion duration and subsequently lower peak pressure rise rate compared to HCCI and PPC. The RCCI operation has highest indicated thermal efficiency but it has lower combustion efficiency. It is suggested that higher efficiency is due to the lower heat transfer losses stemming from the lower maximum PRR [218]. This study also perturbed the intake conditions (intake temperature and pressure) from the baseline to analyse the sensitivity of combustion phasing and its control. It was noted that baseline combustion phasing can be easily recovered through small changes in the global fuel reactivity in RCCI strategy. However, in PCC strategy, the baseline combustion phasing was unrecoverable when combustion becomes advanced due to variations in intake conditions. On the contrary, retarded combustion phasing (due to variation in intake condition) is recoverable by appropriately increasing the level of fuel stratification.

Another study compared the RCCI and GCI strategy at high load conditions at 1300 rpm for optimized engine operating conditions [214]. In this study, RCCI strategy uses dual direct injection of high and low reactivity fuels (Fig. 2.46a). A non-parametric regression analysis tool, known as component selection and smoothing operator (COSSO) for multivariate smoothing splines, is used to choose a single optimum operating point for comparison of RCCI and GCI strategies [219]. The COSSO fits a response surface model to the genetic algorithm data from all the runs combined. The combustion and emission characteristics are compared to optimal engine operating condition. The overall combustion characteristics are very similar with the close to TDC injection initiating combustion for both strategies. The main differences observed in these two (RCCI and GCI) strategies are soot emissions. The difference in soot formation in RCCI and GCI is due to the state of local equivalence ratio and temperature in the cylinder during combustion. Figure 2.49 depicts the variation of local φ and temperature at various crank angle positions after start of combustion (ASOC) on φ-T map. The NO_x and soot formation regions in φ-T map are identified from HCCI simulations of mixtures of gasoline and air from the study [220].

The combustion is delayed in GCI mode due to longer ignition delay as gasoline has lower cetane index (high octane fuel). In RCCI strategy, combustion occurs during injection event due to shorter ignition delay of diesel fuel, which results in higher local φ and more soot formation. In RCCI case, at 10° ASOC (close to CA_{50}), the final diesel fuel injection is ~60% complete and local φ reached above 3 (Fig. 2.49). In GCI case, the final gasoline injection is finished at 7°CA, and post combustion mixing has reduced the peak local φ to ~1.2. The higher local φ during combustion leads to comparatively higher soot formation in RCCI strategy. However, at this expense, RCCI strategy has comparatively better control on combustion phasing due to shorter ignition delay. The peak local temperatures achieved in the RCCI and GCI cases are 2242 K and 2056 K, respectively, due to utilization of high EGR levels. Thus, lower NO_x emissions are obtained in both cases due to low peak temperatures.

Sensitivity to operating conditions in RCCI and GCI strategy is also evaluated in the same study [214]. Figure 2.50 depicts the sensitivity of outputs (IMEP, gross indicated efficiency (GIE), CA_{50} and peak pressure rise rate (PPRR)) to changes in inputs (EGR, intake pressure and temperature, direct injection (DI) mass) for RCCI and GCI strategies.

Figure 2.50 shows that both RCCI and GCI strategies are very sensitive to EGR fluctuations at this high load condition. In contrast to this observation, a sensitivity study at a medium load condition found the RCCI strategy to be insensitive to EGR [221]. Figure 2.50 shows that the GCI strategy is more sensitive to EGR than the RCCI strategy (Fig. 2.50). The global φ of the GCI strategy is close to stoichiometric (0.98), which makes it to be very sensitive to oxygen concentration. Intake valve closing temperature and pressure also affect the combustion phasing

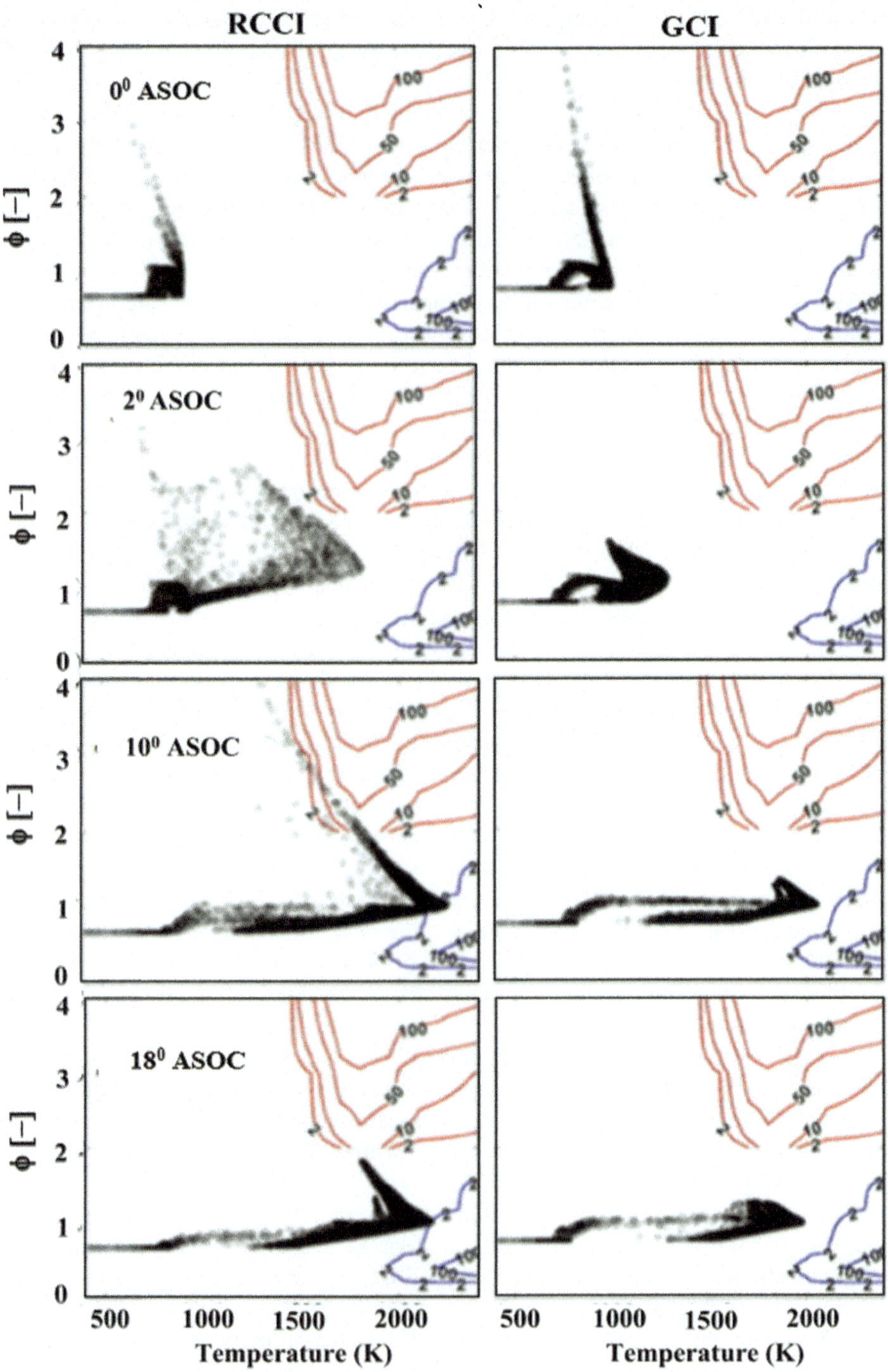

Fig. 2.49 Φ-T map of the charge at various crank angle positions for the optimum RCCI and gasoline compression ignition (GCI) combustion strategies [214]

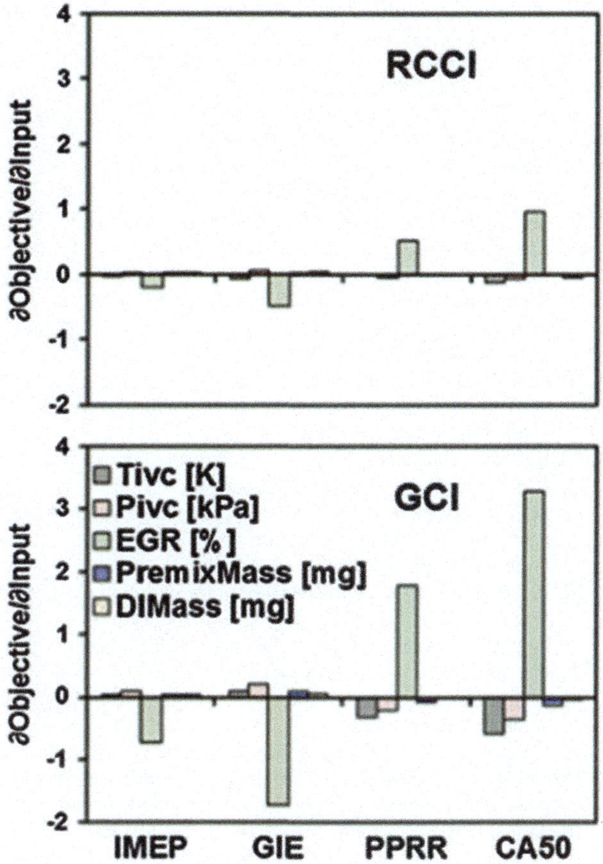

Fig. 2.50 Sensitivity of outputs to changes in inputs for RCCI and GCI strategies [214]

considerably for the GCI strategy, whereas the RCCI strategy is relatively insensitive to these fluctuations.

The characteristics of dual fuel RCCI and single fuel PPCI show their potential for higher thermal efficiency along with reduction of NO_x and soot emission to an ultralow level. Detailed combustion, performance and emission characteristics of HCCI, RCCI and PPC strategies are discussed in Chaps. 6, 7 and 8, respectively.

References

1. Zheng M, Reader GT, Hawley JG (2004) Diesel engine exhaust gas recirculation—a review on advanced and novel concepts. Energy Convers Manage 45(6):883–900
2. Majewski WA, Khair MK (2006) Diesel emissions and their control, vol. 303. SAE technical paper

3. Johnson TV (2002) Diesel emission control: 2001 in review (No. 2002-01-0285). SAE technical paper
4. Dec JE (2009) Advanced compression-ignition engines—understanding the in-cylinder processes. Proc Combust Inst 32(2):2727–2742
5. Dec JE (1997) A conceptual model of DI diesel combustion based on laser-sheet imaging (No. 970873). SAE technical paper
6. Dempsey AB, Curran SJ, Wagner RM (2016) A perspective on the range of gasoline compression ignition combustion strategies for high engine efficiency and low NO_x and soot emissions: effects of in-cylinder fuel stratification. Int J Engine Res 17(8):897–917
7. Lewander M (2011) Characterization and control of multi-cylinder partially premixed combustion. PhD thesis, Lund University. ISBN 978-91-7473-148-4
8. Kitamura T, Ito T, Senda J, Fujimoto H (2002) Mechanism of smokeless diesel combustion with oxygenated fuels based on the dependence of the equivalence ration and temperature on soot particle formation. Int J Engine Res 3(4):223–248
9. Kim D, Ekoto I, Colban WF, Miles PC (2008) In-cylinder CO and UHC imaging in a light-duty diesel engine during PPCI low-temperature combustion. SAE Int J Fuels Lubr 1:933–956. (2008-01-1602)
10. Dec JE, Canaan RE (1998) PLIF imaging of NO formation in a DI diesel engine (No. 980147). SAE technical paper
11. Som S, Aggarwal S (2009) Modeling diesel spray flame lift-off using detailed chemistry and a new primary breakup model. In: 47th AIAA aerospace sciences meeting including the new horizons forum and aerospace exposition, p 666
12. Pickett LM, Siebers DL (2006) Soot formation in diesel fuel jets near the lift-off length. Int J Eng Res 7(2):103–130
13. Pickett LM, Siebers DL, Idicheria CA (2005) Relationship between ignition processes and the lift-off length of diesel fuel jets (No. 2005-01-3843). SAE technical paper
14. Chartier C, Aronsson U, Andersson O, Egnell R, Collin R, Seyfried H et al (2009) Analysis of smokeless spray combustion in a heavy-duty diesel engine by combined simultaneous optical diagnostics (No. 2009-01-1353). SAE technical paper
15. Naber JD, Siebers DL (1996).Effects of gas density and vaporization on penetration and dispersion of diesel sprays (No. 960034). SAE technical paper
16. Akihama K, Takatori Y, Inagaki K, Sasaki S, Dean AM (2001). Mechanism of the smokeless rich diesel combustion by reducing temperature (No. 2001-01-0655). SAE technical paper
17. Musculus MP, Miles PC, Pickett LM (2013) Conceptual models for partially premixed low-temperature diesel combustion. Prog Energy Combust Sci 39(2):246–283
18. Dobbins RA (2002) Soot inception temperature and the carbonization rate of precursor particles. Combust Flame 130(3):204–214
19. Huestis E, Erickson PA, Musculus MP (2007) In-cylinder and exhaust soot in low-temperature combustion using a wide-range of EGR in a heavy-duty diesel engine (No. 2007-01-4017). SAE technical paper
20. Colban WF, Miles PC, Oh S (2007) Effect of intake pressure on performance and emissions in an automotive diesel engine operating in low temperature combustion regimes (No. 2007-01-4063). SAE technical paper
21. Dec JE (2015) Advanced compression ignition combustion for high efficiency and ultralow NO_x and soot. In: Encyclopedia of automotive engineering, vol 1. Wiley, Chichester, pp 267–306
22. Richter M, Engström J, Franke A, Aldén M, Hultqvist A, Johansson B (2000) The influence of charge inhomogeneity on the HCCI combustion process (No. 2000-01-2868). SAE technical paper
23. Dec JE, Hwang W, Sjöberg M (2006) An investigation of thermal stratification in HCCI engines using chemiluminescence imaging (No. 2006-01-1518). SAE technical paper

24. Hultqvist A, Christensen M, Johansson B, Richter M, Nygren J, Hult J, Aldjn M (2002) The HCCI combustion process in a single cycle-speed fuel tracer LIF and chemiluminescence imaging (No. 2002-01-0424). SAE technical paper
25. Johansson B (2007) Homogeneous charge compression ignition: the future of IC engines? Int J Veh Des 44(1–2):1–19
26. Battin-Leclerc F (2008) Detailed chemical kinetic models for the low-temperature combustion of hydrocarbons with application to gasoline and diesel fuel surrogates. Prog Energy Combust Sci 34(4):440–498
27. Saxena S, Bedoya ID (2013) Fundamental phenomena affecting low temperature combustion and HCCI engines, high load limits and strategies for extending these limits. Prog Energy Combust Sci 39(5):457–488
28. Erlandsson O (2002) Early Swedish hot-bulb engines-efficiency and performance compared to contemporary gasoline and diesel engines (No. 2002-01-0115). SAE technical paper
29. Gussak LA, Turkish MC, Siegla DC (1975) High chemical activity of incomplete combustion products and a method of prechamber torch ignition for avalanche activation of combustion in internal combustion engines (No. 750890). SAE technical paper
30. Onishi S, Jo SH, Shoda K, Do Jo P, Kato S (1979) Active thermo-atmosphere combustion (ATAC)-a new combustion process for internal combustion engines (No. 790501). SAE technical paper
31. Noguchi M, Tanaka Y, Tanaka T, Takeuchi Y (1979) A study on gasoline engine combustion by observation of intermediate reactive products during combustion (No. 790840). SAE technical paper
32. Iida N (1994) Combustion analysis of methanol-fueled active thermo-atmosphere combustion (ATAC) engine using a spectroscopic observation (No. 940684). SAE technical paper
33. Duret P, Venturi S (1996) Automotive calibration of the IAPAC fluid dynamically controlled two-stroke combustion process (No. 960363). SAE technical paper
34. Oguma H, Ichikura T, Iida N 1997 A study on adaptability of alternative fuels for lean burn two-stroke ATAC engine (No. 972097). SAE technical paper
35. Yamaguchi J (1997) Honda readies activated radical combustion two-stroke engine for production motorcycle. Automot Eng 105(1):90–92
36. Ishibashi Y, Asai M (1998) A low pressure pneumatic direct injection two-stroke engine by activated radical combustion concept (No. 980757). SAE technical paper
37. Najt PM, Foster DE (1983) Compression-ignited homogeneous charge combustion (No. 830264). SAE technical paper
38. Thring RH (1989) Homogeneous-charge compression-ignition (HCCI) engines (No. 892068). SAE technical paper
39. Christensen M, Hultqvist A, Johansson B (1999) Demonstrating the multi fuel capability of a homogeneous charge compression ignition engine with variable compression ratio (No. 1999-01-3679). SAE technical paper
40. Yao M, Zheng Z, Liu H (2009) Progress and recent trends in homogeneous charge compression ignition (HCCI) engines. Prog Energy Combust Sci 35(5):398–437
41. Lu X, Han D, Huang Z (2011) Fuel design and management for the control of advanced compression-ignition combustion modes. Prog Energy Combust Sci 37(6):741–783
42. Lavy J, Dabadie JC, Angelberger C, Duret P, Willand J, Juretzka A et al (2000) Innovative ultra-low NO_x controlled auto-ignition combustion process for gasoline engines: the 4-SPACE project (No. 2000-01-1837). SAE technical paper
43. Maurya RK, Agarwal AK (2011) Experimental study of combustion and emission characteristics of ethanol fuelled port injected homogeneous charge compression ignition (HCCI) combustion engine. Appl Energy 88(4):1169–1180
44. Stanglmaier RH, Roberts CE (1999) Homogeneous charge compression ignition (HCCI): benefits, compromises, and future engine applications (No. 1999-01-3682). SAE technical paper

45. Affens WA, Sheinson RS (1980) Autoignition: the importance of the cool flame in the two-stage-process. Loss Prev 13:83–88
46. Machrafi H (2010) HCCI combustion chemistry, reduced kinetic mechanisms and controlling strategies. In: Lackner M, Winter F, Agarwal AK (eds) Handbook of combustion, New technologies, vol 5. WILEY-VCH Verlag GmbH & Co. KGaA, Weinheim
47. Machrafi H (2008) Experimental validation of a kinetic multi-component mechanism in a wide HCCI engine operating range for mixtures of n-heptane, iso-octane and toluene: influence of EGR parameters. Energy Convers Manag 49(11):2956–2965
48. Tanaka S, Ayala F, Keck JC (2003) A reduced chemical kinetic model for HCCI combustion of primary reference fuels in a rapid compression machine. Combust Flame 133(4):467–481
49. Hwang W, Dec J, Sjöberg M (2008) Spectroscopic and chemical-kinetic analysis of the phases of HCCI autoignition and combustion for single-and two-stage ignition fuels. Combust Flame 154(3):387–409
50. Sjöberg M, Dec JE (2007) Comparing late-cycle autoignition stability for single-and two-stage ignition fuels in HCCI engines. Proc Combust Inst 31(2):2895–2902
51. Westbrook CK, Pitz WJ, Curran HJ (2007) Auto-ignition and chemical kinetic mechanisms of HCCI combustion. In: HCCI and CAI engines for the automotive industry. Woodhead Publishing Limited, Cambridge, England pp 433–445
52. Zheng J, Yang W, Miller DL, Cernansky NP (2001) Prediction of pre-ignition reactivity and ignition delay for HCCI using a reduced chemical kinetic model (No. 2001-01-1025). SAE technical paper
53. Sjöberg M, Dec JE (2007) EGR and intake boost for managing HCCI low-temperature heat release over wide ranges of engine speed (No. 2007-01-0051). SAE technical paper
54. Serinyel Z, Le Moyne L, Guibert P (2007) Homogeneous charge compression ignition as an alternative combustion mode for the future of internal combustion engines. Int J Veh Des 44 (1–2):20–40
55. Zhao F, Asmus TW, Assanis DN, Dec JE, Eng JA Najt PM (2003) Homogeneous charge compression ignition (HCCI) engines: key research and development issues. SAE PT-94
56. Silke EJ, Pitz WJ, Westbrook CK, Sjöberg M, Dec JE (2008) Understanding the chemical effects of increased boost pressure under HCCI conditions. SAE Int J Fuels Lubr 1:12–25. (2008-01-0019)
57. Curran HJ, Gaffuri P, Pitz WJ, Westbrook CK (1998) A comprehensive modeling study of n-heptane oxidation. Combust Flame 114(1):149–177
58. Luong MB, Yu GH, Lu T, Chung SH, Yoo CS (2015) Direct numerical simulations of ignition of a lean n-heptane/air mixture with temperature and composition inhomogeneities relevant to HCCI and SCCI combustion. Combust Flame 162(12):4566–4585
59. Coskun G, Jonsson M, Bood J, Tunér M, Algotsson M, Li B et al (2015) Analysis of in-cylinder H2O2 and HO2 distributions in an HCCI engine–comparison of laser-diagnostic results with CFD and SRM simulations. Combust Flame 162(9):3131–3139
60. Burns VR, Rapp LA, Koehl WJ, Benson JD, Hochhauser AM, Knepper JC, et al (1995) Gasoline reformulation and vehicle technology effects on emissions-auto/oil air quality improvement research program (No. 952509). SAE technical paper
61. Machrafi H, Cavadias S (2008) Three-stage autoignition of gasoline in an HCCI engine: an experimental and chemical kinetic modeling investigation. Combust Flame 155(4):557–570
62. Vuilleumier D, Kozarac D, Mehl M, Saxena S, Pitz WJ, Dibble RW et al (2014) Intermediate temperature heat release in an HCCI engine fueled by ethanol/n-heptane mixtures: an experimental and modeling study. Combust Flame 161(3):680–695
63. Dec JE, Yang Y (2010) Boosted HCCI for high power without engine knock and with ultra-low NO_x emissions-using conventional gasoline. SAE Int J Engines 3:750–767. (2010-01-1086)
64. Yang J (2007) Four-stroke gasoline HCCI engines with thermal management. In: Zhao H (ed) HCCI and CAI engines for the automotive industry. Woodhead Publishing, Cambridge, England ISBN:978-1-84569-128-8

65. Sun R, Thomas R, Gray CL (2004). An HCCI engine: power plant for a hybrid vehicle (No. 2004-01-0933). SAE technical paper
66. Shahbakhti M (2009) Modeling and experimental study of an HCCI engine for combustion timing control. PhD thesis, University of Alberta
67. Borland M, Zhao F (2002) Application of secondary air injection for simultaneously reducing converter-in emissions and improving catalyst light-off performance (No. 2002-01-2803). SAE technical paper
68. Bauer H, Haldenwanger HG, Hirth P, Brück R (1999) Thermal management of close coupled catalysts (No. 1999-01-1231). SAE technical paper
69. Konstantinidis PA, Koltsakis GC, Stamatelos AM (1997) Computer aided assessment and optimization of catalyst fast light-off techniques. Proc Inst Mech Eng D J Automob Eng 211 (1):21–37
70. Maurya RK, Agarwal AK (2014) Experimental investigations of performance, combustion and emission characteristics of ethanol and methanol fueled HCCI engine. Fuel Process Technol 126:30–48
71. Dec JE, Sjöberg M (2004). Isolating the effects of fuel chemistry on combustion phasing in an HCCI engine and the potential of fuel stratification for ignition control (No. 2004-01-0557). SAE technical paper
72. Dec JE, Yang Y, Dronniou N (2011) Boosted HCCI-controlling pressure-rise rates for performance improvements using partial fuel stratification with conventional gasoline. SAE Int J Engines 4:1169–1189. (2011-01-0897)
73. Yang Y, Dec JE, Dronniou N, Sjöberg M (2011) Tailoring HCCI heat-release rates with partial fuel stratification: comparison of two-stage and single-stage-ignition fuels. Proc Combust Inst 33(2):3047–3055
74. Yang Y, Dec JE, Dronniou N, Sjöberg M, Cannella W (2011) Partial fuel stratification to control HCCI heat release rates: fuel composition and other factors affecting pre-ignition reactions of two-stage ignition fuels. SAE Int J Engines 4:1903–1920. (2011-01-1359)
75. Buda F, Bounaceur R, Warth V, Glaude PA, Fournet R, Battin-Leclerc F (2005) Progress toward a unified detailed kinetic model for the autoignition of alkanes from C 4 to C 10 between 600 and 1200 K. Combust Flame 142(1):170–186
76. Ribaucour M, Minetti R, Sochet LR, Curran HJ, Pitz WJ, Westbrook CK (2000) Ignition of isomers of pentane: an experimental and kinetic modeling study. Proc Combust Inst 28 (2):1671–1678
77. Aceves SM, Flowers DL, Dibble RW, Babajimopoulos A (2007) Overview of modeling techniques and their application to HCCI/CAI engines. In: HCCI and CAI engines for the automotive industry. pp 456–474
78. Sjöberg M, Dec JE (2006) Smoothing HCCI heat-release rates using partial fuel stratification with two-stage ignition fuels (No. 2006-01-0629). SAE technical paper
79. Sjoberg M, Dec J (2010) Ethanol autoignition characteristics and HCCI performance for wide ranges of engine speed, load and boost. SAE Int J Engines 3(1):84–106
80. Sjöberg M, Dec JE (2008) Influence of fuel autoignition reactivity on the high-load limits of HCCI engines. SAE Int J Engines 1:39–58. (2008-01-0054)
81. Martinez-Frias J, Aceves SM, Flowers D, Smith JR, Dibble R (2000) HCCI engine control by thermal management (No. 2000-01-2869). SAE technical paper
82. Chang J, Filipi Z, Assanis D, Kuo TW, Najt P, Rask R (2005) Characterizing the thermal sensitivity of a gasoline homogeneous charge compression ignition engine with measurements of instantaneous wall temperature and heat flux. Int J Engine Res 6(4):289–310
83. Hu T, Liu S, Zhou L, Li W (2006) Effects of compression ratio on performance, combustion, and emission characteristics of an HCCI engine. Proc Inst Mech Eng D J Automob Eng 220 (5):637–645
84. Christensen M, Johansson B (2000) Supercharged homogeneous charge compression ignition (HCCI) with exhaust gas recirculation and pilot fuel (No. 2000-01-1835). SAE technical paper

85. Sjöberg M, Dec JE (2003) Combined effects of fuel-type and engine speed on intake temperature requirements and completeness of bulk-gas reactions for HCCI combustion (No. 2003-01-3173). SAE technical paper
86. Maurya RK, Agarwal AK (2015) Combustion and emission characterization of n-butanol fueled HCCI engine. J Energy Res Technol 137(1):011101
87. Zhao H, Peng Z, Williams J, Ladommatos N (2001) Understanding the effects of recycled burnt gases on the controlled autoignition (CAI) combustion in four-stroke gasoline engines (No. 2001-01-3607). SAE technical paper
88. Cairns A, Blaxill H (2005) The effects of combined internal and external exhaust gas recirculation on gasoline controlled auto-ignition (No. 2005-01-0133). SAE technical paper
89. Sjöberg M, Dec JE, Hwang W (2007) Thermodynamic and chemical effects of EGR and its constituents on HCCI autoignition (No. 2007-01-0207). SAE technical paper
90. Cooper B, Jackson N, Penny I, Truscott T, Rawlins D, Seabrook J (2006) Advanced development techniques for delivering low emissions diesel engines. Proc THIESEL 2006:267–280
91. Dubreuil A, Foucher F, Mounaïm-Rousselle C (2006) Effect of EGR chemical components and intake temperature on HCCI combustion development (No. 2006-32-0044). SAE technical paper
92. Machrafi H, Cavadias S, Guibert P (2008) An experimental and numerical investigation on the influence of external gas recirculation on the HCCI autoignition process in an engine: thermal, diluting, and chemical effects. Combust Flame 155(3):476–489
93. Lemel M, Hultqvist A, Vressner A, Nordgren H, Persson H, Johansson B (2005) Quantification of the formaldehyde emissions from different HCCI engines running on a range of fuels (No. 2005-01-3724). SAE technical paper
94. Hasan AO, Abu-jrai A, Ala'a H, Tsolakis A, Xu H (2016) Formaldehyde, acetaldehyde and other aldehyde emissions from HCCI/SI gasoline engine equipped with prototype catalyst. Fuel 175:249–256
95. Risberg P, Johansson D, Andrae J, Kalghatgi G, Björnbom P, Ångström HE (2006) The influence of NO on the combustion phasing in an HCCI engine (No. 2006-01-0416). SAE technical paper
96. Machrafi H, Guibert P, Cavadias S (2008) HCCI engine modeling and experimental investigations–part 2: the composition of a NO-PRF interaction mechanism and the influence of NO in EGR on auto-ignition. Combust Sci Technol 180(7):1245–1262
97. Sang W (2014) Knock mitigation on boosted controlled auto-ignition engines with fuel stratifications and exhaust gas recirculation. PhD thesis, Massachusetts Institute of Technology
98. Sjöberg M, Dec JE (2009) Influence of EGR quality and Unmixedness on the high-load limits of HCCI engines. SAE Int J Engines 2:492–510. (2009-01-0666)
99. Sjöberg M, Dec JE, Cernansky NP (2005). Potential of thermal stratification and combustion retard for reducing pressure-rise rates in HCCI engines, based on multi-zone modeling and experiments (No. 2005-01-0113). SAE technical paper
100. Sjöberg M, Dec JE (2005) Effects of engine speed, fueling rate, and combustion phasing on the thermal stratification required to limit HCCI knocking intensity (No. 2005-01-2125). SAE technical paper
101. Dec JE, Hwang W (2009) Characterizing the development of thermal stratification in an HCCI engine using planar-imaging thermometry. SAE Int J Engines 2:421–438. (2009-01-0650)
102. Reuss DL, Sick V (2005) Inhomogeneities in HCCI combustion: an imaging study (No. 2005-01-2122). SAE technical paper
103. Schmitt M, Boulouchos K (2016) Role of the intake generated thermal stratification on the temperature distribution at top dead center of the compression stroke. Int J Engine Res 17 (8):836–845

104. Kakuho A, Nagamine M, Amenomori Y, Urushihara T, Itoh T (2006) In-cylinder temperature distribution measurement and its application to HCCI combustion (No. 2006-01-1202). SAE technical paper
105. Krasselt J, Foster DE, Ghandhi J, Herold R, Reuss D, Najt P (2009) Investigations into the effects of thermal and compositional stratification on HCCI combustion–part I: metal engine results (No. 2009-01-1105). SAE technical paper
106. Herold RE, Krasselt JM, Foster DE, Ghandhi JB, Reuss DL, Najt PM (2009) Investigations into the effects of thermal and compositional stratification on HCCI combustion–part II: optical engine results. SAE Int J Engines 2:1034–1053. (2009-01-1106)
107. Sjöberg M, Dec JE, Babajimopoulos A, Assanis DN (2004) Comparing enhanced natural thermal stratification against retarded combustion phasing for smoothing of HCCI heat-release rates (No. 2004-01-2994). SAE technical paper
108. Yang DB, Wang Z, Wang JX, Shuai SJ (2011) Experimental study of fuel stratification for HCCI high load extension. Appl Energy 88(9):2949–2954
109. Hwang W, Dec JE, Sjöberg M (2007). Fuel stratification for low-load HCCI combustion: performance & fuel-PLIF measurements (No. 2007-01-4130). SAE technical paper
110. Dec JE, Yang Y, Dernotte J, Ji C (2015) Effects of gasoline reactivity and ethanol content on boosted, premixed and partially stratified low-temperature gasoline combustion (LTGC). SAE Int J Engines 8:935–955. (2015-01-0813)
111. Lavoie GA, Martz J, Wooldridge M, Assanis D (2010) A multi-mode combustion diagram for spark assisted compression ignition. Combust Flame 157(6):1106–1110
112. Koopmans L, Denbratt I (2001) A four stroke camless engine, operated in homogeneous charge compression ignition mode with commercial gasoline (No. 2001-01-3610). SAE technical paper
113. Reuss DL, Kuo TW, Silvas G, Natarajan V, Sick V (2008) Experimental metrics for identifying origins of combustion variability during spark-assisted compression ignition. Int J Engine Res 9(5):409–434
114. Urushihara T, Yamaguchi K, Yoshizawa K, Itoh T (2005) A study of a gasoline-fueled compression ignition engine∼ expansion of HCCI operation range using SI combustion as a trigger of compression ignition∼ (No. 2005-01-0180). SAE technical paper
115. Persson H, Hultqvist A, Johansson B, Remón A (2007) Investigation of the early flame development in spark assisted HCCI combustion using high speed chemiluminescence imaging (No. 2007-01-0212). SAE technical paper
116. Huang Y, Sung CJ, Eng JA (2004) Dilution limits of n-butane/air mixtures under conditions relevant to HCCI combustion. Combust Flame 136(4):457–466
117. Persson H, Pfeiffer R, Hultqvist A, Johansson B, Ström H (2005) Cylinder-to-cylinder and cycle-to-cycle variations at HCCI operation with trapped residuals (No. 2005-01-0130). SAE technical paper
118. Middleton RJ, Olesky LKM, Lavoie GA, Wooldridge MS, Assanis DN, Martz JB (2015) The effect of spark timing and negative valve overlap on spark assisted compression ignition combustion heat release rate. Proc Combust Inst 35(3):3117–3124
119. Olesky LM, Martz JB, Lavoie GA, Vavra J, Assanis DN, Babajimopoulos A (2013) The effects of spark timing, unburned gas temperature, and negative valve overlap on the rates of stoichiometric spark assisted compression ignition combustion. Appl Energy 105:407–417
120. Olesky LM, Lavoie GA, Assanis DN, Wooldridge MS, Martz JB (2014) The effects of diluent composition on the rates of HCCI and spark assisted compression ignition combustion. Appl Energy 124:186–198
121. Zigler BT, Keros PE, Helleberg KB, Fatouraie M, Assanis D, Wooldridge MS (2011) An experimental investigation of the sensitivity of the ignition and combustion properties of a single-cylinder research engine to spark-assisted HCCI. Int J Engine Res 12(4):353–375
122. Manofsky L, Vavra J, Assanis DN, Babajimopoulos A (2011) Bridging the gap between HCCI and SI: spark-assisted compression ignition (No. 2011-01-1179). SAE technical paper

123. Martz JB, Middleton RJ, Lavoie GA, Babajimopoulos A, Assanis DN (2011) A computational study and correlation of premixed isooctane–air laminar reaction front properties under spark ignited and spark assisted compression ignition engine conditions. Combust Flame 158 (6):1089–1096
124. Quader AA (1976) What limits lean operation in spark ignition engines-flame initiation or propagation? (No. 760760). SAE technical paper
125. Middleton RJ, Martz JB, Lavoie GA, Babajimopoulos A, Assanis DN (2012) A computational study and correlation of premixed isooctane air laminar reaction fronts diluted with EGR. Combust Flame 159(10):3146–3157
126. Yun H, Wermuth N, Najt P (2011) High load HCCI operation using different valving strategies in a naturally-aspirated gasoline HCCI engine. SAE Int J Engines 4:1190–1201. (2011-01-0899)
127. Persson H (2008) Spark assisted compression ignition, SACI. PhD thesis, Lund University, SE-22100 Lund. ISBN 978-91-628-7578-7
128. Wang Z, Wang JX, Shuai SJ, Tian GH, An X, Ma QJ (2006) Study of the effect of spark ignition on gasoline HCCI combustion. Proc Inst Mech Eng D J Automob Eng 220 (6):817–825
129. Hyvönen J, Haraldsson G, Johansson B (2005) Operating conditions using spark assisted HCCI combustion during combustion mode transfer to SI in a multi-cylinder VCR-HCCI engine (No. 2005-01-0109). SAE technical paper
130. Lawler B, Mamalis S, Joshi S, Lacey J, Guralp O, Najt P, Filipi Z (2017) Understanding the effect of operating conditions on thermal stratification and heat release in a homogeneous charge compression ignition engine. Appl Therm Eng 112:392–402
131. Lawler B, Splitter D, Szybist J, Kaul B (2017) Thermally stratified compression ignition: a new advanced low temperature combustion mode with load flexibility. Appl Energy 189:122–132
132. Christensen M, Johansson B (1999) Homogeneous charge compression ignition with water injection (No. 1999-01-0182). SAE technical paper
133. Ogawa H, Miyamoto N, Kaneko N, Ando H (2003) Combustion control and operating range expansion with direct injection of reaction suppressors in a premixed DME HCCI engine (No. 2003-01-0746). SAE technical paper
134. Hyvönen J, Wilhelmsson C, Johansson B (2006) The effect of displacement on air-diluted multi-cylinder HCCI engine performance (No. 2006-01-0205). SAE technical paper
135. Manente V (2010) Gasoline partially premixed combustion: an advanced internal combustion engine concept aimed to high efficiency, low emissions and low acoustic noise in the whole load range. PhD thesis, Lund University, SE 22100 Lund Sweden. ISBN 978-91-628-8144-3
136. Hardy WL, Reitz RD (2006) A study of the effects of high EGR, high equivalence ratio, and mixing time on emissions levels in a heavy-duty diesel engine for PCCI combustion (No. 2006-01-0026). SAE technical paper
137. Kimura S, Aoki O, Kitahara Y, Aiyoshizawa E (2001) Ultra-clean combustion technology combining a low-temperature and premixed combustion concept for meeting future emission standards (No. 2001-01-0200). SAE technical paper
138. Wimmer A, Eichlseder H, Klell M, Figer G (2006) Potential of HCCI concepts for DI diesel engines. Int J Veh Des 41(1–4):32–48
139. Takeda Y, Keiichi N, Keiichi N (1996) Emission characteristics of premixed lean diesel combustion with extremely early staged fuel injection (No. 961163). SAE technical paper
140. Akagawa H, Miyamoto T, Harada A, Sasaki S, Shimazaki N, Hashizume T, Tsujimura K (1999) Approaches to solve problems of the premixed lean diesel combustion (No. 1999-01-0183). SAE technical paper
141. Hashizume T, Miyamoto T, Hisashi A, Tsujimura K (1998) Combustion and emission characteristics of multiple stage diesel combustion (No. 980505). SAE technical paper
142. Iwabuchi Y, Kawai K, Shoji T, Takeda Y (1999) Trial of new concept diesel combustion system-premixed compression-ignited combustion (No. 1999-01-0185). SAE technical paper

143. Yanagihara H, Sato Y, Mizuta JI (1997) A study of DI diesel combustion under uniform higher-dispersed mixture formation. JSAE Rev 18(4):361–367
144. Walter B, Gatellier B (2002) Development of the high power NADI™ concept using dual mode diesel combustion to achieve zero NO_x and particulate emissions (No. 2002-01-1744). SAE technical paper
145. Singh S, Reitz RD, Musculus MPB, Lachaux T (2007) Validation of engine combustion models against detailed in-cylinder optical diagnostics data for a heavy-duty compression-ignition engine. Int J Engine Res 8(1):97–126
146. Genzale CL, Reitz RD, Musculus MP (2008) Effects of piston bowl geometry on mixture development and late-injection low-temperature combustion in a heavy-duty diesel engine. SAE Int J Engines 1:913–937. (2008-01-1330)
147. Noehre C, Andersson M, Johansson B, Hultqvist A (2006) Characterization of partially premixed combustion (No. 2006-01-3412). SAE technical paper
148. Johansson B (2005) High-load partially premixed combustion in a heavy-duty diesel engine. In: Diesel engine emissions reduction (DEER) conference. Chicago, 21–25, 2005
149. Sang W, Cheng WK, Maria A (2014) The nature of heat release in gasoline PPCI engines (No. 2014-01-1295). SAE technical paper
150. Rose KD, Ariztegui J, Cracknell RF, Dubois T, Hamje HDC, Pellegrini L et al (2013) Exploring a gasoline compression ignition (GCI) engine concept (No. 2013-01-0911). SAE technical paper
151. Goyal H, Kook S, Hawkes E, Chan QN, Padala S, Ikeda Y. (2017) Influence of engine speed on gasoline compression ignition (GCI) combustion in a single-cylinder light-duty diesel engine (No. 2017-01-0742). SAE technical paper
152. Sellnau M, Moore W, Sinnamon J, Hoyer K, Foster M, Husted H (2015) GDCI multi-cylinder engine for high fuel efficiency and low emissions. SAE Int J Engines 8:775–790. (2015-01-0834)
153. Kalghatgi GT (2005) Auto-ignition quality of practical fuels and implications for fuel requirements of future SI and HCCI engines (No. 2005-01-0239). SAE technical paper
154. Sjöberg M, Edling LO, Eliassen T, Magnusson L, Ångström HE (2002) GDI HCCI: effects of injection timing and air swirl on fuel stratification, combustion and emissions formation (No. 2002-01-0106). SAE technical paper
155. Aroonsrisopon T, Werner P, Waldman JO, Sohm V, Foster DE, Morikawa T, Iida M (2004) Expanding the HCCI operation with the charge stratification (No. 2004-01-1756). SAE technical paper
156. Dempsey AB, Curran S, Wagner R, Cannella W (2015) Effect of premixed fuel preparation for partially premixed combustion with a low octane gasoline on a light-duty multicylinder compression ignition engine. J Eng Gas Turbines Power 137(11):111506
157. Loeper P, Ra Y, Adams C, Foster DE, Ghandhi J, Andrie M et al (2013) Experimental investigation of light-medium load operating sensitivity in a gasoline compression ignition (GCI) light-duty diesel engine (No. 2013-01-0896). SAE technical paper
158. Ra Y, Loeper P, Andrie M, Krieger R, Foster DE, Reitz RD, Durrett R (2012) Gasoline DICI engine operation in the LTC regime using triple-pulse injection. SAE Int J Engines 5:1109–1132. (2012-01-1131)
159. Sellnau MC, Sinnamon J, Hoyer K, Kim J, Cavotta M, Husted H (2013) Part-load operation of gasoline direct-injection compression ignition (GDCI) engine (No. 2013-01-0272). SAE technical paper
160. Sellnau MC, Sinnamon J, Hoyer K, Husted H (2012) Full-time gasoline direct-injection compression ignition (GDCI) for high efficiency and low NO_x and PM. SAE Int J Engines 5:300–314. (2012-01-0384)
161. Sellnau M, Foster M, Hoyer K, Moore W, Sinnamon J, Husted H (2014) Development of a gasoline direct injection compression ignition (GDCI) engine. SAE Int J Engines 7:835–851. (2014-01-1300)

162. Tanov S, Collin R, Johansson B, Tuner M (2014) Combustion stratification with partially premixed combustion, PPC, using NVO and split injection in a LD-diesel engine. SAE Int J Engines 7:1911–1919. (2014-01-2677)
163. Dempsey AB, Reitz RD (2011) Computational optimization of a heavy-duty compression ignition engine fueled with conventional gasoline. SAE Int J Engines 4:338–359. (2011-01-0356)
164. Kalghatgi GT, Risberg P, Ångström HE (2006) Advantages of fuels with high resistance to auto-ignition in late-injection, low-temperature, compression ignition combustion (No. 2006-01-3385). SAE technical paper
165. Manente V, Johansson B, Tunestal P (2009) Partially premixed combustion at high load using gasoline and ethanol, a comparison with diesel (No. 2009-01-0944). SAE technical paper
166. Manente V, Zander CG, Johansson B, Tunestal P, Cannella W (2010) An advanced internal combustion engine concept for low emissions and high efficiency from idle to max load using gasoline partially premixed combustion (No. 2010-01-2198). SAE technical paper
167. Chang J, Kalghatgi G, Amer A, Viollet Y (2012) Enabling high efficiency direct injection engine with naphtha fuel through partially premixed charge compression ignition combustion (No. 2012-01-0677). SAE technical paper
168. Adhikary BD, Reitz RD, Ciatti S, Kolodziej C (2014) Computational investigation of low load operation in a light-duty gasoline direct injection compression ignition [GDICI] engine using single-injection strategy (No. 2014-01-1297). SAE technical paper
169. Kolodziej CP, Ciatti S, Vuilleumier D, Adhikary BD, Reitz RD (2014) Extension of the lower load limit of gasoline compression ignition with 87 AKI gasoline by injection timing and pressure (No. 2014-01-1302). SAE technical paper
170. Kalghatgi GT, Hildingsson L, Harrison AJ, Johansson B (2011) Autoignition quality of gasoline fuels in partially premixed combustion in diesel engines. Proc Combust Inst 33 (2):3015–3021
171. Manente V, Johansson B, Tunestal P, Cannella W (2009) Effects of different type of gasoline fuels on heavy duty partially premixed combustion. SAE Int J Engines 2:71–88. (2009-01-2668)
172. Manente V, Tunestal P, Johansson B, Cannella WJ (2010) Effects of ethanol and different type of gasoline fuels on partially premixed combustion from low to high load (No. 2010-01-0871). SAE technical paper
173. Chen Y, Wolk B, Mehl M, Cheng WK, Chen JY, Dibble RW (2017) Development of a reduced chemical mechanism targeted for a 5-component gasoline surrogate: a case study on the heat release nature in a GCI engine. Combust Flame 178:268–276
174. Solaka H, Tuner M, Johansson B (2013) Analysis of surrogate fuels effect on ignition delay and low temperature reaction during partially premixed combustion (No. 2013-01-0903). SAE technical paper
175. Kokjohn SL (2012) Reactivity controlled compression ignition (RCCI) combustion. PhD thesis, University of Wisconsin-Madison
176. Bessonette PW, Schleyer CH, Duffy KP, Hardy WL, Liechty MP. (2007) Effects of fuel property changes on heavy-duty HCCI combustion (No. 2007-01-0191). SAE technical paper
177. Inagaki K, Fuyuto T, Nishikawa K, Nakakita K, Sakata I (2006) Dual-fuel PCI combustion controlled by in-cylinder stratification of ignitability (No. 2006-01-0028). SAE technical paper
178. Zeldovich YB (1980) Regime classification of an exothermic reaction with nonuniform initial conditions. Combust Flame 39(2):211–214
179. Chen JH, Hawkes ER, Sankaran R, Mason SD, Im HG (2006) Direct numerical simulation of ignition front propagation in a constant volume with temperature inhomogeneities: I. Fundamental analysis and diagnostics. Combust Flame 145(1):128–144

180. Reitz RD, Duraisamy G (2015) Review of high efficiency and clean reactivity controlled compression ignition (RCCI) combustion in internal combustion engines. Prog Energy Combust Sci 46:12–71
181. Li J, Yang W, Zhou D (2017) Review on the management of RCCI engines. Renew Sust Energ Rev 69:65–79
182. Eichmeier JU, Reitz RD, Rutland C (2014) A zero-dimensional phenomenological model for RCCI combustion using reaction kinetics. SAE Int J Engines 7:106–119. (2014-01-1074)
183. Kokjohn S, Reitz RD, Splitter D, Musculus M (2012) Investigation of fuel reactivity stratification for controlling PCI heat-release rates using high-speed chemiluminescence imaging and fuel tracer fluorescence. SAE Int J Engines 5:248–269. (2012-01-0375)
184. DelVescovo DA (2016) The effects of fuel stratification and heat release rate shaping in reactivity controlled compression ignition (RCCI) combustion. PhD thesis, University of Wisconsin-Madison
185. Kokjohn SL, Musculus MP, Reitz RD (2015) Evaluating temperature and fuel stratification for heat-release rate control in a reactivity-controlled compression-ignition engine using optical diagnostics and chemical kinetics modeling. Combust Flame 162(6):2729–2742
186. Kokjohn SL, Reitz RD (2010) Characterization of dual-fuel PCCI combustion in a light-duty engine. In: 20th international multidimensional engine modeling user's group meeting, Detroit, MI, April 12, 2010
187. Paykani A, Kakaee AH, Rahnama P, Reitz RD (2016) Progress and recent trends in reactivity-controlled compression ignition engines. Int J Engine Res 17(5):481–524
188. Ladommatos N, Parsi M, Knowles A (1996) The effect of fuel cetane improver on diesel pollutant emissions. Fuel 75(1):8–14
189. Hosseini V, Neill WS, Guo H, Chippior WL, Fairbridge C, Mitchell K (2011) Effects of different cetane number enhancement strategies on HCCI combustion and emissions. Int J Engine Res 12(2):89–108
190. Nieman DE, Dempsey AB, Reitz RD (2012) Heavy-duty RCCI operation using natural gas and diesel. SAE Int J Engines 5:270–285. (2012-01-0379)
191. Wang H, DelVescovo D, Yao M, Reitz RD (2015) Numerical study of RCCI and HCCI combustion processes using gasoline, diesel, iso-butanol and DTBP cetane improver. SAE Int J Engines 8:831–845. (2015-01-0850)
192. Dempsey AB, Adhikary BD, Viswanathan S, Reitz RD (2012) Reactivity controlled compression ignition using premixed hydrated ethanol and direct injection diesel. J Eng Gas Turbines Power 134(8):082806
193. Dempsey AB, Walker NR, Reitz RD (2013) Effect of piston bowl geometry on dual fuel reactivity controlled compression ignition (RCCI) in a light-duty engine operated with gasoline/diesel and methanol/diesel. SAE Int J Engines 6:78–100. (2013-01-0264)
194. Li Y, Jia M, Liu Y, Xie M (2013) Numerical study on the combustion and emission characteristics of a methanol/diesel reactivity controlled compression ignition (RCCI) engine. Appl Energy 106:184–197
195. Benajes J, Molina S, García A, Monsalve-Serrano J (2015) Effects of direct injection timing and blending ratio on RCCI combustion with different low reactivity fuels. Energy Convers Manag 99:193–209
196. Hanson R, Curran S, Wagner R, Reitz R (2013) D.. Effects of biofuel blends on RCCI combustion in a light-duty, multi-cylinder diesel engine. SAE Int J Engines 6:488–503. (2013-01-1653)
197. Gharehghani A, Hosseini R, Mirsalim M, Jazayeri SA, Yusaf T (2015) An experimental study on reactivity controlled compression ignition engine fueled with biodiesel/natural gas. Energy 89:558–567
198. Splitter D, Reitz RD, Hanson R (2010) High efficiency, low emissions RCCI combustion by use of a fuel additive. SAE Int J Fuels Lubr 3:742–756. (2010-01-2167)

199. Hanson R, Kokjohn S, Splitter D, Reitz R (2011) D.. Fuel effects on reactivity controlled compression ignition (RCCI) combustion at low load. SAE Int J Engines 4:394–411. (2011-01-0361)
200. Dempsey AB, Walker NR, Reitz RD (2013) Effect of cetane improvers on gasoline, ethanol, and methanol reactivity and the implications for RCCI combustion. SAE Int J Fuels Lubr 6:170–187. (2013-01-1678)
201. Ma S, Zheng Z, Liu H, Zhang Q, Yao M (2013) Experimental investigation of the effects of diesel injection strategy on gasoline/diesel dual-fuel combustion. Appl Energy 109:202–212
202. Benajes J, García A, Monsalve-Serrano J, Boronat V (2017) Achieving clean and efficient engine operation up to full load by combining optimized RCCI and dual-fuel diesel-gasoline combustion strategies. Energy Convers Manag 136:142–151
203. Walker NR, Dempsey AB, Andrie MJ, Reitz RD (2013) Use of low-pressure direct-injection for reactivity controlled compression ignition (RCCI) light-duty engine operation. SAE Int J Engines 6:1222–1237. (2013-01-1605)
204. Splitter D, Wissink M, Kokjohn S, Reitz RD (2012) Effect of compression ratio and piston geometry on RCCI load limits and efficiency (No. 2012-01-0383). SAE technical paper
205. Li J, Yang WM, Zhou DZ (2016) Modeling study on the effect of piston bowl geometries in a gasoline/biodiesel fueled RCCI engine at high speed. Energy Convers Manag 112:359–368
206. Hanson R, Curran S, Wagner R, Kokjohn S, Splitter D, Reitz RD (2012) Piston bowl optimization for RCCI combustion in a light-duty multi-cylinder engine. SAE Int J Engines 5:286–299. (2012-01-0380)
207. Wu Y, Reitz RD (2015) Effects of exhaust gas recirculation and boost pressure on reactivity controlled compression ignition engine at high load operating conditions. J Energy Resour Technol 137(3):032210
208. Yu C, Wang J, Yu W, Liu J, Gao D (2013) Research on low temperature combustion of homogeneous charge induced ignition (HCII) in a light-duty diesel engine. In: Proceedings of the FISITA 2012 world automotive congress, Springer Berlin Heidelberg, pp 195–204
209. Li Y, Jia M, Chang Y, Fan W, Xie M, Wang T (2015) Evaluation of the necessity of exhaust gas recirculation employment for a methanol/diesel reactivity controlled compression ignition engine operated at medium loads. Energy Convers Manag 101:40–51
210. Wissink M, Reitz R (2016) Exploring the role of reactivity gradients in direct dual fuel stratification. SAE Int J Engines 9:1036–1048. (2016-01-0774)
211. Wissink M, Reitz RD (2015) Direct dual fuel stratification, a path to combine the benefits of RCCI and PPC. SAE Int J Engines 8:878–889. (2015-01-0856)
212. Splitter D, Hanson R, Kokjohn S, Reitz RD (2011) Reactivity controlled compression ignition (RCCI) heavy-duty engine operation at mid-and high-loads with conventional and alternative fuels (No. 2011-01-0363). SAE technical paper
213. Dempsey AB, Reitz RD (2011) Computational optimization of reactivity controlled compression ignition in a heavy-duty engine with ultra low compression ratio. SAE Int J Engines 4(2):2222–2239
214. Kavuri C, Paz J, Kokjohn SL (2016) A comparison of reactivity controlled compression ignition (RCCI) and gasoline compression ignition (GCI) strategies at high load, low speed conditions. Energy Convers Manag 127:324–341
215. Kokjohn SL, Hanson RM, Splitter DA, Reitz RD (2011) Fuel reactivity controlled compression ignition (RCCI): a pathway to controlled high-efficiency clean combustion. Int J Engine Res 12(3):209–226
216. Hendricks TL, Splitter DA, Ghandhi JB (2014) Experimental investigation of piston heat transfer under conventional diesel and reactivity-controlled compression ignition combustion regimes. Int J Engine Res 15(6):684–705
217. Li Y, Jia M, Chang Y, Kokjohn SL, Reitz RD (2016) Thermodynamic energy and exergy analysis of three different engine combustion regimes. Appl Energy 180:849–858

218. Dempsey AB, Walker NR, Gingrich E, Reitz RD (2014) Comparison of low temperature combustion strategies for advanced compression ignition engines with a focus on controllability. Combust Sci Technol 186(2):210–241
219. Lin Y, Zhang HH (2003) Component selection and smoothing in smoothing spline analysis of variance models. Institute of Statistics Mimeo Series 2556, NUCS
220. Kavuri C, Tiry M, Paz J, Kokjohn SL (2016) Experimental and computational investigation of soot production from a premixed compression ignition engine using a load extension injection. Int J Engine Res. https://doi.org/10.1177/1468087416650073
221. Klos D, Kokjohn SL (2015) Investigation of the sources of combustion instability in low-temperature combustion engines using response surface models. Int J Engine Res 16 (3):419–440

Chapter 3
LTC Fuel Quality Requirements

Abstract In low temperature combustion (LTC) engines, premixed fuel-air mixture is created in the cylinder, and combustion starts by auto-ignition due to compression of the fuel–air mixture during compression stroke. The LTC process involves various physical processes (atomization, evaporation and mixing) and complex chemical reactions occurring in the cylinder. Fuel properties and fuel composition play an important role in all the physical and chemical processes involved in LTC process. Autoignition depends on the evolution of cylinder pressure and temperature with time and autoignition chemistry of the fuel–air mixture. Autoignition chemistry depends on the fuel composition and fuel-air mixture quality (equivalence ratio). This chapter discusses the autoignition characteristics (autoignition chemistry, impact of fuel molecular structure on autoignition, fuel autoignition quality), fuel effects on autoignition and several fuel indices developed for LTC engines. Fuel design and fuel properties/quality required for LTC engines are also discussed in the present chapter.

Keywords Fuel quality • Octane number • Octane index • HCCI number • Fuel design • LTC • RON • Autoignition

3.1 Autoignition Characteristics

In the LTC engine, the fuel and air are premixed (either fully or partially) before the start of combustion. Combustion starts by autoignition at several locations in the combustion chamber due to pressure and temperature rise during compression stroke of engine. The fuel autoignition quality plays an important role in combustion process as the charge is auto-ignited in the cylinder. Different theoretical, empirical and experimental methods exist for describing the autoignition quality of fuels. In theoretical approach autoignition quality is described using chemical kinetics of autoignition process by different types of reaction mechanisms (detailed or reduced mechanism). Empirical models use basic equations derived from experimental data to describe the autoignition quality of fuels in combustion process. In the experimental approach, a test engine or a constant volume chamber is used for comparison of autoignition quality of various fuels. The standard methods for

R.K. Maurya, *Characteristics and Control of Low Temperature Combustion Engines*, Mechanical Engineering Series, https://doi.org/10.1007/978-3-319-68508-3_3

quantifying autoignition properties of fuels are RON (research octane number) and MON (motor octane number) for gasoline-like fuels and cetane number (CN) for diesel-like fuels. The autoignition quality of diesel fuel is estimated by CN, which is measured for a given fuel by comparing ignition characteristics with reference fuel in a CFR (corporate fuel research) cetane engine using ASTM D613 test method [1]. The reference fuel is prepared by blending of n-cetane, n-hexadecane (CN = 100) and heptamethyl nonane (CN = 15). The CN of reference fuel having x% volume n-cetane is given by $CN = x + 0.15(100 - x)$. The test fuel is assigned the CN of reference fuel which has the same ignition quality [1]. ASTM methods D2699 and D2700 are typically used to measure RON and MON, respectively, in gasoline-like fuels. RON and MON are estimated on a standardized CFR (corporate fuel research) engine using two sets of operating conditions. During the estimation of RON, CFR engine is operated at 600 rpm speed, inlet temperature of 49 °C and fixed spark timing, while in MON estimation, engine speed is 900 rpm, the inlet temperature is 149 °C and spark timing is adjusted for maximum knock [2]. The mixture of n-heptane and iso-octane is used as reference fuel for octane number measurement. The RON is often higher than MON around 5–10 units for practical hydrocarbons. The difference in RON and MON has led to different definitions of octane number worldwide. For example, only RON is used in the European Union, and in the United States, the average of RON and MON is typically used [2]. The fuel sensitivity (S) is defined as the difference between the RON and MON value ($S = RON-MON$), which is a measure of the difference in knock tendency for a fuel at different operating conditions [1]. In modern engines, charge condition in the cylinder after compression is different from the RON and especially MON cases. Therefore, suitability of RON and MON needs to be reconsidered. Modern SI engines are "beyond RON" because at a given pressure, the temperature of the unburned mixture is lower than the RON test. Alternatively, the pressure at a given temperature is higher in modern engines than RON test. All the approaches used to improve the SI engine efficiency such as direct fuel injection, higher compression ratio and turbocharging are pushing modern engines further beyond RON [1, 2]. The HCCI engine can be operated well beyond RON and MON. The HCCI engine experiments helped to understand the fuel effects on autoignition with change in operating conditions. The autoignition characteristics of different fuels in LTC engines are discussed in the next subsections.

3.1.1 Autoignition Chemistry

The autoignition process is investigated at a fundamental level by modelling the kinetics of the chemical reactions involved. The ignition process of a commercial transportation hydrocarbon fuel involves hundreds of chemical species participating in thousands of reactions, which makes the whole process very complex. Autoignition occurs when energy released by chain reactions produces exponential increase in temperature of charge in the cylinder. Autoignition starts with initiation

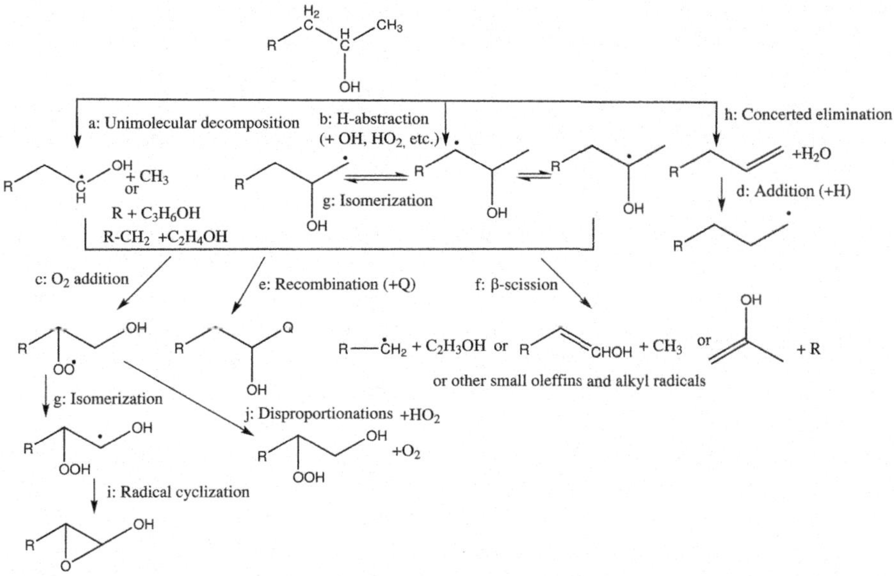

Fig. 3.1 Simplified reaction scheme for hydrocarbon auto-ignition [3]

reactions which generates the radicals from stable species. Chain-branching reactions multiply the number of radicals, which leads to further reaction progress. Chemical reactions involved in the autoignition process can be categorized into ten main types of reactions. Figure 3.1 depicts the summary of simplified reaction scheme of hydrocarbon oxidation by presenting important reactions.

Boot et al. [3] presented the summary of ten main categories of reactions involved in the auto-ignition process as (i) unimolecular decomposition; (ii) hydrogen (H) atom abstraction; (iii) O_2 addition reactions; (iv) other addition reactions of radicals such as H, OH or HO_2, where radicals attach themselves to unsaturated bond sites; (v) recombination reactions; (vi) β-scission; (vii) isomerization; (viii) concerted elimination, two substituents are removed from a molecule in one-step mechanism, forming an unsaturated bond; (ix) radical cyclization; and (x) disproportionation, where two radicals react to produce two different nonradical products. Kinetics of elementary reactions involved in low temperature autoignition process is discussed in detail in the review study [4]. In HCCI combustion engine, the hydrocarbon oxidation process can be divided into low temperature heat release (LTHR), intermediate temperature heat release (ITHR) and high temperature heat release (HTHR) reactions. In Sect. 2.2.2 of Chap. 2, the autoignition reactions involved in HCCI combustion are discussed in detail. Figure 2.4 (Chap. 2) depicts the LTHR, ITHR and HTHR zones as well as single- and two-stage ignition fuels in HCCI engines. Typically fuels containing significant amount of n-paraffins show LTHR characteristics during premixed compression igntion combustion.

Ideally comprehensive chemical models of autoignition process must use all the relevant chemical reactions. However, "comprehensive" schemes cannot be perfectly accurate in the prediction of autoignition process due to uncertainties in the reaction rate constants and their temperature and pressure dependency [1, 5]. Typically detailed reactions mechanisms consist of thousands of reactions. Many practical applications used reduced reaction mechanisms to describe the combustion process [6–8]. Reduced reaction mechanisms are developed by selecting reactions of critical importance for the particular application. Autoignition reaction mechanisms applicable to internal combustion engines are developed for a very few pure compounds. Detailed chemical kinetic models for the oxidation of hydrocarbon fuels are provided in the study [5]. The chemical kinetic schemes are developed for PRFs (primary reference fuels), which contain n-heptane and iso-octane. Conventional gasoline and diesel contain hundreds of species, which have different combustion chemistry than PRFs. Therefore, simplified surrogate fuels are required for representing practical fuels [9]. A surrogate fuel is defined as a fuel composed of a smaller number of pure species, which has the same combustion and emission characteristics as that of targeted practical fuel [1]. The chemical kinetic model of such surrogate fuels is being developed [10–13]. The full chemical kinetic model for surrogate fuels is very large to be used in engine computational models (five-component surrogate has 1328 species and 5825 reactions) [10]. Thus, reduced reaction mechanism is required for autoignition analysis. Empirical autoignition model and experiments can help to understand the autoignition process. Empirical models are discussed in Sect. 3.1.3. Fuel molecular structure also plays an important role in autoignition process, which is discussed in Sect. 3.1.2.

3.1.2 Impact of Fuel Molecular Structure

The chemical composition of commercially available fuel is dependent on the crude oil, refinery processes of oil refineries and season. The HCCI combustion characteristics are very sensitive to the fuel composition, and a fuel standard for advanced combustion is required [14]. Fuel molecular structure affects the auto-ignition characteristics in LTC engines especially in LTHR and ITHR. A typical commercially available gasoline consists of 54.2% total paraffins (10.8% normal paraffins and 43.4% iso-paraffins), 33.6% aromatics, 8.6% olefins, 2.9% naphthenes and 0.88% benzene [1]. A typical diesel contains 44% total paraffins, 29% naphthenes and 25.6% aromatics. Different category of fuel component has different autoignition characteristics. Fuel molecular structure of fuel plays an important role in autoignition characteristics of fuel-air mixture.

The overall autoignition reaction rate is mainly governed by chain-branching reactions (Sect. 2.2.2 of Chap. 2). The chain-branching reaction rate is a function of molecular structure of fuel, pressure and temperature of reactor, equivalence ratio (φ) of mixture and initial concentration of fuel and air [3]. A sufficiently large radical pool is required to autoignite the fuel. At specific initial conditions in the cylinder, the overall autoignition reaction rate is governed by fuel molecular structure. Initial reactions are governed by the type of C-H bond (primary,

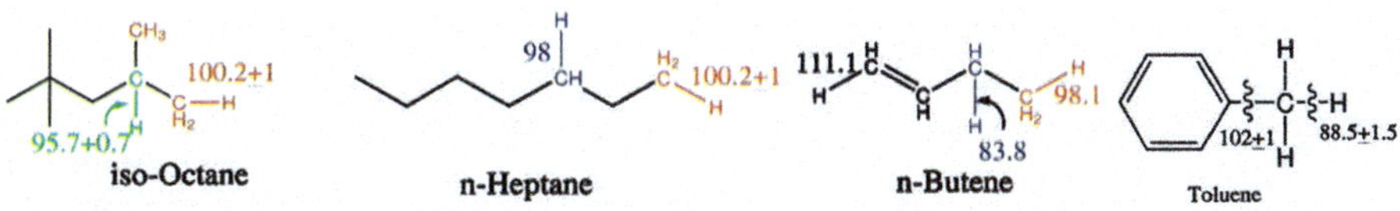

Fig. 3.2 The C–H bond dissociation energy (kcal/mol) for different hydrocarbon fuel component [3, 15]

secondary and tertiary), location of C-H bond in the fuel carbon chain and number of C–H bonds in the fuel molecular structure. Intermediate reactions are governed by hydroperoxyl radical (RO_2) chemistry (propagation and branching reactions). Fuel molecular structure plays an important role in hydroperoxyl radical chemistry in LTHR regime. Terminal reactions are governed by the number and reactivity of intermediate species (termination), which are dependent on fuel structure [3]. The decomposition reactions start with weakest bonds (lowest bond dissociation energy) on the molecular structure of the fuel. Figure 3.2 depicts the C–H bond dissociation energy for different hydrocarbons for different types of C–H bonds (primary, secondary and tertiary bonds). Initiation reactions are followed by H atom abstraction reaction. The bond dissociation energy of C–H bond decreases in the order of primary > secondary > tertiary C–H bond (Fig. 3.2). The molecules having a higher number of primary C–H bonds are more resistant to autoignition reactions. The primary reference fuel n-heptane contains 10 secondary and 6 primary C–H bonds, and iso-octane contains 15 primary, 2 secondary and 1 tertiary C–H bonds. Thus, iso-octane has higher octane number (RON = 100), and it is difficult to autoignite [3].

A study developed an equation to calculate the RON, based on the number of carbon atoms in particular molecular structure by multiple linear regression using data of 58 hydrocarbons [14]. Developed equation for RON calculation is presented in Eq. (3.1):

$$\mathrm{RON}_{\mathrm{calc}} = (-9.13C_2 - 46.0C_d) + (-4.63C_n - 28.0C_{2n}) + 2.30C_b + 107 \quad (3.1)$$

where C_2 is the carbon number of secondary carbon atoms in straight chain, C_d is the dummy variable (if the number of straight chain carbon atoms is higher than 6, $C_d = 1.0$, and if it is less than or equal to 6, $C_d = 0$), C_n is the number of carbon atoms in the naphthene ring, C_{2n} is the number of secondary carbon atoms in the side chains of the naphthene ring and C_b is the number of carbon atoms in the aromatic ring. The calculated RON value correlates well ($R^2 = 0.90$) with experimental data for paraffins, olefins, naphthenes and aromatics. This equation clearly illustrates the importance of molecular structure on the autoignition characteristics. The ignitability of the paraffins increases (RON decreases) with increase in the length of straight chain in the chemical structure (increases rapidly with carbon chain length higher than 6) (see Fig. 3.3 and Eq. 3.1). The ignitability of the hydrocarbon increases with increase in the naphthene ring size and increase in carbon atom number in the side chains of naphthenes, which leads to rapid increase

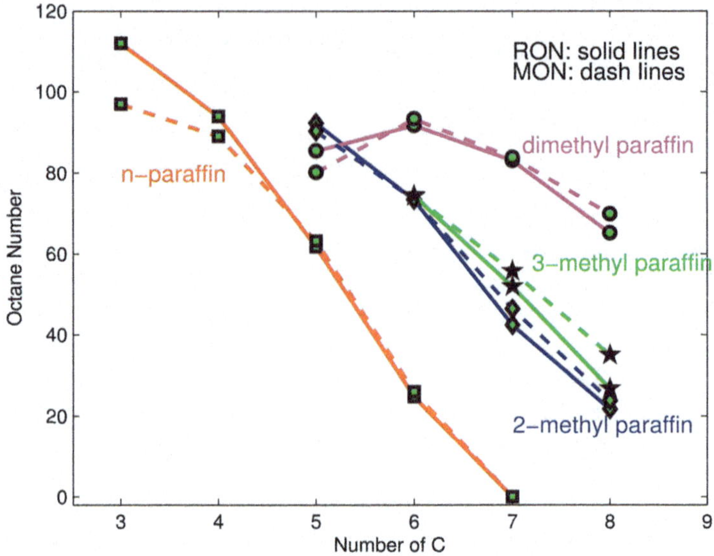

Fig. 3.3 Impact of paraffin chain length and branching on RON and MON [3, 16]

in ignitability or decrease in RON (Eq. 3.1). The ignitability of the hydrocarbon decreases with the addition of aromatics [14].

Figure 3.3 shows that the octane number of fuel decreases with increasing number of carbon atoms in straight chain. This is explained by alkyl radical isomerization reactions at low to intermediate temperatures. The RO_2 radicals (reaction R3, Chap. 2) are very important intermediates in autoignition process. Both alkyl and peroxyl radicals undergo intra-H atom transfer via a transition state ring to produce QOOH, which dominates low temperature chain branching. The isomerization ring strain energy is highly dependent on its size, and larger rings have lower strain energies [3]. The longer average chain length hydrocarbons have more options (more possible site for isomerization reactions) for low energy barrier isomerization reactions and thus higher overall reaction rates in autoignition process leading to higher ignitability or lower octane number (Fig. 3.3). The highly branched paraffins have comparatively lower overall autoignition reaction rates due to drop in average chain length of every branch and higher primary C–H bond (comparatively stronger bond) located at the extremities of molecular structure. This results in lower ignitability and higher RON (Fig. 3.3).

The presence of double bonds and oxygen groups in the fuel affects the possible reaction pathways due to the differences in bond dissociation energies. Additionally, these groups can affect the radical isomerization reactions also. Significant quantity of olefins are present in marketed gasoline. Olefins have a higher RON than paraffins with the same carbon (C) number (Fig. 3.4). The olefinic C bond provides site for addition reactions involving O, H, OH or HO_2 radicals, which initiates the divergence from the paraffinic reaction pathways [3]. The position of

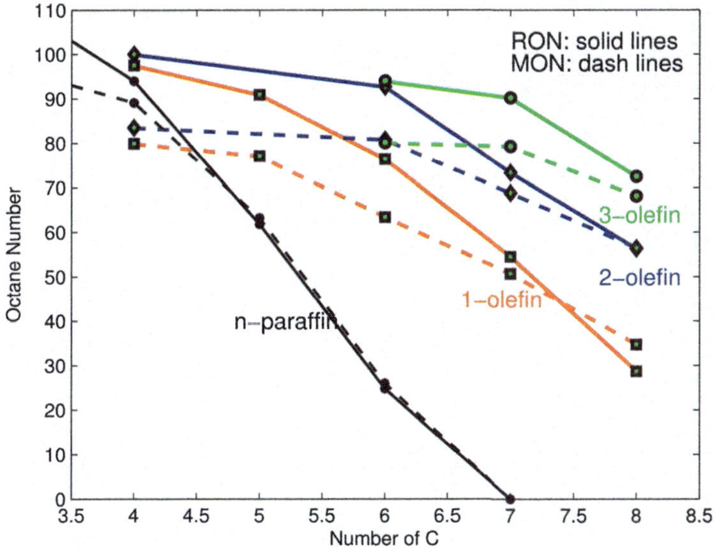

Fig. 3.4 Impact of carbon number and double bond site on olefin reactivity [3, 16]

the double bond in the molecular structure has a significant effect on olefinic autoignition process in the low to intermediate temperature regime [17]. Long-chained olefins having double bond on the extremity tend to show paraffinic-like NTC characteristics [3, 17]. Figure 3.4 also depicts that 3-olefin has higher octane number than 1-olefin.

Almost one-third of gasoline contains aromatics consisting 11.7% toluene, 6.93% m-xylene, 4.04% o-xylene, 3.04% ethyl benzene, 2.58% p-xylene, 1.65% methyl toluene and 1% benzene [1]. Thus, autoignition chemistry of benzene and toluene is important and studied by various researchers [18–22]. Figure 3.5 shows the RON of different aromatic compounds with different number of methyl chains and number of methyl branches. The figure shows that longer side chain aromatics (n-butylbenzene) have lower RON values [14]. The highly stable π-bond in the benzene ring makes H atom abstraction from aromatics difficult. Thus, most aromatics are highly resistant to auto-ignition reactions at low to intermediate temperatures [18]. Benzyl radicals are also resonance stable, and lead to radical–radical reactions in an oxidative environment, which produce even more stable bi-benzyl species. This tendency leads aromatics to both lower propensity for further decomposition and formation of soot during combustion [23]. A more complete discussion on the impact of molecular structure on autoignition characteristics of hydrocarbons can be found in the recent study [3].

Effect of engine operating conditions such as intake air temperature, intake oxygen content and engine speed on ignitability of different hydrocarbon is investigated in HCCI engine [14]. Figure 3.6 shows the effect of intake temperature and engine speed on hydrocarbon ignitability (HI) index of paraffins (n-pentane and

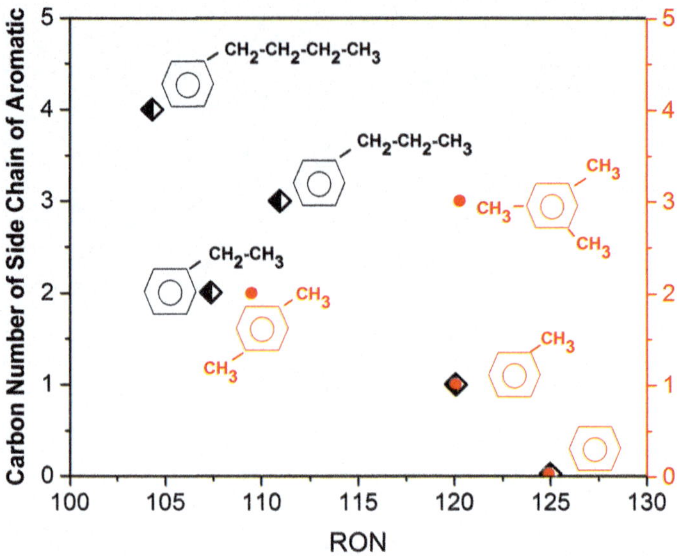

Fig. 3.5 RON versus carbon number of the side chain in aromatics and number of methyl branches in aromatics (Adapted from [14])

iso-octane), naphthenes (cyclohexane and ethyl cyclohexane) and aromatics (toluene, m-xylene and o-xylene) in HCCI engine. The hydrocarbon ignitability index (HI index) is defined as the autoignition characteristic relative to the n-heptane and isooctane scale, based on by octane number measured by the CFR engine [14]. Figure 3.6 shows that the HI indices of the naphthenes increased, and those of the aromatics decreased with the increase in intake air temperature. This observation suggests that the ignitability of the naphthenes decreases and the ignitability of aromatics increases with the increase in intake air temperature. A study suggested that the side chain of naphthenes is responsible for the ignitability improvement, and its effect is stronger at higher intake air temperatures [14]. Oxidation of cyclohexane occurs via two reaction processes (cyclohexane radical reaction and an open-ring reaction process), which are dependent on the charge temperature. The reaction proceeds by cyclohexane radical process at lower low intake air temperature condition, and the LTHR occurs. The naphthene ring opens at the high intake air temperature condition and reactions proceed via open-ring reaction pathway. This reaction pathway does not show LTHR reactions. The two reaction pathways at different temperature lead to different ignitability at higher intake temperature in HCCI engine [14].

In aromatics the chemical ignitability depends on the position of the methyl branch and the side chain length (Fig. 3.5). Ignitability of o-xylene is found higher everywhere in comparison to m-xylene (Fig. 3.6). Differences in the reaction process of o-xylene, m-xylene and p-xylene have been reported in the study [24]. The dehydration reaction in the o-xylene molecule occurs more easily than with m-xylene and p-xylene. This is due to the two methyl groups on fuel molecule

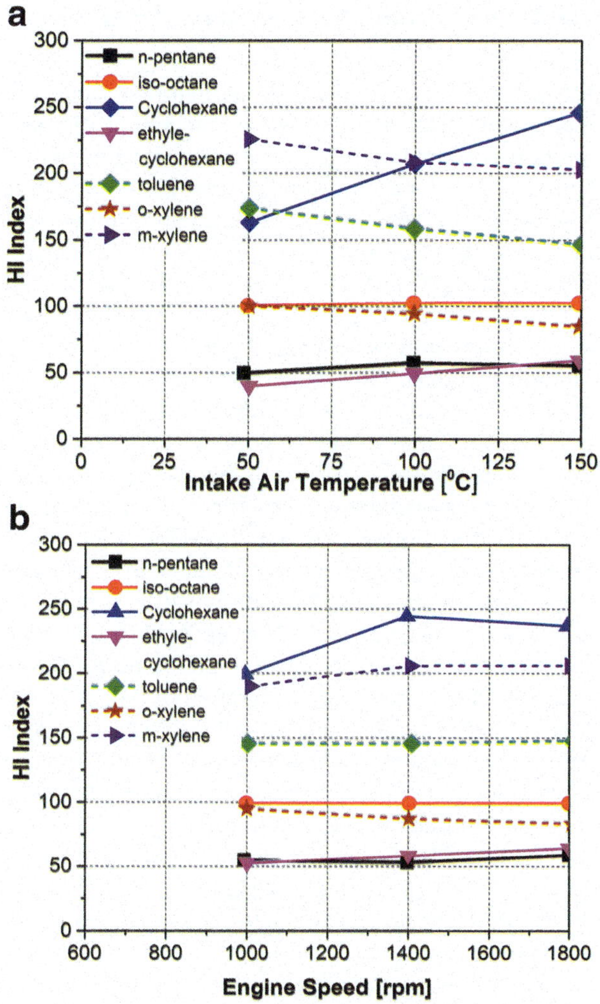

Fig. 3.6 Effect of intake air temperature and engine speed on hydrocarbon ignitability (HI) index in HCCI engine (Adapted from [14])

structure (two methyl groups at the consecutive carbon of the benzene ring) that helps in transition state ring formation during isomerization reaction of o-xylene. This might be a reason of higher ignitability of o-xylene than m-xylene [14].

Figure 3.6b also depicts that the relative ignitability (the HI index) of the naphthenes and aromatics decreases with the increase in engine speed in HCCI engine. With the increase in engine speed, the value of LTHR decreases, and HTHR occurs in the expansion stroke (delayed combustion phasing) particularly for cyclohexane and aromatics with short side chains (toluene, m-xylene, and ethyl-benzene) [14]. At higher engine speed, lower time available for LTHR reactions that decreases the amount of LTHR and affects the autoignition characteristics of fuel showing LTHR.

3.1.3 Empirical Auto-ignition Modelling

Empirical approach to describe autoignition is considered as in-between method of theoretical chemical kinetic-based modelling approach and the experimental auto-ignition test methods [25]. The chemical kinetics of HCCI combustion is very similar to the kinetics of knock (end-gas autoignition) in spark ignition engines. Livengood and Wu introduced a method (empirical correlation) to predict the knock conditions for various fuels [26]. This method is also known as knock integral method (KIM) [27]. A study reviewed the five different types of models (based on different complexity and required input data) used to predict the autoignition timings in HCCI engines [28]. Auto-ignition models range from multidimensional CFD models and multizone models to simple control-oriented models. The simplest model uses a threshold temperature to estimate the start of ignition, but it fails to predict combustion phasing at different operating conditions due to dependency of oxidation reaction reactions on species concentration along with charge temperature in the cylinder [29]. The Shell model is also used to predict HCCI ignition timing [30]. Shell model is a lumped chemical kinetic model using only five representative species in the eight generic reactions and only applicable for single-component fuels. Shell model accurately estimates ignition timing over temperature and engine speed sweep, but is less accurate with load variation [28]. A two-step Arrhenius-type reaction rate-based model [31] is used in various studies [32, 33]. In this model the integration of the Arrhenius global reaction rate for the fuel is tracked until it reaches a threshold value defined from experimental data. This model requires instantaneous fuel and oxygen concentrations as well as in-cylinder gas temperature for predicting autoignition, which is impractical for on-board ignition timing control [28]. The KIM is another category of model used for estimating HCCI ignition timing in control-oriented modelling [26]. The KIM is based on an exponential correlation (function of in-cylinder gas pressure and temperature) to predict the auto-ignition characteristics of a homogeneous fuel–air mixture [34].

The KIM is a common empirical method used in the estimation of HCCI ignition timings. The basis of the correlation stems from the ignition delay of various fuels using a rapid compression testing machine [26]. The resulting empirical relationship in the general form is provided in Eq. (3.2):

$$\tau = Ae^{\left({}^{b}/_{T}\right)p^{n}} \tag{3.2}$$

where τ is the ignition delay (or induction time), p is the charge pressure as a function of time, T is the charge temperature as a function of time and A, b and n are empirical constants. Ignition delay or induction time is defined as the time between the end of compression and autoignition (marked by heat release) in rapid compression machine [1]. The relatively more reactive mixture has lower ignition delay (τ).

Livengood and Wu proposed the functional relation between the concentration ratio, $(x)/(x)_c$, of the significant species in the autoignition reaction and the relative time, t/τ. The $(x)_c$ is the critical concentration ratio, which is the concentration of the species at the end of the reaction. The ignition correlation of Livengood and Wu in terms of the crank angle instead of time (as typical engine combustion characteristics measured on crank angle basis) is shown in Eq. (3.3) [28]:

$$\frac{x}{x_c} = \int_{\theta_0=0}^{\theta=\theta_e} \frac{1}{\omega\tau} d\theta = \int_{\theta_0=0}^{\theta=\theta_e} \frac{1}{A\omega e^{\left(b/T\right)} p^n d\theta = 1.0} \tag{3.3}$$

where θ_e is the crank position at which autoignition or knock occurs and θ_0 is the initial crank position where integration starts. The engine speed (ω) is signified in revolutions per minute (rpm), the pressure in kPa and the temperature in K.

In this model, instantaneous pressure and temperature of cylinder charge are required to estimate the ignition timings. The measurement of in-cylinder pressure on engine is difficult and expensive which limits the utilization of this model for real-time control. A study [28, 35] presented modified knock integral model (MKIM) for estimation of HCCI ignition timings using intake valve closing (IVC) pressure and temperature, equivalence ratio and instantaneous volume. The MKIM is presented by Eq. (3.4):

$$\int_{\theta_{\text{ivc}}}^{\theta_{\text{soc}}} \frac{\phi^B}{A\omega \exp\left(\frac{C\left(P_{\text{ivc}} v_c^{k_c}\right)^D}{T_{\text{ivc}} v_c^{k_c - 1}}\right)} d\theta = 1.0 \tag{3.4}$$

where θ is the engine crank angle and B, C and D are constant parameters. k_c represents the average polytropic coefficient. The value of the expression being integrated increases as the point of SOC is approached. A and v_c are determined by

$$v_c(\theta) = \frac{V_{\text{ivc}}}{V(\theta)}, A = E_1 \text{EGR} + E_2 \tag{3.5}$$

where E_1 and E_2 are constant parameters and the cylinder volume, V (θ), is calculated at any crank angle using slider crank mechanism.

A study investigated the validity of correlation proposed by Livengood and Wu in HCCI engines for various fuels with practical interest (hydrogen, methane, ethanol and n-heptane) [36]. The study concluded that the method predicts ignition timings quite accurately for the fuels (hydrogen, methane and ethanol) with single-stage heat release (no LTHR). However, KIM underpredicts the ignition time when ignition takes place after the TDC for all three cases because KIM cannot account for slowing down the oxidation chemistry due to cooling effect of the expansion. The study also showed that the KIM method fails to accurately predict the ignition timing for two-stage heat release (LTHR and HTHR) fuel (n-heptane), and it overpredicts the ignition timing. Another study describes the ignition delay as a combination of a low temperature regime and high temperature regime induction time, and each stage is described by an Arrhenius equation [37]. The study showed that this is a convenient and flexible method for describing the ignition delay of a wide range of fuels including linear paraffins, iso-paraffins, olefins, aromatics and alcohols.

3.1.4 Fuel Effects on Autoignition in LTC Engines

Gasoline and diesel fuel contain hundreds of components and each component has its specific autoignition temperature. The variations in the temperature and fuel composition make autoignition control very difficult in HCCI combustion. Understanding of fuel effects on autoignition process is required to describe and predict the autoignition process in HCCI combustion. Autoignition temperature and LTHR are closely coupled with fuel autoignition qualities in HCCI engines. Gasoline-like fuels with an octane number lower than 83 show LTHR in HCCI combustion [38]. Diesel-like fuels or high cetane number fuels typically exhibit LTHR in HCCI combustion, and LTHR disappears for fuels having cetane number less than 34 (CN < 34) [39]. Fuel structure is strongly related to LTHR (Sect. 3.1.2), and small variation in chemical composition can change the HCCI combustion characteristics [40].

Figure 3.7 illustrates the effect of PRF fuel composition on LTHR rate in HCCI combustion. The LTHR decreases with increase in PRF octane number (increase in

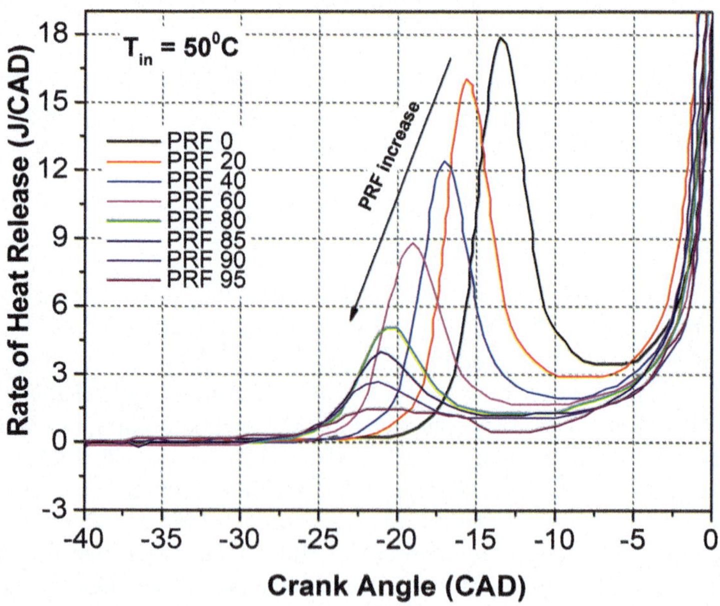

Fig. 3.7 Effect of PRF fuel composition (different n-heptane concentration) on LTHR in HCCI combustion (Adapted from [41])

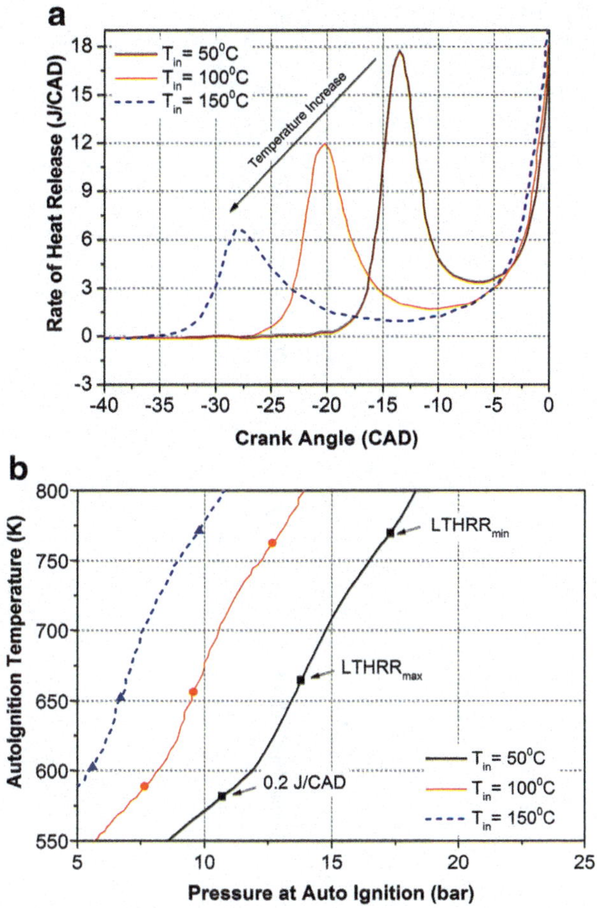

Fig. 3.8 Effect of intake temperature on (**a**) LTHR and (**b**) autoignition temperature for PRF0 (n-heptane) in HCCI combustion (Adapted from [41])

iso-octane content). The LTHR depends on concentration of straight chain paraffins (Sect. 3.1.2), and therefore as n-heptane concentration in fuel decreases (increase in PRF octane number), the amount of LTHR decreases.

Figure 3.8 shows the effect of intake temperature on LTHR and autoignition temperature in HCCI engine using PRF0 (pure n-heptane). The LTHR is higher at lower intake air temperature and decreases with increase in inlet temperature. The main reaction containing alkyl–peroxyl radical formation (reaction R3; Chap. 2) is a reversible reaction and responsible for LTHR. This reaction is highly dependent on temperature and an increase in temperature shifts the reaction towards the reactant side. Therefore, increasing inlet temperature leads to lower amount of LTHR (Fig. 3.8). At lower inlet temperatures, higher compression ratio is used to achieve a constant combustion phasing, and higher pressure is known to increase the LTHR [41]. For higher octane PRF, LTHR diminishes at higher intake temperatures, which leads to delayed start of combustion. Figure 3.8b also depicts that the maximum value of low temperature heat release rate ($LTHRR_{max}$) occurs at almost

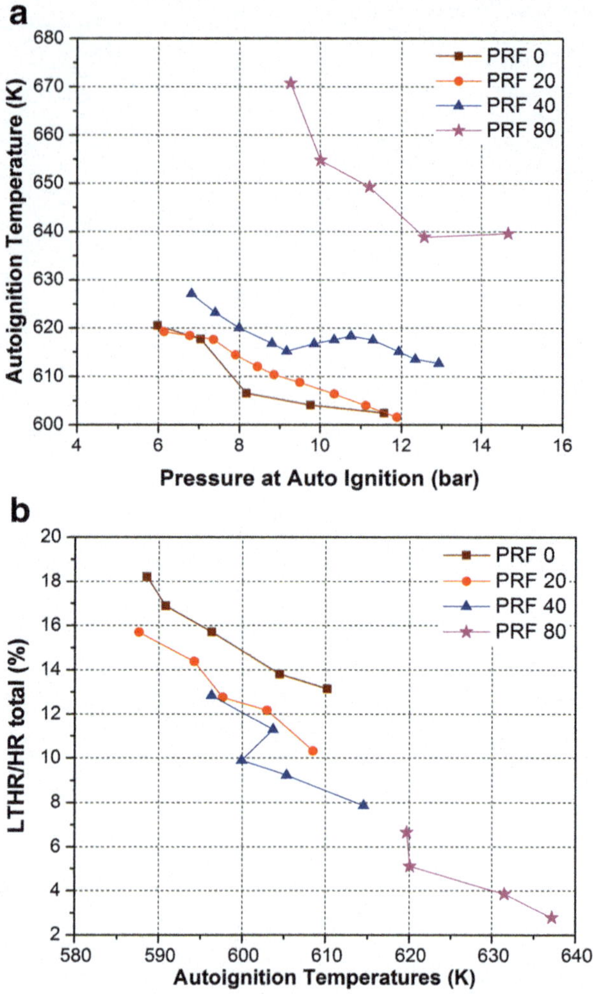

Fig. 3.9 Effect of PRF fuel composition on (**a**) autoignition temperature, and (**b**) LTHR in HCCI engine (Adapted from [41])

same temperature for each test condition. In this figure, autoignition (start of LTHR) is defined as the point where heat release rate of 0.2 J/CAD has reached. The LTHR starts decreasing after reaching $LTHRR_{max}$ in each operating conditions. At lower intake temperature, autoignition temperature is also lower (Fig. 3.8b).

Figure 3.9 shows the effect of PRF composition on autoignition temperature and LTHR fraction. The LTHR fraction is shown in the figure as percentage of total heat release in the cylinder. Different points for each test fuel in Fig. 3.9 correspond to the different inlet air temperatures. Autoignition temperature increases with increase in octane number of the fuel. Fuels having extensive LTHR show similar autoignition characteristics, and autoignition starts at almost the same temperature regardless of the cylinder pressure. The n-heptane (PRF0) auto ignites around

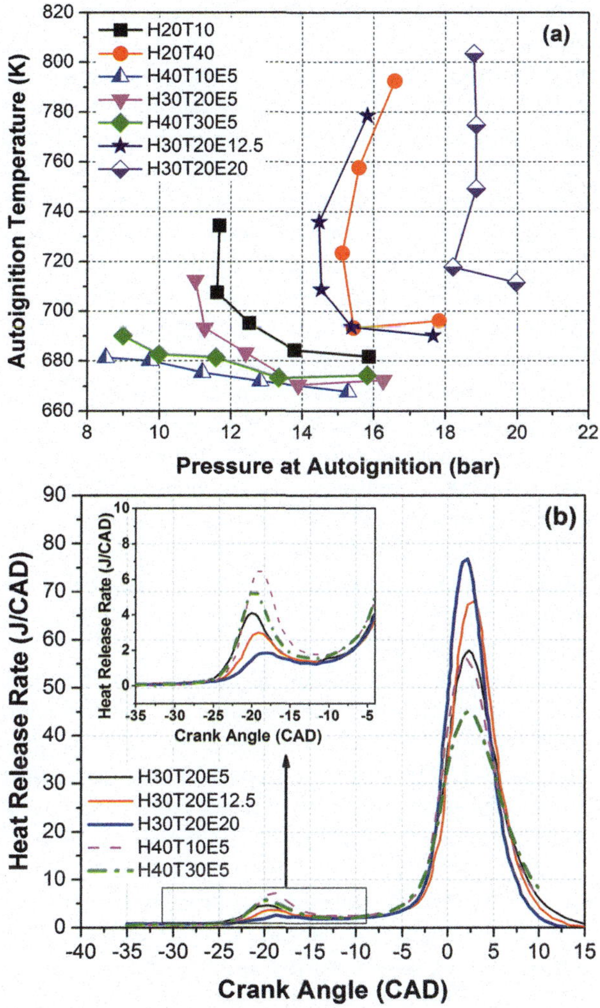

Fig. 3.10 Effect of TERF fuel composition on (**a**) autoignition temperature, and (**b**) LTHR in HCCI combustion (Adapted from [42, 43])

570 K, with a weak trend that comparatively higher autoignition temperature is required for lower cylinder pressure. With decrease in LTHR, autoignition temperature rises quickly [41]. Figure 3.9b depicts almost linear correlation between the amount of LTHR and autoignition temperature. All the test operating conditions having detectable LTHR (>1% and with a $LTHR_{max}$) ignited at temperatures below 650 K (Fig. 3.9b).

Figure 3.10 illustrates the effect of toluene and ethanol content on autoignition temperature and LTHR in HCCI combustion. Five data points on the curve (Fig. 3.10a) represent the five intake temperature for a particular fuel. Fuels designated as HxTyEz contains x% of n-heptane, y% of toluene, z% of ethanol and (100-x-y-z) % of iso-octane. Ethanol and toluene are known to reduce LTHR reactions by consuming radicals during LTHR regime [44, 45]. Generally, autoignition temperature increases with increase in intake temperature for all the test fuels, but the level of change in autoignition temperature is different for different fuels (Fig. 3.10a). The curve for toluene ethanol reference fuels (TERF) is similar to PRFs (Fig. 3.9a). Toluene addition in the fuel increases the auto-ignition temperature (see case H20T10 and H20T40), which can be seen as upward shift in the graphs as toluene content increases. An increase in autoignition temperature is observed mainly at higher intake temperatures due to the lower LTHR. Autoignition temperature differs slightly at lowest intake temperatures. Fuels depicting LTHR at all five intake temperatures, autoignition occurs at similar charge temperatures. Fuels with little or no LTHR have comparatively larger difference in autoignition temperatures. Radical quenching effect of ethanol can be seen by comparing the amount of LTHR for three fuels with a constant amount of n-heptane (30 vol.%) and toluene (20 vol.%) but varying amounts of ethanol (5, 12.5, 20 vol.%) (Fig. 3.10b). Toluene quenching effect is also depicted in Fig. 3.10b by comparing two fuels having constant amount of n-heptane (40 vol.%) and ethanol (5 vol.%) but varying amounts of toluene (10, 30 vol.%). The lower LTHR for the particular fuel with a higher concentration of ethanol or toluene shows the quenching effect. The figure shows that toluene has the weaker LTHR quenching effect than ethanol, despite the higher RON of toluene than ethanol (120 compared to 109) [43].

Gasoline partially premixed combustion (PPC) is another LTC strategy that demonstrated higher operating range and better combustion control, while keeping the benefits of HCCI combustion (Sect. 2.5.2 of Chap. 2). In PPC mode, ignition delay and LTHR have an important role in ignition characteristics. Premixed combustion phase in PPC needs mixing of fuel and air before ignition starts. The desired premixing is achieved by longer ignition delay. During ignition delay period, physical processes (droplet evaporation, mixing of fuel with air and heating due to compression) and chemical process (formation of radicals and intermediates) occur in the engine cylinder [27]. Effect of fuel properties on PPC relates through charge cooling effect during evaporation, mixing process and fuel ignitability (reactivity) [46]. At constant combustion phasing, ignition delay is correlated with fuel RON (Fig. 3.11). Figure 3.11 shows that the ignition delay increases with an increase in RON value for different start of combustion (SOC) operation in PPC mode because fuel reactivity decreases with increase in RON. The ignition delay is found to decrease with retarding SOC, and minimum around SOC at TDC position (0 CAD). Until SOC reaches TDC position, ignition delay decreases due to higher pressure and temperature of air during compression. Air pressure and temperature decrease for SOC after TDC, which also leads to higher ignition delay.

Duration of LTHR reaction depends on the ignition delay. Prolonged ignition delay gave an increased duration of LTHR in PPC engine. Figure 3.12 shows the variations of ignition delay, LTHR duration and LTHR fraction with engine load at constant combustion phasing of 6 CAD after TDC and 1500 rpm. Low load limit achieved with stable combustion for low and high octane fuels is also indicated in Fig. 3.12. The low load limit is defined as the point where ambient intake

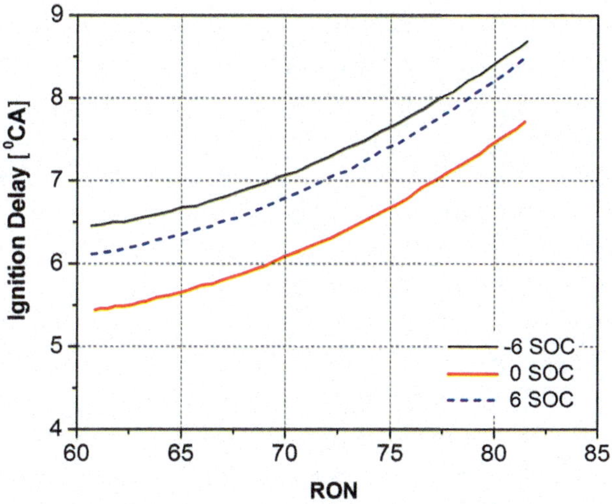

Fig. 3.11 Predicted ignition delay as a function of RON in PPC engine at 8 bar IMEP and 1500 rpm (Adapted from [47])

temperature is not enough for stable combustion. The lowest stable load achieved with higher octane number fuel is higher in comparison to lower octane number in PPC. Justification of trend can be found in Sect. 2.5 of Chap. 2. Ignition delay period decreases with increase in engine load for all fuels above the low load limits due to the low temperature at start of injection (SOI). The SOI timings are adjusted to achieve the constant combustion phasing for all the test conditions. Fuel is injected at lower cylinder pressure-temperature at early (advanced) SOI timings. Ignition delay decreased again when λ increases (lower IMEP), in case of below the low load limit conditions. Apparently, the higher oxygen fraction has a stronger impact on the ignition delay than the decrease in charge temperature [48]. LTHR reaction rate depends on both fuel composition and gas temperature in the combustion chamber. Above the low load limit, duration of LTHR decreases with an increase in engine load. Fraction of LTHR is a function of engine load (Fig. 3.12b), and LTR fraction increased with decrease in engine load because of reduction in cylinder temperature. The fraction of the LTHR decreased with increased load from about 5% at 2 bar IMEP to about 3% at 8 bar IMEP [48].

Figure 3.13 shows the effect of inlet oxygen content and fuel composition on ignition delay and LTHR fraction (also referred as low temperature reaction (LTR) fraction in few studies) in PPC combustion. Ignition delay depends on the injection parameters (injection quantity, injection timing and injection pressure) and engine operating parameters (air charge conditions, load, engine speed, combustion chamber geometry and swirl rate) for a particular fuel. Figure 3.13a shows the variations of ignition delay at constant engine speed and load conditions for constant combustion phasing, where fuel injection timing and duration are adjusted to achieve the load (IMEP 8 bar) and combustion phasing (3 CAD aTDC). Experiment is conducted at absolute inlet pressure of 2.8 bar. Ignition delay is longer for higher RON fuel due to high resistance to autoignition, and distinct difference in ignition delay is observed proportional to difference in RON for each fuel. At

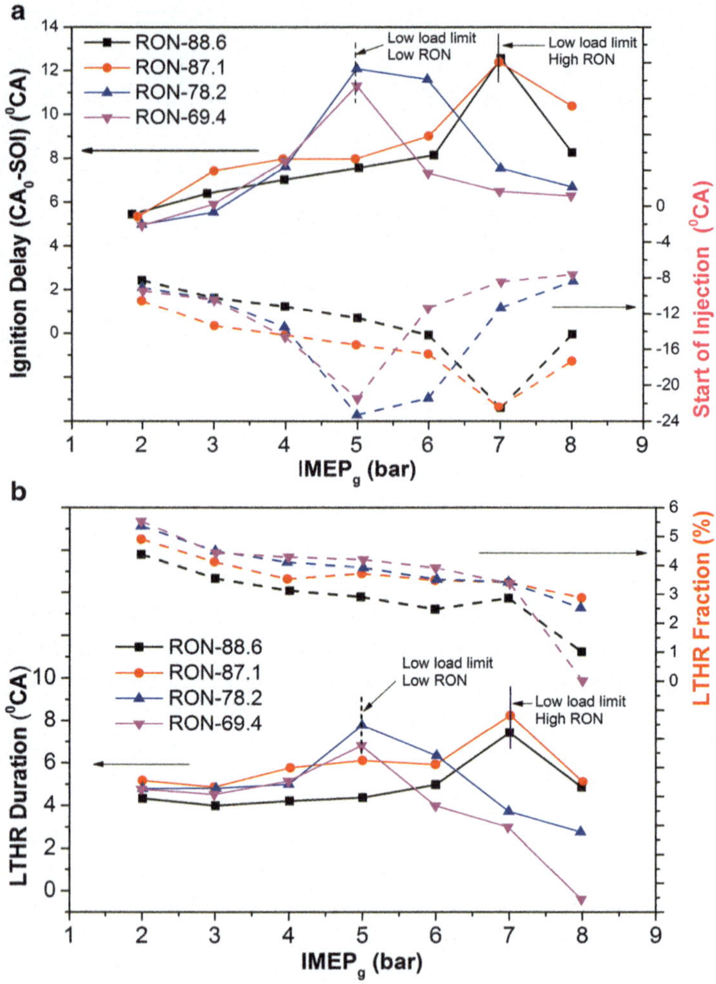

Fig. 3.12 Variation of ignition delay, LTHR duration and LTHR fraction with engine load in PPC at constant combustion phasing of 6 CAD after TDC (Adapted from [48])

lower oxygen concentration, more ambient gas needs to mix with fuel for release of same energy. Therefore, additional time is required for the mixing and hence ignition delay is higher.

In HCCI combustion, LTHR increases with increase in n-heptane content and decreases with increase in iso-octane content [49, 50]. In PPC combustion, LTHR increases with decrease in n-heptane content and increase in EGR rate. This trend is opposite to the trend observed in HCCI combustion. The SOI is advanced to achieve constant combustion phasing for high RON fuels, and advancing SOI increases LTHR phase due to lower cylinder temperature and pressure. The inlet oxygen concentration has stronger impact on LTHR than fuel composition for PRF fuel (Fig. 3.13a) [47].

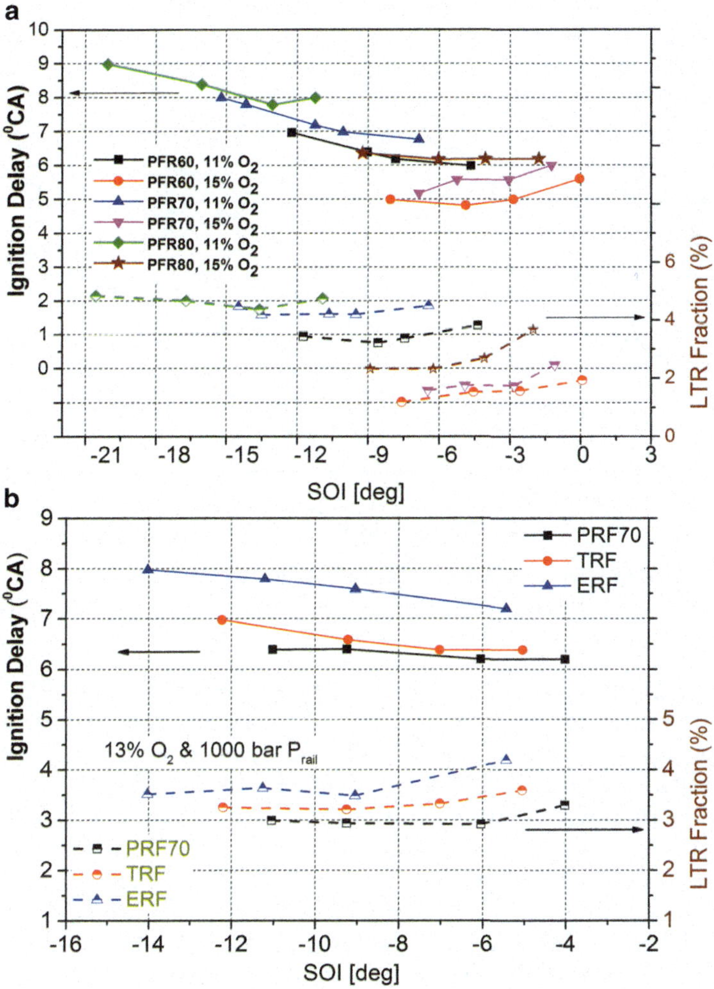

Fig. 3.13 Ignition delay and LTHR as a function of SOI for different (**a**) PRFs and oxygen content, (**b**) ERF and TRF (Adapted from [47])

Figure 3.13b depicts the effect of fuel composition (especially fuel containing aromatic and alcohol) on ignition delays and LTHR phase in PPC. A base fuel PRF70 is compared to TRF (toluene reference fuel) and ERF (ethanol reference fuel). The n-heptane concentration is the same for all the fuel (30% by volume) fractions, and iso-octane in PRF70 is replaced by 15% toluene for TRF fuel and by 10% for ERF fuel. To maintain the same combustion phasing, SOI is advanced for TRF by one degree. Ignition delay is higher for TRF and ERF fuel in comparison to PRF70 as octane number is higher for these fuels. Ignition delay is more influenced by ethanol than toluene. Ethanol and toluene have amplified the LTHR regime in PPC mode which is the opposite of HCCI mode trend (Fig. 3.13b) [47]. However, ethanol has higher impact on LTHR than toluene in PPC combustion.

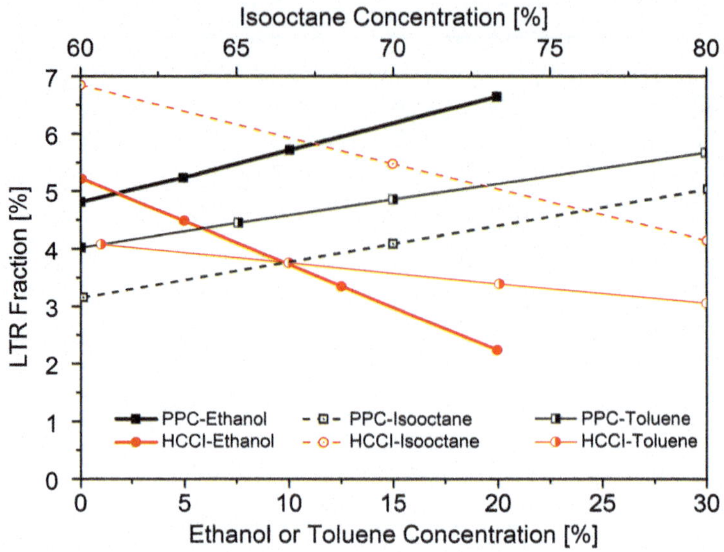

Fig. 3.14 LTHR fraction as a function of ethanol, toluene and iso-octane concentration in HCCI and PPC combustion (Adapted from [51])

Figure 3.14 compared the LTHR (or LTR) fraction in HCCI and PPC combustion at various fuel compositions (different ethanol, toluene and iso-octane concentrations in fuel) at constant engine load. The figure shows that LTHR fraction increases with increase in ethanol concentration with a slope of 0.08 in PPC combustion. However, LTHR fraction decreases with a slope of −0.1 with increase in ethanol content in HCCI combustion mode. Similarly, LTHR fraction increases with increase in toluene in PPC and decreases in HCCI combustion mode. For both PPC and HCCI combustion, the steepness of the slopes is similar but in reverse direction. Figure 3.14 also depicts that ethanol has a higher effect on LTHR fraction than toluene. Similar trend is also observed for iso-octane fraction.

In summary, fuel effect on LTHR in PPC is reversed as compared to HCCI mode. It is suggested that chemical fuel effects dominate over physical effects for port-injected HCCI. However, the processes involved in LTHR regime are more complex in PPC mode than in HCCI mode. The strong relation between LTHR and ignition delay in PPC mode shows that the ignition quality plays a key role in LTHR regime [51].

3.1.5 Autoignition Quality Test Limitations

The cetane number (CN) and octane number (ON) are the main parameters used to characterize chemical properties of fuel in conventional CI and SI engines, respectively. The HCCI operating conditions cover both the ignition quality (CN) and anti-knock quality (ON) at both lean (low load) and rich (high load) limit conditions. Thus, conventional CN or ON may not be optimal parameter for fuel characterization in HCCI engines. Studies suggested that HCCI ignition characteristics cannot be estimated by conventional measurement of autoignition quality of fuels (RON, MON or CN) [52].

The compression ratio required for auto-ignition of different kinds of paraffin does not rank with ON [53]. Issues related to RON and MON for characterizing the auto-ignition behaviour in HCCI engine are further investigated [43, 54–56]. These studies concluded that RON and MON methods do not describe pressure fluctuations from auto-ignition in the end gas, but it rather describes the rate of change of pressure prior to knocking. Additionally, the pressure at a given temperature is higher in modern engines than RON test.

No correlation is found between combustion phasing (CA_{50}) and RON or MON in HCCI combustion with negative valve overlap by testing 15 different fuels with high octane numbers [57]. Another study showed that fuel composition affects the ignition delay for fuels with same RON or MON, and thus neither RON nor MON is a complete predicator of autoignition quality in HCCI engines [58]. Therefore a new index is required for estimating fuel performance in HCCI combustion engines.

3.2 LTC Fuel Index

In HCCI engine, fuel composition strongly affects the combustion phasing and magnitude of heat release rate (HRR). Fuel with high olefin/aromatic content can reduce the maximum HRR and cylinder pressure as much as 20% and shift the maximum HRR position up to 10 CAD in HCCI engine operated at similar RON and MON fuels with different composition [58]. A study showed that the marketed gasoline has variations in aromatic content ranges from 4% to 54% and olefinic fraction ranges from 0% to 30% [59]. Fuel composition variability makes HCCI engine control more challenging. Fuel composition effect on combustion characteristics and variations in fuel composition in pump fuels leads to the development of newer methodology for prediction autoignition characteristics in LTC engine. Next subsections present the several fuel index developed to describe the autoignition characteristics of LTC engine fuel.

3.2.1 Octane Index

To describe the autoignition characteristics of HCCI combustion, Kalghatgi et al. developed a parameter called octane index (OI), which better correlates the combustion phasing [1, 60]. The octane index is defined by Eq. (3.6):

$$\mathrm{OI} = (1 - K).\mathrm{RON} + K.\mathrm{MON} = \mathrm{RON} - K.S \tag{3.6}$$

The value of parameter K depends on engine operating conditions (value can be positive, zero or negative), and S is fuel sensitivity (S=RON−MON). Fuel sensitivity depicts the response of a particular fuel at higher intake temperatures for knock tendency relative to iso-octane and n-heptane [61]. Fuel sensitivity is zero for PRFs and, hence, OI for PRF is equal to RON. The value of K is 0 and 1 for RON and MON test conditions, respectively. The higher OI of fuel means more resistant to autoignition and leads to retarded combustion phasing in HCCI engine.

The OI represents the combustion characteristics of the fuel. Combustion phasing (CA_{50}) is proposed to be a linear function of RON and MON. The K value is estimated by regression analysis [1]:

$$\mathrm{CA50} = a\mathrm{RON} + b\mathrm{MON} + c \tag{3.7}$$

$$\mathrm{CA50} = c + (a + b)\mathrm{OI} \tag{3.8}$$

where $\mathrm{OI} = ((a/(a+b))/\mathrm{RON} + ((b/(a+b))/\mathrm{MON})$ and K is defined as $b/(a+b)$.

The K value strongly depends on the temperature at 15 bar cylinder pressure (T_{comp15}) and increases with T_{comp15} [62]. T_{comp15} is a function of physical state in the cylinder (based on engine speed, intake temperature, etc.), and K is dependent on engine operating condition, but not on the fuel properties. A predictive equation for K as a function of T_{comp15} and the relative air/fuel ratio (λ) is obtained and presented in Eq. (3.9):

$$K = 0.00497^{*}T_{\mathrm{comp15}} - 0.15^{*}\lambda - 3.67 \tag{3.9}$$

At every engine operating condition, there exists an ideal fuel with $\mathrm{OI} = \mathrm{OI}_0$ where combustion phasing (CA_{50}) is at TDC. OI_0 increases with increase of compression pressure and temperature at TDC position. OI_0 value reduces with increasing λ and engine speed. OI_0 can be recognized as engine's fuel requirement.

The OI is a good indicator of autoignition characteristics for most of the PRF blends and surrogates composed of a few pure chemical components. However, OI does not correlate well for gasoline fuels containing ethanol, naphthenes and aromatics [61, 63, 64]. These species are typically found in commercial gasoline worldwide.

A recent study improved the OI and proposed a new OI model (JKZ model) [61]. The modified OI is shown in Eq. (3.10):

$$\begin{aligned}\mathrm{OI}_{\mathrm{modified}} = {} & \mathrm{RON} - \mathrm{K}'.\mathrm{S} + \kappa(\mathrm{Aromatics})^2(\mathrm{Olefines} + \mathrm{Saturates}) \\ & + \varepsilon(\mathrm{Aromatics.Ethanol})\end{aligned} \tag{3.10}$$

The two new parameters (κ and ε) are used in conjunction with K′ defined previously in order to keep the same convention as Kalghatgi's OI. In this model, combustion phasing (CA_{50}) is related to RON and MON value of fuel by Eq. (3.11):

$$CA_{50} = c' + a'\text{RON} + b'\text{MON} + d\,(\text{Aromatics})^2\,(\text{Olefins} + \text{Saturates}) + e(\text{Aromatics.Ethanol}) \quad (3.11)$$

In Eq. (3.11), aromatics, olefins, saturates and ethanol represent the volume fraction of each respective component in the fuel. The parameters K′, κ and ε are calculated as

$$k = b'/(a' + b'); \kappa = d/(a' + b'); \varepsilon = e/(a' + b');$$

Modified OI is more capable of capturing the auto-ignition characteristics of the refinery stream fuels for all the intake temperature conditions, especially low intake temperatures.

3.2.2 CAI/HCCI Index

Shibata and Urushihara [65] developed a HCCI index based on MON and amounts of five components (n-paraffin (nP), iso-paraffin (iP), olefin (O), aromatics (A) and oxygenates (OX)). The Shibata–Urushihara HCCI index is given in Eq. (3.12):

$$\text{S-U HCCI Index (abs)} = m.\text{MON} + a.(\text{nP}) + b.(\text{iP}) + c.(\text{O}) + d.(\text{A}) + e.(\text{OX}) + Y \quad (3.12)$$

All fuel component concentration in Eq. (3.12) is the percent by volume, and m, a, b, c, d, e and Y are temperature-dependent constants. The value of coefficient "e" depends upon the specific oxygenate in the fuel formulation (ethanol, MTBE or ETBE). The SU–HCCI index corresponds to the crank angle timing for 20% completion of the HTHR.

The same study also developed HCCI index in terms of RON, which is more widely used index, and HCCI index expression is presented in Eq. (3.13) with a separate set of coefficients [65]:

$$\text{S-U HCCI Index (abs)} = r.\text{RON} + a'.(\text{n-p}) + b'.(\text{i-p}) + c'.(\text{O}) + d'.(\text{A}) + e'.(\text{OX}) + Y' \quad (3.13)$$

Studies with different gasoline composition suggest that SU–HCCI index poorly predicts the autoignition characteristics of HCCI combustion [64, 66].

Another group proposed a HCCI index based on the fuel's tendency to decrease or increase the HCCI operating region [67]. This index compares the operating area (in HCCI mode) achieved with the given fuel to the HCCI operating area achieved with the reference fuel. Figure 3.15 illustrates the definition of HCCI index. Moreover, a good HCCI fuel must not deteriorate the full load performance. A modified HCCI index is proposed and presented by Eqs. (3.15) and (3.16) to have

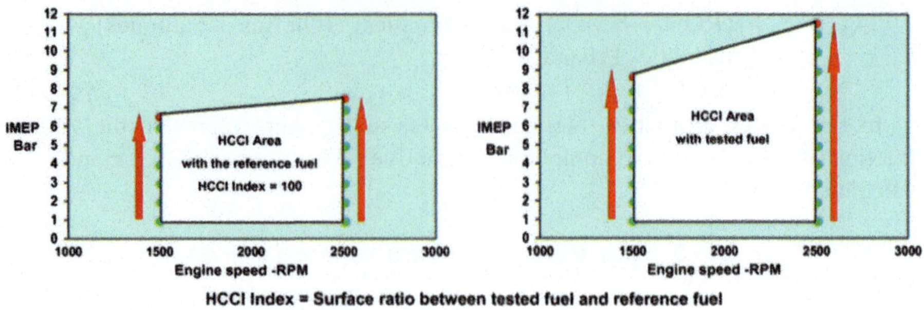

Fig. 3.15 HCCI index definition [67]

higher importance to full load operation in conventional mode, which eliminates the improvement allowed in HCCI mode:

$$\text{HCCI index} = \frac{\text{Tested Fuel HCCI Surface}}{\text{Ref.Fuel HCCI Surface}} \tag{3.14}$$

$$\begin{aligned}\text{Mod.HCCI index}(1) = 0.8 \times \frac{\text{Tested Fuel HCCI Surface}}{\text{Ref.Fuel HCCI Surface}} \times 100 + 0.2 \\ \times \frac{\text{Max Power Tested Fuel}}{\text{Max Power Ref.Fuel}} \times 100\end{aligned} \tag{3.15}$$

$$\begin{aligned}\text{Mod.HCCI index}(2) = \frac{\text{Tested Fuel HCCI Surface}}{\text{Ref.Fuel HCCI Surface}} \\ \times \frac{\text{Max Power Tested Fuel}}{\text{Max Power Ref.Fuel}} \times 100\end{aligned} \tag{3.16}$$

A similar procedure is also used to evaluate the fuel effect on the controlled autoignition (CAI) combustion [68, 69]. The CAI index is used to precisely quantify the effect of fuel on four CAI zones defined as low-speed/low load, low-speed/high load, high-speed/high load, and high-speed/low load zones. The HCCI index is used to compare HCCI operating region achieved from the tested fuel with that achieved from the reference fuel.

3.2.3 *Lund–Chevron HCCI Number*

Lund–Chevron HCCI number is developed to fulfill the need of a HCCI number analogous to that of RON and MON for SI engines. To estimate HCCI number, first a set of reference curves of the compression ratio (CR) is generated for combustion phasing (CA_{50}) at 3° aTDC using PRFs at five different intake temperatures (50–150 °C) [66]. A quadratic function of CR with octane number is obtained by fitting quadratic relation in PRF data (PRF60–100) as shown in Eq. (3.17):

$$\mathrm{CR} = a.x^2 + b.x + c \tag{3.17}$$

where "x" is the volume % of iso-octane in the primary reference fuels used to develop the correlation. The constants a, b and c obtained by curve fitting and their values are different for different intake temperatures. For particular fuel and engine operating conditions of interest, the autoignition CR is plotted vs. RON, along with the reference quadratic curve for the PRFs at the same conditions. The HCCI number of fuel is assigned equal to the RON of the PRF that had the same CR. The quadratic equation can be rewritten for calculating HCCI number from CR by Eq. (3.18) [43]:

$$\text{HCCI number} = -\frac{b}{2.a} + \sqrt{\left(\frac{b}{2.a}\right)^2 - \left(\frac{c - \mathrm{CR}}{a}\right)} \tag{3.18}$$

In this manner, HCCI numbers can be determined for each test fuel at engine operating condition of interest. The Lund–Chevron HCCI number is found to be closely coupled with the amount of low temperature reactions [66].

3.2.4 LTC Fuel Performance Index

The LTC index is a new metric developed for ranking the appropriateness of fuels in LTC engines based on the fraction of potential fuel savings obtained in the vehicle driving cycle [70]. The LTC index can be utilized to rank LTC fuels and predict their future performance. This index can also be used to recognize attractive fuels for LTC engines or assist in the development of new fuels for LTC engines.

The LTC fuel index is defined as the percentage of fuel savings obtained by HCCI operation and depicted by Eq. (3.19):

$$\begin{aligned} I_{\mathrm{LTC}} &= \frac{m_{fs,\,\mathrm{SI+HCCI}}}{m_{fs,\,\mathrm{HCCI}}} \times 100\% \\ &= \frac{m_{f,\,\mathrm{SI\epsilon HCCI}}}{m_{f,\mathrm{SI}}} \times 100\% \end{aligned} \tag{3.19}$$

where $m_{fs,\mathrm{SI+HCCI}}$ is the fuel consumed over the hybrid SI/HCCI operation, $m_{fs,\mathrm{HCCI}}$ is the fuel savings for full HCCI operation, $m_{f,\mathrm{SI\epsilon HCCI}}$ is the fuel consumed during the baseline SI operation over the possible HCCI operating range and $m_{f,\mathrm{SI}}$ is the baseline SI fuel consumption over the entire driving cycle. The LTC index (I_{LTC}) can also be understood as the mass-weighted percentage of discrete viable HCCI operating points [70]. The LTC fuel index combines information about the fuel operating region (the viable HCCI combustion engine loads and speeds) with the operating conditions required during engine operation at a particular real driving cycle.

3.3 LTC Fuel Design

The HCCI engine combustion is typically controlled by the reaction kinetics of different species present in fuel (Sect. 2.2.2 of Chap. 2). The chemical kinetics is governed by physical and chemical properties of fuel along with their spatial and time histories in the cylinder, which are dependent on engine design and operating parameters.

Generally, high cetane number fuels tend to extend lean-burn region in LTC engine due to higher fuel reactivity. Higher reactivity of high cetane fuels limits the high load operation at richer mixture due to advanced combustion phasing and excessive HRR. To overcome the problem of higher reactivity at higher engine loads, higher octane number (lower reactivity) fuels are used in LTC engines. Autoignition-resistant quality of high octane fuels prevents advanced combustion phasing at higher engine loads. The higher octane fuels face the challenge at lower load conditions as engine is operated on leaner fuel–air mixtures. The lean engine operation encounters misfire or partial burn cycles at lower intake temperature and load operations. Hence, cold start and lean-burn challenges are present in HCCI combustion using high octane number fuels. Additionally, various studies depicted that conventional fuels (diesel/gasoline) or pure single-component fuels are not able to fulfil the HCCI combustion requirements over a wide range of engine operating conditions [52]. Therefore, it is required to make fuels more suitable for HCCI combustion by reforming or designing the fuel's physical–chemical properties, fuel components and composition along with fuel molecular structure. Figure 3.16 illustrates a proposed HCCI fuel design concept.

The main challenge in HCCI engine is the control the ignition timings and HRR, to avoid the excessive noise and exhaust emissions. To achieve HCCI combustion over a wide range of engine operating conditions, reconfiguration of fuel properties and compositions is required for improving the adaptability and compatibility of various fuels [52]. All the approaches such as fuel additives, dual fuel engine operation, fuel blending and optimized fuel composition in real time can be utilized to redesign chemical properties of the charge based on HCCI engine load (Fig. 3.16).

Stratified charge compression ignition (SCCI) is proposed to enlarge the operating region of HCCI combustion engine (Sects. 2.4, 2.5 and 2.6 of Chap. 2). Typically, moderate inhomogeneities (temperature, air–fuel ratio and fuel composition) are required to obtain wider engine operating range, a smooth heat release, ignition timing controllability and lower emissions in LTC engines. Flexible control of combustion phasing using temperature stratification, fuel concentration stratification and fuel composition stratification can be utilized to achieve higher operating range, while maintaining higher maximum thermal efficiency and lower exhaust emissions. This target can be achieved by a combination of fuel design and fuel injection strategies. Figure 3.17 demonstrates the basic principle of fuel design and management in SCCI combustion engines.

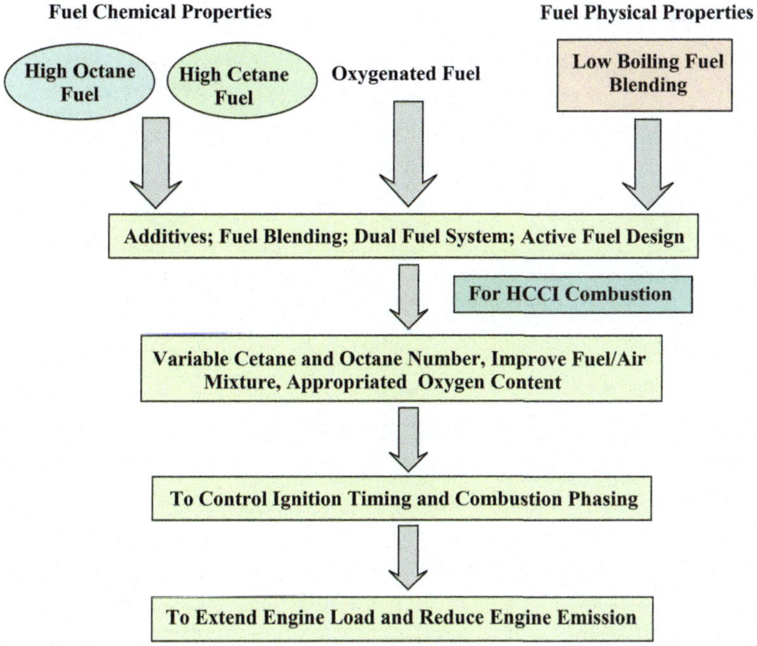

Fig. 3.16 Fuel design principles for HCCI combustion [52]

Figure 3.17 presents that in LTC engines, physical and chemical properties of fuels as well as fuel composition need to be managed actively on real-time basis depending on engine operating ranges. In advanced compression ignition engines, fuel concentration and composition stratification can be created using fuel injection strategies (PFI and DI) like in reactivity-controlled compression ignition (RCCI) and PPC engines. Test fuels can be selected with opposing chemical properties for port and direct injection, which can moderate the spatial local fuel–air composition distribution. Utilization of well-regulated controlled EGR is an effective method to control the HRR by reduction in combustion temperature [52].

3.4 Fuel Requirement in LTC Mode

Ideally any fuel can be used in HCCI engine by choosing the right design and operating conditions for autoignition of fuel–air mixture. In premixed charge compression ignition such as HCCI combustion, fully premixed fuel–air mixture is required before the start of combustion. The fuel must be volatile enough to allow sufficient mixing of fuel with air such that fuel-rich regions in the charge can be avoided to accomplish lower soot and NOx emissions. The simple method for homogeneous/premixed charge preparation is to inject the fuel in intake manifold

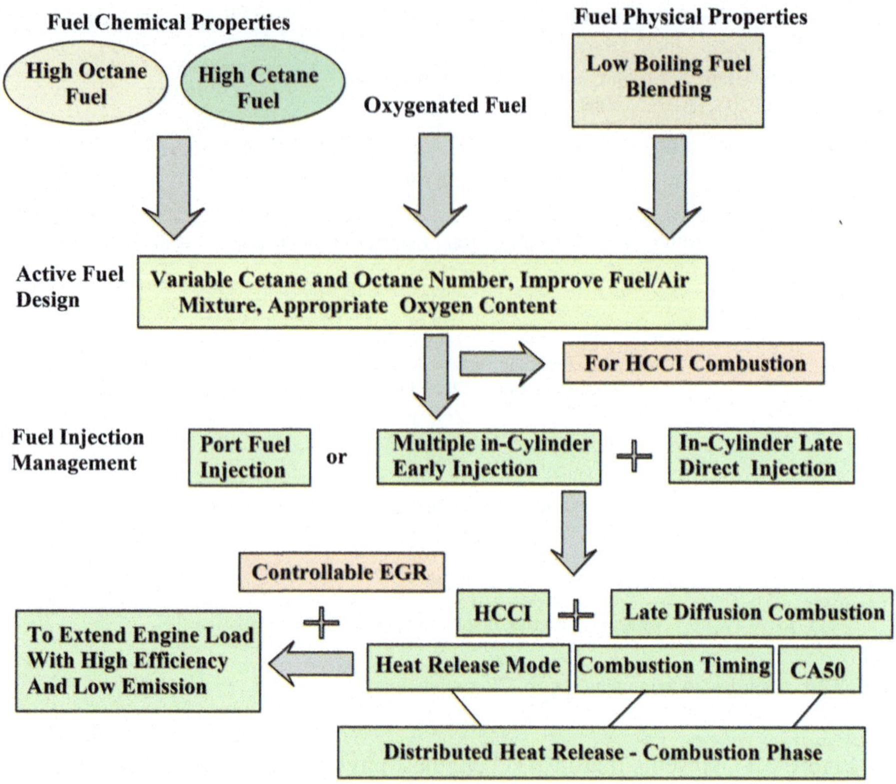

Fig. 3.17 SCCI combustion fuel design and management concept [52]

by port fuel injection (PFI) system. Fuel boiling point range plays an important role for using PFI system. The PFI works sufficiently well with gasoline-like fuels with boiling point up to 180 °C [2]. Gaseous fuels do not have issues related to mixing with port injection. Thus, most of the available high octane fuels can be used by PFI in LTC engines.

Fuels with lower volatility can use direct injection (DI) mixture preparation strategies by using high-pressure injection, multiple injection and improved injector design. Diesel has a high initial boiling point, and thus it will not evaporate and mix with air without intake preheating during PFI in the manifold. Diesel fuel is very difficult to use in HCCI engines using PFI system. A combination of low compression ratio and high intake temperature is required to use diesel in HCCI engine [38]. Minimum 100 ° C intake temperature is required to use high-boiling-point fuel (diesel). Thus, it is concluded that the high-boiling-point diesel-like fuels should not use PFI, and use direct injection system to obtain proper atomization and mixing. To prepare sufficiently premixed charge, early direct injection at comparatively lower injection pressure is used to obtain premixed charge in LTC engines.

Fuel autoignition quality is of critical importance in LTC engine due to the nature of ignition. The autoignition characteristics of fuels must satisfy the varying requirement of the engines over a wide range of operating conditions. On investigating wide ignition quality fuels in HCCI engines, it is summarized that ideal HCCI ignition quality for dedicated HCCI engines is in the range of CN 35–45 (approximately 70–50 RON, respectively) [25, 62, 71, 72]. Other LTC strategies, such as RCCI combustion, uses both high and low reactivity fuels.

The PPC combustion mode can be achieved with diesel fuel as well as high octane gasoline-like fuels [73]. At lower engine loads, diesel is a well-suited fuel, and at higher engine load, gasoline-like fuels are more suitable fuel for PPC mode operation. Ideally ON of PPC fuel should be a function of engine load. A study suggested a simple thumb rule for optimum fuel ON as a function of engine load. The optimum ON is five times the load in indicated mean effective pressure (IMEP). This indicates that the fuel can be very similar to diesel at idle engine operating conditions, and at full engine load conditions, an ON of 100 or higher is required for the desired performance of the engine.

The charge preparation and autoignition control strategies in LTC engine for different automotive fuels are discussed in Chaps. 4 and 5, respectively.

References

1. Kalghatgi G (2014) Fuel/engine interactions, SAE International, Warrendale, ISBN 978-0-7680-6458-2
2. Boot M (2016) Biofuels from lignocellulosic biomass: innovations beyond bioethanol. Wiley, Weinheim
3. Boot MD, Tian M, Hensen EJ, Sarathy SM (2017) Impact of fuel molecular structure on auto-ignition behavior–design rules for future high performance gasolines. Prog Energy Combust Sci 60:1–25
4. Zádor J, Taatjes CA, Fernandes RX (2011) Kinetics of elementary reactions in low-temperature autoignition chemistry. Prog Energy Combust Sci 37(4):371–421
5. Simmie JM (2003) Detailed chemical kinetic models for the combustion of hydrocarbon fuels. Prog Energy Combust Sci 29(6):599–634
6. Griffiths JF (1995) Reduced kinetic models and their application to practical combustion systems. Prog Energy Combust Sci 21(1):25–107
7. Maurya RK, Akhil N (2016) Numerical investigation of ethanol fuelled HCCI engine using stochastic reactor model. Part 1: development of a new reduced ethanol oxidation mechanism. Energy Convers Manag 118:44–54
8. Maurya RK, Akhil N (2017) Development of a new reduced hydrogen combustion mechanism with NO x and parametric study of hydrogen HCCI combustion using stochastic reactor model. Energy Convers Manag 132:65–81
9. Farrell JT, Cernansky NP, Dryer FL, Law CK, Friend DG, Hergart CA et al (2007) Development of an experimental database and kinetic models for surrogate diesel fuels (No. 2007-01-0201). SAE technical paper
10. Naik CV, Pitz WJ, Westbrook CK, Sjöberg M, Dec JE, Orme J et al (2005) Detailed chemical kinetic modeling of surrogate fuels for gasoline and application to an HCCI engine (No. 2005-01-3741). SAE technical paper

11. Pitz WJ, Mueller CJ (2011) Recent progress in the development of diesel surrogate fuels. Prog Energy Combust Sci 37(3):330–350
12. Battin-Leclerc F (2008) Detailed chemical kinetic models for the low-temperature combustion of hydrocarbons with application to gasoline and diesel fuel surrogates. Prog Energy Combust Sci 34(4):440–498
13. Puduppakkam KV, Naik CV, Wang C, Meeks E (2010) Validation studies of a detailed kinetics mechanism for diesel and gasoline surrogate fuels (No. 2010-01-0545). SAE technical paper
14. Shibata G, Kawaguchi R, Yoshida S, Ogawa H (2014) Molecular structure of hydrocarbons and auto-ignition characteristics of HCCI engines. SAE Int J Fuels Lubr 7(2014-32-0003):1050–1061
15. Luo. Y (2003) Handbook of bond dissociation energies in organic compounds. CRC Press, Boca Raton
16. Knocking characteristics of pure hydrocarbons, STP225, American Society for Testing Materials (1958). ASTM International, Philadelphia
17. Vanhove G, Ribaucour M, Minetti R (2005) On the influence of the position of the double bond on the low-temperature chemistry of hexenes. Proc Combust Inst 30(1):1065–1072
18. Brezinsky K (1986) The high-temperature oxidation of aromatic hydrocarbons. Prog Energy Combust Sci 12(1):1–24
19. Saggese C, Frassoldati A, Cuoci A, Faravelli T, Ranzi EA (2013) Wide range kinetic modeling study of pyrolysis and oxidation of benzene. Combust Flame 160(7):1168–1190
20. Sakai Y, Inamura T, Ogura T, Koshi M, Pitz WJ (2007) Detailed kinetic modeling of toluene combustion over a wide range of temperature and pressure (No. 2007-01-1885). SAE technical paper
21. Zhao L, Cheng Z, Ye L, Zhang F, Zhang L, Qi F, Li Y (2015) Experimental and kinetic modeling study of premixed o-xylene flames. Proc Combust Inst 35(2):1745–1752
22. Battin-Leclerc F, Warth V, Bounaceur R, Husson B, Herbinet O, Glaude PA (2015) The oxidation of large alkylbenzenes: an experimental and modeling study. Proc Combust Inst 35 (1):349–356
23. Metcalfe WK, Dooley S, Dryer FL (2011) Comprehensive detailed chemical kinetic modeling study of toluene oxidation. Energy Fuel 25(11):4915–4936
24. Murakami Y (2009) Progress in elementary reaction kinetics in combustion by quantum chemical models. J Combustion Soc Jpn 51(157):192–199
25. Risberg P (2006) Describing the auto-ignition quality of fuels in HCCI engines. PhD thesis, Royal Institute of Technology, S-100 44 Stockholm. ISRN/KTH/MMK/R-06/07-SE
26. Livengood JC, Wu PC (1955) Correlation of autoignition phenomena in internal combustion engines and rapid compression machines. In Symposium (international) on combustion (vol 5, no 1, pp 347–356). Elsevier
27. Heywood JB (1988) Internal combustion engine fundamentals. McGrawHill, New York
28. Swan K, Shahbakhti M, Koch CR (2006) Predicting start of combustion using a modified knock integral method for an HCCI engine (No. 2006-01-1086). SAE technical paper
29. Shaver GM, Gerdes JC, Jain P, Caton PA, Edwards CF (2003) Modeling for control of HCCI engines. In American control conference, 2003. Proceedings of the 2003 (vol 1, pp 749–754). IEEE
30. Halstead MP, Kirsch LJ, Quinn CP (1977) The autoignition of hydrocarbon fuels at high temperatures and pressures – fitting of a mathematical model. Combust Flame 30:45–60
31. Turns SR (2000) An introduction to combustion: concepts and applications, 2nd edn. McGraw-Hill, Boston
32. Shaver GM, Roelle M, Gerdes JC (2005) Tackling the transition: decoupled control of combustion timing and work output in residual-affected HCCI engines. American control conference proceedings, pp 3871–3876. Portland, OR, USA

33. Rausen DJ, Stefanopoulou AG, Kang J-M, Eng JA, Kuo W (2004) A mean-value model for control of homogeneous charge compression ignition (HCCI) engines. American control conference, pp 125–131
34. Souder JS, Mehresh P, Hedrick JK Dibble RW (2004) A multi-cylinder HCCI engine model for control. In ASME 2004 International Mechanical Engineering Congress and Exposition. American Society of Mechanical Engineers, pp. 307–316
35. Shahbakhti M (2009) Modeling and experimental study of an HCCI engine for combustion timing control. PhD thesis, University of Alberta
36. Hernández JJ, Lapuerta M, Sanz-Argent J (2014) Autoignition prediction capability of the Livengood–Wu correlation applied to fuels of commercial interest. International Journal of Engine Research 15(7):817–829
37. Yates AD, Swarts A, Viljoen CL (2005) Correlating auto-ignition delays and knock-limited spark-advance data for different types of fuel (No. 2005-01-2083). SAE technical paper
38. Christensen M, Hultqvist A, Johansson B (1999) Demonstrating the multi fuel capability of a homogeneous charge compression ignition engine with variable compression ratio (No. 1999-01-3679). SAE technical paper
39. Bunting BG, Wildman CB, Szybist JP, Lewis S, Storey J (2007) Fuel chemistry and cetane effects on diesel homogeneous charge compression ignition performance, combustion, and emissions. Int J Engine Res 8(1):15–27
40. Shibata G, Oyama K, Urushihara T, Nakano T (2004) The effect of fuel properties on low and high temperature heat release and resulting performance of an HCCI engine (No. 2004-01-0553). SAE technical paper
41. Truedsson I, Tuner M, Johansson B, Cannella W (2012) Pressure sensitivity of HCCI auto-ignition temperature for primary reference fuels. SAE Int J Engines 5(2012-01-1128):1089–1108
42. Truedsson I, Tuner M, Johansson B, Cannella W (2013) Pressure sensitivity of HCCI auto-ignition temperature for gasoline surrogate fuels (No. 2013-01-1669). SAE technical paper
43. Truedsson I (2014) The HCCI fuel number – measuring and describing auto-ignition for HCCI combustion engines. PhD thesis, Lund University. ISBN 978-91-7473-949-7
44. Andrae JC, Head RAHCCI (2009) Experiments with gasoline surrogate fuels modeled by a semidetailed chemical kinetic model. Combust Flame 156(4):842–851
45. Mehl M, Pitz W, Sarathy M, Yang Y, Dec JE (2012) Detailed kinetic modeling of conventional gasoline at highly boosted conditions and the associated intermediate temperature heat release (No. 2012-01-1109). SAE technical paper
46. Solaka H (2014) Impact of fuel properties on partially premixed combustion. PhD thesis, Lund University, Sweden
47. Solaka H, Tuner M, Johansson B (2013) Analysis of surrogate fuels effect on ignition delay and low temperature reaction during partially premixed combustion (No. 2013-01-0903). SAE technical paper
48. Solaka H, Aronsson U, Tuner M, Johansson B (2012) Investigation of partially premixed combustion characteristics in low load range with regards to fuel octane number in a light-duty diesel engine (No. 2012-01-0684). SAE technical paper
49. Shibata G, Oyama K, Urushihara T, Nakano T (2005) Correlation of low temperature heat release with fuel composition and HCCI engine combustion (No. 2005-01-0138). SAE technical paper
50. Tanaka S, Ayala F, Keck JC, Heywood JB (2003) Two-stage ignition in HCCI combustion and HCCI control by fuels and additives. Combust Flame 132(1):219–239
51. Aronsson HS, Truedsson I, Tuner M, Johansson B, Cannella W (2014) Comparison of fuel effects on low temperature reactions in PPC and HCCI combustion (No. 2014-01-2679). SAE technical paper
52. Lu X, Han D, Huang Z (2011) Fuel design and management for the control of advanced compression-ignition combustion modes. Prog Energy Combust Sci 37(6):741–783

53. Leppard WR (1990) The chemical origin of fuel octane sensitivity (No. 902137). SAE technical paper
54. Yates A, Swarts A, Viljoen C (2003) An investigation of anomalies identified within the ASTM research and motor octane scales (No. 2003-01-1772). SAE technical paper. doi:https://doi.org/10.4271/2003-01-1772.
55. Swarts A, Yates A, Viljoen C, Coetzer R (2005) A further study of inconsistencies between autoignition and knock intensity in the CFR octane rating engine (No. 2005-01-2081). SAE technical paper
56. Perumal M, Floweday G (2011) An investigation of cascading autoignition and octane number using a multi-zone model of the CFR engine. SAE Int J Engines 4(2011-01-0850):976–997
57. Koopmans L, Strömberg E, Denbratt I (2004) The influence of PRF and commercial fuels with high octane number on the auto-ignition timing of an engine operated in HCCI combustion mode with negative valve overlap (No. 2004-01-1967). SAE technical paper
58. Farrell JT, Bunting BG (2006) Fuel composition effects at constant RON and MON in an HCCI engine operated with negative valve overlap (No. 2006-01-3275). SAE technical paper
59. Angelos JP, Andreae MM, Green WH, Cheng WK, Kenney T, Xu Y (2007) Effects of variations in market gasoline properties on HCCI load limits (No. 2007-01-1859). SAE technical paper
60. Kalghatgi G, Risberg P, Ångstrom HE (2003) A method of defining ignition quality of fuels in HCCI engines (No. 2003-01-1816). SAE technical paper
61. Lacey J, Kameshwaran K, Sathasivam S, Filipi Z, Cannella W, Fuentes-Afflick PA (2017) Effects of refinery stream gasoline property variation on the auto-ignition quality of a fuel and homogeneous charge compression ignition combustion. Int J Engine Res 18(3):226–239
62. Risberg P, Kalghatgi G, Ångstrom HE (2003) Auto-ignition quality of gasoline-like fuels in HCCI engines (No. 2003-01-3215). SAE technical paper
63. Liu H, Yao M, Zhang B, Zheng Z (2009) Influence of fuel and operating conditions on combustion characteristics of a homogeneous charge compression ignition engine. Energy Fuel 23(3):1422–1430
64. Rapp VH, Cannella WJ, Chen JY, Dibble RW (2013) Predicting fuel performance for future HCCI engines. Combust Sci Technol 185(5):735–748
65. Shibata G, Urushihara T (2007) Auto-ignition characteristics of hydrocarbons and development of HCCI fuel index (No. 2007-01-0220). SAE technical paper
66. Truedsson I, Cannella W, Johansson B, Tuner M (2014) Development of new test method for evaluating HCCI fuel performance (No. 2014-01-2667). SAE technical paper
67. Starck L, Lecointe B, Forti L, Jeuland N (2010) Impact of fuel characteristics on HCCI combustion: performances and emissions. Fuel 89(10):3069–3077
68. Jeuland N, Montagne X, Duret P (2003) Engine and fuel related issues of gasoline CAI (Controlled Auto-Ignition) combustion (No. 2003-01-1856). SAE technical paper
69. Jeuland N, Montagne X, Duret P (2004) New HCCI/CAI combustion process development: methodology for determination of relevant fuel parameters. Oil Gas Sci Technol – Rev. IFP 59 (6):571–579
70. Niemeyer KE, Daly SR, Cannella WJ, Hagen CL (2015) Investigation of the LTC fuel performance index for oxygenated reference fuel blends. Fuel 155:14–24
71. Risberg P, Kalghatgi G, Ångstrom HE, Wåhlin F (2005) Auto-ignition quality of diesel-like fuels in HCCI engines (No. 2005-01-2127). SAE technical paper
72. chun Hou Y, cai Lu X, lin Zu L, bin Ji L, Huang Z (2006) Effect of high-octane oxygenated fuels on n-heptane-fueled HCCI combustion. Energy Fuel 20(4):1425–1433
73. Manente V, Johansson B, Tunestal P (2009) Partially premixed combustion at high load using gasoline and ethanol, a comparison with diesel (No. 2009-01-0944). SAE technical paper

Chapter 4
Premixed Charge Preparation Strategies

Abstract Mixing of fuel and oxidiser (air) for combustion is termed as charge preparation. Creation of premixed charge is the key feature of low temperature combustion (LTC) engines. Quality of fuel–air mixture governs the combustion process and its rate. Premixed charge is required prior to the start of combustion in LTC engines. Depending on LTC strategy, different quality of premixed charge (degree of homogeneity) is required for higher thermal efficiency and control of the combustion rate. The premixed charge can be created by injecting fuel outside the engine cylinder (external charge preparation) or inside the engine cylinder (internal charge preparation), depending on fuel properties and combustion strategy. This chapter presents the summary of premixing techniques used for charge preparation in LTC engines. Charge preparation strategies for well-mixed charge (more often external charge preparation) and partially stratified charge (internal charge preparation using direct injection) are discussed in detailed for gasoline-like fuels as well as diesel-like fuels. Dual fuel charge preparation (using high and low reactivity fuels) is typically used in RCCI combustion is also discussed in the present chapter.

Keywords PFI • Direct injection • Dual fuel • Spray • Air–fuel mixture • Fuel stratification • PFS • Equivalence ratio • DI • Dual injection • HCCI • LTC • RCCI

4.1 External Charge Preparation

Quality of charge governs the combustion and emissions characteristics of the engine. In HCCI combustion engine, as the name suggests, well-mixed homogeneous fuel–air mixture is required. Creating purely homogeneous mixture of fuel and air without any fuel concentration or thermal inhomogeneity is not possible in real engines. In real engines, some degree of inhomogeneity is always present in the fuel–air mixture. However, fairly well-mixed charge is essential in HCCI combustion to produce lower particulate emissions. Recent studies have shown that inhomogeneities can be used to increase the engine operating loads of HCCI combustion (see Chap. 7). In HCCI combustion engine, a well-mixed charge of highly diluted fuel–air mixture is required to achieve reasonable combustion rate and lower engine emissions. Several strategies of charge preparation are used for premixed charge

R.K. Maurya, *Characteristics and Control of Low Temperature Combustion Engines*, Mechanical Engineering Series, https://doi.org/10.1007/978-3-319-68508-3_4

preparation based on fuel properties and combustion control strategies. Charge preparation strategies in all LTC modes can be categorized into three main types: (i) external charge preparation (fuel and air are mixed before entering in the engine cylinder), (ii) internal charge preparation (fuel is mixed with air inside the engine cylinder) and (iii) dual fuel charge preparation (low reactivity and high reactivity fuels use both external and internal charge preparation). All of these charge preparation strategies have different degree of mixture homogeneity due to the different amount of time available for mixing fuel with air.

External charge preparation is typically used for producing well-mixed fuel–air mixture (close to homogeneous mixture) before the start of combustion. Strategy for creating premixed charge also depends on volatility of the fuel. Next subsections present the discussion of external charge preparation for volatile (gasoline-like fuels) and non-volatile fuels (diesel-like fuels).

4.1.1 Gasoline-Like Fuels

Volatile fuels evaporate easily to form fuel vapour at relatively lower temperatures (sometimes near ambient temperature). Preparation of premixed charge outside the cylinder is easier with easily evaporating fuels having low boiling temperatures. Typically, port fuel injection (PFI) or external fuel vaporizers are used for premixed charge preparation in HCCI engines. Figure 4.1 illustrates the external charge preparation strategies used for creating well-mixed fuel–air mixture in HCCI engines. External fuel vaporizer is typically used to create well-premixed charge, where fuel is injected into a separate chamber and fuel vapour mixes with air in the chamber [1, 2]. Fuel vaporizer is well ahead of intake valve such that fuel vapours have enough time to mix with air prior to entering in the engine cylinder. A study used a premixed fuelling system (as shown in Fig. 4.1a) consisting of a gasoline-type direct injector (GDI) installed in an electrically heated fuel-vaporizing chamber. Appropriate plumbing is done to ensure thorough premixing of fuel with the air upstream of the intake plenum [2]. This type of strategy creates fairly well-mixed charge with relatively lower inhomogeneity. To investigate the effect of mixing of fuel vapour on charge homogeneity in HCCI combustion, another study delivered a pre-vaporized fuel in upstream of the intake valve by a U-shaped capillary tube (2.3 mm inside diameter) using a gaseous fuel injector (Fig. 4.1b) [1]. This type of arrangement for premixed charge preparation does not have liquid film on port walls or intake valve, which creates additional inhomogeneity in the charge. Depending on fuel injection timings (with respect to intake valve timings), fuel unmixedness creates uneven distribution of fuel in the cylinder in comparison to well-mixed charge created by fuel vaporizer [1]. The PFI system (Fig. 4.1c) is the most commonly used external charge preparation method in HCCI engines using different gasoline-like fuels [3–5].

In PFI engines, intake port of each cylinder has fuel injector installed in the manifold. This type of fuel injection system is relatively inexpensive and performs robustly, except at low engine operating temperatures where poor fuel vaporization

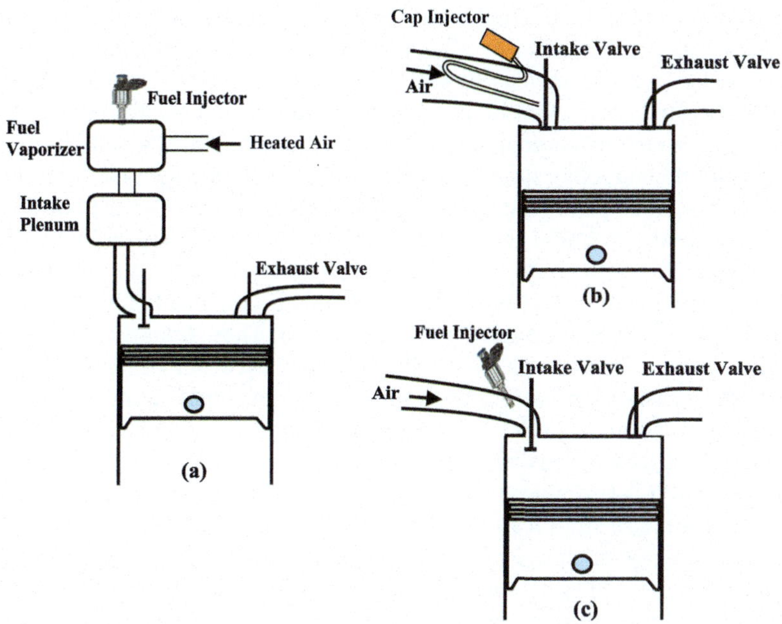

Fig. 4.1 Schematic of different external charge preparation strategies in well-mixed HCCI combustion engine

is a concern. In PFI system, fuel is injected at constant pressure usually in the range of 3–5 bar. Depending on injector characteristics (nozzle hole diameter, injection pressure, etc.), fuel atomization takes place with a droplet sauter mean diameter (SMD) in the range of 120–200 μm [6]. In PFI engine, fuel is injected into the intake port in late exhaust or early intake stroke. Then some of the fuel droplets would hit and adhere onto the intake port wall or intake valves, while some others would float in the intake port, a part of which would evaporate before being taken into the cylinder [7]. In the next cycle, adhered fuel would evaporate, and floating droplets and vapour remnants in intake port would also flow into cylinder. Therefore, in-cylinder mixture distribution can possibly be different in the next cycle. Depending on the quantity of fuel injected per cycle, it is implausible that enough time is available before inlet valve opening (IVO) for complete mixture preparation in the port before entering in the cylinder even in fully warmed engine conditions. In case of the lower temperature of intake port/ valve and inadequate time available for complete evaporation, a fraction of fuel enters in the cylinder as liquid rivulet and droplets, while some residual remains at the surface of intake port. This may lead to various consequences such as imprecise fuel metering, incomplete combustion and soot formation due to incomplete droplet evaporation and worse time response during transients [6]. However, charge preparation in HCCI combustion engine using PFI of gasoline-like fuels does not show visible smoke [3, 8] due to enough premixing of fuel. The PFI injection is more suitable for premixed charge preparation of highly volatile liquid fuels and gaseous fuels. In case of low volatility diesel-like fuels, PFI results in inhomogeneous mixture formation due to poor evaporation as well as increased wall impingement, which leads to higher HC and CO emissions along with higher fuel consumption and oil dilution [9, 10].

The PFI system design generally consists of three features: injection drop size, injection timing and spray targeting. These design parameters govern the level of charge preparation quality achieved by the system. Less than the ideal charge preparation results in liquid fuel presence and mixture inhomogeneity that affects engine behaviours such as air–fuel control, emissions and stability [11]. In transient engine operating conditions, air–fuel ratio control is complicated by the equilibration of fuel films within the engine port and cylinder. These films can be quite large, particularly under cold engine conditions [12]. The existence of large fuel films in the cylinder as well as mixture stratification has serious implications on emissions, particularly hydrocarbons (HCs). The open valve injection (OVI) timing and valve targeting minimized A/F excursions by reducing port surface wetting. The close valve injection (CVI) timing is preferred over OVI timing for minimizing hydrocarbons and mixture stratification. The minimization of surface wetting with OVI timing resulted in poor charge preparation as the drop sizes are too large to vaporize and mix effectively [11]. A study identified four transport mechanisms of liquid fuel into the engine cylinder during the intake stroke: (i) first forward flow atomization, (ii) spray contribution, (iii) high-speed intake flow transport and (iv) fuel film squeezing [13]. First forward flow atomization is the atomization of liquid fuel accumulated in the intake valve seat vicinity by the forward motion of the intake air at the beginning of the intake process (after equalization of cylinder and intake pressure) as the air strips liquid off the wall [14]. Injection spray contribution is the fuel transport into the cylinder due to injection towards the open intake valve with OVI. The fuel film squeezing is the squeezing of any fuel film accumulated on the valve seat at the end of the intake process as the intake valve closes. Study [13] concluded that increased film atomization occurs in areas which feature high magnitudes of intake air flow velocity due to the shape of the intake port. In areas where flow recirculation occurs, liquid fuel film flow is more prominent resulting in liquid fuel droplet production due to fuel film squeezing. With CVI, these film-driven transport mechanisms are observed longer during warm-up than with OVI due to more significant wall film development. With OVI the dominant liquid fuel transport mechanism is the direct transport of liquid fuel from the injector into the cylinder due to injection contribution [13].

Two types of fuel adhesion occur during PFI injection: (i) fuel adhesion on port wall and (ii) fuel adhesion on valves. Liquid film on port needs to be reduced as it affects the particulate formation and emissions. Fuel adhesion on the port wall does not generate particle number emission directly. However, it would affect the estimation of in-cylinder air–fuel ratio in the following several cycles because liquid film on intake port wall would evaporate and enter into the cylinder, whose amount is not predictable [7]. Measurement of particle number in homogeneous charge spark ignition engine showed that particle number concentration is approximately constant for equivalence ratio below 1.4, and particle number emission

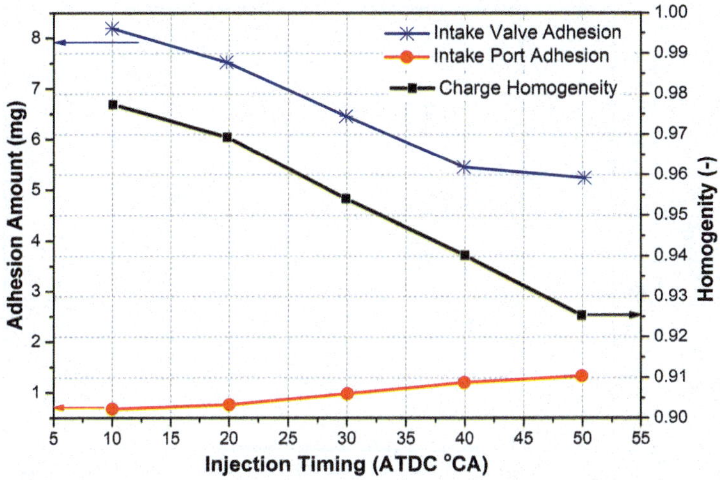

Fig. 4.2 Maximum fuel adhesion quantity on intake port and valves and in-cylinder mixture homogeneity for the different end of PFI injection timings (Adapted from [7])

steeply increased at equivalence ratios more than 1.45 [15]. Therefore, it is believed that 1.4–1.5 is the critical equivalence ratio for PM generation in premixed gasoline combustion. Thus, liquid fuel adhesion needs to be controlled in the engine.

Figure 4.2 shows the maximum fuel adhesion quantity on intake port and valves for the different end of injection (EOI) timings in PFI system using OVI. The figure depicts that advance injection timing is helpful to reduce fuel adhesion on intake port wall because the injection timing is more harmonized with valve opening time and fuel adhesion could be avoided [7]. Figure 4.2 also depicts the in-cylinder mixture homogeneity variance at ignition timing for different EOI timings. Mixture homogeneity improves with advanced injection timings, and it relates colsely with fuel adhesion on valves (Fig. 4.2). Maximum adhesion quantity of fuels on valves increases with advanced injection timing. Fuel adhesion on valves takes relatively long time to evaporate, which would be a continuous source of fuel supply in an even flow rate. However, all the fuel adhesion on valves would evaporate and flow into cylinder before IVC in the same cycle. Thus, early injection timings have relatively more homogeneous mixture in cylinder [7].

Effect of fuel spray impact and the interactions between successive injections on air–fuel mixture preparation in PFI engine is investigated for HCCI combustion [16]. Figure 4.3 depicts the overall mass fraction of evaporated fuel as a function of the overall heat flux removed from the wall at different temperatures. The fuel vaporization rate deteriorates with the fuel injection frequency due to the interaction between successive spray impacts, while its significance decreases as the surface temperature reaches to its overall Nukiyama [16]. Interaction between consecutive fuel injections is expected to have significant effect on charge preparation in HCCI combustion engine. Additionally, deterioration of the charge quality can be avoided at high loads only if the heat transfer regime of impacting droplets is

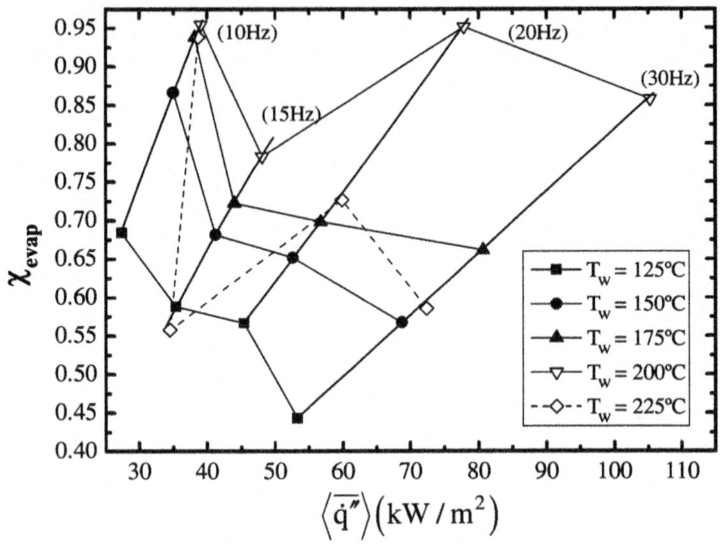

Fig. 4.3 Mass fraction of evaporated fuel as a function of overall heat flux for constant surface temperatures [16]

close to the overall critical heat flux (Fig. 4.3), while in the nucleate/boiling or above the overall critical heat flux, the interaction between consecutive fuel injection reduces the fuel evaporation [16].

The dominant soot particle production mechanism is often described through the existence of liquid fuel pools on the mentioned in-cylinder areas, where pool fires are originated. Removal or attachment of liquid film on the cylinder head-intake port surroundings is found to be highly influenced by the mode of in-cylinder liquid film transport and the characteristics of the flow. Liquid film transport mechanisms are highly affected by higher engine speed, early valve timing and narrow spray pattern with OVI strategy, leading to the higher amount of fuel deposits on those zones [17]. Thus, inhomogeneity in charge preparation is important during combustion.

Figure 4.4 shows the degree of heterogeneity of the fuel mass fraction, and temperature field in a port injected HCCI engine at intake BDC and combustion TDC positions. The heterogeneity is computed using 3D CFD calculation and assuming perfect mixture at intake valve opening (no simulation in the intake port). Two heterogeneity scales are selected (+/−25% and +/−5%) to investigate the level of homogeneity. The scale +/−25% is selected because it matches with the sensitivity of the optical diagnostic techniques typically used for fuel-air mixture investigation, which indicates homogeneous mixture close to TDC position (Fig. 4.4). The +/−5% scale is more suitable for homogeneity visualization near TDC [18]. At BDC position, the best scale is the +/−25%, which suggests that there is a difference of around 50% of the average value between the highest and

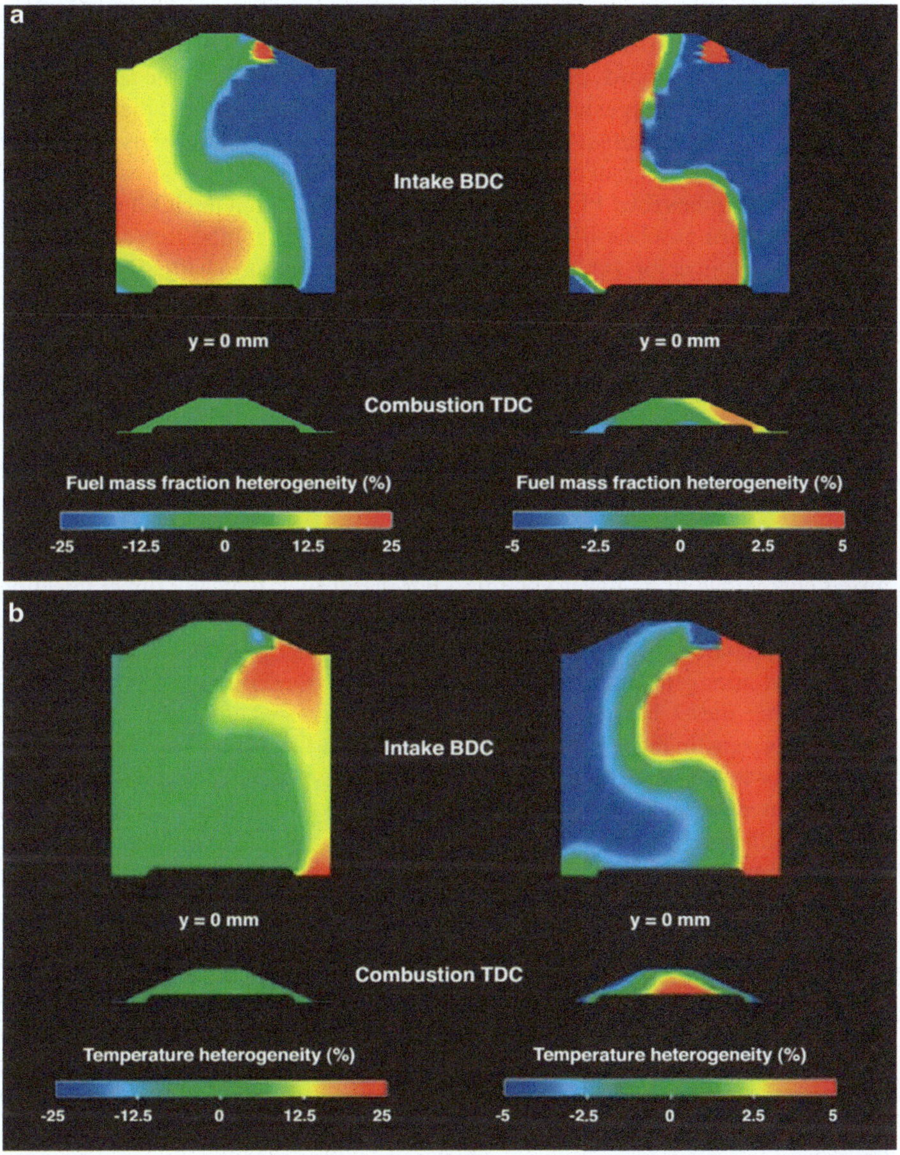

Fig. 4.4 Degree of heterogeneity of the (**a**) fuel mass fraction and (**b**) temperature field in a PFI engine [18]

lowest local values. These heterogeneity fields in combustion chamber demonstrate that the homogeneity is not absolute. The fuel mass fraction is higher below the intake valve at TDC position in the symmetry plane of the combustion chamber. Figure 4.4 shows that local equivalence air–fuel ratio at compression TDC is higher below intake valves, where autoignition can be favourable first. Highest local temperature is found in the centre of the combustion chamber due to higher heat losses near the piston because of much lower squish height. Region below the intake valve facing towards exhaust valve can be the most favourable auto-ignition region due to the heterogeneity of charge [18]. It can also be noted that while this result is by assuming perfect mixture from the port, the inhomogeneity of mixture entering in the cylinder can create a different distribution.

Numerous studies used PFI strategy for charge preparation in HCCI combustion using gasoline or gasoline-like fuels. Some initial HCCI combustion studies have also used PFI injector for creating homogeneous mixture with diesel [9, 19, 20]. Diesel PFI has difficulties due to its higher boiling range resulting in poor evaporation. Next section discusses the external charge preparation using diesel-like fuels.

4.1.2 *Diesel-Like Fuels*

Premixed charge combustion using diesel encounters challenge of preparing well-mixed charge due to its lower volatility. For diesel fuel, typically direct injection (internal charge preparation) is suggested to avoid the evaporation and mixing issues. To prepare the homogeneous mixture with diesel fuel outside the cylinder, the external arrangement is required to vaporize the diesel. External charge preparation using diesel has been explored using air-assisted injectors and external diesel vaporizers. In the air-assisted injection system, fuel is injected using gasoline port injector inside an air-assisted cap, where it mixes with the externally supplied air [21]. In external diesel vaporizer, fuel is injected in the externally heated chamber, which creates diesel vapours [22, 23]. Figure 4.5 illustrates the schematic diagram of external charge preparation using diesel vaporizer.

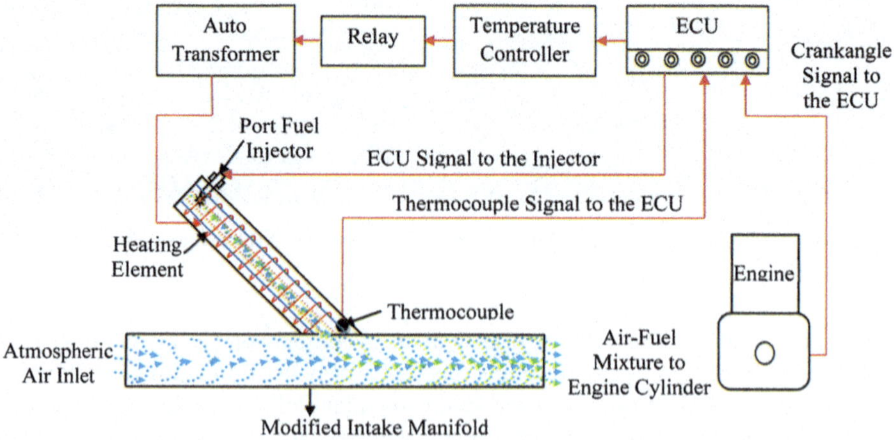

Fig. 4.5 Schematic diagram of electronically controlled diesel vaporizer system for HCCI combustion [23]

The fuel vaporizer system is used in creating vapour of diesel fuel. The intake manifold of the engine is modified to house the fuel vaporizer, and the diesel vapours are injected in air present in the intake manifold [23]. The diesel vapours follow the intake air motion very well, and the droplets have high evaporation rate because of very high surface to volume ratio. Vaporized diesel fuel disperses uniformly in the surrounding airstream creating fairly well-mixed air–fuel mixture. In this type of system, additional requirement of energy and temperature control system is required. However, fuel injection system is less demanding and it operates at lower fuel injection pressure. Due to additional complexity of the system, internal charge preparation by direct fuel injection at appropriate timings is recommended for diesel-like fuels.

4.2 Internal Charge Preparation

Internal charge preparation means fuel and air are mixed inside the engine cylinder. In this strategy, fuel is directly injected in the cylinder. Quality of premixed charge obtained depends on fuel injection characteristics and fuel properties. Fuel injection timings, number of injections and fuel injection pressure are the main fuel injection characteristics, which governed the premixed charge preparation quality and degree of stratification for a particular fuel. Fuel injection timings can be broadly categorized as early and late direct injection. Early fuel injection timing refers to the fuel injection early in the compression stroke. Late direct injection timings typically refer to the fuel injection timings close to combustion TDC of engine cycles. Major concern in this strategy is that the fuel injection ends before autoignition starts to achieve premixed charge combustion in the cylinder. Other important concern during direct injection is to avoid fuel impingement on the cylinder walls by controlling fuel spray penetration length. Internal charge preparation strategy also offers direct control of the combustion timings. Fuel injection strategy also depends on fuel properties. Charge preparation strategy with gasoline- and diesel-like fuels is presented in next subsections.

4.2.1 Gasoline-Like Fuels

Presently most of the low temperature combustion research is focused on gasoline-like fuels. Internal charge preparation strategies using direct injection (DI) produce premixed stratified fuel–air mixture. Low temperature combustion strategies using gasoline direct injection can be divided into three main categories based on stratification of charge: (i) partial fuel stratification (PFS), (ii) moderate fuel stratification (MFS) and (iii) heavy fuel stratification (HFS) [24]. Figure 4.6 illustrates the fuel injection strategies and corresponding charge quality based on stratification (PFS, MFS and HFS) in low temperature gasoline compression

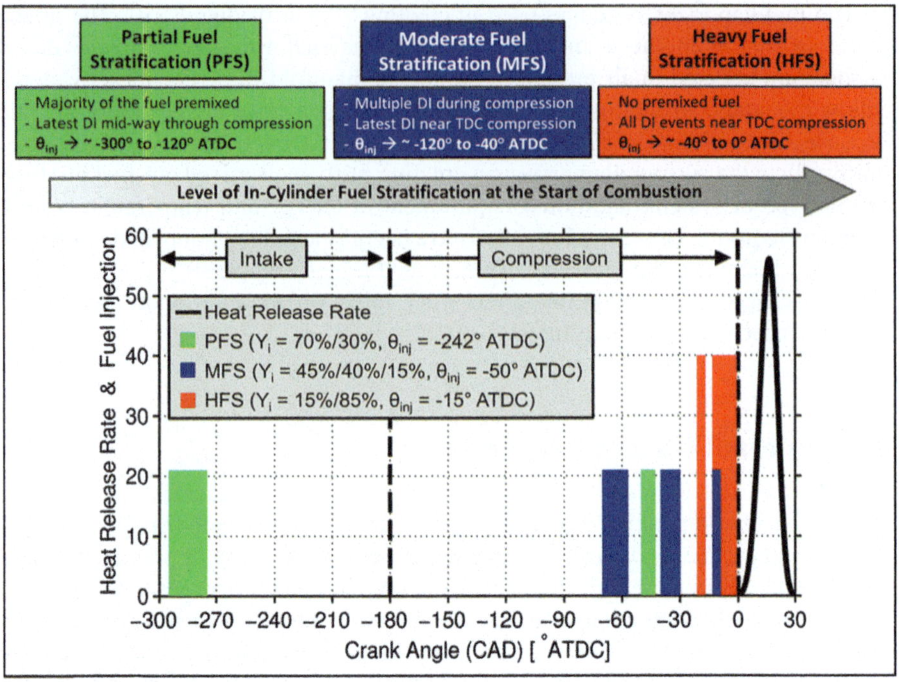

Fig. 4.6 Description of different fuel injection strategies and charge qualities based on stratification in low temperature combustion using gasoline [24]

ignition (LTGCI) engines. In Fig. 4.6, fuel injection bar width is proportional to the quantity of injected fuel, and height of the bar is proportional to corresponding fuel injection pressure.

In the HFS strategy, relatively higher fuel injection pressure is used to create the air–fuel mixture. In some of the fuel injection strategies, multiple fuel injection is also used (Fig. 4.6). In case of multiple injection strategy, injection centroid is used as metric to evaluate the fuel stratification category. The injection centroid (θ_{inj}) is mass weighted average injection timing, and it is calculated by Eq. (4.1) [24].

$$\theta_{\text{inj}} = \frac{\sum_{i=1}^{N_{\text{inj}}} Y_i \text{SOI}_i}{\sum_{i=1}^{N_{\text{inj}}} Y_i} = \sum_{i=1}^{N_{\text{inj}}} Y_i \text{SOI}_i \tag{4.1}$$

where N_{inj} is the number of fuel injection events, Y_i is fuel mass fraction in injection even i and SOI_i is the start of injection (SOI) in injection event i. This metric is used to determine the level of fuel stratification as shown in Fig. 4.6.

The PFS strategy has lowest level of stratification, and most of the fuel is injected early in the intake stroke or by port fuel injection. The PFS strategy is typically used to increase the high load limit (while maintaining ultralow NO_x and soot emissions of HCCI combustion) by reducing heat release rate (HRR) due to sequential autoignition of partially stratified charge [2, 25–29]. For reduction of HRR by fuel stratification, autoignition of fuel must be sensitive to variations in the local fuel–air equivalence ratio (ϕ), such that the ϕ stratification in the cylinder generates a staged heat release event due to sequential autoignition occurring down the ϕ gradient [26]. In the PFS strategy, fuel is fully premixed and a fraction (up to about 20%) of fuel is directly injected into the latter part of the compression stroke using a GDI injector to create the mixture stratification. This PFS strategy is selected because it provides adequate fuel stratification and the premixed fuel retains a sufficiently high ϕ in the leanest parts of the charge to achieve good combustion efficiency [27]. The large premixed portion also promises good air utilization throughout the combustion chamber. This fuel injection strategy is also termed as standard PFS (std-PFS), where 80–95% of the fuel is fully premixed and rest of the fuel directly injected at 35–60 CAD before the combustion TDC. It is important to note that the injection timings of the late DI fuel are still relatively early, $\geq$35° CA before TDC. These SOI timings ensure that fuel injection is complete before the start of combustion. Therefore, the late DI fuel undergoes substantial mixing with the air before the onset of the sequential autoignition process [29]. In the std-PFS, the intake temperature must be kept sufficiently high to prevent fuel condensation in the intake system.

The DI fuelling removes fuel condensation issue in the intake system, and it allows the intake temperature close to ambient temperature. Therefore, another strategy termed as early DI PFS (E-DI PFS) is used to increase the engine load for a given intake boost, i.e. increasing the load-to-boost ratio, compared to premixed fuelling (see Chap. 7). In this strategy all the fuel is injected directly into the cylinder during intake stroke. In the DI fuelling, injection even in early intake stroke (60° CA after intake TDC), the charge mixture is not completely uniform by the time combustion occurs, and the resulting mixture distribution works well for PFS [2]. This strategy is able to increase the thermal efficiency of engine. To improve the fuel stratification further, dual direct injection (D-DI) PFS strategy is proposed [29]. In the D-DI PFS strategy, majority of fuel is injected as E-DI PFS (60° CA a TDC intake), and a fraction of fuel injected in the later part of compression stroke (more than $\geq$35° CA before TDC) to achieve a sequential auto-ignition combustion mode similar to std.-PFS. The D-DI strategy can slightly improve the thermal efficiency in comparison to E-DI PFS strategy.

The MFS strategy increases the fuel stratification in comparison to PFS by reducing the premixed quantity of fuel. MFS strategy typically injects all the fuel directly in the cylinder during compression stroke. Additionally, the MFS strategy typically has a fuel injection event near TDC to trigger the combustion [30–35]. A low temperature partially premixed combustion process in MFS regimes is also termed as gasoline direct injection compression ignition (GDCI) in various studies [31, 32, 34, 35]. In this strategy, gasoline is injected at typical GDI fuel injection pressure by multiple injections, and it rapidly vaporizes and mixes with air. In GDCI, the engine has a shallow pent roof quiescent combustion chamber and central-mounted injector. To facilitate the injection-controlled fuel stratification, quiescent chamber is selected, and swirl, tumble and squish are minimized so that excessive flow is not able to destroy the stratification produced by injection process [34]. The piston and bowl shape are also matched with the injection system and spray characteristics. Multiple late direct injections are found to be beneficial for this LTC strategy [30]. The basic idea is to form combustible mixtures in different regions of the cylinder which have their different characteristic air–fuel ratio, ignition delay time and temperature history (by fuel stratification). The stratification in combustion chamber leads to long enough combustion duration for reducing the combustion noise and fast enough heat release to achieve higher thermal efficiency [35]. It is found that with three injections, the desired efficiency and emission can be achieved in GDCI. Typically first injection event takes place in the middle of compression stroke at relatively lower pressure and temperature and delivers fuel above the piston. The second injection typically targets the lower outer bowl of the piston. The third injection event takes places close to TDC, which leads to low spray penetration, and this fuel occupies the inner bowl region. In this way, this strategy creates a partially stratified charge in the cylinder.

Fuel stratification in HFS strategy is higher than MFS. A low temperature premixed combustion process in HFS regime is also termed as partially premixed combustion (PPC) in various studies [36–39]. In PPC strategies, typically gasoline-like fuel (higher octane number than diesel) is used. In this strategy, single and multiple fuel injection is used to create the fuel–air mixture. Figure 4.7 illustrates the difference in the temporal evolution of liquid length in gasoline and diesel fuel spray at SOI 14°CAD bTDC under vaporizing but non-reacting conditions. The images are presented from the SOI to the stabilized zone. Figure 4.7 depicts that there are significant differences in terms of liquid length in gasoline and diesel spray, and gasoline spray penetration is relatively shorter than diesel.

Injection timing has significant role in spray formation process because gas density and temperature are different for different SOI timings. Fuel spray characteristics depend on the gas density and temperature in the combustion chamber. Figure 4.8 shows the liquid length of diesel and gasoline spray for different injection timings relevant to PPC strategy. Figure 4.8 depicts that stabilized value of the liquid length is shorter with gasoline fuel in comparison to diesel fuel and it is 1.8–2.4 times shorter than diesel liquid length. The liquid length increases with advanced fuel injection timings. Simplified one-dimensional model to estimate the liquid length (LL) is given by Eq. (4.2), where D_0 is the nozzle diameter, ρ_f is the fuel density, ρ_g is the gas density and $Y_{f,\mathrm{evap}}$ is the vaporized fuel mass fraction [40].

$$\mathrm{LL} = k \cdot D_0 \cdot \sqrt{\frac{\rho_f}{\rho_g}} \cdot \frac{1}{Y_{f,\mathrm{evap}}} \quad (4.2)$$

For advanced SOI timings, liquid length is higher mainly due to the lower chamber temperature attained (in the range of SOI showed). The changes in temperature have a great influence on vaporized fuel mass fraction, and it is

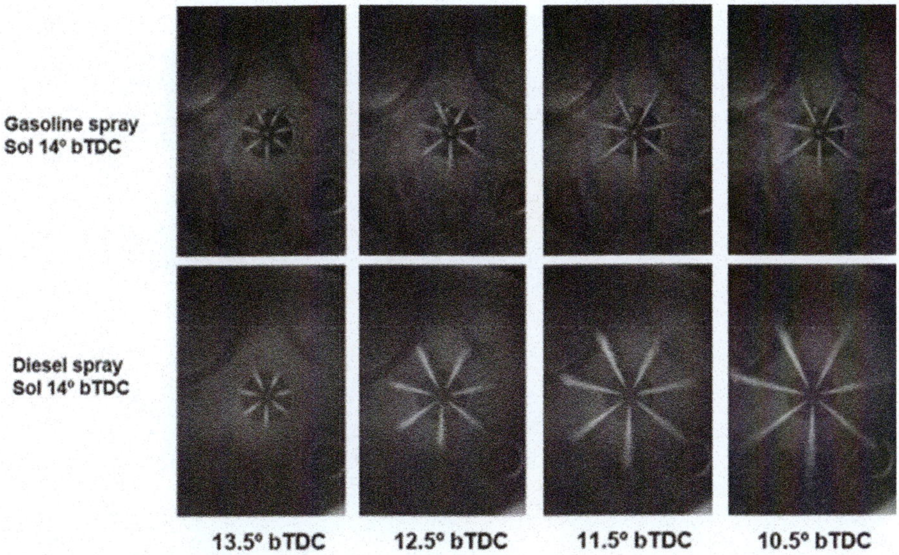

Fig. 4.7 Temporal evolution of the liquid length for gasoline and diesel fuel spray at same fuel injection timings [36]

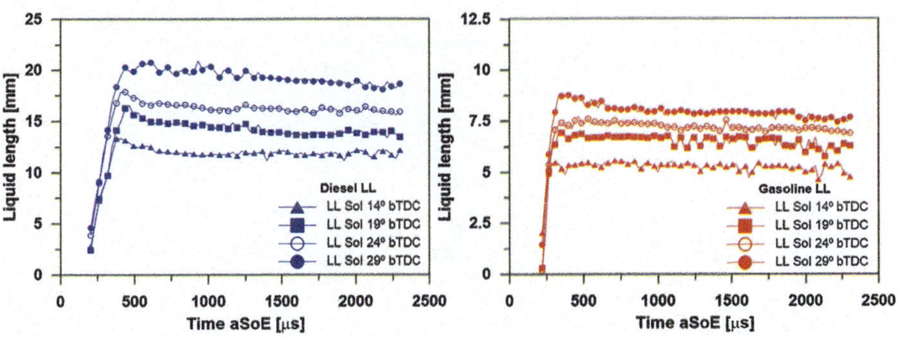

Fig. 4.8 Liquid length of diesel and gasoline spray for different fuel injection timings [36]

found to be the key parameter governing the differences in liquid length for gasoline and diesel under stabilized conditions [36].

Figure 4.9 shows the fuel–air equivalence ratio distribution after the end of injection (EOI) in PPC engine for SOI timing 25° bTDC using gasoline. Air–fuel ratio is estimated using average of ten selected single-shot fuel-tracer PLIF images at particular crank positions. Figure 4.9 shows that for most of the laser sheet region, the equivalence ratio is below 4 with some exceptions in the core of jets. The fuel jets interact with the cylinder wall and, with the swirl of intake airflow, create three local parcels with $\Phi > 1.0$ (at −5° CAD). The vaporized fuel continues to mix with the local air, and, finally, two local zones are created at 5° (marked by

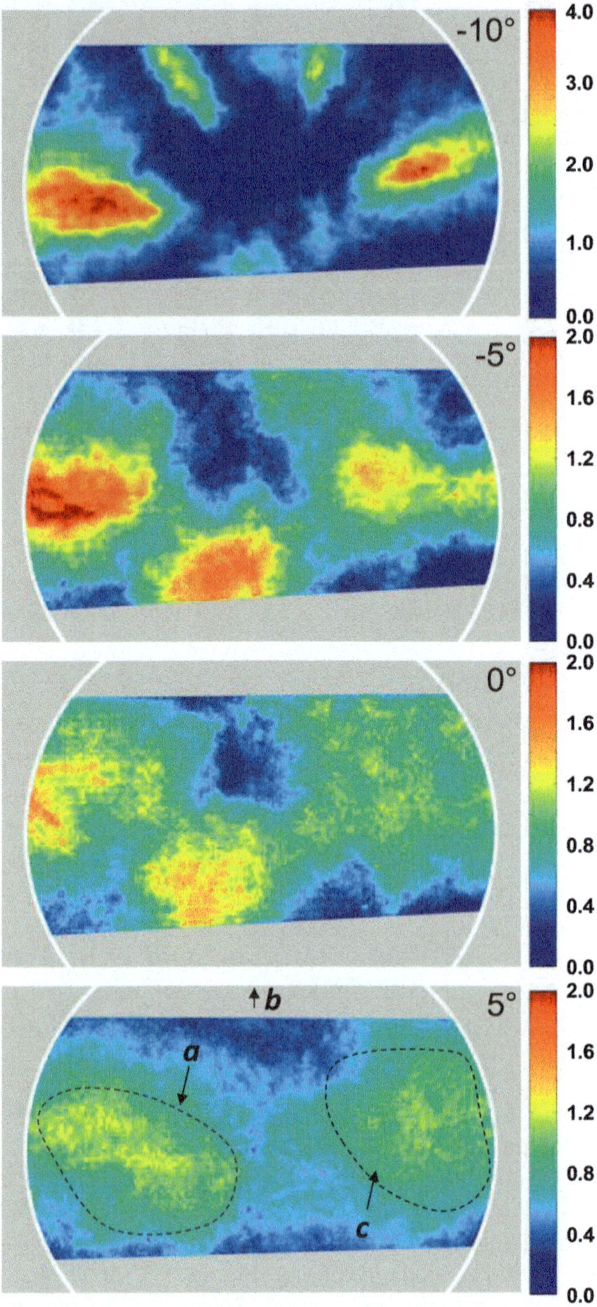

Fig. 4.9 Fuel–air equivalence ratio distribution after the end of fuel injection in PPC engine [37]

"a" and "c") with Φ in the range of 0.8–1.2. The study also showed that the earlier SOI timings (90° bTDC) provide a further leaner charge in the diagnostic region at 5° CAD [37].

To achieve higher engine load and lower emission in PPC engine using gasoline, a fuel injection strategy consisting two injection events is developed [41]. The first injection event occurs early in the compression stroke (at 60° before TDC), and it is fixed for each load condition. This injection event is used to create a homogeneous mixture, while the second injection event is close to TDC to create fuel stratification, which triggers the PPC combustion process. The quantity of fuel in the first injection is independent of the load, and it is a function only of compression ratio, reactivity of the fuel and EGR level [39]. The engine load and combustion phasing are controlled through the second injection.

Figure 4.10 shows the fuel percentage in the first injection and mixing period as function of IMEP in PPC engine. Fuels designated as G.ON x/y in the Fig. 4.10 represents gasoline octane number, where 'x' is RON, and 'y' is MON. Figure 4.10 depicts the amount of fuel in the first injection depends on the compression ratio and octane number of the fuel. To avoid the oxidation reactions in the fuel injected during first injection, high amount of EGR is used. For higher reactivity fuel, the pilot quantity of fuel is zero and only main injection takes place (Fig. 4.10a). At lower compression ratio, higher amount of pilot fuel can be used due to relatively lower compression temperature. At lower compression ratio, higher engine load can be obtained by burning higher quantity of fuel for same combustion noise limits. Mixing period is relatively shorter at lower compression ratio (Fig. 4.10b). Mixing period is defined as the crank angles between the end of the main injection (EOIm) and 10% of total heat release (CA10). At high load, the mixing period decreases together with the fuel octane number (Fig. 4.10b).

A study also investigated the effects of injection timing on fluid flow characteristics of PPC engine [42]. The study demonstrated that during injection, the turbulence level noticeably increases by 2–4 times depending on the injection timing. The vortex inside the piston bowl, driven by the spray, persists during the ignition delay period. After the end of injection, turbulence level and mean velocity decrease, which decreases the mixing efficiency until the start of combustion (SOC). Later injection timings in PPC combustion cause higher mean velocity and turbulence level inside the piston bowl and consequently higher mixing efficiency [42].

4.2.2 *Diesel-Like Fuels*

The level of homogenization realizable with diesel fuel for the charge preparation is far away from that in a conventional SI engine. In fact, a study suggested that a homogeneous stratification (rather than a homogeneous mixture) should be more suitable term for lean HCCI diesel combustion [43]. It means that charge in the cylinder contains homogeneously distributed small clouds of rich mixture in the

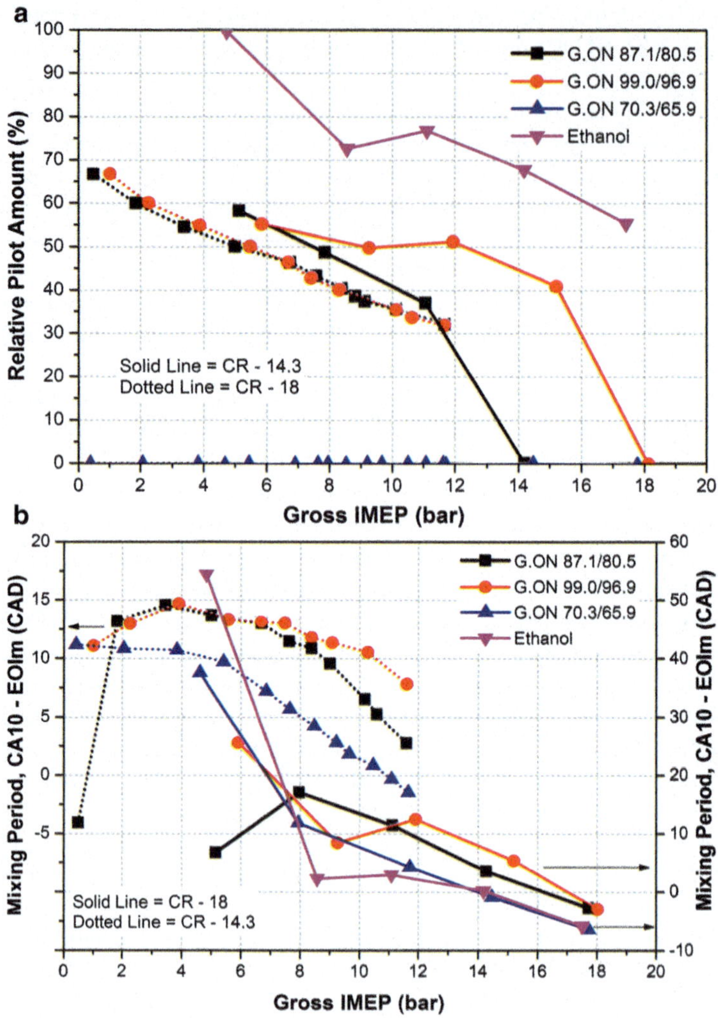

Fig. 4.10 (**a**) Fuel percentage in the first (pilot) injection and (**b**) mixing period as function of IMEP in PPC combustion (Adapted from [38, 39])

cool air and oxidation reactions must progress concurrently in every cloud, in such a way that combustion temperature is relatively lower and a lean combustion occurs. Premixed charge compression ignition (PCCI) combustion using diesel fuel is discussed in Sect. 2.5.1 of Chap. 2. Typically, charge preparation (fuel–air mixing) using diesel is termed as internal charge preparation because diesel is mixed with air inside the cylinder. Premixed charge preparation strategies using diesel can be broadly categorized into early direct injection and late direct injection techniques. In early direct injection PCCI combustion, various strategies exist such as premixed lean diesel combustion (PREDIC), multiple stage diesel combustion (MULDIC),

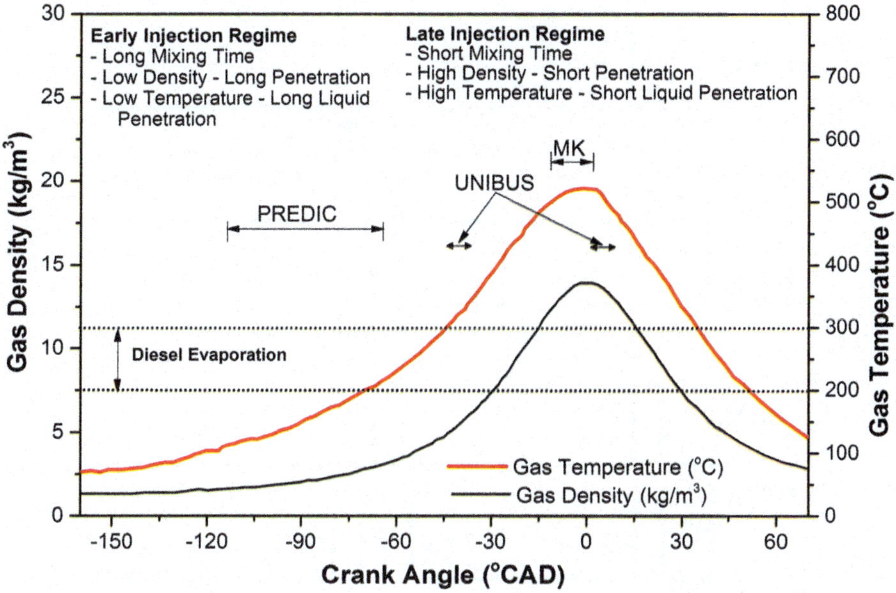

Fig. 4.11 In-cylinder gas density and temperature as function of crank position in CI engine (Adapted from [45])

homogeneous charge intelligent multiple injection combustion (HiMICS), premixed compression ignition (PCI), uniform bulky combustion system (UNIBUS) and narrow angle direct injection (NADI). In late injection premixed charge combustion, strategies such as modulated kinetics (MK), homogeneous charge late injection (HCLI) and the highly premixed late injection (HPLI) are investigated.

Some major issues exist in the formation of homogeneous air–fuel mixture in the engine cylinder using diesel fuel [44]. To create homogeneous mixture, relatively higher amount of time is required for homogenization, which can be achieved by well-advanced fuel injection. During advanced fuel injection, air density and temperature are lower, which leads to poor atomization and evaporation. To avoid the fuel impingement on cylinder wall, higher fuel injection pressure is not possible. Other main concern is lower vaporization of diesel droplets due to the low volatility of diesel fuel. Figure 4.11 shows the variation of gas density and temperature as function of crank position along with diesel vaporization region. In early direct injection strategy, fuel is injected early in the compression stroke (well advanced compared to conventional diesel combustion). During early injection timings, air temperature and density are relatively lower, which increases the penetration length of diesel spray. However, during late injection strategy, fuel is injected in higher gas temperature and density resulting into faster evaporation and lower liquid penetration length. Typically used injection timing for early and late injection strategies is shown in Fig. 4.11.

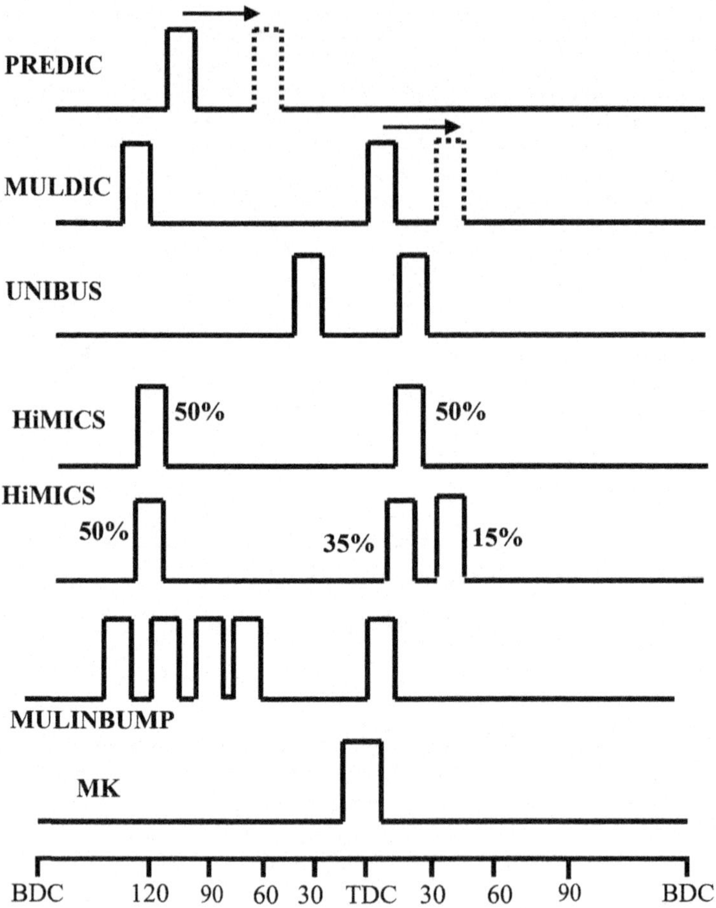

Fig. 4.12 Fuel injection strategies in different lean premixed diesel combustion strategies (Adapted from [46–50])

Figure 4.12 demonstrates the typical fuel injection timings in different lean premixed diesel combustion strategies in early and late direct injection regime. The figure depicts that single and multiple direct injection strategies are used to create premixed charge. In early injection strategies, avoiding fuel impingement and cylinder wall wetting is the key challenge due to fuel injection in relatively lower gas density and temperature. To achieve this target, redesign of fuel injection system or geometrical arrangement is essential. Figure 4.12 illustrates the injector arrangement in the cylinder for PREDIC and conventional diesel combustion (CDC) strategies. Two side injectors are installed for PREDIC and MULDIC strategies to create lean premixed charge in the cylinder. Side injector is employed to increase the distance between fuel injector and cylinder wall in comparison with central injection in CDC to avoid fuel impingement on the cylinder wall. Therefore,

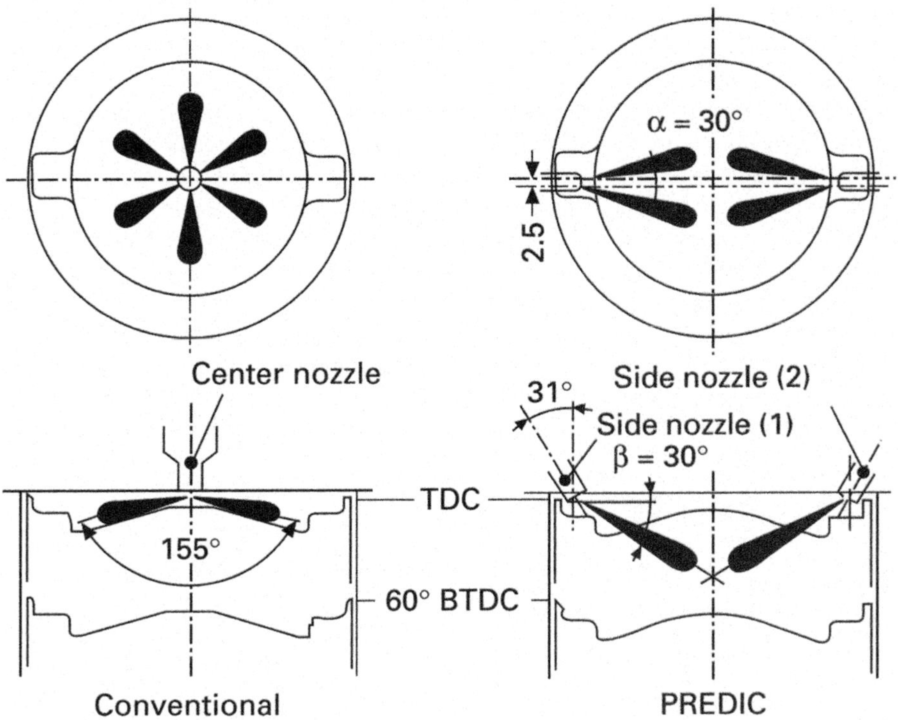

Fig. 4.13 Fuel injector arrangement in PREDIC and CDC strategies [51]

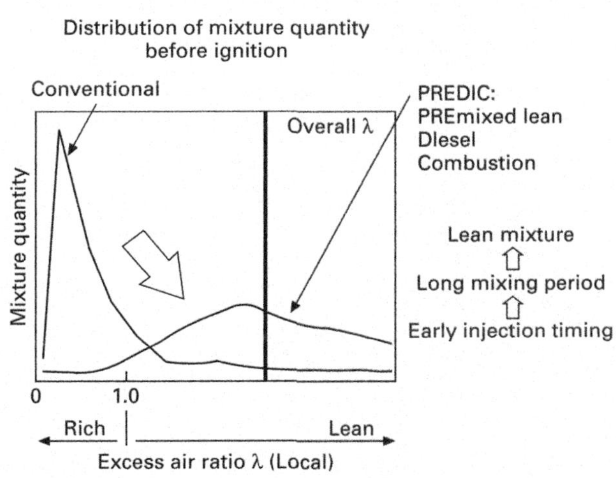

Fig. 4.14 Illustration of air–fuel mixture formation processes and differences in PREDIC and CDC [51]

the side injectors are beneficial for lean premixed fuel–air mixture preparation using diesel fuel [51] (Fig. 4.13).

Figure 4.14 schematically illustrates the differences in air–fuel mixture formation in PREDIC and CDC strategies. Horizontal coordinates represent excess air ratio λ and vertical coordinates the magnitude of fuel–air mixture prior to ignition, and the distribution is determined using numerical analysis. In CDC strategy, fuel is injected close to TDC, which leads to the formation of overly rich mixture immediately after injection prior to ignition. However, in PREDIC strategy, early direct injection (typically 90°BTDC) takes place, which allows sufficient time for evaporation and mixing of fuel. Therefore, overly rich mixture created after injection becomes leaner prior to the start of combustion. Thus, PREDIC strategy has leaner premixed combustion than CDC even with same overall excess air ratio (λ).

Different strategies in early direct injection regime have been explored for charge preparation (Fig. 4.12). These strategies can be broadly categorized into impinging sprays, hollow cone sprays and ordinary diesel sprays, which are typically used in combination with a narrow included angle and/or multiple injections [53, 54]. Narrow angle direct injection (NADI) strategy is proposed in order to avoid fuel deposition on the cold cylinder liner, which can be achieved by reducing the spray angle [55]. Figure 4.15 illustrates the differences in combustion chamber

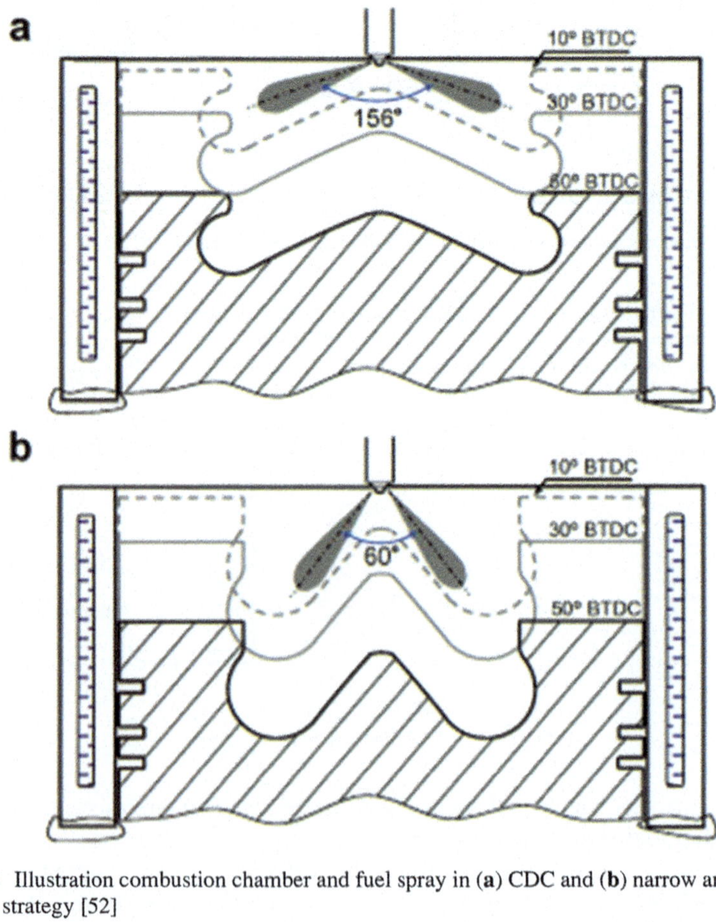

Fig. 4.15 Illustration combustion chamber and fuel spray in (**a**) CDC and (**b**) narrow angle direct injection strategy [52]

and spray in CDC and narrow angle direct injection system at different crank positions. In CDC configuration, spray will miss the piston bowl for early fuel injection timings (Fig. 4.15a). However, in narrow spray cone angle injector, the fuel spray will not miss the piston bowl for early injection timings (Fig. 2.15b) [52].

Modulated kinetic (MK) combustion system is of the widely discussed late direct injecting regime for premixed charge diesel combustion [56]. In this strategy diesel is directly injected into the cylinder near or after TDC. However, heavy amount of cooled EGR, higher swirl and lower compression ratio are used to increase the ignition delay for creating premixed charge. This system employs two mutually independent intake ports, one of which is a helical port for producing an ultra-high swirl ratio and the other is a tangential port for creating a low swirl ratio. Late direct injection of diesel can remove the problems related to fuel wall impingement, but shorter ignition delay of diesel limits this strategy to lower engine loads [57]. The more discussion on early and direct injection strategies can be found in Chap. 2 and in reference [58].

The lean premixed charge combustion using neat diesel in late and early injection regime is able to achieve only low or medium engine loads. Recent studies focus more on using high octane fuels (like gasoline) to achieve higher engine loads in LTC using premixed charge (see Chap. 7). Premixed charge preparation using gasoline-like fuels is already discussed in Sect. 2.1. In LTC regime, diesel injection is done along with low reactivity gasoline-like fuels in dual fuel RCCI combustion. Dual fuel charge preparation strategy is discussed in the next section.

4.3 Dual Fuel Charge Preparation

In dual fuel LTC strategy (such as RCCI), two fuel of different reactivity are used. Detailed discussion on fundamentals of RCCI combustion is given in Sect. 2.6 of Chap. 2. Typically, low reactivity fuel (gasoline-like fuels) is injected in the intake manifold by PFI strategy, and high reactivity fuel is directly injected into the cylinder to create reactivity stratification in the combustion chamber. In recent studies, significant advantage is demonstrated by injecting both high and low reactivity fuels directly in the cylinder (see Chaps. 2 and 7). Charge preparation strategy in dual fuel combustion using single direct injection and dual direct injection is presented in next subsections.

4.3.1 Single Fuel Direct Injection

In RCCI combustion engine, performance and emission characteristics depend on fuel injection parameters such as fuel premixing ratio, fuel injection timings, fuel injection pressure and spray angle of injected fuel. These parameters affect the charge quality and fuel distribution (stratification) in the cylinder. To achieve the

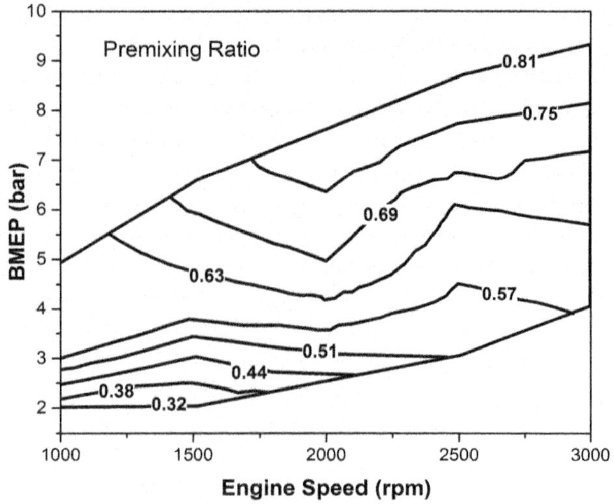

Fig. 4.16 Fuel premixing ratio as a function of engine speed and load in RCCI engine (Adapted from [59])

RCCI combustion, diesel is directly injected in the cylinder by single or dual injection, and gasoline is typically injected into the intake port. Figure 4.16 presents the premixing ratio (ratio of gasoline mass and total fuel mass) as a function of engine speed and load condition. The RCCI operation is achieved by an early single DI pulse of diesel fuel (between 30 and 70° BTDC) and port fuelling of gasoline during closed intake valve conditions [59]. Fuel rail pressure is reduced as diesel start of injection (SOI) timing is advanced to avoid spray impingement on the cylinder walls. Figure 4.16 depicts that the premixed ratio varies with engine speed and load to achieve reasonable burn rate and lower emissions.

At particular premixing ratio, injection parameters such as injection timing, injection pressure and spray angle govern the fuel reactivity distribution in the chamber. Figures 4.17 and 4.18 show the effects of spray angle and SOI timings on the equivalence ratio distribution in the combustion chamber of the RCCI engine operated at 1500 rpm and 4 bar BMEP. Selection of proper spray angle can define how diesel fuel parcels distribute throughout the cylinder and it has significant effect on combustion characteristics of RCCI engine.

Figure 4.17 depicts that for narrower spray angles, diesel fuel parcels are gathered in the piston bowl. For spray angles 25°, 35° and 45°, some fuel droplets are attached to the piston bowl surface because of the spray impingement. This event occurs when the incident angle of the impinging droplets is perpendicular to the piston bowl surface. Diesel parcels are not able to mix with gasoline–air mixture perfectly in the impinging conditions, which leads to lower heat release rate [60]. However, piston bowl impingement is more preferred in comparison to cylinder liner impingement or piston top impingement, because of relatively higher piston bowl temperature than in cylinder walls. Diesel fuel parcels target wall and the squish region for wider spray angles (Fig. 4.17). Thus, autoignition initiates and propagates in/from region leading to complete consumption of the fuel in the squish region. Therefore, wider spray angles have lower unburned hydrocarbon emissions.

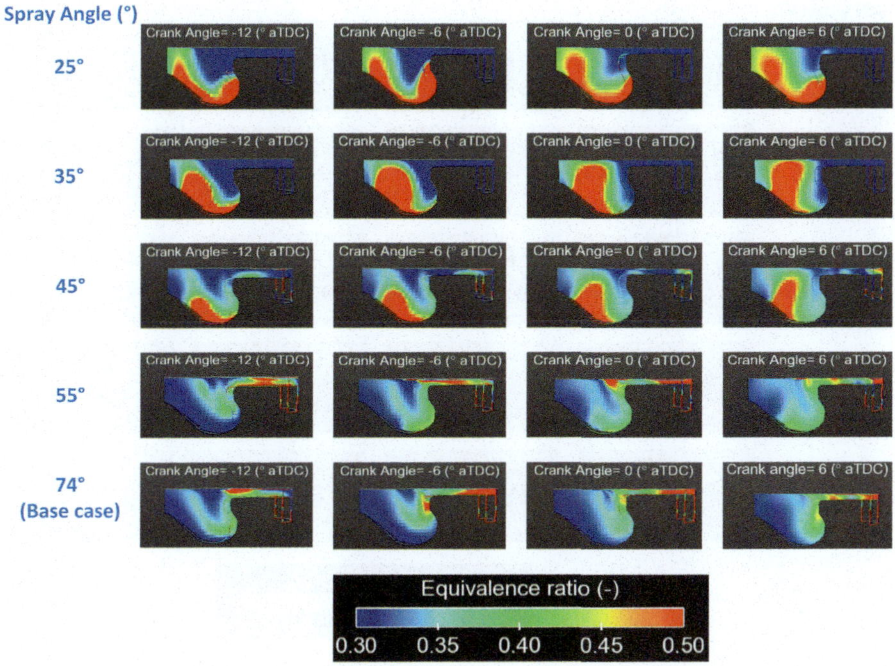

Fig. 4.17 Effects of spray angle on equivalence ratio at the different crank angle position in RCCI engine operated at 1500 rpm, 4 bar BMEP, SOI = −53° aTDC and fuel injection pressure 480 bar [60]

However, homogeneously distributed gasoline burns incompletely near the liner in squish region because high reactivity fuel targets in the piston bowl region.

Mixing time of diesel increases with advanced SOI timings in the compression stroke, which produce a partially premixed mixture of gasoline, diesel and air. Local equivalence ratio gradient reduces, and mixture homogeneity increases with increase in mixing time that can also lead to very low reactivity mixture zones in the cylinder. Therefore, with too advanced SOI timings, RCCI combustion approaches partial burn/ misfire burn conditions. For delayed SOI timing, fuel-rich regions are created due to shorter mixing time, which leads to higher NO_x and soot emissions [60]. Figure 4.18 demonstrates that advanced SOI timing causes more premixed mixture with lower local reactivity gradients, which can be observed by comparing the SOI = −63° aTDC with SOI = −45° aTDC at crank shaft positions −12° and −2° aTDC.

To determine the operating map of RCCI engine with EURO VI limits, an experimental procedure is developed with a well-defined criteria [61, 62]. Figure 4.19 presents the summary of the RCCI engine operation strategy along with fuel injection strategy. Operating limitations used to decide an acceptable RCCI operation: NO_x = 0.4 g/kWh, soot = 0.01 g/kWh, maximum PRR = 15 bar/CAD and P_{max} = 190 bar (see Chap. 7 for more details).

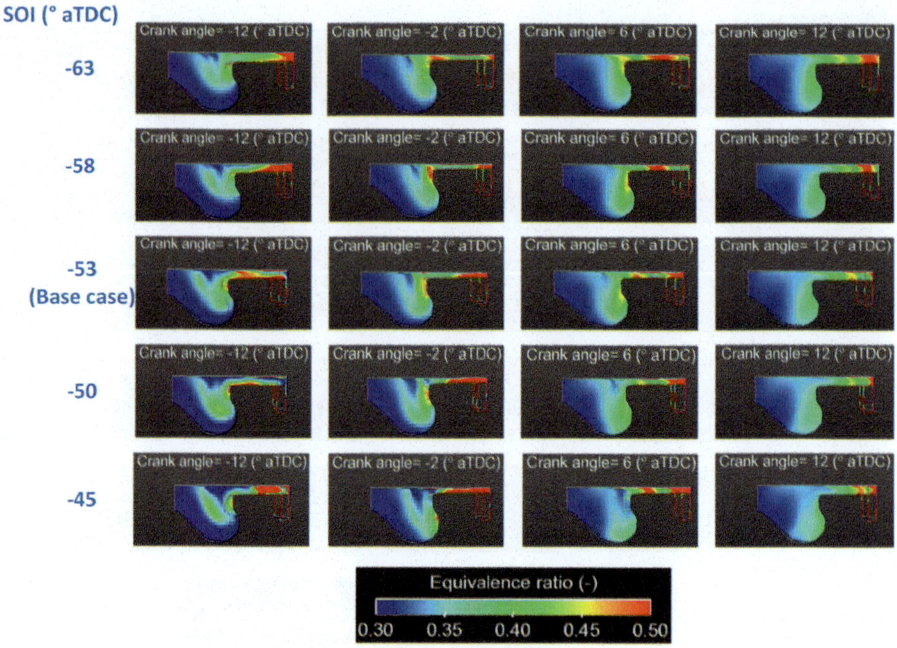

Fig. 4.18 Effects of SOI timings on equivalence ratio at the different crank position in RCCI engine operated at 1500 rpm, 4 bar BMEP, SOI = −53° aTDC and fuel injection pressure 480 bar [60]

Figure 4.19 depicts three key steps in RCCI engine operation: (i) find potential engine settings for stable operation at the desired load, (ii) achieve the EURO VI (NOx and soot) limits by adjusting premixing ratio and injection timings along with EGR and (iii) improve the fuel consumption and minimize the HC and CO emissions while maintaining the EURO VI NOx and soot limits. Fuel injection timings and premixing ratio play an important role in achieving the desired objective.

Typically, SOI timing of diesel in RCCI engine is around 60° before TDC to ensure the premixing of diesel in case of single direct injection [63]. In case of double direct injection RCCI combustion, first injection event occurs around 60° before TDC, and second injection event takes place around 35° before TDC [64]. A study used two different injection settings to operate the engine in fully premixed RCCI and highly premixed RCCI operation (Fig. 2.45 of Chap. 2), where the dual direct injection is used at two different injection timings [65]. Fuel injection pressure affects the atomization, vaporization and dispersion of the diesel injected liquid fuel. Higher fuel injection pressure leads to rapid vaporization and mixing process, which results into more homogeneous mixture [66]. However, droplet diameter decreases with increase in fuel injection pressure, and less penetration is achieved due to inertial effect of injected fuel droplets leading to relatively lower fuel distribution. At lower injection pressure, higher droplet diameter is achieved, which leads to higher flexibility for fuel distribution throughout the cylinder [60]. In

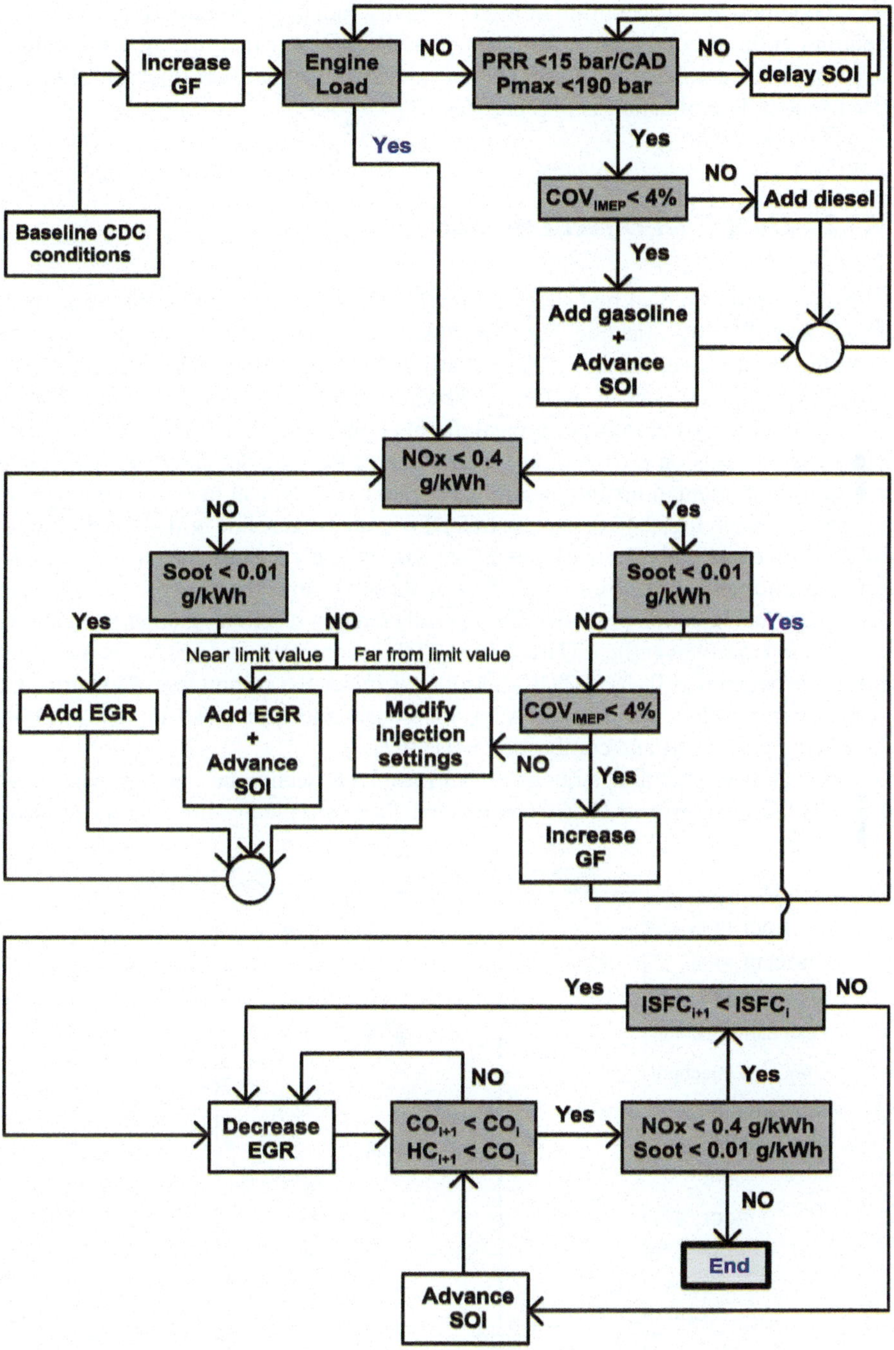

Fig. 4.19 RCCI engine operation strategy to determine operating map with EURO VI limits [61]

RCCI combustion, distribution of high reactivity fuel plays an important role in the determination of combustion characteristics. Thus, lower fuel injection pressure is used in RCCI combustion in comparison to CDC. Typical fuel injection pressure used in RCCI combustion is 500 bar [63, 64].

4.3.2 Dual Fuel Direct Injection

The direct injection dual fuel stratification (DDFS) strategy combines the benefits of RCCI and PPC by injecting both gasoline and diesel directly into the cylinder, enabling control over the in-cylinder distribution of both fuels. The DDFS strategy provides additional control on combustion for engine operation with conventional fuels and compression ratios of typical diesel engines. Section 2.6.4 of Chap. 2 provides more details on DDFS strategy. Drawing of the cylinder showing the injector locations and orientations in typical DDFS along with typical injection strategies in DDFS is given in Fig. 2.46 (Chap. 2). In DDFS strategy, gasoline is injected during the intake stroke and it is effectively premixed. Gasoline is typically injected at 1000 bar fuel injection pressure. Similar to RCCI, diesel is injected 40–60° before TDC (at typical fuel injection pressure 500 bar) and creates a reactivity gradient that allows for precise control over the LTHR. After LTHR is complete, and just as the main heat release is beginning (0–10° BTDC), gasoline is injected a second time and burns in a diffusion-controlled manner, allowing for considerable increase in load without increasing peak HRR and combustion noise [67].

Typical fuel injection strategy in various LTC techniques is summarized in Fig. 4.20. In well-mixed HCCI combustion, fuel is typically injected into intake)

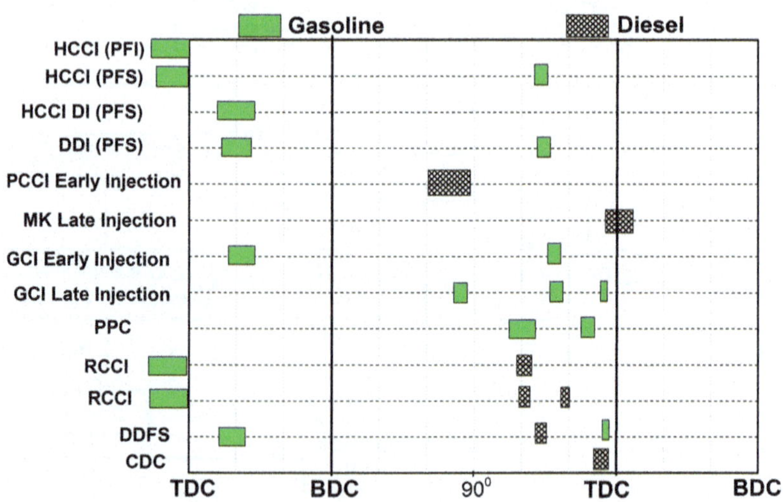

Fig. 4.20 Fuel injection strategies in different LTC techniques (not to scale) (Adapted from [68, 69])

manifold by PFI system. In the HCCI combustion using PFS, most of the fuel is injected into the manifold and a direct injection event occurs in the range of 60°–35° before TDC. To create the partial stratification in HCCI combustion, single direct injection (HCCI-DI) and dual direct injection (D-DI) take place. In HCCI-DI PFS strategy, all the fuel is injected around 60° after intake TDC. However, in D-DI-PFS, first direction injection event occurs at 60° after intake TDC, while the second injection occurs 35° before TDC.

Premixed diesel injection strategies are also described in Fig. 4.12. Gasoline compression ignition (GCI) has two strategies of early and late injection. In RCCI combustion, gasoline is injected into the intake port, and diesel is directly injected in the cylinder. Diesel is injected by single or double direct injection as shown in Fig. 4.20. In DDFS strategy, both gasoline and diesel fuel are directly injected in the cylinder to create the desired level of fuel stratification and premixing.

References

1. Herold RE, Foster DE, Ghandhi JB, Iverson RJ, Eng JA, Najt PM (2007) Fuel unmixedness effects in a gasoline homogeneous charge compression ignition engine. Int J Engine Res 8 (3):241–257
2. Dec JE, Dernotte J, Ji C (2017) Increasing the load range, load-to-boost ratio, and efficiency of low-temperature gasoline combustion (LTGC) engines. SAE Int J Engines 10(2017-01-0731)
3. Maurya RK, Agarwal AK (2015) Combustion and emission characterization of n-butanol fueled HCCI engine. J Energy Resour Technol 137(1):011101
4. Maurya RK, Agarwal AK (2011) Experimental study of combustion and emission characteristics of ethanol fuelled port injected homogeneous charge compression ignition (HCCI) combustion engine. Appl Energy 88(4):1169–1180
5. Maurya RK, Agarwal AK (2009) Experimental investigation of the effect of the intake air temperature and mixture quality on the combustion of a methanol-and gasoline-fuelled homogeneous charge compression ignition engine. Proc Inst Mech Eng D J Automob Eng 223(11):1445–1458
6. Fiengo G, di Gaeta A, Palladino A, Giglio V (2013) Basic concepts on GDI systems. In: Common rail system for GDI engines. Springer, London, pp 17–33
7. Yu J, Abe M, Sukegawa Y (2017) A numerical method to simulate intake-port fuel distribution in PFI engine and its application (No. 2017-01-0565). SAE technical paper
8. Maurya RK, Agarwal AK (2014) Experimental investigations of performance, combustion and emission characteristics of ethanol and methanol fueled HCCI engine. Fuel Process Technol 126:30–48
9. Ryan TW, Callahan TJ (1996) Homogeneous charge compression ignition of diesel fuel (No. 961160). SAE technical paper
10. Suzuki H, Koike N, Ishii H, Odaka M (1997) Exhaust purification of diesel engines by homogeneous charge with compression ignition part 1: experimental investigation of combustion and exhaust emission behavior under pre-mixed homogeneous charge compression ignition method (No. 970313). SAE technical paper
11. McGee J, Curtis EW, Russ S, Lavoie G (2000) The effects of port fuel injection timing and targeting on fuel preparation relative to a pre-vaporized system (No. 2000-01-2834). SAE technical paper
12. Imatake N, Saito K, Morishima S, Kudo S, Ohhata A (1997) Quantitative analysis of fuel behavior in port-injection gasoline engines (No. 971639). SAE technical paper

13. Meyer R, Yilmaz E, Heywood JB (1998) Liquid fuel flow in the vicinity of the intake valve of a port-injected SI engine (No. 982471). SAE technical paper
14. Meyer R, Heywood JB (1997) Liquid fuel transport mechanisms into the cylinder of a firing port-injected SI engine during start up (No. 970865). SAE technical paper
15. Sukegawa Y, Oryoji K (2015) 3D modeling of particulate matter from spark ignition engines. SAE Int J Engines 8(2015-01-0391):419–425
16. Panão MRO, Moreira ALN (2007) Interpreting the influence of fuel spray impact on mixture preparation for HCCI combustion with port-fuel injection. Proc Combust Inst 31 (2):2205–2213
17. Claret J, Lauer T, Bobicic N, Posselt A, Schlerfer J (2017) Impact of the injection and gas exchange on the particle emission of a spark ignited engine with port fuel injection (No. 2017-01-0652). SAE technical paper
18. Knop V, Jay S (2006) Latest developments in gasoline auto-ignition modelling applied to an optical CAI (Tm) engine. Oil Gas Sci Technol 61(1):121–137
19. Gray AWB, Ryan TW (1997) Homogeneous charge compression ignition (HCCI) of diesel fuel (No. 971676). SAE technical paper
20. Christensen M, Hultqvist A, Johansson B (1999) Demonstrating the multi fuel capability of a homogeneous charge compression ignition engine with variable compression ratio (No. 1999-01-3679). SAE technical paper
21. Simescu S, Fiveland SB, Dodge LG (2003) An experimental investigation of PCCI-DI combustion and emissions in a heavy-duty diesel engine (No. 2003-01-0345). SAE technical paper
22. Ganesh D, Nagarajan G (2010) Homogeneous charge compression ignition (HCCI) combustion of diesel fuel with external mixture formation. Energy 35(1):148–157
23. Ganesh D, Nagarajan G, Ganesan S (2014) Experimental investigation of homogeneous charge compression ignition combustion of biodiesel fuel with external mixture formation in a CI engine. Environ Sci Technol 48(5):3039–3046
24. Dempsey AB, Curran SJ, Wagner RM (2016) A perspective on the range of gasoline compression ignition combustion strategies for high engine efficiency and low NOx and soot emissions: effects of in-cylinder fuel stratification. Int J Engine Res 17(8):897–917
25. Sjöberg M, Edling LO, Eliassen T, Magnusson L, Ångström HE (2002) GDI HCCI: effects of injection timing and air swirl on fuel stratification, combustion and emissions formation (No. 2002-01-0106). SAE technical paper
26. Dec JE, Yang Y, Dronniou N (2011) Boosted HCCI-controlling pressure-rise rates for performance improvements using partial fuel stratification with conventional gasoline. SAE Int J Engines 4(2011-01-0897):1169–1189
27. Sjöberg M, Dec JE (2006) Smoothing HCCI heat-release rates using partial fuel stratification with two-stage ignition fuels (No. 2006-01-0629). SAE technical paper
28. Yang Y, Dec JE, Dronniou N, Sjöberg M (2011) Tailoring HCCI heat-release rates with partial fuel stratification: comparison of two-stage and single-stage-ignition fuels. Proc Combust Inst 33(2):3047–3055
29. Dernotte J, Dec J, Ji C (2017) Efficiency improvement of boosted low-temperature gasoline combustion engines (LTGC) using a double direct-injection strategy (No. 2017-01-0728). SAE technical paper
30. Sellnau M, Sinnamon J, Hoyer K, Husted H (2011) Gasoline direct injection compression ignition (GDCI)-diesel-like efficiency with low CO2 emissions. SAE Int J Engines 4(2011-01-1386):2010–2022
31. Ra Y, Loeper P, Andrie M, Krieger R, Foster DE, Reitz RD, Durrett R (2012) Gasoline DICI engine operation in the LTC regime using triple-pulse injection. SAE Int J Engines 5(2012-01-1131):1109–1132
32. Sellnau MC, Sinnamon J, Hoyer K, Kim J, Cavotta M, Husted H (2013) Part-load operation of gasoline direct-injection compression ignition (GDCI) engine (No. 2013-01-0272). SAE technical paper

33. Ciatti S, Subramanian SN (2011) An experimental investigation of low-octane gasoline in diesel engines. J Eng Gas Turbines Power 133(9):092802
34. Sellnau M, Moore W, Sinnamon J, Hoyer K, Foster M, Husted H (2015) GDCI multi-cylinder engine for high fuel efficiency and low emissions. SAE Int J Engines 8(2015-01-0834):775–790
35. Sellnau MC, Sinnamon J, Hoyer K, Husted H (2012) Full-time gasoline direct-injection compression ignition (GDCI) for high efficiency and low NOx and PM. SAE Int J Engines 5 (2012-01-0384):300–314
36. López JJ, García-Oliver JM, García A, Domenech V (2014) Gasoline effects on spray characteristics, mixing and auto-ignition processes in a CI engine under partially premixed combustion conditions. Appl Therm Eng 70(1):996–1006
37. Tang Q, Liu H, Li M, Yao M, Li Z (2017) Study on ignition and flame development in gasoline partially premixed combustion using multiple optical diagnostics. Combust Flame 177:98–108
38. Manente V, Tunestal P, Johansson B, Cannella WJ (2010) Effects of ethanol and different type of gasoline fuels on partially premixed combustion from low to high load (No. 2010-01-0871). SAE technical paper
39. Manente V, Johansson B, Tunestal P, Cannella W (2009) Effects of different type of gasoline fuels on heavy duty partially premixed combustion. SAE Int J Engines 2(2009-01-2668):71–88
40. Desantes JM, Pastor JV, Payri R, Pastor JM (2005) Experimental characterization of internal nozzle flow and diesel spray behavior. Part II: Evaporative conditions. Atomization Sprays 15 (5):517–544
41. Manente V, Johansson B, Tunestal P (2009) Partially premixed combustion at high load using gasoline and ethanol, a comparison with diesel (No. 2009-01-0944). SAE technical paper
42. Najafabadi MI, Tanov S, Wang H, Somers B, Johansson B, Dam N (2017) Effects of injection timing on fluid flow characteristics of partially premixed combustion based on high-speed particle image velocimetry. SAE Int J Engines 10 (2017–01-0744):1443–1453
43. Yanagihara H (2001) Ignition timing control at Toyota "Unibus" combustion system. In: Proceedings of a new generation of engine combustion processes for the future, Editions Technip, Peris, France pp 35–42
44. Pastor J, Luján J, Molina S, García JM (2007) Overview of HCCI diesel engines. In: HCCI and CAI engines for the automotive industry. Woodhead Publishing Limited. Cambridge, England
45. Wåhlin F (2007) Experimental investigation of impinging diesel sprays for HCCI combustion. Doctoral dissertation, KTH
46. Takeda Y, Keiichi N, Keiichi N (1996) Emission characteristics of premixed lean diesel combustion with extremely early staged fuel injection (No. 961163). SAE technical paper
47. Hasegawa R, Yanagihara H (2003) HCCI combustion in DI diesel engine (No. 2003-01-0745). SAE technical paper
48. Hashizume T, Miyamoto T, Hisashi A, Tsujimura K (1998) Combustion and emission characteristics of multiple stage diesel combustion (No. 980505). SAE technical paper
49. Yokota H, Kudo Y, Nakajima H, Kakegawa T, Suzuki T (1997) A new concept for low emission diesel combustion (No. 970891). SAE technical paper
50. Su W, Lin T, Pei Y (2003) A compound technology for HCCI combustion in a DI diesel engine based on the multi-pulse injection and the BUMP combustion chamber (No. 2003-01-0741). SAE technical paper
51. Aoyagi Y (2007) HCCI combustion with early and multiple injections in the heavy-duty diesel engine. In: HCCI and CAI engines for the automotive industry. Woodhead Publishing Limited. Cambridge, England
52. Kim MY, Lee CS (2007) Effect of a narrow fuel spray angle and a dual injection configuration on the improvement of exhaust emissions in a HCCI diesel engine. Fuel 86(17):2871–2880
53. Agarwal AK, Singh AP, Maurya RK (2017) Evolution, challenges and path forward for low temperature combustion engines. Prog Energy Combust Sci 61:1–56

54. Gan S, Ng HK, Pang KM (2011) Homogeneous charge compression ignition (HCCI) combustion: implementation and effects on pollutants in direct injection diesel engines. Appl Energy 88(3):559–567
55. Walter B, Gatellier B (2003) Near zero NOx emissions and high fuel efficiency diesel engine: the NADITM concept using dual mode combustion. Oil Gas Sci Technol 58(1):101–114
56. Kimura S, Aoki O, Ogawa H, Muranaka S, Enomoto Y (1999) New combustion concept for ultra-clean and high-efficiency small DI diesel engines (No. 1999-01-3681). SAE technical paper
57. Stanglmaier RH, Roberts CE (1999) Homogeneous charge compression ignition (HCCI): benefits, compromises, and future engine applications (No. 1999-01-3682). SAE technical paper
58. Yao M, Zheng Z, Liu H (2009) Progress and recent trends in homogeneous charge compression ignition (HCCI) engines. Prog Energy Combust Sci 35(5):398–437
59. Curran S, Gao Z, Wagner R (2014) Reactivity controlled compression ignition drive cycle emissions and fuel economy estimations using vehicle systems simulations with E30 and ULSD. SAE Int J Engines 7(2014-01-1324):902–912
60. Nazemi M, Shahbakhti M (2016) Modeling and analysis of fuel injection parameters for combustion and performance of an RCCI engine. Appl Energy 165:135–150
61. Benajes J, García A, Monsalve-Serrano J, Balloul I, Pradel G (2016) An assessment of the dual-mode reactivity controlled compression ignition/conventional diesel combustion capabilities in a EURO VI medium-duty diesel engine fueled with an intermediate ethanol-gasoline blend and biodiesel. Energy Convers Manag 123:381–391
62. Benajes J, Pastor JV, García A, Monsalve-Serrano J (2015) The potential of RCCI concept to meet EURO VI NOx limitation and ultra-low soot emissions in a heavy-duty engine over the whole engine map. Fuel 159:952–961
63. Curran SJ, Hanson RM, Wagner RM (2012) Reactivity controlled compression ignition combustion on a multi-cylinder light-duty diesel engine. Int J Engine Res 13(3):216–225
64. Splitter D, Wissink M, DelVescovo D, Reitz RD (2013) RCCI engine operation towards 60% thermal efficiency (No. 2013-01-0279). SAE technical paper
65. Benajes J, García A, Monsalve-Serrano J, Boronat V (2017) Achieving clean and efficient engine operation up to full load by combining optimized RCCI and dual-fuel diesel-gasoline combustion strategies. Energy Convers Manag 136:142–151
66. Hariprasad T (2013) Effect of injection pressure on performance of dual fuel diesel engine (No. 2013-01-2887). SAE technical paper
67. Wissink M, Reitz RD (2015) Direct dual fuel stratification, a path to combine the benefits of RCCI and PPC. SAE Int J Engines 8(2015-01-0856):878–889
68. Zhang Y (2017) Comparisons of particulate size distributions from multiple combustion strategies. Doctoral dissertation, The University of Wisconsin-Madison
69. Ohsawa K, Kamimoto T (2008) Advanced diesel combustion. In: Flow and combustion in reciprocating engines. Springer, Berlin/Heidelberg, pp 353–380

Chapter 5
Combustion Control Variables and Strategies

Abstract Low temperature combustion (LTC) engines need different enabling technologies depending on the fuel and strategy used to achieve combustion of the premixed fuel–air mixture. Controlling the combustion rate is one of the major challenges in LTC engines, particularly in HCCI combustion engine. To achieve higher thermal efficiency, the desired phasing of combustion timings is essential even at moderate combustion rates. Present chapter describes the combustion control variables and control strategies used for LTC engines. Various methods demonstrated to control the LTC engines can be categorized in to two main strategies: (i) altering pressure–temperature history and (ii) altering fuel reactivity of the charge. Temperature history of the charge in the cylinder can be altered by several parameters such as intake conditions (temperature and pressure), EGR, variable valve timing (VVT), variable compression ratio (VCR), water injection, supercharging and fuel injection strategies. Fuel reactivity of charge in the cylinder can be altered by various parameters such as equivalence ratio (Φ), fuel stratification, fuel additives, ozone additions and dual fuel. All these combustion control strategies are discussed for utilizing gasoline-like fuels in HCCI, PPC and RCCI combustion mode engines.

Keywords Combustion control • HCCI • EGR • PPC • NVO • Dual fuel • Additives • VVT • VCR • RCCI • LTC

5.1 Altering Time Temperature History

Time–temperature history of the charge in the cylinder determines the combustion timings, which essentially governs the combustion rate and thermal efficiency. In different LTC strategies, charge temperature history is controlled to achieve desired combustion phasing and extend the engine operating load. Temperature and pressure history of charge in the cylinder is correlated by gas equations. The time–temperature history of the charge is affected by several engine operating parameters discussed in the following subsections.

R.K. Maurya, *Characteristics and Control of Low Temperature Combustion Engines*, Mechanical Engineering Series, https://doi.org/10.1007/978-3-319-68508-3_5

5.1.1 Intake Thermal Management

Appropriate thermodynamic and chemical in-cylinder conditions are required in a HCCI combustion close to TDC position to achieve the autoignition at desired combustion phasing. Additionally, the air–fuel mixture in HCCI combustion needs to be sufficiently dilute to keep combustion rates and maximum charge temperature low enough to achieve acceptable ringing intensity (see Chap. 7) and NO_x emissions. Fuel typically used in HCCI combustion is gasoline-like fuels, which have higher autoignition temperatures. Intake preheating of charge is often used to enable the autoignition and phase the HCCI combustion at adequate crank position. Basic idea is to increase the temperature of charge at intake valve closing (IVC), where compression of charge starts. This temperature can be increased by increasing intake temperature or trapping the hot residual gases in the cylinder by negative valve overlap (NVO). Since in HCCI combustion engine is mainly kinetically controlled, it has the sensitivity to the charge temperature. Several techniques are used to increase the initial temperature of the charge at compression start position such as electrical preheating, exhaust heat exchanger, glow plugs etc. [1–4].

Figure 5.1 shows the required inlet temperature at IVC position as a function of engine speed and load in naturally aspirated HCCI engine using ethanol for different compression ratios. Higher intake temperature advances the combustion phasing, which increases the pressure rise rate drastically and sometimes above acceptable limits. Lower intake temperature retards the combustion phasing, which may leads to partial burn or misfire due to low combustion temperature. The lower temperature at retarded combustion phasing leads to the lower combustion efficiency. Operating limits shown in Fig. 5.1 used the acceptable pressure rise rate <5 MPa/ms and combustion efficiency >85%. At particular engine load (IMEP), the temperature shown in Fig. 5.1 corresponds to most efficient inlet temperature (T_{IVC} with highest thermal efficiency). Figure depicts that higher T_{IVC} is required at lower IMEP for every engine speed at given compression ratio. Higher inlet temperature required is due to leaner mixture operation (lower reactivity) at lower IMEP conditions. Richer engine operating conditions require relatively lower intake temperature. For higher engine load conditions, comparatively lower T_{IVC} is required at higher compression ratio (Fig. 5.1). The higher compression ratio increases the compression temperature required for autoignition. Outside the operating range (Fig. 5.1), at higher intake temperature, the volumetric and thermal efficiency of the engine is very low. Intake temperature is required to be adjusted along with change in engine speed and load conditions as shown in Fig. 5.1.

Intake temperature control is usually thought as too slow response parameter for closed-loop control of transient HCCI combustion. The most direct implementation of intake temperature control is an electric heating element in the intake air path. The thermal inertia of the heating element itself causes a time constant of the order of seconds, which is indeed too slow for closed-loop control of transient HCCI combustion (as engine cycle typically completes in few milliseconds) [6]. A more sophisticated solution to this problem is fast thermal management (FTM) [7]. Fast thermal management can be used to control the temperature of the mixture at the

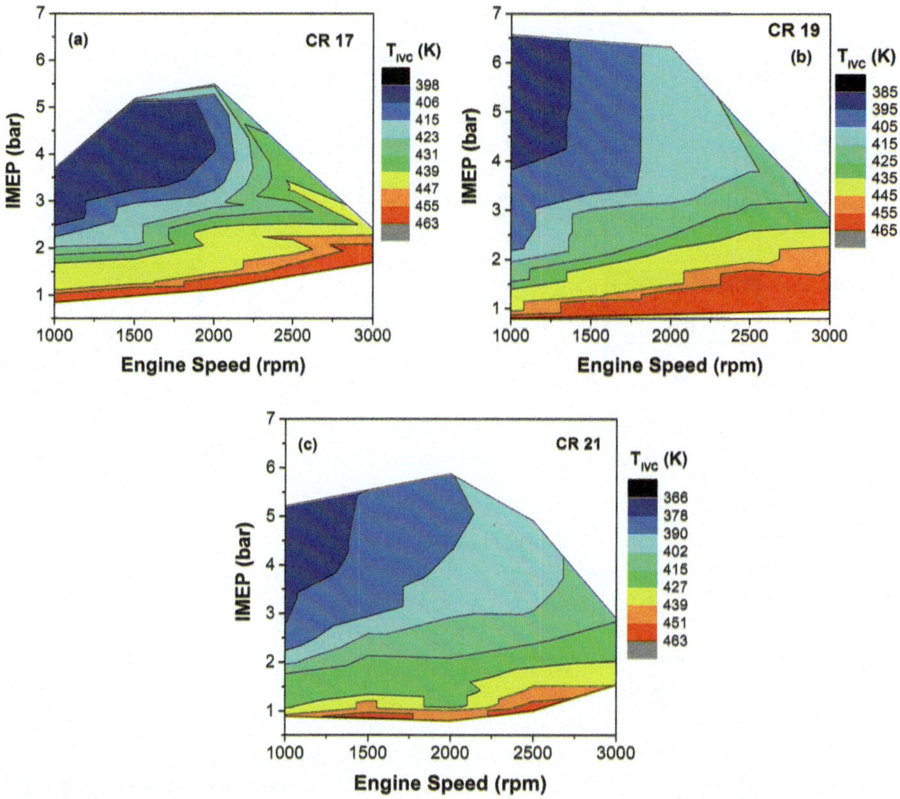

Fig. 5.1 Engine T_{IVC} (inlet temperature) as function of engine load and speed in ethanol HCCI combustion at different compression ratios [5]

beginning of compression stroke. The inlet temperature strongly affects the combustion phasing. It is demonstrated that thermal management could be used for HCCI control in a wide range of operating conditions [8]. Figure 5.2 shows the schematic of FTM system used for control of HCCI combustion. To achieve fast thermal management control of the inlet temperature, a source of cold ambient air and a source of hot air (heated electrically or by exhaust heat recovery) is used. By controlling the valves of the hot and cold airflows, cycle-to-cycle control of the inlet temperature can be achieved. The benefit of this method is that it does not require major engine modifications or use of fuel additives. The major drawback of FTM is that the engine system has to be fitted with a heater (electrical or exhaust heat exchanger) [10]. The mixing of hot can cold air can be inside or outside the cylinder (Fig. 5.2) by independently controlling the valves.

To avoid the electric heater for intake preheating in HCCI combustion, heat exchangers are used to extract the heat from engine exhaust. Figure 5.3 illustrates the system used for extracting heat from engine exhaust and coolant to preheat the intake air. At lower engine loads, the difference between exhaust gas temperature

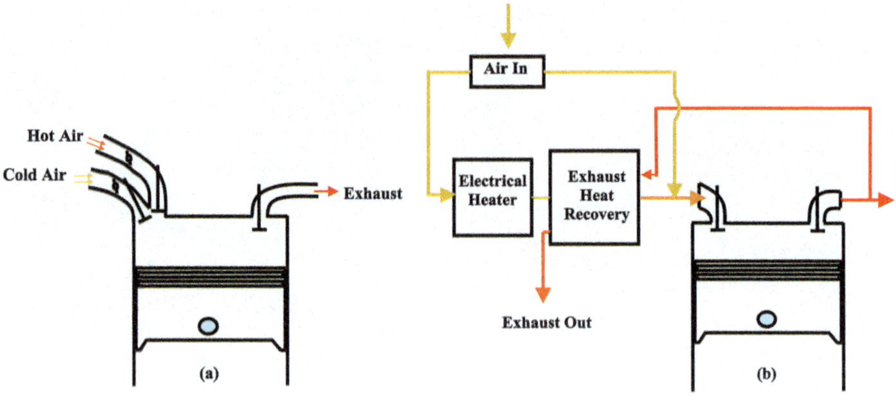

Fig. 5.2 (**a**) Schematic of heated and cold air arrangement in engine with two intake valve (Adapted from [9]), and (**b**) schematic of FTM system (Adapted from [7])

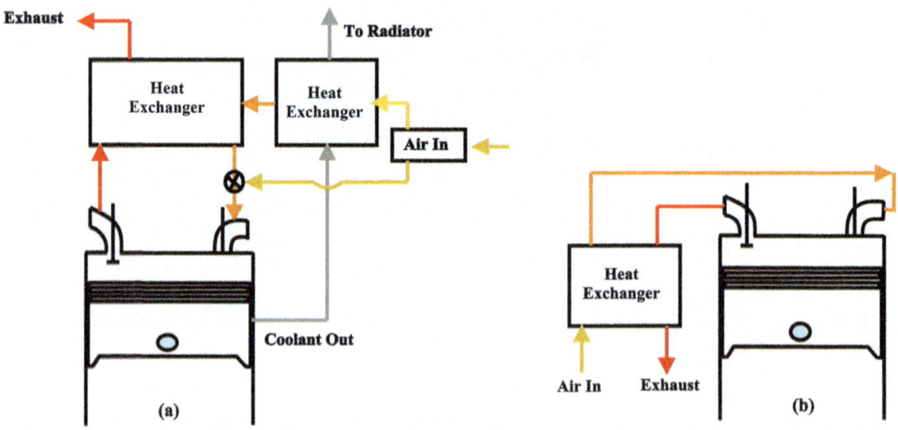

Fig. 5.3 Schematic of air preheating system by using heat exchanger taking heat input from (**a**) both exhaust and coolant (Adapted from [11]), and (**b**) exhaust (Adapted from [4])

and intake air temperature can become insufficient for intake preheating. Intake throttling is necessary to compensate for this temperature difference. At lower engine load, exhaust temperature is also lower, and typically, higher intake temperature requirement is also higher at low engine loads. Figure 5.4 shows the interaction between inlet air temperature, combustion phasing and exhaust gas temperature in HCCI combustion with exhaust heat recovery. Higher inlet temperature leads to advanced combustion phasing, which results in lower exhaust temperature. Lower exhaust temperature leads to the decrease in intake temperature resulting into delayed combustion timings. Delayed combustion timings then increase the exhaust temperatures (Fig. 5.4). The exhaust heat recovery-based system shown in Fig. 5.3 is not typically operated during engine start-up [4, 12]. In general idea of using heat exchanger may have difficulty in adjusting the intake

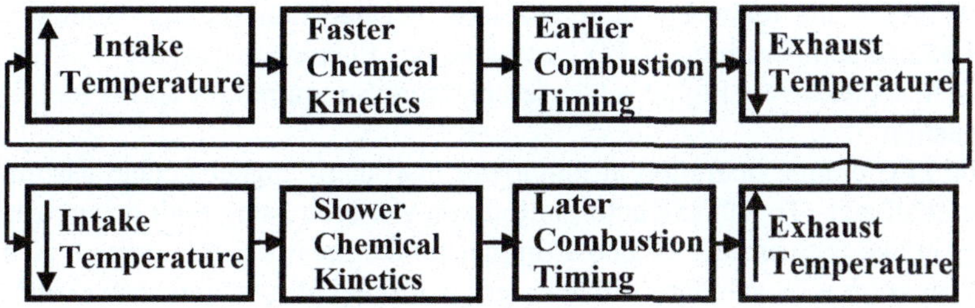

Fig. 5.4 Interaction between inlet air temperature, combustion phasing and exhaust gas temperature in HCCI combustion with exhaust heat recovery [12]

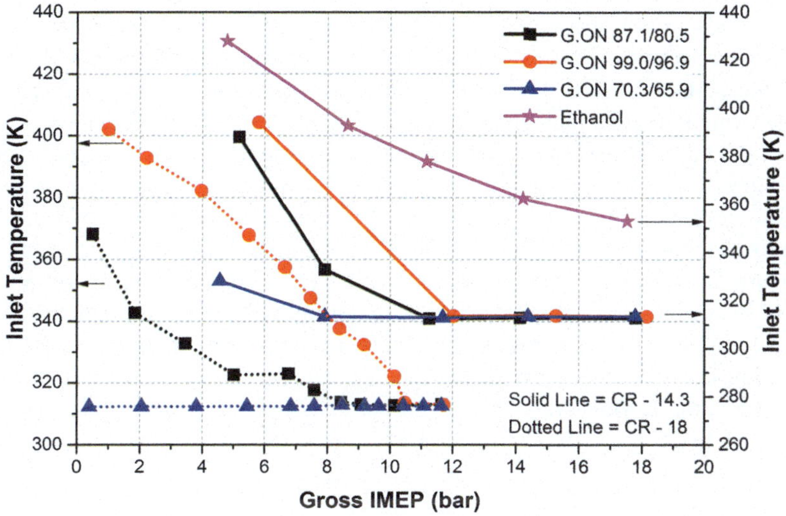

Fig. 5.5 Intake temperature as a function of engine load and fuel type in PPC engine for different compression ratios (Adapted from [13, 14])

temperature rapid enough to meet the control requirement during transient operation due to the thermal inertia of heat process. The FTM is presented as an acceptable alternative of exhaust heat recovery method.

Intake temperature requirement depends on the engine load as well as combustion strategy. Figure 5.5 shows the intake temperature requirements in PPC engine using different gasoline-like fuels at 1300 rpm for two compression ratio. Fuels designated as G.ON x/y represents gasoline octane number, where 'x' is RON, and 'y' is MON. Figure depicts that intake temperature requirements decrease with increase in engine loads. To achieve higher engine loads, typically higher intake pressure is employed. Gasoline becomes Φ sensitive at higher intake pressure (see Chap. 2), and stratification can be used to increase the load range. In PPC engine, heavy stratification of gasoline is used, and thus, lower intake temperature is required at higher engine load. At higher engine load (>10 bar IMEP), intake temperature requirement is close to ambient temperature (Fig. 5.5). The lower octane fuel is the relatively more reactive fuel and thus requires comparatively lower intake temperature for PPC combustion. Figure also depicts that at lower engine loads relatively higher intake temperature is required for all the fuels in PPC combustion.

The PPC engine operation at idle/lower load using higher octane gasoline is a major challenge. To achieve autoignition at lower load using high octane gasoline, higher intake temperature or boost (higher intake pressure) is required. Available boost pressure with standard turbocharger is limited at low engine load and speed conditions. Higher intake temperature requirement at lower engine load is also illustrated in Fig. 5.5. The usefulness of negative valve overlap (NVO) is investigated in PPC engine at lower engine load conditions [15]. The purpose of NVO is to trap hot residual gases in the cylinder for increasing the charge temperature to enable auto-ignition while using high octane gasoline. At lower engine load, exhaust temperature is also relatively lower, and thus, trapping residual gases of lower temperature results into relatively lower charge temperature. To increase the charge temperature for facilitating autoignition, higher amount of residual gases by larger NVO is required. Increased NVO also increases the EGR (internal) fraction in the cylinder. Higher amount of EGR leads to lower charge temperature by increasing specific heat of the charge [15]. Hence, there exists optimum NVO duration that can be used to operate the PPC engine at lower load condition.

Figure 5.6 shows the combustion instability (represented by standard deviation of IMEP) and ignition delay variation at different NVO operating conditions with and without the use of glow plug. Glow plug is used in combination with NVO, and glow plug is continuously on during the experiment. Figure 5.6 depicts that glow plug operation lowers standard deviation in IMEP (improves the combustion

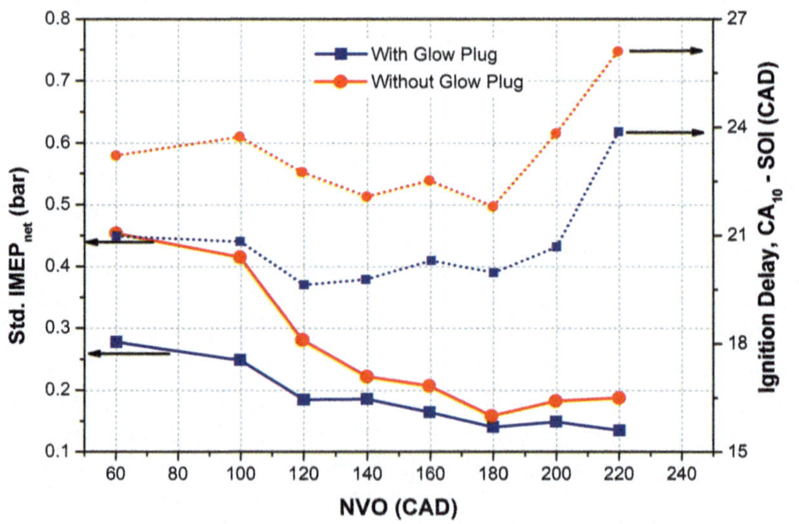

Fig. 5.6 Standard deviation of IMEP and ignition delay variations (with and without glow plug) in PPC engine using gasoline (Adapted from [15])

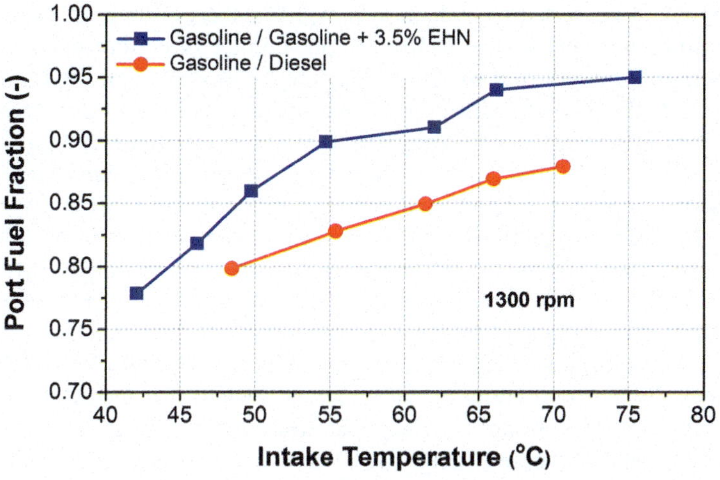

Fig. 5.7 Intake temperature requirement in RCCI engine with port fuel fraction (Adapted from [16])

instability) and reduces the ignition delay due to additional increase in charge temperature by glow plug. Study [15] showed that the lowest ignition delay is observed from approximately 120 CAD up to 200 CAD NVO (corresponding residual gas fraction is 20–45%). The NVO duration of 180–200 CAD (40–45% residual gas fraction) resulted into shortest combustion duration in PPC mode. This is the result of several factors such fuel, EGR and temperature distribution in addition to global charge temperature. Considering a limiting value of standard deviation on IMEP for minimum achievable load, the low load limit of PPC engine can be extended from 3.2 bar IMEP down to 2.2 bar IMEP by increasing NVO from 60 CAD to 180 CAD [59].

The RCCI combustion engine requires relatively lower intake temperature than other LTC strategies. In RCCI combustion, typically a high reactivity fuel is injected to enable the combustion in the chamber. The intake temperature requirement depends on the amount of low reactivity fuel injected in the cylinder. Figure 5.7 shows the variation of intake temperature with port injected fuel fraction in RCCI engine operation with gasoline/diesel and gasoline/gasoline +EHN. Intake temperature requirement increases with increase in the low reactivity fuel to enhance the autoignition. Similarly, intake temperature requirement is higher in single fuel RCCI with additives (EHN) as overall low reactivity fuel increases.

5.1.2 Exhaust Gas Recirculation

Exhaust gas recirculation (EGR) is a well-established method for facilitating control over ignition and combustion phasing, which affects the combustion performance in CI engines [17]. The EGR is broadly divided in two category (internal and external EGR) based on induction method in the cylinder. In external EGR (eEGR), a faction of exhaust gases from engine tailpipe is recirculated back into the intake manifold by typically passing through an EGR cooler. However, in case of internal EGR (iEGR), hot residual gases are trapped in the cylinder by changing the valve timings. The effect of EGR in HCCI engine is discussed in Sect. 2.2.4.5 of Chap. 2. The use of EGR changes the charge composition and charge mixture properties as the thermodynamic and chemical properties of EGR are significantly different than air. EGR typically consists of significant amount of complete combustion products (CO_2 and H_2O) and a large variety of incomplete combustion products, some of which is trace quantity. The EGR as a diluent has the potential to significantly alter and affect combustion characteristics including burn rates, ignition timing, combustion and engine thermal efficiencies. Effect of EGR and its constituents on ignition timing for gasoline and PRF80 fuel in HCCI combustion is shown in Fig. 2.22 (Chap. 2). Chemical effect of incomplete combustion production on the start of combustion in HCCI combustion is also shown in Fig. 6.29b (Chap. 6). The EGR is also used to increase the engine operating load range of HCCI combustion (Fig. 2.23 of Chap. 2).

In internal EGR, trapping of hot residuals from the previous cycle is achieved by the using flexible valve trains. Desired amount of internal EGR is obtained by adjusting the valve lift and valve timings (typically NVO) depending on engine operating conditions. In NVO strategy, the exhaust valve is closed early and a fraction of exhaust gas trapped in the cylinder undergo “recompression” due motion of piston towards TDC. The trapped residual gases subsequently mix with the cooler fresh incoming charge. Mixing of the fresh charge with the hot residual gases leads to overall increase in the charge temperature, which facilitate to achieve autoignition in HCCI combustion using high octane fuels [18]. In conventional engines, in contrast to NVO, the positive valve overlap (PVO) is typically used where both the exhaust and intake valves are open simultaneously for a short time period around the intake TDC position. Figure 5.8 illustrates the PVO and NVO valve lift profiles along with corresponding typical cylinder pressure trace. The PVO shown in figure has negligible overlap between intake and exhaust events, but it can vary depending on engine speed. Figure 5.8 also shows that advance in exhaust valve closing (EVC) timing is complemented by a corresponding and equal retard in intake valve open (IVO) timing in case of the NVO, and this strategy is called “symmetric” NVO. This strategy is employed to minimize the pumping losses related to recompression.

Another strategy for internal EGR is known as exhaust reinduction or rebreathing. In exhaust reinduction, fraction of exhaust gases from the previous engine cycle are inducted back into the cylinder during the intake stroke along with fresh air by a short exhaust valve event during the intake stroke. A study demonstrated the viability of reinduction HCCI and its potential for effective combustion phasing control [20]. The sensible energy of exhaust inducted in the cylinder is utilized to initiate the HCCI combustion. However, one of the main challenges with reinduction HCCI is the cylinder-to-cylinder coupling, which occurs by mixing of exhaust from different cylinders in the exhaust manifold and mixed exhaust is

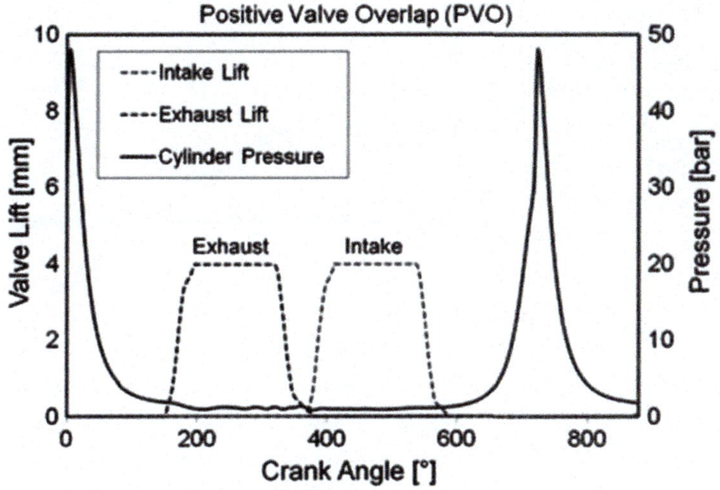

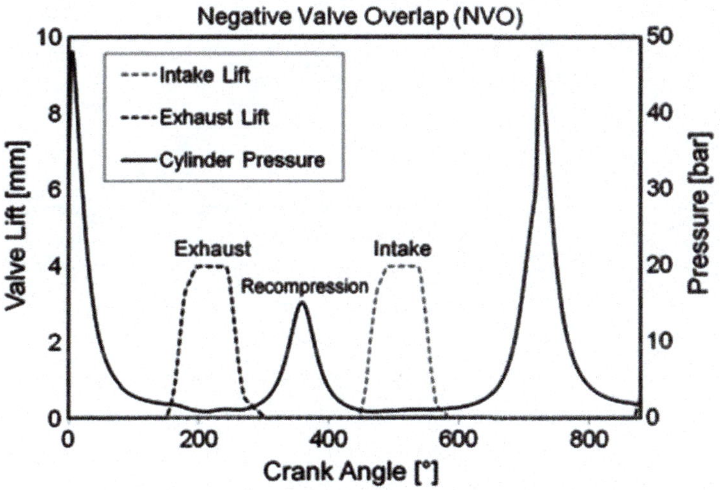

Fig. 5.8 Illustration of PVO and NVO valve lift profiles along with typical cylinder pressure curve [19]

reinducted into the cylinders. This mixing couples the contents from one cylinder to the contents of the other cylinders of the engine [21]. A model and control strategies are developed to handle the cylinder-to-cylinder coupling concerns accompanied with exhaust reinduction HCCI [22]. However, exhaust recompression (NVO strategy) does not have cylinder to cylinder coupling issues altogether.

The HCCI combustion is significantly affected by the homogeneity of EGR. The HCCI engine tends to tolerate more heterogeneously distributed EGR. For the same EGR level, power out can be increased by heterogeneous EGR as it lowers the pressure rise rate due to thermal stratification. Inhomogeneity among fresh charge and residual gas is favourable condition because temperature in EGR rich regions of

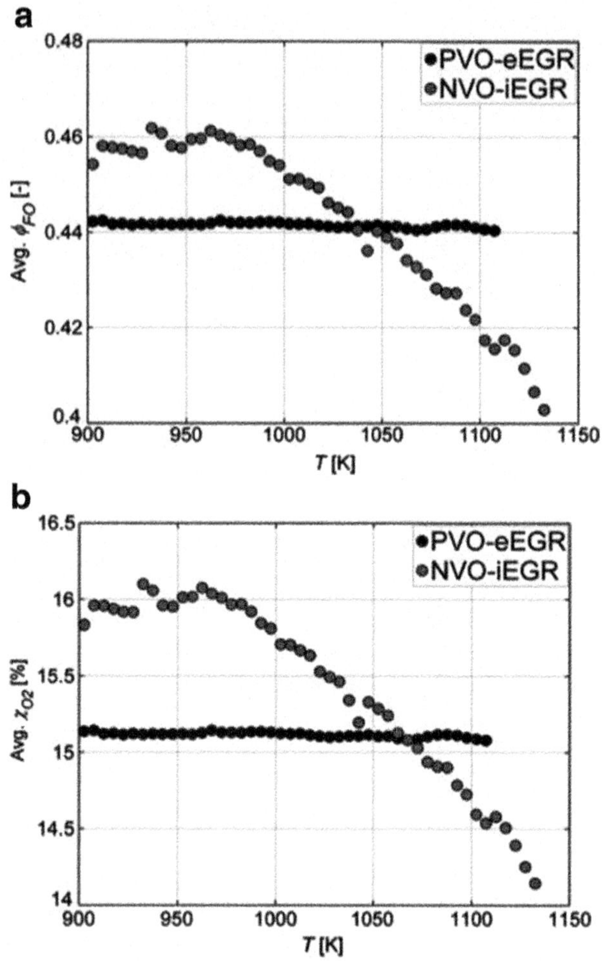

Fig. 5.9 Pre-ignition (10° CA bTDC) distribution of the average Φ_{FO} and χ_{o2} within a zone versus charge temperature (T) for PVO-eEGR and NVO-iEGR [19]

the chamber is higher than those attained in completely homogeneous in-cylinder charge. The small inhomogeneity could be beneficial in reducing heat release rate (see Sect. 2.2.4 of Chap. 2) [23].

Figure 5.9 illustrates the mean values of fuel oxygen equivalence ratio (Φ_{FO}) and oxygen mole fraction (χ_{o2}) as a function of charge temperature for understanding the relation between the thermal and compositional stratification for internal and external EGR conditions. This is determined by computing the average values of Φ_{FO} and χ_{o2} of all CFD (computational fluid dynamics) cells within a particular temperature range. The figure depicts that there is insignificant variation in composition (Φ_{FO} and χ_{o2}) with respect to temperature in the PVO case with external EGR. However, in case the of NVO internal EGR, both Φ_{FO} and χ_{o2} are negatively correlated to the stratification in temperature. The hotter zones have a lower Φ_{FO} and lower χ_{o2} than the colder regions, which have higher fuel to oxygen equivalence ratios and less residual dilution [19].

A study showed that the composition of the EGR gases is different from that of the engine exhaust gases due to an evolution inside the EGR pipe resulting from several mechanisms: fragmentation, conglomeration and chemical reactions for unburned hydrocarbons, oxygenates and NO_x [24]. A weak evolution of CO, CO_2 and O_2 concentration is observed in the EGR pipe. Conglomeration of some unburned hydrocarbons is found, which could not be interesting to control the combustion because PAHs are not very reactive and present in the EGR gases in very low concentration. Presence of aldehydes and ketones could impact HCCI combustion phasing because of their high reactivity [24].

EGR has thermal and chemical effect on HCCI combustion. However, EGR is used in HCCI combustion mainly for thermal effect to reduce the charge temperature in the cylinder. The EGR dilution also reduces the oxygen content in the charge, and at higher engine loads, oxygen deficiency limits can also be encountered (see Chap. 7). At higher engine loads, very high amount of EGR is required to control the heat release rate. A PCCI combustion system used 54% EGR to retard the combustion phasing and improve the obtained IMEP [25]. Another premixed charge combustion study used high EGR rates up to 68% to effectively control the start of combustion [26]. However, higher EGR amount decreases the thermodynamic efficiency, and it also creates problems in engine transient response and temperature-stability characteristics. Current research on LTC strategies are trying to reduce the quantity of EGR required to control the heat release rate at higher engine loads.

Figure 5.10 presents the EGR flow rate as a function of engine load and fuel type in PPC engine for different compression ratios. At higher compression ratio (18), EGR flow rate around 50% is required for higher engine load, and low octane fuel has lower requirement of EGR. At lower compression ratio, lower octane gasoline

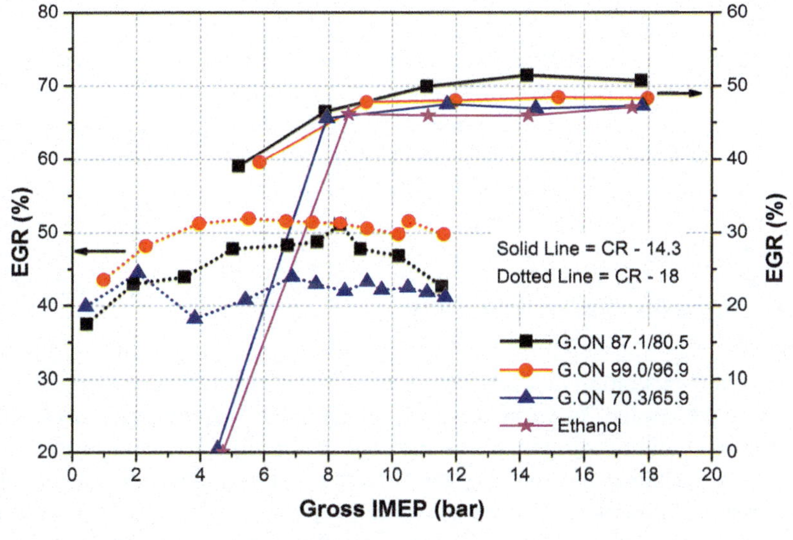

Fig. 5.10 EGR rate as a function of engine load and fuel type in PPC engine for different compression ratios (Adapted from [13, 14])

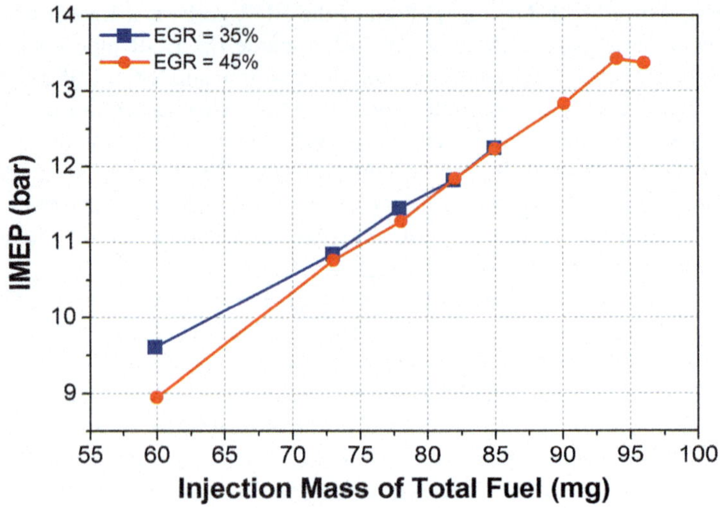

Fig. 5.11 IMEP as a function of direct injection fuel mass (at fixed premixed ratio 85%) in RCCI engine for two EGR flow rates (Adapted from [27])

and ethanol have lower EGR requirement for low load conditions. The EGR requirement depends on fuel, compression ratio, injection timings and intake pressure used for PPC combustion. The EGR rate shown in Fig. 5.10 is for injection strategy (two direct injection per cycle) developed in reference [13].

The RCCI combustion engine requires relatively lower EGR rate because the combustion rate is controlled by varying the reactivity of charge by using two fuels with significantly different reactivity. Figure 5.11 shows the IMEP achieved with two different EGR flow rates at fixed premixing ratio of 85%. Figure 5.11 depicts that more than 12 bar IMEP can be achieved with 35% EGR flow rate, while around 50% EGR rate is required in PPC combustion for the same IMEP (Fig. 5.10). Typically, in the optimized RCCI operation conditions require the EGR assistance, particularly at higher engine load to control the pressure rise rate, and up to 50% may be required depending on engine load and fuel.

5.1.3 Variable Valve Actuation

Variable valve actuation (VVA) offers a very fast method of varying the breathing process of the engine. In variable valve actuation strategy, mainly variable valve timing (VVT) and/or variable valve lift (VVL) is used to (i) control the residual gases in the cylinder and (ii) vary the effective compression ratio of the engine. Compression ratio control can be achieved over a wide range of HCCI operating conditions as the compression ratio strongly affects the combustion phasing. The HCCI engine has typically higher compression ratio and can obtain lower compression ratios by delaying the IVC during the compression stroke. For HCCI control purposes, two major methodologies, residual gas control and effective compression ratio control, are used very often [6]. A fully flexible VVA also offers the advantage of changing the temperature and composition of charge by rebreathing the residual gases from the previous cycle into the cylinder. With a full VVA system, the timing for the IVO, IVC, exhaust valve opening (EVO) and EVC can be changed to any desired timings [10].

Effect of different VVA strategies is investigated in premixed diesel combustion for their performance and cost/complexity [28]. Study compared several configurations such as VVTe (variable valve timing exhaust), VVTei (variable valve timing exhaust and intake), VVLei (variable valve lift exhaust and intake), DEe (double event of exhaust valves during intake phase) and SEe (single event of exhaust valve during intake phase) using VVA in premixed diesel combustion (Fig. 5.12). First three strategies (VVTe, VVTei and VVLei) are exhaust recompression strategy, and last two (DEe and SEe) are exhaust reinduction strategy to control the residual gases in the cylinder. In VVTe configuration, only the timing of exhaust valve is advanced while maintaining the same lift and duration, to trap the residual gases. In case of VVTei, valve timing is adjusted for both intake and exhaust valve symmetrically. In case of VVLei, IVC and EVO timings are kept fixed, while IVO and EVC timings are adjusted along with valve lifts. Reduction in effective compression ratio is avoided by keeping the fixed timing of IVC. Fixed EVO timing remove very early opening of the exhaust valve, which avoids the reduction in the expansion phase of the engine cycle. SEe and DEe are the two configurations used for reinduction of exhaust gases in

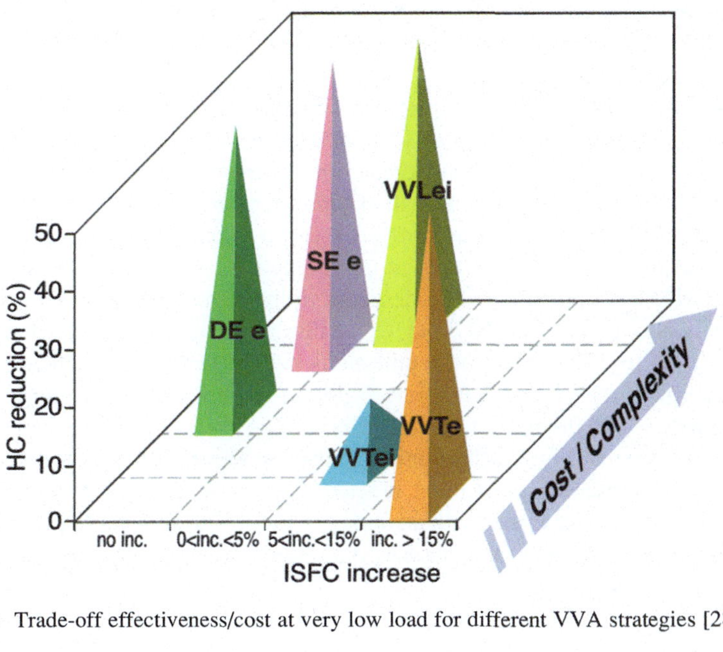

Fig. 5.12 Trade-off effectiveness/cost at very low load for different VVA strategies [28]

the cylinder by reopening of single and both (double) exhaust valves, respectively, during the intake stroke of the engine cycle [28].

Utilization of internal EGR reduces the CO and HC emissions at lower engine loads by increasing the charge temperature. Figure 5.12 compares different VVA configurations with respect to effectiveness in reducing HC emissions versus increase in indicated specific fuel consumption (ISFC) along with cost/complexity of the system with reference to fixed valve actuation (commercial configuration). The VVTe configuration is effective in decreasing HC emissions (50%) and relatively lower complexity, but it increases the ISFC (>15%) because of the early opening of exhaust valve leading to reduction in expansion work. The VVTei is slightly better in ISFC but relatively lower HC reduction in this strategy. In case of reinduction strategy, there is no reduction in the expansion phase. The double event of exhaust (DEe) during intake appears to provide the best trade-off between reduction in HC emission, ISFC increase and cost of the tested configuration [28]. A more complete discussion can be found in the original study.

Technological limitations of air path (compressor behaviour at the high-pressure ratio and low airflow, EGR rate vs. turbine energy, EGR cooling capacity and circuit permeability) and combustion limitation (torque reserve at high equivalence ratio, particulate emissions and combustion noise) limit the engine load range of highly premixed combustion [28]. Compression ratio of engine affects the charge temperature, and thus, NO_x and noise emissions are higher at the higher compression ratio. Reduction in compression ratio leads to increase in engine load range. The effective compression ratio can be reduced by varying the IVC timings because effective compression ratio of the engine is directly related to the cylinder volume at IVC. The compression affects only the cylinder volume at IVC and not the entire engine displacement volume. Figure 5.13 shows the summary of trade-off observed in reduction of effective compression ratio using VVA by presenting positive as well as negative impacts.

The reduction in effective compression ratio must have a positive effect on NOx emission due to decrease in pressure and temperature in the engine cylinder. However, a major effect is the decrease in volumetric efficiency resulting into decrease in EGR rate at particular intake pressure and equivalence ratio, which leads to higher NOx. Similar trade-off is observed for smoke emissions. Reduction in compression ratio increases the ignition delay thus more premixed combustion leads to reduction in smoke, while decrease in the volumetric efficiency produces opposite result. In case of HC and CO emissions, the combustion temperature increase leads to decrease in these emissions but the lower effective compression ratio also lower the charge temperature during injection phase leading to lower vaporization, which creates a negative effect.

Different types of VVA technologies are used in the engine such as cam phasing, piezoelectric, hydraulic, permanent magnet, electromagnetic and thier combinations [29]. A summary of the main advantages and disadvantages of different VVA technologies are provided in Table 5.1. A more complete discussion can be found in the original study [29].

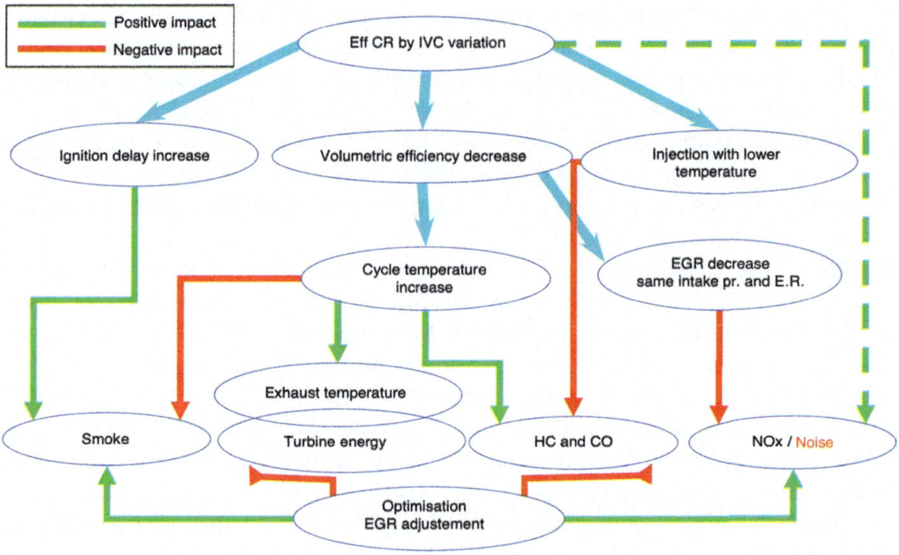

Fig. 5.13 Summary of trade-off observed in reduction of effective compression ratio using VVA strategy [28]

Table 5.1 Advantages and disadvantages of different VVA technologies [29]

Technology	Advantages	Disadvantages
Cam phasing	Reliability; variable valve lift	Cycle-by-cycle actuation only at low engine speed
Piezoelectric	Fully variable timing	Small displacements; complicated control
Hydraulic	Fully variable timing; soft landing; variable valve lift	Space requirements; temperature sensitivity; precise/expensive valves; high-power consumption
Permanent magnet and motors	Fully variable timing	Low force generation; weight; low valve speeds
Electromagnetic	Fully variable timing; space requirements	Additional electronics; soft landing needed; weight

5.1.4 Variable Compression Ratio

Compression ratio has very strong influence on intake temperature requirement for control of combustion timings in HCCI combustion engine. The variable compression ratio (VCR) can be used to control combustion timings by varying the compression ratio, which leads to variation in the charge temperature after compression. The fuel–air mixture auto-ignites at early crank angle position at a higher compression ratio due to higher charge temperature. VCR can be achieved by several methods. One method is to tilt the upper part of the engine block, the mono head. A hydraulic motor, controlled by an electronic valve turns the eccentric shaft that changes the engine configuration to vary the compression ratio [30]. Another method is to mount a plunger in the cylinder head whose position can be varied to change the compression ratio [31]. The main drawback of VCR system is that it presently does not allow individual cylinder control that is necessary to obtain good combustion timing control. The VCR systems are also expensive and complex [10].

The compression ratio is a key structural parameter that affects the engine efficiency. The LTC engine has typically higher efficiency, and relatively lower compression ratio is used to increase the load range of LTC engine (see Figs. 5.5 and 5.10). In order to extend the engine operating load, a RCCI engine is optimized at a very low compression ratio of 11.7 [32]. Another RCCI combustion study investigated two different compression ratios of 14 and 17 at different engine speeds [33]. At lower compression ratio, NOx emission reduced and unburned hydrocarbon emissions are increased in RCCI engine. At compression ratio 14, longer ignition delay, combustion duration and also lower maximum HRR are achieved. Therefore, compression ratio 14 is preferred at high load due to the limitation on peak pressure rise rate of the engine [33, 34]. A study also investigated the RCCI combustion by varying effective compression ratio by changing IVC timings [35]. The shape of heat release curve varies with IVC timings. The IVC timings can be optimized depending on fuel injection timings and fuel premixing ratio at particular engine speed and load conditions.

5.1.5 Water Injection

HCCI combustion control has been demonstrated using water injection. Water injection in the port is investigated in HCCI engine, and study indicated that evaporative cooling in the intake manifold (particularly at higher intake temperatures) offers the control of ignition timings by controlling the intake valve closing temperature [36]. The study confirmed that it is possible to control the ignition timing in a range of engine operating conditions with the use of water injection but at price of an increase in the already higher emissions of unburned hydrocarbons and carbon monoxide. Atomization, vaporization and distribution of injected water are the key parameters for system optimization [37].

A study used water injection to control diesel HCCI combustion timing and demonstrated that water injection could reduce the heat release rate significantly [38]. Direct water injection (using diesel fuel injector) is used to extend the load range of premixed compression ignition [39]. The study showed that the combustion suppression effect increases with increase in water injection quantity, and excessive water injection leads to higher THC emission along with lower thermal efficiency. Although the reaction suppressing effect of water injection enhances with advanced timings, an optimum water injection timing occurs. With more advanced timings, water vaporization is delayed, which decreases the reaction suppressing effect. Thus, amount of water injection must be limited to the minimum required for sufficient suppression of oxidation reactions at excessively advanced combustion timings [23].

A new strategy called thermally stratified compression ignition (TSCI) is proposed to artificially create the thermal stratification in the cylinder. In this combustion mode, water is directly injected in the cylinder to control the mean temperature along with temperature distribution in the cylinder, which offers cycle-to-cycle control of start of the combustion and HRR in premixed compression ignition [40]. More detailed description of TSCI strategy is provided in Sect. 2.4 of Chap. 2.

5.1.6 Boosting

Boosting (increasing intake air pressure) is considered as an effective technique to increase the engine operating load and extend the operational range of air–fuel ratio in HCCI combustion mode. However, boosting is accompanied by a higher peak cylinder pressure, which may lead to reach peak pressure limit of the engine. Higher intake pressure can enable an increase of the overall dilution of fuel–air mixture and affect the combustion characteristics through chemical effects related to the higher pressure of combustion chamber. At particular fuelling rate, relatively larger amount of inducted mass at higher intake pressure leads to higher level of charge dilution leads to reduction in pressure rise rates in HCCI combustion. For a fixed pressure rise rate limit, relatively higher amount of fuel can be burned in the cylinder and thus the maximum load limit increases via intake boosting. The higher load limit achieved in different LTC strategies is discussed in Sect. 7.1.2 of Chap. 7.

A very higher intake pressure (up to 3.6 bar) is required to achieve HCCI engine operating load (IMEP ~ 20 bar) comparable to conventional engines (Fig. 7.6 of Chap. 7). Figure 5.14 illustrates the intake pressure requirement as a function of engine load in PPC strategy at two different compression ratios. The figure depicts that very high intake pressure (~3.6 bar) is required at higher engine load. This level of intake boost is not practical with current turbochargers installed on marketed engines. Thus, boost requirement needs to be reduced by further research. Figure 5.14 also demonstrates that at relatively lower compression ratio, higher engine load can be achieved in PPC engine.

Turbocharger/supercharger providing the intake boost pressure needs to be adequately sized and matched to the requirements of LTC engine to achieve particular engine load and efficiency. Recent studies [41, 42] demonstrated the partial fuel stratification method to reduce the boost pressure requirement in low temperature gasoline combustion (LTGC) engines. Different direct injection strategies can be utilized to create the fuel stratification, which reduces the boost to load ratio in LTC engines [41, 42].

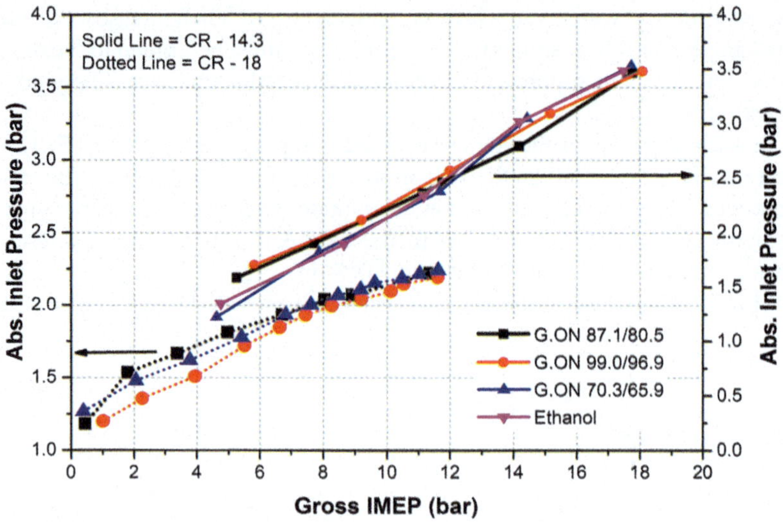

Fig. 5.14 Absolute inlet pressure as a function of engine load and fuel type in PPC engine for different compression ratios (Adapted from [13, 14])

5.1.7 In-Cylinder Injection Strategies

The local charge temperature and the air–fuel ratio are the key parameters controlling the initiation of HCCI combustion process. The direct injection of fuel into the cylinder provides the potential to control the combustion process by altering the local fuel concentration by varying the injection timings. The gas temperature is also altered through the charge cooling from fuel evaporation. Early fuel injection timings provide sufficient time for fuel vaporization and mixing with the air to create a premixed fuel–air mixture. The late pilot injection in the cylinder during compression stroke can control the combustion phasing by creating fuel stratification (increasing the local fuel concentration in some regions) in the cylinder [23]. Additionally, rich mixture decreases the ratio of specific heat and thus amount of disposable compression heating in the charge. Consequently mixture has to be further compressed to reach the autoignition temperature. The capability of split injection can combine these two functions.

Figure 5.15 illustrates the consecutive flame images of the gasoline auto-ignition by single and double injections (at three pilot injection timings of −20, −16 and −12 aTDC) with fixed main injection timing at TDC and fuel injection pressure at 400 bar, intake pressure 1.4 bar at 1200 rpm. Heat release rate and the normalized spatially integrated natural luminosity (SINL) are shown to investigate the correlation between the two (Fig. 5.15b). The natural luminosity of a flame generally consists of the chemiluminescence (emitted from the electronically excited gaseous species including CH, C_2 and formaldehyde) and the soot incandescence [43]. Figure 5.15 depicts that the single injection gasoline compression ignition (GCI) has a

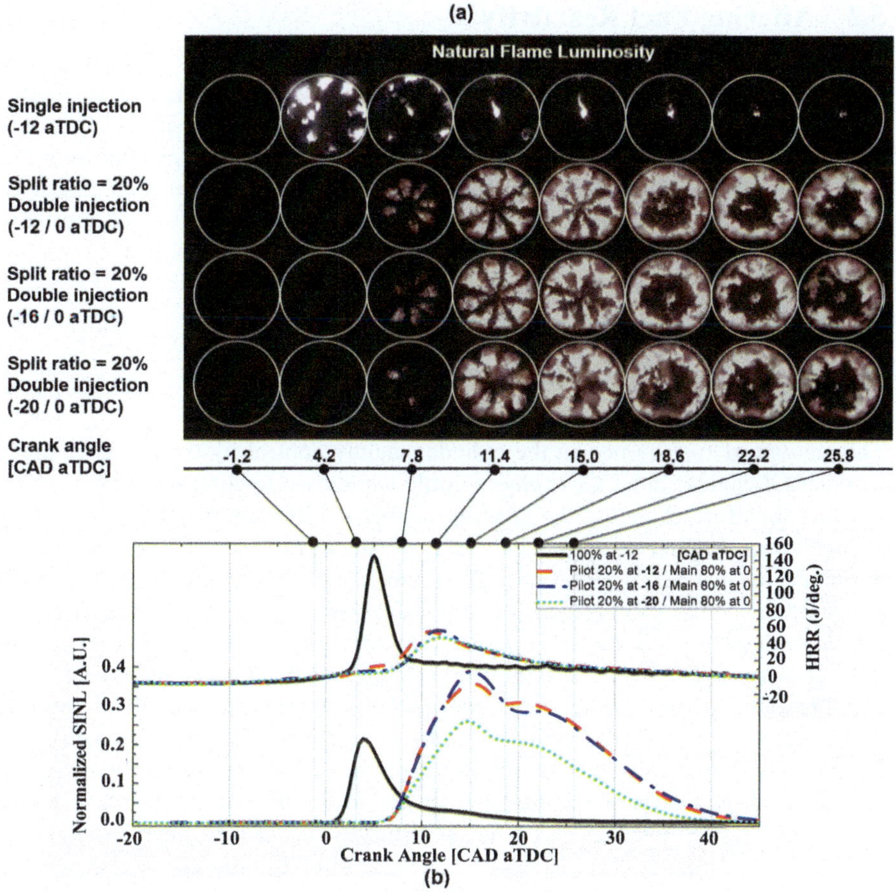

Fig. 5.15 Natural flame luminosity and heat release rate in gasoline combustion with single and double injection at 1200 rpm with total fuel injection quantity of 20 mg/stroke [43]

weak natural luminosity by the chemiluminescence from the premixed flame, while the double injection GCIs showed a strong natural luminosity similar to the diesel-like diffusion flame of the main injection by the soot incandescence. Double fuel injection creates a relatively more stratified charge in the cylinder leading to higher soot formation. Figure 5.15 also depicts that double injection strategy reduces the heat release rate of GCI combustion.

Fuel injection timings and number of injection govern the fuel stratification in the combustion chamber and thus heat release rate in LTC combustion. In case of diesel-like fuels, typically direct fuel injection strategy is used to achieve premixed diesel combustion. However, both port and direct fuel injection strategies are used in gasoline premixed compression ignition engines. Detailed charge preparation strategies using gasoline and diesel are discussed in Chap. 4. Comparison of fuel injection strategies in different LTC modes is provided in Figs. 4.12 and 4.20 (Chap. 4).

5.2 Altering Fuel Reactivity

Altering reactivity (autoignition characteristics) of fuel–air mixture is another strategy to control the combustion rate and combustion phasing in LTC engines. Mixture autoignition properties are affected by fuel types, fuel concentration, blending of two or more fuels, fuel additives, residual rate and residual reactivity, boosting etc. In this section, LTC controlled by altering the fuel reactivity using different strategies is presented.

5.2.1 *Fuel–Air Equivalence Ratio*

The amount of fuel burned in the cylinder mainly control by the power required from HCCI engine. In HCCI engine, throttle is not used to control the inducted air quantity therefore changing the quantity of fuel at different engine loads leads to change in equivalence ratio. Influence of equivalence ratio on HCCI combustion is discussed in Sect. 2.2.4.1 of Chap. 2. Higher equivalence ratio (richer mixture) advances the combustion phasing at fixed inlet conditions. Overall fuel–air equivalence ratio (Φ) of inducted charge governs the engine load in LTC engines. The Φ-sensitivity (autoignition reactivity) of charge also depends on intake boost pressure of engine (see Fig. 2.15 of Chap. 2). For example, gasoline is less sensitive fuel at naturally aspirated conditions, and fuel reactivity changes at higher intake pressure. With Φ-sensitive fuels, stratification of fuel can be used to increase the engine operating load for fixed pressure rise rate limits. Several LTC strategies such as PFS HCCI, PPC and RCCI uses fuel stratification in the cylinder by using direct fuel injection. Internal charge preparation in these LTC strategies is described in Sect. 4.2 of Chap. 4.

5.2.2 *In-Cylinder Fuel Stratification*

Fuel stratification in the cylinder is created by direct injection of fuel. Fuel stratification is one of the main methods to control the combustion rate by increasing the combustion duration. Fuel stratification is typically used to increase the engine operating load of LTC engines (see Sect. 7.1.2 of Chap. 7). A recent study demonstrated the direct injection of gasoline (by dual pulse) increases the thermal efficiency of partially stratified HCCI combustion [42]. PPC combustion also creates heavy stratification in the cylinder by using fuel injection system of diesel engine for gasoline-like fuels. In RCCI combustion, fuel stratification as well as reactivity stratification is created in the engine cylinder. Fuel injection strategy is an important variable among many factors in improving the performance of LTC engines. In RCCI combustion, injection strategy (number of pulse and fuel injection timing) affects the fuel and reactivity stratification in the cylinder. The number of fuel injection pulses in the injection strategy can affect the high reactivity fuel distribution and consequently shape the reactivity distribution in the combustion chamber. Single, double and triple fuel injection pulses have been investigated in different RCCI studies [34]. Fuel injection timing of high reactivity fuel is most important variable which affects the mixing process of fuel spray with mixture of air and low reactivity fuels. This mixing defines the reactivity distribution in the cylinder which affects the combustion characteristics in RCCI engine. Early injection of high reactivity fuel leads the combustion more prone to reactivity controlled, and relatively late injection results in combustion more similar to mixing/diffusion controlled. In direct injection dual fuel stratification (DDFS) strategy, both the high and low reactivity fuels are directly injected into the cylinder to create the stratified charge in the cylinder. Fuel injection strategies of different LTC modes with different stratification levels are presented in Fig. 4.20 (Chap. 4).

5.2.3 Dual Fuel

Fuel autoignition property variation can be used as a method for HCCI combustion control. The combustion phasing can be controlled by varying the fuel properties, and thus operating range can also be expanded. A study investigated the effect of fuel properties on low and high temperature heat release of HCCI combustion and its performance [44]. Fuel composition affects the low temperature heat release (LTHR) and high temperature heat release (HTHR) values. The effect of LTHR on HCCI combustion is discussed in Chap. 2. Fuel stratification in the cylinder leads to retarded combustion phasing, and it provides an additional actuator for HCCI combustion control [45]. However, very large stratification may lead to unstable combustion.

Dual fuel method can be used to actively vary the fuel octane number. The idea of using dual fuels is to use two fuels with different auto-ignition reactivity. The dual fuel system has a main fuel with a high octane number and a secondary fuel with low octane number (high cetane number) [46, 47]. This feature can then be used to control the combustion phasing in HCCI as blending the two fuels at different fuel ratio changes the auto ignition properties. Injection of low octane fuel in higher quantity leads to earlier autoignition in the cylinder. For advanced combustion phasing, the secondary fuel quantity is increased. For individual cylinder control of the combustion phasing, each cylinder must have two injectors for two different reactivity fuels. The benefit with dual fuels operation is that it provides an accurate control without any large engine modifications and cost. Only additional injectors are required. One demerit of this system is the requirement of two fuel tanks and its refuelling. However, secondary fuel consumption is very low and could be refuelled only in the maintenance intervals [10]. Typically, this system is demonstrated with fuel injection in the port by external blending of the fuels.

The RCCI combustion is another strategy, where in-cylinder blending of two different reactivity fuels occurs. With in-cylinder blending of fuels, fuel stratification as well as reactivity stratification can be controlled in this strategy. Fuel ratio of low and high reactivity fuel is one of the variables that affect the reactivity stratification in the combustion chamber. Reactivity in the cylinder can be varied by changing the fuel ratio, and consequently it affects the ignition delay. Ignition delay affects the mixing time of fuel and thus homogeneity of the charge in the cylinder. The low reactivity fuel ratio up to 90% is used in RCCI engines, and most of the studies showed that combustion phasing is retarded with the increase of low reactivity fuel. This is because the blending of low reactivity fuel increases the ignition delay and consequently retarding the combustion phasing [34].

5.2.4 Fuel Additives

Selection of appropriate fuel is a key aspect of LTC engine design and development. Autoignition characteristics and fuel volatility are the key variables in the fuel selection. In order to easily create premixed charge, fuel needs to have high volatility characteristics. Chemically, single-stage heat release fuels have lower sensitivity to variations in engine load and speed. Lower sensitivity to variations makes the requirements on HCCI control system easier over a wide range of operating conditions. To achieve high fuel conversion efficiency, fuel autoignition temperature is critical parameter for selection of an optimum geometrical compression ratio of the engine [23]. Required fuel in HCCI combustion needs to satisfy the requirements over wide range of engine loads (low to high load). At lower engine loads, low octane fuel is required to facilitate the autoignition, while at higher engine loads relatively higher octane fuel is required to avoid the knocking during combustion. This opposing requirement leads to difficulty in HCCI engine development, and there exists no universal fuel that satisfies the specific requirement of HCCI combustion engine. The optimal fuel selection is influenced by the combustion control strategies used in LTC engine and engine operating conditions. Thus, LTC engines can be operated on any fuel, but adaptation is required either fuel to specific engine design or engine to a specific fuel or even a specific engine operating condition [23].

Autoignition in HCCI combustion is mainly controlled by chemical kinetics, which is sensitive to several parameters including fuel properties, oxygen and EGR concentration, charge temperature and inlet pressure. In naturally aspirated conditions, even though HCCI combustion can be achieved using regular gasoline, a high intake temperature or high amount of hot residual gas is required for autoignition because of low autoignition reactivity of gasoline [1, 48, 49]. Due to high intake temperature requirements, charge densities are much lower, which limits the high load HCCI engine operation. This problem can be significantly reduced by using low octane gasoline in HCCI engine [50]. Furthermore, the combustion phasing can be retarded farther with good stability for a more reactive fuel, which allows higher charge mass fuel–air equivalence ratio (ϕ_m) operation without combustion knock. This leads to significant increase in higher engine load limit of HCCI combustion. Even though low octane gasoline is more suitable for naturally aspirated HCCI engines, these fuels are not readily available in the market. Moreover, the high reactivity of low octane gasoline can turn into a difficulty for intake-boosted operation due to very high reactivity above particular boost level. To avoid the knocking in this condition, large amount of EGR is required, and maximum load can become limited by a lack of oxygen at higher engine load operation [50]. To meet the fuel requirements in HCCI combustion, several fuel additives are added to regular low reactivity fuels. Fuel additives used in HCCI combustion include 2-ethylhexyl nitrate (EHN), di-tertiary butyl peroxide (DTBP), hydrogen (H_2), hydrogen peroxide (H_2O_2), formaldehyde (CH_2O) and ozone (O_3) [49, 51–56]. A study showed that hydrogen addition in DME (dimethyl ether) fuelled HCCI combustion could mitigates low temperature reactions and delay its occurrence by consuming OH radicals. Induction of hydrogen intensifies high temperature reaction of DME, and oxidation reaction process becomes fast due to the active nature of hydrogen in high temperature oxidation. Due to reduced low temperature reactions, high temperature reactions of DME is also retarded by hydrogen addition [53].

The EHN and DTBP are typically used to improve the cetane number (CN) of diesel fuels. These additives are also used to enhance the autoignition of gasoline-like fuels in HCCI combustion [49, 51]. Figure 5.16 illustrates the effect of addition of EHN and DTBP on intake temperature requirement in HCCI combustion engine. Figure 5.16a depicts the intake temperature requirement decreases with increase in concentration of additives to maintain same start of combustion (CA_{10}) position. Typically, high intake temperature ($T_{in} \sim 140$ °C) is required for fuels with low HCCI reactivity, such as regular gasoline and E10 at naturally aspirated engine operating conditions. By adding small amount of EHN (0.15%) or DTBP (0.35%), the tin drops drastically from 144 °C to 95 °C, and it can be further reduced to 60 °C using 0.4% EHN (Fig. 5.16a). The figure also shows that, DTBP has a smaller effect on T_{in} compared with EHN for a particular addition percentage. Furthermore, a nonlinear decrease in T_{in} is found for both EHN and DTBP that means the decrease of T_{in} is greater at low additive concentration (<0.25%) and the effect becomes more moderate as the additive percentage increases [49]. At higher intake pressure reactivity is higher at even lower addition percentage. Figure 5.16b illustrates the dependence of combustion (CA_{50}) on intake temperature for different concentrations of DTBP in 25% DEE-ethanol blends and ethanol. Figure 5.16b depicts that intake temperature requirement reduces at given combustion phasing with increase in DTBP concentration.

Figure 5.17a illustrates the effect of EHN on heat release rate (HRR) at low and intermediate temperature range at constant CA_{10} (366° CA) position. The HRR curves have been normalized by the total amount of detected heat release (THR) in order to eliminate differences between HRR curve caused solely by a difference in the total amount of fuel injected per cycle as T_{in} is varied. The figure depicts that low temperature heat release is significantly enhanced by increased additive

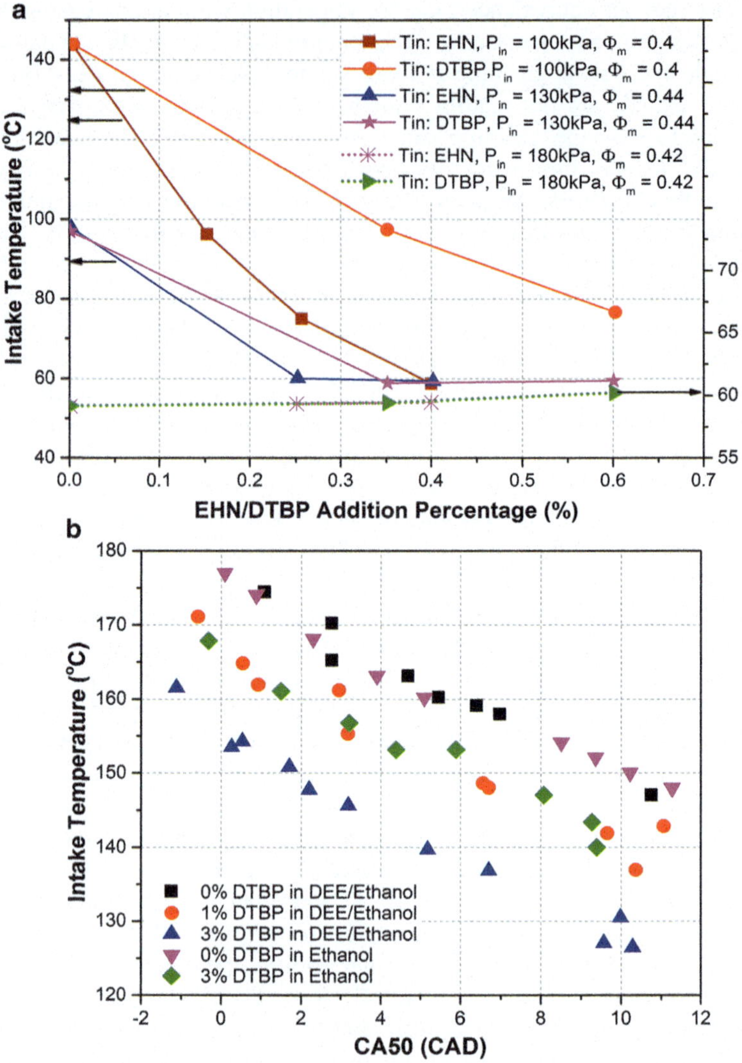

Fig. 5.16 (**a**) Intake temperature requirement as function of additive concentration at different intake pressure in HCCI combustion (Adapted from [49]), and (**b**) effect of DTBP addition to ethanol and ethanol-DEE blends on dependence between intake temperature and combustion phasing in HCCI combustion (Adapted from [51])

concentration, which leads to reduction in intake temperature requirements. Figure 5.17b shows the ϕ-sensitivity of E10 with 0.4% EHN additive. ϕ-sensitivity is an important parameter in applying partial fuel stratification because it allows the autoignition to occur sequentially down the equivalence ratio gradient, which reduces the HRR by increasing the combustion duration [49]. Figure 5.17b shows that addition of EHN significantly increases the ϕ-sensitivity of E10 at even lower

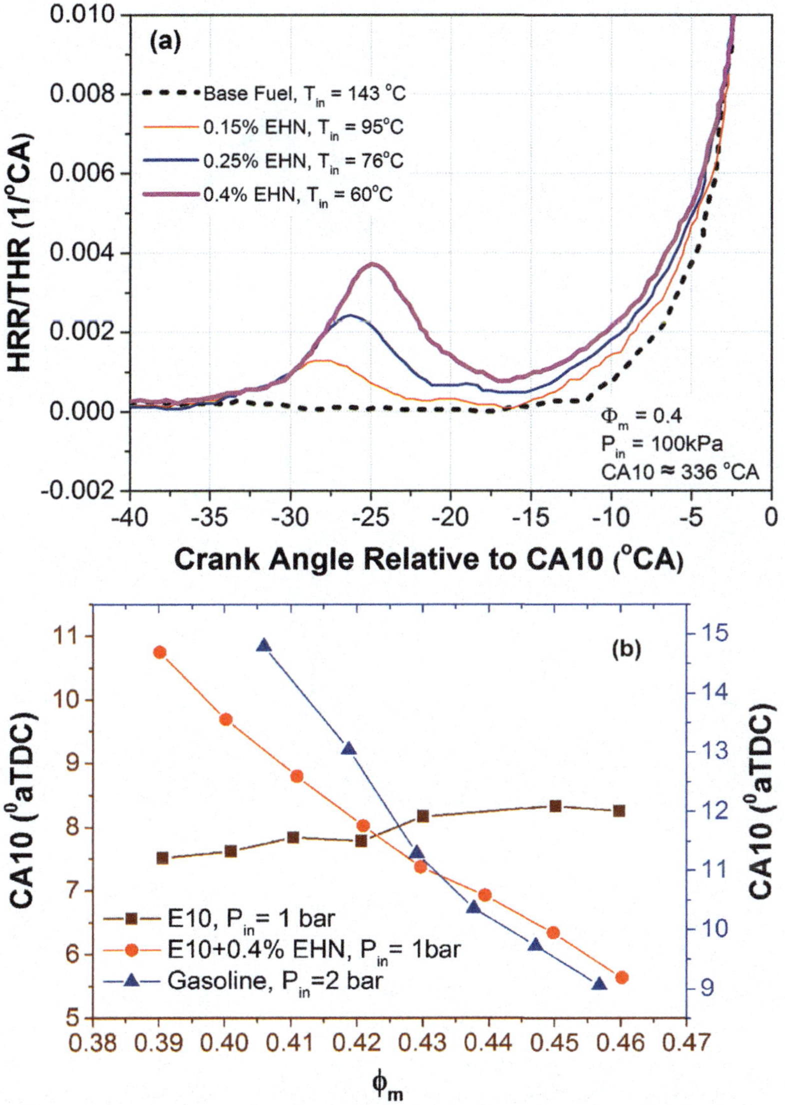

Fig. 5.17 (**a**) Effect of EHN on heat release rate at low and intermediate temperature range, and (**b**) ϕ-sensitivity of CCG-E10 with 0.4% EHN additive in HCCI engine (Adapted from [49, 57])

intake pressure, where E10 is not φ-sensitive without additives. Addition of EHN produces the effect similar to ϕ-sensitivity of gasoline at 2.0 bar intake pressure. It is expected that partial fuel stratification will work well for EHN additive added fuels to improve the thermal efficiency of the engine. Increasing the intake boost pressure also enhances the autoignition. Gasoline having additives can become very reactive as boost level is increased, requiring a large amount of EGR to prevent

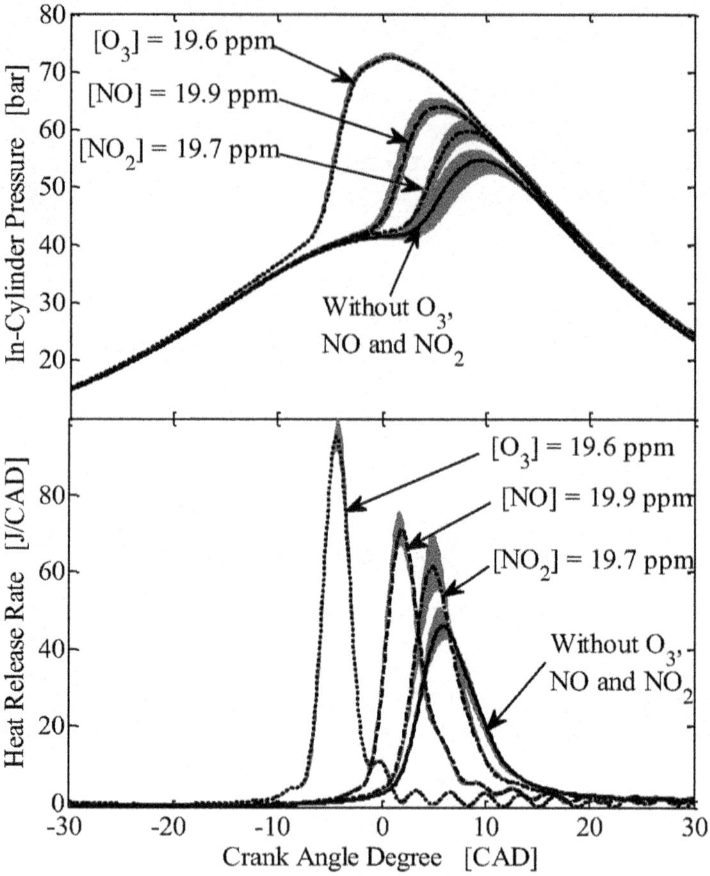

Fig. 5.18 Cylinder pressure and HRR variations with 20 ppm injection of NO, NO_2 and O_3 species in HCCI combustion [58]

overly advanced autoignition timings. These high EGR levels reduce the amount of oxygen in the charge, and the maximum load can become limited by oxygen availability before it reaches the knock/stability limit [49].

The reactivity of the charge can be modified by injecting oxidizing species such as nitric oxide (NO), nitrogen oxide (NO_2) and ozone (O_3) in the intake manifold. Figure 5.18 shows the cylinder pressure and HRR variations by injecting 20 ppm of NO, NO_2 and O_3 species in HCCI combustion. NO and NO_2 are the species that can even present in EGR operated HCCI combustion. These species can advance the combustion phasing of HCCI combustion (see Fig. 6.29a in Chap. 6). Figure 5.18 also depicts that all NO, NO_2 and O_3 species advances the combustion phasing of HCCI combustion and ozone is the most potent oxidizer.

A new strategy to control HCCI combustion is evolved, which is dominated by chemical kinetics, use oxidizing chemical species to initiate fuel oxidation. Ozone

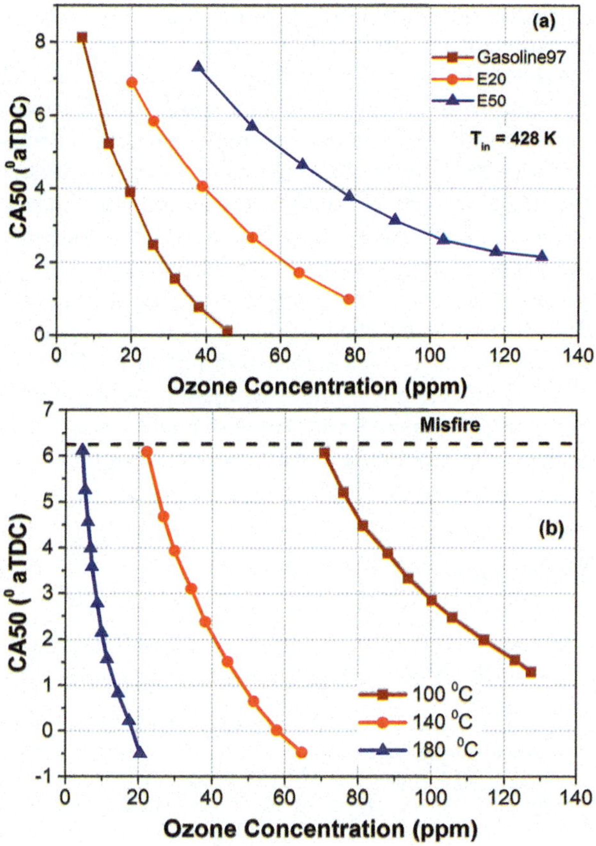

Fig. 5.19 Combustion phasing as function of intake ozone concentration for (**a**) different fuels (Adapted from [55]), and (**b**) different intake temperature using iso-octane (Adapted from [56]) in HCCI combustion

seems to be one of the most promising species. Present development of increasingly smaller ozonizers may lead to their installation on vehicles. This kind of ozonizer in HCCI engine can be used to control combustion phasing and promote HCCI engine application in future vehicles similar to conventional engines [56]. It is proposed to use ozone as oxidizer to modify the molecular structure of the fuel and to control the combustion timing inside the cylinder or to seed the intake of engines with some few ppm of ozone in order to enhance HCCI combustion [56, 59].

Figure 5.19 shows the combustion phasing as function of intake ozone concentration for different fuels and different intake temperature using iso-octane in HCCI combustion. The figure depicts that combustion phasing is advanced with increase of ozone concentration for all studied fuels and intake temperatures. Ethanol blends require higher ozone concentration than gasoline to maintain the same combustion phasing due to lower reactivity (higher octane number) of ethanol. The slope of the curves can be interpreted as the ozone seeding sensitivity, and a steeper slope corresponds to a higher sensitivity. Late combustion phasing at lower intake temperatures and lower ozone concentrations is limited around 7° aTDC or by unstable combustion or misfire (Fig. 5.19).

Figure 5.19b depicts that at 180 °C inlet temperature, less than 20 ppm of ozone is sufficient to vary the CA_{50} by 6 CAD and for the same combustion phasing variation around 120 ppm of ozone is required at 100 °C intake temperature. The decrease in the inlet temperature slows down ozone decomposition because the decomposition of ozone is strongly affected by temperature. Therefore, there are fewer O-atoms formed and oxidation of the fuel by ozone is reduced. This explains the high amount of ozone seeding in the intake manifold is necessary for maintaining the same CA_{50}. The sensitivity of the CA_{50} is reduced at lower intake temperatures [56]. Figure 5.19b also illustrates the intake temperature reduction potential of ozone. The study showed that with sufficient ozone concentration, the cool flame occurs even in the case of iso-octane HCCI combustion [56]. Low temperature heat release is also observed in small quantities for the gasoline and E20 fuels at the lower intake temperatures, and it increases with increase in ozone addition [55]. Increase in low temperature heat release results into lower intake temperature requirement in HCCI combustion.

References

1. Maurya RK, Agarwal AK (2014) Experimental investigations of performance, combustion and emission characteristics of ethanol and methanol fueled HCCI engine. Fuel Process Technol 126:30–48
2. Maurya RK, Agarwal AK (2014) Effect of intake air temperature and air–fuel ratio on particulates in gasoline and n-butanol fueled homogeneous charge compression ignition engine. Int J Engine Res 15(7):789–804
3. Annen K, Stickler D, Woodroffe J (2006) Glow plug-assisted HCCI combustion in a miniature internal combustion engine (MICE) generator. In: 44th AIAA Aerospace sciences meeting and exhibit. American Institute of Aeronautics and Astronautics, Reston, VA, p 1349
4. Saxena S, Vuilleumier D, Kozarac D, Krieck M, Dibble R, Aceves S (2014) Optimal operating conditions for wet ethanol in a HCCI engine using exhaust gas heat recovery. Appl Energy 116:269–277
5. Maurya RK, Akhil N (2016) Numerical investigation of ethanol fuelled HCCI engine using stochastic reactor model. Part 2: Parametric study of performance and emissions characteristics using new reduced ethanol oxidation mechanism. Energy Convers Manag 121:55–70
6. Tunestål P, Johansson B (2007) HCCI control. In: CAI and HCCI engines for the automotive industry. Woodhead Publishing Limited. Cambridge, England, pp 164–184
7. Haraldsson G, Tunestål P, Johansson B, Hyvönen J (2004) HCCI closed-loop combustion control using fast thermal management (No. 2004-01-0943). SAE technical paper
8. Martinez-Frias J, Aceves SM, Flowers D, Smith JR, Dibble R (2000) HCCI engine control by thermal management (No. 2000-01-2869). SAE technical paper
9. Lee D, Stefanopoulou AG, Makkapati S, Janković M (2010) Modeling and control of a heated air intake homogeneous charge compression ignition (HCCI) engine. In: American Control Conference (ACC). Baltimore, MD, USA. IEEE, pp 3817–3823
10. Bengtsson J (2004) Closed-loop control of HCCI engine dynamics. PhD theses. Lund University, Sweden
11. Yang J, Culp T, Kenney T (2002) Development of a gasoline engine system using HCCI technology-the concept and the test results (No. 2002-01-2832). SAE technical paper
12. Saxena S, Schneider S, Aceves S, Dibble R (2012) Wet ethanol in HCCI engines with exhaust heat recovery to improve the energy balance of ethanol fuels. Appl Energy 98:448–457

13. Manente V, Johansson B, Tunestal P, Cannella W (2009) Effects of different type of gasoline fuels on heavy duty partially premixed combustion. SAE Int J Engines 2(2009-01-2668):71–88
14. Manente V, Tunestal P, Johansson B, Cannella WJ (2010) Effects of ethanol and different type of gasoline fuels on partially premixed combustion from low to high load (No. 2010-01-0871). SAE technical paper
15. Borgqvist P, Tuner M, Mello A, Tunestal P, Johansson B (2012) The usefulness of negative valve overlap for gasoline partially premixed combustion, PPC (No. 2012-01-1578). SAE technical paper
16. Hanson R, Kokjohn S, Splitter D, Reitz RD (2011) Fuel effects on reactivity controlled compression ignition (RCCI) combustion at low load. SAE Int J Engines 4(2011-01-0361):394–411
17. Klinkert S (2014) An experimental investigation of the maximum load limit of boosted HCCI combustion in a gasoline engine with negative valve overlap. Doctoral dissertation, The University of Michigan
18. Kodavasal J (2013) Effect of charge preparation strategy on HCCI combustion. Doctoral dissertation, University of Michigan
19. Kodavasal J, Lavoie GA, Assanis DN, Martz JB (2015) The effects of thermal and compositional stratification on the ignition and duration of homogeneous charge compression ignition combustion. Combust Flame 162(2):451–461
20. Caton PA, Simon AJ, Gerdes JC, Edwards CF (2003) Residual-effected homogeneous charge compression ignition at a low compression ratio using exhaust reinduction. Int J Engine Res 4 (3):163–177
21. Jungkunz AF (2013) Actuation strategies for cycle-to-cycle control of homogeneous charge compression ignition combustion engines. Doctoral dissertation, Stanford University
22. Kulkarni AM, Adi GH, Shaver GM (2007) Modeling cylinder-to-cylinder coupling in multi-cylinder HCCI engines incorporating reinduction. In: ASME 2007 International Mechanical Engineering Congress and Exposition. American Society of Mechanical Engineers, pp 1597–1604
23. Zhao F, Asmus TW, Assanis DN, Dec JE, Eng JA, Najt PM (2003) Homogeneous charge compression ignition (HCCI) engines: key research and development issues PT-94. Progress in technology, vol 94. SAE International. Warrendale, Pennsylvania
24. Piperel A, Montagne X, Dagaut P (2007) HCCI engine combustion control using EGR: gas composition evolution and consequences on combustion processes (No. 2007-24-0087). SAE technical paper
25. Kanda T, Hakozaki T, Uchimoto T, Hatano J, Kitayama N, Sono H (2005) PCCI operation with early injection of conventional diesel fuel (No. 2005-01-0378). SAE technical paper
26. Boyarski NJ, Reitz RD (2006) Premixed compression ignition (PCI) combustion with modeling-generated piston bowl geometry in a diesel engine (No. 2006-01-0198). SAE technical paper
27. Liu H, Wang X, Zheng Z, Gu J, Wang H, Yao M (2014) Experimental and simulation investigation of the combustion characteristics and emissions using n-butanol/biodiesel dual-fuel injection on a diesel engine. Energy 74:741–752
28. Walter B, Pacaud P, Gatellier B (2008) Variable valve actuation systems for homogeneous diesel combustion: how interesting are they? Oil Gas Sci Technol-Revue de l'IFP 63 (4):517–534
29. Mashkournia M (2012) Electromagnetic variable valve timing on a single cylinder engine in HCCI and SI. MS thesis, University of Alberta
30. Haraldsson G, Tunestål P, Johansson B, Hyvönen J (2002) HCCI combustion phasing in a multi cylinder engine using variable compression ratio (No. 2002-01-2858). SAE technical paper
31. Christensen M, Hultqvist A, Johansson B (1999) Demonstrating the multi fuel capability of a homogeneous charge compression ignition engine with variable compression ratio (No. 1999-01-3679). SAE technical paper

32. Dempsey AB, Reitz RD (2011) Computational optimization of reactivity controlled compression ignition in a heavy-duty engine with ultra low compression ratio. SAE Int J Engines 4 (2):2222–2239
33. Jia Z, Denbratt I (2015) Experimental investigation of natural gas-diesel dual-fuel RCCI in a heavy-duty engine. SAE Int J Engines 8(2015-01-0838):797–807
34. Li J, Yang W, Zhou D (2017) Review on the management of RCCI engines. Renew Sust Energ Rev 69:65–79
35. Hanson RM, Kokjohn SL, Splitter DA, Reitz RD (2010) An experimental investigation of fuel reactivity controlled PCCI combustion in a heavy-duty engine. SAE Int J Engines 3(2010-01-0864):700–716
36. Christensen M, Johansson B (1999) Homogeneous charge compression ignition with water injection (No. 1999-01-0182). SAE technical paper
37. Iwashiro Y, Tsurushima T, Nishijima Y, Asaumi Y, Aoyagi Y (2002) Fuel consumption improvement and operation range expansion in HCCI by direct water injection (No. 2002-01-0105). SAE technical paper
38. Kaneko N, Ando H, Ogawa H, Miyamoto N (2002) Expansion of the operating range with in-cylinder water injection in a premixed charge compression ignition engine (No. 2002-01-1743). SAE technical paper
39. Ogawa H, Miyamoto N, Kaneko N, Ando H (2003) Combustion control and operating range expansion with direct injection of reaction suppressors in a premixed DME HCCI engine (No. 2003-01-0746). SAE technical paper
40. Lawler B, Splitter D, Szybist J, Kaul B (2017) Thermally stratified compression ignition: a new advanced low temperature combustion mode with load flexibility. Appl Energy 189:122–132
41. Dec JE, Dernotte J, Ji C (2017) Increasing the load range, load-to-boost ratio, and efficiency of low-temperature gasoline combustion (LTGC) engines. SAE Int J Engines 10 (2017-01-0731):1256–1274
42. Dernotte J, Dec J, Ji C (2017) Efficiency improvement of boosted low-temperature gasoline combustion engines (LTGC) using a double direct-injection strategy (No. 2017-01-0728). SAE technical paper
43. Kim D, Bae C (2017) Application of double-injection strategy on gasoline compression ignition engine under low load condition. Fuel 203:792–801
44. Shibata G, Oyama K, Urushihara T, Nakano T (2004) The effect of fuel properties on low and high temperature heat release and resulting performance of an HCCI engine (No. 2004-01-0553). SAE technical paper
45. Dec JE, Sjöberg M (2004) Isolating the effects of fuel chemistry on combustion phasing in an HCCI engine and the potential of fuel stratification for ignition control (No. 2004-01-0557). SAE technical paper
46. Olsson JO, Tunestål P, Johansson B (2001) Closed-loop control of an HCCI engine (No. 2001-01-1031). SAE technical paper
47. Maurya RK, Agarwal AK (2013) Experimental investigation of close-loop control of HCCI engine using dual fuel approach (No. 2013-01-1675). SAE technical paper
48. Dec JE, Yang Y (2010) Boosted HCCI for high power without engine knock and with ultra-low NOx emissions-using conventional gasoline. SAE Int J Engines 3(2010-01-1086):750–767
49. Ji C, Dec JE, Dernotte J, Cannella W (2014) Effect of ignition improvers on the combustion performance of regular-grade E10 gasoline in an HCCI engine. SAE Int J Engines 7(2014-01-1282):790–806
50. Yang Y, Dec JE, Dronniou N, Cannella W (2012) Boosted HCCI combustion using low-octane gasoline with fully premixed and partially stratified charges. SAE Int J Engines 5(2012-01-1120):1075–1088
51. Mack JH, Dibble RW, Buchholz BA, Flowers DL (2005) The effect of the di-tertiary butyl peroxide (DTBP) additive on HCCI combustion of fuel blends of ethanol and diethyl ether (No. 2005-01-2135). SAE technical paper

52. Noorpoor AR, Ghaffarpour M, Aghsaee M, Hamedani A (2009) Effects of fuel additives on ignition timing of methane fuelled HCCI engine. J Energy Inst 82(1):37–42
53. Hu E, Chen Y, Cheng Y, Meng X, Yu H, Huang Z (2015) Study on the effect of hydrogen addition to dimethyl ether homogeneous charge compression ignition combustion engine. J Renew Sustain Energ 7(6):063121
54. El-Din HA, Elkelawy M, Yu-Sheng Z (2010) HCCI engines combustion of CNG fuel with DME and H 2 additives (No. 2010-01-1473). SAE technical paper
55. Truedsson I, Rousselle C, Foucher F (2017) Ozone seeding effect on the ignition event in HCCI combustion of gasoline-ethanol blends (No. 2017-01-0727). SAE Technical Paper
56. Masurier JB, Foucher F, Dayma G, Mounaïm-Rousselle C, Dagaut P (2013) Towards HCCI control by ozone seeding (No. 2013-24-0049). SAE technical paper
57. Dec JE, Yang Y, Dronniou N (2011) Boosted HCCI-controlling pressure-rise rates for performance improvements using partial fuel stratification with conventional gasoline. SAE Int J Engines 4(2011-01-0897):1169–1189
58. Masurier JB, Foucher F, Dayma G, Dagaut P (2015) Investigation of iso-octane combustion in a homogeneous charge compression ignition engine seeded by ozone, nitric oxide and nitrogen dioxide. Proc Combust Inst 35(3):3125–3132
59. Schönborn A, Hellier P, Aliev AE, Ladommatos N (2010) Ignition control of homogeneous-charge compression ignition (HCCI) combustion through adaptation of the fuel molecular structure by reaction with ozone. Fuel 89(11):3178–3184

Chapter 6
Combustion Characteristics

Abstract Combustion characteristics of an engine determine its performance and emission characteristics. This chapter presents the combustion characteristic of LTC engines using different gasoline-like fuels (ethanol and methanol) vis-à-vis conventional fuels. First, combustion kinetics of ethanol and methanol in premixed charge compression ignition engine is discussed using reaction pathway and sensitivity analysis. Combustion kinetics in LTC engine using hydrocarbon fuels is discussed in Chap. 2. Ignition and heat release characteristics of LTC combustion process are described by analysis of ignition delay, in-cylinder pressure, pressure rise rate, ringing intensity, heat release rate, start of combustion, combustion phasing, combustion duration and combustion efficiency. Effect of different engine operating conditions on combustion characteristics of ethanol and methanol vis-à-vis conventional fuels in HCCI, PPC and RCCI combustion is discussed in the present chapter. Combustion stability and cyclic variation analysis of combustion parameters using statistical and nonlinear dynamic methods are also discussed in the last section of this chapter.

Keywords Combustion • Heat release rate • LTC • HCCI • PPC • RCCI • Ignition • Pressure rise rate • Cyclic variability • Knocking • Ringing • Gasoline • Ethanol • Methanol

6.1 Ignition Characteristics

6.1.1 Chemical Kinetics

In premixed charge compression ignition process, combustion rate is governed mainly by chemical kinetics (Sect. 2.2.2, Chap. 2). All the low temperature combustion (LTC) strategies are mostly kinetically controlled. The LTC process is primarily determined by autoignition chemistry (discussed in Chap. 3). In LTC engines, combustion starts by spontaneous auto-ignition of the fuel–air mixture because of the increase in pressure and temperature in the cylinder from compression work by piston. Auto-ignition process is mainly controlled by some important radical species such as H, OH, HO_2 and the more stable H_2O_2. The species H_2O_2

R.K. Maurya, *Characteristics and Control of Low Temperature Combustion Engines*, Mechanical Engineering Series, https://doi.org/10.1007/978-3-319-68508-3_6

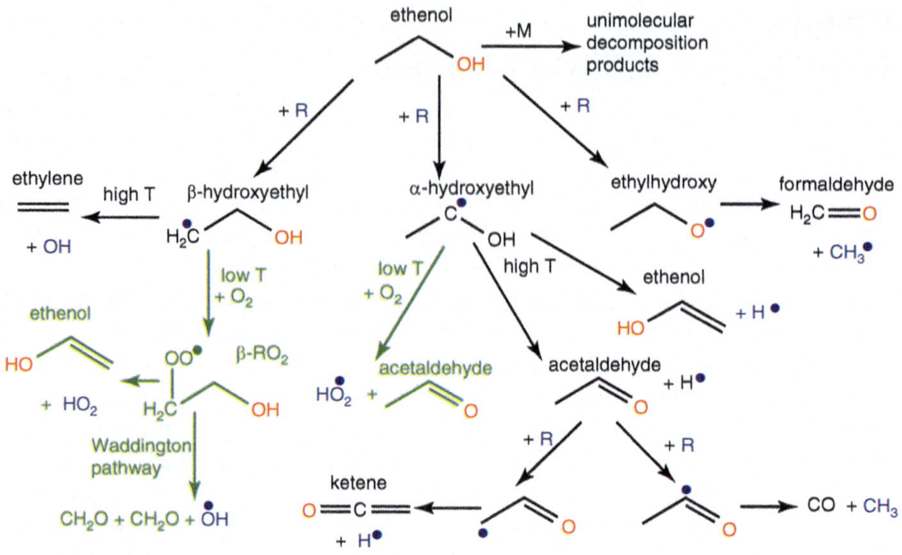

Fig. 6.1 Reaction path diagram for ethanol combustion. *R* indicates OH, HO_2 and *H* radicals in the diagram [2]

and HO_2 are associated with low temperature heat release (LTHR) process (described in Chap. 2). These radicals increase slowly as the engine cycle progresses and reach to a maximum point close to main combustion event [1]. Reactions involved in single- and two-stage heat release are already discussed for hydrocarbon fuels (Sect. 2.2.2 of Chap. 2). In this section, main reactions involved with combustion of ethanol and methanol are discussed for completeness as combustion characteristics of ethanol and methanol along with conventional hydrocarbon fuel are discussed in this book.

Ethanol is the most widely used alcohol in internal combustion engines. Ethanol is used as a neat fuel as well as an octane enhancer and oxygenate in gasoline. Ethanol is also used in various LTC strategies to extend the operating range and reduce the emissions. The combustion chemistry of alcohols is investigated in detail to understand the ignition characteristics in combustion engines [2]. The general reaction path diagram of ethanol combustion is presented in Fig. 6.1. The low temperature pathways are presented in green colour.

Ethanol oxidation starts with hydrogen abstraction reactions by OH, HO_2 and H radicals (denoted as R). At higher combustion temperatures, these radicals undergo β-scission wherein C–C and C–O bonds are favoured, leading to the formation of stable intermediates, including acetaldehyde, formaldehyde and ethylene. The intermediate acetaldehyde is mainly consumed through hydrogen abstraction from the aldehydic carbon, and the consequent radical then decomposes to form CO. Another minor pathway of acetaldehyde decomposition continues through abstraction from the methyl group resulting into ketene formation [2]. At low temperatures (i.e. below 1000 K), the α- and β-hydroxyethyl radicals can form intermediate species by reacting with O_2. The α-hydroxyethyl radical mainly produces acetaldehyde and HO_2 radical. These reactions govern the low temperature ignition characteristics of ethanol–air mixtures, particularly at higher combustion pressures. The low energy barrier for the chemically activated α-hydroxyethyl + O_2 route prevents conventional peroxy stabilization and isomerization routes, which eventually lead to chain branching processes in alkanes (e.g. ethane, propane, etc.) [2, 3]. The β-hydroxyethyl routes are of lower significance in combustion systems because initial hydrogen abstraction reactions by OH radicals strongly prefer α-carbon site at lower temperatures. The lack of low temperature reactivity in ethanol combustion is due to lack of an OH radical chain branching route [2].

A study developed reduced reaction mechanism for neat ethanol HCCI combustion and performed sensitivity analysis to determine the most important reactions in ethanol HCCI combustion [4]. The study conducted sensitivity analysis to quantify the sensitivities of reactions for different combustion parameters such as start of combustion (CA_{10}), maximum cylinder temperature, peak pressure rise rate and combustion duration (CA_{90-10}). Normalized sensitivity coefficients are estimated for every parameter, and the reactions having higher sensitivity coefficients are recognized as highly sensitive reaction to the corresponding parameter. Considering the high sensitivity reactions (in all combustion parameters) as the important reactions for ethanol combustion, important reaction in ethanol oxidation in HCCI combustion is presented in Appendix 1 (Table A1) along with their kinetic parameters. Sensitivity analysis showed that $HO_2 + C_2H_5OH$ [=] $H_2O_2 + CH_3CHOH$ is the most sensitive reaction in ethanol HCCI combustion [4]. This observation confirms that composition and decomposition of the intermediate species hydrogen peroxide (H_2O_2) and hydroperoxyl radical (HO_2) plays an important role in HCCI combustion process. Sensitivity of oxidation reactions is also dependent on quality of air–fuel mixture (equivalence ratio (φ) or relative air–fuel ratio (λ)). Figure 6.2 presents the variations of normalized sensitivity coefficients for CA_{10} at different λ. The figure shows that leanest mixture ($\lambda = 6$) has the highest normalized sensitivity coefficients followed by $\lambda = 4$ and 2 for most of the reactions. This shows that leaner fuel has higher sensitivity of the reactions toward start of combustion due to comparatively lower combustion temperature at lean engine operating conditions. Thus, leaner engine operating conditions have higher chances of misfire and limit the lower engine load of HCCI engine.

In naturally aspirated HCCI combustion, neat ethanol exhibits single-stage heat release, i.e. only high temperature heat release (HTHR) [5]. The LTHR depends on both fuel and engine operating conditions. The LTHR characteristics typically observed at higher intake pressure and lower engine speeds (Sects. 2.2.4.1 and 2.2.4.4 of Chap. 2). In HCCI combustion, gasoline exhibits single-stage heat release (only HTHR) at lower intake pressures and moderate engine speeds. At higher intake pressure (boosted engine operation), gasoline starts showing LTHR. Ethanol addition to gasoline (gasoline–ethanol blends) increases the required intake pressure, where gasoline starts exhibiting LTHR. Ethanol in gasoline blend acts as a radical sink and leads to both delaying ignition and reducing LTHR magnitude [6]. Another study showed that intermediate temperature heat release (ITHR) can

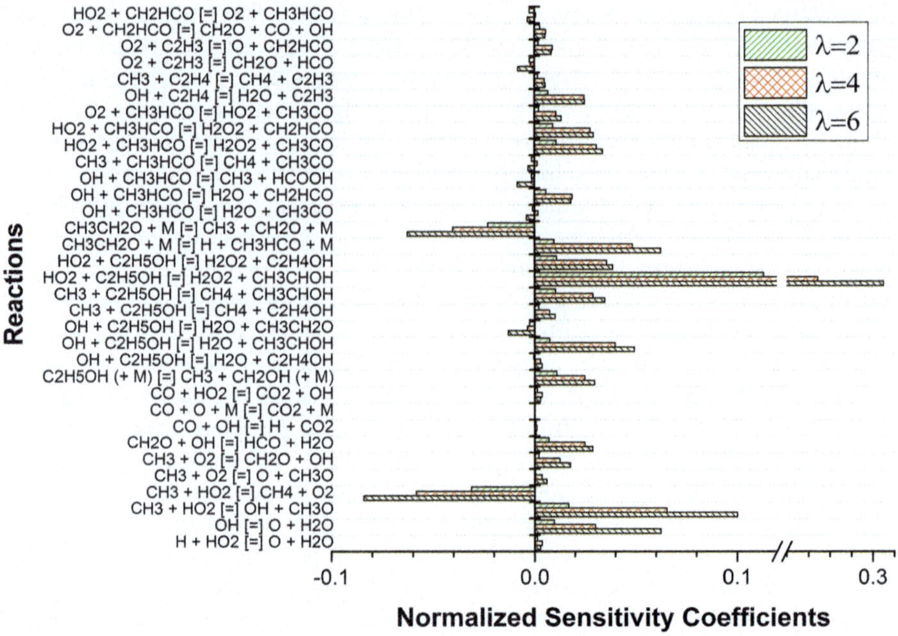

Fig. 6.2 Variations of CA_{10} sensitivity of ethanol oxidation reactions at different λ in HCCI combustion [4]

be introduced in fuels not exhibiting ITHR (such as ethanol) by adding highly reactive fuel (such as n-heptane) that shows LTHR [7]. At higher intake pressures, it is found that ITHR is governed by both ethanol and n-heptane oxidation reactions in HCCI combustion (Fig. 2.10).

Methanol is the shortest chain alcohol, and its combustion mechanism offers a basis for the combustion of larger alcohols. The recombination of methyl and hydroxyl radicals leads to the formation of methanol during hydrocarbon combustion, and several intermediate species important to methane combustion (e.g. CH_2O, CH_3O, CH_2OH and HCO) are also relevant to methanol combustion [2]. Figure 6.3 presents the general reaction path diagram for methanol combustion. The relative importance of the different pathways varies with temperature and stoichiometry [8]. Methanol decomposition starts with hydrogen abstraction by OH, HO_2 and H radicals. The branching ratio of hydrogen abstraction by OH radical is strongly favourable to the production of CH_2OH with CH_3O being a minor route. Formaldehyde formation takes place by both these radical intermediates, and it controls the reactivity of the system. Hydrogen abstraction from formaldehyde (HCHO) creates formyl (HCO) radical, eventually resulting into the formation of CO and CO_2 via the sequence $CH_2O \rightarrow HCO \rightarrow CO \rightarrow CO_2$ [2, 8]. Another study also indicated that methanol is unreactive below 800 K temperature and no intermediate or product species are formed. Above 800 K combustion temperature, methanol starts reacting to produce formaldehyde that subsequently oxidized to form CO and later CO_2 [2, 9].

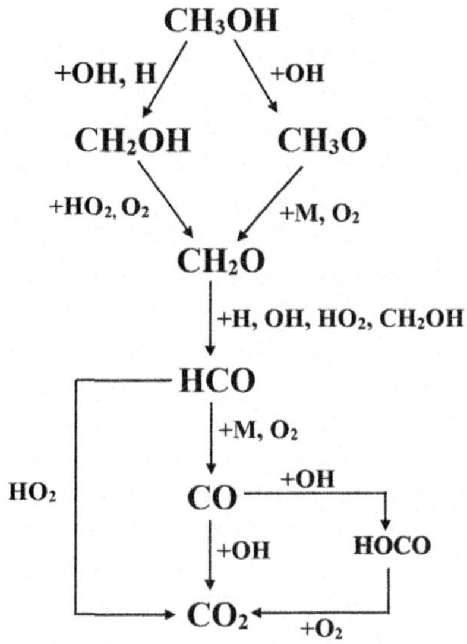

Fig. 6.3 Reaction path diagram for methanol combustion [8]

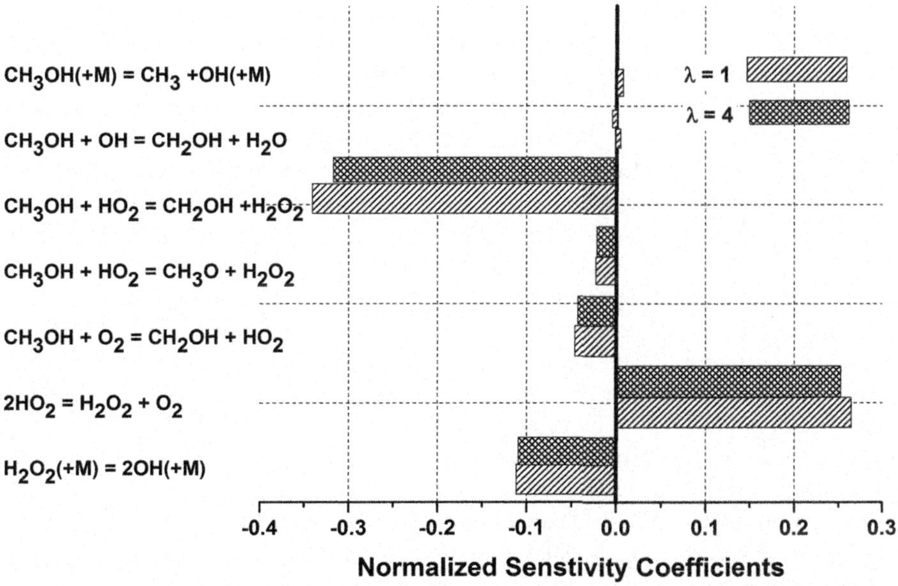

Fig. 6.4 Normalized sensitivity coefficients for the predicted ignition delay with respect to the rate constants of the reactions for methanol at $\lambda = 1$ and 4 [8]

Figure 6.4 depicts a sensitivity analysis for the ignition delay for methanol in rapid compression machine (RCM). The highest negative sensitivity coefficients (promoting ignition) are found for $CH_3OH + HO_2 \leftrightarrow CH_2OH + H_2O_2$, $CH_3OH + O_2 \leftrightarrow$CH2OH $+ HO_2$ and $H_2O_2(+M) \leftrightarrow OH + OH(+M)$, whereas recombination of HO_2 radicals, $HO_2 + HO_2 \leftrightarrow H_2O_2 + O_2$, inhibits ignition of methanol. Study showed that flow reactor predictions using methanol is also sensitive to the same subset of reactions as ignition delay [8]. A more complete and detailed discussion on combustion chemistry of alcohol can be found in reference [2].

Most attractive feature of alcohols for utilization in internal combustion engine applications is their auto-ignition propensity (i.e. knock resistance, octane rating). Methanol and ethanol have higher octane number than gasoline and thus used for improving the octane number of SI engine fuels to mitigate knocking and engine operation at higher compression ratios. Another characteristics of alcohols is their high octane sensitivity (S = RON − MON) (for details on octane sensitivity, see Chap. 3). This unique fuel property can have benefits in modern LTC engines and highly boosted direct injection spark ignition engines that prefer high octane sensitivity fuels [2, 10].

6.1.2 Ignition Temperature and Ignition Delay

All the LTC strategies are typically kinetically controlled process, and combustion starts with autoignition of fuel–air mixture in the combustion chamber. Chemical kinetics plays an important role in ignition characteristics and estimation of ignition delay. Ignition delay is an important parameter that affects mixing and combustion characteristics of fuel–air mixture in the engine cylinder. Ignition delay is affected by several operating parameters such as intake temperature and pressure, exhaust gas recirculation (EGR), fuel properties and fuel composition. In different LTC strategies, ignition occurs at different pressure and temperature conditions depending on engine operating conditions. Figure 6.5 illustrates the different regimes of ignition process that are important for autoignition of fuel in premixed compression ignition. The temperature region is divided into three main regions, namely, low temperature (LT), negative temperature coefficient (NTC) region and high temperature (HT) region, and a time scale is associated with each region (Fig. 6.5). In NTC region, a flat or opposite trend in ignition delay is observed with charge temperature. Reaction involved for appearance of NTC region is discussed in Sect. 2.2.2 of Chap. 2. Figure 6.5 depicts that in the LT region (τ_1) and HT (τ_3) region ignition delay trends show linear behaviour. Significant deviation in ignition delay is observed only in the intermediate temperature range.

A phenomenological auto-ignition tool called "3-Arrhenius model" is used to predict low and high temperature ignition delays of HCCI combustion processes. In two-stage heat release conditions, the low and medium temperature reactions occur sequentially. Therefore, they are directly added to one another. The high temperature reactions lead to a parallel single-stage ignition path. This is why it is an independent term. The overall ignition delay for the full temperature range can then be modelled as shown in Eq. (6.1) [11, 12].

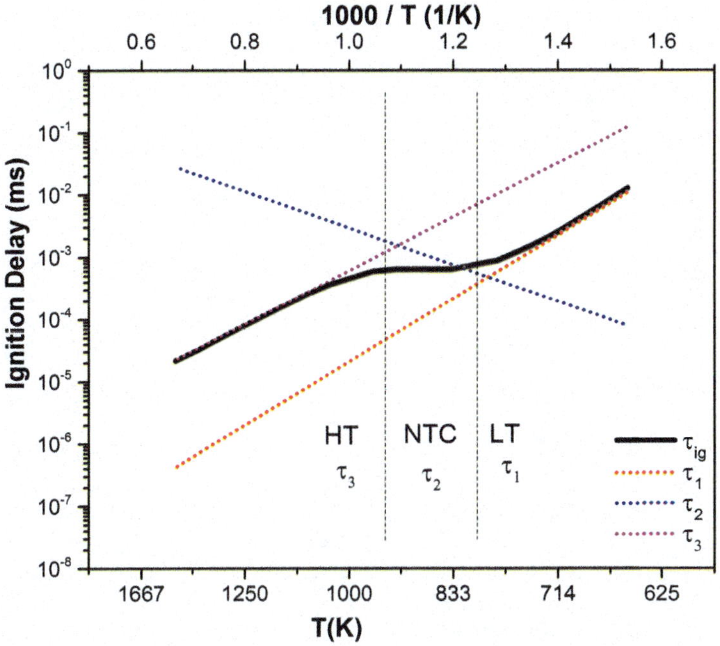

Fig. 6.5 Illustration of different regimes of ignition delay to explain overall ignition delay based on time scales τ_1,τ_2 and τ_3 [11]

$$\frac{1}{\tau} = \frac{1}{\tau_1 + \tau_2} + \frac{1}{\tau_3} \tag{6.1}$$

where individual time scales in different regions are estimated from an Arrhenius-type correlation given by Eq. (6.2) [11].

$$\tau_i = A_i \left(\frac{p}{p_{\text{ref}}}\right)^{\beta_i} T^{b_i} \exp\left(\frac{T_{A,i}}{T}\right) \phi^{\gamma_i} \quad \text{For } i = 1, 3 \tag{6.2}$$

where A_i, β_i, b_i, $T_{A,i}$ and γ_i are the kinetic model parameters and φ is equivalence ratio, which is introduced to describe φ dependency at intermediate temperature range.

Equivalence ratio affects the ignition delay in the intermediate temperature range and has a lower dependence in low and high temperature range. Figure 6.6 depicts the effect of equivalence ratio on ignition delay for n-heptane and ethanol. Figure 6.6a shows that NTC behaviour is observed for richer mixture ($\varphi = 0.4$), while leaner mixture does not show the NTC regime. No φ dependence is observed for pre-ignition delay times (shown with half open symbol) for pre-ignition happening at $T < 900$ K (Fig. 6.6a). Ethanol is single-stage heat release fuel; therefore, no NTC regime is observed (Fig. 6.6b). Another study conducted on n-heptane–air mixture also confirmed weak dependence of ignition delay on equivalence ratio

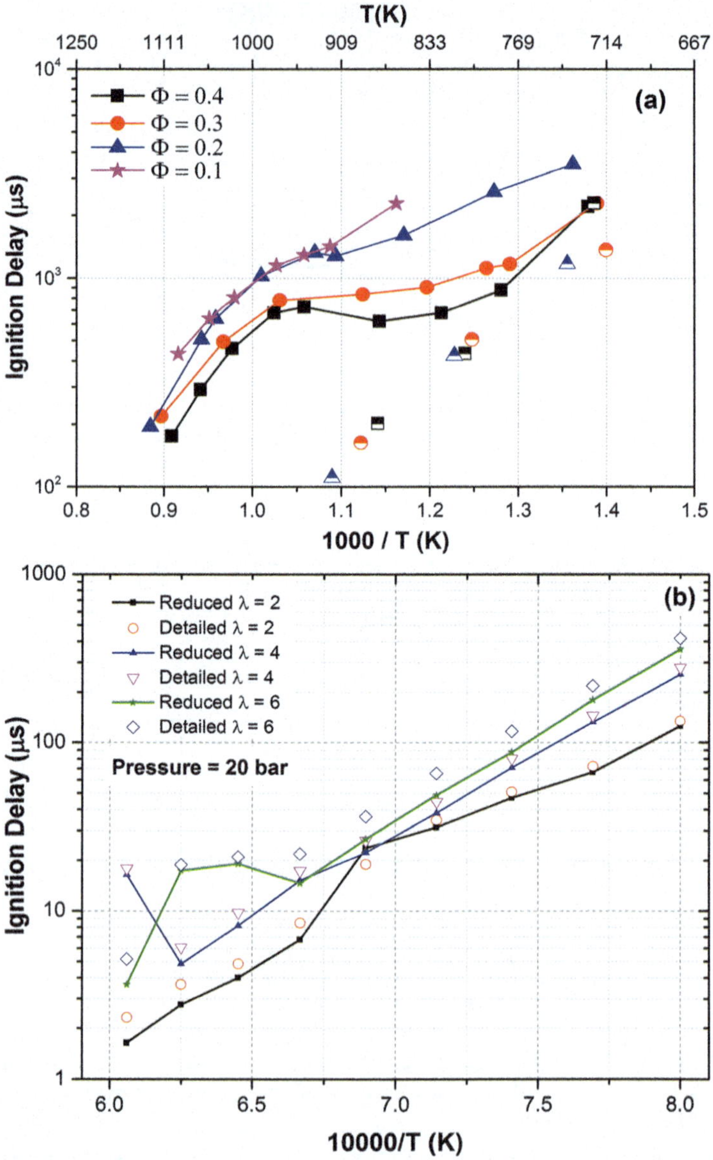

Fig. 6.6 Variations of ignition delay with φ for (**a**) n-heptane at 50 bar [13] and (**b**) ethanol at 20 bar (Adapted from [4])

outside of the intermediate temperature range [14]. Thus, during engine experiments, φ sensitivity can be utilized to increase the local ignition delay by concentration gradient (fuel stratification) in the charge (see Sects. 2.2.4.1 and 2.2.4.5 of Chap. 2).

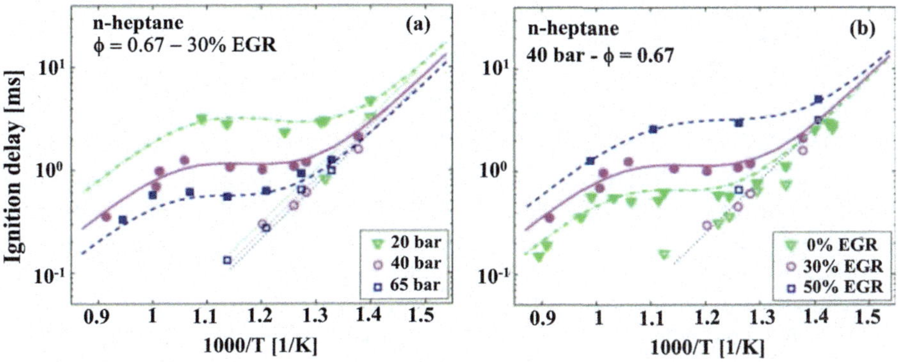

Fig. 6.7 Variations of ignition delay with (**a**) combustion pressure and (**b**) EGR for n-heptane [11]

Other essential parameters which affect the ignition delay in LTC engines are pressure and EGR. The EGR is one of the important parameters used in PPC and RCCI engines at higher load conditions. It is well known that ignition process for several fuels can be pressure sensitive (e.g. long-chain fully saturated alkanes). Figure 6.7 shows the effect of pressure and EGR on ignition delay for n-heptane fuel. Figure 6.7a shows the ignition delay for second-stage heat release decreases with increase in pressure as expected. Figure 6.7a also depicts that pressure has a prominent effect on the ignition delay in the intermediate and subsequent high temperature regimes. It can also be noticed that first-stage reactions mainly depend on only temperature and not dependent on pressure (Fig. 6.7a). These observations illustrate that fuel reactivity is likely to increase for a given mixture at higher pressure. Another study showed that higher intake pressure has significant effect on gasoline reactivity in the intermediate temperature range, while similar high octane PRF fuels and ethanol do not show pressure dependency on fuel reactivity [15]. In RCCI combustion engines, the fuel pressure sensitivity can be of more importance at higher engine loads and efficiency conditions. High-pressure conditions typically exist at higher engine loads. Similarly, high pressure (dilution) is often used to reduce the emissions and increase the engine efficiency. Therefore, higher engine load conditions may enhance pressure-sensitive fuel reactivity more than the desired. The increased reactivity may lead to lower operational window or performance; or engine may need to operate at comparatively lower compression ratios [16].

At higher engine loads in LTC engines, higher amount of EGR is often used. Ignition delay increases with increase in EGR fraction (Fig. 6.7b). Experimental data also indicated the flattening of NTC regime with increase in EGR fraction, and at higher EGR (e.g. 50%), NTC behaviour seems to supress completely [11]. To account the effect of EGR dilution and charge dilution on ignition delay, ignition delay model (Eq. 6.2) is extended as shown in Eq. (6.3) [11].

$$\tau_i = A_i \left(\frac{p}{p_{\text{ref}}}\right)^{\beta_i} T^{b_i} \exp\left(\frac{T_{A,i}}{T}\right) \phi^{\gamma_i} \left(\frac{[N_2]}{[O_2]}\right)^{d_i} \exp\left(\frac{[N_2]e_i}{[O_2]T}\right) \quad \text{For } i = 1,3 \qquad (6.3)$$

In the extended equation, two additional terms (an exponential weighted factor and a simple power law term) are introduced along with two new model parameters, d_i and e_i. The exponential term accounts for altered temperature dependence, whereas the power law term reduces or prolongs the ignition delay with the variation in EGR fraction [11]. The same study also indicated that at heavy EGR conditions, behaviour of reactive fuels (such n-heptane) is similar to gasoline, which has lower tendency to exhibit NTC and more tendency for ITHR.

The HCCI combustion is initiated by a spontaneous auto-ignition of the charge, when temperature and pressure of cylinder exceed a certain level. The temperature and pressure at the point of auto-ignition is time dependent. Longer available time for autoignition has a decreasing effect on the auto-ignition temperature and pressure. Typically in HCCI combustion, autoignition position is estimated from heat release curve because there is no direct control of ignition timing in HCCI engine. To estimate the accurate value of autoignition point, a tuned heat release model is essential (discussed in Sect. 6.2.1). Definition of autoignition position varies in different studies. A study defined the ignition position as one crank angle degree (CAD) before the crank angle (CA) position of 1% heat release [17]. Another study defined the autoignition position as CA position when value of heat release rate exceeds 0.2 J/CAD [18]. This study also used 1% burn position as ignition point, and ignition temperature has similar trend that observed with threshold heat release rate value (0.2 J/CAD) estimation.

Figure 6.8 shows the variations of ignition temperature and required intake temperature for autoignition of ethanol, methanol and iso-octane in HCCI combustion. Ignition temperature is calculated at 1 CAD before the 1% heat release position. The figure shows that ignition temperature is dependent on λ but with lower sensitivity (curve is flat as mixture becomes leaner). Both ethanol and methanol have lower ignition temperatures than iso-octane, although they have higher octane number (ON) than iso-octane. This observation indicates that the ON is not the only measure of the fuel's auto-ignition temperature. Other factors such as heat of vaporization (charge cooling) also affect the autoignition temperature. However, study found that for hydrocarbons with similar heat of vaporization and molecule size, the ON is quite good measure of the ignition temperature [17]. Typically, ethanol requires the lowest intake temperature and increases with methanol and iso-octane (Fig. 6.8). The requirement of inlet temperature increases with λ because of lower reactivity of lean fuel mixture, lower residual temperature at leaner mixture and lower cylinder wall temperature at lean engine operation [17].

Figure 6.9 shows the effect of ethanol composition in oxygenated reference fuels on autoignition temperature and LTHR for HCCI combustion. Autoignition position is defined as CA position where heat release rate exceeds 0.2 J/CAD. Five symbol marks in the graph indicated five different intake air temperatures. Reference fuels consist of n-heptane (H) and ethanol (E) and the remaining is iso-octane. Volume percentage of fuel composition is indicated with preceding number. Fuels

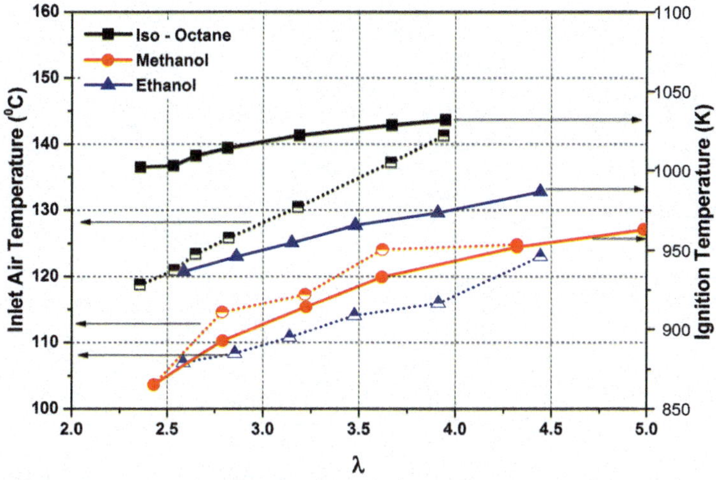

Fig. 6.8 Variations of ignition temperature and required intake temperature as function of λ for ethanol, methanol and iso-octane (Adapted from [17])

designated as HxTyEz contains x% of n-heptane, y% of toluene, z% of ethanol and (100-x-y-z)% of iso-octane. In fuels containing constant n-heptane (H30E5, H30E10 and H30E20), autoignition temperature increases with increase in ethanol content (Fig. 6.9a). It can also be noticed that autoignition temperature increases more significantly at higher intake temperatures because at higher intake temperature LTHR decreases. The LTHR fraction decreases with increase in ethanol fraction and became zero for 60% ethanol (Fig. 6.9b). Ethanol acts as sink for radicals responsible for LTHR (Sect. 6.1.1). Similar trend is also observed for fuel containing only heptane and ethanol (H40E60, H50E50 and H60E40) that autoignition temperature increases with increase in ethanol content. Although these fuels have higher content of reactive fuel (n-heptane), they have higher autoignition temperatures (shown with solid lines) in comparison to fuel having lower content of heptane (30%) along with ethanol and iso-octane (shown with dashed lines). It can also be observed from Fig. 6.9 that fuels containing lower amount of LTHR have higher autoignition temperatures. The fuel effects on autoignition characteristics in LTC engines are discussed in Sect. 3.1.4 of Chap. 3.

6.2 Heat Release Characteristics

6.2.1 Heat Release Estimation

Heat release is typically used to analyse the combustion characteristics in internal combustion engines. Heat release information cannot be measured directly from engine cylinder but generally calculated by processing of measured cylinder

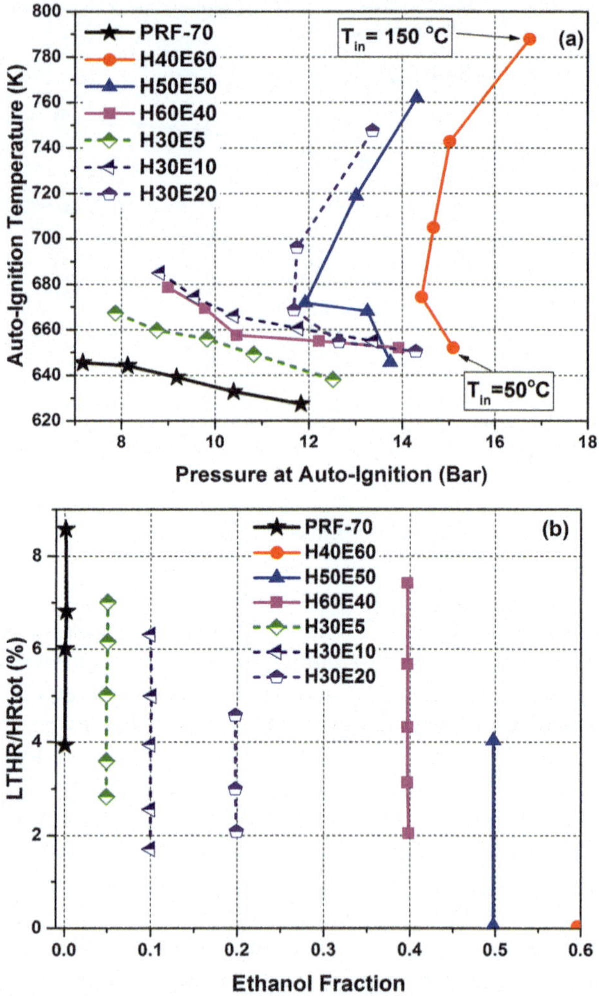

Fig. 6.9 Effect of ethanol concentration on (**a**) autoignition temperature (**b**) LTHR in HCCI engine (Adapted from [18])

pressure data. Processing of in-cylinder pressure data can produce valuable informations that can be utilized for further development, optimization and calibration of combustion engines. Informations derived from cylinder pressure data processing related to engine combustion, performance and emissions are summarized in Table A2 (Appendix 1) along with their signal processing methods. In-cylinder pressure data is typically measured from piezoelectric pressure sensor flush mounted on cylinder head of the engine. To measure the cylinder pressure data on crank angle basis, crank angle encoder is used. Schematic diagram of typical in-cylinder pressure measurement system in HCCI engine is shown in Figs. A1 and A2 of Appendix 1. Piezoelectric pressure sensor typically provides the charge that is converted in voltage by charge amplifier. The analog voltage signal is converted into digital and stored by high speed data acquisition system. Four-step pressure signal processing consisting of (i) absolute pressure correction, (ii) phasing w.r.t. crank angle, (iii) cycle averaging and (iv) smoothening (filtering the noise in signal) is used to obtain sufficiency accurate cylinder pressure data [19]. The cylinder pressure is further analysed to estimate combustion characteristics.

Cyclic variations in measured pressure data require averaging of numerous successive cycles for accurate calculation heat release from cylinder pressure data. Averaging of measured data can eliminate the point-to-point variations because of random signal noise, but it cannot remove systematic errors in the signal. The mean input parameters (mass of fuel and air, engine speed, etc.) are commonly used to calculate heat release rate (HRR). Thus, average in-cylinder pressure data is required for accurate representation of combustion cycle. To estimate the sufficient number of cycles for the average data in HCCI combustion, a standard deviation-based method is proposed [19–21]. Averaging of cycles is not possible for combustion analysis on cycle-to-cycle basis or during transient operation of engine. Therefore, filtering of measured cylinder pressure signal is necessary as it is affected by various factors such as signal noise, engine operating condition, etc. Signal noise in the measured pressure signal is present from various sources such as (i) conversion of pressure to electrical signal in the sensor (because of thermal effects, lack of linearity in sensor, sensor resonance, etc.), (ii) signal transmission (bad connections, electrical effects, etc.), (iii) analog to digital conversion and (iv) combustion chamber resonance [21]. Cyclic variations can occur even at steady-state conditions due to variations in the amount of fuel and air supplied, fuel and temperature stratification, air motion in the cylinder, etc. [19].

Figure 6.10 shows the variations in the standard deviation envelope for different numbers of cycles in cylinder pressure for without and with filer case in HCCI combustion. To evaluate the standard deviation in the figure, 3000 consecutive engine cycles are used. For different numbers of cycles (n), average cycle with function of CA position is obtained using the equation $\bar{X}_{n,j}(\theta) = \frac{1}{n}\sum_{i=j}^{j+n} X_i(\theta)$, where X_i is pressure signal during ith cycle and θ is the CA position. For each chosen number of cycles, a set of $N-n+1$ mean signals $\bar{X}_{n,j}(\theta)$ can be estimated. For each set, standard deviation as a function of CA position can be evaluated as $\sigma_n(\theta) = \sigma_n\{\bar{X}_{n,j}(\theta)\}$. Minimum and maximum value of standard deviation is used for standard deviation envelope (shaded region with a colour) for different numbers of cycles (Fig. 6.10). The figure shows that the value of standard deviation decreases with increase in the number of cycles for both the cases (with and without signal filter). The area between the maximum and minimum standard deviation curve shrinks with the increase in the number of cycles. The figure also shows that the maximum variations in the cylinder pressure are observed during combustion period close to top dead centre (TDC). Filtering the cylinder pressure data using Savitzky–Golay filter reduces the maximum value of standard deviation as well as the area between the maximum and minimum standard deviation curve significantly. This suggests that filtering the signal and selection of optimal number cycle is required for accurate heat release analysis.

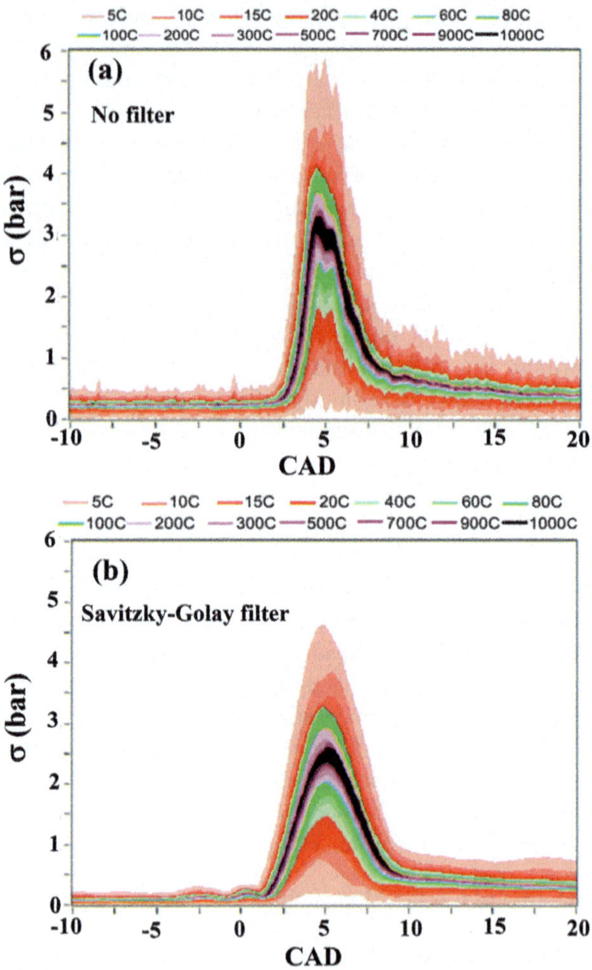

Fig. 6.10 Variations of standard deviation in average cycles for different numbers of cycles chosen to average for measured cylinder pressure signal with and without filter at 1200 rpm and $\lambda = 2.1$ in HCCI engine [19]

Calculation of standard deviation for crank angle basis is also conducted for pressure rise rate (PRR) and HRR [19]. Figure 6.11 shows the variations in the maximum difference between the maximum and minimum envelope curve $\{(\sigma_{max}-\sigma_{min})_{max}\}$ of the standard deviation during entire cycles with selected number of cycles for various HCCI engine operating conditions. Figures show that maximum of $(\sigma_{max}-\sigma_{min})_{max}$ decreases quickly with the increase in the number of averaged cycles and by using digital signal filters to eliminate high-frequency signal noise. Figure 6.11 also depicts that there are a number of cycles afterward; additional number of averaging cycles does not reduce the value of $(\sigma_{max}-\sigma_{min})_{max}$. For randomly selected threshold value of $(\sigma_{max}-\sigma_{min})_{max}$ at 5 J/CAD to evaluate optimum number of cycles based on HRR, then without filtering the optimum number of cycles is more than 1000 cycles (Fig. 6.11a). For same threshold value of (5 J/CAD), the optimum number of cycles is 300, 800 and 600 using

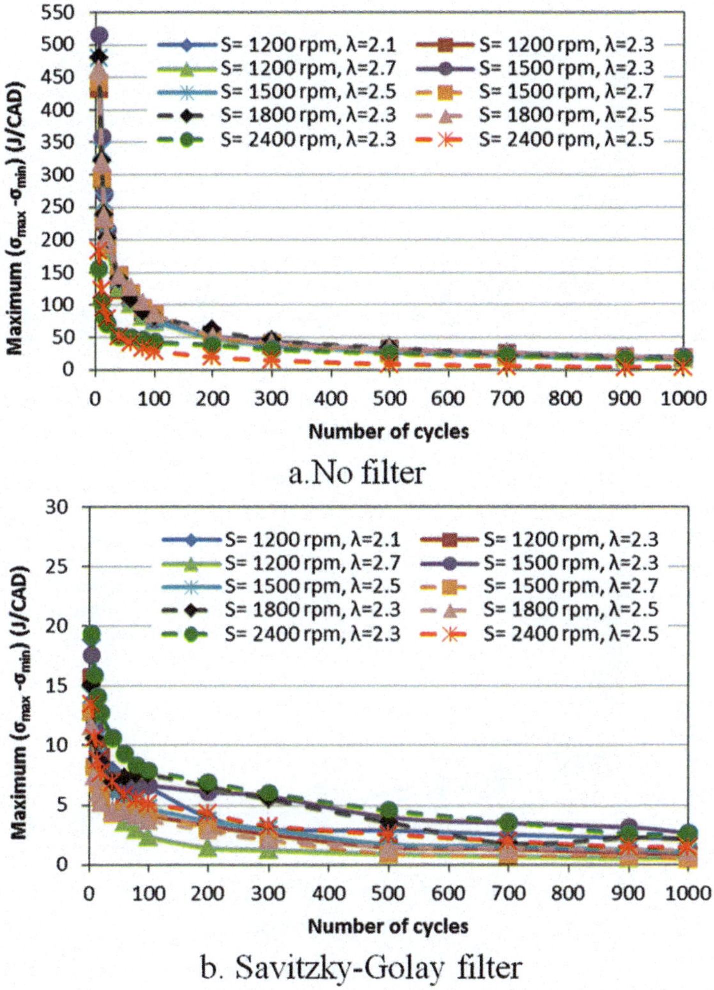

Fig. 6.11 Variations of the maximum difference between the maximum and minimum envelope of the standard deviation with number of cycles selected for averaging for rate of heat release under different load/speed conditions of the HCCI engine. (**a**) No filter, (**b**) Savitzky–Golay filter, (**c**) zero phase filter and (**d**) Butterworth filter [19]

Savitzky–Golay, zero-phase and Butterworth filter, respectively, for all the test conditions (Fig. 6.11b–d). The optimum number of cycles is found to be lower for leaner engine operating conditions. Figure 6.11 clearly indicates that the optimum number of cycles required for heat release analysis is dependent on the engine operating conditions as well as higher-frequency signal noise removal technique [19]. A more complete discussion using cylinder pressure, PRR and HRR to estimate optimal number of cycle is provided in study [19].

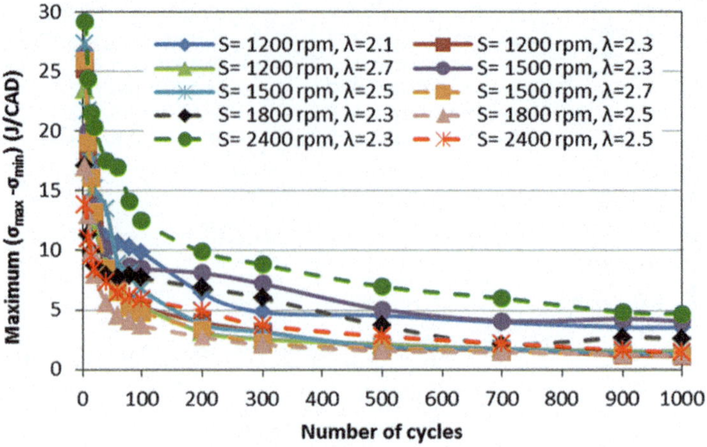

c. Zero Phase filter

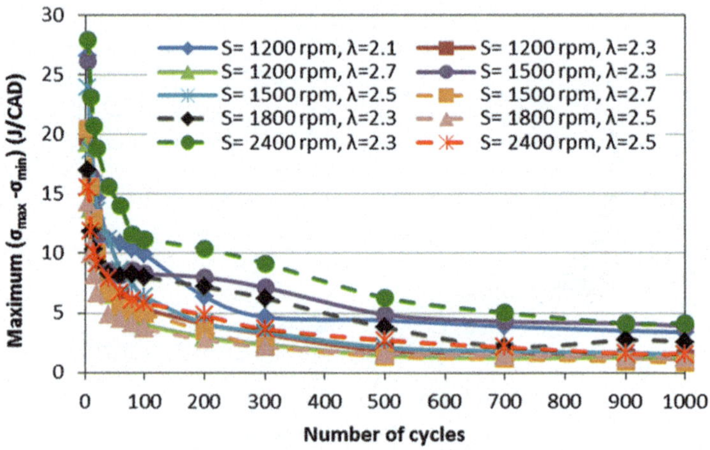

d. Butterworth filter

Fig. 6.11 (continued)

The heat release rate is calculated from measured cylinder pressure by well-known Eq. (6.4). The detailed derivation of equation along with assumption taken to derive the equations is presented in section "Heat Release Rate Analysis" of Appendix 2.

$$\frac{\partial Q}{\partial \theta} = \frac{\gamma}{\gamma - 1} p \frac{dV}{d\theta} + \frac{1}{\gamma - 1} V \frac{\partial p}{\partial \theta} + \frac{\partial Q_w}{\partial \theta} + \frac{\partial Q_{\text{Crevice}}}{\partial \theta} \tag{6.4}$$

Several input parameters are used to accurately calculate the HRR such as manifold pressure for pegging the pressure data, phasing between cylinder pressure and crank angle position, compression ratio for instantaneous volume calculation and heat transfer coefficients for wall heat transfer calculations. In order to have correct evaluation of HRR, the input parameters need to be tuned such that motoring cylinder pressure has zero value of HRR. Figure 6.12a shows the well-tuned motoring heat release curve (HRR is close to zero value) by tuning several parameters such as intake air pressure, intake air temperature, phasing between the acquired pressure and CA position, compression ratio and heat transfer coefficient. Fuel is not injected in the cylinder during motoring; therefore, no heat release is expected for correctly tuned input parameters. In this tuning procedure for input parameters, offline iterative processing of motored cylinder pressure data is used [22].

Correctly tuned values of input parameters obtained from the analysis of motored cylinder pressures can be used for the analysis of fired cycles also. Figure 6.12b presents the typical pressure and HRR in HCCI combustion at 1500 rpm and $\lambda = 2.1$ for well-tuned conditions. The figure shows that the HRR is almost zero at locations other than combustion duration. The HRR calculated using accurately tuned input parameters can be used for ignition position estimation in HCCI engines (Sect. 6.1.2).

Error in input parameters can cause significant errors in estimation of combustion and performance parameters. A summary of amount of errors that can be introduced during calculations of various performance- and combustion-related parameters in HCCI combustion by measurement errors is presented in Table 6.1.

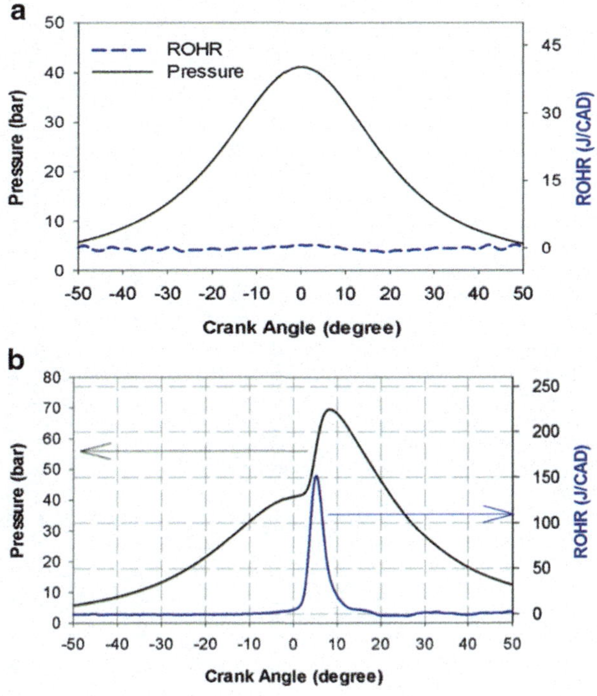

Fig. 6.12 In-cylinder pressure and rate of heat release (ROHR) curves for (**a**) motoring cycle and (**b**) a HCCI firing cycle in well-tuned parameter condition [22].

Table 6.1 Effect measurement errors on HCCI engine combustion and performance parameters [22]

	IMEP (%)	PMEP (%)	HRR_{max} (%)	CA_{50} (CAD)	CD (CAD)	CAD HRR_{max}	Total HR (%)	T_{max} (%)
TDC +1(CAD) ↑	8.87–11.66↑	1.23–0.8↑	4.58–7.80↑	1.0–0.5↑	0.167–0.5↑	1.166↑	9.88–10.81↑	3.21–3.94↑
r_c + 1↑	NE	NE	6.90–3.36↓	0.33–1.0↑	5.16–4.67↑	–	0.99–1.10↓	4.51–3.45↓
IP −0.1 (bar) ↓	–	0.02↓	4.48–3.24↑	0.16–0.33↓	0.50–1.16↓	0.166↑	3.87–2.53↑	10.68↑
IT +10 °C↑	NE	NE	1.02–0.78↑	–	0.333↓	–	0.63–0.67↑	2.54↑
HT + 15% ↑	NE	NE	0.16–0.41↑	–	0.5–0.66↑	–	2.17–2.52↑	–

NE no effect, r_c compression ratio, *IT* intake temperature, *IP* intake pressure, *HT* heat transfer, *TDC* top dead centre, *HR* heat release, *CD* combustion duration, *PMEP* pumping mean effective pressure, ↑ increase, ↓ decrease

The calculation of different parameters (from measured cylinder pressure data) such as indicated mean effective pressure (IMEP), maximum mean gas temperature (T_{max}), combustion duration, etc. is presented in Appendix 2. Table 6.1 shows that TDC phasing has maximum effect on IMEP and total heat release estimation. Table 6.1 shows that measurement error of 1 CAD in TDC position can cause an error in the range of 8.87–11.66% and 9.88–10.81% in IMEP and total heat release, respectively, depending on engine operating conditions. Other parameters have comparatively lower effect on heat release calculations.

6.2.2 Heat Release Rate in LTC Engines

Single zone heat release model is typically used for calculating heat release rate (HRR) from measured in-cylinder pressure data (details provided in Appendix 2). The HRR measures the rate of conversion of chemical energy of fuel into thermal energy during combustion in the cylinder. The HRR directly affects pressure rise rate (PRR) and, accordingly, the torque produced in the cycle. The HCCI engines have comparatively higher HRR because of simultaneous autoignition of entire premixed charge in the cylinder. Figures 2.12 and 2.18 (Chap. 2) show the typical HRR variations at different equivalence ratios and intake temperatures for gasoline HCCI combustion in naturally aspirated engine conditions. Single-stage HRR curve is observed for gasoline, and HRR increases rapidly with increase in intake temperature and with richer fuel–air mixtures. Figure 6.13 shows the HCCI combustion HRR for ethanol and methanol for different relative air–fuel ratio (λ) at 1200 rpm. It can be observed from the figure that with increase in λ (leaner mixture), the HRR and PRR decrease along with increase in duration of heat release. Ethanol and methanol both have single-stage heat release process as they are low autoignition reactivity high octane fuels. The figure shows that initially fuel–air mixture starts burning at comparatively lower combustion rate and increases rapidly to a very high value around halfway through the combustion event. After achieving to maximum point, combustion rate decreases to zero in almost identical manner as it increased resulting in to almost symmetrical HRR curve. HCCI combustion is primarily controlled by chemical kinetics (Chap. 2) and is directly affected by λ and thermodynamic state of the charge in the cylinder. Figure 6.13 depicts that highest HRR is obtained at richest charge operating condition (lowest λ) and the lowest HRR is found for the leanest mixture (highest λ) at constant inlet air temperature.

Heat release and autoignition process in HCCI combustion process also depend on fuel characteristics as well as engine operating conditions. High reactivity (lower octane number) fuels typically exhibit two-stage heat release (LTHR and HTHR), and lower reactivity fuels exhibit single-stage heat release (HTHR) in HCCI combustion process. Reactions involved in the LTHR and HTHR are discussed in Sect. 2.2.2 of Chap. 2. Figure 6.14 shows the variations in LHTR at different engine operating conditions and ethanol blends in HCCI combustion. The figure shows that magnitude of LTHR decreases with increase in intake temperature and engine

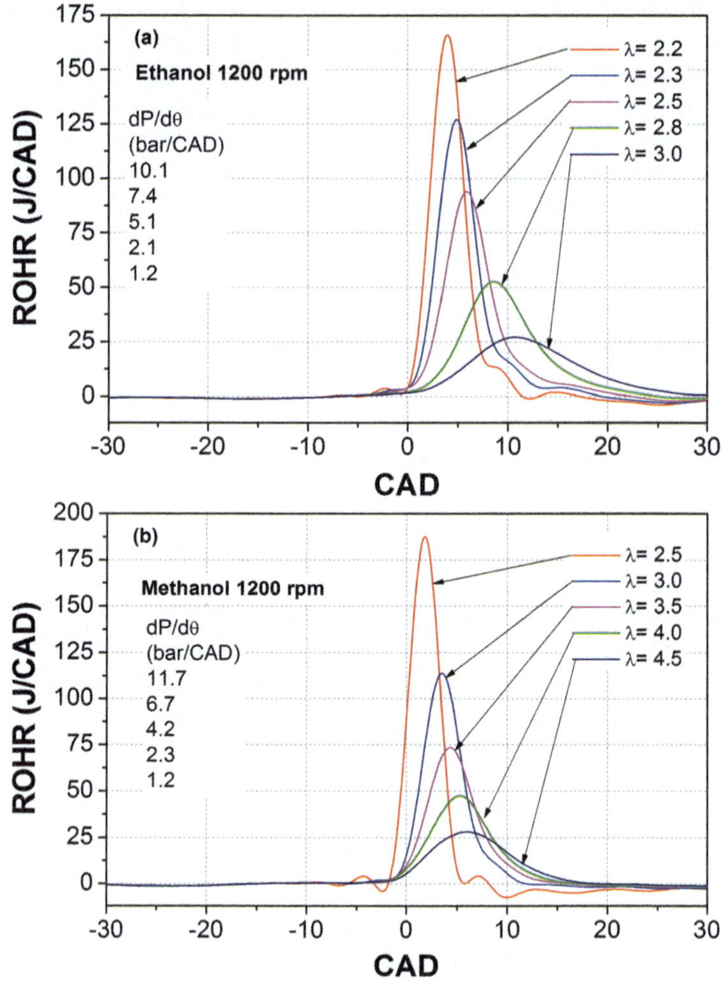

Fig. 6.13 The HCCI combustion rate of heat release (ROHR) curves for (**a**) ethanol and (**b**) methanol for different λ at 1200 rpm (Adapted from [24])

speed. The LTHR is a fairly slow oxidation reactions that occurs actively in 760–880 K temperature range. There is less time for the LTHR to take place before 880 K as the engine speed is increased, which leads to lower LTHR at higher engine speed. Inhibiting effect of both increasing engine speed and intake temperature on low temperature chemistry is observed (Fig. 6.14). The LTHR is also get advanced by increasing intake temperature and engine speed. The ethanol addition reduces the LTHR drastically due to radical (OH and HO_2 radicals) scavenging nature of ethanol. Ethanol addition fully suppresses the LTHR at higher engine speed and intake temperature (E15, 900 rpm, 149ºC in Fig. 6.14). Effectiveness of ethanol is reduced with increasing speed and intake temperature in suppressing LTHR

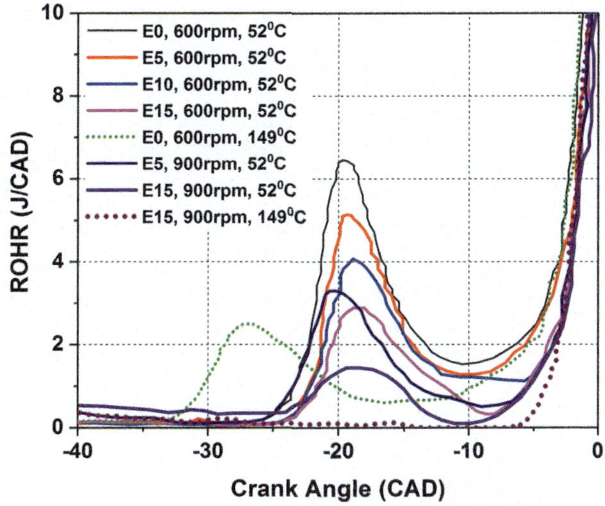

Fig. 6.14 The HCCI combustion LTHR variations at different engine operating conditions and ethanol blends (Adapted from [23])

(Fig. 6.14). Ethanol addition also retards the LTHR onset position, leading to competing mechanism, with lower autoignition tendency from higher ethanol concentration [23]. The factors jointly also explain the nonlinearity in octane boosting tendency or HCCI number of ethanol.

Intake air pressure significantly affects the HRR characteristics in HCCI combustion (Sect. 2.2.4.3 of Chap. 2). Autoignition reactivity of gasoline changes at higher intake pressure conditions (Fig. 2.15). Figure 6.15 shows the effect of boost pressure on HRR of different ethanol blends in HCCI combustion. In figure 6.15, the HRR is represented as ratio of HRR to the total heat release (THR) in the cylinder. The figure shows that at lower intake pressure (1 bar), there are gasoline and ethanol blends showing only HTHR. At higher intake temperatures, ITHR appears before HTHR for gasoline and lower blends of ethanol. Neat ethanol operation does not exhibit the ITHR at higher intake pressure (Fig. 6.15). The reactions involved in ITHR of gasoline are discussed in Sect. 2.2.2 of Chap. 2. This ability of having ITHR of fuels helps in retarding combustion phasing with acceptable level of cyclic variations (combustion instability limits). The HCCI engine needs to be run at retarded combustion phasing to achieve higher engine loads. In typical gasoline and gasoline-like fuels, the stability with combustion phasing retard is related to the rate of temperature rise leading up to the onset of the main combustion (i.e. just prior to hot ignition) [25]. To maintain a strong rate of temperature rise at retarded combustion phasing after TDC, sufficient heat release by early autoignition reactions is required to overcome the expansion cooling produced by the piston motion. This requirement is fulfilled by ITHR reactions, and the fuels showing ITHR reactions can achieve higher combustion phasing retard and higher engine loads. Typically, ITHR occurs at temperatures below those of hot ignition but above those of the more well-known LTHR, which can be observed on crank angle basis in Fig. 6.15.

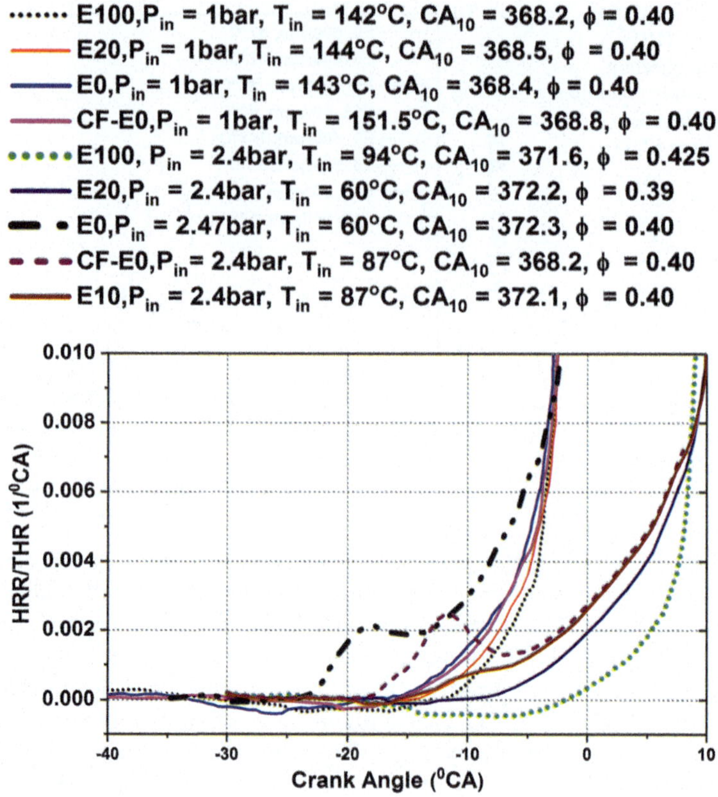

Fig. 6.15 Effect of intake pressure on ITHR in HCCI combustion for different ethanol-blended fuels (Adapted from [25])

The partial fuel stratification (PFS) is another strategy to achieve higher engine loads in LTC regime (see Sect. 2.2.4.6 of Chap. 2). In PFS, some portion of fuel is injected directly in the cylinder to create the stratification, which lowers the pressure rise rate due to sequential autoignition of the charge. Figure 6.16 shows the effect of single and dual direct injection on HRR in HCCI combustion different amount of direct fuel injection quantity. In single fuel injection case, fuel is injected at start of injection (SOI) timing 60° CA from start of intake stroke. This injecting timing is found to have higher thermal efficiency. In case of dual direct fuel injection case, injection timing is held constant at 60° and 305° CA position from intake TDC for constant combustion phasing (CA_{50}) at 372ºCA. The charge equivalence ratio (φ_m) is constant for all the test. Hence, fuel fraction of first injection proportionally decreases, when second injection amount is increased (0 to 35%). Constant combustion phasing is maintained by adjusting amount of EGR. Experiments are conducted at intake pressure of 2.4 bar as fuel is found to be strongly φ-sensitive at this boost pressure.

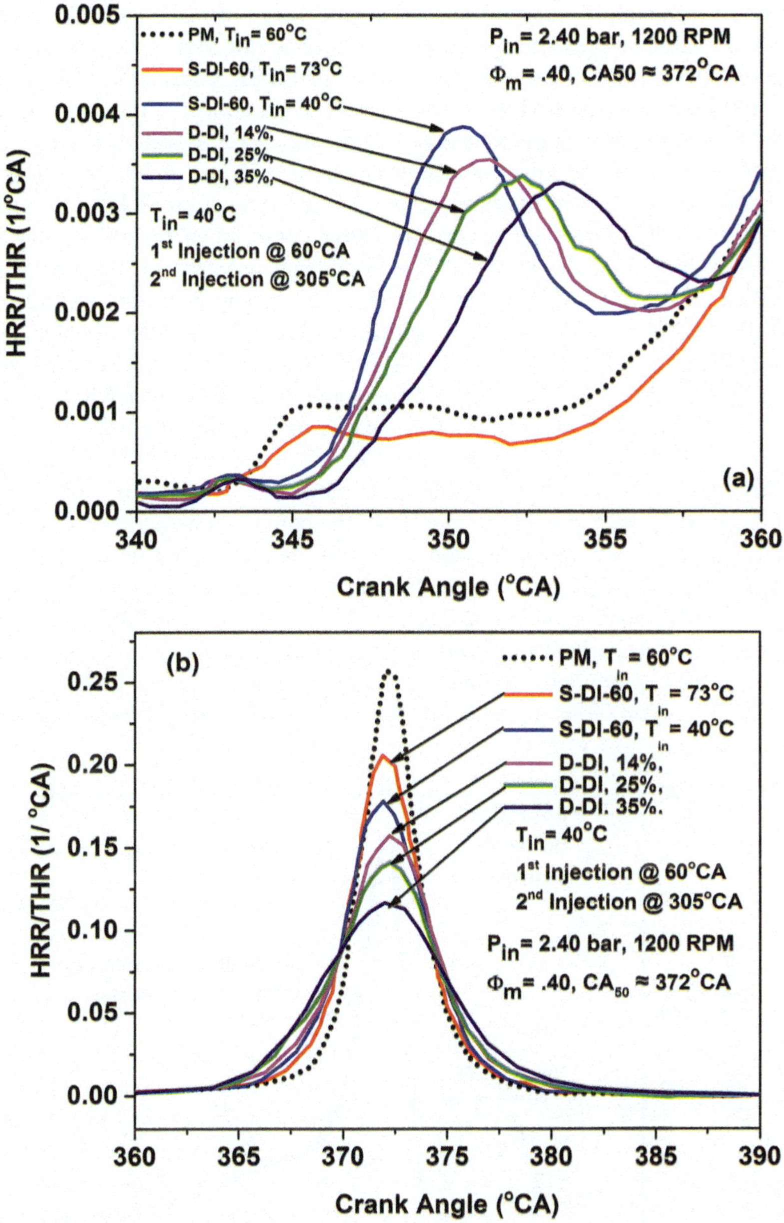

Fig. 6.16 The effect of single and dual direct injection on HRR at constant combustion phasing in HCCI engine at 1200 rpm (Adapted from [26])

Figure 6.16 shows that increasing the amount of fuel stratification significantly reduces the peak HRR and increases the combustion duration at constant combustion phasing. This occurs because the increased stratification greatly enhances the amount of sequential autoignition of the charge, which leads to a slower combustion-induced pressure rise along the entire combustion event [26]. Although the amount of LTHR is very small, the overall LTHR consistently reduces with increased late-DI fraction (Fig. 6.16a). The LTHR almost diminishes at higher intake temperature for premixed and single injection case, which indicates that fuel becomes φ sensitive at lower intake temperature. Figure 6.16 also shows that compared to the premixed fuelling, the single direct injection (S-DI) also provides a reduction in the PRR and a corresponding decrease in the peak HRR. This strongly confirms that the S-DI, despite being injected early in the intake stroke to create a fairly well-mixed mixture, produces a moderate amount of fuel stratification that reacts sequentially and lengthens the combustion duration. The stratification comparatively increases in dual direct injection. Study concluded that the double DI is a very effective fuelling method to reduce the peak HRR and increase the burn duration or to advance the CA_{50} while maintaining the PRR constant.

Premixed combustion using gasoline-like fuels with high fuel stratification is typically known as partially premixed combustion (PPC) (see Sect. 2.5.2 of Chap. 2). In PPC, fuel is directly injected in the combustion chamber, and a highly stratified premixed charge is burned. The PPC is considered as the combustion process in between HCCI and conventional diesel combustion. Figure 6.17 shows the typical HRR curve showing different combustion phases in PPC. The PPC heat release can be categorized into four phases: ignition delay (ID), low temperature reaction (LTR), premixed combustion phase and late mixing controlled phase (Fig. 6.17). Ignition delay phase is the period between start of fuel injection (SOI)

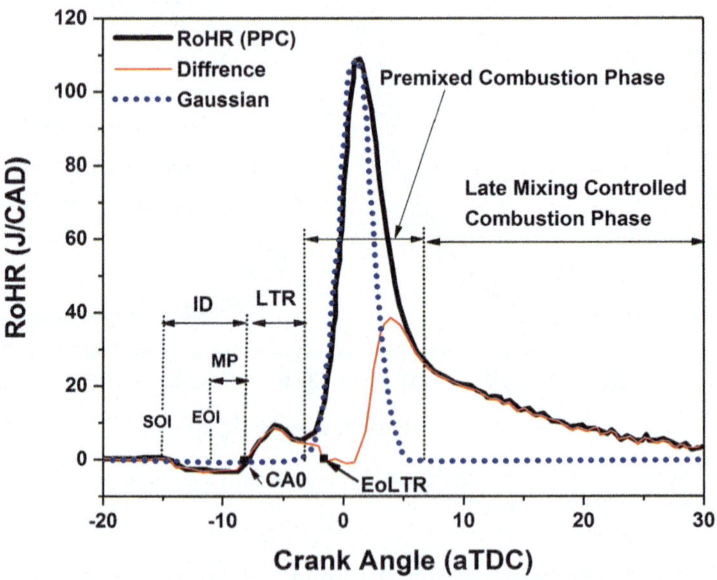

Fig. 6.17 Typical heat release rate curve indicating different combustion phasing in PPC engine (Adapted from [27]) (*MP* mixing period, *EoLTR* end of low temperature reactions)

and start of combustion (SOC). Typically the end of injection (EOI) is before the start of combustion. Start of combustion can be defined as the position where HRR becomes positive (CA0). In ignition delay period, most of the fuel is injected and evaporation takes place. This phase is important in PPC as it influences the premixed combustion phase [27]. Low temperature reaction phase is the combustion phase with low reaction temperature, where several reactions occur simultaneously and lead to the main combustion. The LTR is depicted by small peak before the main premixed heat release. The premixed combustion phase is characterized by a rapid combustion that occurs after LTR phase. This phase is mainly controlled by chemical kinetics, and reaction rate depends on temperature and composition of mixture. In the late mixing controlled combustion phase, combustion rate is mainly controlled by fuel vapour–air mixing process. The heat release continues at lower rate into the expansion stroke. The low heat release rate is because of slower chemical reaction process in this phase due to lower cylinder temperature during expansion stroke. The separation of these two phases is difficult because the part of the premixed phase combustion is limited by mixing and part of the late combustion phase depends on slow reactions rather than mixing [28]. A Gaussian profile is used to separate these two phases in a well-defined manner. Gaussian profile is fitted to the rising flank of the premixed peak, between end of LTR (EoLTR) and the actual peak, and the integrated area of the profile is used as a measure of the premixed reactions [27, 28].

Another LTC strategy called reactivity controlled combustion (RCCI) uses partially stratified charge using two fuels of different reactivity (see Sect. 2.6 of Chap. 2). In RCCI combustion, low reactivity fuel is typically premixed (port injection), and high reactivity fuel is directly injected to create fuel concentration and reactivity stratification in the cylinder. The HRR in RCCI combustion depends on fuel premixing ratio, injection timings, fuel composition and intake charge conditions. Figure 6.18 shows the effect of fuel premixing ratio on the HRR for gasoline/diesel and methanol/diesel RCCI combustion. Figure 6.18a depicts the effect of gasoline/diesel premixing ratio on the HRR, and it shows that the combustion of neat diesel (i.e. 100% diesel) demonstrates the cool flame reactions and premixed combustion phase is followed by diffusion combustion phase. In this duration, fuel injection and combustion process are overlapped. At the end of diesel injection, there is a stage of late burning process. With increase in the gasoline/diesel premixing ratio (up to 50% gasoline and 50% diesel), peak of heat release rate increases because more charge is burned during premixed combustion phase. Higher premixing is achieved due to longer ignition delay caused by lower global fuel reactivity. Therefore, diesel has more time to mix with premixed air and gasoline charge, which results in more fraction of charge burn during premixing combustion phase and shows higher peak HRR [29]. At this condition, SOC was observed before the end of diesel injection (Fig. 6.18). Hence, richer local equivalence ratio was observed at the SOC. With further increase in the gasoline/diesel premixing ratio, combustion characteristics change. In lower fuel premixing ratios, lower global fuel reactivity leads to longer ignition delay. In addition to this, lower injection duration of diesel is still able to mix diesel fuel with premixed air and

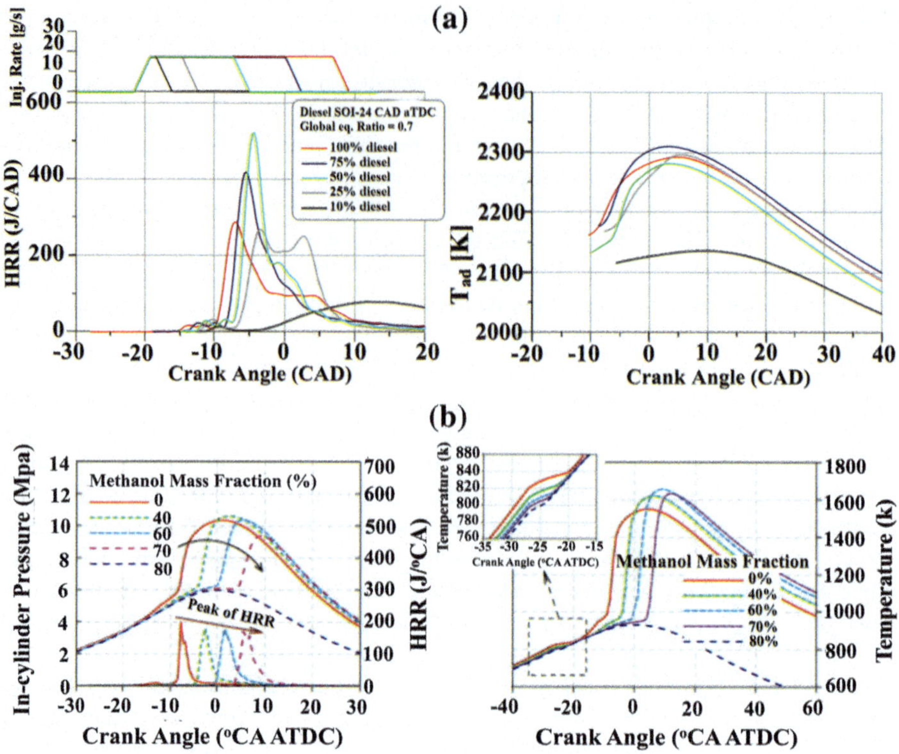

Fig. 6.18 Effect of fuel premixing ratio on the HRR in RCCI combustion mode for (**a**) gasoline/diesel [29] and (**b**) methanol/diesel [30]

gasoline before the auto-ignition starts. However, with further increase in the gasoline/diesel premixing ratio (50/50% to 25/75%), combustion is found with two peaks in heat release rate curve. The first peak of HRR curve is lower as compared to lower premixing ratio cases. The first peak appears due to the combustion of partially mixed injected diesel fuel and its corresponding premixed air and gasoline charge. After this stage, stratified charge of air, gasoline and diesel burns due to the rise in the temperature and pressure from previous combustion stage [29]. Moreover, with further increase in the gasoline/diesel premixing ratio (25/75% up to 10/90%), poor combustion was observed. It is already discussed that with increase in the premixing ratio, global reactivity of the charge is reduced. In addition with further increase in premixing ratio (i.e. 90% gasoline and 10% diesel), diesel injection duration is too small, and energy given by diesel is not sufficient to ignite the premixed air and gasoline. Combined effect of these two factors leads to increase of the ignition delay (coupled with the fact that the injection is ended much before the SOC), and overmixing of air and gasoline with diesel results in poor charge stratification (leanest local φ) leading to poor combustion efficiency.

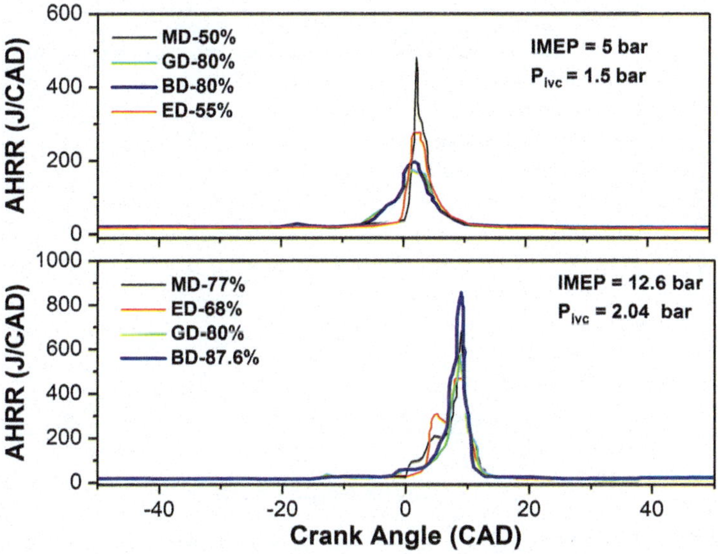

Fig. 6.19 Comparison of heat release rate at different engine operating conditions for different in-cylinder blending fuels (Adapted from [31])

Figure 6.18b shows the effect of methanol fraction on HRR in RCCI combustion. The figure reveals that ignition timing is retarded and peak HRR decreases with increase in the methanol fraction. This is due to fuel reactivity distribution (lower global fuel reactivity) in the combustion chamber. With increase in the methanol fraction, ignition delay increases due to two main reasons. Firstly, lower cetane number of methanol delays the ignition timings. Second, higher amounts of methanol injection in the manifold lead to lower temperature rise rate in the cylinder due to higher specific heat of methanol [30].

Figure 6.19 shows the HRR at constant combustion phasing for different in-cylinder blending fuels (i.e. gasoline/diesel (GD), n-butanol/diesel (BD), ethanol/diesel (ED) and methanol/diesel (MD)). The percentage value in legend of figure shows the preming ratio. The figure reveals that higher fraction of diesel is required (or lower fuel premixing ratio) for maintaining the constant combustion phasing with methanol/diesel and ethanol/diesel as compared to gasoline/diesel and n-butanol/diesel fuels. It attributes to the combined effect of combustion chemistry, higher enthalpy vaporization of methanol and ethanol as compared to n-butanol and gasoline, as well as the lower global fuel reactivity (high octane number of methanol and ethanol) [31]. Figure 6.19 also depicts that at 5 bar IMEP, shorter combustion duration and higher pressure rise rate were obtained with methanol/diesel and ethanol/diesel as compared to gasoline/diesel and n-butanol/diesel RCCI combustion. It might be because of higher amount of diesel injection in case of MD and ED RCCI combustion. However, this trend changes at higher engine load (12.6 bar IMEP). In case of 12.6 bar IMEP, diesel fuel quantity with various tested fuel combinations is almost similar. Combustion of such stratified charge with

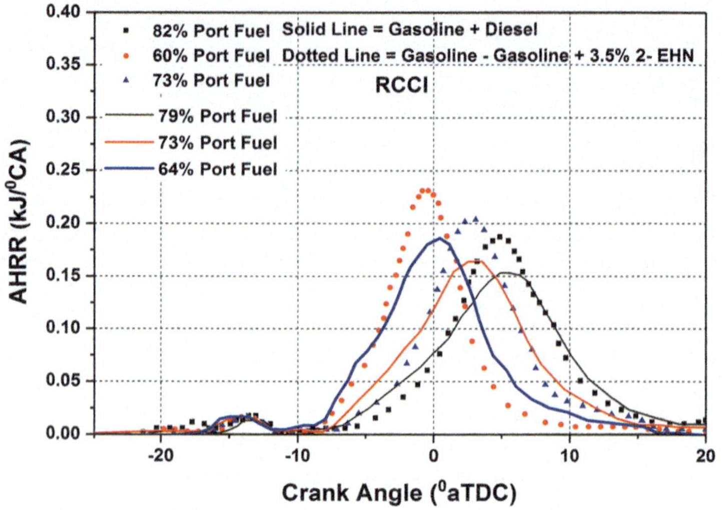

Fig. 6.20 Effect of premixing ratio on HRR ingasoline/diesel and gasoline/gasoline+2-ethylhexyl nitrate RCCI combustion at 1300 rpm (Adapted from [33])

higher temperature and overall equivalence ratio results in typically similar combustion duration and ignition timing.

The RCCI engine can also be operated on a single fuel by adding small amount of cetane improver in directly injected fuel (Sect. 2.6 of Chap. 2). A study demonstrates the single fuel RCCI operation at medium engine load conditions and found the similar results with gasoline/gasoline + DTBP as with gasoline/diesel RCCI engine [32]. Another study investigated the single fuel RCCI operation at lower load condition with gasoline/gasoline +2-ethylhexyl nitrate (EHN) [33]. Similar to the dual fuel RCCI, combustion phasing is easily controlled by adjustments to the ratio of more to less reactive fuel. Figure 6.20 illustrates the effect of gasoline/diesel and gasoline/gasoline +2-EHN (3.5% by volume) premixing ratio on HRR at 4.5 bar IMEP and 1300 rpm. The figure reveals that single fuel RCCI operation shows much faster HTHR than dual fuel RCCI operation. Faster HTHR can be possibly the result of higher fraction of port injected fuel, or 2-EHN added gasoline in the direct injection stream might have greater reactivity than that of diesel fuel [33]. Significant difference is also noted for LTHR in single and dual fuel RCCI. Gasoline is a high octane fuel and typically does not exhibit two-stage heat release. The comparatively advanced LTHR reaction (comparison to diesel) was observed because of the decomposition of 2-EHN.

Direct fuel injection timing has significant effect on fuel gradient distribution, which affects the combustion characteristics of RCCI engine. Figure 6.21a shows the effect of diesel injection timing on HRR for 50% methanol/diesel premixing ratio in RCCI engine. The figure shows that the heat release advances with advanced SOI timings from 7 CAD bTDC to 27 CAD bTDC. With further advance in SOI timing, combustion retards (Fig. 6.21a). Retard of combustion at more

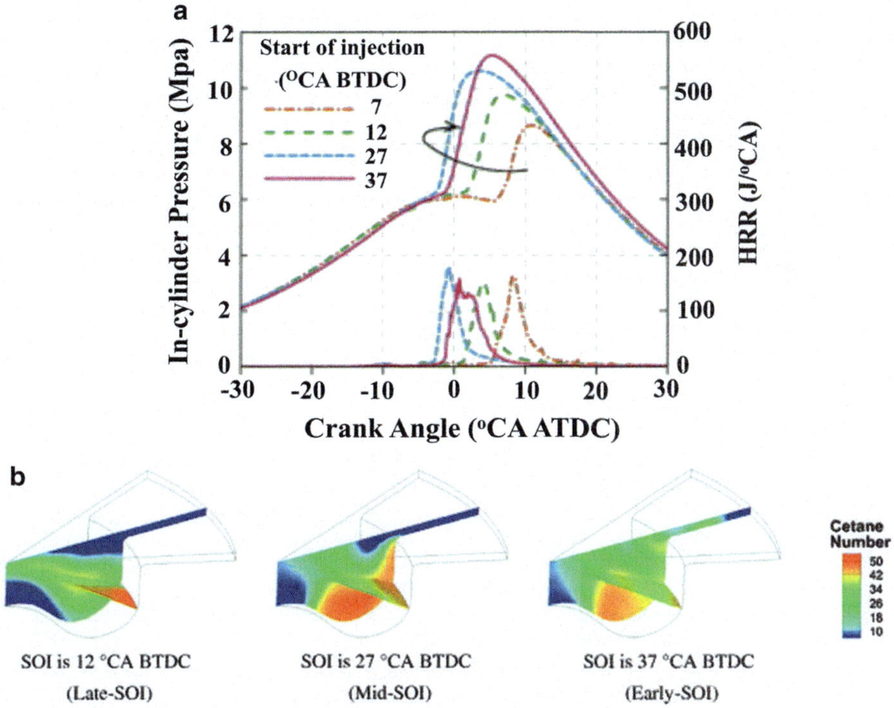

Fig. 6.21 (**a**) Effect of injection timing on HRR in methanol/diesel RCCI combustion [30] and (**b**) cetane number distribution for different SOI at 2 °CA bTDC [30]

advanced SOI can be explained by cetane number distribution in the chamber. Figure 6.21b shows the cetane number (CN) distribution for three different SOI timings, i.e. 12 CAD bTDC, 27 CAD bTDC and 37 CAD bTDC. The figure shows that the CN values are higher in comparatively larger region for 27 CAD bTDC, and in relatively smaller region for 12 CAD bTDC. Higher cetane number value indicates the high reactivity region, therefore, it leads to earlier ignition timings. However, suitable SOI timing may ensure a region of higher reactivity (higher CN) of charge either early SOI (leads to wide distributions of diesel) or late SOI (not enough time for atomization of diesel) [30]. In addition to this, higher specific heat of charge with advanced SOI timings reduces the temperature rise rate of charge, which results in lower ignition ability. Therefore, auto-ignition occurs firstly with mid-SOI (27 CAD bTDC); too advanced or too retarded SOI can lead to the delayed ignition timing.

Initial conditions of charge significantly affect the HRR in premixed LTC engines. The effect of intake temperature and pressure on HCCI combustion is discussed in Sect. 2.2.4.3 of Chap. 2. The HRR in HCCI combustion increases with increase in intake temperature, and combustion phasing advances with higher intake temperature (Fig. 2.18). Figure 6.22 shows the effect of intake temperature for dual fuel (gasoline/diesel (G/D)) RCCI and single fuel (gasoline/ gasoline (G/G)

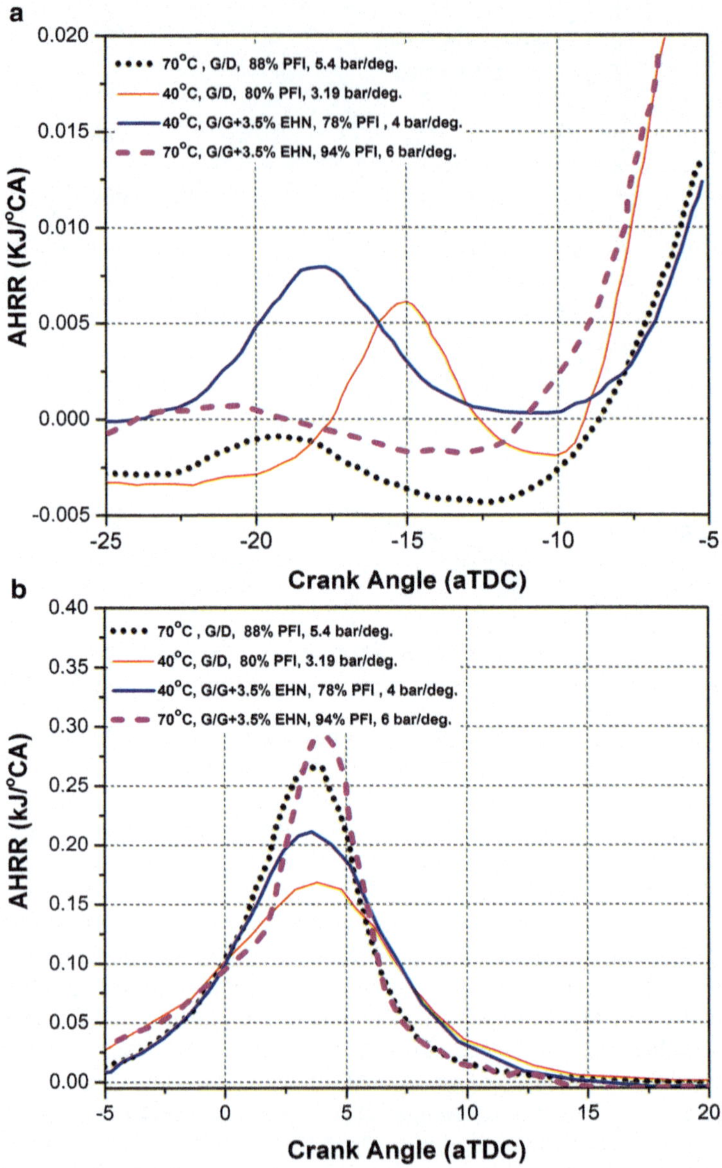

Fig. 6.22 Effect of intake temperature on HRR in gasoline/diesel (G/D) and gasoline/gasoline (G/G) +2-EHN RCCI combustion at 4.5 bar IMEP and1300 rpm (Adapted from [33])

+2-EHN) RCCI engine. The figure reveals that RCCI engine could be operated at similar combustion phasing for both the single and dual fuel strategies. Shorter combustion duration is obtained with increase in the intake temperature for both the strategies (Fig. 6.22). It attributes to higher fraction of port injected fuel (used for maintaining the constant combustion phasing) which leads to burn more premixed charge. Figure 6.22a also reveals that peak of LTHR decreases with increase in intake temperature for both the strategies as expected at higher temperature. Study concluded that controlling the premixed PFI fuel fraction is able to accommodate large variations in intake temperature by maintaining constant combustion phasing. Combustion phasing is also able to be easily controlled at desired CA_{50} timings at fixed intake conditions by varying the PFI fuel fraction for both fuel injection strategies [33].

6.2.3 Combustion Phasing and Duration

Combustion phasing and combustion duration affect the performance and emissions characteristics of engine. At a particular boost level (intake pressure), the engine efficiency can be maximized if the combustion is correctly phased and it has the correct burn duration [34]. Too early combustion phasing or too short combustion duration typically leads to very high combustion temperature, which results into more heat transfer loss. Conversely, late combustion phasing or long combustion duration leads to a decrease in effective expansion ratio, thus higher exhaust losses and lower engine efficiency. Typically all the combustion phasing parameters are calculated from heat release analysis of measured cylinder pressure signal (see details in Appendix 2). Crank angle (CA) position at which 10% heat release (CA_{10}) occurs in the cylinder is typically used as an estimation of SOC. Combustion duration (CA_{90-10}) is generally defined as the period between CA positions of 90% and 10% heat release. Combustion duration is a measure of overall combustion rate, i.e. how fast or slow combustion process is happening in the cycle. Crank angle position for 50% heat release (CA_{50}) is typically used as combustion phasing in LTC engines as it affects the performance of engine.

Figures 6.23 shows the variations of CA_{10} in ethanol- and methanol-fuelled HCCI combustion for different λ at different intake air temperatures. The figure shows that on increasing inlet air temperature, CA_{10} advances at any λ because of increase in speed of reaction kinetics leading to earlier auto-ignition due to higher combustion chamber temperature resulting from higher initial charge temperature. This trend is observed for both ethanol and methanol. Auto-ignition of leaner mixture requires higher inlet temperature for both fuels to achieve particular CA_{10} position. This requirement of higher intake air temperature arises due to (i) comparatively lower charge reactivity of leaner mixture, (ii) lower temperature of residuals at leaner engine operation and (iii) mean cylinder wall temperature that is also lower for leaner mixture engine operations [35]. Thus, leaner mixtures have retarded CA_{10} position at a given intake air temperature as longer time is required to set off the autoignition reactions at leaner engine operating conditions. Figure 6.23 also shows that at particular intake temperature, CA_{10} is less sensitive to λ at higher intake air temperature as compared to lower intake air temperature (curves become horizontal). Methanol shows earlier start of combustion (CA_{10}) position at constant

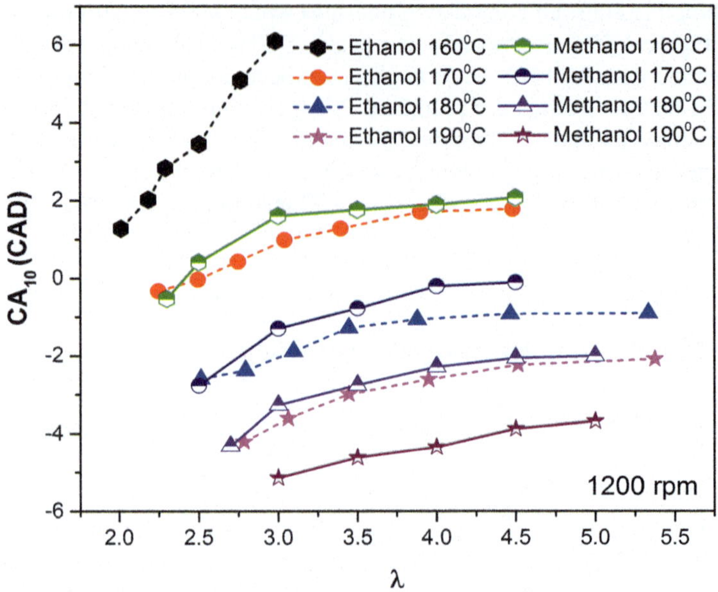

Fig. 6.23 Variation of CA_{10} with λ in HCCI combustion for ethanol and methanol at 1200 rpm (Adapted from [35])

λ and intake temperature in comparison to ethanol. This is possibly due to higher oxygen content in methanol and autoignition chemistry of methanol.

Figure 6.24 illustrates the correlation of ignition timing (CA_{10}) with CA_{50} and CA_{90-10} using gasoline and ethanol in HCCI combustion. Figure 6.24a shows that at particular intake air temperature, ignition timings are strongly coupled to the combustion phasing (CA_{50}) for both the fuels. Polynomial second-order regression curve drawn through the experimental data points captures the trends in behaviour of combustion phasing with R^2 value 0.99. Figure 6.24b shows that combustion duration (CA_{90-10}) increases drastically with retarded SOC (CA_{10}) timings. Third-order polynomial regression curve fit in the experimental data points captures the trend in behaviour of combustion duration. The combustion duration increases due to lower combustion rate resulting from lower combustion temperature. In case of late CA_{10} positions, gas expansion (due to piston movement) before transition to high temperature oxidation further lowers the charge temperature for early part of heat release and makes initial combustion rate comparatively lower. Initially slow combustion rate leads to lower temperature rise rate, and lower peak temperature is achieved in cylinder because of further volume expansion. Heat transfer is also higher near TDC position, which further widens the temperature distribution within charge with time contributing in the extension of the combustion event [35, 36].

Figure 6.25 shows the variation of CA_{10} in HCCI operating range using gasoline and ethanol at 1200 rpm. The HCCI operating range is defined by higher and lower load limit boundary. Knock region shown in the figure represents the engine

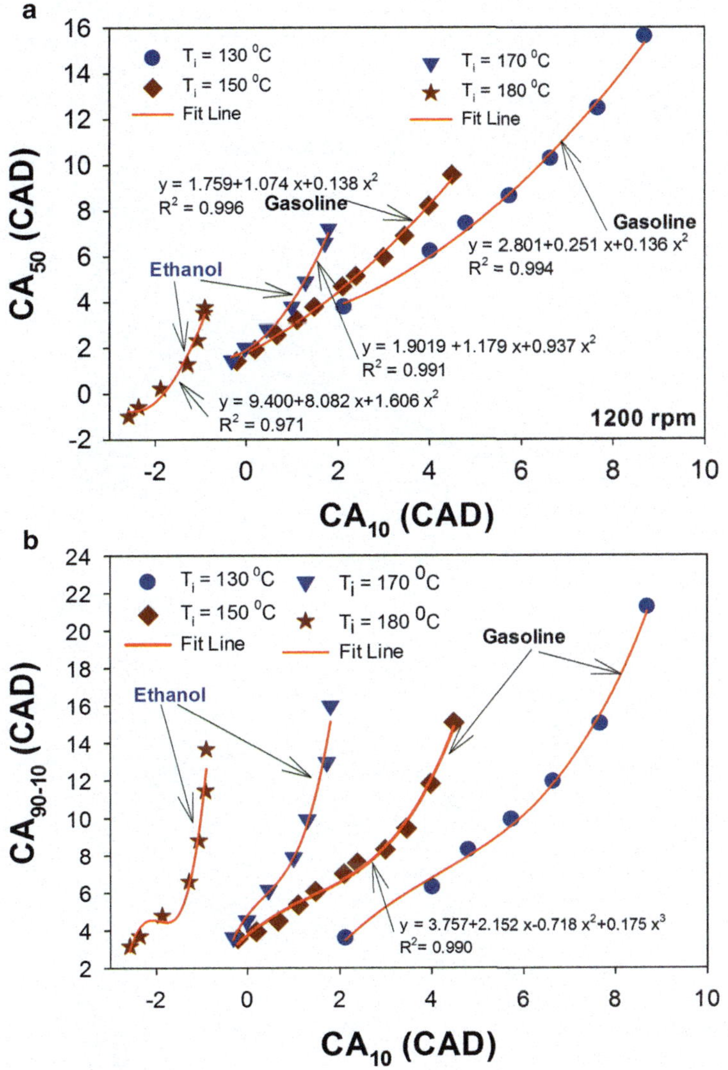

Fig. 6.24 Correlation of CA_{10} with (**a**) CA_{50} and (**b**) CA_{90-10} in HCCI combustion for gasoline and ethanol (Adapted from [24, 35])

operating conditions having ringing intensity (RI) >6 MW/m^2. Misfire region represents the engine operating conditions where coefficient of variation in IMEP (COV_{IMEP}) >3.5%. Thus, engine operating conditions having RI <6 MW/m^2 and COV_{IMEP} <3.5% are considered in HCCI operating range. At constant λ, increase in intake temperature advances the CA_{10} positions. Figure 6.25 shows that contour lines of CA_{10} are closer to vertical for ethanol HCCI conditions, which suggests that ignition timing (CA_{10}) is primarily dependent on intake air temperature.

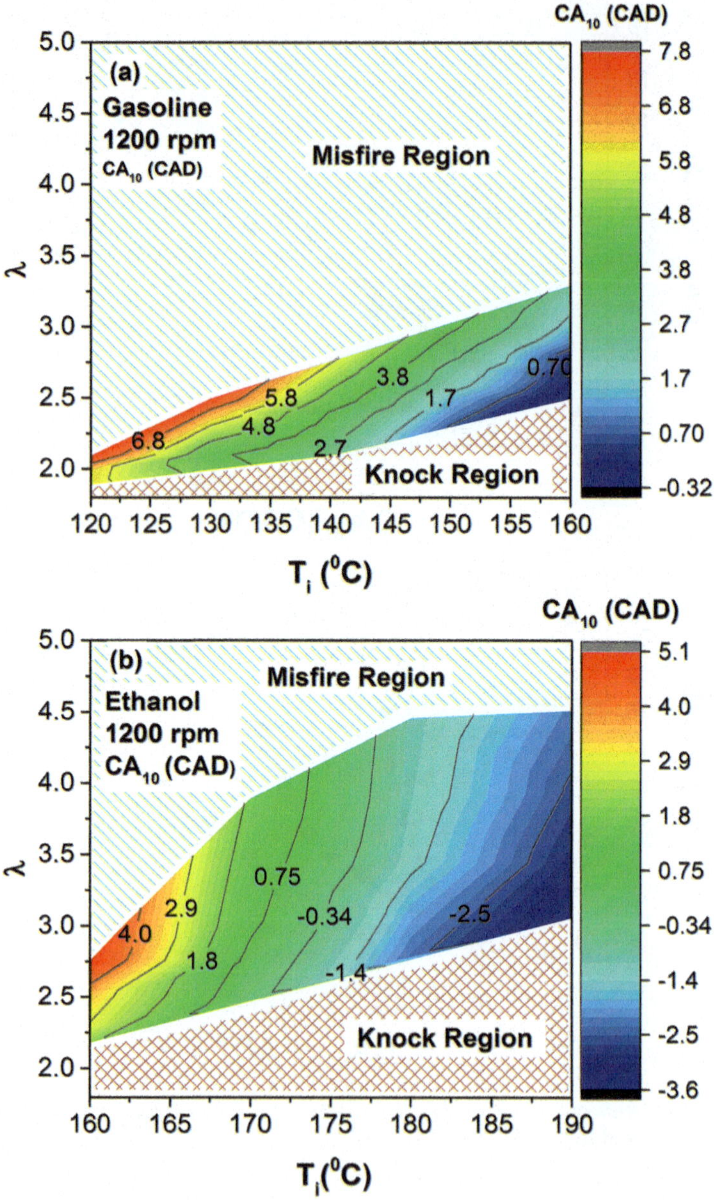

Fig. 6.25 Variation of CA_{10} in HCCI operating range for gasoline and ethanol (Adapted from [35])

Dependency of ethanol CA_{10} position on λ is very weak. The contour lines of CA_{10} position in gasoline HCCI are comparatively flatter as compared to ethanol indicating that CA_{10} position has greater dependency on λ as compared to ethanol.

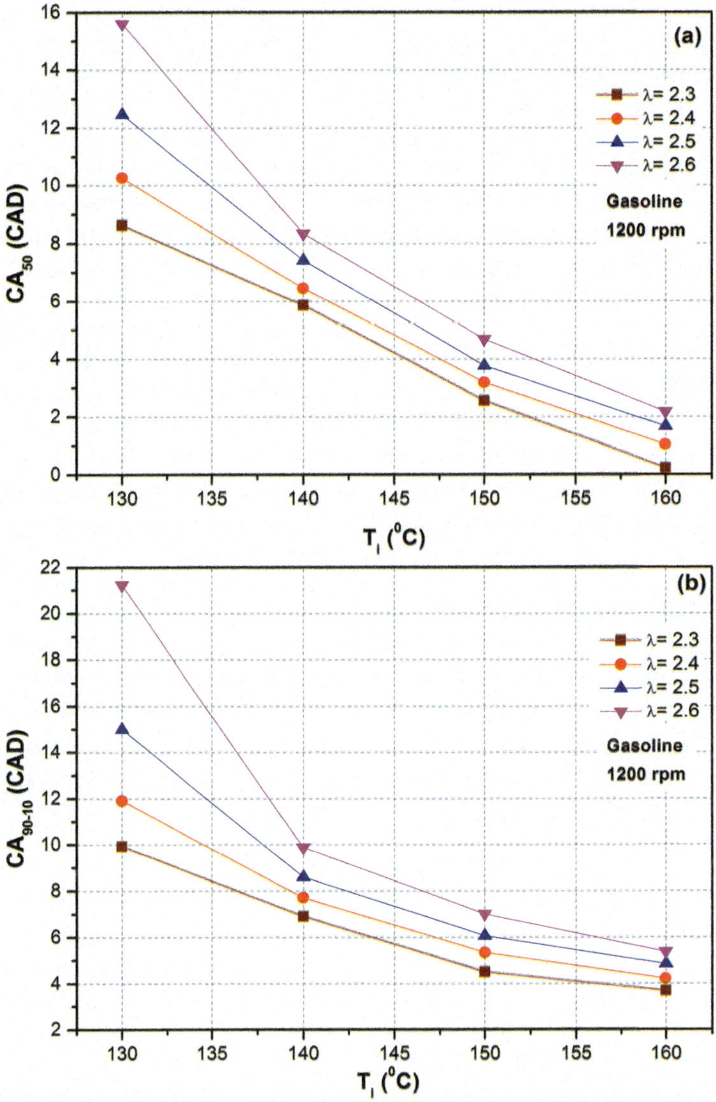

Fig. 6.26 Effect of intake air temperature on CA_{50} and CA_{90-10} at different λ at 1200 rpm in gasoline HCCI combustion [24]

Figure 6.26 shows the effect of intake air temperature (T_i) on CA_{50} and CA_{90-10} in gasoline HCCI at 1200 rpm for different λ. For each λ, combustion phasing (CA_{50}) advances with increase in intake air temperature due to earlier ignition by higher combustion chamber temperature caused by higher initial charge temperature. Combustion phasing is very sensitive to intake air temperature. Figure 6.26a shows that at $\lambda = 2.6$, increase of 30 °C intake air temperature advances the CA_{50}

by 14 CAD. High sensitivity of combustion phasing for intake air temperature at delayed phasing can be described as follows. Auto-ignition occurring at delayed timings in expansion stroke becomes more sensitive to charge temperature at TDC because volume expansion due to piston movement tends to counteract the temperature rise by auto-ignition process. Intake air temperature directly affects the charge temperature in the cylinder, therefore leading to change in combustion phasing. This change in combustion phasing changes the wall temperature, which affects the heat transfer rates and feeds back into charge temperature at TDC [36].

Figure 6.26b shows that on increasing intake air temperature, combustion duration decreases for all values of λ. It can be noticed that leaner mixtures have higher combustion durations at a given intake temperature. Intake air temperature directly affects the combustion phasing (Fig. 6.26a). Combustion phasing is one of the main factors which affects the combustion duration. As the combustion phasing retards, the combustion duration is longer since the maximum combustion temperature decreases. Several factors contribute to lower maximum temperature at delayed combustion phasing. These are as follows: (i) delayed CA_{50} is the result of lower initial charge temperature, which leads to lower cylinder temperature, (ii) delayed CA_{50} gives lower in-cylinder pressures resulting in lower maximum temperature and (iii) delayed CA_{50} results in lower combustion efficiency, which leads to lesser heat release in the cylinder, therefore lower maximum in-cylinder temperatures [17]. Combustion duration in HCCI combustion is shorter than conventional combustion modes. At higher intake air temperature, combustion duration is extremely short, as short as 3 CAD for rich mixtures (Fig. 6.26b). This very fast combustion leads to knocking combustion and the engine operation becomes extremely noisy. At lower intake temperature and leaner mixtures, combustion duration is longer. Very long combustion duration may lead to very higher cyclic variations or partial burn/misfire cycle.

Figure 6.27 shows the variation of combustion duration (CA_{90-10}) in HCCI operating range using gasoline and ethanol at 1200 rpm. Engine operating conditions having RI <6 MW/m^2 and COV_{IMEP} $<3.5\%$ are considered in HCCI operating range. The figure shows that the CA_{90-10} decreases as intake air temperature increases at constant λ. The CA_{90-10} increases as mixture becomes leaner for both fuels. Figure 6.27 also exhibits that contour lines of CA_{90-10} have higher slope for ethanol as compared to gasoline, which indicates that CA_{90-10} is more dependent on intake air temperature in ethanol HCCI operation as compared to gasoline HCCI. The maximum CA_{90-10} obtained in HCCI operating range is less than 15 CAD for both fuels at 1200 rpm.

Combustion duration in HCCI combustion is also dependent on engine compression ratio. Increasing the compression ratio leads to higher cylinder pressure and hence higher combustion temperature, which causes shorter combustion duration. Another, factor contributing to shorter combustion duration is increased species concentration as the pressure level is higher. Increased species concentration leads to faster combustion chemistry [17]. Typically, EGR is used as diluent to reduce the peak HRR in HCCI combustion. The EGR affects the combustion duration in HCCI combustion. The total effect of EGR on combustion duration is

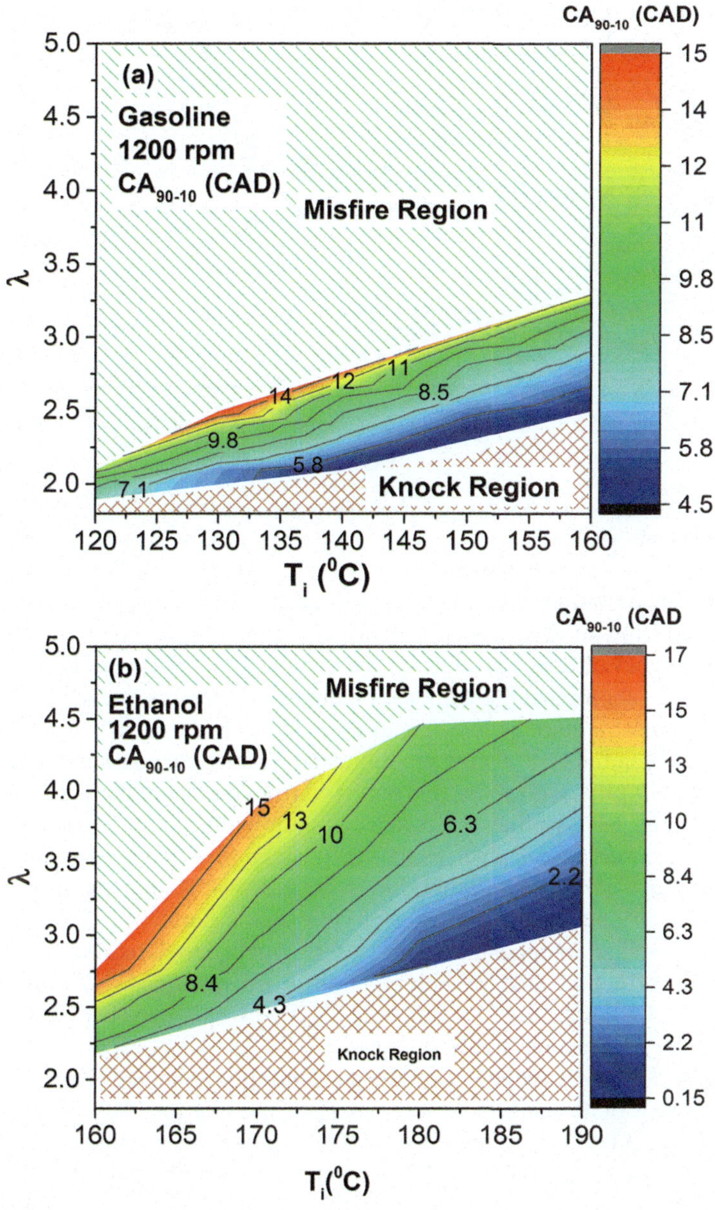

Fig. 6.27 Variation of combustion duration in HCCI operating range for gasoline and ethanol at 1200 rpm (Adapted from [35])

contributed by four main factors: (i) dilution effect, burned gases in EGR lowers the oxygen concentration; (ii) heat capacity effect, due to higher specific heat capacity of combustion products in EGR; (iii) chemical effect, chemical reactions involving

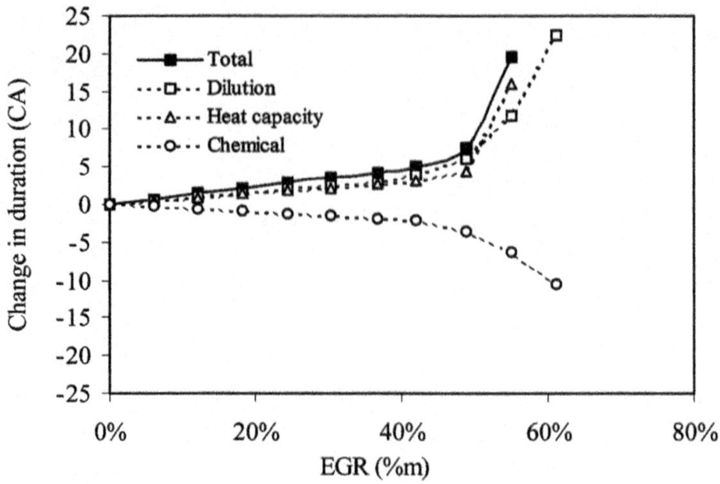

Fig. 6.28 Contribution of different types of effects of isothermal EGR on the combustion duration in HCCI combustion [37]

species of the burned gases; and (iv) stratification effect, recirculated burned gases are stratified from the fuel–air mixture [37, 38]. Figure 6.28 shows the contribution of different types of effect of EGR on the combustion duration in HCCI engine.

Figure 6.28 shows that the combustion duration increases linearly with the EFR fraction up to 50%. Further increase in EGR fraction above 50% leads to a much faster increase in the combustion duration, and incomplete combustion appears for more than EGR 60% utilization. It is observed from the figure that dilution and heat capacity effects have similar effect in slowing down the combustion rate in HCCI engine. This observation suggests that the lower combustion temperature (because of the heat capacity effect) and the lower oxygen availability (the dilution effect) are mainly contributing in extension of combustion duration [37]. However, the chemical effect has an opposing effect on the combustion duration (Fig. 6.28). The acceleration of combustion by participation of trace species (presents in burned gases) in HCCI combustion (see Sect. 2.2.4.5 of Chap. 2). This chemical effect decreases the combustion duration.

Figure 6.29a shows the effect of NO concentration in intake charge (which can be there during EGR operation) on combustion phasing for PRF (84% iso-octane/ 16% n-heptane blend), TRF (65% toluene/ 35% n-heptane mixture) and gasoline in HCCI engine. At intake pressure 2.0 bar, main combustion phasing of the gasoline and the TRF have similar behaviour with increase in NO concentration, and the CA_{50} advances in the entire concentration range tested. The CA_{50} of the gasoline HCCI combustion is however earlier in comparison to TRF. Figure 6.29a also depicts that the effect of NO is strongest at low concentrations, which is the concentration range most relevant to HCCI engines. The CA_{50} for the PRF fuel advances also sharply at low NO concentrations, and at higher concentrations starts retarding at boosted intake condition (2.0 bar). At intake pressure 1.0 bar, gasoline

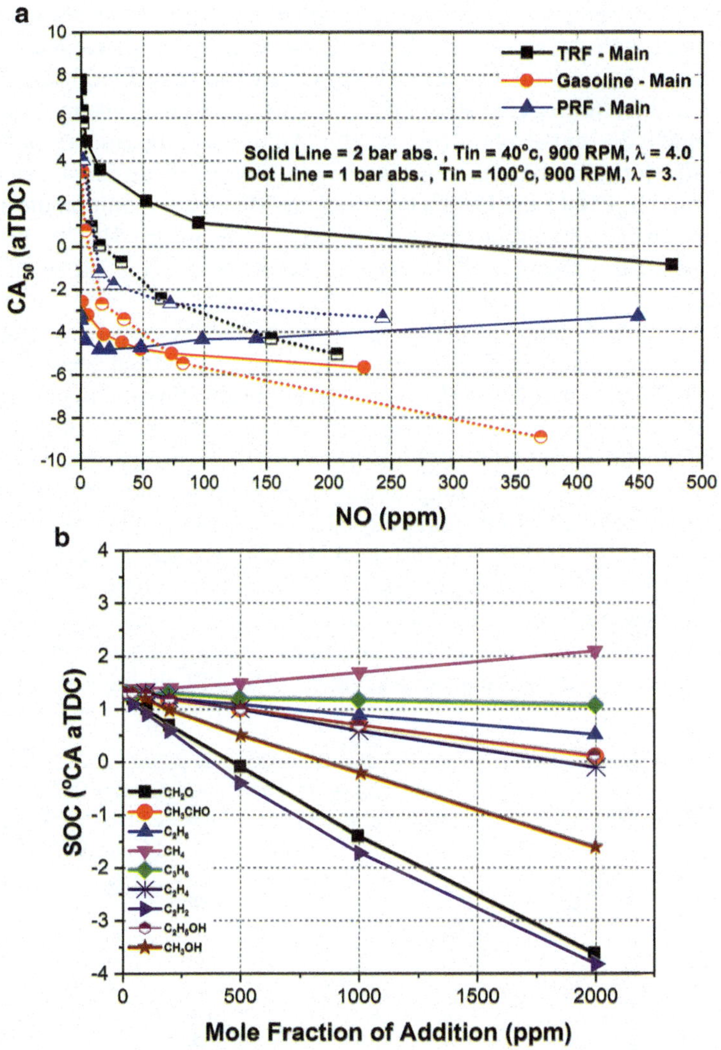

Fig. 6.29 (**a**) Effect of NO concentration in intake charge on CA_{50} for PRF, TRF and gasoline in HCCI engine (Adapted from [39]) and (**b**) effect on incomplete oxidation products on SOC with different mole fractions (Adapted from [40])

CA_{50} behaves more like the TRF fuel than the PRF fuel with NO concentration, and CA_{50} advances in the entire concentration range tested for both the gasoline and the TRF fuel. This is also the case for the PRF fuel but not to the same extent. At lower intake pressure also, the NO addition has the strongest impact at lower concentrations. The advance combustion phasing effect at low NO concentrations and low intake pressure (1.0 bar) is even bigger than in the higher intake pressure (2.0 bar) for TRF. The effect of NO in intake charge is more marked for the TRF and the gasoline in comparison to the PRF fuel [39].

Figure 6.29b shows the effect of incomplete oxidation products induction in intake charge with different mole fractions on SOC of HCCI engine. The main incomplete oxidation products in residual gas that can have significant effects on the HCCI ignition timings are short-chain alkanes, short-chain alkenes, short-chain alkynes, aldehydes and alcohols (Fig. 6.29b). All of these species except CH_4 have the potential for advancing the HCCI ignition timings to different extents. It is found that C_2H_2, CH_2O and CH_3OH enhance the HCCI auto-ignition most without considering their real concentrations, and CH_4 can retard the HCCI auto-ignition slightly. Detailed discussion with reaction mechanism is provided in the original study [40].

Typically in LTC engines, port fuel injection (PFI) is used for preparation of premixed charge. The PFI timings also affect the combustion phasing in HCCI engines. Figure 6.30 shows the effect of SOI timings of PFI system on combustion phasing at different amounts of fuelling (21, 25 and 29 mg/cycle) and different intake temperatures (150, 160, 170 and 180 °C). To investigate the effect of PFI timing on the combustion phasing, different SOI timings (5, 70, 210, 500 and 600 CAD) are selected in open and closed intake valve conditions. Intake valve opening duration (−10 to 198 CAD) is shown with shaded box below the graph (Fig. 6.30). Different injection timings affect fuel mixing differently, and various degrees of fuel inhomogeneity are caused by different fuel injection timings [41]. During closed intake valve fuel injection, majority of fuel spray strikes the base of the intake passage and intake valves. The PFI timing has a significant

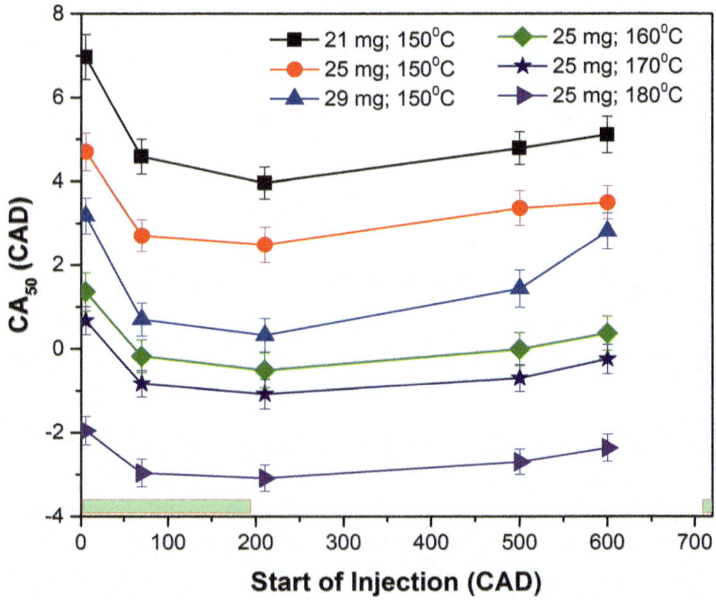

Fig. 6.30 CA_{50} variation for varying SOI timings of PFI in methanol HCCI combustion at 1500 rpm (Adapted from [43])

influence on the mixture formation, and it depends on several factors such as characteristics of fuel spray in the intake manifold, homogeneity of the mixture at the time of ignition, cold or hot engine operation, etc. More details regarding the mixture preparation during closed and open valve injection timing and flow of fuel–air mixture with its effect on HC emissions are discussed in [42].

Figure 6.30 shows that on increasing the amount of fuel in the cylinder, the CA_{50} position advances because richer fuel–air mixture auto-ignites early due to higher mixture reactivity. At constant fuelling, combustion timing advances with increase in intake air temperature (Fig. 6.30) as explained for Fig. 6.26. The CA_{50} position depends on the λ and inlet air temperature. At particular fuelling and intake air temperature, the CA_{50} position depends on the fuel injection timing. Figure 6.30 depicts that for closed valve timing, CA_{50} position is earlier as compared to open valve injection due to greater homogeneity of mixture formation during closed valve injection timing and more time available for mixing of fuel and air.

In well-mixed HCCI combustion, the HRR is controlled by chemical kinetic rate, and amount of thermal stratification exits in the cylinder. Thermal stratification leads to sequential autoignition which results into slow combustion process and lowers the PRR. Thus, higher amount of fuel can be burned or more advanced CA_{50} can be achieved without knock (see Sect. 2.2.4.6 of Chap. 2). Since the control of thermal stratification is difficult, typically fuel stratification is used to burn higher amount of fuel to achieve higher engine load in HCCI engine. Some amount of fuel is directly injected in the cylinder to create the fuel stratification, and the rest of fuel is premixed by port fuel injection or using separate mixing tank. The higher equivalence ratio region created by direct injection (DI) fraction autoignites faster, which leads to advanced ignition. The equivalence ratio of leanest region is reduced to comparatively lower premixed fraction, and this leaner region autoignites slowly. The overall effect of fuel stratification is the increase in combustion duration and reduction in peak HRR. The combustion duration increases with increase in amount of DI fraction of fuel (Fig. 2.27).

Figure 6.31 shows the variation in CA_{50} for well-premixed case and partial fuel stratification (PFS) case for different charge equivalence ratios and IMEP. At all the test points, ringing intensity is <5 MW/m^2. Figure 6.31a shows that at particular equivalence ratio, CA_{50} is advanced in case of PFS. Well-mixed combustion phasing needs to be delayed to keep PRR below the limit. The reduction in HRR by PFS allowed the CA_{50} to be advanced significantly for the same load (Fig. 6.31b), which leads to higher thermal efficiency. For the constant CA_{50} retard, the PFS strategy can achieve higher engine load, while maintaining the ringing intensity limits. Figure 6.31b also compares standard PFS and early DI PFS strategy. In the standard PFS, 80–95% of the fuel is premixed, and the remaining fuel is directly injected in the latter part of the compression stroke, typically at 300–325° CA, to create the stratification in the cylinder. In early DI PFS, all of the fuel is supplied by direct injection early in the intake stroke (60° CA), and due to incomplete mixing, resulting mixture behaves similar effect that produced by standard PFS (Fig. 6.31b) [25]. However, early DI PFS has the advantage that lower intake temperature (T_{in} <60 °C) can be used without concern about premixed

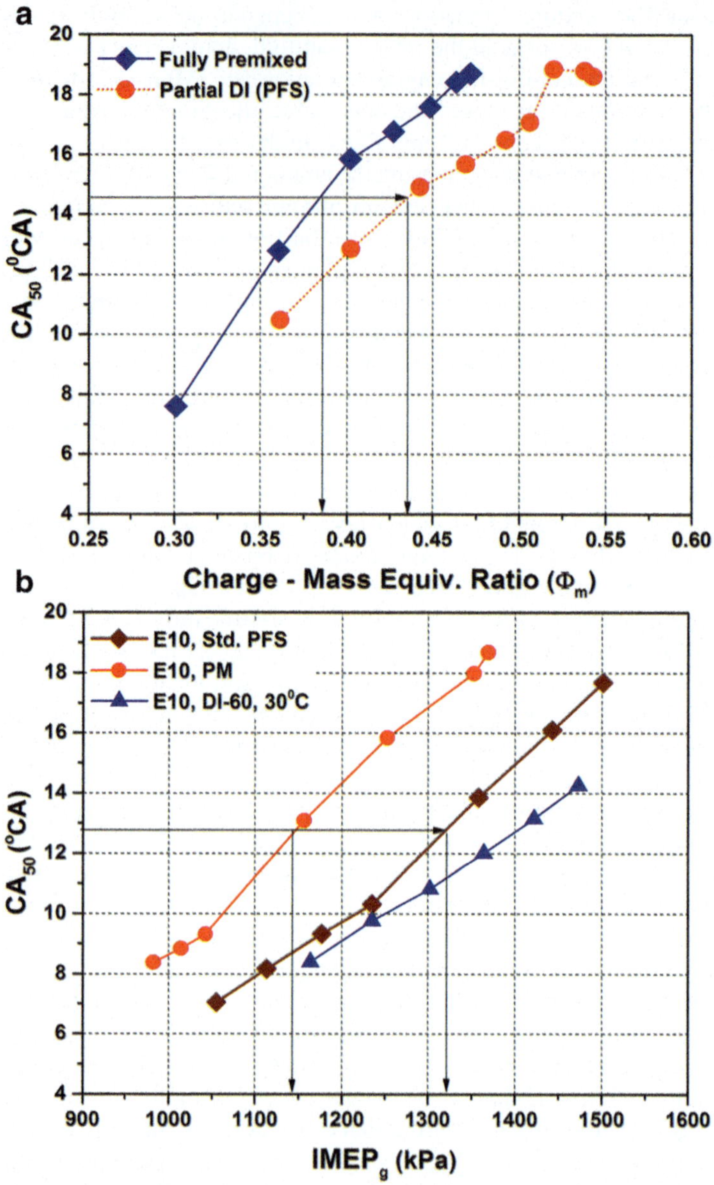

Fig. 6.31 CA_{50} variations for well premixed and PFS with (**a**) charge equivalence ratio (Adapted from [44]) and (**b**) IMEP (Adapted from [25])

fuel condensing in the intake system. Lower intake temperatures can be advantageous to increase the thermal efficiency. Lower intake temperature significantly reduces the amount of EGR required to retard CA_{50} to maintain RI $\approx$ 5 MW/m^2 [25]. Lower temperature also results into lower combustion temperature for the

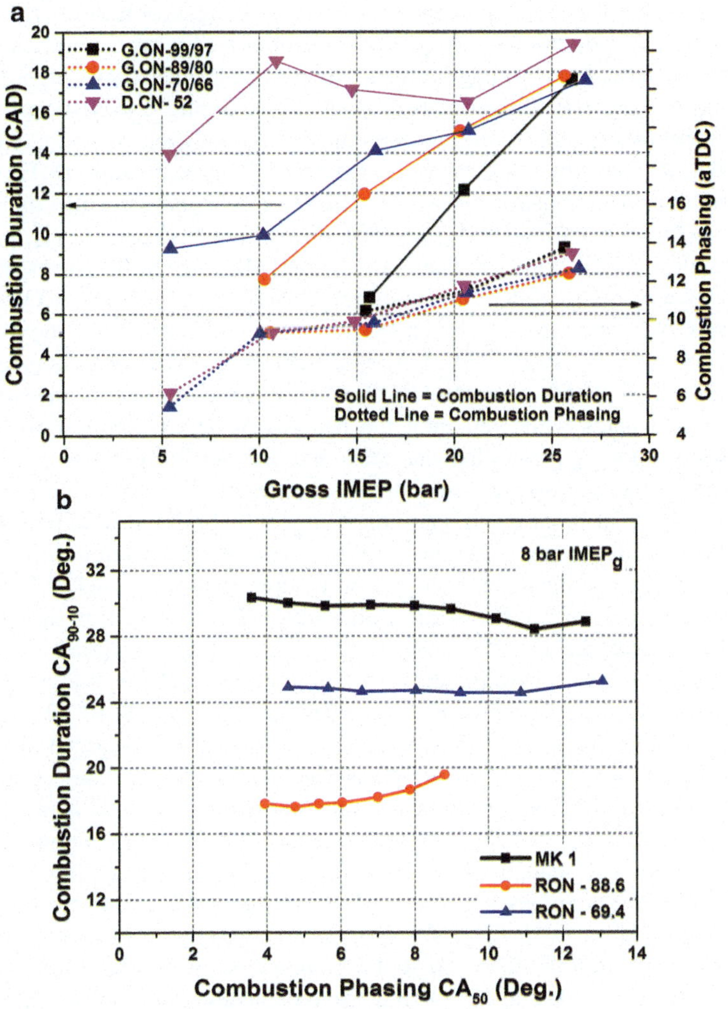

Fig. 6.32 Variation of combustion duration with (**a**) IMEP (Adapted from [45]) and (**b**) combustion phasing (CA_{50}) (Adapted from [28]) for different octane number fuels in PPC strategy

same load. Both the reduced EGR and the lower combustion temperatures act to increase the ratio of specific heat (γ) of the combustion products leading to more work to be extracted during the expansion stroke.

Figure 6.32 shows the variations of CA_{50} and combustion duration PPC strategy, which used heavy stratification in cylinder. Fuels designated as G.ON x/y in the Fig. 6.32a represents gasoline octane number, where 'x' is RON, and 'y' is MON. Figure 6.32a shows that combustion phasing retards with increase in engine load to keep the maximum PRR within the high load limit. Fuels with higher octane number can be used at higher loads and cannot be operated at lower engine loads.

Figure 6.32a also shows that combustion duration is longer for lower octane number fuels. Combustion duration is comparatively shorter for higher octane gasoline-like fuels, and it increases with engine load. Higher octane fuels have higher ignition delay and mixing is comparatively higher. Thus, during combustion they have shorter combustion period due to simultaneous combustion of premixed charge. Figure 6.32b shows the CA_{90-10} as a function of CA_{50} at constant IMEP of 8 bar. The variations in combustion phasing is achieved by changing the injection timings. The figure shows that diesel (MK1 Swedish diesel) fuel has longer CA_{90-10} than the gasoline fuels in PPC strategy. The CA_{90-10} for MK1 and for the lowest RON fuel is insensitive to CA_{50}, while it increased for the other gasoline fuels (RON 88) with retarding CA_{50}. The CA_{90-10} is longest for MK1 and shortest for the high RON fuels [28]. This is because of rapid premixed combustion in higher octane fuels.

The dual fuel RCCI is another LTC strategy, which is widely investigated to higher engine loads while retaining the benefits of HCCI combustion. In RCCI combustion mode, ignition characteristics and combustion phasing can be easily controlled by fuel premixing ratio and SOI timings of high reactivity fuel (diesel). Lower fuel premixing ratio and retard diesel injection timing lead to increase in the inhomogeneity of charge inside the cylinder due to shorter mixing time. Combustion of highly inhomogeneous charge results into higher NO_x and PM emissions. An excessively advanced diesel injection timing leads to very lean distribution of diesel fuel inside the cylinder that results in slower combustion. Figure 6.33a represents the effect of gasoline/diesel premixing ratio and diesel injection timing on CA_{50} for two different amounts of injected fuel per cycle in RCCI engine. The figure shows that combustion phasing (CA_{50}) advances with retarded diesel injection timings for each test fraction of gasoline. Retarded diesel injection timings produce more stratified charge, which results in higher local equivalence ratio and higher local reactivity of charge in some regions. Higher local reactivity and local equivalence ratio of charge reduce the ignition delay, and autoignition starts early showing advanced combustion phasing [33]. Figure 6.33b shows the variation of combustion duration with CA_{50} for different premixing ratios of gasoline/diesel (G/D) and methanol/diesel (M/D) RCCI combustion. It can be noticed from the figure that for both G/D and M/D RCCI, the combustion duration is shortened with increased premixed ratio (R_p) at $T_{IVC} = 400$ K. However, ringing intensity reduces with increased premixing ratio because of retarded CA_{50} with increase in premixing ratio [46]. Additionally, the combustion rate is significantly affected by intake valve closing temperature [47], and combustion duration shortens with higher intake value closing temperature due to the accelerated combustion rate (Fig. 6.33b). Figure 6.33b also depicts that combustion duration of M/D RCCI is much shorter than G/D RCCI for particular premixing ratio and intake temperature, despite of late combustion phasing in M/D RCCI. Shorter combustion duration in M/D RCCI is mainly due to the differences in homogeneity of the fuel distribution and the different fuel chemical properties. Comparatively more retarded CA_{50} in M/D RCCI has more time for mixing diesel in the cylinder to form a more homogeneous fuel–air mixture before main heat release. This leads to comparatively smaller reactivity gradient in M/D RCCI, and the combustion occurs at multiple hot ignition sites leading to shorter combustion duration [46].

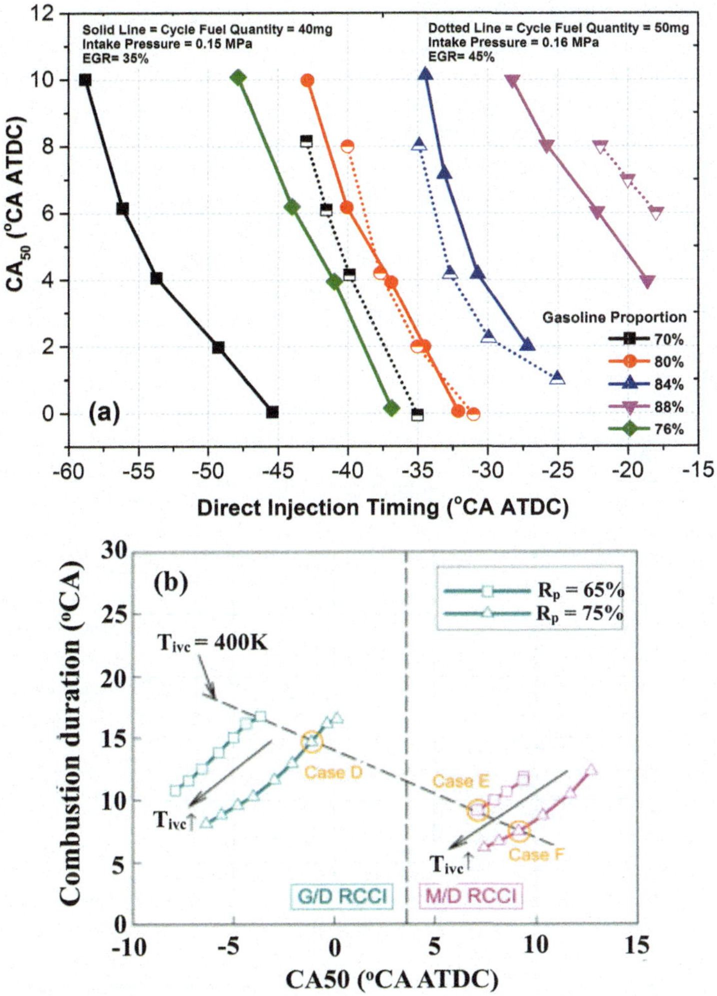

Fig. 6.33 (**a**) Effect of gasoline fuel fraction and diesel injection timing on CA_{50} (Adapted from [48]) and (**b**) combustion duration as function of CA_{50} [46] in RCCI combustion

Figure 6.34 depicts the effect of ignition timing and ignition delay on the burning duration of dimethyl ether (DME)–ethanol RCCI combustion. The burning duration between CA_{10} and CA_{50} represents the premixed combustion duration, while the diffusion burning duration is represented by CA_{50}–CA_{90} in Fig. 6.34. The figure reveals that retarded ignition timings lead to the increase of the premixed combustion duration while reducing the mixing controlled diffusion combustion duration. At fixed ignition timing, increase in the ethanol fraction reduces the premixed combustion duration and enhances the mixing controlled combustion duration. It attributes to increase in ignition delay period due to the lower cetane number of

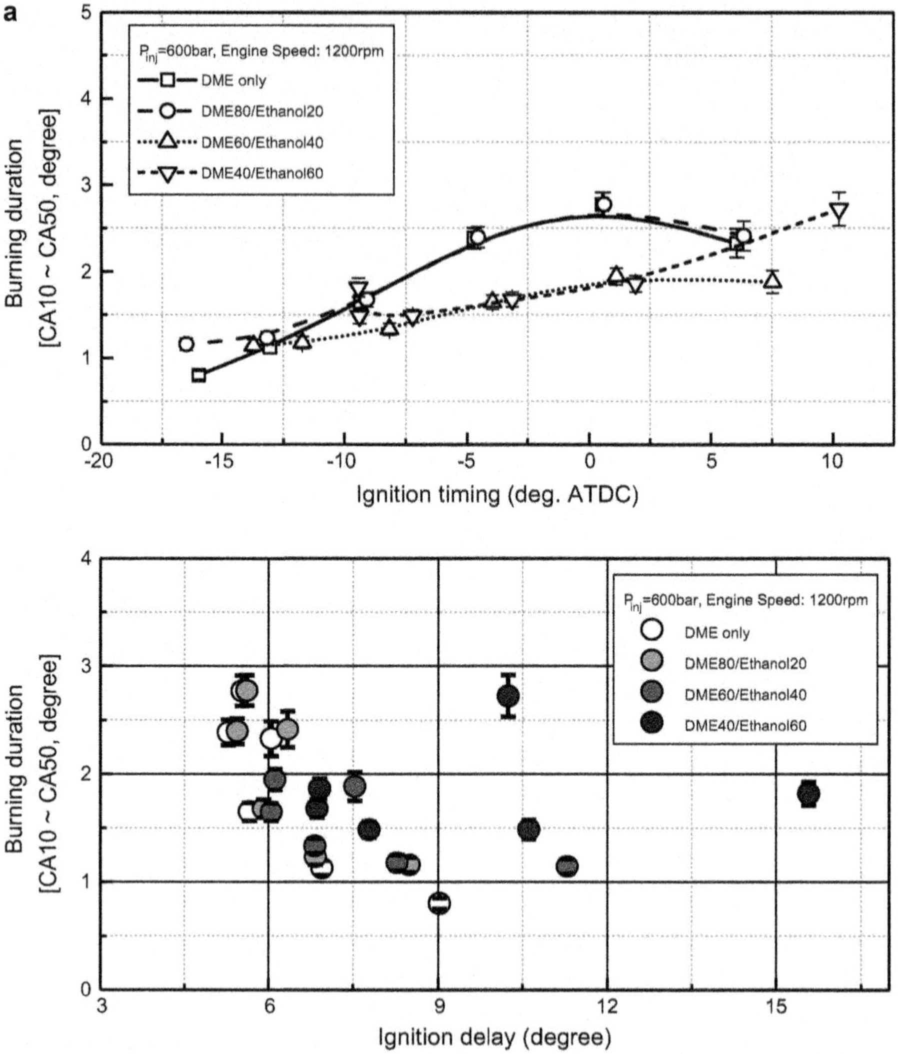

Fig. 6.34 Effect of ignition timing and ignition delay on (**a**) premixed burning duration (CA_{10}–CA_{50}) and (**b**) diffusion burning duration (CA_{50}–CA_{90}) in DME–ethanol RCCI combustion [49]

ethanol fuel. Longer ignition delay allows more time for premixing of charge. Therefore, autoignition of premixed charge shortens the premixed combustion duration [49]. Study reported that with an increase in the ethanol concentration, relationship between premixed combustion duration (i.e. CA_{10}–CA_{50}) and ignition delay has smaller slope, which indicates that ignition delay period has weak much significant effect on the CA_{10}–CA_{50}. Moreover, slope between mixing controlled combustion (i.e. CA50–CA90) and ignition delay period reduces with an increase in

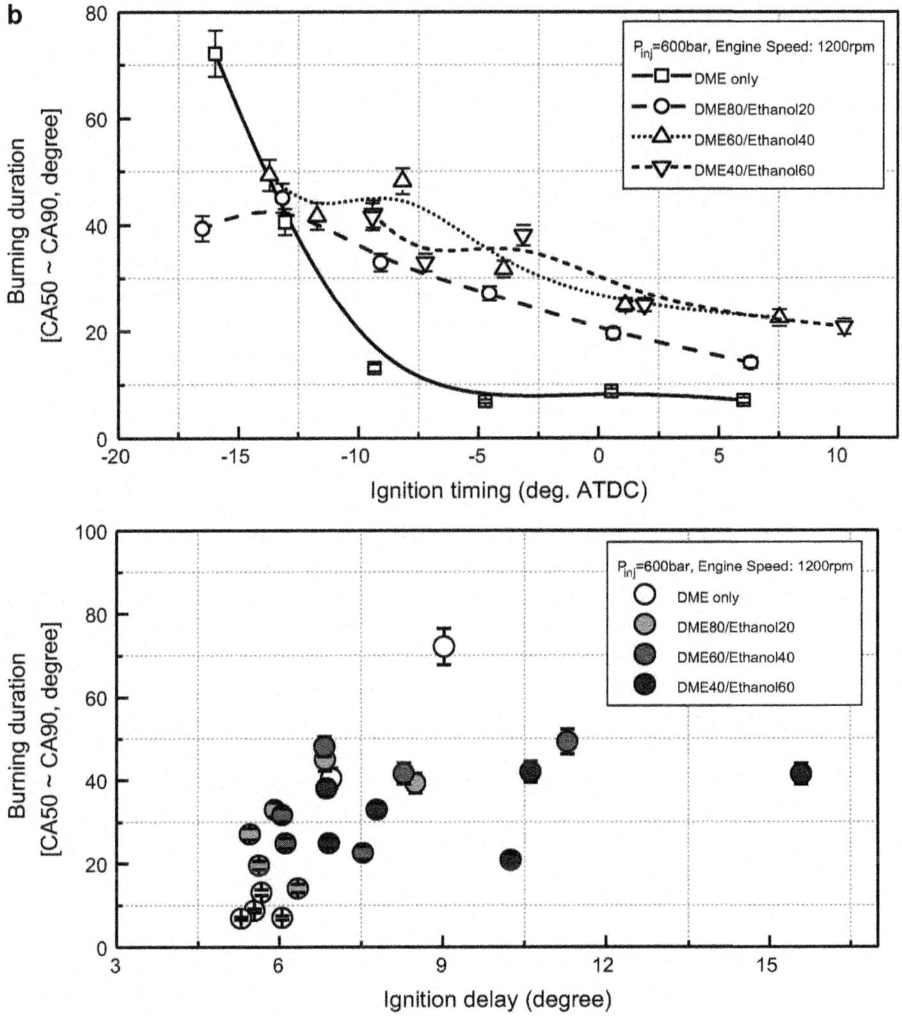

Fig. 6.34 (continued)

the ethanol fraction. Hence, it is concluded that increased ethanol concentration hinders in combustion progression [49].

Figure 6.35a shows the effect of initial pressure (P), methanol fraction (MF), initial temperature (T), start of injection (SOI) and exhaust gas recirculation rate (EGR) on combustion phasing (CA_{50}) variations in RCCI engine. The figure indicates that CA_{50} monotonically retarded or advanced with the variations in engine operating parameters (T, P, EGR and MF) except SOI. These parameters have more sensitive effect on combustion phasing, and ignition timing can be controlled using these parameters. However, more sensitive parameters need to be controlled precisely in order to minimize the cycle-to-cycle variations [44]. Study suggested that

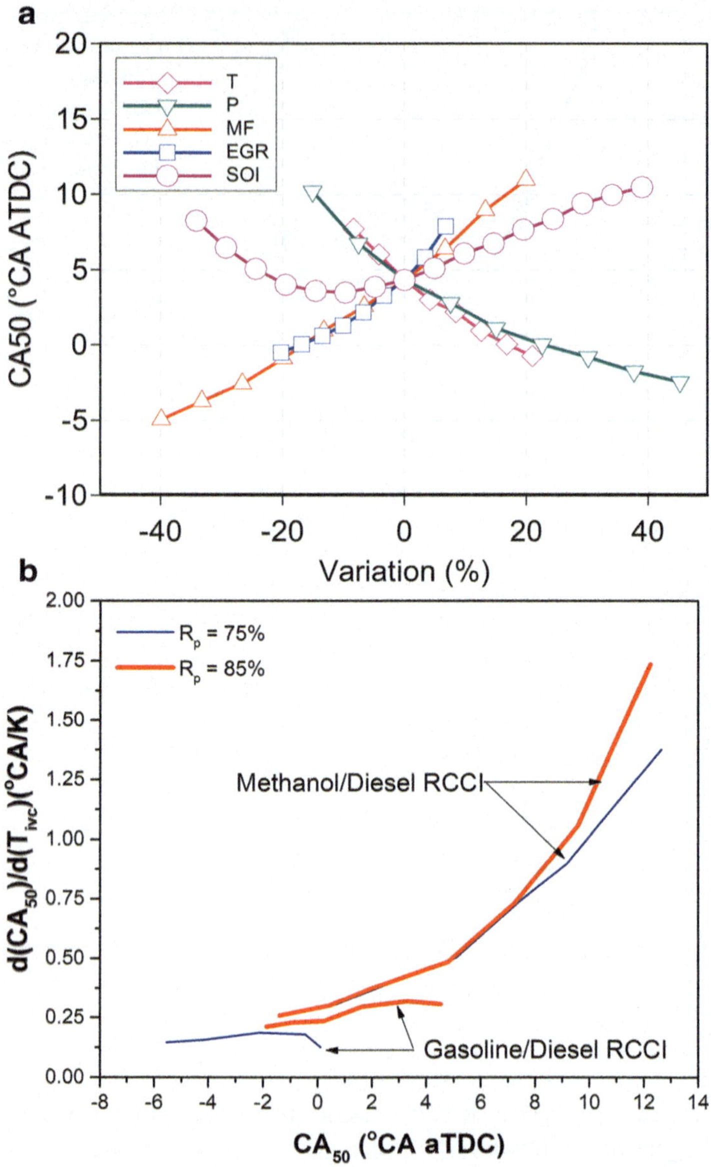

Fig. 6.35 (**a**) The CA_{50} variations with the changes of the different operating parameters [50]. (**b**) Sensitivity of CA_{50} to the variation of intake valve closing temperature for different fuels as a function of CA_{50} in RCCI combustion (Adapted from [46])

adjusting the initial pressure is the best option to achieve appropriate combustion phasing with minimum impact on other parameters [50]. However, intake pressure adjustment in practical transient engine operation is challenging. Adjusting the SOI can also be an option to achieve appropriate combustion phasing, when CA_{50} is later than 4.3 CAD aTDC (Fig. 6.35a). The EGR and initial temperature adjustment to achieve desired combustion phasing also affect the fuel consumption and emissions. Thus, it is indicated from Fig. 6.35a that varying the methanol fraction can be effective and feasible strategy to control the combustion phasing in RCCI combustion.

Figure 6.35b shows the sensitivity of combustion phasing with intake temperature for gasoline/diesel (G/D) and methanol/diesel (M/D) RCCI combustion at different premixing ratios. The figure shows that at same premixing ratio and CA_{50}, M/D RCCI has comparatively larger sensitivity ($d(CA_{50})/d(T_{IVC})$) in comparison to G/D RCCI. The larger sensitivity is attributed to lower LTHR, which leads to larger sensitivity of in-cylinder temperature to the variation of T_{IVC} in M/D RCCI [46]. The higher sensitivity found at higher premixing ratio for both fuels at fixed CA_{50} is also due to lower LTHR. Figure 6.35b also depicts that sensitivity of combustion phasing with intake temperature increases at retarded combustion phasing. For CA_{50} is earlier than 5° CA ATDC, the sensitivity remains constant for different combustion phasing (Fig. 6.35b). The sensitivity increases rapidly (nonlinear relationship is found between CA_{50} and T_{IVC}) around TDC and is found very high for further retarded CA_{50} than 5° CA ATDC. This is due to significant cooling effect of piston expansion on in-cylinder temperature before autoignition [36]. Since M/D RCCI has comparatively larger retarded combustion phasing (Fig. 6.33b), therefore more sensitivity towards intake temperature is observed.

6.3 Combustion Efficiency

Combustion efficiency is a measure of combustion quality (how well fuel is burned) of fuel–air mixture in the cylinder. Combustion efficiency affects the thermal efficiency of engine as well as engine emissions. Combustion efficiency is generally calculated from measured exhaust emission or calculated heat release using measured cylinder pressure (equations are provided in Appendix 2). Combustion efficiency of LTC engine is affected by various engine operating parameters such as equivalence ratio, EGR, combustion phasing, fuel quality, fuel injection parameters, intake pressure and temperature, etc. Figure 6.36 shows the effect of combustion phasing (CA_{50}) and λ on combustion efficiency in HCCI engine at 1200 rpm. Combustion efficiency values presented in the figure are calculated as ratio of the total heat release to total energy supplied (see Appendix 2). Figure 6.36a shows that higher intake air temperature and close to TDC CA_{50} have higher combustion efficiency due to higher combustion temperatures. The combustion efficiency is very poor for very late combustion phasing. During late combustion phasing combustion, volume of combustion chamber is increased, and the combustion temperature decreases. Especially at the end of combustion, the temperature becomes too low for complete oxidation of the fuel; therefore, a large amount of unburned hydrocarbon (HC) and carbon monoxide (CO) is emitted in the exhaust,

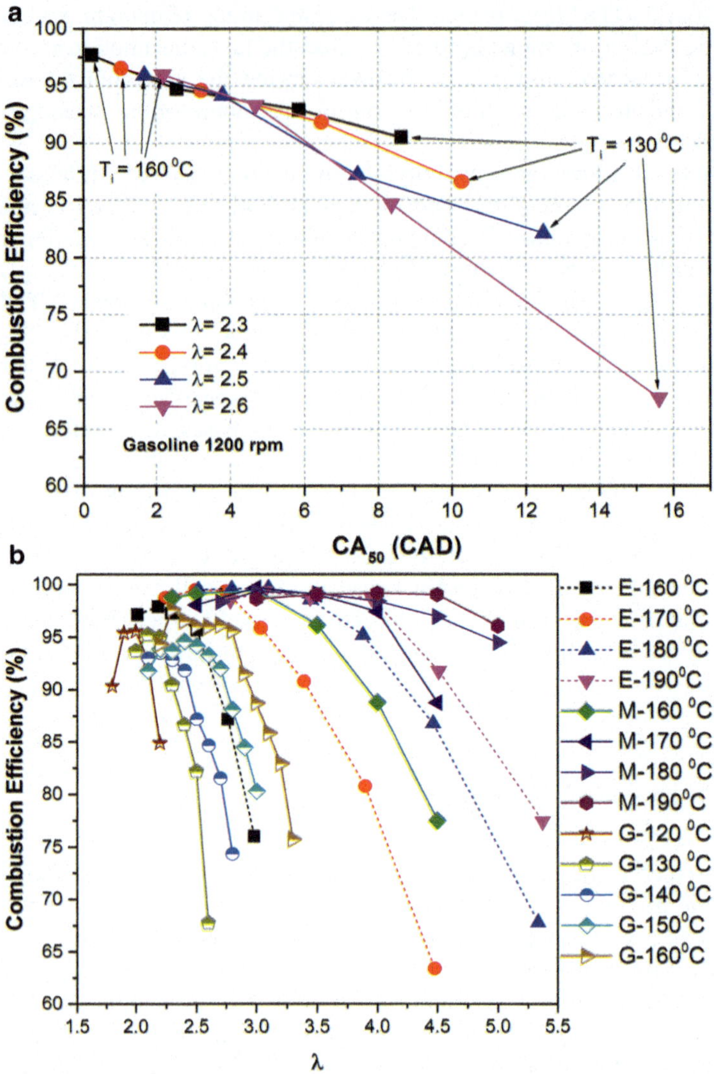

Fig. 6.36 (**a**) Effect of combustion phasing on combustion efficiency [24] and (**b**) effect of λ on combustion efficiency for gasoline, ethanol and methanol at different intake air temperatures (Adapted from [35]) in HCCI combustion

which also leads to lower thermal efficiency. This effect of increased volume during combustion becomes stronger with increased compression ratio, because the change in relative volume around TDC increases with higher compression ratio [51].

Figure 6.36b shows the variations of combustion efficiency with λ using gasoline, ethanol and methanol at different intake air temperatures. In figure legend, numbers after the name of fuels gasoline (G), ethanol (E) and methanol

(E) represent intake air tempearture. Higher combustion efficiency is achieved with reasonably rich mixtures. Combustion efficiency increases for engine operation at higher inlet charge temperatures and richer mixture (lower λ) due to higher combustion temperature and combustion rate at these operating conditions. Moreover, CA_{50} is advanced with higher intake temperatures that leads to more complete fuel oxidation before the volume expansion reduces temperatures to a level, which is very low to complete the oxidation reactions. In very advanced CA_{50}, PRR increases drastically and knocking starts. During very high PRR (with very rich and high inlet temperature), combustion efficiency decreases (gasoline $\lambda = 1.9$; Fig. 6.36b). Combustion efficiency for ethanol and methanol fuel is comparatively higher than gasoline for fairly rich mixture and higher intake air temperatures.

Figure 6.37 shows the variations of combustion efficiency with engine load and boost pressure using ethanol and iso-octane in HCCI combustion. Combustion efficiency is calculated from measured unburned species in exhaust gases. Figure 6.37a shows that at lower engine load combustion efficiency is very poor for both fuels and it improves with increase in intake temperature. At a particular engine load, combustion efficiency of ethanol is found to be higher in comparison to iso-octane. At lower engine loads (leaner mixtures) and lower intake temperatures, the limit for stable combustion after autoignition is reached. Very small amount of heat is released from the combustion, which results in severe quenching [52] and poor combustion efficiency. Figure 6.37b shows that the combustion efficiency increases with engine load for both iso-octane and ethanol due to the use of a richer mixture and hence higher temperature in the cylinder. In supercharge conditions, combustion efficiency increases due to the increase in cylinder pressure and temperature. The highest combustion efficiency is yielded with ethanol at higher compression ratio. Typical maximum combustion efficiency at higher engine loads is found around 95%, which is lower than the typical combustion efficiency in conventional engines. This is possibly due to diluted low temperature combustion in HCCI engines.

Figure 6.38 shows the variations of combustion efficiency with different topland widths of piston and with different fuels in HCCI engines. Figure 6.38a shows the combustion efficiency as a function of λ at different engine speeds. With the smallest topland width, the combustion efficiency decreases slightly with increased λ. At $\lambda = 2.5$, the case with the smallest topland width shows the lowest combustion efficiency. At $\lambda = 4.5$ the result is reversed. Topland width 1.3 mm shows the comparatively lower combustion efficiency at all the speeds. More fuel is trapped in the topland and not participating in the normal combustion process [52]. Study also concluded that opening up the topland to 2.8 mm, HC is reduced by over 50% in some of the test cases. Figure 6.38b shows the combustion efficiency variations with compression ratio for different fuels. Fuels designated as HxEy contains x% of n-heptane, y% of ethanol and (100-x-y) % of iso-octane. Each fuel is tested at five different inlet air temperature levels from 50° to 150° C and combustion phasing of CA_{50} at $3 \pm 1°$ after TDC is maintained by varying the compression ratio at an equivalence ratio of 0.33. The figure shows that in PRF fuels, combustion efficiency decreases as CR increases (corresponding to a lower inlet air temperature). At

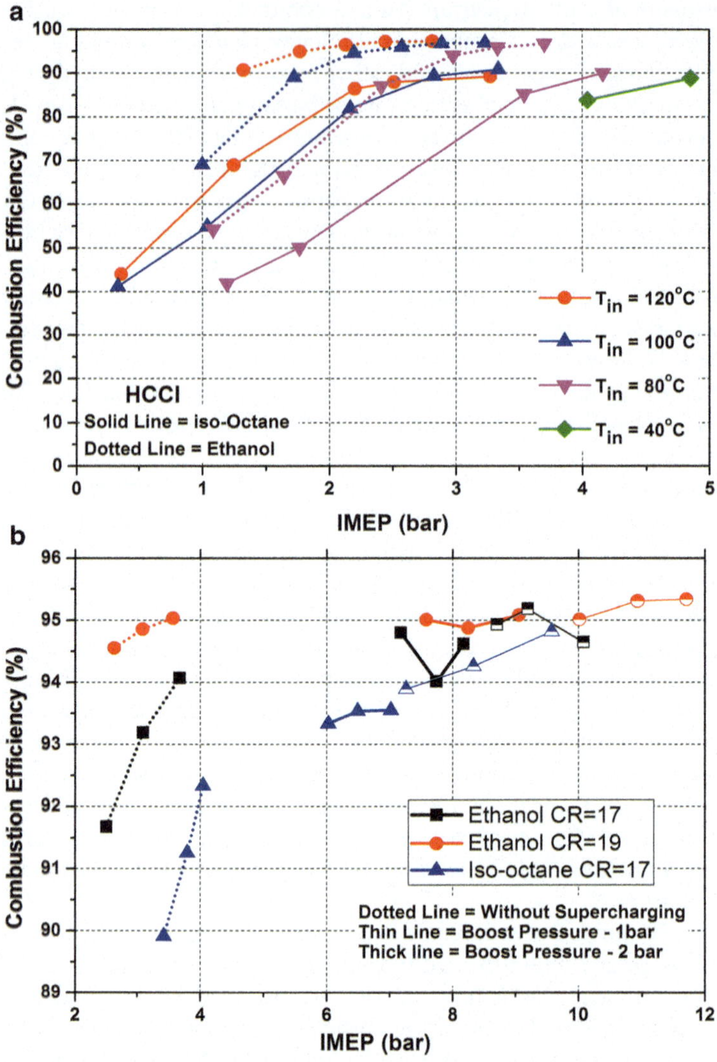

Fig. 6.37 (**a**) Variations of combustion efficiency with IMEP at different intake temperatures (Adapted from [52]) and (**b**) effect of boost pressure on combustion efficiency using ethanol (Adapted from [53]) and iso-octane in HCCI combustion

particular compression ratio, a higher octane number fuel has higher combustion efficiency than a fuel of a comparatively lower octane number. A higher toluene concentration (corresponding to a higher octane rating) results in to higher combustion efficiencies even though a higher compression ratio is used. For all test fuels, combustion efficiency decreases as compression ratio increases (Fig. 6.38b) [53].

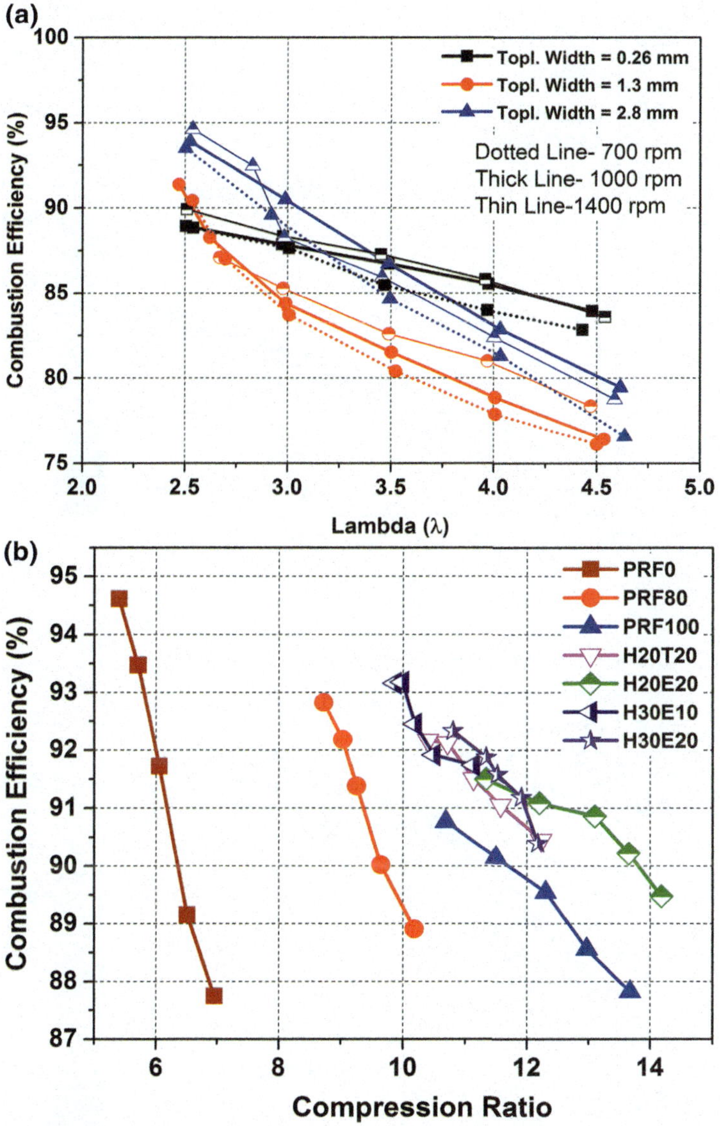

Fig. 6.38 (**a**) Effect of piston topland width on combustion efficiency at different operating conditions [54] and (**b**) effect of different fuels on combustion efficiency at constant combustion phasing and 600 rpm (Adapted from [55]) in HCCI combustion

Figure 6.39 shows the variations of combustion efficiency in HCCI operating range using gasoline and ethanol. Engine operating conditions having RI <6 MW/m^2 and COV_{IMEP} <3.5% are considered in HCCI operating range. The figure shows that the combustion efficiency is higher for richer mixture and high inlet air temperature. Combustion efficiency is highest close to knock boundary and

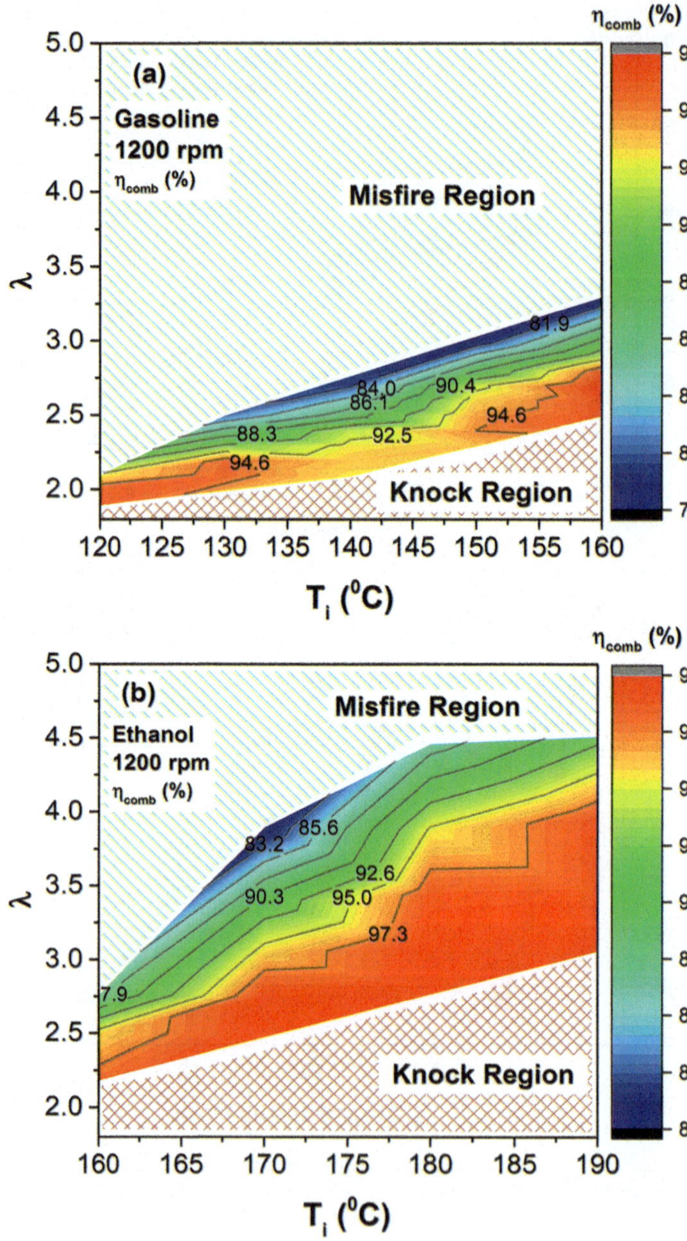

Fig. 6.39 Variation of combustion efficiency in HCCI operating range for (**a**) gasoline and (**b**) ethanol in HCCI combustion at 1200 rpm (Adapted from [24])

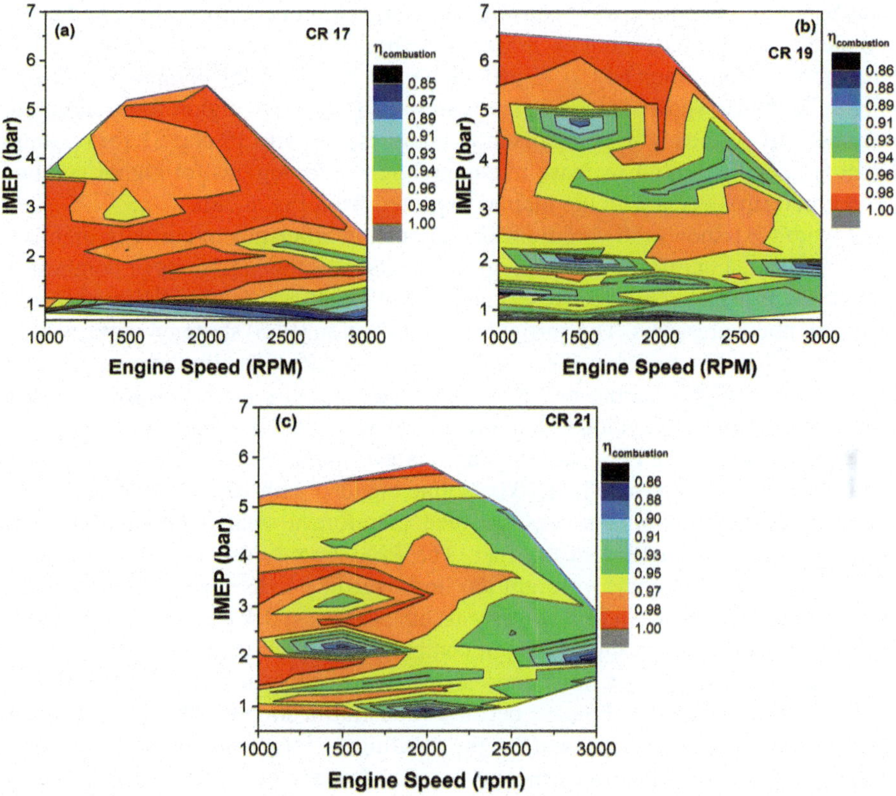

Fig. 6.40 Combustion efficiency map over wide range of engine speed at compression ratios of (**a**) 17:1, (**b**) 19:1 and (**c**) 21:1 for ethanol-fuelled HCCI combustion [5]

lowest close to misfire boundary for both fuels. Contour lines of constant combustion efficiency are inclined indicating the dependency of combustion efficiency on intake air temperature and air–fuel ratio. Combustion efficiency in HCCI operating range is higher for ethanol in comparison to gasoline (Fig. 6.39).

Figure 6.40 shows the combustion efficiency map of a wide range of engine speed at different compression ratios for ethanol-fuelled HCCI engine in naturally aspirated conditions. The HCCI operating region at any engine speed is defined as engine operating conditions having peak PRR <5 MPa/ms and combustion efficiency >85%. At each engine speed, there are several combinations of intake temperature, and λ is satisfying this criterion. For each λ, intake temperature at which thermal efficiency is maximum is used to develop the combustion efficiency map at each speed. Figure 6.40 shows that the combustion efficiency decreases at higher engine speed. Time available for combustion decreases with increase in engine speed, and to increase combustion rate, higher inlet temperature is required. Combustion phasing is also dependent on intake temperature in HCCI engine, and advanced combustion phasing results in higher PRR. With the higher limits of PRR and based on the combustion phasing and combustion time, available combustion efficiency is achieved, and it decreases with engine speed [5]. It can also be noticed that combustion efficiency is higher at higher loads for all compression ratios. Figure 6.40 also shows that compression ratio 17 has large region for around 99% combustion efficiency in comparison to compression ratio 19 and 21 for optimum inlet temperature. With increase in compression ratio, regions for higher combustion efficiency decrease in the map as expected from Fig. 6.38b.

Figure 6.41 shows the variations in combustion efficiency in partially premixed combustion using different fuels. Figure 6.41a shows the variations of combustion efficiency in well-premixed (HCCI) and partial DI (PFS) case for different fuels at various intake pressures. The partial fuel stratification (PFS) strategy works well for high φ sensitivity fuel (see Sect. 2.2.4.1 of Chap. 2). Gasoline shows higher φ sensitivity at higher intake pressure, and hence gasoline data is shown at 2 bar intake pressure. Combustion efficiency is similar for PFS and premixed strategy in case of gasoline, but PFS strategy can achieve higher engine loads due to fuel stratification (Fig. 6.41a). The E10 (gasoline blended with 10% ethanol) fuel is not φ sensitive at lower intake pressure, and therefore a small amount of reactive chemical (EHN) is added in the fuel. The PFS strategy can be used to achieve higher thermal efficiency with ignition improver like EHN. The combustion efficiency is slightly lower in PFS strategy at some cases. At 0.4% EHN and 1.3 bar, the combustion efficiency rapidly falls in comparison to premixed case at higher load because of the high load limit approached due to oxygen limit in case of PFS. Oxygen limit occurs in PFS case because fuel stratification by PFS creates locally rich zones in the cylinder, and as the fuelling is increased to the limit, these regions (rich zones) have insufficient oxygen, leading to incomplete combustion even though there is still oxygen available in other regions of the cylinder [56]. Hence, although overall amount of intake oxygen is same for well premixed and PFS, the combustion efficiency of PFS falls much more quickly near the oxygen limit leading to reduction in the thermal efficiency of PFS compared to premixed even with CA_{50} still being more advanced. This disadvantage is removed with 0.15%EHN concentration at same intake pressure (1.3 bar). Due to the lower reactivity of the fuel with 0.15%EHN compared to fuel with 0.4%EHN, less EGR is required, and more oxygen is available [56], and in this case PFS shows similar trend as the fuel with 0.4%EHN at 1 bar intake pressure.

Figure 6.41b shows the variations of combustion efficiency with engine load for various gasoline (different research octane number) fuels at two compression ratios (17.1 and 14.3). In PPC strategy, high octane fuels are directly injected, and to lower the PRR rate, very high amount of EGR is used. Figure 6.41b shows that despite the use of 50% of EGR in PPC strategy, the combustion efficiency is always above 98% for engine loads higher than 7 bar IMEP. In this strategy, it is possible to burn the fuel–air mixture within the temperature range of 1500 K and 2000 K with suitable combination of λ and EGR. Thus, low NO_x and high combustion efficiency are achieved between 4 and 18 bar gross IMEP for different gasoline-like fuels. Ethanol (higher octane fuels) shows comparatively lower combustion efficiency at lower engine loads in PPC strategy (Fig. 6.41b).

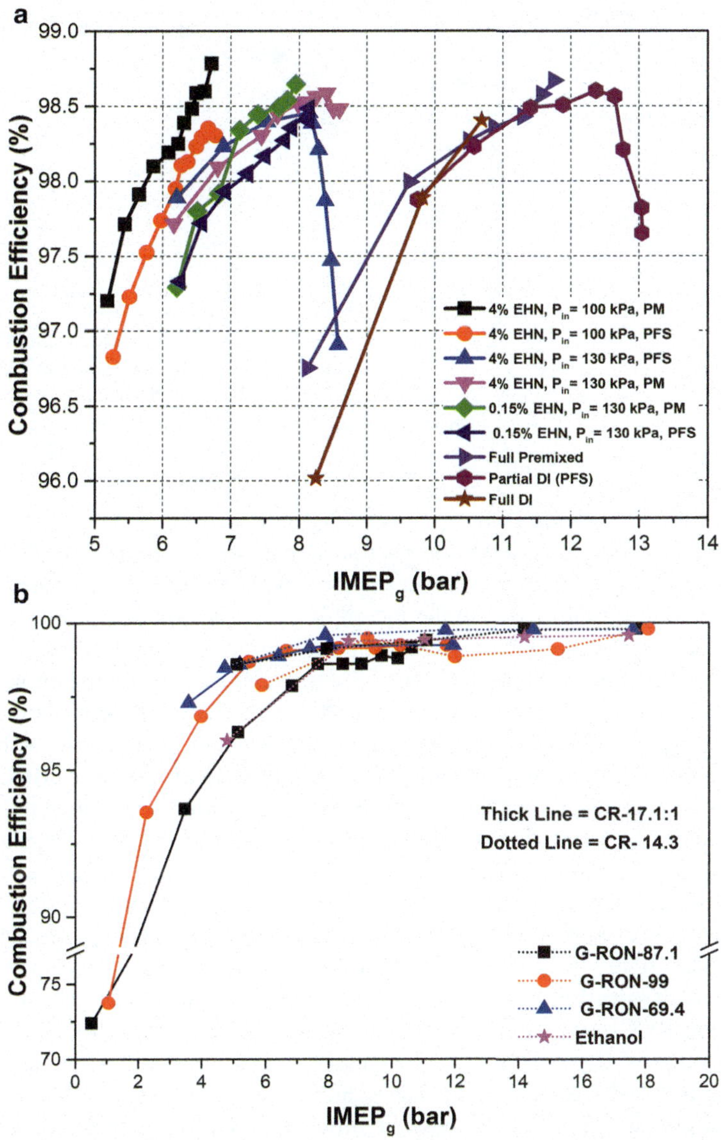

Fig. 6.41 (**a**) Variations of efficiency in PFS and well-mixed strategy in HCCI combustion (Adapted from [44, 56]) and (**b**) combustion efficiency as function of engine load in PPC strategy (Adapted from [34, 57])

Figure 6.42 shows the variations of combustion efficiency with engine load for gasoline/diesel (G/D) and ethanol-blended gasoline (20%, 85%)/diesel (E20/D; E85/D) in RCCI strategy vis-à-vis conventional diesel combustion (CDC). Combustion efficiency is a function of engine load, and it increases with an increase in

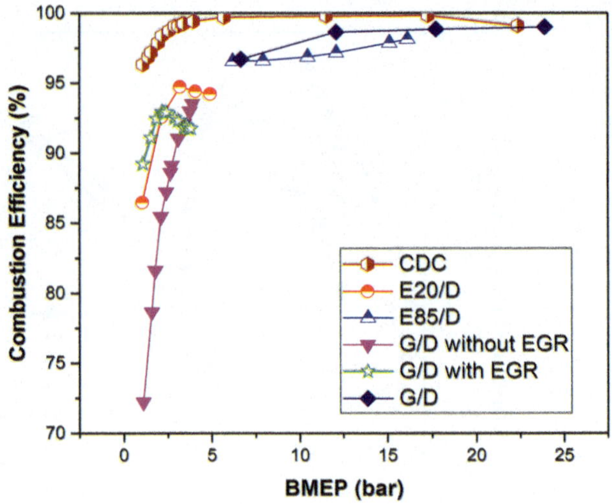

Fig. 6.42 Variations of combustion efficiency with engine load for different fuels in RCCI combustion (Adapted from [58–60])

the engine load due to higher combustion temperature. Figure 6.42 reveals that RCCI combustion strategy has lower combustion efficiency as compared to CDC strategy, which is mainly due to the incomplete combustion of low reactivity fuel. Figure 6.42 also depicts that RCCI engine operation with EGR has a higher combustion efficiency at lower load as compared to RCCI operation without EGR. The intake charge temperature is higher with EGR, which leads to increase the oxidation rate. Therefore, higher combustion efficiency is obtained with EGR at lower load condition. Comparatively lower combustion efficiency of RCCI operation leads to increase the unburned hydrocarbon emissions as compared to CDC.

In-cylinder fuel blending and fuel injection timing affect the combustion efficiency of RCCI combustion as it affects the local equivalence ratio and charge stratification in the combustion chamber. Figure 6.43a represents the effect of diesel main injection timing and in-cylinder fuel blending ratio on the combustion efficiency. The figure reveals that combustion efficiency slightly improves with retarded main injection timings and reaches about 97.5%. With retard main injection timing RCCI operation, combustion phasing reaches close to TDC position, which increases the combustion temperature. Higher combustion temperature enhances the fuel oxidation rate that reduces the HC as well as CO emissions, which results into higher combustion efficiency. However, higher combustion temperature promotes the NO formation reactions, which results in higher NO_x emissions also. Additionally, ringing intensity is also higher for combustion phasing close to TDC but still below the limit of 5 MW/m^2. With too advanced fuel injection timing (i.e. −45 to −50 CAD), overmixing of fuel and air may occur that results into overleaning of charge, and poor combustion efficiency is achieved. However, with these injection timings (i.e. −45 to −50 CAD), NO_x emissions are below the EURO VI limits [47]. Study also concluded that to keep NO_x and ringing intensity below the acceptable limits, advanced fuel injection timing of diesel is necessary in RCCI combustion.

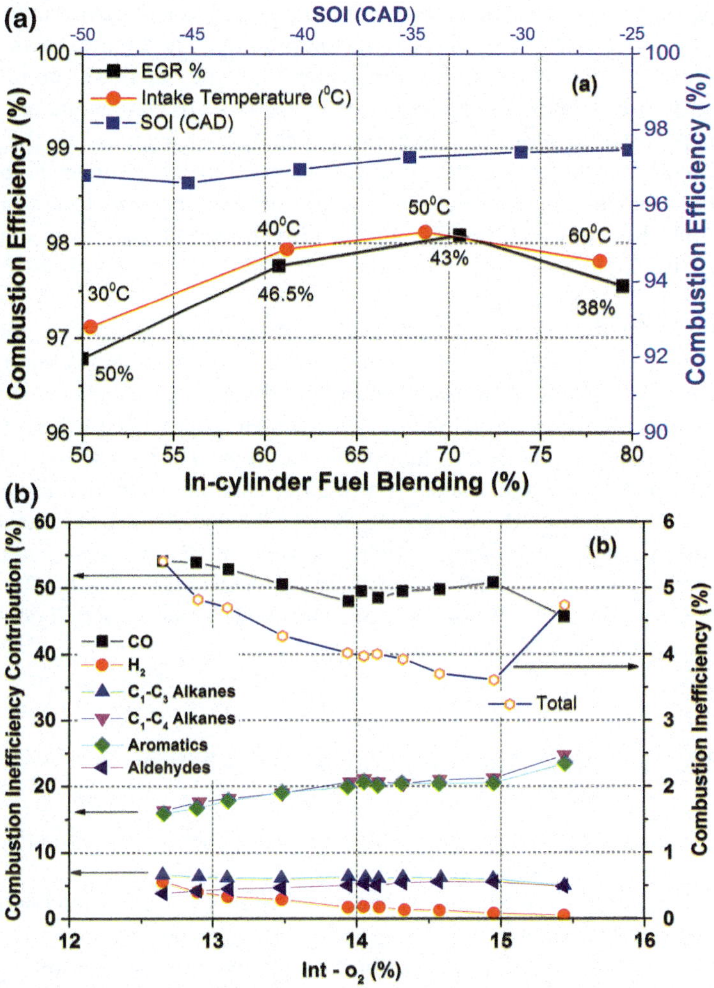

Fig. 6.43 (**a**) Effect of injection timing and fuel premixing ratio on the combustion efficiency (Adapted from [47]). (**b**) Contributors to combustion inefficiency (Adapted from [61]) in RCCI engine

Figure 6.43a also suggests that by increasing the oxygen concentration (or reducing the exhaust gas recirculation rate), combustion efficiency improves due to increased HC and CO oxidation rate [47]. With the reduction in EGR rate or an increase in the in-cylinder fuel blending ratio, combustion efficiency improves (Fig. 6.43a). Reduction in EGR rate results in progressive advance combustion phasing, which leads to increase the combustion temperature. Maximum achieved combustion efficiency is 98.2% which is obtained with EGR rate of 43% and 70.8% in-cylinder fuel blending ratio (Fig. 6.43a). Combined effect of in-cylinder fuel blending and intake temperature on combustion efficiency is also presented in

Fig. 6.43a. Increase in the intake temperature is required with increase in gasoline fraction to maintain the constant combustion phasing. The figure shows that with increase in the intake temperature, combustion efficiency increases up to a certain fraction of gasoline. Higher temperature of air enhances oxidation rate of the higher octane fuel (such as gasoline). However, with further increase in the gasoline fraction, overall reactivity of charge inside the combustion chamber is lower, and combustion rate is slower. Thus, it increases the combustion losses and lower combustion efficiency is observed (Fig. 6.43a). Maximum achieved combustion efficiency is 98.1% at intake temperature of 50.2 ° C and gasoline fraction of 68.7%.

Figure 6.43b illustrates the contribution of different species in combustion inefficiency in gasoline/diesel RCCI combustion. The figure shows that CO is the highest contributor (~50%) of combustion inefficiency and it remains fairly constant over the entire range of intake oxygen. The HCs and OHCs have a combined contribution almost equivalent to that of the CO (Fig. 6.43b). The C4–C7 fraction and the aromatics are the major contributors in combustion inefficiency that indicate the presence of unburned fuel in the RCCI exhaust. At lower intake oxygen concentrations, the trends depict slightly lower impact of HCs and OHCs in agreement with the slightly higher contribution of CO and H_2 (Fig. 6.43b). Increase in the unburned fuel and consequent heavy HC contribution for RCCI could be due to the entrapment of the gasoline–air mixture in the crevices [61].

6.4 Pressure Rise Rate and Combustion Noise

The HCCI combustion is by nature flameless volumetric (entire cylinder) heat release process, which occurs in a very short time period in comparison to conventional engine combustion (SI and CI) process. The HCCI combustion completion is limited by chemical kinetics in well-mixed mixture, while in conventional engine combustion process, its completion is limited by flame propagation/mixing process governed by heat and mass transfer. The HCCI combustion mode is similar to the knock phenomenon in conventional SI engine because of its nearly simultaneous autoignition of charge in the cylinder. Thus, high load operating range of HCCI engine is limited by high combustion rate (due to autoignition of entire well-mixed charge in the cylinder) resulting in high pressure rise rates (PRR) and heavy knocking. The two major concerns arising from very high PRR and knocking are unacceptable levels of combustion noise and accelerated wear on the mechanical components of the engine. Typically, a threshold value of PRR is used to limit the high load operation of HCCI engine. Another parameter named as "ringing intensity" (RI) (equation is provided in Appendix 2) is also used to better represent the knock intensity of HCCI combustion [62]. The RI indicates the wave energy of pressure oscillation caused by knocking during HCCI combustion.

Figure 6.44a presents the variations of RI with λ and intake air temperature for gasoline, ethanol and methanol. In figure legend, numbers after the name of fuels

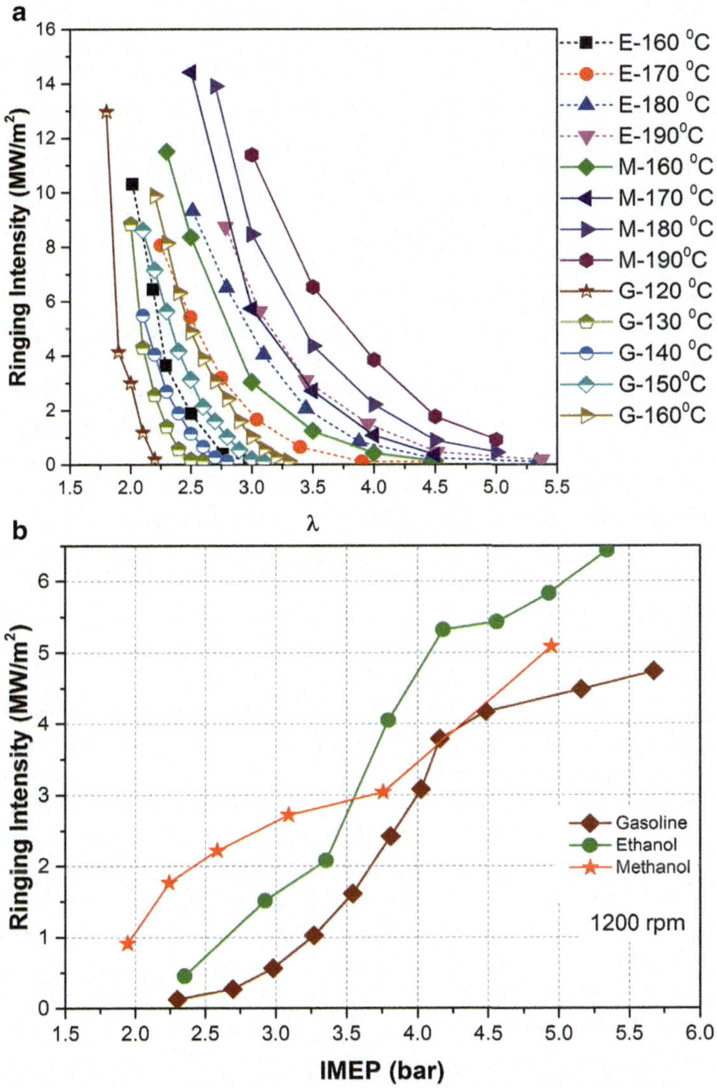

Fig. 6.44 (**a**) Variations of ringing intensity with λ and Ti using different fuels (Adapted from [35]). (**b**) Variations of ringing intensity with engine load for gasoline, ethanol and methanol [35] in HCCI engine

gasoline (G), ethanol (E) and methanol (E) represent intake air tempearture (T_i). The figure shows that RI increases drastically at each intake air temperature as mixture becomes richer for both the fuels. When mixture becomes richer, combustion chamber temperature increases due to advanced combustion phasing resulting in faster combustion rate, which leads to higher PRR. On increasing the intake temperature at any λ, RI increases. The RI values depend on intake temperature at any λ for a particular fuel. Intake air temperature is typically increased to improve the combustion efficiency. Therefore, these two factors act opposite to each other, and optimum intake air temperature needs to be selected for each λ.

Figure 6.44b shows the RI variations with engine load (IMEP) at combustion phasing corresponding to maximum efficiency (by varying intake air temperature) for gasoline, ethanol and methanol. The figure shows that RI increases with engine load for each fuel because richer air–fuel mixture is used at higher engine load. Figure 6.44b also indicates that ethanol and methanol have higher RI than gasoline as intake air temperature used to auto-ignite the fuel is higher due to higher octane number of ethanol and methanol. At higher intake temperatures, combustion phasing is also more advanced for ethanol and methanol as compared to gasoline at optimum efficiency. Another possible reason is at higher intake temperatures, comparatively lower natural thermal stratification for autoignition, which results into comparatively lower sequential autoignition (due to natural thermal stratification) leading to smaller combustion duration and higher RI in case of ethanol and methanol.

Figure 6.45a shows the effect of CA_{50} on the RI at different intake air temperatures using gasoline and ethanol at 1200 rpm. The figure depicts that RI is very high at advanced CA_{50} for both fuels. At retarded combustion phasing, RI is comparatively much lower. Thus, combustion phasing retard is one method to control the RI in HCCI combustion. At retarded combustion phasing, sensitivity of RI (slope of curve in Fig. 6.45a) with CA_{50} is lower, and it increases drastically for advanced combustion phasing near TDC. Third-order polynomial regression curve fits with experimental data and captures the trend in RI variations with CA_{50} for gasoline and ethanol with a coefficient of determination (R^2) of over 0.99.

Figure 6.45b shows the correlation of RI with maximum PRR (PRR_{max}). In this figure, RI is plotted against PRR_{max} for the λ and intake air temperature sweep data using gasoline, ethanol, methanol and butanol at 1200 and 2400 rpm. The figure shows that RI strongly correlates with PRR_{max}. Quadratic regression of experimental data shows that RI dependence with PRR_{max} is quadratic. The sensitivity of RI (slope of curve in Fig. 6.45b) with PRR_{max} depends on engine speed and it is higher at higher engine speed. The figure also indicates that depending on engine operating conditions, RI or PRR can be very high and unacceptable, which limits the high load engine operation. To decrease the PRR or RI at higher engine load, it is important to understand the phenomenon occurring during knocking combustion. Based on understanding of HCCI knock, different methods can be developed for HCCI knock mitigation. Next few sections describe the HCCI knock, knock mitigation methods, knock metrics and combustion noise in detail.

6.4.1 HCCI Knock

In HCCI combustion, well mixture charge is burned in the cylinder within a time period in the order of 0.1 ms, while combustion duration of conventional engines

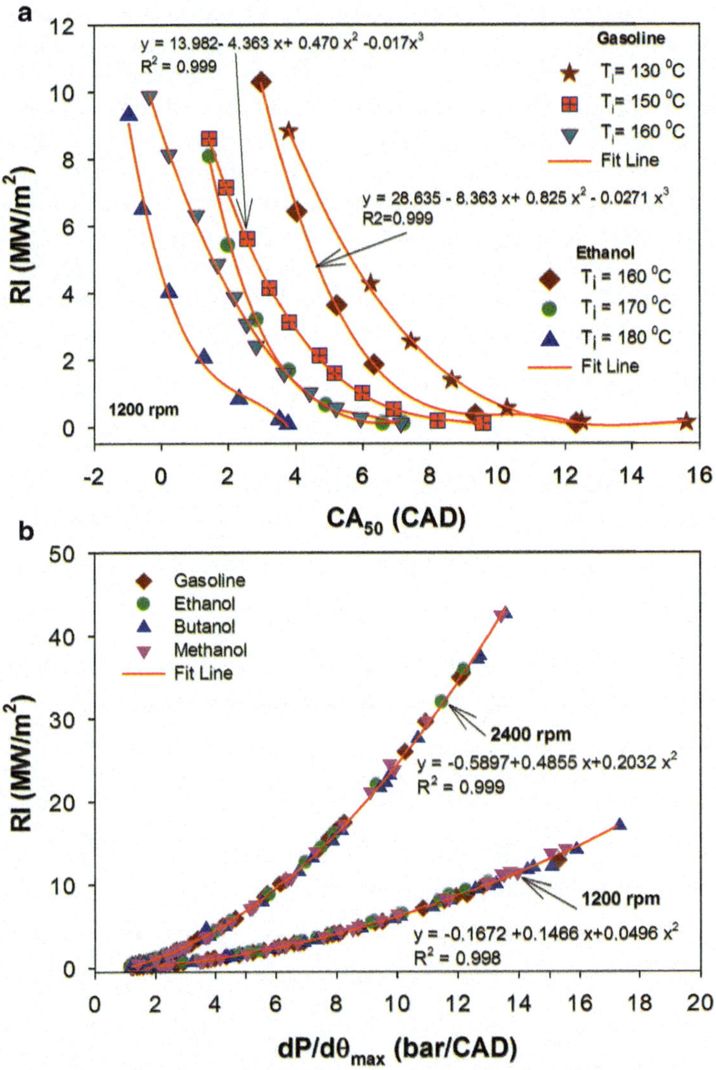

Fig. 6.45 (**a**) Effect of combustion phasing on ringing intensity for gasoline and ethanol HCCI combustion (Adapted from [35]). (**b**) Correlation of RI with PRR_{max} using different fuels at 1200 and 2400 rpm (Adapted from [24]) in HCCI engine

occurs in tens of milliseconds [63, 64]. Comparatively shorter combustion duration leads to very high PRR in HCCI combustion. The HCCI engines are prone to pressure oscillation even though audible knock is not detected due to the presence of pressure wave in combustion chamber [62, 65]. The knocking phenomenon in HCCI combustion can be explained by understanding the initiation of pressure wave in the cylinder. It is well understood that there exist fuel and thermal inhomogeneities (parcels with different temperatures and equivalence ratios) in the HCCI engine cylinder, although charge is well mixed and named as homogeneous mixture (see Sect. 2.2.4.6 of Chap. 2). Combustion occurs sequentially due to fuel and thermal inhomogeneities in the cylinder. Andreae et al. developed a criterion for initiating a pressure wave in HCCI engine, when there is a volumetric heat release rate of $\dot{q}$ locally [65]. If the heat release is over a sphere of radius r and volume $V(=4\pi r^3/3)$, to maintain uniform pressure, the expansion due to the rise in gas temperature should be provided for by the volumetric expansion of the sphere. Assuming constant ideal gas properties, for uniform pressure in the cylinder, the equation $P\Delta V = \text{m}R\Delta T$ holds, and from energy conservation, equation $m\,C_p\,\Delta T = \dot{q}\,V\Delta t$ can be written for time Δt. Combining these equations, volumetric rate of change can be written as:

$$\frac{\Delta V}{\Delta t} = \frac{\gamma}{\gamma - 1}\frac{\dot{q}V}{P} \tag{6.5}$$

For acoustic expansion with speed of sound "u", the volumetric change can also be written as:

$$\frac{\Delta V}{\Delta t} = 4\pi r^2 u \tag{6.6}$$

The acoustic expansion would not be able to accommodate the thermal expansion to maintain the uniform pressure in the cylinder when [65]:

$$\dot{q} > \frac{3\gamma}{\gamma - 1}\frac{P}{(r/u)} \tag{6.7}$$

Similar criteria are also developed in another study by Yelvington et al. [66] that knock would not be expected when:

$$\beta = \frac{L_c(\gamma - 1)}{\gamma P}\frac{\dot{q}}{u} \leq 1 \tag{6.8}$$

throughout the cylinder, at all times. γ is ratio of specific heat (C_p/C_v) and L_c is a characteristic length scale (volume = area) of the inhomogeneities in the HCCI cylinder (taken as 0.1bore).

From this analysis, it is found that the parameters such as local ambient pressure in the cylinder (P), mixture composition (γ), temperature in the cylinder (affecting speed of sound $u = \sqrt{\gamma RT}$) and charge stratification (characteristic length of combustion) affect the threshold for pressure wave formation [63]. At higher pressure in cylinder (increasing the boost), higher HRR can be accommodated without pressure oscillations. The more stratified charge (smaller characteristic length) and higher EGR (lower γ) can burn more amount of fuel without pressure oscillations which leads to extension of higher load with same RI limit.

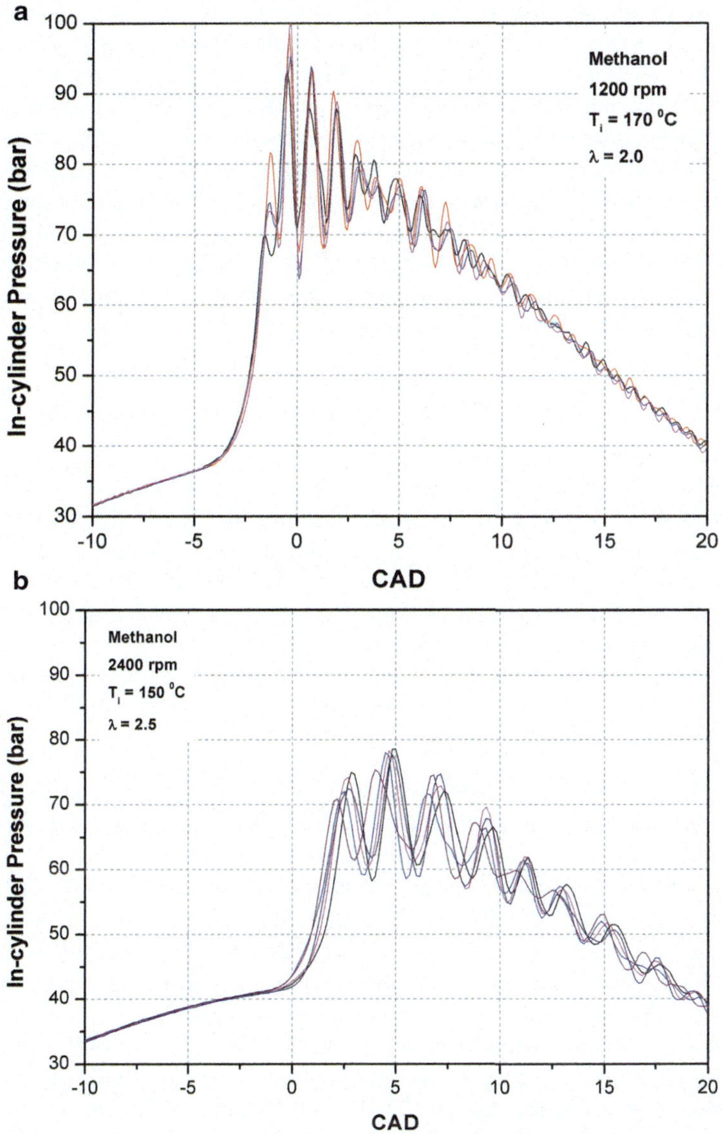

Fig. 6.46 Experimental in-cylinder pressure trace for five knocking combustion cycles for (**a**) $\lambda = 2.0$ and 1200 rpm and (**b**) $\lambda = 2.5$ and 2400 rpm in methanol HCCI combustion [24]

The pressure oscillations can be severe (excessive PRR) at very rich fuel mixtures in HCCI combustion. Figure 6.46 illustrates the pressure oscillations in HCCI combustion for five consecutive knocking combustion cycles at two different engine operating conditions. In-cylinder pressure data is acquired at 86.4 kHz for both the operating conditions to observe the pressure wave phenomena in closer detail.

Pressure oscillations observed in Fig. 6.46 are due to very fast combustion in a given parcel of fuel–air mixture, and these local overpressures lead to resonant pressure waves. For slow enough combustion rate in a fuel–air parcel, the parcel expands and maintains its pressure nearly equal to that of the rest of the gas in the cylinder. These pressure oscillations are not expected in combustion of a perfectly homogeneous mixture (no thermal and fuel stratification). These oscillations occur due to inhomogeneities present in the cylinder in real HCCI engine. These pressure oscillations have originated in the cylinder which leads to the excitation of engine structure and eventually results in acoustic radiation perceived as combustion noise.

The resonant frequencies of a cylinder can be determined analytically by solving the wave equation presented by Eq. (6.9) [62].

$$\frac{\partial^2 P}{\partial t^2} = u^2 \nabla^2 P \tag{6.9}$$

where P is the cylinder pressure and u is the speed of sound ($u = \sqrt{\gamma RT}$). γ is the ratio of specific heats, R is the ideal gas constant and T is the temperature.

The frequencies at which the engine knocks are evaluated by the acoustic vibration modes specific to the combustion chamber. For a simple cylindrical combustion chamber, the vibration mode frequencies can be calculated analytically using C. S. Draper's acoustic pressure wave formula [67].

$$f_{m,n} = \frac{\rho_{m,n} u}{\pi B} \tag{6.10}$$

where $f_{m,n}$ = specific vibration frequency for mode m,n [Hz], $\rho_{m,n}$ = vibration mode factor, u = local speed of sound, B = bore of the engine, m = circumferential mode number and n = radial oscillation mode number.

The combustion chamber temperature affects the acoustic velocity "u", as well as the ratio of specific heats (γ). The in-cylinder temperature is required for calculation of γ and the speed of sound. Using Draper's equation, mean theoretical frequencies for each vibration mode are calculated in order to compare them with the measured frequencies from the pressure traces. Figure 6.47a shows the theoretically calculated acoustic mode frequencies as function of crank angle position for a typical fired HCCI engine cycle. It is observed that the expected frequencies of different modes are in the range of 4–15 kHz. Figure 6.47b presents the power spectrum of pressure oscillations present in knocking cycle, which illustrates the frequencies present in the cylinder pressure oscillations. It is observed that three frequencies mainly present are 5.6 kHz, 6.9 kHz and 8.4 kHz. Peak for third mode is very weak in some of the engine cycles in frequency domain plot. None of these frequencies are multiples of each other suggesting that no overtone frequencies exist [24]. Frequencies are corresponding to modes (1, 0), (2, 0) and (0, 1) termed as first, second and third mode of pressure oscillations (Fig. 6.47b). The exact value of frequencies depends on the combustion chamber geometry and charge conditions [65]. The charge conditions can even shift the frequency corresponding to peak power spectrum amplitude.

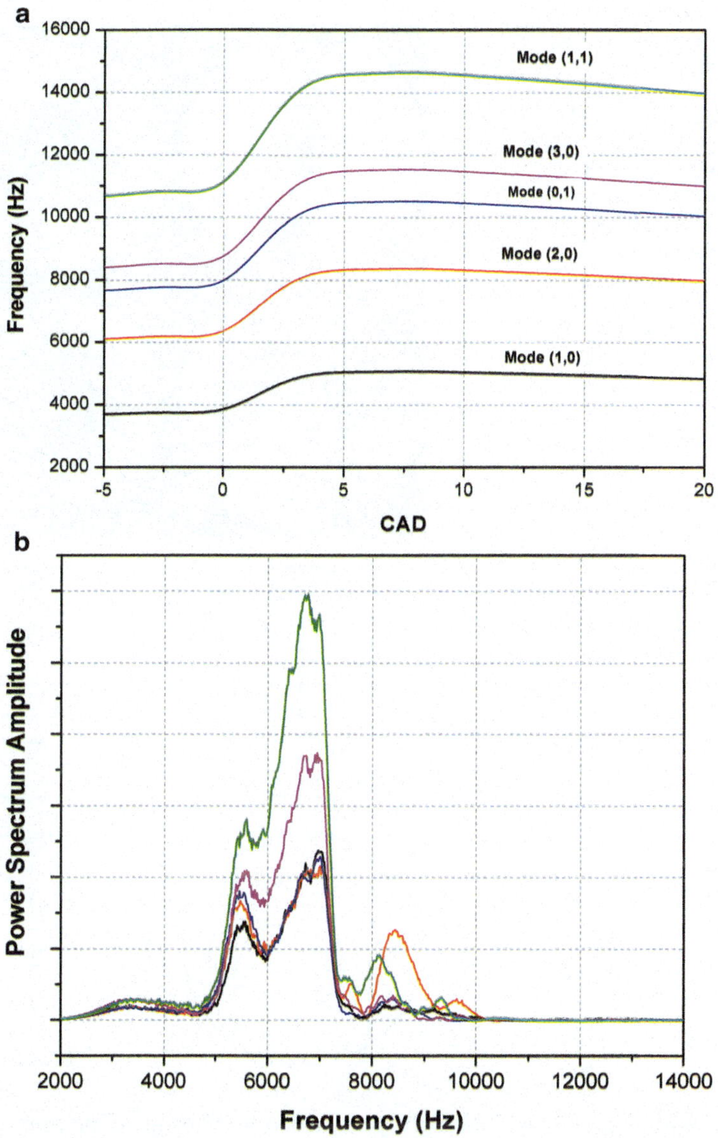

Fig. 6.47 (**a**) Theoretically estimated acoustic mode frequencies as function of CAD for a typical fired HCCI engine cycle [24]. (**b**) Frequency domain plot for four different knocking HCCI combustion cycles [24]

Another study found that the lower frequency (4–8 kHz) knock dominates over higher frequencies (8–20 kHz) in HCCI engines [62], which is also observed in Fig. 6.47b. The type of acoustic modes and the frequency of the pressure waves are dependent on the local speed of sound and the combustion chamber geometry.

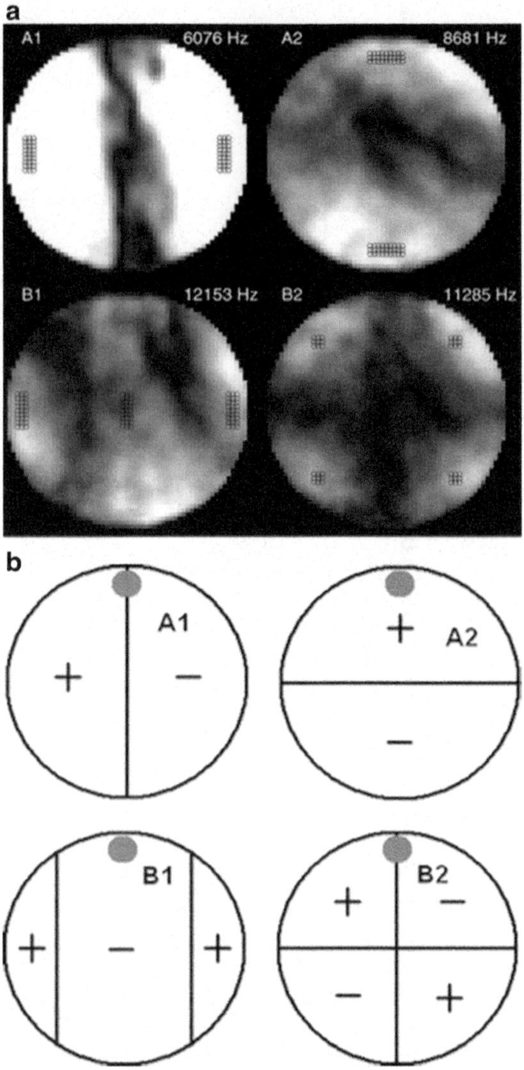

Fig. 6.48 (**a**) Spatially resolved amplitudes of four different frequencies extracted from the high-speed video sequence showing the acoustic modes in the cylinder in HCCI combustion. A brighter colour indicates higher amplitudes [68]. (**b**) Schematic of the four acoustic modes measured with high-speed video. Lines represent the nodes [68]

A study used image-intensified high-speed video to investigate the pressure oscillations, or acoustic modes, in the HCCI combustion chamber by developing an image analysis procedure which extracts the different acoustic modes [68]. Figure 6.48 illustrates the amplitude of different frequency revealing acoustic modes. Acoustic modes obtained from image analysis show good agreement with modes estimated by measurements of in-cylinder pressure data. In Fig. 6.48a, the nodes and antinodes of the four acoustic modes are apparent and illustrated schematically in Fig. 6.48b. This particular combustion cycle is selected as an example as it is one of the few cycles where all four modes occurred [68]. Typically, the first circumferential mode (A1) is the dominant mode in HCCI combustion.

Andreae et al. [65] investigated the knock behaviour of a HCCI engine by measuring the sound radiation from the engine and the cylinder pressure data. Prominent PSD (power spectral density) peaks are observed at the 1st mode of cylinder gas oscillation for both the cylinder pressure and the audible signal. However, the development of the pressure wave in time is very different than that of the audible wave within a cycle. No direct interdependence is shown at the same time between the amplitude of the acoustic radiation and that of the cylinder pressure wave. Therefore, study concluded that the audible wave is not the result of the transmission of the in-cylinder pressure wave through the engine structure. It was rather, the sound radiation from the structural vibration excited by the first few waves of the cylinder gas vibration. After the vibrational excitation by first few waves, the cylinder gas vibration and the structural vibration are essentially decoupled. Thus, the amplitudes of the structural acoustic wave and pressure wave, at the same time in the subsequent development in one cycle, are independent [63].

Net energy transfer from the combustion gases to the engine structure is essential that causes the outer surface of the engine to vibrate, in order to radiate sound waves from the engine. Therefore, the difference between "quiet" combustion and "noisy" combustion is not the frequency content of the pressure data itself, but the energy contained within the pressure pulsations at the frequencies that the engine structure is sensitive [62]. Eng [62] proposed a quantity defining the total acoustic energy flux I (watt/m^2) related with the pressure oscillation in the cylinder, by decomposing the waveform into a series of harmonic waves and adding up the wave intensity of all of them. Acoustic engine flux is given by Eq. (6.11).

$$I = \frac{1}{2\gamma}\frac{\Delta P^2}{P}\sqrt{\gamma RT} = \frac{1}{2\gamma}\frac{\sqrt{\gamma RT}}{P}\sum_n (\Delta P_n)^2 \tag{6.11}$$

where ΔP is the pressure pulsation amplitude, ΔP_n the amplitude of the nth harmonic wave from the Fourier decomposition, R the universal gas constant and P and T the pressure and temperature, respectively.

To estimate the amplitude of pressure pulsation, Eng [62] considered that amplitude of pressure pulsation is directly proportional to maximum pressure rise rate in the cylinder. Thus,

$$\Delta P = \beta \frac{dP}{dt}_{max} \tag{6.12}$$

where β is a scale factor determined from the experimental data and $(dP/dt)_{max}$ is the maximum PRR in kPa/ms.

The value for β can be rationally determined from the experimentally measured data by analysing the maximum amplitude of the pressure pulsations over a range of operating conditions. Eng [62] found that the amplitude of the pressure pulsations is typically on the order of 5% of the maximum PRR over a range of conditions and accordingly the coefficient β is set to 0.05. The values of β can be different for engine of different bore size and different compression ratios [69]. Andreae [65] also proposed similar relation between the amplitude of pressure pulsation and maximum PRR (Eq. (6.13)).

$$\Delta P \approx \frac{1}{4f}\frac{\mathrm{d}P}{\mathrm{d}t}_{\max} \tag{6.13}$$

where f is the oscillation frequency which can be approximately calculated from the 1st mode oscillation frequencies. This correlation can be utilized to analyse and evaluate the engine knock tendency with only measured cylinder pressure data without resolving the acoustic waves.

6.4.2 Knock Metrics and High Load Limit

Accurate measurement of a knocking combustion process is essential to understand and predict knock in combustion engines. In published literature several methods have been investigated for accurate knock detection in engines. The accuracy of knock onset is higher for measurement using an optical approach, while other methods have limited accuracy due to measurement of secondary affect triggered by the rapid autoignition process. However, the optical measurement of knock has limited use due to lower durability of the measurement system. Knock detection methods other than optical method can be divided into four categories based on the type of data used for knock analysis as (i) in-cylinder pressure, (ii) engine block vibration, (iii) ion current signal and (iv) rate of heat transfer [70]. The most widely used method for knock detection is cylinder pressure-based methods because of relatively high accuracy and durability of cylinder pressure measurement systems. To ensure the reliability of acquiring a pressure oscillation, the pressure sensor should be installed near the cylinder wall and be flush mounted to avoid cavity resonance [71]. Cylinder pressure sensor mounting position also affects the accuracy of data to be used for knock detection purpose. Typically knock onset (KO) and knock intensity (KI) are the main knock properties. The knock onset indicates the timing of the start of the pressure oscillations (hence, indirectly a measure of autoignition time). The knock intensity is an indication of the severity of the pressure oscillations [72]. The knock intensity can be computed on data either in the time or frequency domain; all frequency-domain calculations are conducted on the power spectrum (also known as the power spectral density). The rapid variations in pressure signal due to the knock event can be detected by analysing either filtered in-cylinder pressure and its derivatives or the HRR and its derivatives. Finally, KI metrics can be based on a single value (a maximum or minimum value of a quantity) such as RI, MAPO (maximum amplitude of pressure oscillations), MPRR etc, or an averaged or integrated value such as IMP (integral of modulus of pressure oscillations), AEFD (average energy in frequency domain) etc [70].

Based on the understanding of the energy associated with pressure oscillation in HCCI engine, Eng [62] proposed a metric called ringing intensity (RI) by using maximum pressure ($P_{\max}$), maximum temperature ($T_{\max}$) and maximum PRR in the cylinder as:

$$\mathrm{RI} = \frac{\sqrt{\gamma R T_{\max}}}{2\gamma P_{\max}}\left[\beta\left(\frac{\mathrm{d}P}{\mathrm{d}t}\right)_{\max}\right]^2 \tag{6.14}$$

Another widely used parameter is maximum pressure rise rate in the combustion cycle. The relation between RI and maximum PRR value in HCCI engine is shown in Fig. 6.45b. The variation of RI with different engine operating conditions is presented in Fig. 6.45a.

Typically three main metrics based on cylinder pressure are used for analysis at high engine load operation. The commonly used metrics are maximum pressure rise rate (MPRR), RI and combustion noise level (CNL). The crank angle-based MPRR (i.e. bar/°CA) is perhaps the most commonly used metric used as limiting criterion for the harshness of HCCI combustion. Presently it is well known that acceptable ringing/knock limit (i.e. the onset of knock) for HCCI combustion is not correctly detected by setting a fixed criterion on the maximum allowable PRR ($\mathrm{d}P/\mathrm{d}\theta$) at different engine speeds and boost pressures [15, 62, 69, 73–75]. At higher intake pressure (boost level), it is found that engine can withstand higher PRR without creating an undue increase in the characteristic audible knocking sound. This observation is thought to be the result of the increased dampening effect of the pressure waves by the higher in-cylinder charge density [62, 69]. The immediate utility of MPRR is due to its ease of computation from in-cylinder pressure data and can also be calculated in real time. The MPRR metric is directly applicable to the issue of engine durability, but its usefulness in describing combustion noise is not universal [76]. MPRR can be used to analyse relative changes in combustion noise between tests on similar engine platforms operating under similar conditions. The RI metric directly addresses the issue of engine durability, as it is an estimate of the intensity of in-cylinder pressure waves, but it shares with MPRR the limitation that it does not include the effects of frequency distribution and attenuation and therefore is not a consistently good prediction for combustion noise level [76].

The RI is a knock metric based on the acoustic energy flux of the pressure wave, which is computed by relating the amplitude of pressure pulsation (ΔP) and the global properties of charge in the cylinder. Experimental data shows that relating ΔP to MPRR by Eq. (6.12) is unsatisfactory because a proportional relationship with a constant β value cannot be established over wide range of engine operating conditions [75]. The ΔP–MPRR relationship depends on the engine operating condition (such as boost pressure) and on the extent of charge stratification in the cylinder. Due to large degree of inaccuracy in prediction of pressure oscillations (when only MPRR is used), modified relationship between ΔP and MPRR is presented [75]. By applying the conservation of energy to the burning mass inside engine cylinder, which might build up pressure wave, the relation between local PRR and HRR can be given by Eq. (6.15).

$$\dot{P} = (\gamma - 1)\dot{q} - \gamma P \frac{\dot{V}}{V} \tag{6.15}$$

Here $\dot{q}$ is the volumetric HRR and V is the volume of the charge that produces the heat release.

Integrating over a small time Δt that leads to the pressure rise is given by Eq. (6.16).

$$\Delta P = (\gamma - 1)\dot{q}\,\Delta t - \gamma P \frac{\Delta V}{V} \tag{6.16}$$

Equation (6.16) shows that pressure rise magnitude is influenced by the heat release (1st term) as well as by work needed for expansion against the ambient pressure (2nd term) inside engine cylinder. Hence, higher intake pressure leading to higher in-cylinder ambient pressure, reduces the pressure oscillation amplitude for the same amount of fuel burned. Also with more stratified charge (smaller heat release space size), the pressure rise under the same extent heat release should be smaller as well [63]. The volumetric expansion of combustible mixture parcel is limited by acoustic speed, so to relate the ΔP with *MPRR*, an offset term is introduced to consider this effect [63, 75].

$$\Delta P = \beta\,(\mathrm{MPRR} - \mathrm{MPRR_{offset}}) \tag{6.17}$$

where

$$\mathrm{MPRR_{offset}} = \gamma P \frac{\Delta V}{V \Delta V} = \gamma P \frac{C * S * \Delta t}{V \Delta t} = \gamma P \frac{C}{L} \sim P \frac{\sqrt{T}}{L}$$
$$= (\mathrm{MPRR_{offset}})_0 \frac{P_{\max}}{P_0} \sqrt{\frac{T_{\max}}{T_0} \frac{L}{L_0}} \tag{6.18}$$

where S is the surface area of the parcel, $L = V/S$ standing for the characteristic length of the parcel and $(MPRR_{offset})_0$, P_0, T_0, L_0 the reference values determined by the specific combustion conditions (intake pressure, temperature, fuel stratification extent, etc.). The correlation of ΔP shows great consistency with the modified relationship. However, authors suggested additional data using different engines and engine configurations is required to solidify the relationship [75].

Maximum load achieved in HCCI engines is constrained by mainly ringing/ knocking limits. There are other operating constraints which limit the achievable maximum engine load such as misfire limit, cyclic variation limits, oxygen availability limits, maximum cylinder pressure limits and efficiency and emission limits [77]. Table 6.2 presents the different cylinder pressure-based knock metrics used for knock limit (high load limit) in HCCI engines. The RI is mainly used as knock criteria in low temperature combustion engines. Figure 6.49 shows the variations of RI in HCCI operating range using gasoline and ethanol in naturally aspirated conditions. Engine operating conditions having RI <6 MW/m^2 (knock region)

Table 6.2 Different in-cylinder pressure-based methods for characterization of knock in HCCI engines [78]

Knock criteria	Limits
MPRR $(dP/d\theta)_{max}$	2 bar/CAD @3000 rpm [79] 6–8 bar/CAD@1500 rpm [80] 6–8 bar/CAD @2000 rpm [81] 10 bar/CAD @1500 rpm [82] 12 bar/CAD @ 1000 rpm [53]
MPRR $(dP/dt)_{max}$	5 MPa/ms [65]
Ringing intensity (RI)	2 MW/m^2 [62] 5 MW/m^2 [83] 6 MW/m^2 [84]
Pressure oscillations	0.5 bar oscillations; 10% of cycles [85]
Wavelet packet transform	Wavelet coefficients [86]

and COV_{IMEP} <3.5% (misfire region) are considered in HCCI operating range. RI is observed to be higher for richer mixture and high inlet air temperature as explained earlier. RI is highest close to the knock boundary and lowest close to the misfire boundary for both test fuels. Contour lines of constant RI are inclined representing the dependency of RI on intake air temperature and air–fuel ratio. It can also be noticed that RI in HCCI operating range is higher for ethanol compared to gasoline due to different combustion phasing.

Misfire of engine cycle creates a further constraint upon maximum load. Misfire limit occurs in the case where insufficient energy is available for the fuel to reach its HTHR (high temperature heat release) regime. EGR is an effective strategy used to suppress the MPRR at higher engine load. However, EGR displaces air so that for a given value of manifold air pressure and negative valve overlap (NVO) setting, the load decreases. If the NVO is adjusted to keep the amount of burned gas (residual and EGR) constant, the temperature of the charge in the cylinder decreases with increase in EGR. In this case, at particular condition, the charge temperature can be insufficient to sustain HCCI operation and misfire limit occurs [87]. Delayed combustion phasing is another important strategy used to reduce the excessive PRR at higher loads in HCCI engine. At particular engine operating condition, such a combustion phasing delay required that cycle misfires because of quenching effect due to charge expansion with downwards piston motion. Reactions quench due to insufficient temperatures and unable to reach HTHR. This constrains the high load limit by misfire [77].

The maximum cylinder pressure limit also constrains the achievable high load in HCCI engine. This limit arises mainly due to design considerations specifically related to specific engine decided by high pressure capability of engine components and typically provided by engine manufacturer. This limit can also occur in a number of ways such as high boost pressure and high equivalence ratio engine operation, excessive ringing in engine and early combustion phasing operation [77]. Cyclic variability limits and their causes are discussed in Sect. 6.5 of this chapter. The oxygen availability limit typically comes into picture where higher

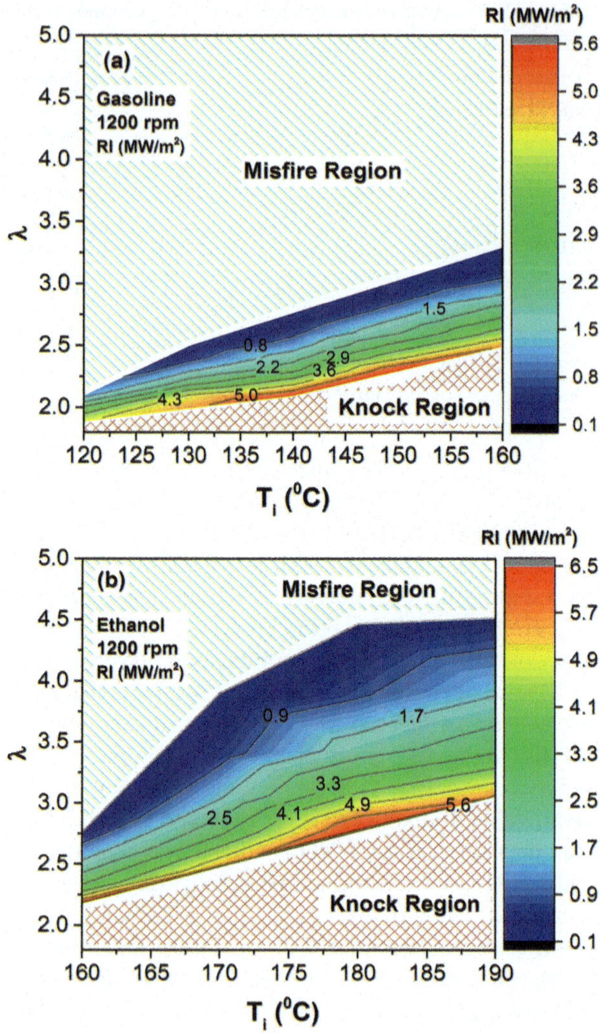

Fig. 6.49 Variation of RI in HCCI operating range using gasoline and ethanol at 1200 rpm (Adapted from [24])

amount of EGR is used as charge diluent. The HCCI engine operation with high reactivity fuels requires the use of external (cooled) EGR to avoid the excessive PRR [88]. Cooled EGR offers a dilution effect and also reduces the charge reactivity at higher intake pressures, which is used to achieve higher engine load. Efficiency and emission limits further constrain the higher engine load in HCCI combustion. At very early or very late combustion phasing, power out and efficiency of HCCI engine is lower. Unburned hydrocarbon and CO emissions are higher in HCCI combustion, when engine is operated on leaner mixture or highly delayed combustion phasing due to lower combustion temperature. The NO_x emissions are higher in HCCI combustion when operated on richer mixture or highly advanced combustion phasing due to higher combustion temperature. These operating conditions limit the high load engine operations. In well-mixed HCCI combustion, normally ringing limit appears before the NO_x limit [77].

The fundamental process affecting HCCI high load operation can be divided into five main categories: (1) chemical kinetics and fuel properties, (2) initial charge conditions (pressure, temperature and φ), (3) inhomogeneities (fuel and thermal stratification), (4) combustion phasing and (5) heat transfer from the cylinder. Chemical kinetics is dependent on fuel characteristics. HCCI fuels are categorized into single- and dual-stage heat release fuel. The φ sensitivity of fuel affects the high load operation of HCCI engine (see Sect. 2.2.4.1 of Chap. 2). Initial charge conditions mainly affecting the high load are initial temperature, initial pressure (boost pressure) and residual fraction or EGR fraction in the charge. Inhomogeneities of fuel or temperature in the combustion chamber play an important role in high load engine operation in HCCI combustion (see Sect. 2.2.4.6 of Chap. 2). Combustion phasing affects the overall performance of HCCI engines. The combustion phasing is affected by several engine operating parameters. Detailed discussion of each of these parameters affecting high load engine operation is provided in the reference [77]. The parameters affecting the fundamental phenomenon which governs the high load limit can be utilized to control the PRR or mitigate the knocking process in HCCI engines. Methods for controlling the HCCI knock are discussed in the next subsection.

6.4.3 Controlling Pressure Rise Rate

As discussed in Sect. 6.4.1, the HCCI engine knock is governed by the abrupt pressure rise initiated by local rapid heat release in different parcels in the combustion chamber. The relationship between the global and the local PRR is also discussed in previous Sects. 6.4.1 and 6.4.2. Reduction in global PRR is required to decrease the knock propensity of HCCI combustion. Thus, assessment of PRR dependency on various engine operating parameters is essential before suggesting the control methods. Wildman et al. developed methods for thermodynamic assessment of MPRR [89]. A simple thermodynamic model is used for the analysis. The PRR can be obtained as Eq. (6.19) by energy balance.

$$\dot{p} = (\gamma - 1)\left[\dot{q} - \frac{\dot{Q}_L}{V}\right] - \left[\gamma P \frac{\dot{V}}{V}\right] \tag{6.19}$$

To assess the maximum pressure rise rate at the high load limit, the heat loss term $\dot{Q}_L/V$ and the volumetric expansion term are small compared to the volumetric heat release rate $\dot{q}$ term. Therefore:

$$\dot{p} \approx (\gamma - 1)\dot{q} \tag{6.20}$$

The volumetric heat release rate at crank angle may be written as in Eq. (6.21).

$$\dot{q}\,(\theta) = \frac{\mathrm{LHV}\left(\frac{m_f}{V(\theta)}\right)}{\tau_{\mathrm{reaction}}} \tag{6.21}$$

where V(θ) is the cylinder volume and τ_{reaction} is the chemical reaction time scale. Thus, MPRR can be written as:

$$\mathrm{MPRR} = (\gamma - 1)\left[\frac{m_f \mathrm{LHV}}{\tau_{\mathrm{reaction}}}\right]_{\max} \tag{6.22}$$

Study found fairly good correlation by using the 10–90% burn duration as an estimate for τ_{reaction} and the volume at CA_{50} as estimate for the charge volume at the MPRR point [89]. To analyse the effect of engine load on MPRR, Eq. (6.22) can be written in terms of NIMEP (net indicated mean effective pressure) and indicated fuel conversion efficiency ($\eta_{f,i}$) by Eq. (6.23).

$$\mathrm{MPRR} = (\gamma - 1)\left[\frac{\mathrm{NIMEP}}{\eta_{f,i}}\,\frac{V_D}{V(\theta^*)}\right]\frac{1}{\tau_{\mathrm{reaction}}} \tag{6.23}$$

where $V(\theta^*)$ is the cylinder volume at the MPRR point and V_D is the displacement volume. In order to achieve high engine load (IMEP) using higher intake pressure, MPRR will increase proportionally if other parameters remain the same. At high charge density (higher boost pressure), τ_{reaction} will decrease (reaction becomes faster), and combustion will also advance due to lower ignition delay leading to lower $V(\theta^*)$. These factors lead to higher increase in *MPRR* at boosted engine operation. The MPRR can be reduced by delaying combustion phasing and increasing the $V(\theta^*)$ value.

Scaringe et al. investigated the dependency of τ_{reaction} on thermochemical parameters by regression of experimental data [87]. Correlation found in the study is presented in Eq. (6.24).

$$\frac{1}{\tau_{\mathrm{reaction}}} = 3.32 \cdot X_{O_2}^{1.88} \cdot X_{\mathrm{fuel}}^{0.35} \cdot P_{10\mathrm{BTC}}^{1.92} \cdot \exp\left(\frac{-2560}{T_{10\mathrm{BTC}}}\right) \tag{6.24}$$
$$\left(\tau_{\mathrm{reaction}} \text{in ms}, P \text{ in bar}, T \text{ in } {}^0\mathrm{K}\right)$$

The X_{O_2} is mole fraction of oxygen concentration and X_{fuel} is fuel mole fraction. The compression pressure and temperature at 10° before-top-centre ($P_{10\mathrm{BTC}}$ and $T_{10\mathrm{BTC}}$) are selected as the thermodynamic variables because up to this crank angle position, there is no significant heat release. The temperature and pressure ($P_{10\mathrm{BTC}}$ and $T_{10\mathrm{BTC}}$) also encompassed the effects of the trapped enthalpy of the residual, the intake air temperature and the heat loss [87]. This correlation gives satisfactory ($R^2 = 0.71$) agreement with experimental data. It is conceived that charge stratification also contribute in the data scatter and correlation will improve if stratification parameters can be included in the analysis. It is difficult to find the precise quantitative measures of stratification to assess the stratification factors. Study used variables such as $(T_{exh} - T_{in})$, $x_r(1 - x_r)$ and $(1 - x_{EGR})$, to consider the stratification for the regression [87]. The variable $(T_{exh} - T_{in})$ represents the temperature non-uniformity due to the mixing between the hot trapped residual and the relatively cool intake fluid. T_{exh} is the exhaust temperatue and T_{in} is inlet charge temperature. The $x_r(1 - x_r)$ represents the concentration non-uniformity due to the mixing between the residual and the fresh charge, and $(1 - x_{EGR})$ represents the concentration non-uniformity due to the mixing between the EGR and the fresh charge. X_{EGR} is the mole fraction of exhaust gas recirculation in unburned charge, and X_r is mole fraction of residual gas in unburned charge.The modified correlation found is presented in Eq. (6.25).

$$\begin{aligned} \frac{1}{\tau_{reaction}} &= 4.93 \cdot X_{O_2}^{0.11} \cdot X_{fuel}^{1.38} \cdot P_{10BTC}^{2.55} \cdot \exp\left(\frac{-893}{T_{10BTC}}\right) \\ &(T_{exh} - T_{in})^{-1.63} [X_r(1 - X_r)]^{-3.73} (1 - X_{EGR})^{3.46} \\ &\tau_{reaction} \text{in ms}, P \text{ in bar}, T \text{ in } {}^0\text{K} \end{aligned} \quad (6.25)$$

Though the stratification variables are a very simplified representation of stratification, the agreement with experimental data improved significantly ($R^2 = 0.87$). This observation indicates the significant role of stratification in characteristic time responsible for MPRR in HCCI combustion.

A study investigated the combustion limits of HCCI engines and showed that MPRR is dependent on φ, P_{in} (intake pressure), CA_{50} and S (fuel sensitivity) [90]. From multiple linear regression of experimental data, best linear correlation is represented by Eq. (6.26).

$$\text{MPRR} = 36.7[\varphi] + 2.86[P_{in}] - 0.344[\text{CA}_{50}] - 0.053[S] - 8.45; \quad (6.26)$$

where MPRR is expressed in bar/degree, P_{in} in bar, S in fuel sensitivity (RON − MON) and CA_{50} in CAD [90]. From Eq. (6.26), it is observed that MPRR value increases with richer mixtures (higher φ) and boost pressure (high P_{in}) and decreases with increasing CA_{50} and fuel sensitivity, if these parameters are varied individually while keeping other parameters constant.

Another study selected other parameters to investigate the variations of MPRR in naturally aspirated HCCI engine using gasoline, ethanol, butanol and methanol as fuel [24]. Parameters selected to characterize the MPRR variations are energy injected per cycle (J/cycle) (magnitude parameter), CA_{50} (CAD aTDC) (location parameter) and combustion duration (CA_{90-10}) (CAD) (dispersion parameter). The correlation found to work well for the experimental data is given by Eq. (6.27).

$$\text{MPRR} = a + b(\text{CA}_{50}) + c\,(\text{CA}_{90-10}) + \text{d (Energy/cycle)} \quad (6.27)$$

where a, b, c and d are constants, which were determined using regression of the experimental data. Empirical correlation found has good coefficient of determination (R^2) value (>0.91). The values of empirical constants are given in Table 6.3.

Table 6.3 Values of empirical constant for calculation of MPRR by Eq. 6.27

Speed	a	b	c	d	R^2
1200	−5.3282	−0.8593	0.1872	0.0249	0.95
2400	−5.6838	−0.7656	0.0575	0.0305	0.91

This correlation is used to predict the MPRR values, and the results indicate that the simple correlation captures the trend of MPRR variations and presents good agreement with the experimental data for the majority of operating conditions. An uncertainty of 2 bar/CAD is found in the predictions using this correlation. The uncertainty is evaluated by 2σ, where σ is the standard deviation of residual errors. This means that the true value, with 95% confidence, lies within 2σ of the estimated value. This correlation suggests that the main parameters affecting the $ROPR_{max}$ in the HCCI combustion engine are fuel energy injected per cycle, combustion phasing and combustion duration.

Based on the discussion, main strategies to control pressure rise rate are discussed in next subsections.

6.4.3.1 Extension of Combustion Duration

The relationship between MPRR and reaction characteristic time is represented by Eq. (6.22). The value of m_f.LHV in the equation cannot be varied at constant engine load to reduce the MPRR. Volume of engine cylinder at MPRR position ($V(\theta)$) is affected by combustion phasing but its effect will be very limited. The ratio of specific heat (γ) is not a directly controlled parameter, but it is affected by fuel concentration and temperature in the cylinder [63]. The most potential variable in Eq. (6.22) for reduction in MPRR is $\tau_{reaction}$. A fairly good correlation is found for MPRR by using the 10–90% combustion duration as an estimate for $\tau_{reaction}$ [89]. This observation suggests that extending the combustion duration can result into reduction in MPRR in HCCI combustion. Experimental studies confirmed that increase in combustion duration leads to reduction in MPRR values. Figure 6.50 demonstrates the effect of combustion duration and maximum mean cylinder temperature (T_{max}) on the MPRR in HCCI combustion using gasoline, methanol, ethanol and butanol. The figure shows that MPRR increases drastically with increase in maximum in-cylinder temperature and smaller combustion duration. For shorter combustion duration, combustion rate is high and combustion temperature is also high. Correlation of combustion duration and maximum mean cylinder temperature with MPRR is stronger than linear correlation. It is found that quadratic regression of experimental data has coefficient of determination (R^2) of more than 0.90 [24].

6.4.3.2 Thermal and Fuel Stratification

Charge inhomogeneity in the engine cylinder (thermal and/or fuel stratification) affects the combustion characteristics of the HCCI engines (see Sect. 2.2.4.6 of Chap. 2). In HCCI engines, the actual overall reaction time is fundamentally

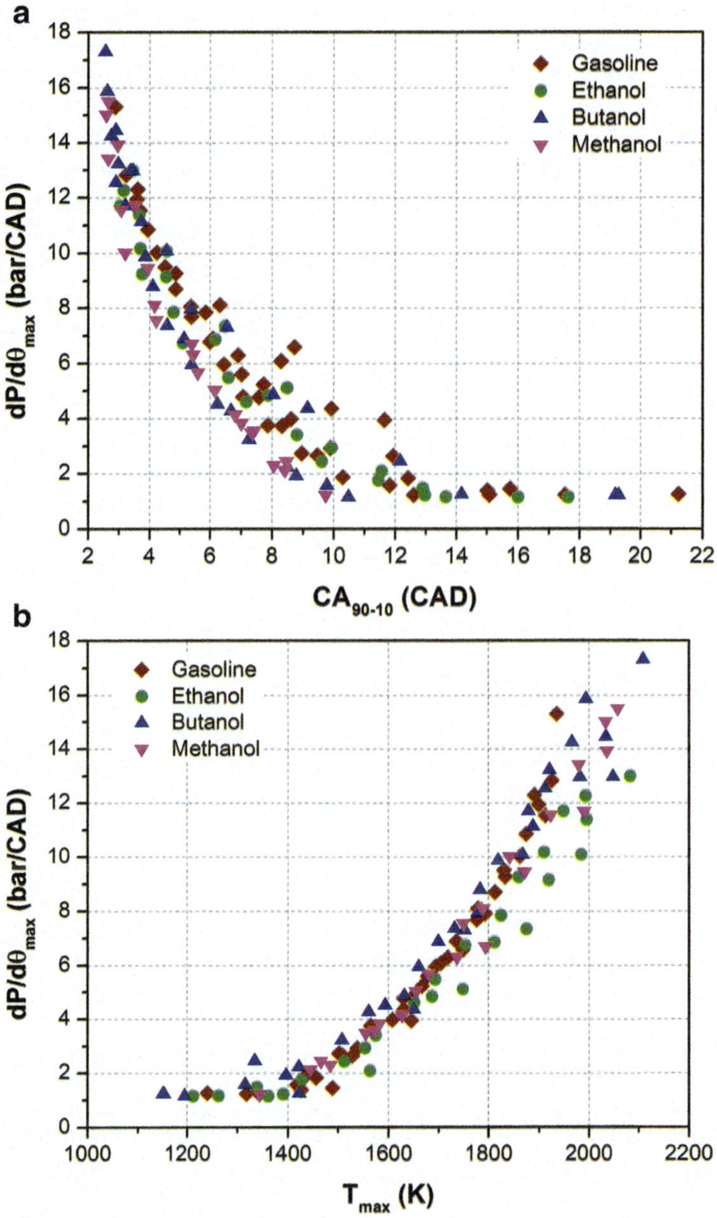

Fig. 6.50 Effect of (**a**) combustion duration and (**b**) maximum mean in-cylinder temperature on MPRR using different fuels at 1200 rpm in HCCI combustion [24]

governed by the charge inhomogeneity. With non-uniform temperature and fuel concentration distribution in the cylinder, different mixture parcels in the combustion chamber ignite at different times sequentially. Ignition delay difference among the mixture parcels mainly governs the time scale of the entire combustion process [63]. Thus, stratification (temperature and fuel) can be used to reduce the HCCI engine combustion intensity and avoid the knocking issues. However, the ignition delay of the mixture parcels needs to be sufficiently sensitive to the difference of temperature and/or concentration of the fuel to use stratification as an effective strategy for MPRR reduction. Sang [63] developed simple analytical equation to illustrate the effect of stratification on heat release rate in premixed combustion. Study considered the one-dimensional temperature and fuel concentration distribution of charge in x direction to simplify the equation. Equilibrium pressure $P = P(t')$ (t' is a time point within the period of consideration) and all the other conditions assumed to be fixed across the cylinder. Then ignition delay of a mixture parcel can be determined as in Eq. (6.28) [63].

$$\tau = \tau\left(T(x,t'), \rho_f(x,t'), P(t')\right) \tag{6.28}$$

Based on Livengood–Wu approach [91], the ignition of the mixture parcel takes place when the integration (Eq. 6.29) reaches to 1.

$$I(x,t) = \int_{t_0}^{t} \frac{\mathrm{d}t'}{\tau\left(T(x,t'), \rho_f(x,t'), P(t')\right)} \tag{6.29}$$

Assume a portion of charge with I >1 has already ignited by time t. For the sequential ignition of the charge due to the temperature gradient, the next mass element Δm to be ignited in the time step Δt is given by Eq. (6.30).

$$\Delta m = \rho_f(x,t') * \Delta x \tag{6.30}$$

where Δx is the increment in space dimension over which the ignition criterion will be reached in the next time step Δt and Δx is given by Eq. (6.31).

$$\Delta x = \Delta t \frac{\left(\frac{\partial I}{\partial t}\right)_{t,I=1}}{\left(\frac{\partial I}{\partial x}\right)_{t,I=1}} \tag{6.31}$$

By using Eqs. 6.28, 6.29, 6.30 and 6.31, heat release rate can be written as Eq. (6.32).

$$\dot{q}(x^*,t) = \mathrm{LHV}\frac{\Delta m}{\Delta t} = \mathrm{LHV}\frac{\rho_f(x^*,t)}{\tau(x^*,t)\left[\int_{t_0}^{t} \mathrm{d}t' \left(\frac{\mathrm{d}\left(1/\tau\right)}{\mathrm{d}\rho_f}\frac{\mathrm{d}\rho_f}{\mathrm{d}x} + \frac{\mathrm{d}\left(1/\tau\right)}{\mathrm{d}T}\frac{\mathrm{d}T}{\mathrm{d}x}\right)\right]_{x^*,t}} \tag{6.32}$$

Equation (6.32) provides the coupling between the temperature sensitivity of the ignition delay ($\mathrm{d}(1/\tau)/\mathrm{d}T$) and the temperature gradient ($\mathrm{d}T/\mathrm{d}x$), the concentration sensitivity of the ignition delay ($d(1/\tau)/d\rho_f$) and the concentration gradient ($d\rho_f/dx$) and its influence on the local heat release rate. Additionally, Eq. (6.32) also depicts that the actual value of the ignition delay of the parcel under consideration and the local fuel concentration also impacts directly to the heat release process [63]. Dilution by EGR extends the ignition delay and fuel concentration can also be reduced in HCCI combustion. These effects justify the use of EGR in avoiding knock besides its modification to in-cylinder temperature and specific heat ratio.

It is difficult to produce thermal stratification in the cylinder to control the MPRR in HCCI engines. Typically fuel stratification is used to control the PRR and achieve higher engine loads. Two main strategies PPC and RCCI in LTC regimes use fuel stratification using single and dual fuel, respectively. Figure 6.51a shows the effect of fuel stratification on MPRR at 1200 rpm in HCCI engine at constant combustion phasing, where fuel stratification is created using direct injection of fuel (EOI 30° and 50° bTDC). Figure 6.51a shows that the MPRR decreases almost linearly with increase in proportion of fuel in the late direct injection. Increase in stratification in fuel leads to decrease in MPRR and increase in combustion duration (Fig. 6.51a). Figure 6.51b shows the variation of MPRR with engine load (IMEP) in PPC engine for various gasoline with different octane number (RON/MON indicated in the figure). The PPC strategy uses very heavy stratification of fuels, and all the fuel is directly injected in the cylinder. Large amount of EGR along with high boost pressure is used in PPC engine to lower the MPRR [45]. The figure shows that MPRR is function of load as well as fuel type and it increases with engine load. Even though values above 15 bar/CAD might look like high for heavy duty applications, however, highly boosted operation (up to 3.7 bar intake pressure for data points in Fig. 6.51b) has to be considered. It is well known that higher boost pressure leads to an increase in MPRR without resulting into higher acoustic noise [45].

Equation (6.32) shows that the concentration gradient and concentration sensitivity with ignition delay affect the HRR in premixed combustion. In RCCI engine, along with concentration gradient reactivity, gradient also exists, which leads to ignition delay distribution in the cylinder (Fig. 2.42 of Chap. 2). Figure 6.52a shows the effect of gasoline/diesel (G/D) premixing ratio and diesel injection timing on MPRR for two different injected fuel quantity. The figure indicates that the MPRR decreases with an advanced fuel injection timings or increased gasoline fraction. This study also observed the similar trend for the in-cylinder mean temperature curve, as MPRR is directly affected by the in-cylinder mean temperature [48]. Delaying the fuel direct injection timing with fixed gasoline proportion or lowering the gasoline proportion with constant fuel injection timing can cause an enhancement of the charge stratification.

Figure 6.52b shows the effect of engine load on the MPRR for various fuels at constant combustion phasing. Higher MPRR is observed for methanol/diesel and ethanol/diesel RCCI operation as compared to gasoline/diesel RCCI combustion. Study [31] reported that in case of methanol/diesel and ethanol/diesel RCCI operation, higher fraction of direct injected fuel (i.e. diesel) is required to maintain the constant combustion phasing. It is attributed to lower fuel reactivity and higher

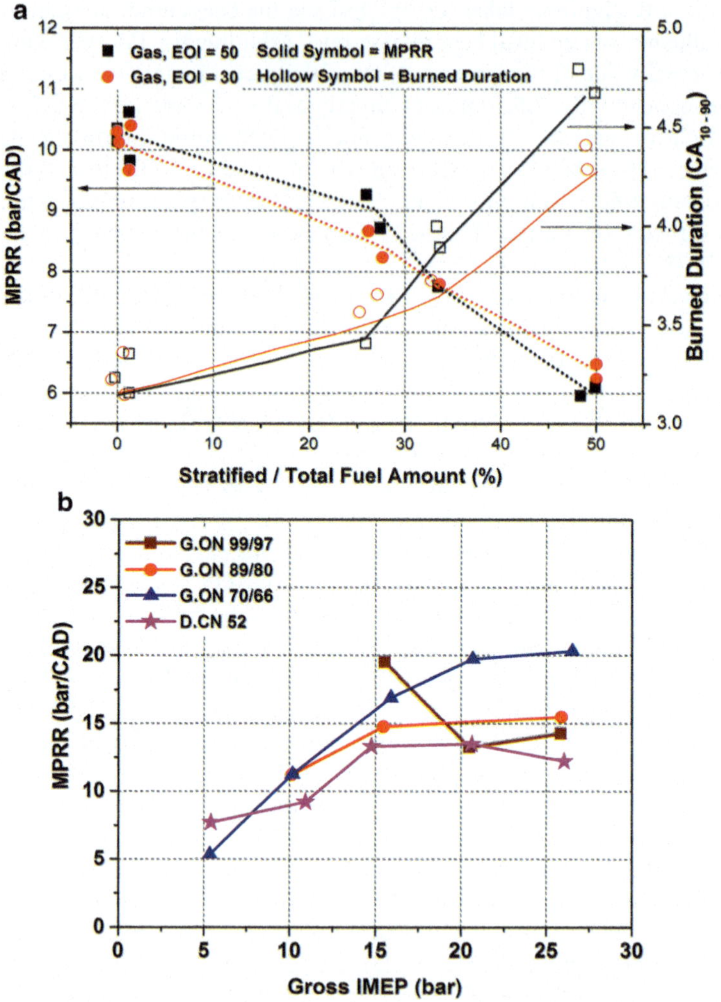

Fig. 6.51 (**a**) MPRR variations with amount of fuel stratification in HCCI engine for gasoline (Adapted from [92]). (**b**) Variation of MPRR with engine load for different octane fuels in PPC engine at 1250 rpm (Adapted from [45])

heat of vaporization of methanol and ethanol. Higher fraction of injected diesel fuel is responsible for higher MPRR as it increases the reactivity of mixture and combustion phasing is advanced.

Figure 6.53 shows the effect of CA_{50} on the RI in RCCI engine by varying different engine operating parameters. Figure 6.53a shows the effect of CA_{50} on RI by varying, i.e. initial pressure (*P*), methanol fraction (*MF*), initial temperature (*T*), start of injection (*SOI*) and exhaust gas recirculation rate (*EGR*). The change in CA_{50} by varying these parameters (increasing or decreasing from base value) is

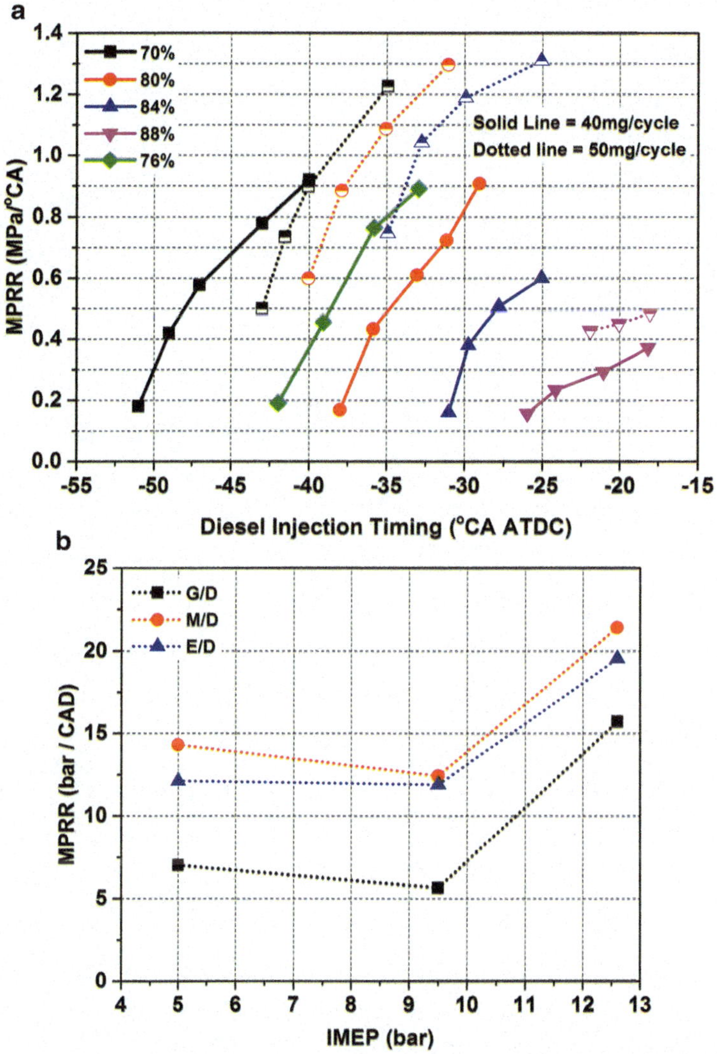

Fig. 6.52 (**a**) Effect of premixing ratio and diesel injection timing on MPRR in RCCI engine (Adapted from [48]). (**b**) Variation of MPRR with engine operating load for gasoline/diesel (G/D), methanol/diesel (M/D) and ethanol/diesel (E/D) RCCI operation (Adapted from [31])

shown in Fig. 6.35a. Figure 6.53 shows the *RI* reduces with delayed CA_{50} (i.e. after 4.3 CAD ATDC) for all the parameters. The higher *RI* with the advanced CA_{50} is due to increased maximum combustion temperature and pressure, fast burning rate with short combustion duration [31]. The figure shows that *RI* monotonically increases with the increased initial temperature or the reduced EGR rate (CA_{50} advances: see also Fig. 6.35a). Lower EGR rate and higher initial temperature significantly increase the combustion temperature along with advanced combustion

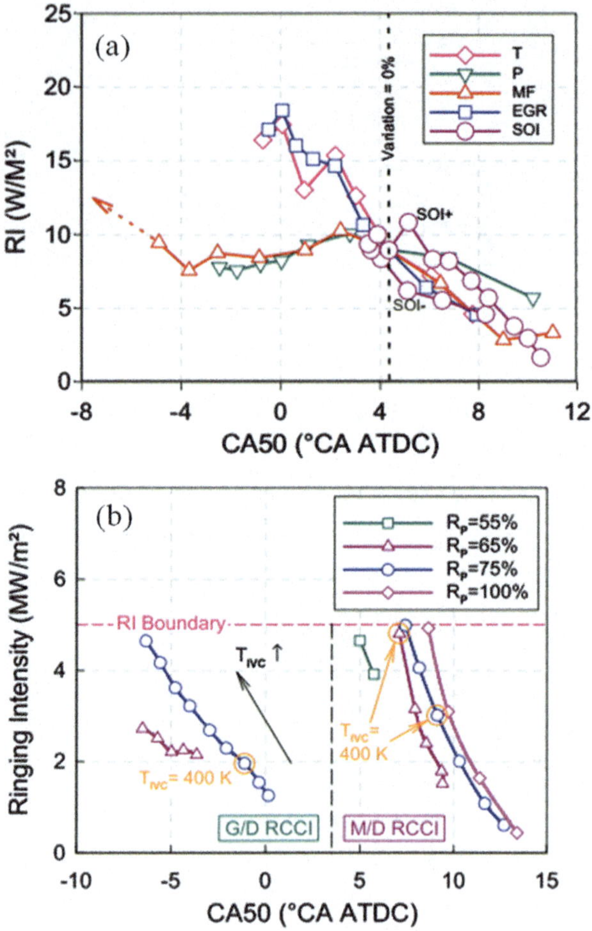

Fig. 6.53 (**a**) Variation of combustion phasing (CA_{50}) with ringing intensity (RI) for methanol/diesel RCCI engine (Adapted from [50]). (**b**) Comparison of ringing intensity versus combustion phasing (CA_{50}) variations for various premixing ratio of gasoline/diesel and methanol/diesel RCCI engine (Adapted from [46])

phasing, and combustion duration is substantially shortened, which results into increase of *RI*. Both methanol fraction and initial pressure depict the similar characteristics on RI as CA_{50} is retarded from −3.69°CA ATDC. *RI* remains a negligible change and then decreases significantly with the increased methanol fraction or the decreased initial pressure (Fig. 6.53b). For large methanol fraction, combustion phasing is significantly retarded to lower reactivity of methanol. Subsequently, the long ignition delay and the nearly homogeneous fuel distribution because of higher methanol fraction result into a HCCI-like (only premixed) combustion. This leads to high single peak HRR and short combustion duration. In case of relatively lower methanol fraction, the high reactivity region created by the large amount of the injected diesel contributes to the early combustion phasing. The short ignition delay and more injected diesel lead to more inhomogeneous distribution of φ in the cylinder, and the HRR curve shows two peaks correspond to the premixed combustion of diesel and the mixing controlled combustion of diesel together with the premixed combustion of methanol, respectively. Hence, the overall peak of HRR is decreased with the relatively lower methanol fraction, and the combustion duration is prolonged. When the CA_{50} is between −3.69° CA ATDC and 4.3°CA ATDC (the cases with low methanol fraction), the low peak of HRR counteracts the high combustion temperature and pressure resulting from the early CA_{50}. Therefore, RI remains almost constant with the variation of methanol fraction during this combustion phasing.

Figure 6.53b shows the variation of ringing intensity with CA_{50} for various premixing ratio (R_p) of gasoline/diesel (G/D) and methanol/diesel (M/D) RCCI engine. The figure depicts that for methanol/diesel RCCI, the RI can be significantly reduced by increasing premixing ratio at engine operation on $T_{IVC} = 400$ K (case circled in Fig. 6.53b). The RI of methanol/diesel RCCI is slightly higher than that of gasoline/diesel at the premixed ratio of 75% and T_{IVC} of 400 K, although the CA_{50} of methanol/diesel RCCI is rather delayed. The *RI* is extremely high for methanol/diesel RCCI at advanced CA_{50} before TDC. Therefore, in order to avoid excessive RI (>5 MW/m^2), the acceptable CA_{50} of methanol/diesel RCCI should be considerably later than that of gasoline/diesel RCCI [46].

Figure 6.54 represents the allowable operating range of combustion phasing at various fuel premixing ratio for gasoline/diesel and methanol/diesel RCCI operation at IMEP = 4.7 bar and speed of 1900 rpm. The operating range is selected on the basis of acceptable RI, equivalent indicated specific fuel consumption (EISFC) and Euro VI NO_x limits. The figure reveals that to keep the RCCI operation for both the fuels within the acceptable limit, retarded combustion phasing is required with increase in the fuel premixing ratio. It means charge homogeneity plays a significant role determination of optimal combustion phasing in RCCI operation. The figure depicts that in methanol/diesel operation, more retard combustion phasing is required to meet the RI limit and later combustion phasing is required to meet EISFC limit as compared to gasoline/diesel RCCI operation due to the faster combustion rate of methanol/diesel RCCI operation. In methanol/diesel RCCI operation, weaker low temperature heat release occurs; thus, it is more sensitive for combustion phasing to the variation in T_{IVC} as compared to gasoline/diesel RCCI operation. Therefore, in methanol/diesel RCCI operation, operating range of T_{IVC} is very narrow as compared to gasoline/diesel RCCI in order to achieve stable operation (Fig. 6.54).

6.4.3.3 Intake Boost and EGR

Boosting the intake air is a widely accepted method to increase the higher engine load operation of the engine. Increasing intake pressure leads to more air inducted in the cylinder, which leads to burn more amount of fuel at the same level of lean engine operation. However, it also renders higher knock propensity due to more charge burned in the cylinder which leads to higher PRR at constant CA_{50}, and boost pressure also enhances the auto-ignition event resulting to higher PRR [15]. Early HCCI experiments demonstrate that IMEP of 14 bar can be achieved

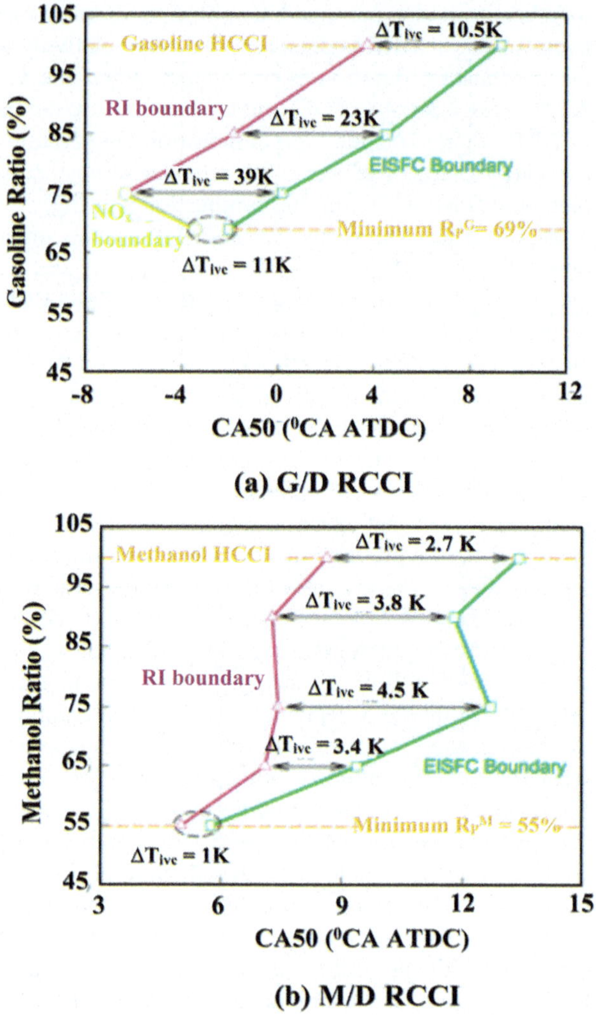

Fig. 6.54 Allowable operating range of combustion phasing at various fuel premixing ratio for gasoline/diesel and methanol/diesel RCCI operation (Adapted from [46])

at an intake pressure of 3 bar (obtained with an external compressor) as compared to 2.38 bar IMEP achieved at naturally aspirated conditions using natural gas as fuel [53]. Advanced combustion phasing by higher intake pressure is adjusted by intake air temperature variations. This study also found that intake boost has different impact on various gasoline-like fuels. Another study also demonstrated that boosted intake pressure could decrease the pressure rise induced combustion noise and results into comparatively wider operating range for turbocharged HCCI engine [74]. Dec et al. showed that increasing the intake pressure using cooled EGR dilution rather than air dilution of charge can significantly extend the achievable higher IMEP [15]. Cooled EGR dilution of charge is more effective because air dilution contains higher availability of oxygen, which advances the combustion phasing, while cooled EGR displaces the excess oxygen resulting into slowdown of the auto-ignition reactions and therefore delayed the combustion phasing. At boost pressure of 3.25 bar, IMEP of 16.34 bar is achieved by increasing the EGR rate until unstable combustion occurs (Fig. 2.23 of Chap. 2). Another study found that higher intake pressure increases the ambient pressure of cylinder and reduces the pressure oscillations and gives lower acoustics energy flux for the same PRR [64]. The reason for this observation is already discussed in Sect. 6.4.2. Sang [63] summarized the effect of boost pressure on HCCI combustion as (i) increased boost pressure engine operation has potential for utilization of higher quantity of fuel per cycle, (ii) higher boost pressure leads to advanced combustion phasing and risk of knock increases and (iii) higher boost pressure lowers the HCCI engine knock propensity from the acoustic perspective.

The EGR is extensively utilized for HCCI combustion control at high engine load and discussed in detail in Sect. 2.2.4.5 of Chap. 2. The use of EGR is not a load increasing strategy in itself, but at higher loads, NO_x emission levels are excessively high even for HCCI combustion. In these conditions, EGR can also be used to achieve stoichiometric conditions to mitigate the NO_x emission by the three-way catalyst. In HCCI combustion, internal EGR (to enhance the initiation of combustion) and cooled external EGR (to reduce HRR and achieve higher load) are typically used [63]. The two types of EGR (internal and external) can also be used simultaneously to lower the knock propensity while maintaining stable combustion [80]. Wildman [93] found that using external EGR instead of internal EGR (NVO trapped residual gas) for charge dilution extended the high load limit. The potential of EGR on controlling the LHTR of two-stage fuels can be used for knock mitigation in HCCI engine [94]. This study showed that the EGR effectively counteracts auto-ignition and lowers the intake temperature control requirements. Thermal and chemical effect of EGR in HCCI combustion is discussed in Sect. 2.2.4.5 of Chap. 2. Chemical effect of trace species on combustion phasing is also discussed in Sect. 6.2.3 of the present chapter. Investigation of main incomplete oxidation products in exhaust shows that short-chain alkanes, short-chain alkenes, short-chain alkynes, aldehydes and alcohols have significant effects on the HCCI auto-ignition [40]. Study demonstrated that all of these incomplete oxidation species (except for CH_4) can advance the HCCI auto-ignition timings to different extents. The C_2H_2, CH_2O and CH_3OH enhance HCCI auto-ignition most without considering their real concentrations in the chamber, and CH_4 can retard the HCCI auto-ignition slightly [40]. The advanced ignition timing may lead to higher PRR in HCCI combustion, which can be adjusted by other parameters. Sang [63] summarized the impact of EGR on HCCI combustion as the EGR has the potential for HRR reduction, as fundamentally inert gas dilution might extend the ignition delay, and effect of EGR can have different trend and sensitivity for different fuels.

Several other strategies have also been investigated for HCCI combustion control. Detail discussed on those strategies can be found in Chap. 5.

6.4.4 Combustion Noise

Typically, the HCCI combustion process is faster in comparison to conventional engines. The HCCI combustion is characterized by a relatively high PRR, which potentially generates two types of combustion noise [73]. First, a "base" noise produced by the "normal" combustion process as a result of the uniform pressure rise associated with the combustion process. This base noise can be similar to CDC and greater than conventional SI engine. Second, combustion noise source for HCCI engines originates from the "knocking" combustion regime at higher engine loads leading to high PRR. At higher PRR, acoustic pressure oscillations are created in cylinder and eventually result in a distinctive, sharp and unpleasant sound [73]. Both types of noise cause the vibrations in the engine structure (block) that are subsequently radiated into the surrounding environment. Several cylinder pressure-based metrics are commonly used to characterize the combustion HCCI, and these metrics are useful for adjusting and optimizing the control variables to maintain the engine operating well and to avoid engine knock as well as excessive base noise. Several studies used MPRR or RI to indicate combustion noise instead of microphone data or using a combustion noise meter that simulates the attenuation characteristics of the engine structure. Combustion noise meters are devices that provide a noise level (in dB) by taking cylinder pressure as the input signal. Filters are applied to the pressure signal to first account for the attenuation due to engine structure and then (optionally) to account for the frequency response of the human hearing system [95].

Combustion noise level (CNL) is a commonly used metric designed to evaluate the intensity of the combustion-induced noise emitted through the engine structure. The cylinder pressure is converted to the frequency domain where it is filtered by a transfer function to account for the noise attenuation by the engine structure and how the transmitted noise is perceived by the human ear [73, 95].

Figure 6.55 shows the variation of CNL with peak pressure rise rate (PPRR), RI, P_{max} and IMEP at steady-state conditions of HCCI engines. The solid lines in the figure represent the regression line fit, and the two horizontal dashed lines represent the misfire and ringing limits. Figure 6.55a, b depicts the quadratic and cubic polynomial regression curve to represent the strong correlation ($R^2 = 0.9$) between CNL with PPRR and RI, respectively. The CNL does not have strong linear relation ($R^2 = 0.68$) with PPRR [96]. Figure 6.55a also shows that the CNL becomes more than 90 d*B* for PPRR >8 bar/CAD and HCCI operation at these high values should be avoided. The higher PPRR engine operation leads engine ringing and excessive heat transfer. The RI is almost constant despite a large variation of CNL from 72 to 84 d*B* in steady-state data points (Fig. 6.55b). Figure 6.55c shows a strong linear correlation between the CNL and P_{max} (R^2 = 0.91). No correlation is found between CNL and IMEP (Fig. 6.55d).

Figure 6.56 depicts the variations of CNL with start of combustion (SOC), CA_{50}, CA_{90} and burn duration (BD) in HCCI combustion. The figure shows that these parameters have moderate leaner correlation with CNL (R^2 ranges from 0.45 to

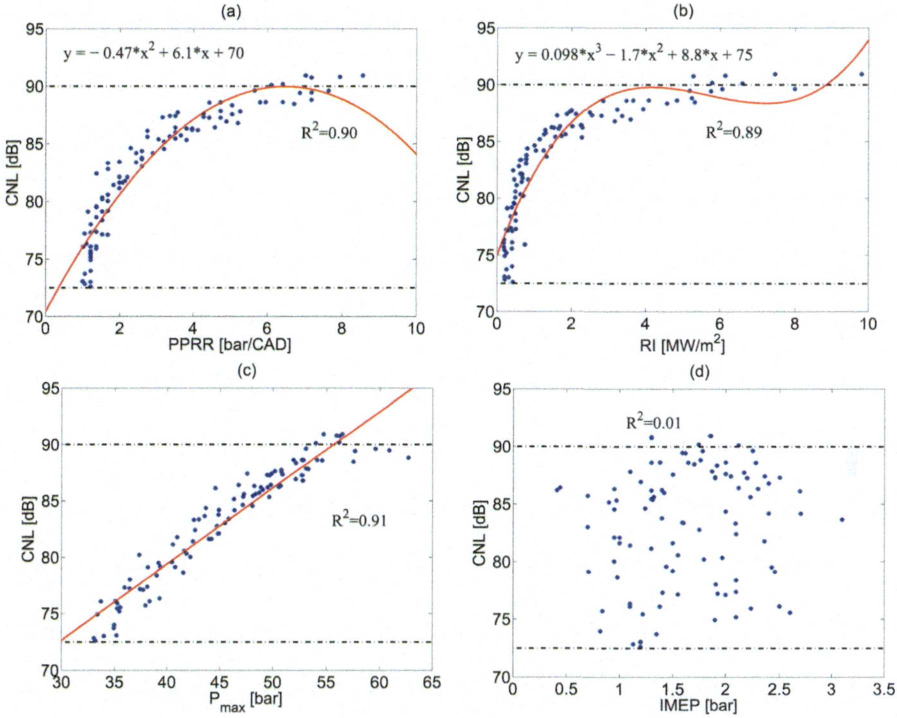

Fig. 6.55 Variation CNL with the PPRR, RI, P_{max} and IMEP at steady-state HCCI combustion [96]

0.55). Advanced combustion timing (i.e. SOC, CA_{50}) results into higher levels of CNL (Fig. 6.56a–b), and retarded combustion timings (CA_{50} >10 CAD aTDC) produce lower CNL values (i.e. CNL <80 dB). Early CA_{50} (CA_{50} <8 CAD aTDC) and shorter BD (BD <12 CAD) test conditions have CNL in noisy region. A correlation with $R^2 = 0.44$ is observed between CNL and BD. Test conditions with 1< SOC <3 CAD aTDC, 4< CA_{50} <6 CAD aTDC, 8< CA_{90} <10 CAD aTDC and 8< BD <10 CAD exhibit the highest propensity for higher CNL (Fig. 6.56) [96]. Retarded combustion phasing and longer burn duration typically lead to lower CNL.

Figure 6.57 shows the relationship between RI/MPRR and CNL in HCCI engine for different operating conditions. Figure 6.57 shows at particular intake pressure (P_{in}), the CNL increases linearly with log(RI) for CNL above 80 d*B*. It is also observed that for any RI (~0.8 to 7 MW/m^2), the CNL increases by ~2.2 d*B* as P_{in} increases from 1.0 to 2.0 bar. The increase in the CNL with RI and boost pressure is attributed to higher magnitude of the zero mode [73]. This relationship between RI and CNL for variations in P_{in} has significant importance from the standpoint of selecting a metric to constrain the operating range of HCCI-like combustion to avoid excessive harshness [73]. For example, CNL threshold of 90 d*B* is selected

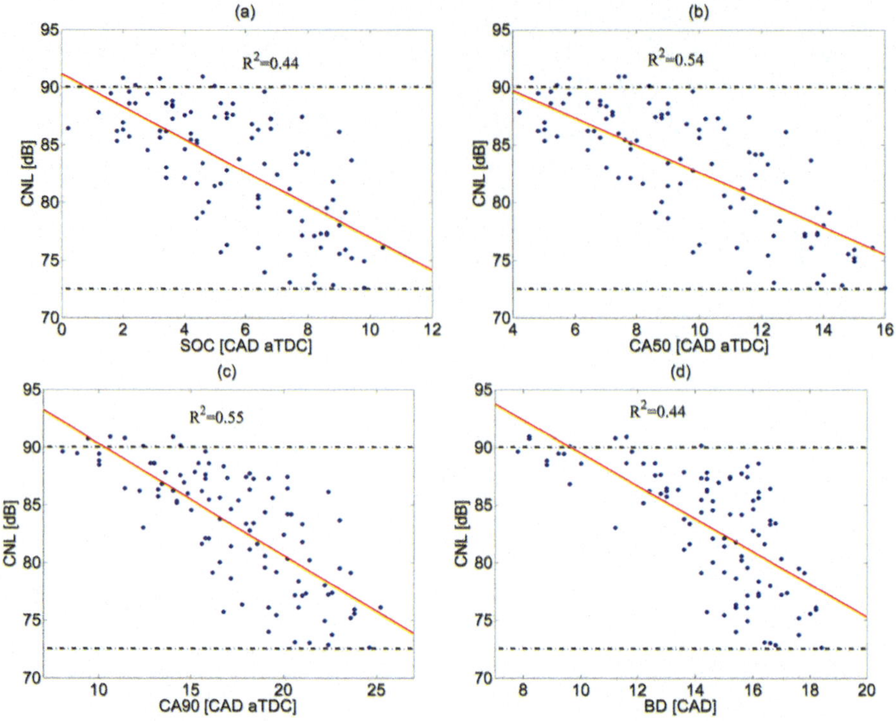

Fig. 6.56 Variation CNL with the SOC, CA_{50}, CA_{90} and burn duration (BD) at steady-state HCCI combustion [96]

(RI = 5 MW/m^2 for P_{in} = 2.0 bar) as knocking limit; then significant knock can occur at P_{in} = 1.0 bar at which RI = 7.5 MW/m^2 (which is severe knocking such as above in which data is not being taken) for the same CNL. Figure 6.57 also shows the effect of engine speed on CNL in HCCI combustion. The CNL significantly increases with increase in engine speed. The figure also presents that the spacing between the data curves at the various engine speeds is unequal, which indicates that the slope of the increase in the CNL with engine speed decreases with the increase in engine speed [73]. This study concluded that CNL is not a useful metric for estimation of the knock propensity of HCCI combustion because the CNL is not sensitive to the power content of the knock/ringing resonant modes, and the onset of knock does not correlate with a fixed value of the CNL. This study also emphasize the fact that the RI and the CNL are designed to provide two distinctly different but complementary measurements. The RI is found to be a better criterion to avoid knocking combustion regimes and the associated characteristic noise, while the CNL provides a valuable metric for estimating the overall loudness of the combustion, irrespective of whether the engine is knocking or not.

Figure 6.58 shows the effect of combustion phasing on sound pressure level (SPL) for CDC, RCCI and HCCI combustion engines. The figure shows the SPL

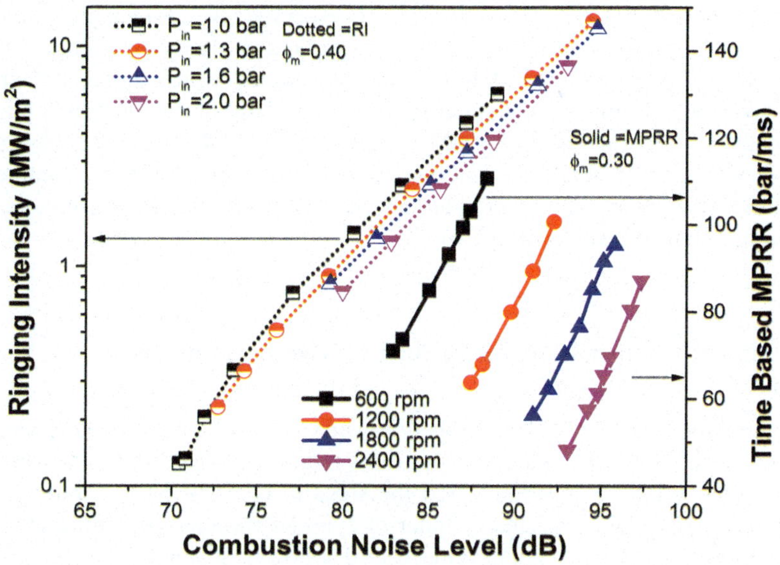

Fig. 6.57 CNL variation with RI and MPRR at different engine operating conditions in HCCI combustion (Adapted from [73])

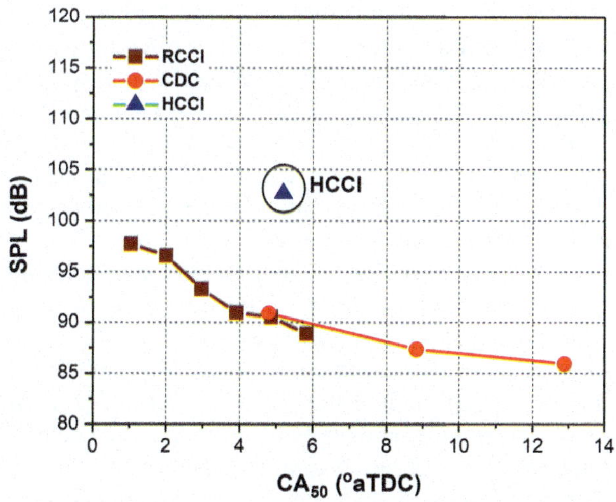

Fig. 6.58 Variation of sound pressure level (SPL) with CA_{50} in RCCI, HCCI and CDC engines [76]

decreases with retard of combustion phasing for both RCCI and CDC. The SPL is lower for RCCI combustion in comparison with HCCI combustion at similar combustion phasing (Fig. 6.58) due to stratified combustion in RCCI engine, and combustion duration is also higher in RCCI combustion. The CDC combustion has comparatively lower SPL due to delayed combustion phasing and larger combustion duration. Study also concluded that magnitude and variation of SPL is generally not proportional to PPRR or RI, especially when changing operating conditions or combustion strategy [76]. Another major observation from the study is that the first two circumferential resonant frequencies (which appear in the ranges of 2–4 kHz and 4–6 kHz) accounted for more than 80% of the integrated power in the acoustic vibrational range for all HCCI, RCCI and CDC cases. This observation indicates that in general, the audible ringing noise produced by compression ignition strategies on that particular engine platform primarily originates from circumferential pressure waves [76].

6.5 Combustion Instability and Cyclic Variations

The engine revolutions and developed power vary from cycle to cycle and from cylinder to cylinder. It means that no two engine cycles are the same similar to fingerprints. Atkins [97] summarized the signs of combustion variability in conventional engines as: cyclic and cylinder-to-cylinder variations in torque and engine speed, engine roughness, compromised torque/power, lower resistance to knock, efficiency losses, higher emissions, compromised dilution tolerance and lower fuel economy. Atkins [97] also listed the causes of combustion variability in conventional engines as mixture motion at the location and time of spark, variations in the amount of air and fuel inducted in each cycle, mixing of the fuel and exhaust residuals, low ignition energy or a small plug gap, fuel preparation (droplet size, cone angle, targeting, swirl), excessive dilution (by exhaust gas recirculation or air) and long burn duration due to poor combustion system hardware design. It has been demonstrated that cyclic variations correlate well with variations in the torque produced by engine; thus, it directly affects the drivability of vehicle. Therefore, it is essential to minimize cycle-to-cycle variations during the engine design phase itself for stable engine operation and optimized performance in terms of emissions and fuel conversion efficiency.

As discussed in Chap. 2, there is no direct control on ignition timings in HCCI engines, and ignition timing is sensitive to equivalence ratio, intake temperature, compression ratio, residual gases/EGR, cylinder wall temperature and chemical kinetics of fuel–air mixture. Li et al. [98] summarized the possible strategies to initiate and maintain stable HCCI combustion as (i) methods to rapidly adjust the intake temperature and pressure, as well as the rate of EGR; (ii) methods to vary the total mass, composition and temperature of the residual gases trapped in the cylinder; (iii) mechanisms to alter the compression ratio during transient operation such as variable valve timing; (iv) methods of optimizing fuel composition in the cylinder using on-board fuel reforming; (v) methods for smooth transition between HCCI and conventional combustion modes; and (vi) methods for rapid air–fuel mixing to prepare premixed charge. Depending on the strategies used to achieve the HCCI combustion and combustion phasing, typically HCCI combustion has higher unburned hydrocarbon and CO emissions, as well as higher cycle-to-cycle and cylinder-to-cylinder variations than expected, especially at lower engine operating loads [98, 99]. High cyclic variations (misfire) typically limits the HCCI operating range at lower engine loads. To extend the engine operating load of HCCI engine, fast switching between HCCI and conventional mode is proposed, which essentially needs cycle-by-cycle control of ignition timing. Developing new methods to effectively control combustion phasing over a wide range of operating conditions and combustion-mode switching needs clear understanding of cyclic variations in HCCI combustion. Combustion phasing control is even more challenging during transient conditions due to cyclic variations in HCCI combustion [100]. Understanding of in-cylinder process of HCCI cyclic variations is required in order to make the combustion more stable to cyclic variations. The sources of cyclic variability and their characteristics in LTC engines are discussed in next subsections in detail.

6.5.1 Source of Cyclic Variability

In HCCI engines, cycle-to-cycle variations are lower in comparison to conventional SI engines, but variations can be very large under particular engine operating conditions [189, 101–102]. Li et al. [98] identified and summarized several factors affecting HCCI combustion and the extent of cyclic variations. These include variations in factors and parameters such as (i) equivalence ratio or fuel flow rate; (ii) intake pressure and temperature; (iii) coolant and lubricating oil temperatures; (iv) mixture inhomogeneity (thermal and fuel stratifications); (v) diluents, i.e. external or internal EGR; (vii) fuel–air mixing and homogeneous mixture formation strategies; (vi) intensity of intake charge motion and bulk turbulence in the cylinder; and (viii) completeness of combustion in the preceding cycle. To control cycle-to-cycle variations in HCCI engines, various actuation strategies are used, which also affect the combustion phasing such as intake air preheating, exhaust reinduction, exhaust recompression, VCR, multi-fuel and multiple injection strategies [101]. Several studies characterized the cycle-to-cycle and cylinder-to-cylinder variations using different fuels [101–109]. Shahbakhti et al. [102] recognized four major potential sources of cycle-to-cycle variations in HCCI engines, namely, (i) thermal stratification, (ii) fuel concentration stratification, (iii) cyclic variations in equivalence ratio of charge and (iv) cyclic variations in diluents (EGR and/or residual gases).

Spatial variations in fuel concentration in the charge exist in the combustion chamber because of the inadequate mixing of fuel, air and residual gas. The fuel injection strategy affects the air–fuel homogeneity at the initial phase of combustion. Typically, port fuel injection (PFI) is used for charge preparation in HCCI engines, which can lead to the mixture composition inhomogeneity [110]. In the fuel concentration stratified (inhomogeneity) case, locally rich zones autoignite earlier with suitable temperature in the cylinder. The effect of fuel stratification is specifically prominent at lower engine loads [111]. Cycle-to-cycle variations in the fuel concentration inhomogeneity of the mixture lead to the cycle-to-cycle variations in HCCI combustion. It is demonstrated that the fuel inhomogeneity has comparatively smaller effect on the cycle-by-cycle spatial variations of HCCI combustion than the temperature inhomogeneity [110].

Temperature inhomogeneity in the charge is caused by heat transfer from different temperature surfaces (piston, valves, cylinder head) and from inadequate mixing of fuel, air and residual gas [112]. Spatial temperature stratification in the combustion chamber can also occur due to the inhomogeneous heat release from the cool flame chemistry [113]. Experimental measurements demonstrated the heterogeneity in combustion of charge always exists with large spatial and temporal variations due to temperature stratification in HCCI engines [111, 113, 114]. Effect of inhomogeneity on HCCI combustion is discussed in Sect. 2.2.4.6 of Chap. 2. It is discussed that thermal stratification in the bulk gas directly determines the maximum HRR in HCCI combustion. Thermal stratification is naturally created during intake and compression stroke. Delayed combustion phasing substantially enhanced the thermal stratification due to more time available for the stratification to develop [36, 114]. Analysis of single-cycle temperature maps (obtained from optical imaging) shows that the thermal stratification existing in the bulk charge of cylinder is not characterized by isolated cold pockets, but is rather due to the penetration of cold structures extending from the walls [114]. These cold structures are found to be not visible in average images, which indicates that they are mainly produced by in-cylinder turbulent flows that vary significantly from cycle to cycle. Investigation on effect of turbulence and initial temperature on inhomogeneity in HCCI combustion showed that increasing temperature inhomogeneity in the bulk flow leads to an earlier ignition of the charge. For the same temperature inhomogeneity, increasing turbulence would retard the ignition process [113]. Enhanced natural thermal stratification can be achieved by lowering the coolant temperature. Lower coolant temperature leads to the increased heat loss which results in to a fuel penalty [36]. For possible additional enhancement of the natural thermal stratification, the heat transfer rates can be further enhanced by increasing the in-cylinder air swirl, which further increases fuel penalty. Cycle-to-cycle variations in the temperature of residual gas can create cyclic thermal stratification in the combustion chamber, which can result into cycle-to-cycle variations in HCCI combustion [115]. A coupling between wall temperature and combustion timing also exists, and cyclic variations in cylinder wall temperature can lead to cyclic variations in combustion timing and vice versa [116].

Cyclic variations in mean air–fuel ratio of inducted charge can vary significantly in PFI engines [117], which is a widely used method of charge preparation in HCCI engines. Cyclic variations in the gas exchange process and incomplete evaporation of the liquid fuel (especially at very lean mixture operation) can lead to cyclic variations in air–fuel ratio [118]. These cyclic variations in air–fuel ratio lead to variations in mixture richness (reactivity) as well as variations in the ratio of specific heats (γ) of the mixture, which creates the differences in the charge temperature [111]. The cyclic variations in mean air–fuel ratio of the charge lead to cyclic variations in HCCI combustion [102].

To control the combustion phasing in HCCI engine, cooled EGR and internally trapped residuals are the two widely used methods among other strategies. The external EGR (EEGR) typically retards the combustion phasing by reduction of combustion temperature and suppression of the autoignition reactions, which is caused by higher specific heat of combustion products and reduction in oxygen concentration in the charge. In contrast, the trapped internal EGR (REGR) from the previous cycle leads to advancing the combustion phasing due to higher temperature of residuals which preheats the incoming charge in cylinder [119, 120]. The presence of hot residual pockets trapped from the previous cycle can further increase the degree of thermal stratification in cylinder due to incomplete mixing, unless the potentially more intense mixing homogenizes the core of the charge very fast [119]. The amount of diluents (internal or external) in the cylinder affects the HCCI combustion [110]. Therefore, cyclic variations in quantity and composition of diluents can lead to cyclic variations in HCCI combustion that can create a variation in the diluent for the next combustion cycle [102]. In case of dilution by EGR, both the thermal and chemical effects play a role in cycle-to-cycle variations in HCCI combustion [119].

Figure 6.59 illustrates the cyclic variations in HCCI combustion for stable and unstable engine operation. The stable operation with a CA_{50} avg. of 13.4° CA

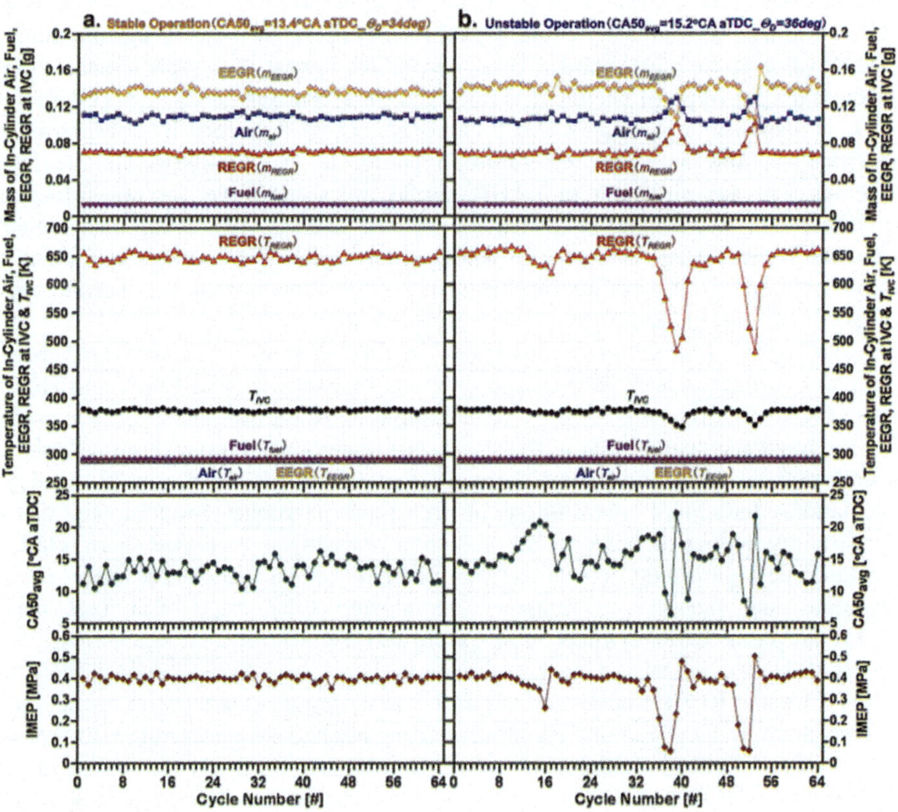

Fig. 6.59 Cyclic variations in HCCI combustion for (**a**) stable operation and (**b**) unstable operation for 64 consecutive cycles [119]

aTDC ($COV_{IMEP} = 3.56\%$) has only small cyclic variations of the mass and temperature of each gas at IVC (m_{air}, m_{fuel}, m_{EEGR}, m_{REGR}, T_{air}, T_{fuel}, T_{EEGR}, T_{REGR}: m and T indicates mass and temperature respectively) and the T_{IVC}. Cyclic variations for the stable operation are equally distributed to values higher and lower than the average resulting into lower cyclic variations in the CA_{50} and the IMEP. In contrast, the unstable operation has various cycles that exhibit a much larger spread around the average for m_{air}, m_{EEGR}, m_{REGR}, T_{REGR} and the resulting T_{IVC} (e.g. cycles no 34–42 and cycles no 49–57 of Fig. 6.59b).

Detailed analysis of IMEP time series (Fig. 6.59b) shows that a partial burn cycle is often followed by another partial burn cycle because a partial burn cycle has lower T_{REGR} that leads to comparatively lower T_{IVC}. Hence, the subsequent combustion cycle after a partial burn cycle has a higher probability for a delayed initiation of combustion. Delayed ignition timing is likely to result into a second partial burn cycle with more reduced IMEP. However, comparison of cyclic variations in IMEP and CA_{50} shows that cyclic variations of IMEP do not match the cyclic variations of CA_{50} well (particularly for the partial burn cycles) [119]. Additionally, Fig. 6.59b shows that immediately after an excessive partial burn cycle (cycle number 38 and 52 having the lowest IMEP), the IMEP recovers towards the average value of IMEP at significantly lower T_{IVC} and highly delayed CA_{50} timings. Particularly, cycle number 53 (Fig. 6.59b) shows the maximum IMEP value at the most retarded CA_{50} despite the lowest T_{IVC}. This indicates that chemical state (type of chemical species and thier concentration) of the charge at IVC is responsible for higher IMEP and it cannot be described by considering a thermal state (pressure and temperature) of charge at IVC. It is shown that the trace species that are produced in LTHR period are responsible for the sufficient enhancement of autoignition which recover the IMEP [119]. Study also argued that the chemical enhancement of autoignition by the trace species recirculated from previous cycle with internal residual gases is greater than the thermal suppression of autoignition by low charge temperature.

The sources of combustion instability in LTC engines are investigated using response surface model (RSM), and data analysis showed that RCCI and HCCI engines have significantly large cyclic variations and have a much higher sensitivity to the input parameters than CDC [121]. An uncertainty propagation analysis is conducted by using the partial derivatives of the RSM and the input standard deviations. Individual contributions of each input parameter are computed using uncertainty propagation technique and then normalized to reveal their relative magnitude. Figure 6.60 depicts the relative effects of each input parameter (temperature and pressure at intake valve closing (T_{ivc}, P_{ivc}), PFI fuel mass (PFIFuelMass), EGR, cylinder temperature (CylTemp), direct injected fuel mass (CRFuelMass)) on IMEP, CA_{50}, gross indicated efficiency (GIE) and peak PRR for the HCCI and RCCI combustion engines. The figure depicts that the cyclic variability are introduced mainly from four sources: trapped gas temperature, PFI mass, DI mass and EGR%. Cylinder surface temperature shows modest effect on cyclic variability in HCCI and RCCI combustion. Combustion chamber pressure at intake valve closing (IVC) has only small contribution to the variation in RCCI combustion, but it is a significant factor in cyclic variation HCCI combustion [121]. The

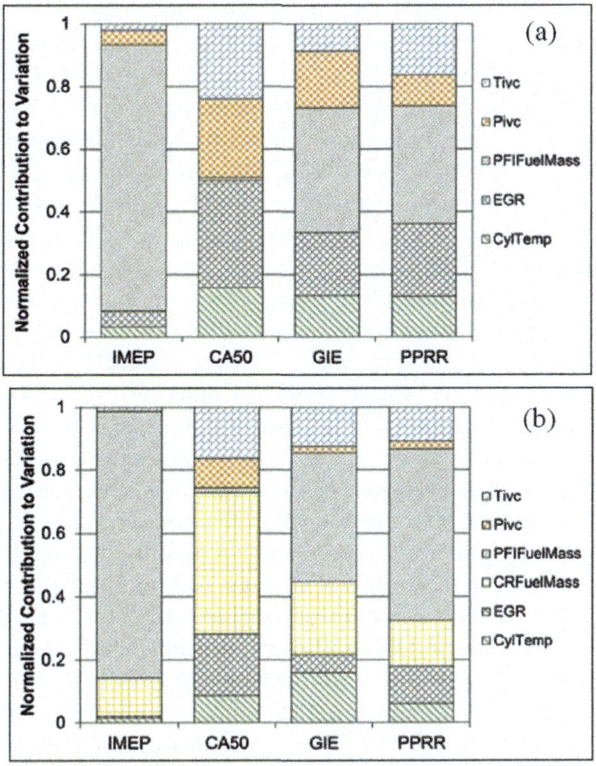

Fig. 6.60 Normalized contribution to variations in output parameters (IMEP, CA_{50}, GIE and PRRR) by various input parameters in (**a**) HCCI and (**b**) RCCI engines [121]

combustion phasing (CA_{50}) in HCCI is influenced by four main inputs: T_{IVC}, P_{IVC}, EGR% and cylinder surface temperature. Even though PFI fuel mass has very strong effect on IMEP and PPRR, it has a minimal effect on CA_{50}. It can also be noticed that none of the four main parameters that affect the HCCI cyclic variability can be controlled on a cycle-to-cycle basis and each parameter can have cycle coupling effects [121]. In RCCI combustion, CA_{50} cyclic variations have three main contributors, namely, T_{IVC}, CR fuel mass and EGR%.

It is concluded that accurate and consistent fuel delivery systems are essential to enable the advantages of LTC modes in production engines. Additionally, it is required to manage the temperature and EGR sensitivity in engines. To control the temperature and EGR sensitivity, development of predictive control strategies is required to adjust fuelling on a cycle-to-cycle basis [121].

6.5.2 Characterization of Cyclic Variability

6.5.2.1 Effect of Operating Parameters and Combustion Modes

Cyclic variations in HCCI combustion are analysed by different statistical and chaotic approaches. Different types of patterns were observed in the cyclic variations of different combustion parameters in HCCI engines. Figure 6.61 illustrates the three distinct patterns in cyclic variations of start of combustion (SOC), P_{max} and IMEP in HCCI combustion. Patterns observed in Fig. 6.61 demonstrate that cyclic variations in HCCI combustion are not always a random (unstructured) event. Figure 6.61a shows the normal variation pattern that occurs more frequently in HCCI combustion [102]. These variations do not follow a definite structured pattern. Figure 6.61b depicts almost periodic pattern, which is oscillating within two limits. This periodic variations can be due to oscillating equivalence ratio of charge supplied to engine [102]. A cyclic variation pattern with several weak/misfired ignitions and some strong ignitions is illustrated in Fig. 6.61c. This operating condition has large cyclic variations in SOC, P_{max} and IMEP. At these operating conditions, values of P_{max} and IMEP are generally very lower at weak/misfire cycles. The misfire/weak ignition cycles produce significant quantity of unburned/partially burned fuel, which is recirculated to the next combustion cycle by residuals. The recirculated partially oxidized fuel from previous misfired cycle leads to enhanced autoignition in the next cycle [36]. This also describes the strong ignition cycles observed after few weak cycles. The weak/misfire pattern and periodic pattern of cyclic variations must be removed from HCCI combustion using appropriate control strategy. The amplitude of cyclic variation in the normal pattern must be minimized in order to control the HCCI ignition timing to a set point. To control the ignition timings of HCCI combustion, characteristics of cyclic variation dynamics need to be explored.

Figure 6.62 depicts the cumulative percentage of occurrence for coefficients of variations in IMEP (COV_{IMEP}) for different ranges of CA_{10}, CA_{50} and combustion duration (CA_{90-10}) for 273 HCCI test conditions obtained by using gasoline, ethanol, methanol and butanol at 1200, 1800 and 2400 rpm for different intake temperatures and air–fuel ratio without using EGR. The cumulative percentage of occurrence is computed by counting the number of test points out of all the test points with CA_{10}, CA_{50} and CA_{90-10} interval which have the same range of cyclic variations. To illustrate the calculations, take a particular case in Fig. 6.62c. The figure shows that 107 test points have CA_{90-10} between 5 and 9 CAD, and 73 out of 107 test points have COV_{IMEP} less than 3.5%; thus, percentage of occurrence for COV_{IMEP} <3.5% is 68.2% for CA_{90-10} occurring between 5 and 9 CAD. Figure 6.62a shows that test points with advanced CA_{10} have lower variations in IMEP in comparison to the delayed CA_{10} positions. For 80% test points with CA_{10} between −5.5 and −2 CAD have COV_{IMEP} less than 3.5%, and for 60% test points having CA_{10} between −2 and 2.5 CAD have COV_{IMEP} lower than 3.5%. Delayed CA_{10} between 2.5 and 9.8 CAD have 53% test points with COV_{IMEP} less than 3.5% (Fig. 6.62a). The effect of combustion phasing on COV_{IMEP} is demonstrated in Fig. 6.62b. The figure shows that the delayed CA_{50} has comparatively larger cyclic variations in IMEP. For CA_{50} between −4.2 and 1 CAD, 90% test points have COV_{IMEP} lower than 3.5%; however, CA_{50} later than 5 CAD has only 44% of test conditions having COV_{IMEP} lower than 3.5%. Delayed CA_{50} curve has lower number of test points in all the ranges of COV_{IMEP} values. There are various factors contributing to higher cyclic variations in for delayed CA_{50} after TDC. These

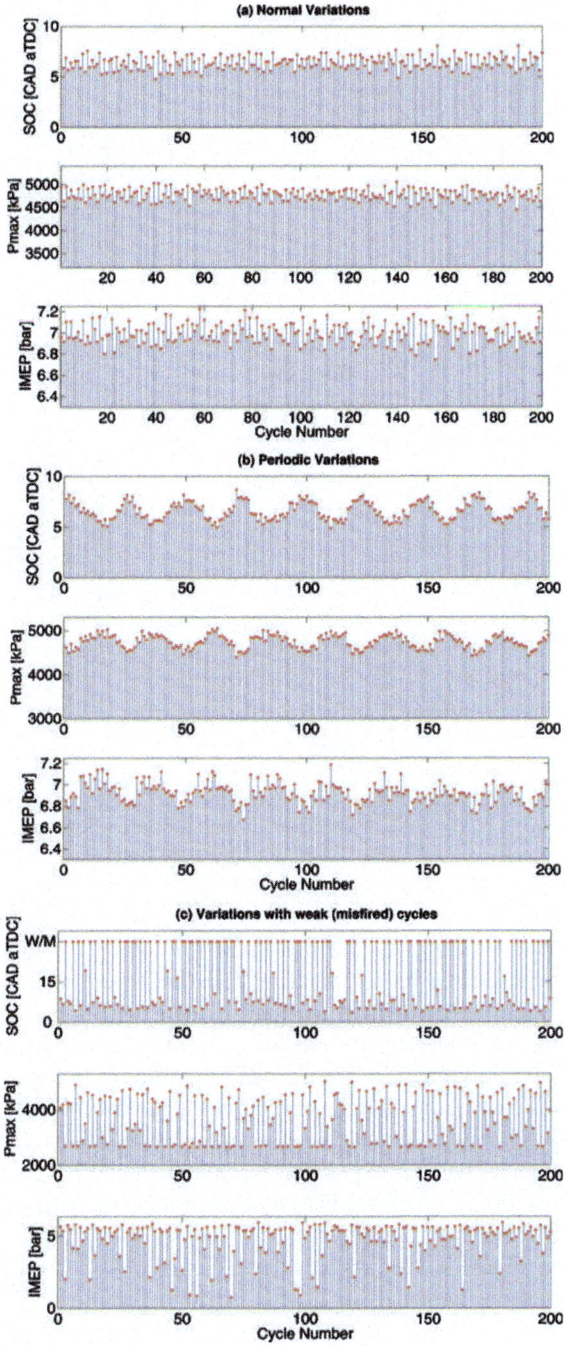

Fig. 6.61 Cyclic variations patterns of IMEP, P_{max} and SOC in HCCI combustion for (**a**) normal variation ($\varphi = 0.40$, $T_i = 121$ °C, $ON = 20$, $P_i = 119$ kPa, 800 rpm), (**b**) periodic variation ($\varphi = 0.36$, $T_i = 133$ °C, $ON = 20$, $P_i = 125$ kPa, 1003 rpm) and (**c**) variations with weak (misfire) cycles ($\varphi = 0.42$, $T_i = 116$ °C, $ON = 40$, $P_i = 124$ kPa, 907 rpm) [102]

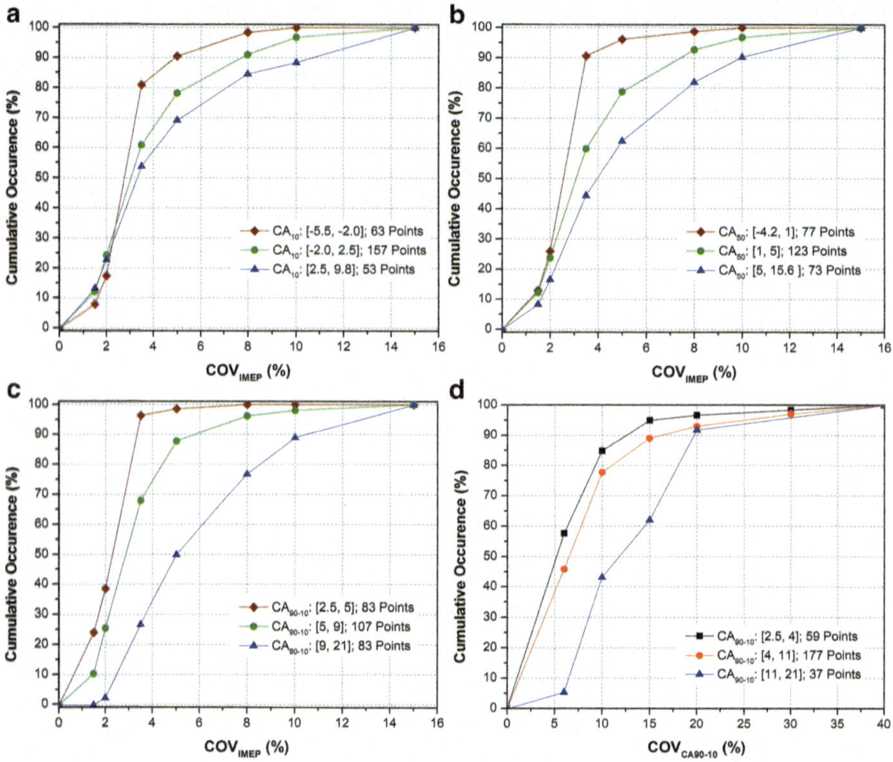

Fig. 6.62 Cumulative occurrence for COV_{IMEP} in HCCI combustion for different (**a**) CA_{10} positions, (**b**) CA_{50} positions and (**c**) combustion duration (CA_{90-10}) and cumulative occurrence for COV of combustion duration (**d**) for different CA_{90-10} [107]

factors include [102]: (a) at delayed CA_{50} timings, volume expansion rate (by piston movement) is higher leading to reduction in charge temperature which counteracts the temperature rise due to combustion [122]; (b) at retarded CA_{50}, natural thermal stratification is enhanced [36]; (c) the combustion phasing becomes increasingly sensitive to fluctuations of the charge temperature, when it is shifted later into the expansion stroke [36]; and (d) by retarding the CA_{50}, both combustion efficiency and thermal efficiency decrease [116, 123] and completeness of combustion also decreases with lower peak combustion temperature [36]. Additionally, higher chamber expansion rate also lowers PRR and P_{max} by delaying the oxidation reactions, which leads to reduction in work output with possibilities of partial burn and misfired cycles.

Figure 4.106c reveals the effect of CA_{90-10} on variation of IMEP, and it shows that the larger CA_{90-10} has higher variation in IMEP. CA_{90-10} between 2.5 and 5 CAD (96% of the test points) has COV_{IMEP} lower than 3.5%; however, CA_{90-10} larger than 9 CAD having only 25% of test points has COV_{IMEP} lower than 3.5%. Longer CA_{90-10} has a lower number of test points in the entire ranges of COV_{IMEP}.

Figure 6.62d depicts the cumulative occurrence for COV of CA_{90-10}. The figure shows that faster HCCI combustion (shorter CA_{90-10}) tends to have smaller cyclic variations in CA_{90-10}. Larger CA_{90-10} curve has lower number of test points for entire ranges of COV of CA_{90-10}.

Figure 6.63 illustrates the variations of IMEP as a function of CA_{50} and CA_{90-10} for gasoline and ethanol at 1200 rpm. Figure shows that larger CA_{90-10} and late

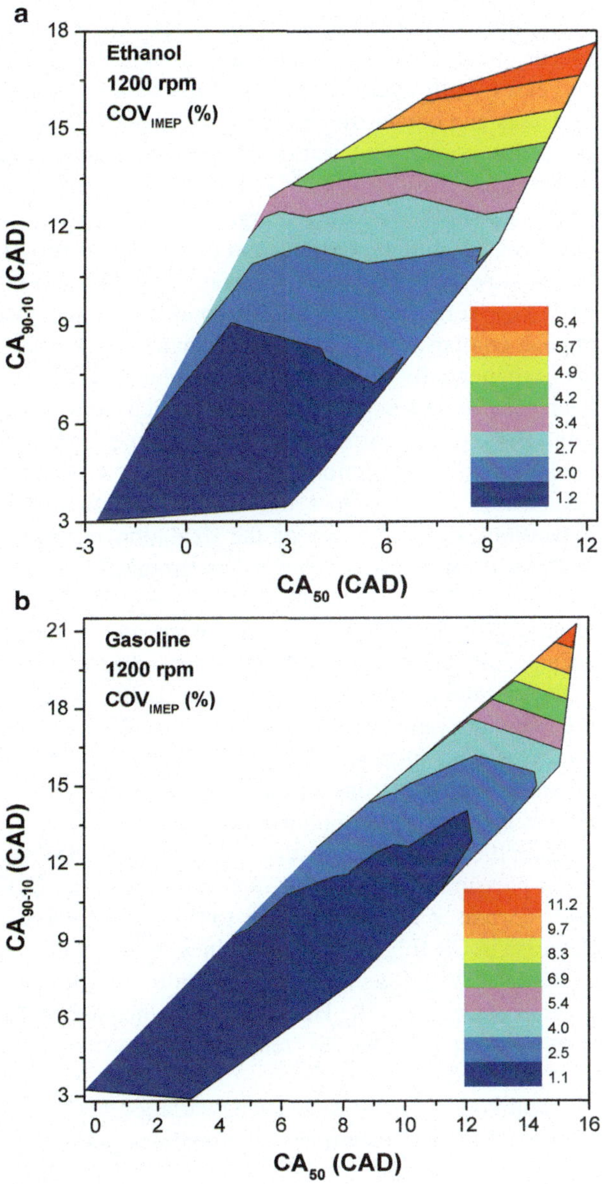

Fig. 6.63 Variations of COV_{IMEP} in HCCI combustion with combustion phasing and combustion duration for (**a**) ethanol and (**b**) gasoline [107]

CA_{50} have higher cyclic variations of IMEP for both the test fuels. It is also observed that contour lines of COV_{IMEP} are almost horizontally inclined for ethanol, which indicates that COV_{IMEP} is more dependent on CA_{90-10} in comparison to CA_{50}.

Figure 6.64 shows the variations of COV of IMEP for different HCCI engine operating conditions (λ, intake temperature (T_i), intake pressure, engine speed, compression ratio (CR) and EGR). Figure 6.64a shows the effect of λ and T_i on COV_{IMEP} for gasoline, and COV_{IMEP} is calculated from 2000 consecutive engine cycle data. At particular T_i, cyclic variations (COV_{IMEP}) increase at both engine speeds as λ increases (leaner mixtures). For a particular mixture, COV_{IMEP} decreases with increase in T_i (Fig. 6.64a) because combustion temperature increases and combustion phasing advances with higher T_i. This observation indicates that combustion stability improves with increasing T_i. For comparatively richer mixture ($\lambda = 1.8$; $T_i = 120$ °C) and very advanced timings (Fig. 6.64a), COV_{IMEP} starts increasing marginally as knocking combustion starts from these operating conditions. Combustion is retarded away from TDC position at higher λ values (lower fuelling rate) and comparatively lower T_i, which leads to very high COV_{IMEP} (Fig. 6.64a). This observation can be justified as follows. In HCCI engine, combustion is governed by chemical kinetics, and oxidation reaction rates are exponentially dependent on combustion temperature. Charge temperature decreases during expansion stroke, which slows down the combustion reaction rate. Thus, cyclic variation in charge temperature at TDC can lead to significant cyclic variation in reaction rate in expansion stroke of engine. An engine cycle with combustion phasing close to TDC is less affected by piston movement during expansion stroke as most of the oxidation reactions occur close to TDC position. Piston speed is also very slow around TDC, and zero at TDC position. For engine cycle with late combustion phasing in expansion stroke is more prone to larger cyclic variations as charge temperature decreases at higher rate (as piston speed is higher) and reactions rates are reduces [107].

Figure 6.64b shows the effect of EGR, engine speed, intake pressure and compression ratio (CR) on COV of IMEP. The figure shows that COV_{IMEP} increases with increase in engine speed (600–1400 rpm), when other operating parameters remain constant. This is because of lower time availability at higher engine speeds for autoignition reactions, which results into retarded combustion phasing [98]. Figure 6.64b also shows that the COV_{IMEP} decreases with increase in boost pressure at constant air–fuel ratio (50). The COV_{IMEP} decreased marginally with decreases in CR from 16 to 11, and it increases drastically with further decrease in CR (Fig. 6.64b) due to incomplete combustion. Similarly, with increase in EGR, COV_{IMEP} decreases as without EGR light/weak knocking is present, which is suppressed by EGR. However, with increase in EGR beyond 55%, COV_{IMEP} increases rapidly due to retarded combustion phasing and partial misfiring (Fig. 6.64b) [98].

Figure 6.65 illustrates the cyclic variation in CA_{50} and SOC with different HCCI engine operating parameters. Figure 6.65a shows the effect of T_i and λ on the variation of standard deviation (σ) of CA_{50} for gasoline. At every T_i, the cyclic

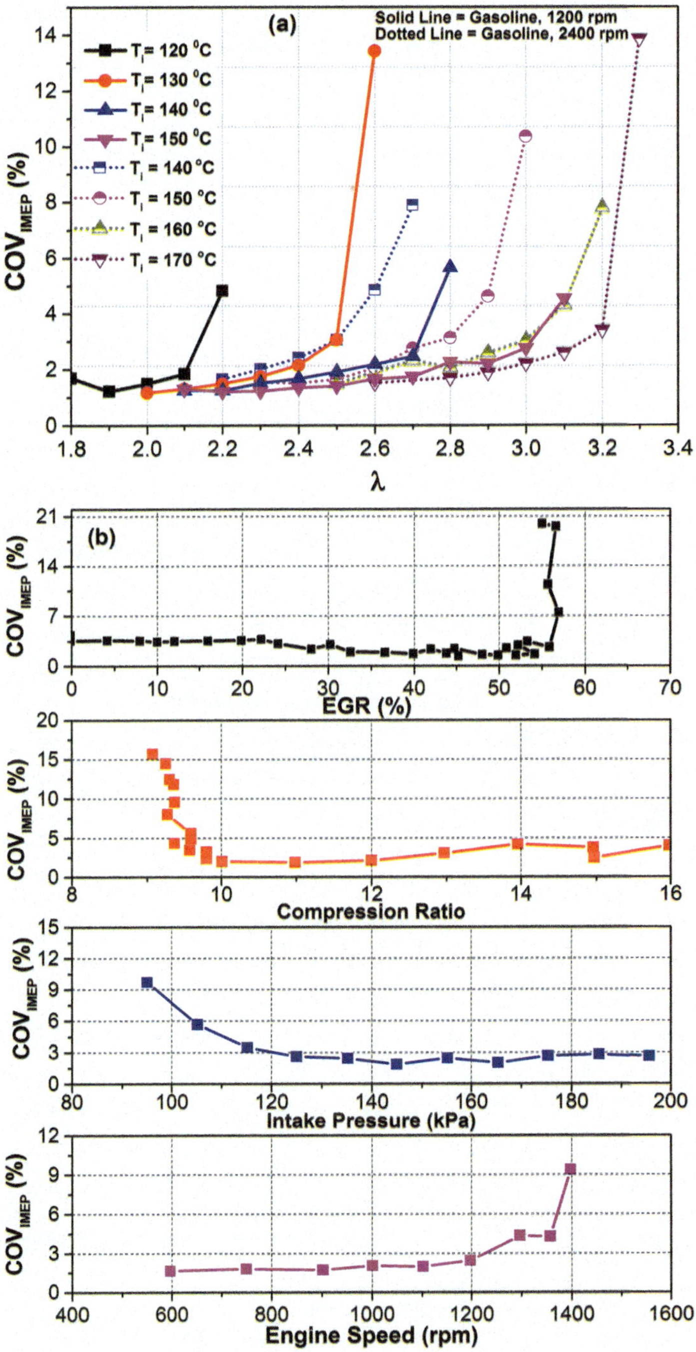

Fig. 6.64 Variations of COV_{IMEP} in HCCI combustion (**a**) at different λ and T_i at 1200 and 2400 rpm (Adapted from [107]) and (**b**) with EGR, compression ratio, intake pressure and engine speed (Adapted from [98])

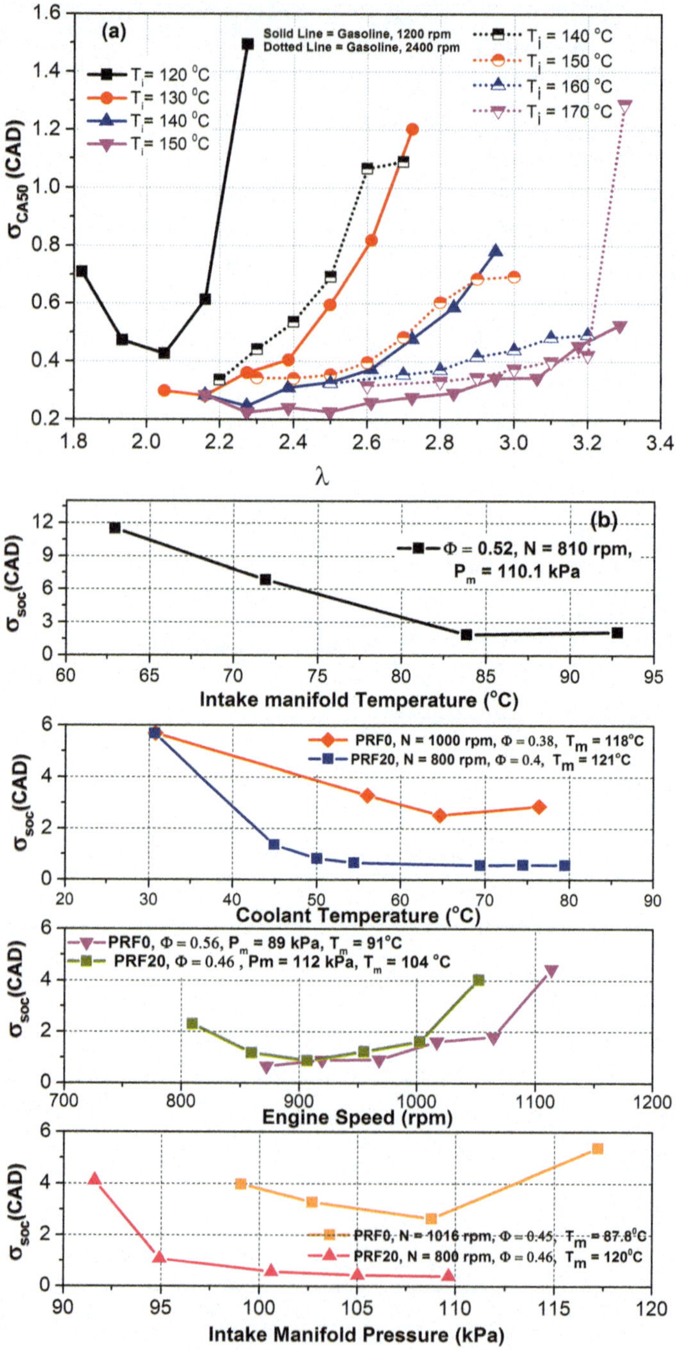

Fig. 6.65 (**a**) Variations of $\sigma_{CA_{50}}$ at different λ and T_i at 1200 and 2400 rpm (Adapted from [107]) and (**b**) variations of standard deviation of SOC at different intake temperatures, intake pressures, engine speeds and coolant temperatures in HCCI combustion (Adapted from [102])

variation ($\sigma_{CA_{50}}$) decreases as the mixture becomes richer (decrease in λ) for both speeds and $\sigma_{CA_{50}}$ decreases with increase in T_i at particular λ (Fig. 6.65a). At very rich fuel–air mixture and advanced combustion phasing ($\lambda = 1.8$; $T_i = 120$ °C), $\sigma_{CA_{50}}$ starts increasing because of knocking combustion in the cylinder. Figure 6.65b shows the cyclic variation of SOC with different HCCI engine operating conditions and SOC is calculated from third derivative of pressure data [102]. Figure 6.65b shows that cyclic variations are higher at lower intake temperatures. This trend can be justified by a combination of three main effects. First, cyclic variations of residual gas temperature lead to cyclic variations in the core charge temperature which results into variations in autoignition process [36]. Higher T_i increases the charge temperature, and thus similar residual gas temperature variations affect to a lesser degree. Second, SOC is advanced near TDC at higher T_i and thus lower cyclic variations is expected. Third, combustion duration is longer at lower T_i, and longer combustion duration is expected to have higher cyclic variations (Figs. 6.62 and 6.63).

Cyclic variation of SOC increases with a reduction in the coolant temperature (Fig. 6.65b). Lower cooling temperature enhances the wall heat transfer, and cold thermal boundary in the cylinder expands, where partial burning takes place. At lower coolant temperature, thermal stratification in the combustion chamber increases, which makes the HCCI combustion more sensitive towards the variations of wall temperature [102, 124]. Additionally, variations in the heat transfer increase the chances of higher cyclic variations at lower coolant temperature [102]. Study further indicated that cyclic variations of SOC are more sensitive to variations in the intake temperature in comparison to the variation in coolant temperature, while cyclic variation of combustion duration shows higher sensitivity to the coolant temperature in comparison to the intake temperature [102]. This study also showed that effect of both engine speed and intake pressure on the cyclic variations of SOC is heavily depended on the location of SOC with respect to TDC. Increasing intake pressure and decreasing engine speed lead to the advance SOC, and cyclic variation of SOC decreases (Fig. 6.65b). Cyclic variation of SOC is affected mainly by a combination of three factors: (i) location of SOC, (ii) combustion duration and (iii) secondary effects (thermal and compositional inhomogeneities) [102]. The operating parameters affecting these factors will affect the cyclic variations in SOC.

Figure 6.66 shows the cyclic variation of CA_{50} in RCCI combustion at different engine operating conditions. Figure 6.66a shows the variations of standard deviation of CA_{50} for different diesel mass fractions in RCCI combustion for gasoline/diesel (G/D) and methanol/diesel (M/D) RCCI combustion. The zero diesel fraction represents the HCCI conditions using gasoline and methanol, respectively. Figure 6.66a shows that the cyclic variations in M/D RCCI are significantly higher than G/D RCCI due to lower reactivity of methanol. The RCCI combustion strategy with a small quantity of diesel injection drastically lowers cyclic variations as compared to pure gasoline or pure methanol HCCI combustion. This is particularly observed at constant T_{ivc} as at $T_{ivc} = 405$ K, standard deviation of CA_{50} decreases from 2.5 to 0.8° CA only by injecting 20% diesel in G/D RCCI combustion

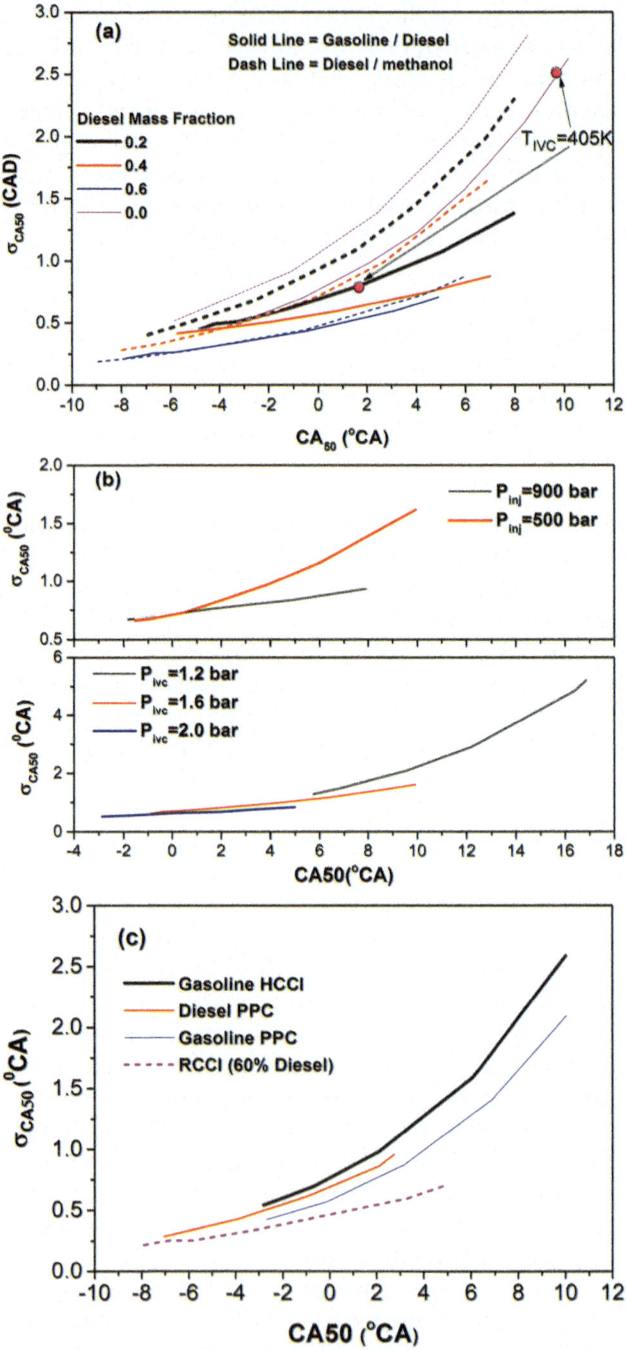

Fig. 6.66 (**a**) Effect of diesel mass fraction on variation of CA_{50} in G/D and M/D RCCI combustion, (**b**) effect of injection pressure and intake pressure variation of CA_{50} in G/D RCCI combustion and (**c**) effect of combustion modes on variation of CA_{50} at IMEP 5.9 bar and 1900 rpm (Adapted from [125])

(Fig. 6.66a). This is attributed to advanced combustion phasing with addition of diesel and increased reactivity of mixture. However, the advantage of diesel injection on reduction of cyclic variability for M/D is not as notable as that for G/D RCCI combustion [125].

Figure 6.66b shows the effect of intake pressure and injection pressure on cyclic variations of RCCI combustion. Cyclic variations in CA_{50} decrease with increase in intake pressure (Fig. 6.66b). The LTHR process is strongly enhanced by increase in boost pressure. Stronger LTHR in the cylinder leads to the higher charge temperature rise rate before the occurrence of HTHR. Higher charge temperature rise rate before HTHR compensates for variations in charge temperature [126], which results into lower sensitivity of CA_{50} to variation in compression temperature. This leads to lower cyclic variations in combustion phasing [125]. Figure 6.66b also demonstrates that combustion stability improves with increase in the injection pressure of direct injection diesel in RCCI combustion. Figure 6.66c compares cyclic variations in combustion phasing for the different combustion modes. Fuel is directly injected with same fuel injection strategy for diesel and gasoline PPC. To obtain desired combustion phasing, T_{ivc} is adjusted for different combustion modes. For the same CA_{50}, variations of CA_{50} are lower in gasoline PPC in comparison to gasoline HCCI (Fig. 6.66c) due to locally rich gasoline mixture in PPC engine (by direct injection of gasoline leading to fuel stratification), which lowers the sensitivity of ignition timing to variations in T_{ivc} [125]. This leads to lower cyclic variations in PPC engine. Cyclic variations of CA_{50} of diesel PPC are higher than that of gasoline PPC as same CA_{50}. In order to keep constant CA_{50}, extremely low T_{ivc} should be used for diesel PPC. The RCCI combustion (with 60% diesel) exhibits the slowest variations of CA_{50} among all the combustion modes (Fig. 6.66c). The moderate T_{ivc} and the appropriate fuel reactivity are the major factors for the lower cyclic variations in G/D RCCI combustion compared to diesel PPC and gasoline HCCI combustion [125].

6.5.2.2 Return Maps

The return maps are often used as qualitative tool to evaluate the inherent deterministic interaction cycles of combustion events [127, 128]. Return map is a tool to identify the probable interaction between different engine combustion parameters of a particular combustion cycle and its next consecutive combustion cycle. In random time series of combustion events, consecutive combustion cycles are not interrelated, and the return map depicts an unstructured cloud of data points collected around a fixed point. In case of deterministic coupling between combustion events, the return map depicts more structures (as a function on the graph depending on correlation) such as dispersed data points about a diagonal line [129]. Return maps plot pairs of successive time series values of combustion parameters versus each other (crank angle position for maximum pressure (θ_{Pmax}) and heat release value for cycle i versus cycle $i + 1$ are plotted in Fig. 6.67). Each cycle data point relates to the next cycle through the general statistical picture of the

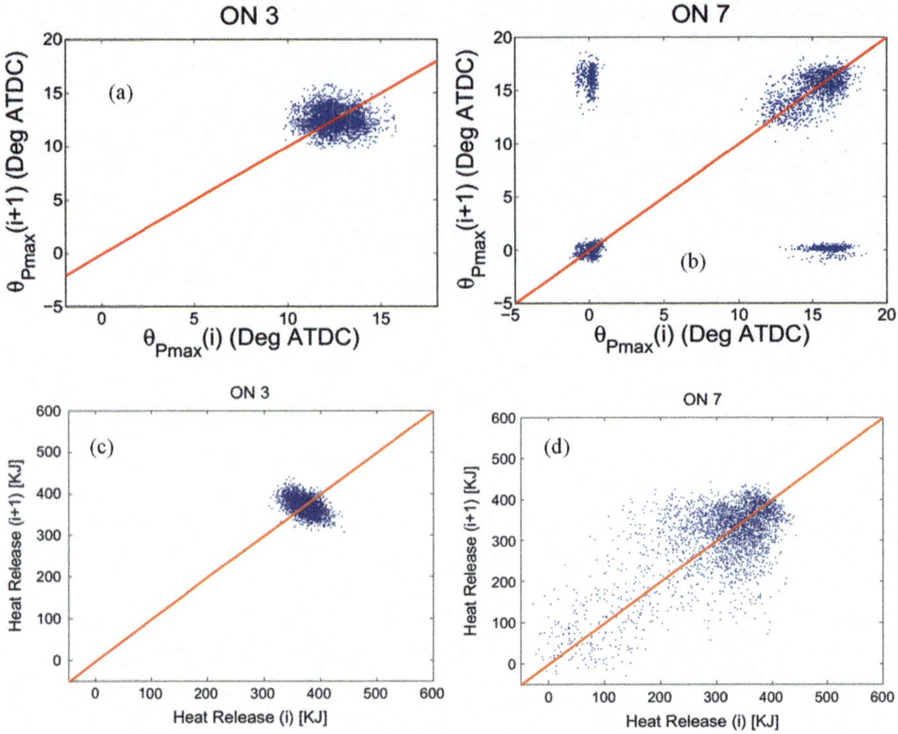

Fig. 6.67 Return map of (**a**) $\theta_{P_{max}}$ using ON3, (**b**) $\theta_{P_{max}}$ using ON7, (**c**) heat release using ON3 and (**d**) heat release using ON7 in HCCI combustion at 1000 rpm and manifold temperature 38 °C (Adapted from [127, 131])

whole cycle interrelation using return map plots [130]. Figure 6.67 shows the return maps for HCCI combustion parameters using two different PRF fuels (ON 3 and 7) using $\theta_{P_{max}}$ and heat release. Return maps result for $\theta_{P_{max}}$, and heat release for octane number (ON) 3 shows an unstructured cluster of circular data gathered around a fixed point (Fig. 6.67a, c), which suggests the stochastic variation in combustion parameter, which indicates the combustion being stable. At higher octane fuel (ON7), fixed concentrated points destabilize in certain direction of return map, and data points scatter over diagonal line (Fig. 6.67b, d). The structured patterns noticed in the data can be attributed to the deterministic coupling between consecutive combustion cycles [127, 131].

Figure 6.68 shows the return map of CA_{50} for different λ and T_i in HCCI combustion at 1500 rpm. The return maps in Fig. 6.68a show a circular cloud of combustion phasing for leaner mixtures (showing stochastic component), and the combustion phasing distribution becomes flat along the diagonal line for rich mixture. It is difficult to find correlation by this graph for rich mixture as skewed distributions can also produce dispersed data, which makes the interpretation of return maps complicated. Additional issue with return maps is the difficulty of

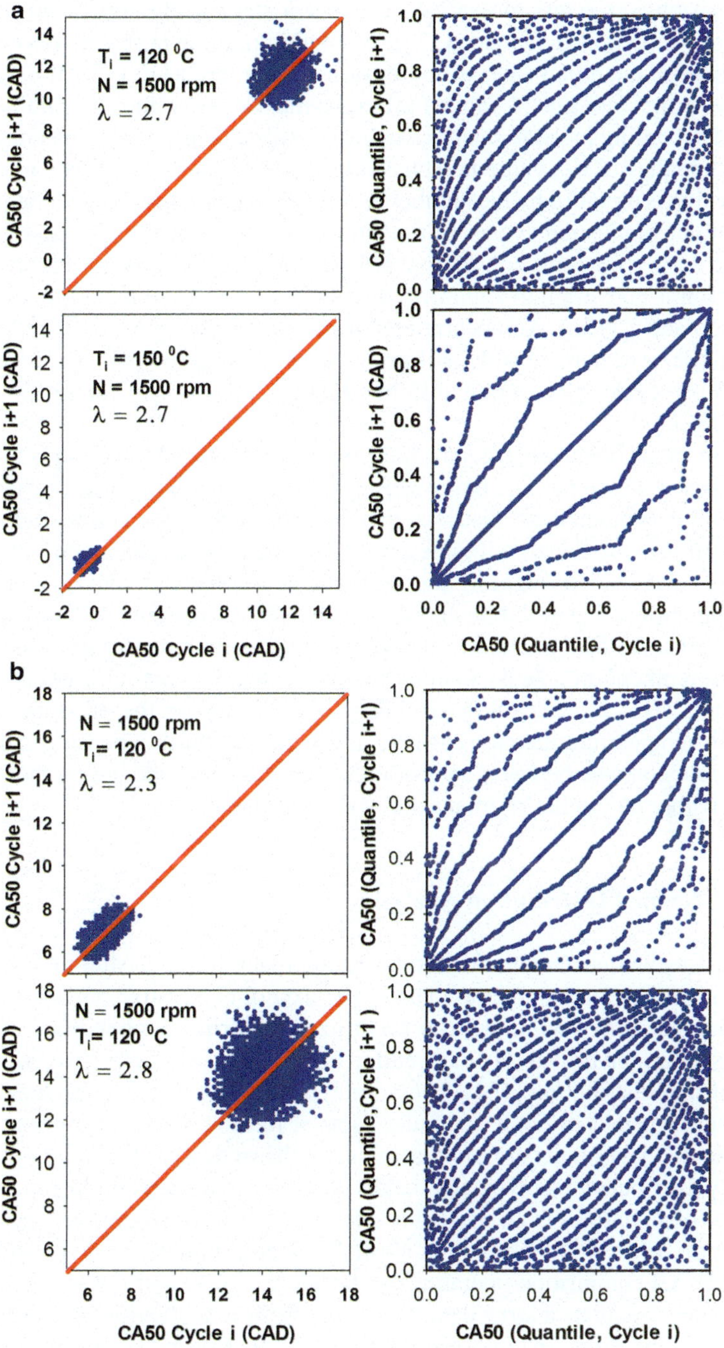

Fig. 6.68 Return maps for CA_{50} with lag of one cycle for different (**a**) λ and (**b**) T_i in HCCI engine at 1500 rpm in HCCI combustion (Adapted from [24, 133])

finding an objective test to evaluate whether a statistically significant degree of correlation is present in the data [132]. A solution to these problems is to construct return maps of quantile values of combustion timing instead of CA_{50} timing values. To convert the data set (N values) into quantile values, values of parameters are replaced according to value such as smallest CA_{50} value is replaced by number $1/N$, the next smallest CA_{50} value is replaced by the number $2/N$ and the largest CA_{50} timing value is replaced by the number N/N, or 1 without disturbing the sequence of consecutive combustion cycles. Entire data set is now converted to values between 0 and 1 during quantile conversion process, and in this process highly skewed distribution is converted to a uniform distribution of data points evenly distributed between 0 and 1 [132]. An uncorrelated time series of combustion parameters has the quantile return map with uniform density of data points, while correlated combustion cycles have non-uniform (higher and lower) densities of data points [132].

Return maps of quantile combustion timings are also shown corresponding to return maps in Fig. 6.68. The figure shows that in leaner mixtures cases, the quantile CA_{50} timing data points are distributed in the entire map space and for rich mixture, data points are distributed with uneven density areas with higher and lower densities. This observation suggests the deterministic behaviour during rich mixture engine operation. Figure 6.68b also shows the similar pattern. At lower intake air temperatures, the data points are distributed in the entire map space, and at higher intake air temperature, the data points are distributed with non-uniform densities of data points. This indicates a deterministic dependency on the previous cycles as intake air temperature increases at very advanced combustion phasing. This deterministic dependence between combstion cycles can be explained by thermal coupling possibly due to hot residuals and wall heat transfer.

6.5.2.3 Normal Distribution Analysis

Statistical techniques are another common approach for analysing the cyclic variations in combustion parameters. The cyclic data of consecutive engine cycles at a particular operating conditions can be utilized to form a probability distribution. The most common probability distribution often used to characterize measured data is the normal distribution [134]. The normal probability plot is a graphical method to evaluate whether or not a particular data set follows a normal distribution. The experimental data points are plotted against a theoretical normal distribution. The normally distributed data typically makes an approximate straight line in the graph. The level of deviations from straight line is considered as deviation from the normal distribution. Figure 6.69 shows the normal probability plots for combustion duration for HCCI combustion operating at knocking, misfire and stable conditions.

Figure 6.69 depicts a large deviation from the normal distribution for the HCCI operating conditions in knocking (RI = 12.9 MW/m^2) and misfire range (COV_{IMEP} = 12.55%). A stretched C shape distribution in knocking range (Fig. 6.69a) and a V shape distribution in misfire range (Fig. 6.69b) is

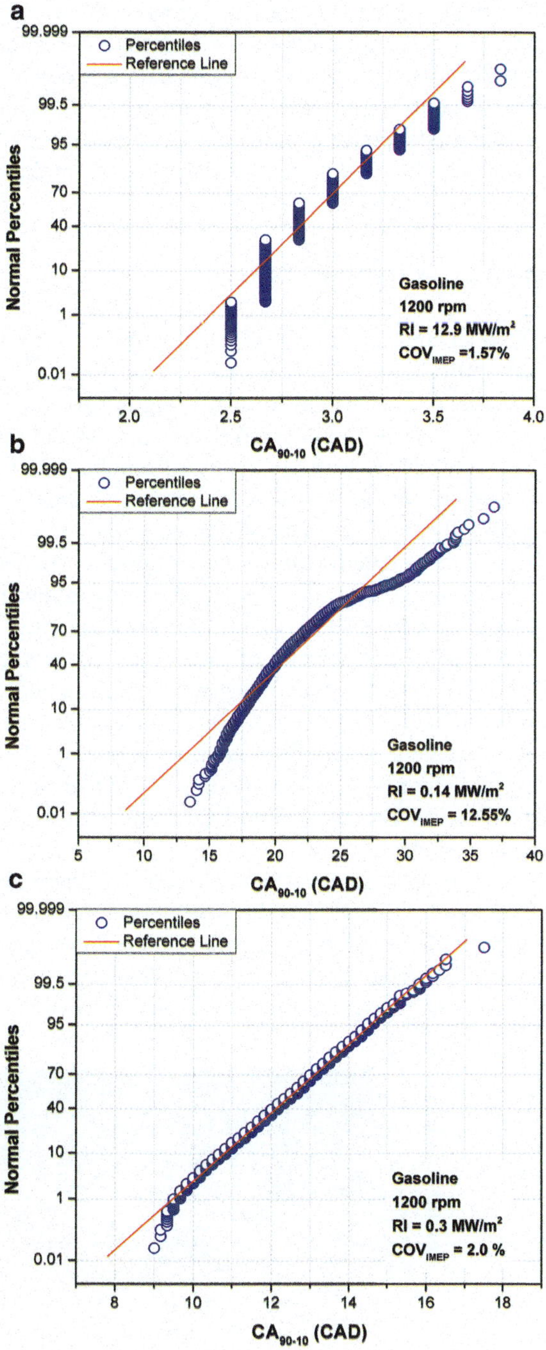

Fig. 6.69 Normal probability plots for combustion duration (**a**) knocking, (**b**) misfire and (**c**) normal stable conditions in HCCI engine (Adapted from [107])

observed in normal probability plots. [107]. In the stable HCCI condition, distribution is close to normal distribution line (Fig. 6.69c). Assuming a normal distribution for the random component of the data set, deviations from the normal distribution are indicative of possible deterministic components in the data [134]. This suggests deterministic patterns in the CA_{90-10} data during knocking and misfire conditions, which can be used to control purposes by modelling and prediction.

6.5.2.4 Wavelet Analysis

Wavelet analysis is a powerful method for identifying the dominant modes of variability in a time series and to describe how these modes vary in time. Hence, it is particularly useful for analysing transient and intermittent processes. Wavelet analysis has been successfully used to determine the combustion variability in conventional as well as advanced combustion modes [135–139]. In the wavelet spectrum, few patterns persist over hundreds of engine combustion cycles at nominally stable conditions, showing long-term structure in the engine dynamics, especially coupled with its control systems and environment, in ways not normally seen using methods that favour short time scales [140]. Figure 6.70 shows wavelet power spectra (WPS) of the heat release time series in HCCI combustion. A complex Morlet wavelet of order 6 as the mother wavelet is used to analyse the time series. In this case, the scale is approximately equal to the Fourier period. The WPS is calculated using a continuous wavelet transform (CWT). The horizontal axis depicts the number of engine cycles, and the vertical axis shows the period in cycles. The region below the thin U-shaped curve represents the cone of influence (COI) and only outside the COI result is reliable [135]. Colours denote the WPS power levels.

Figure 6.70 depicts the following small-amplitude, low-frequency periodicities: (i) a band around the 430-cycle period persisting between 625 and 2250 cycles, (ii) a band around the 315-cycle period spanning approximately 1500–2300 cycles and (iii) a band around the 110-cycle period ranging from 1650 to 2050 cycles [135]. The persistent structure is noticed over very long time scales, so long that

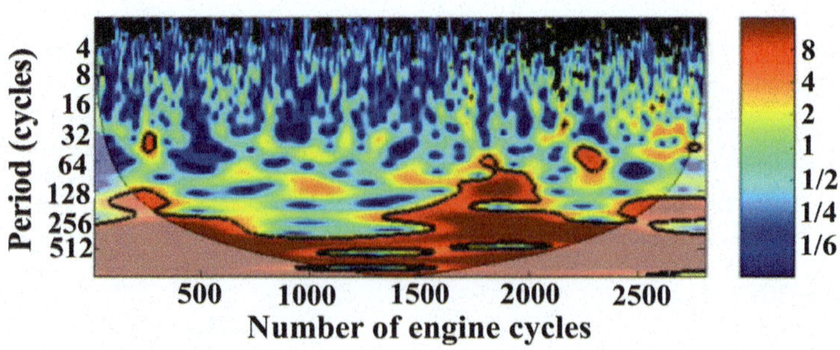

Fig. 6.70 Wavelet power spectra (WPS) of the heat release time series in HCCI combustion (Adapted from [135])

they most likely correspond more to the effects of long-term system variations such as temperature changes or slow chemical reservoir oscillations and less on internal cylinder feedback mechanisms [140].

6.5.2.5 Symbol Sequence Statistics

The nonlinear dynamics and chaos theory have been used to develop several mathematical techniques for deriving important inferences from the time-series data. A very useful method for time-series data analysis is known as "symbol sequence statistic method". This method gives important insights into the behaviour noticed for different combustion parameters in an engine operation. The presence of determinism in cyclic variation implies that intelligent control of the system could be a potential approach to extend the limits of engine operation significantly [78]. Intelligent controllers can effectively use the deterministic information for engine control, and small variation in control input parameters/actuators can shift back the engine to stable operating conditions [141]. Symbol-statistics predictive approach is used for controlling cyclic variations in a HCCI engine [131]. Symbolization of engine combustion data can very useful and effective in identifying the cyclic pattern in the combustion parameters for noisy measured data and data having measurement errors. Symbolization can also evaluate the deterministic effect of previous combustion cycle accurately for properly selected partition number. Symbol sequence technique can also act as data compression strategy, and leads to fast data processing at the time of data logging. Due to these characteristics, it can act as effective tool for real-time control and on-board diagnostics of HCCI engine [142].

Symbolization is a beneficial technique to distinguish the patterns in series of engine cycle-based combustion data. Symbolization transforms the combustion data into discrete bins that are denoted by a particular symbol. Figure 6.71 illustrates the symbolization method using binary partition and calculation of relative frequency of occurrence of particular symbol using IMEP data for 1500 consecutive HCCI engine cycles. The IMEP data time series is separated into two equiprobable partitions (which means equal number of data points in every partition), and each partition is represented by symbol 0 and 1 (Fig. 6.70a). This is the simplest partition to convert the data into a binary series of 1500 length. However, more than two partitions can be used for any data, but higher number of partition may contain higher noise. Relative frequency of particular sequence length is calculated from symbolized time series. Figure 6.70b shows the relative frequency (estimated from symbolic time series) of occurrence of symbols for sequence length 8. For binary partition, total 2^8 or 256 possible combination of sequence using eight sequence lengths by symbol 0 and 1, and all the 256 sequence numbers are presented in decimal format (shown on x-axis of Fig. 6.70b). Relative frequency is the ratio of number of occurrences to the total number of possible occurrence for that sequence length. This relative frequency is plotted on y-axis (Fig. 6.70b) for each 256 numbers. Relative frequency of occurrence of exactly random sequences is equal

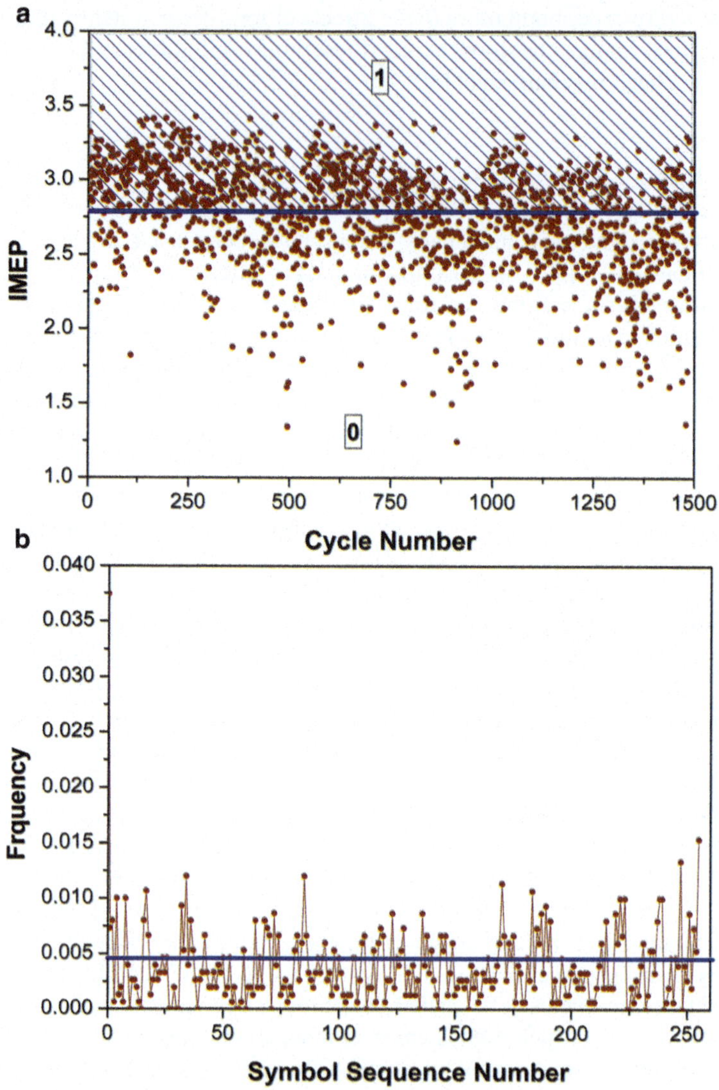

Fig. 6.71 Illustration of (**a**) binary partition symbolization and (**b**) symbol sequence histogram for sequence length "8" at $\lambda = 2.6$ for IMEP in HCCI engine (Adapted from [143])

because of equiprobable partitioning. For a sequence length (Y) and number of partition (X), the random frequency (F_r) of occurrence can be calculated as $F_r = (1/X)^Y$. Random frequency estimated for purely random data is plotted as thick line in the figure. Relative frequency values significantly above the random frequency line show deterministic structure and cyclic correlation in experimental data series [142]. Relative frequency peaks above baseline random frequency (Fig. 6.70b) depict the repeating deterministic events in time series. Figure 6.70b shows that sequence numbers have higher frequency than background random frequency but few sequence numbers have significantly higher and noteworthy frequency. For example, numbers 8, 17, 26, 34, 85, 170, 238, etc. (their corresponding numbers in binary 00001000, 00010001, 00011010, 00100010, 01010101, 10101010, 11101110, respectively) have relatively higher frequency.

In symbol sequence method, choosing optimal sequence length is required to find significant determinism in the data. A modified form of Shannon entropy is used to quantify the deviation of symbol statistics from randomness [141]. Shannon entropy value "1" indicates truly random data with no deterministic structure. Shannon entropy values lower than "1" indicates inherent deterministic structure and correlation between sequential points. Shannon entropy can be utilized to find the optimal sequence length for calculating symbol sequence histogram for estimation of deterministic structure in combustion data. The modified Shannon entropy H_s is calculated by Eq. (6.33).

$$H_s = \frac{1}{\log n_{seq}} \sum_k p_k \log p_k \tag{6.33}$$

where p_k is the probability with which sequence "k" occurs and n_{seq} is total number of sequence with non-zero probability. Figure 6.72a depicts the variations in modified Shannon entropy calculated using IMEP data series at different λ for different sequence lengths using binary partition. Highest presence of determinism in data series is represented by lowest value of Shannon entropy at particular sequence length. Sequence length with minimum Shannon entropy can be considered as the optimal sequence length. Figure 6.72a shows that Shannon entropy depends on sequence length and λ. Minimum Shannon entropy is around 8 sequence length for most of the test conditions using binary partition (Fig. 6.72a). This observation indicates that for effective control strategy, controller must take eight previous cycles into consideration, while using binary partition for experimental data series.

In symbol sequence method, optimal combination of number of partition and sequence length is required for effective control strategy. To estimate the optimal combination of number of partition and sequence length, a matrix of values ranging 2–10 is evaluated (Fig. 6.72b). Figure 6.72b shows that minimum entropy is for partitions "8–10" at sequence length "3". For binary partition, optimal sequence length is around "8". Higher number of sequence length need more memory of previous engine cycles. Due to memory issue of controller and better insight in pattern, higher number of partitions can be selected. If controller uses octal partition, then three previous cycles should be taken into consideration.

Figure 6.73 shows the symbol sequence histograms of CA_{50} for different λ and intake air temperature using octal partition and sequence length of 3 at 1500 rpm in HCCI combustion. Figure 6.73b shows that frequency of sequence number increases more above the red line as mixture becomes richer (lower λ), which indicates the inherent deterministic structures as mixtures become richer. The sequence code occurring more frequently with higher frequency are 0, 8, 16, 64,

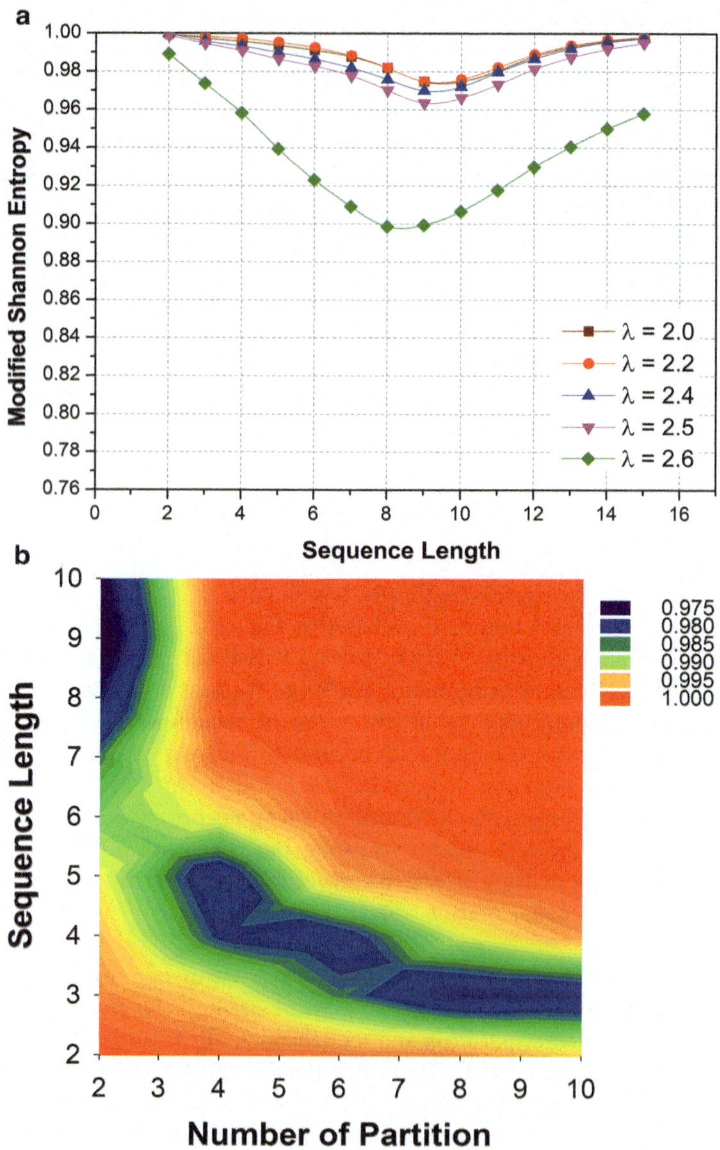

Fig. 6.72 (**a**) Modified Shannon entropy variation using binary partition for IMEP and (**b**) variation of modified Shannon entropy with sequence length and number of partition for IMEP at $\lambda = 2.0$ in HCCI engine (Adapted from [143])

128, 276 and 511. Figure 6.73a depicts the symbol sequence histograms of CA_{50} for different intake air temperatures. The frequency of sequence number increases more above the random frequency line with increase in the intake air temperature. This suggests the increase in deterministic behaviour with increase in temperature at constant λ.

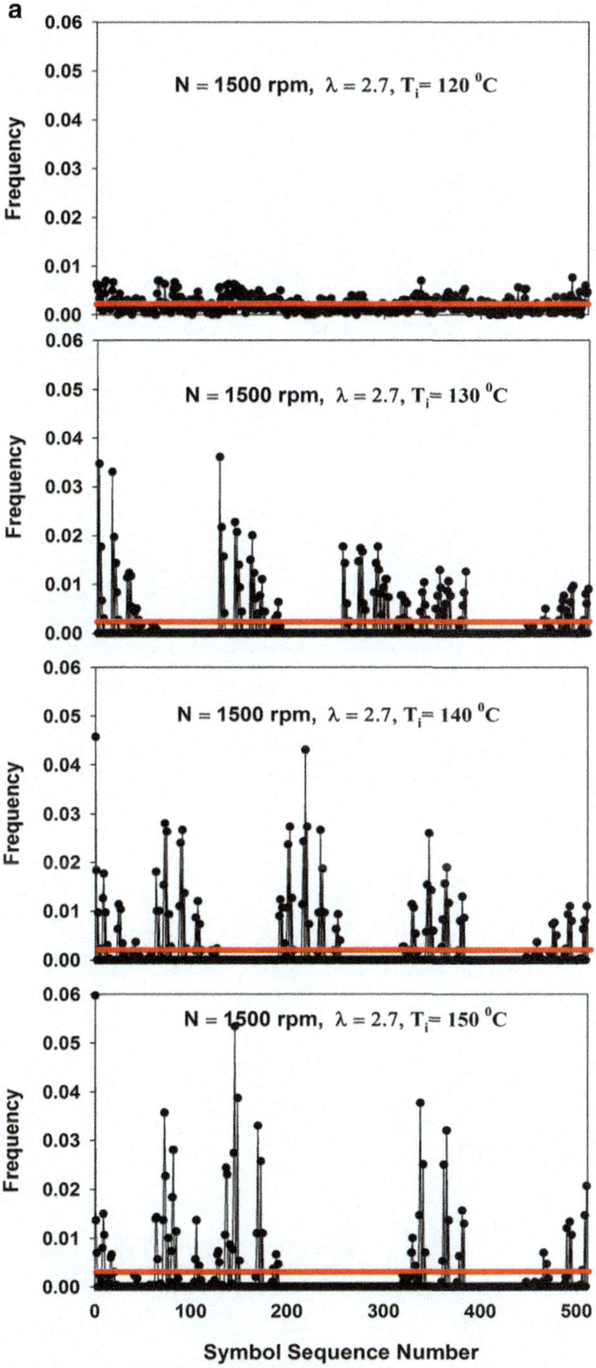

Fig. 6.73 Symbol sequence histograms of CA_{50} in HCCI combustion at 1500 rpm using octal partition and sequence length 3 for different (**a**) intake temperatures and (**b**) λ (Adapted from [24, 133])

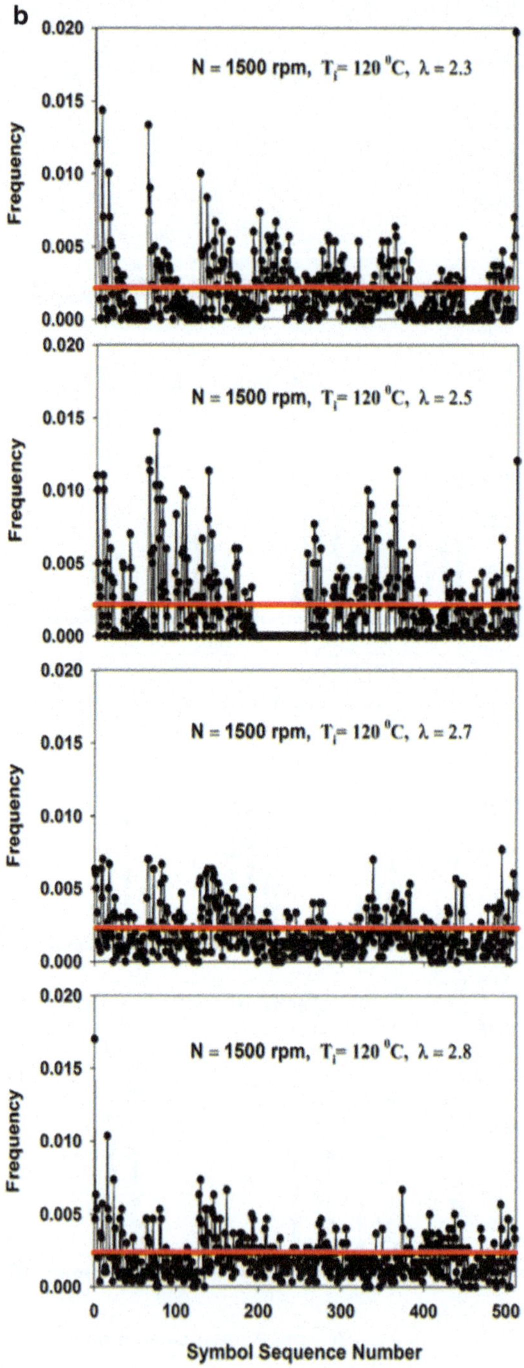

Fig. 6.73 (continued)

6.5.3 Sensing and Control

Understanding the cyclic variations of combustion timing with varying HCCI operating conditions is a necessary step for effectively controlling of the engine. For effective control of HCCI engine, first step is to determine a comprehensive ignition timing metric that is applicable over wide range of operating conditions especially in partial burn regions [127]. The complex dynamics of HCCI combustion in the partial burn operating region needs control for avoiding misfire [144]. For developing control strategy to avoid misfire/partial burn region, understanding of the HCCI engine characteristics during partial burn region is required. In-cylinder pressure [145], ionization current [146] and crankshaft angular speed [147] signal from engine are typically used for partial burn cycle recognition. Cylinder pressure-based methods are more accurate and widely used in engine research.

A study compares two combustion timing parameters ($\theta_{P_{max}}$ and CA_{50}) as online method in order to find and determine accurate and robust method for HCCI combustion timing, which can be used over a wide range of operating conditions for both normal and partial burn operating conditions [148]. In this analysis, 329 different engine operating conditions are used. Metric $\theta_{P_{max}}$ is used as ignition timing parameter because it is easy to calculate and requires minimum computational resources [149]. Figure 6.74 compares the cyclic variations of ignition timing for two criteria of $\theta_{P_{max}}$ and CA_{50} in HCCI combustion. The figure shows that the cyclic variation of $\theta_{P_{max}}$ is higher than CA_{50} which means $\theta_{P_{max}}$ has higher sensitivity. In particular $\theta_{P_{max}}$ registered one cycle of early ignition timing when misfire occurs at cycle 44, while CA_{50} does not able to capture. Validity of these

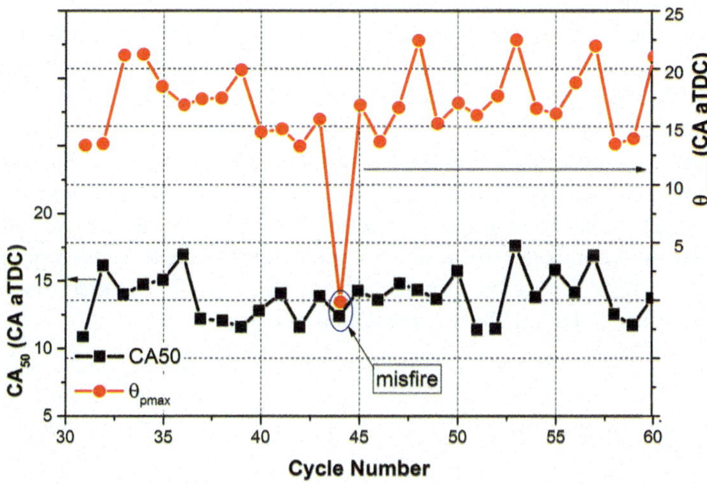

Fig. 6.74 $\theta_{P_{max}}$ and CA_{50} of consecutive cycles for HCCI combustion for PRF 0, speed = 1000 rpm, T_i = 100 °C, ϕ = 0.57, EGR = 0% and $T_{coolant}$ = 75 °C (Adapted from [127, 148])

criteria is further confirmed by analysing correlations between θ_{Pmax} and P_{max}, CA_{50} and IMEP. The $\theta_{P_{max}}$ is found to be a good ignition timing criteria to differentiate between normal and misfire operation in HCCI engine [148]. The CA_{50} is found to be a poor measure of cyclic variations, when the standard deviation of CA_{50} is greater than 2 CAD.

It is essentially required to be able to correctly identify partial burn operating conditions in order to develop a robust ignition timing methods. The COV_{IMEP} is typically used as measure of cyclic variability of engine parameters [150], and to determine the partial burn with IMEP, several criteria have been developed [126, 150, 151]. For example, a standard deviation of IMEP more than 2% indicates the appearance of partial burn and misfire cycles [126]. Using these criteria, a study found that COV_{IMEP} is not a reliable parameter in recognizing high cyclic variation, and it is found that several conditions with higher COV_{IMEP} have few or zero partial burn cycles [150]. A partial burn cycle is defined as when the total heat release is less than 90% of previous cycle [126]. Using these criteria, another study defined HCCI operating conditions as partial burn operating conditions when it has more than 14% partial burn cycles [151] and misfire operating conditions, when it contains more than 30% partial burn cycles [152]. Investigation of cyclic variations in CA_1, CA_{10}, CA_{50} and burn duration (BD) to identify misfire operating conditions found that CA_1 (crank angle position of 1% heat release) and CA_{50} are not effective metrics for HCCI misfire recognition [152]. However, cyclic variations of CA_{10} and BD are effective measures of misfire recognition, and it can be used as alternative metric to COV_{IMEP} for identifying normal, partial burn and misfire HCCI operating conditions. A study compares the several crank angle-based methods (CA_{50}, overall CA_{10}, CA_{10} for main heat release, third derivative of pressure with single and two limits) for the estimation of SOC in stable and unstable engine operating conditions [150]. This study found that criterion CA_{10} based on the main stage heat release is effective for all engine operating conditions.

Several actuation strategies that affect combustion timings can be used to control the HCCI cyclic variations such as intake air preheating, variable compression ratio, EGR, multi-fuel and multiple injection strategies (detailed discussion in Chap. 5). A study developed a control strategy based on symbol-statistics predictive approach for controlling cyclic combustion timing variations [131]. Another study developed a simple controller, designed using a linearized model, which improves the system's disturbance rejection characteristics by transforming the negative eigenvalue into a positive one, and controller reduces both the magnitude and frequency of oscillations in combustion timing on a cycle-to-cycle basis [153]. Detailed discussion on control strategies and algorithms is presented in Chap. 9.

References

1. Aceves SM, Flowers DL, Westbrook CK, Smith JR, Pitz W, Dibble R et al (2000) A multi-zone model for prediction of HCCI combustion and emissions (No. 2000-01-0327). SAE technical paper. doi:https://doi.org/10.4271/2000-01-0327

2. Sarathy SM, Oßwald P, Hansen N, Kohse-Höinghaus K (2014) Alcohol combustion chemistry. Prog Energy Combust Sci 44:40–102
3. Da Silva G, Bozzelli JW, Liang L, Farrell JT (2009) Ethanol oxidation: kinetics of the α-hydroxyethyl radical+ O2 reaction. Chem A Eur J 113(31):8923–8933
4. Maurya RK, Akhil N (2016) Numerical investigation of ethanol fuelled HCCI engine using stochastic reactor model. Part 1: development of a new reduced ethanol oxidation mechanism. Energy Convers Manag 118:44–54
5. Maurya RK, Akhil N (2016) Numerical investigation of ethanol fuelled HCCI engine using stochastic reactor model. Part 2: parametric study of performance and emissions characteristics using new reduced ethanol oxidation mechanism. Energy Convers Manag 121:55–70
6. Vuilleumier DM (2016) The effect of ethanol addition to gasoline on low- and intermediate temperature heat release under boosted conditions in kinetically controlled engines. PhD thesis, University of California, Berkeley
7. Vuilleumier D, Kozarac D, Mehl M, Saxena S, Pitz WJ, Dibble RW, Chen JY, Sarathy SM (2014) Intermediate temperature heat release in an HCCI engine fueled by ethanol/n-heptane mixtures: an experimental and modeling study. Combust Flame 161(3):680–695
8. Aranda V, Christensen JM, Alzueta MU, Glarborg P, Gersen S, Gao Y, Marshall P (2013) Experimental and kinetic modeling study of methanol ignition and oxidation at high pressure. In J Chem Kinet 45(5):283–294
9. Dayma G, Ali KH, Dagaut P (2007) Experimental and detailed kinetic modeling study of the high pressure oxidation of methanol sensitized by nitric oxide and nitrogen dioxide. Proc Combust Inst 31(1):411–418
10. Amer A, Babiker H, Chang J, Kalghatgi G, Adomeit P, Brassat A, Günther M (2012) Fuel effects on knock in a highly boosted direct injection spark ignition engine. SAE Int J Fuels Lubr 5:1048–1065. (2012-01-1634)
11. Vandersickel A, Hartmann M, Vogel K, Wright YM, Fikri M, Starke R, Schulz C, Boulouchos K (2012) The autoignition of practical fuels at HCCI conditions: high-pressure shock tube experiments and phenomenological modeling. Fuel 93:492–501
12. Blomberg CK, Mitakos D, Bardi M, Boulouchos K, Wright YM, Vandersickel A (2016) Extension of the phenomenological 3-Arrhenius auto-ignition model for six surrogate automotive fuels. SAE Int J Engines 9:1544–1558. (2016-01-0755)
13. Herzler J, Jerig L, Roth P (2005) Shock tube study of the ignition of lean n-heptane/air mixtures at intermediate temperatures and high pressures. Proc Combust Inst 30 (1):1147–1153
14. Ciezki HK, Adomeit G (1993) Shock-tube investigation of self-ignition of n-heptane-air mixtures under engine relevant conditions. Combust Flame 93(4):421–433
15. Dec JE, Yang Y (2010) Boosted HCCI for high power without engine knock and with ultra-low NOx emissions-using conventional gasoline. SAE Int J Engines 3:750–767. (2010-01-1086)
16. Splitter DA (2012) High efficiency RCCI combustion. PhD thesis, University of Wisconsin-Madison
17. Christensen M (2002) HCCI combustion-engine operation and emission characteristics. PhD thesis, Lund University, Sweden, ISBN 91-628-5424-0
18. Truedsson I, Tuner M, Johansson B, Cannella W (2013) Pressure sensitivity of HCCI auto-ignition temperature for oxygenated reference fuels. J Eng Gas Turbines Power 135 (7):072801
19. Maurya RK, Pal DD, Agarwal AK (2013) Digital signal processing of cylinder pressure data for combustion diagnostics of HCCI engine. Mech Syst Signal Process 36(1):95–109
20. Maurya RK (2016) Estimation of optimum number of cycles for combustion analysis using measured in-cylinder pressure signal in conventional CI engine. Measurement 94:19–25
21. Payri F, Broatch A, Tormos B, Marant V (2005) New methodology for in-cylinder pressure analysis in direct injection diesel engines—application to combustion noise. Meas Sci Technol 16(2):540

22. Maurya RK, Agarwal AK (2013) Investigations on the effect of measurement errors on estimated combustion and performance parameters in HCCI combustion engine. Measurement 46(1):80–88
23. Singh E, Waqas M, Johansson B, Sarathy M (2017) Simulating HCCI blending octane number of primary reference fuel with ethanol (No. 2017-01-0734). SAE technical paper
24. Maurya RK (2012) Performance, emissions and combustion characterization and close loop control of HCCI engine employing gasoline like fuels. PhD thesis, Indian Institute of Technology, Kanpur
25. Dec JE, Yang Y, Dernotte J, Ji C (2015) Effects of gasoline reactivity and ethanol content on boosted, premixed and partially stratified low-temperature gasoline combustion (LTGC). SAE Int J Engines 8:935–955. (2015-01-0813)
26. Dernotte J, Dec J, Ji C (2017) Efficiency improvement of boosted low-temperature gasoline combustion engines (LTGC) using a double direct-injection strategy (No. 2017-01-0728). SAE technical paper
27. Solaka H (2014) Impact of fuel properties on partially premixed combustion. PhD thesis, Lund University, Sweden
28. Solaka H, Tunér M, Johansson B (2012) Investigation on the impact of fuel properties on partially premixed combustion characteristics in a light duty diesel engine. In: ASME 2012 internal combustion engine division spring technical conference. American Society of Mechanical Engineers, p 335–345, 2012, May
29. Benajes J, Molina S, García A, Belarte E, Vanvolsem M (2014) An investigation on RCCI combustion in a heavy duty diesel engine using in-cylinder blending of diesel and gasoline fuels. Appl Therm Eng 63(1):66–76
30. Li Y, Jia M, Liu Y, Xie M (2013) Numerical study on the combustion and emission characteristics of a methanol/diesel reactivity controlled compression ignition (RCCI) engine. Appl Energy 106:184–197
31. Zou X, Wang H, Zheng Z, Reitz R, Yao M (2016) Numerical study of the RCCI combustion processes fuelled with methanol, ethanol, n-butanol and diesel (No. 2016-01-0777). SAE technical paper
32. Splitter D, Reitz RD, Hanson R (2010) High efficiency, low emissions RCCI combustion by use of a fuel additive. SAE Int J Fuels Lubr 3:742–756. (2010-01-2167)
33. Hanson R, Kokjohn S, Splitter D, Reitz RD (2011) Fuel effects on reactivity controlled compression ignition (RCCI) combustion at low load. SAE Int J Engines 4:394–411. (2011-01-0361)
34. Manente V (2010) Gasoline partially premixed combustion-an advanced internal combustion engine concept aimed to high efficiency, low emissions and low acoustic noise in the whole load range. PhD thesis, Lund University, Sweden
35. Maurya RK, Agarwal AK (2014) Experimental investigations of performance, combustion and emission characteristics of ethanol and methanol fueled HCCI engine. Fuel Process Technol 126:30–48
36. Sjöberg M, Dec JE, Babajimopoulos A, Assanis DN (2004) Comparing enhanced natural thermal stratification against retarded combustion phasing for smoothing of HCCI heat-release rates (No. 2004-01-2994). SAE technical paper
37. Zhao H, Xie H, Peng Z (2005) Effect of recycled burned gases on homogeneous charge compression ignition combustion. Combust Sci Technol 177(10):1863–1882
38. Zhao H, Peng Z, Williams J, Ladommatos N (2001) Understanding the effects of recycled burnt gases on the controlled autoignition (CAI) combustion in four-stroke gasoline engines (No. 2001-01-3607). SAE technical paper
39. Risberg P, Johansson D, Andrae J, Kalghatgi G, Björnbom P, Ångström HE (2006) The influence of NO on the combustion phasing in an HCCI engine (No. 2006-01-0416). SAE technical paper

40. Xie H, Lu J, Chen T, Li L, Li C, Zhao H (2014) Chemical effects of the incomplete-oxidation products in residual gas on the gasoline HCCI auto-ignition. Combust Sci Technol 186 (3):273–296
41. Herold RE, Foster DE, Ghandhi JB, Iverson RJ, Eng JA, Najt PM (2007) Fuel unmixedness effects in a gasoline homogeneous charge compression ignition engine. Int J Engine Res 8 (3):241–257
42. Robert Bosch GmbH (2006) Gasoline engine management, 3rd edn. Wiley. Chichester, England ISBN:978-0-470-05757-5
43. Maurya RK, Agarwal AK (2011) Effect of start of injection on the particulate emission from methanol fuelled HCCI engine. SAE Int J Fuels Lubr 4:204–222. (2011-01-2408)
44. Dec JE, Yang Y, Dronniou N (2011) Boosted HCCI-controlling pressure-rise rates for performance improvements using partial fuel stratification with conventional gasoline. SAE Int J Engines 4:1169–1189. (2011-01-0897)
45. Manente V, Zander CG, Johansson B, Tunestal P, Cannella W (2010) An advanced internal combustion engine concept for low emissions and high efficiency from idle to max load using gasoline partially premixed combustion (No. 2010-01-2198). SAE technical paper
46. Li Y, Jia M, Chang Y, Xie M, Reitz RD (2016) Towards a comprehensive understanding of the influence of fuel properties on the combustion characteristics of a RCCI (reactivity controlled compression ignition) engine. Energy 99:69–82
47. Desantes JM, Benajes J, García A, Monsalve-Serrano J (2014) The role of the in-cylinder gas temperature and oxygen concentration over low load reactivity controlled compression ignition combustion efficiency. Energy 78:854–868
48. Wang Y, Yao M, Li T, Zhang W, Zheng Z (2016) A parametric study for enabling reactivity controlled compression ignition (RCCI) operation in diesel engines at various engine loads. Appl Energy 175:389–402
49. Park SH, Shin D, Park J (2016) Effect of ethanol fraction on the combustion and emission characteristics of a dimethyl ether-ethanol dual-fuel reactivity controlled compression ignition engine. Appl Energy 182:243–252
50. Li Y, Jia M, Chang Y, Liu Y, Xie M, Wang T, Zhou L (2014) Parametric study and optimization of a RCCI (reactivity controlled compression ignition) engine fueled with methanol and diesel. Energy 65:319–332
51. Christensen M, Johansson B (1999) Homogeneous charge compression ignition with water injection (No. 1999-01-0182). SAE technical paper
52. Christensen M, Johansson B, Einewall P (1997) Homogeneous charge compression ignition (HCCI) using isooctane, ethanol and natural gas-a comparison with spark ignition operation (No. 972874). SAE technical paper
53. Christensen M, Johansson B, Amnéus P, Mauss F (1998) Supercharged homogeneous charge compression ignition (No. 980787). SAE technical paper
54. Christensen M, Johansson B, Hultqvist A (2001) The effect of piston topland geometry on emissions of unburned hydrocarbons from a homogeneous charge compression ignition (HCCI) engine (No. 2001-01-1893). SAE technical paper
55. Truedsson I, Tuner M, Johansson B, Cannella W (2013) Emission formation study of HCCI combustion with gasoline surrogate fuels (No. 2013-01-2626). SAE technical paper
56. Ji C, Dec JE, Dernotte J, Cannella W (2014) Effect of ignition improvers on the combustion performance of regular-grade E10 gasoline in an HCCI engine. SAE Int J Engines 7:790–806. (2014-01-1282)
57. Manente V, Tunestal P, Johansson B, Cannella WJ (2010) Effects of ethanol and different type of gasoline fuels on partially premixed combustion from low to high load (No. 2010-01-0871). SAE technical paper
58. Molina S, García A, Pastor JM, Belarte E, Balloul I (2015) Operating range extension of RCCI combustion concept from low to full load in a heavy-duty engine. Appl Energy 143:211–227

59. Hanson RM (2014) Experimental investigation of transient RCCI combustion in a light duty diesel engine, PhD thesis, University of Wisconsin-Madison
60. Benajes J, Molina S, García A, Monsalve-Serrano J (2015) Effects of direct injection timing and blending ratio on RCCI combustion with different low reactivity fuels. Energy Convers Manag 99:193–209
61. Dev S, Divekar P, Xie K, Han X, Chen X, Zheng M (2015) A study of combustion inefficiency in diesel low temperature combustion and gasoline–diesel RCCI via detailed emission measurement. J Eng Gas Turbines Power 137(12):121501
62. Eng JA (2002) Characterization of pressure waves in HCCI combustion (No. 2002-01-2859). SAE technical paper
63. Sang W (2013) Knock mitigation on boosted controlled auto-ignition engines with fuel stratification and exhaust gas recycling. Doctoral dissertation, Massachusetts Institute of Technology
64. Maria AG (2012) On fuel selection in controlled auto-ignition engines: the link between intake conditions, chemical kinetics, and stratification. Doctoral dissertation, Massachusetts Institute of Technology
65. Andreae MM, Cheng WK, Kenney T, Yang J (2007) On HCCI engine knock (No. 2007-01-1858). SAE technical paper
66. Yelvington PE, Rallo MBI, Liput S, Tester JW, Green WH, Yang J (2004) Prediction of performance maps for homogeneous-charge compression-ignition engines. Combust Sci Technol 176(8):1243–1282
67. Draper CS (1938) Pressure waves accompanying detonation in the internal combustion engine. J Aeronaut Sci 5(6):219–226
68. Dahl D, Andersson M, Denbratt I (2011) The origin of pressure waves in high load HCCI combustion: a high-speed video analysis. Combust Sci Technol 183(11):1266–1281
69. Hyvönen J, Haraldsson G, Johansson B (2005) Operating conditions using spark assisted HCCI combustion during combustion mode transfer to SI in a multi-cylinder VCR-HCCI engine (No. 2005-01-0109). SAE technical paper
70. Kim KS (2015) Study of engine knock using a Monte Carlo method. Doctoral dissertation, The University of Wisconsin-Madison
71. Brunt MF, Pond CR, Biundo J (1998) Gasoline engine knock analysis using cylinder pressure data (No. 980896). SAE technical paper
72. Shahlari AJ (2016) An investigation of the knock measurement. Doctoral dissertation, The University of Wisconsin-Madison
73. Dernotte J, Dec JE, Ji C (2014) Investigation of the sources of combustion noise in HCCI engines. SAE Int J Engines 7:730–761. (2014-01-1272)
74. Johansson T, Johansson B, Tunestål P, Aulin H (2009) HCCI operating range in a turbo-charged multi cylinder engine with VVT and spray-guided DI (No. 2009-01-0494). SAE technical paper
75. Maria A, Cheng WK, Kar K, Cannella W (2013) Understanding knock metric for controlled auto-ignition engines. SAE Int J Engines 6:533–540. (2013-01-1658)
76. Wissink M, Wang Z, Splitter D, Shahlari A, Reitz RD (2013) Investigation of pressure oscillation modes and audible noise in RCCI, HCCI, and CDC (No. 2013-01-1652). SAE technical paper
77. Saxena S, Bedoya ID (2013) Fundamental phenomena affecting low temperature combustion and HCCI engines, high load limits and strategies for extending these limits. Prog Energy Combust Sci 39(5):457–488
78. Agarwal AK, Singh AP, Maurya RK (2017) Evolution, challenges and path forward for low temperature combustion engines. Prog Energy Combust Sci 61:1–56
79. Fuerhapter A, Piock WF, Fraidl GK (2003) CSI-controlled auto ignition-the best solution for the fuel consumption-versus emission trade-Off? (No. 2003-01-0754). SAE technical paper
80. Cairns A, Blaxill H (2005) The effects of combined internal and external exhaust gas recirculation on gasoline controlled auto-ignition (No. 2005-01-0133). SAE technical paper

81. Milovanovic N, Blundell D, Pearson RJ, Turner JWG, Chen R (2005) Enlarging the operational range of a gasoline HCCI engine by controlling the coolant temperature (No. 2005-01-015). SAE technical paper
82. Maurya RK, Agarwal AK (2011) Experimental study of combustion and emission characteristics of ethanol fuelled port injected homogeneous charge compression ignition (HCCI) combustion engine. Appl Energy 88(4):1169–1180
83. Yeom K, Bae C (2009) Knock characteristics in liquefied petroleum gas (LPG)– dimethyl ether (DME) and gasoline– DME homogeneous charge compression ignition engines. Energy Fuel 23(4):1956–1964
84. Johansson T, Borgqvist P, Johansson B, Tunestal P, Aulin H (2010) HCCI heat release data for combustion simulation, based on results from a turbocharged multi cylinder engine (No. 2010-01-1490). SAE technical paper
85. Oakley A, Zhao H, Ladommatos N, Ma T (2001) Experimental studies on controlled auto-ignition (CAI) combustion of gasoline in a 4-stroke engine (No. 2001-01-1030). SAE technical paper
86. Hou J, Qiao X, Wang Z, Liu W, Huang Z (2010) Characterization of knocking combustion in HCCI DME engine using wavelet packet transform. Appl Energy 87(4):1239–1246
87. Scaringe RJ, Wildman CB, Cheng WK (2010) On the high load limit of boosted gasoline HCCI engine operating in NVO mode. SAE Int J Engines 3(1):35–45. https://doi.org/10.4271/2010-01-0162
88. Shi L, Cui Y, Deng K, Peng H, Chen Y (2006) Study of low emission homogeneous charge compression ignition (HCCI) engine using combined internal and external exhaust gas recirculation (EGR). Energy 31(14):2665–2676
89. Wildman C, Scaringe RJ, Cheng W (2009) On the maximum pressure rise rate in boosted HCCI operation (No. 2009-01-2727). SAE technical paper
90. Kalghatgi GT, Head RA (2006) Combustion limits and efficiency in a homogeneous charge compression ignition engine. Int J Engine Res 7(3):215–236
91. Livengood JC, Wu PC (1955) Correlation of autoignition phenomena in internal combustion engines and rapid compression machines. Symp (Int) Combust 5(1). Elsevier):347–356
92. Dahl D, Andersson M, Berntsson A, Denbratt I, Koopmans L (2009) Reducing pressure fluctuations at high loads by means of charge stratification in HCCI combustion with negative valve overlap (No. 2009-01-1785). SAE technical paper
93. Wildman CBE (2009) High load limits of the controlled autoignition engine. Doctoral dissertation, Massachusetts Institute of Technology
94. Sjöberg M, Dec JE (2007) EGR and intake boost for managing HCCI low-temperature heat release over wide ranges of engine speed (No. 2007-01-0051). SAE technical paper
95. Shahlari AJ, Hocking C, Kurtz E, Ghandhi J (2013) Comparison of compression ignition engine noise metrics in low-temperature combustion regimes. SAE Int J Engines 6:541–552. (2013-01-1659)
96. Bahri B, Shahbakhti M, Aziz AA (2017) Real-time modeling of ringing in HCCI engines using artificial neural networks. Energy 125:509–518
97. Atkins RD (2009) An introduction to engine testing and development. SAE International, Warrendale. ISBN:978-0-7680-2099-1
98. Li H, Neill WS, Chippior WL (2012) An experimental investigation of HCCI combustion stability using n-heptane. J Energy Resour Technol 134(2):022204
99. Yun H, Kang JM, Chang MF, Najt P (2010) Improvement on cylinder-to-cylinder variation using a cylinder balancing control strategy in gasoline HCCI engines (No. 2010-01-0848). SAE technical paper
100. Shaver GM, Roelle MJ, Gerdes JC (2006) Modeling cycle-to-cycle dynamics and mode transition in HCCI engines with variable valve actuation. Control Eng Pract 14(3):213–222
101. Jungkunz AF (2013) Actuation strategies for cycle-to-cycle control of homogeneous charge compression ignition combustion engines. Doctoral dissertation, Stanford University

102. Shahbakhti M, Koch CR (2008) Characterizing the cyclic variability of ignition timing in a homogeneous charge compression ignition engine fuelled with n-heptane/iso-octane blend fuels. Int J Engine Res 9(5):361–397
103. Koopmans L, Backlund O, Denbratt I (2002) Cycle to cycle variations: their influence on cycle resolved gas temperature and unburned hydrocarbons from a camless gasoline compression ignition engine (No. 2002-01-0110). SAE technical paper
104. Maurya RK, Agarwal AK (2009) Experimental investigation of cycle-by-cycle variations in CAI/HCCI combustion of gasoline and methanol fuelled engine (No. 2009-01-1345). SAE technical paper
105. Maurya RK, Agarwal AK (2011) Experimental investigation on the effect of intake air temperature and air–fuel ratio on cycle-to-cycle variations of HCCI combustion and performance parameters. Appl Energy 88(4):1153–1163
106. Maurya RK, Agarwal AK (2012) Statistical analysis of the cyclic variations of heat release parameters in HCCI combustion of methanol and gasoline. Appl Energy 89(1):228–236
107. Maurya RK, Agarwal AK (2013) Experimental investigation of cyclic variations in HCCI combustion parameters for gasoline like fuels using statistical methods. Appl Energy 111:310–323
108. Lü X, Ji L, Ma J, Huang Z (2007) Combustion stabilities and cycle-by-cycle variations of n-heptane homogeneous charge compression ignition combustion. Energy Fuel 21 (3):1468–1473
109. Persson H, Pfeiffer R, Hultqvist A, Johansson B, Ström H (2005) Cylinder-to-cylinder and cycle-to-cycle variations at HCCI operation with trapped residuals (No. 2005-01-0130). SAE technical paper
110. Richter M, Engström J, Franke A, Aldén M, Hultqvist A, Johansson B (2000) The influence of charge inhomogeneity on the HCCI combustion process (No. 2000-01-2868). SAE technical paper
111. Aleiferis PG, Charalambides AG, Hardalupas Y, Taylor AMKP, Urata Y (2006) Autoignition initiation and development of n-heptane HCCI combustion assisted by inlet air heating, internal EGR or spark discharge: an optical investigation (No. 2006-01-3273). SAE technical paper
112. Pan J, Sheppard CGW, Tindall A, Berzins M, Pennington SV, Ware JM (1998) End gas inhomogeneity, autoignition and knock (No. 982616). SAE technical paper
113. Yu R, Bai XS, Lehtiniemi H, Ahmed SS, Mauss F, Richter M, et al (2006) Effect of turbulence and initial temperature inhomogeneity on homogeneous charge compression ignition combustion (No. 2006-01-3318). SAE technical paper
114. Dronniou N, Dec JE (2012) Investigating the development of thermal stratification from the near-wall regions to the bulk-gas in an HCCI engine with planar imaging thermometry. SAE Int J Engines 5:1046–1074. (2012-01-1111)
115. Chiang CJ, Stefanopoulou AG (2007) Stability analysis in homogeneous charge compression ignition (HCCI) engines with high dilution. IEEE Trans Control Syst Technol 15(2):209–219
116. Olsson JO, Tunestål P, Johansson B, Fiveland S, Agama R, Willi M, Assanis DN (2002) Compression ratio influence on maximum load of a natural gas fueled HCCI engine (No. 2002-01-0111). SAE technical paper
117. Grünefeld G, Beushausen V, Andresen P, Hentschel W (1994) A major origin of cyclic energy conversion variations in SI engines: cycle-by-cycle variations of the equivalence ratio and residual gas of the initial charge (No. 941880). SAE technical paper
118. Liu C, Karim GA, Sohrabi A, Xiao F (2006) Combustion and cyclic variation for lean mixture operation. In: Proceedings of ASME ICED 2006 spring technical conference. ASME paper no. ICES2006-01-1564, Aachen
119. Jung D, Iida N (2017) Thermal and chemical effects of the in-cylinder charge at IVC on cycle-to-cycle variations of DME HCCI combustion with combustion-phasing retard by external and rebreathed EGR. Appl Therm Eng 113:132–149

120. Borgqvist P, Tunestal P, Johansson B (2013) Comparison of negative valve overlap (NVO) and rebreathing valve strategies on a gasoline PPC engine at low load and idle operating conditions. SAE Int J Engines 6:366–378. (2013-01-0902)
121. Klos D, Kokjohn SL (2015) Investigation of the sources of combustion instability in low-temperature combustion engines using response surface models. Int J Engine Res 16 (3):419–440
122. Sjöberg M, Dec JE, Cernansky NP (2005) Potential of thermal stratification and combustion retard for reducing pressure-rise rates in HCCI engines, based on multi-zone modeling and experiments (No. 2005-01-0113). SAE technical paper
123. Guo H, Neill WS, Chippior W, Li H, Taylor JD (2010) An experimental and modeling study of HCCI combustion using n-heptane. J Eng Gas Turbines Power 132(2):022801
124. Chang J, Filipi Z, Assanis D, Kuo TW, Najt P, Rask R (2005) Characterizing the thermal sensitivity of a gasoline homogeneous charge compression ignition engine with measurements of instantaneous wall temperature and heat flux. Int J Engine Res 6(4):289–310
125. Jia M, Dempsey AB, Wang H, Li Y, Reitz RD (2015) Numerical simulation of cyclic variability in reactivity-controlled compression ignition combustion with a focus on the initial temperature at intake valve closing. Int J Engine Res 16(3):441–460
126. Sjöberg M, Dec JE (2007) Comparing late-cycle autoignition stability for single-and two-stage ignition fuels in HCCI engines. Proc Combust Inst 31(2):2895–2902
127. Ghazimirsaied A (2012) Extending HCCI low load operation using Chaos prediction and feedback control. Doctoral dissertation, University of Alberta
128. Daw CS, Finney CEA, Green JB, Kennel MB, Thomas JF, Connolly FT (1996) A simple model for cyclic variations in a spark-ignition engine (No. 962086). SAE technical paper
129. Daw CS, Wagner RM, Edwards KD, Green JB (2007) Understanding the transition between conventional spark-ignited combustion and HCCI in a gasoline engine. Proc Combust Inst 31 (2):2887–2894
130. Green JB, Daw CS, Armfield JS, Finney CEA, Wagner RM, Drallmeier JA et al (1999) Time irreversibility and comparison of cyclic-variability models (No. 1999-01-0221). SAE technical paper
131. Ghazimirsaied A, Koch CR (2012) Controlling cyclic combustion timing variations using a symbol-statistics predictive approach in an HCCI engine. Appl Energy 92:133–146
132. Scholl D, Russ S (1999) Air-fuel ratio dependence of random and deterministic cyclic variability in a spark-ignited engine (No. 1999-01-3513). SAE technical paper
133. Maurya RK, Agarwal AK (2011) Experimental investigation on intake air temperature and air-fuel ratio dependence of random and deterministic cyclic variability in a homogeneous charge compression ignition engine (No. 2011-01-1183). SAE technical paper
134. Chang J, Filipi Z, Assanis D, Kuo T, Najt P, Rask R (2008) NIST/SEMATECH e-Handbook of statistical methods. http://www.itl.nist.gov/div898/handbook
135. Sen AK, Litak G, Edwards KD, Finney CE, Daw CS, Wagner RM (2011) Characteristics of cyclic heat release variability in the transition from spark ignition to HCCI in a gasoline engine. Appl Energy 88(5):1649–1655
136. Maurya RK, Nekkanti A (2016) Combustion instability analysis using wavelets in conventional diesel engine. In: Mathematical concepts and applications in mechanical engineering and mechatronics, IGI Global, Hershey, Pennsylvania, USA. p 390
137. Hunicz J (2014) On cyclic variability in a residual effected HCCI engine with direct gasoline injection during negative valve overlap. Math Probl Eng 2014:1–11
138. Maurya RK, Saxena MR (2016) Investigation of effect of butanol addition on cyclic variability in a diesel engine using wavelets. In: The international symposium on intelligent systems technologies and applications. Springer International Publishing, Cham, Switzerland. p 965–976
139. Sen AK, Longwic R, Litak G, Górski K (2008) Analysis of cycle-to-cycle pressure oscillations in a diesel engine. Mech Syst Signal Process 22(2):362–373

140. Finney CE, Kaul BC, Daw CS, Wagner RM, Edwards KD, Green JB Jr (2015) Invited review: a review of deterministic effects in cyclic variability of internal combustion engines. Int J Engine Res 16(3):366–378
141. Kaul BC, Vance JB, Drallmeier JA, Sarangapani J (2009) A method for predicting performance improvements with effective cycle-to-cycle control of highly dilute spark ignition engine combustion. Proc Inst Mech Eng Part D J Automob Eng 223(3):423–438
142. Finney CEA, Green JB, Daw CS (1998) Symbolic time-series analysis of engine combustion measurements (No. 980624). SAE technical paper
143. Maurya RK (2017) Experimental investigation of cyclic variability in HCCI engine using symbol sequence analysis. Submitted. Under review
144. Knierim KL, Park S, Ahmed J, Kojic A, Orlandini I, Kulzer A (2008). Simulation of misfire and strategies for misfire recovery of gasoline HCCI. In: American Control Conference, IEEE, p 3947–3952, June 2008
145. Haskara I, Mianzo L (2001) Real-time cylinder pressure and indicated torque estimation via second order sliding modes. In Proceedings of the 2001 American Control Conference, IEEE, vol 5, p 3324–3328
146. VanDyne EA, Burckmyer CL, Wahl AM, Funaioli AE (2000) Misfire detection from ionization feedback utilizing the Smartfire® plasma ignition technology (No. 2000-01-1377). SAE technical paper
147. Bue FL, Stefano AD, Giaconia C, Pipitone E (2007) Misfire detection system based on the measure of crankshaft angular velocity. In: Proceeding of the 11th annual AMAA conference, Berlin, March 2007
148. Ghazimirsaied A, Shahbakhti M, Koch CR (2011) Ignition timing criteria for partial burn operation in an HCCI engine. In: Proceedings of CI/CS conference, Winnipeg, Canada, May 2011
149. Asad U, Zheng M (2008) Fast heat release characterization of a diesel engine. Int J Therm Sci 47(12):1688–1700
150. Ghazimirsaied A, Shahbakhti M, Koch CR (2010, January). Comparison of crankangle based ignition timing methods on an HCCI engine. In ASME 2010 internal combustion engine division fall technical conference. American Society of Mechanical Engineers, p 379–390
151. Ghazimirsaied A, Shahbakhti M, Koch CR (2010). Recognizing partial burn operation in an HCCI engine. In: 2010 Combustion Institute-Canadian Section (CICS) spring technical conference, 9–12, May 2010
152. Ghazimirsaied A, Shahbakhti M, Koch CR (2009) Partial-burn crank angle limit criteria comparison on an experimental HCCI engine. In: Proceeding of Combustion Institute-Canadian Section spring technical meeting, University of Montreal, Quebec, p 11–13, May 2009
153. Jungkunz AF, Ravi N, Liao HH, Erlien SM, Gerdes JC (2015) An analytical method for reducing combustion instability in homogeneous charge compression ignition engines through cycle-to-cycle control. Int J Engine Res 16(3):485–500

Chapter 7
Performance Characteristics

Abstract Low temperature combustion (LTC) engines use emerging combustion strategies that offer the potential for higher fuel conversion efficiency along with ultralow NO_x and PM emissions. This chapter presents the performance characteristics of LTC engine employing gasoline-like fuels. The LTC operating range is typically constrained by several limiting factors such as combustion noise, combustion instability, maximum cylinder pressure, oxygen availability, excessive reactivity and emission limits. Present chapter describes the operating range (in terms of engine speed and load) achieved by different LTC strategies (HCCI, PPC and RCCI) within operating limitations. Performance parameters such as thermal efficiency, fuel consumption and exhaust gas temperature are discussed for utilization of different fuels in LTC engines. Experimental data showed that high octane fuels can be best utilized in compression ignition engines with higher fuel conversion efficiency (~ 57% indicated thermal efficiency). Maximum indicated mean effective pressure (IMEP) up to 20 bar with well-mixed HCCI combustion and IMEP up to 25 bar in PPC strategy can be achieved, while keeping the NO_x and soot level below Euro VI limits and ringing intensity (RI) in acceptable level. The dual-fuel RCCI combustion also showed more than 58% indicated thermal efficiency.

Keywords LTC • Ethanol • Efficiency • Gasoline • IMEP • RCCI • PPC • HCCI • Operating range • Methanol • Ringing intensity • Knock • Misfire • Cyclic variations

7.1 LTC Operating Range

Automotive vehicles operate over a wide range of speed and load conditions along with transient operating conditions. Therefore, it is essential to operate the engines used in the vehicles over a significantly large range of engine speed and load conditions (considering the gear ratio of transmission system of vehicle). In fully premixed LTC strategies (such as HCCI), engine load range obtained is comparatively very low due to leaner air–fuel mixture (dilution by air or EGR) operation in order to control the combustion rate and NO_x emissions. In LTC strategies, maximum

R.K. Maurya, *Characteristics and Control of Low Temperature Combustion Engines*, Mechanical Engineering Series, https://doi.org/10.1007/978-3-319-68508-3_7

achievable engine load is limited by several constraints depending on the approach used to achieve combustion in the engine cylinder. The LTC engine operating limitations and LTC operating range are discussed in the following sections.

7.1.1 Operating Limitations

7.1.1.1 Ringing and Combustion Noise Limits

In the well-premixed charge compression ignition (like HCCI), very high heat release rate occurs due to simultaneous autoignition of the entire charge in the cylinder. High heat release characteristics of HCCI combustion result into very high-pressure rise rate. Thus, induced combustion noise from HCCI engine can be very high. Maximum achievable engine load is often limited by very high combustion rate. To operate the engine at higher load, comparatively larger quantity of fuel is burned in the cylinder, which increases and intensified the HCCI combustion rate and progressively leads to unacceptable noise due to severe knocking in the cylinder. These phenomena can cause engine damage during longer operation, and also lead to higher NOx emission. Therefore, higher HCCI engine load limits are often determined by knocking combustion limits. Detailed discussion on HCCI engine knock is given in Sect. 6.4.1 of Chap. 6. Pressure rise rate in the combustion chamber provides a measure of combustion noise. Maximum acceptable threshold limit of pressure rise rate is not able to capture the real combustion noise in some of HCCI engine operating conditions such as higher intake pressure operating conditions [1].

Analysis of the relation between combustion-generated pressure waves and the audible noise in HCCI engine showed that combustion-generated pressure oscillations lead to vibration of engine structure, which produces audible noise, and the phenomenon is known as ringing [2]. Ringing process in HCCI is basically different from knocking which occurs in spark ignition engine due to autoignition of end-gas unburned charge. In HCCI combustion, autoignition occurs uniformly in the entire combustion chamber. Pressure waves created by both knocking and ringing can be captured by cylinder pressure sensor installed in the combustion chamber. Typically, most of the acoustic wave energy during ringing is in the frequency range of 5–6 kHz while in case of knocking more energy in the frequency range of 8–25 kHz [3]. Pressure oscillation amplitude in HCCI combustion is larger, and it could be an order of magnitude higher in comparison to SI engine combustion knocking. Engine cylinder liner can attenuate the pressure wave energies less than 8 kHz in the significant amount [2], which creates the difficulty in ringing detection by externally mounted sensors.

The correlation for ringing intensity is given by Eq. 6.12 in Chap. 6. The ringing intensity (RI) considers the peak pressure rise rate, peak in-cylinder pressure, maximum temperature, and engine speed for calculation, and thus, it is a more generic metric than pressure rise rate. Detailed discussion on knock metric is provided in Sect. 6.4.2 of Chap. 6. Table 6.2 describes the various cylinder pressure-based metrics used for defining HCCI high load limit along with their limiting value. Acceptable RI value mostly used is less than 5 MW/m^2 [4–6] and in some of the studies 5 MW/m^2 [4–6].

Peak pressure rise rate is another method to quantify the harshness of HCCI combustion. A time-based peak pressure rise rate threshold is proposed, and recommended value of peak pressure rise rate is less than 5 MPa/ms for normal HCCI combustion [7]. This study concluded that sound emitted from the structural vibration is excited by the first few pressure waves of cylinder gas vibration and it was not due to the transmission of the cylinder pressure wave through the engine structure. Peak pressure rise rate threshold value of 5 MPa/ms is used to define the high load limit in HCCI engine in recent studies [8, 9]. In terms of crank angle position, maximum acceptable peak pressure rise rate is often used in the range of 5–12 bar/CAD depending on the engine speed.

7.1.1.2 Combustion Instability Limits

In combustion stability limits, two types of constraints can be defined as misfire limit and cyclic variability limits. Misfire cycle appears when fuel is not able to reach high temperature heat release process due to insufficient conditions required in the cylinder for autoignition. Large amount of unburned fuel emitted and very small amount of power or no power are produced in misfire or partial burn cycles. Figure 7.1 demonstrates the partial burn and misfire cycles in an unstable HCCI engine operating conditions. Section 6.5.3 of Chap. 6 provides the criteria for determination of partial burn and misfire cycles. Typically, a combustion cycle is defined as partial burn cycle if it has less than 90% total heat release of the previous cycle [10]. Using these criterion, another study defined HCCI operating conditions as partial burn operating conditions if it consists of more than 14% partial burn cycles and misfire operating conditions when it contains more than 30% partial burn cycles [11].

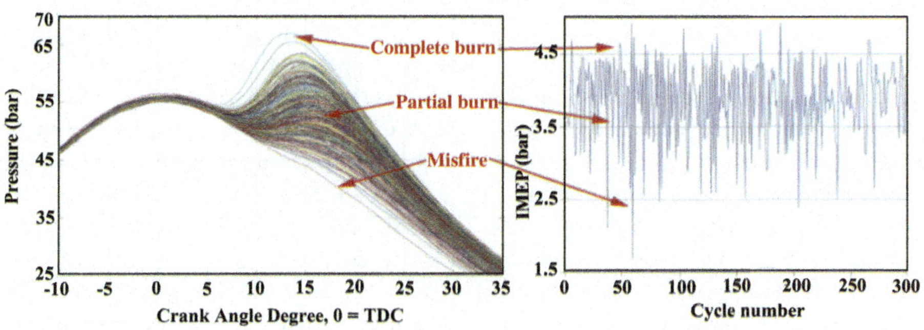

Fig. 7.1 Cylinder pressure and IMEP variations in HCCI combustion at an unstable engine operating condition for 300 consecutive engine cycles [3]

Misfire operating condition puts a constraint on maximum achievable engine load in HCCI combustion. To avoid the excessive ringing, HCCI engine is operated at retarded combustion phasing at higher engine loads in order to burn higher amount of fuel. In some of the test conditions, it is required to delay the combustion phasing (late in the expansion stroke) close to misfire point [3]. At this point, misfire occurs due to a quenching of oxidation reactions before the high temperature heat release. Quenching of oxidation reactions occurs due to low temperature of charge in cylinder because of cylinder volume expansion by movement of piston [12].

The relative pressure fluctuations in the intake manifold are expected to decrease in the HCCI engine due to its unthrottled operation leading to lower cylinder temperature and turbulence fluctuations. The spark ignition combustion is very sensitive to the local charge conditions such as air–fuel ratio, turbulence and mean velocity because ignition starts from a small kernel created between the spark plug electrodes. However, HCCI combustion is much less sensitive to the local charge conditions because ignition occurs in multiple ignition kernels, which results in lower cyclic variations in combustion parameters [13]. However, cyclic variations can be quite large in some of the operating conditions particularly at late combustion phasing operations. Retarding the combustion is more difficult at higher engine speeds as time window for proper combustion is smaller. Detailed characterization of cyclic variability in LTC is provided in Sect. 6.5.2 of Chap. 6. Researchers found that there is lower risk of misfire for the engine operating conditions where coefficient of variation (COV) of IMEP is below 3.5% [1, 13, 14]. At lower engine loads near idle operation, combustion is limited by the standard deviation (STD) of IMEP instead of COV, and the limit is 15 kPa of IMEP for stable combustion [13, 14]. In some of the studies, stable engine load is defined as the operating conditions where standard deviation of IMEP is below 2% [15].

7.1.1.3 Emission Limits

Engine operating conditions with excessively high emissions or very lower overall efficiency should be avoided. The engine control system design may consider these operating limitations as part of its control strategy [3]. Unburned HC and CO emissions are higher in HCCI engine for engine operating conditions leading to lower combustion temperature like lean mixture operation or excessively delayed combustion timing (detailed discussion is provided in Sects 8.2 and 8.3 of Chap. 8). Soot emissions are generally ultralow in HCCI combustion; however it can be important for other LTC strategies, where fuel stratification is used to extend the engine operating load range such as PPC and RCCI. Soot emission limits are also required to define in HCCI combustion where direct injection is used for premixed charge preparation. A HCCI combustion study used filter smoke number (FSN) limit for maximum soot level which is 0.1 FSN for a single operating point and 0.05 FSN for the average results from a complete speed and load range test [1, 13]. Detailed soot emission characteristics are discussed in Sect. 8.4.1 of Chap. 8. At richer fuel–air mixture or advanced combustion phasing operating conditions, NO_x emissions are higher in LTC combustion (detailed discussion can be found in Sect. 8.1 of Chap. 8). Typically in HCCI engines (well-mixed homogeneous charge), the ringing limit is encountered before the NO_x limit [3].

7.1.1.4 Maximum Cylinder Pressure Limits

Maximum cylinder pressure limit is normally a design consideration with respect to stress produced in various engine components. These limits are specific to engine and typically provided by the engine manufacturer. There is risk for structural damage of the engine when this design limit is exceeded at particular engine operating condition. The maximum cylinder pressure limits are evaluated by the high-pressure operation capacity of several engine components such as the piston rings and head gasket that can fail at extreme pressure in the cylinder [3]. The maximum cylinder pressure can become a limiting factor in several engine operating conditions such as (1) high intake pressure along with high equivalence ratios, (2) excessive ringing, (3) very advanced combustion phasing and (4) more fuel present in a given cycle due to previous misfiring cycle [3].

7.1.1.5 Excessive Reactivity Limits

High load operation in HCCI engine can be limited by excessive reactivity of fuels due to autoignition of charge before the desired combustion phasing. Advanced combustion phasing may lead to very high-pressure rise rate and high peak cylinder pressure. To achieve autoignition in HCCI engines, several techniques are used such as higher compression ratios, higher intake temperature and pressure, retaining hot residual gas in cylinder or using higher reactivity fuels. Fuels having significant amount of low temperature heat release (LTHR) or intermediate temperature heat release (ITHR) reach more easily to excessive reactivity limits because these fuels can autoignite before desired ignition timing as high reactivity fuels have requirement of low intake temperature [3]. Detailed discussion on single- and two-stage heat release fuels along with their autoignition chemistry is provided in Sect. 2.2.2 of Chap. 2. Single-stage heat release fuels like ethanol, methanol and iso-octane, having little or no LTHR and ITHR, require higher intake temperature and pressure for autoignition and thus may never face the excessive reactivity limit.

Fuels like gasoline show less reactivity (like single-stage fuel with no LTHR or ITHR) in naturally aspirated conditions or lower intake pressures. At higher intake pressure, gasoline shows higher reactivity and more Φ sensitivity (Fig. 2.15; Chap. 2). Therefore, gasoline requires lower intake temperature for autoignition at desired combustion phasing in HCCI engine at higher intake pressure. Figure 7.2 shows the intake temperature requirement as function of combustion phasing for different intake pressure using gasoline. Figure 7.2 depicts that the requirement of higher intake temperature is reduced to achieve a particular combustion phasing at

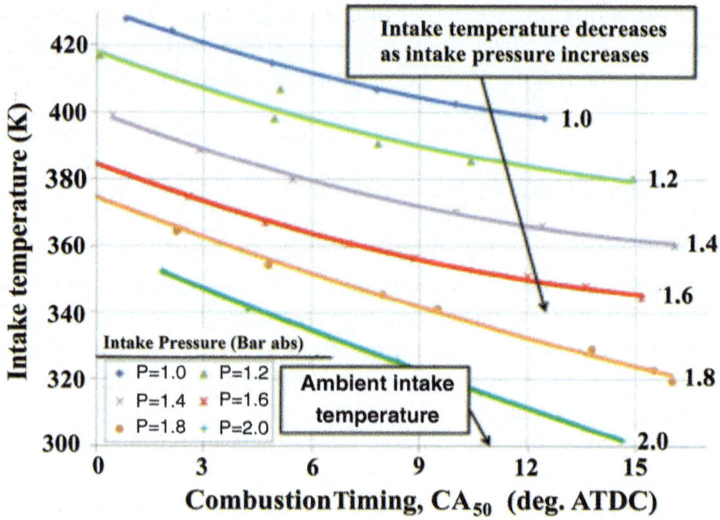

Fig. 7.2 Intake temperature requirement as a function of combustion phasing for different intake pressure at $\Phi = 0.45$ at 1800 rpm [3]

higher intake pressure. At higher intake temperature, the excessive reactivity limit can be encountered for these conditions.

At lower engine speeds, high reactivity fuels with LTHR can have intake temperature requirement below ambient temperature, which is impractical operating condition [16]. EGR can be used to address this issue and manage the amount of LTHR with engine speed. The EGR is found to an effective method for controlling the amount of LTHR over a wide range of engine speeds in HCCI combustion [16, 17]. Practically, intake temperature can be used to control the combustion phasing up to the air temperature (typically 60 °C) on the exit side of compressor in turbocharger [18]. The excessive reactivity limits are typically encountered for two-stage heat release fuels. In case of single-stage fuels, generally other limits appear before the excessive reactivity limits.

7.1.1.6 Oxygen Availability Limits

Generally, oxygen availability limit is encountered in LTC for the conditions using EGR as diluent to control the pressure rise rate. At sufficient intake boost, the charge becomes overly reactive, and EGR is required to prevent overly advanced combustion phasing. The amount of EGR required increases with boost pressure and with engine load for a given boost. As a result, at higher boost pressures, the amount of EGR can become so high that it limits the oxygen available for combustion. Thus, the maximum load that can be achieved at a given intake pressure with high EGR operation is limited by oxygen availability [5]. Studies demonstrated higher power out in HCCI engine by achieving IMEP up to 16.34 bar with gasoline [18] and 18.1 bar with E10 [6], by increasing the boost pressure and using EGR to retard the combustion phasing. This strategy is used up to the oxygen availability limits, where further fuel addition does not lead to increased power output due to incomplete combustion by lack of oxygen in the engine cylinder [3]. For some particular engine, maximum cylinder pressure limit may be encountered before the onset of oxygen availability limits in order to achieve higher engine load in HCCI combustion.

7.1.2 LTC Operating Range

Power output of a HCCI combustion engine depends on several factors at a particular engine operating conditions. Engine load/power output is typically expressed as indicated mean effective pressure (IMEP) because it allows the comparison of different engine sizes and technologies. Main parameters affecting IMEP of HCCI engine are expressed in Eq. (7.1) [19]:

$$\mathrm{IMEP} = f\left(\phi, P_{\mathrm{in}}, \mathrm{CA}_{50}, N, \dot{Q}_{\mathrm{out}}, \mathrm{fuel}, \%\mathrm{EGR}\right) \tag{7.1}$$

where ϕ is the equivalence ratio, P_{in} is the intake charge pressure, CA_{50} is the combustion phasing, N is the engine speed, Q_{out} is the rate of cylinder wall heat loss, fuel is the type of fuel used and %EGR is the fraction of exhaust gas recirculation included in the intake charge.

Figure 7.3a shows the effect of combustion phasing (CA_{50}) on IMEP for different relative air fuel ratio (λ) at 1200 rpm using gasoline in a HCCI engine. Combustion phasing is advanced with increasing the intake air temperature (Fig. 7.3a) because the charge reaches to autoignition temperature early by compression at higher intake air temperature. Figure 7.3a depicts that IMEP decreases as combustion phasing is retarded for particular λ. For retarded combustion phasing, combustion temperature is lower, and higher amount of unburned fuel is emitted due to lower combustion efficiency. For very advanced combustion phasing (before TDC), IMEP may decrease. Figure 7.3b depicts the variation of IMEP with oxygen to carbon ratio (O/C) in the intake charge for different fuels in HCCI combustion. At each test point, combustion phasing is chosen for best indicated thermal efficiency.

The (O/C) ratio is calculated as total oxygen (in air and fuel) (w/w) in the charge to total carbon in the fuel. Ethanol, methanol and butanol contain significant amount of oxygen in the fuel. Figure 7.3b illustrates that IMEP decreases as charge becomes leaner (increase in O/C ratio) due to lower quantity of input energy in the cylinder. At particular operating condition (O/C ratio), ethanol and methanol have higher IMEP compared to gasoline in HCCI engine.

Maximum engine load (IMEP) obtained in HCCI engine is constrained by operating limits discussed in Sect. 7.1.1. Figure 7.4 shows the stable HCCI

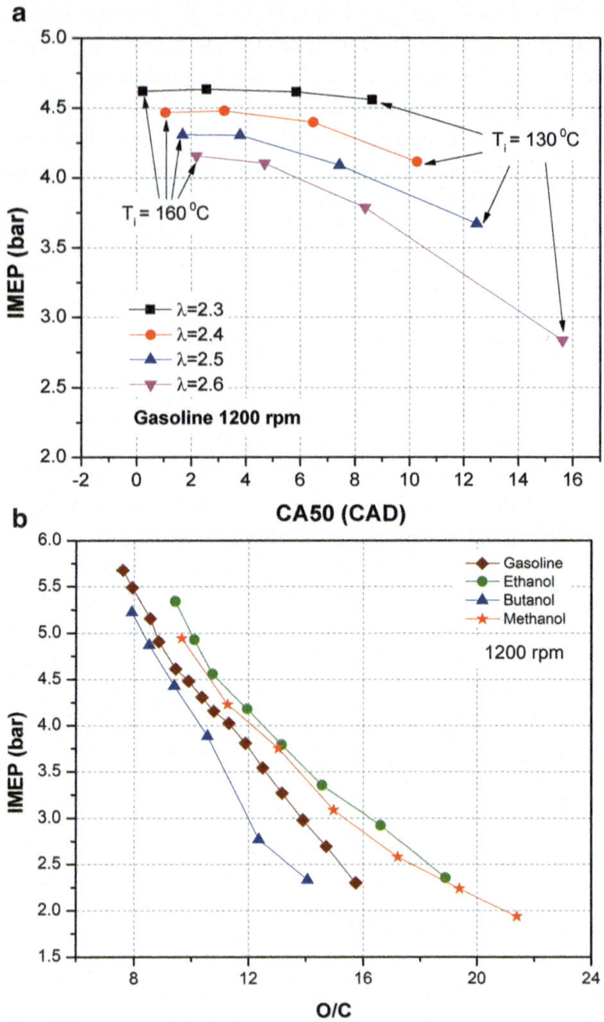

Fig. 7.3 (**a**) IMEP as function of combustion phasing using gasoline, and (**b**) IMEP as function of oxygen to carbon ratio at optimal combustion phasing for different fuels at 1200 rpm in HCCI engine at naturally aspirated conditions (Adapted from [20, 21])

operating range obtained by using RI and cyclic variability limits at higher and lower engine load, respectively, in gasoline-fuelled engine at naturally aspirated conditions. Knock region is defined as engine operating conditions where RI is greater than 6 MW/m^2. Misfire region is defined as engine operating conditions where COV of IMEP is greater than 3.5%. The HCCI operating range constitutes the engine operating points where RI < 6 MW/m^2 and COV_{IMEP} < 3.5% (coloured region in Fig. 7.4). Constant IMEP lines are inclined more horizontally, which indicates that IMEP is mainly dependent on λ. Maximum IMEP is found for lowest λ (richest mixture) and intake temperature (Fig. 7.4) at the apex of wedge-shaped region. The maximum attainable IMEP can be improved by shifting this wedge-shaped region towards the origin of the graph so as to increase the amount of fuel

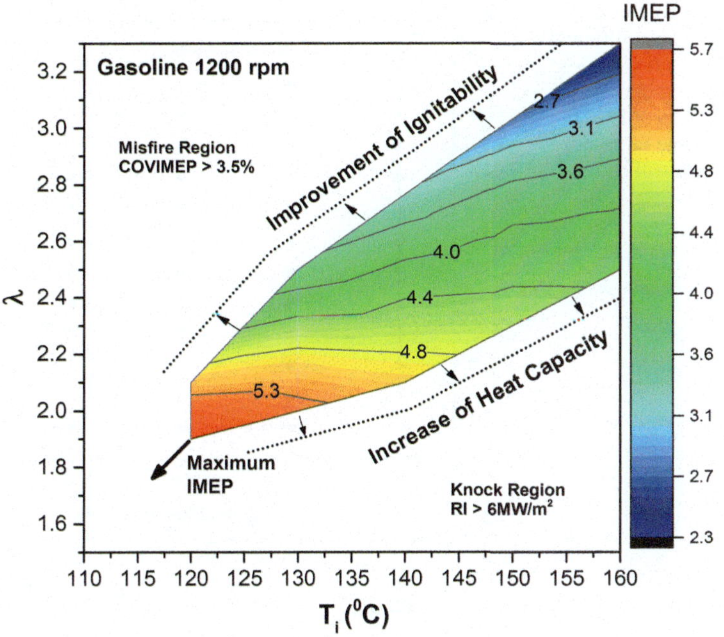

Fig. 7.4 Stable HCCI operating range at 1200 rpm using gasoline (Adapted from [20])

that is supplied at the apex of the wedge. To increase the HCCI operating range, two strategies can be employed (as indicated in Fig. 7.4): (1) increasing the heat capacity of the in-cylinder gas and (2) improving the ignitability of the mixture. For increasing the heat capacity of the in-cylinder gas, supercharging of fresh air or inducting EGR (in the intake manifold) is effective. This method is well known as one of the important ways to increase maximum torque in HCCI engine [22]. In case of the fresh air supercharging, the fuel-air ratio at knocking limit does not decrease because of the additional oxygen in the fresh air though. Several measures can be taken to improve the ignitability of mixture such as utilization of low octane number fuel, fuel reformulation in negative valve overlap, ozone addition in intake air and increasing the compression ratio of the engine [22].

Figure 7.5 shows the HCCI operating range at different engine speeds limited by operating constraints. Figure 7.5a shows the range of λ for which stable HCCI operation can be achieved using ethanol at different engine speed for compression ratio 17, 19 and 21. Operating range is limited by combustion efficiency (>85%) for lower engine load operation, and maximum pressure rise rate (<5 MPa/ms) for higher engine load operation. Engine operating conditions satisfying this criteria are considered in HCCI operating range. For each λ, intake air temperature corresponding to best indicated fuel conversion efficiency is selected. Figure 7.5a depicts that higher engine speed has comparatively shorter range of λ for each compression ratio. There is comparatively less time available for combustion at higher engine speeds because engine cycle completes in shorter duration. To

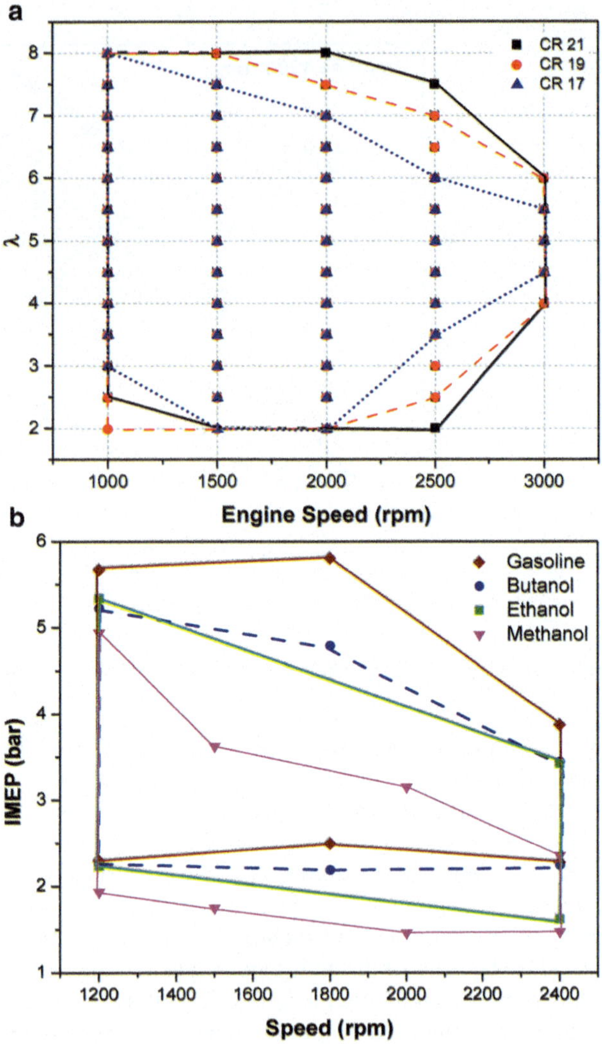

Fig. 7.5 (**a**) HCCI operating range at different compression ratios for ethanol [8] and (**b**) the comparison of HCCI operating range of gasoline, ethanol, methanol and butanol [21]

increase the oxidation rate of reactions, higher intake temperature is required to compensate for shorter combustion duration. Higher intake temperature leading to faster combustion rate leads to faster encounter of RI limits. For constant RI limit, engine has to operate on leaner mixtures at higher intake air temperature, which leads to lower IMEP at higher engine speed (Fig. 7.5b) [8].

Figure 7.5a also depicts that HCCI operating region with respect to λ is larger for higher compression ratio. Combustion temperature increases at higher compression ratio, which can autoignite the leaner mixtures, and thus, leaner range increases more at higher compression ratios (Fig. 7.5a). Figure 7.5b shows the IMEP range obtained at different engine speed for gasoline, ethanol, methanol and butanol. RI limit of 6 MW/m^2 and COV_{IMEP} limit of 3.5% are used to determine the operating range. In these operating conditions, combustion is phased at best indicated thermal efficiency by adjusting the intake air temperature. Maximum IMEP obtained is 5.8 bar for gasoline, which is the typical maximum IMEP that can be achieved with HCCI combustion in naturally aspirated conditions. Maximum IMEP obtained from HCCI combustion is basically dependent on the amount of fuel that can be burned in the cylinder by simultaneously satisfying the operating limitations discussed in Sect. 7.1.1. A study selected three parameters energy input per cycle (J/cycle) (magnitude parameter), P_{max} (bar) (magnitude parameter) and CA_{Pmax} (CAD aTDC) (location parameter) to characterize the IMEP variation in HCCI engine [23]. The correlation which is the best fits to the experimental data is expressed by Eq. (7.2):

$$\text{IMEP} = a + b(\text{Energy}) + c\,(P_{\max}) + d\,(\text{CA}_{\text{Pmax}}) \tag{7.2}$$

where a, b, c and d are constants, which can be calculated by regression analysis of experimental data. Empirical correlation has good coefficient of determination (R^2 > 0.95) [23]. Study also showed that this correlation can predict the IMEP with uncertainty of 0.3 bar in naturally aspirated HCCI combustion engine.

Maximum IMEP achieved in naturally aspirated condition is very low in comparison to conventional SI or CI engine. In automotive vehicles, maximum BMEP is about 12 bar in light-duty spark ignition engines and 18 bar for light-duty compression ignition engines [3, 24]. For heavy-duty compression ignition engines, maximum BMEP can be as high as 25 bar. In naturally aspirated HCCI operation, maximum IMEP obtained is around 5 bar in the typical engine speed range of 1000–1500 rpm, which is about 40% and 30% of the possible loads in spark ignition and diesel engines, respectively, in limited speed range [3]. Initial efforts to employ intake boosting to HCCI engine for higher loads while maintaining high dilution levels lead to only limited success due to excessively high-pressure rise rate and engine knock [25, 26].

Recent studies demonstrated the engine loads of 16 and 18 bar $IMEP_g$ with ultralow NO_x and particulate emissions using well-mixed (WM) HCCI combustion using a regular (87 Anti-Knock Index (AKI)) E0 gasoline and an E10 ethanol blend (10% ethanol in gasoline), respectively [6, 18]. In these studies, the high-pressure rise rate is controlled by retarding combustion phasing through a combination of lower intake temperatures and well-controlled EGR addition. Higher engine load up to 20 bar $IMEP_g$ is achieved using the same approach with the fuel having increased ethanol content to 20% (E20) (93 AKI) [27]. For these high-load well-mixed HCCI studies, NO_x and particulate emissions are more than factor of 10 below US 2010 limits for all boosted conditions. However, the boost pressures required to achieve these high engine loads are fairly high, around 3.2–3.6 bar absolute, which is very difficult for a turbocharger to produce these boost levels at the stated loads in a high-efficiency LTC engine that has a relatively low-enthalpy exhaust stream [5].

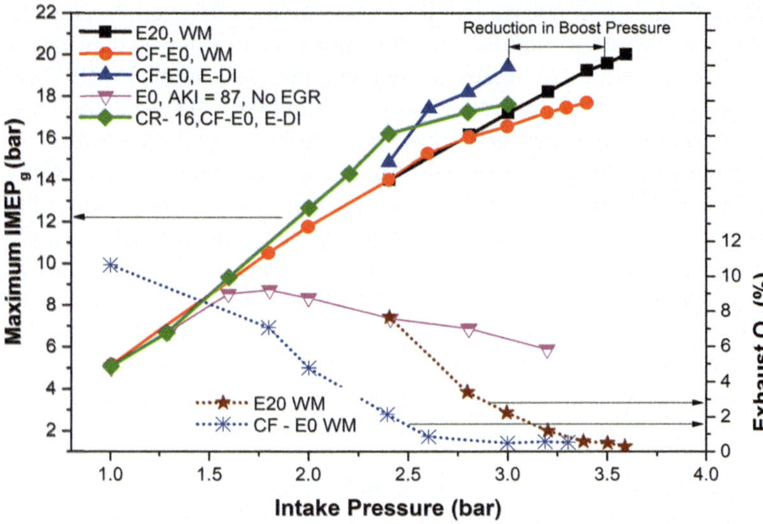

Fig. 7.6 Maximum IMEP as a function of intake boost pressure using different strategies at compression ratio (CR) 14 at 1200 rpm in low-temperature gasoline combustion (LTGC) (Adapted from [5, 27])

Figure 7.6 demonstrates the maximum achieved IMEP as a function of intake boost pressure by using different fuel and fuel injection strategies at compression ratio (CR) 14 at 1200 rpm. Figure 7.6 depicts that maximum IMEP is increased with increase in boost pressure for well-premixed (WM) charge (HCCI) as well as early direct injection (E-DI) operation. All boosted data points are acquired within limit of RI = 5 MW/m^2 to maximize thermal efficiency while avoiding knock during combustion. In engine operation without EGR, maximum IMEP decreases after intake boost pressure (P_{in}) of 1.8 bar (Fig. 7.6) due to higher-pressure rise rate, and RI will be higher if more amount of fuel is burned. The zero ethanol (E0) 87 AKI fuel is the most reactive fuel for intake boosted operation, and EGR addition starts with the P_{in} = 1.8 bar as evident from the sharp decrease in exhaust O_2 concentration from P_{in} = 1.6–1.8 bar. After this point, increase in intake pressure increases the requirement of EGR to prevent overly advanced combustion timing [5].

At P_{in} = 2.6 bar using CF-E0 (certification gasoline) WM (well mixed) case, the exhaust O_2 concentration is nearly zero, which means the charge is stoichiometric with a very high level of dilution from the EGR. Therefore, the maximum load for E0 is oxygen limited for $P_{in} \geq 2.6$ bar while knock/stability limited for $P_{in} < 2.6$ bar. Even though for $P_{in} \geq 2.6$ bar combustion is oxygen limited, a maximum load of 16.3 bar IMEP$_g$ is obtained at P_{in} = 3.2 bar, where maximum cylinder pressure limit is encountered. Ethanol addition to E0 gasoline lowers its autoignition reactivity in boosted conditions, shifting to higher pressures (P_{in}) for which the maximum load becomes oxygen limited (Fig. 7.6). For 20% ethanol content (E20), IMEP increases up to 20.0 bar at P_{in} = 3.6 bar, where maximum cylinder pressure limits are reached and simultaneously exhaust O_2 also reaches to zero at the same point (Fig. 7.6).

The partial fuel stratification (PFS) strategy creates controlled non-uniformities in the charge mixture, which is used to increase the high load limit by increasing the burn duration and reducing the peak HRR (see Sect. 2.2.4.6 of Chap. 2). The PFS strategy results in sequential autoignition from the richest regions to the leanest if the fuel's autoignition reactivity is sensitive to the local equivalence ratio (ϕ) within the charge [28]. The PFS strategy also always produces some amount of associated thermal stratification for two reasons. First, DI fuelling involves vaporization cooling which decreases the local temperature proportionally to the increase in ϕ. Second, the larger quantity of fuel lowers the local γ of the richer regions, which in turn decreases their temperature rise with compression [4]. As a result of these two effects, the richest fuel regions are also the coldest, which acts to reduce their autoignition reactivity and thus works against the increased chemical reactivity of these richer regions for ϕ-sensitive fuels. In standard PFS (std-PFS) strategy, major quantity of the fuel (~90%) is well premixed (similar to WM strategy), and the remaining fuel is supplied by a late direct injection, typically around 310 CA (50° CA before TDC compression) to create a non-uniform fuel distribution in the cylinder. To avoid condensation of the premixed fuel, intake temperature (T_{in}) is kept at 60 °C, the same as for well-mixed (WM) strategy [5].

Figure 7.7 shows the comparison of WM and PFS strategy for achieving maximum load as function of intake pressure. For the E0 gasoline, PFS begins to be effective at $P_{in} = 1.8$ bar, as fuel becomes sufficiently ϕ sensitive at this P_{in} [6, 28], and it is very effective for $P_{in} \geq 2.0$ bar, where the fuel is strongly ϕ sensitive [28]. In PFS strategy, higher engine load is achieved (13 bar IMEP at

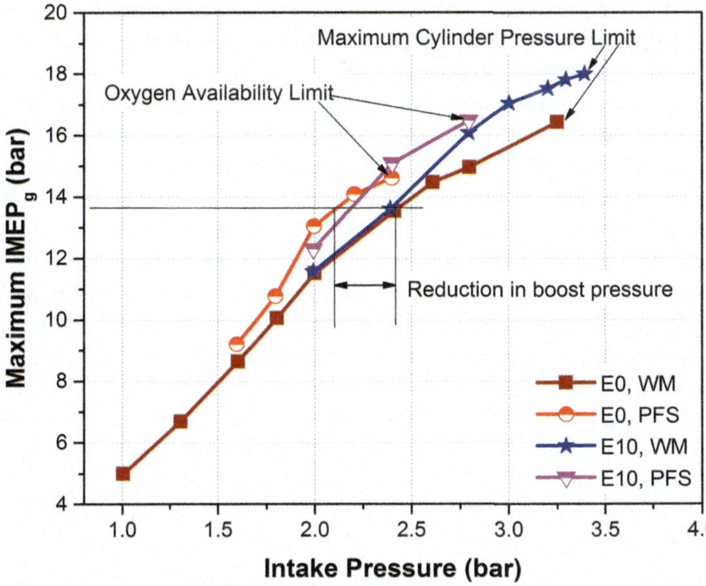

Fig. 7.7 Variation of maximum IMEP as a function of intake pressure for premixed and PFS strategy in HCCI combustion (Adapted from [6, 28])

$P_{in} = 2.0$ bar) in comparison to WM strategy (11.5 bar IMEP at $P_{in} = 2.0$ bar). For boost pressure $2.0 \leq P_{in} < 2.4$ bar, the load at each P_{in} is higher for std-PFS until it was oxygen limited. Since the maximum load is already oxygen limited with premixed fuelling for $P_{in} \geq 2.6$ bar, std-PFS cannot be used to increase the load in this pressure boost range as stratification will create richer mixture than stoichiometric [5]. E10 is less reactive fuel than E0, which makes to achieve for higher engine load, but the trend is similar to E0. Figure 7.7 also depicts that for a constant load, std-PFS strategy requires comparatively lower intake pressure, which means that std-PFS strategy reduces the requirement of boost pressure.

The std-PFS strategy cannot increase the engine load for intake pressures above 2.6 bar as shown in Fig. 7.7 due to oxygen limit constraint. However, E-DI PFS can mitigate this problem, where all the fuel is injected directly into the cylinder in early intake stroke. Direct injection of fuel in E-DI PFS leads to operate the engine at lower intake temperature 30–40 °C without concern about fuel condensation in the intake system, which is 60 °C for well-mixed (WM) strategy. Additionally, in-cylinder vaporization of the DI fuel provides further charge cooling, and it also increases the charge density. Therefore, the charge mass inducted is also higher. In E-DI, start of injection (SOI) of 60 CA after TDC intake is selected. In this case also the charge is not completely mixed and acts similar to PFS. The E-DI PFS strategy leads to a substantial load increase at higher boost pressures (Fig. 7.6). Figure 7.6 also depicts that E-DI strategy can also reduce the requirement of higher boost pressure at high load operation. A more complete discussion on high load increase with std-PFS, E-DI and premixed fuelling can be found in original studies [5, 27, 28].

In order to increase the high load limit, heavy stratification is created using diesel-type direct injectors with both gasoline-like and diesel-like fuels in partially premixed combustion (PPC) strategy [29, 30]. In diesel PPC, NO_x and smoke emissions are higher particularly at higher engine loads. Therefore PPC strategy is focused more on gasoline-like fuels. Figure 7.8 shows the engine load range that can be achieved with PPC strategy using gasoline-like fuels, while keeping NO_x emissions below 0.30 g/kWh and soot below 0.30 filter smoke number (FSN). To achieve this high engine load (25 bar IMEP), intake air pressure up to 3.6 bar (depending on load) and EGR rate up to 50% are used with gasoline-like fuels of different octane numbers. A custom injection strategy is used to operate the PPC engine on gasoline-type fuels in low to high engine load [31]. This injection strategy used two injection pulses in a cycle. During first injection, fuel must not react during the compression stroke, and fuel is injected at −60 aTDC. To create the stratification in the engine cylinder, a second injection takes place near TDC, and it also triggers the combustion. Thus, the combustion timing is governed by the position of the second fuel injection. First fuel injection timing is kept constant for all the load, but fuel quantity depends on the fuel octane number (ON), CR and EGR rate. Figure 7.8a depicts the PPC load range achieved as function of octane number (RON) for unmodified fixed layout of the engine at constant intake temperature of 303 K. Lower load boundary depends on the fuel ON (Fig. 7.8a), and a study suggested that gasoline with 70 octane number is the best fuel [31]. Since

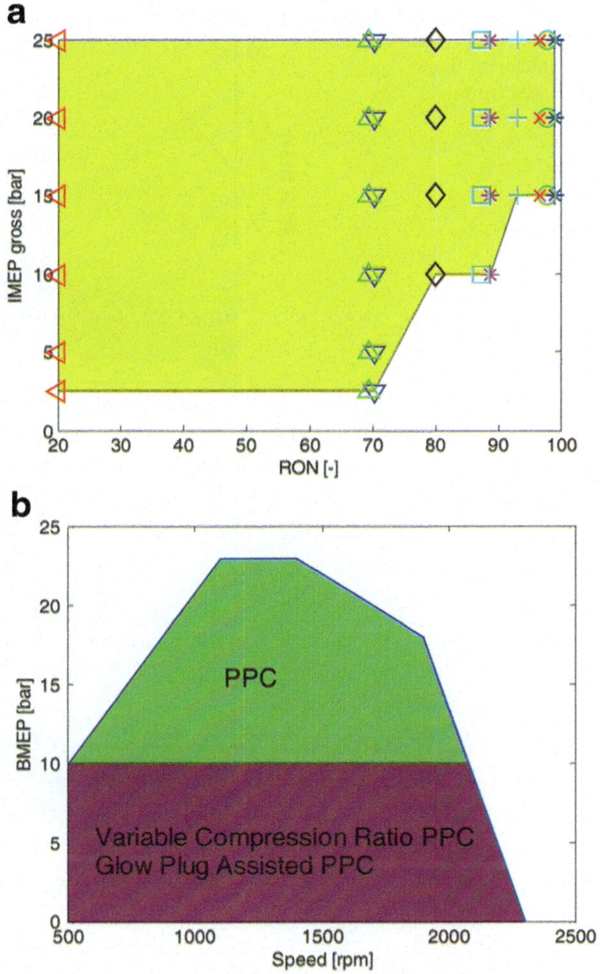

Fig. 7.8 (**a**) Stable operating load range for different types of gasoline fuel in PPC engine and (**b**) **h**ybrid PPC combustion concept using medium- or high-ON gasoline fuels [31]

gasoline of RON 70 is not available in the market, hybrid PPC mode operation is suggested (Fig. 7.8b) for currently available gasoline of higher ON than 70. Engine can be operated in PPC mode at higher engine load, and for lower engine load, variable compression ratio or glow plug-assisted PPC mode needs to be run. Higher octane fuel has difficulty in autoignition and thus requires higher compression ratio or glow plug assistance at lower engine load as boost pressure is also relatively low at lower engine loads.

The RCCI combustion is another LTC strategy, which requires relatively lower EGR rate and comparatively lower fuel injection pressure than PPC while maintaining ultralow NOx and soot emissions. Figure 7.9 shows the RCCI operating range vis-à-vis conventional diesel combustion (CDC) using different fuels. To determine RCCI operating range, a self-imposed constraint of pressure rise rate

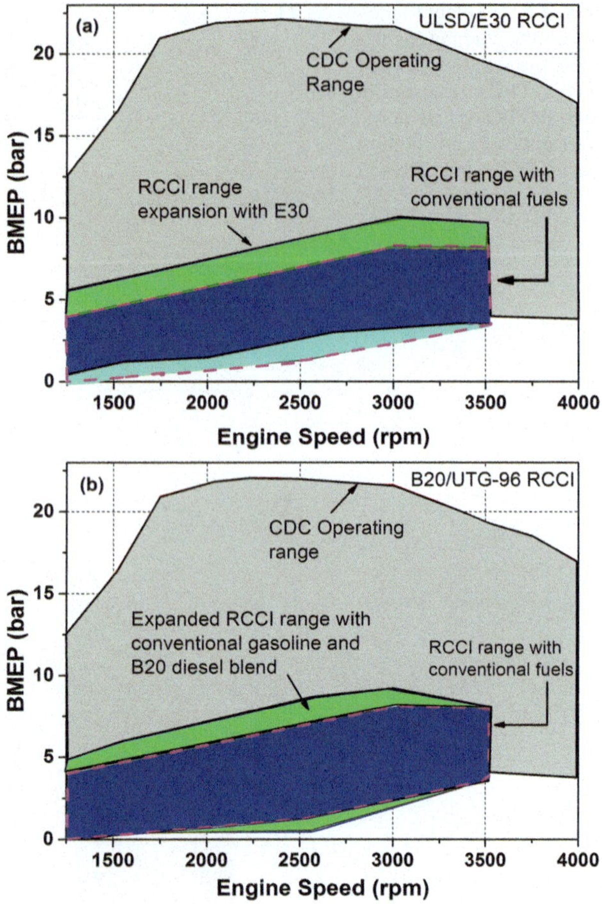

Fig. 7.9 Engine operating range for RCCI combustion with different fuels (Adapted from [32, 33])

(<10 bar/CAD) and CO emissions (<5000 ppm) is used in addition to COV of IMEP of 3%. During RCCI engine operation, an early direct injection of diesel (between 30 and 70 BTDC) and port injection of gasoline at a closed intake valve timing are used. Figure 7.9 depicts that RCCI operating range is comparatively lower than CDC. To achieve these engine load boost pressure up to 1.8 bar is used depending on the engine load. Figure 7.9 shows that RCCI operating range can be expanded by using ethanol and biofuel blends. The maximum BMEP up to 10 bar is obtained in this study. Figure 7.9b indicates that biodiesel/gasoline RCCI obtained higher load range in comparison to the RCCI engine with conventional fuels (gasoline and diesel). The premixing ratio of low to high reactivity fuels could be increased by using biodiesel without compromising the engine stability [32]. The high engine load operation requires higher amount of fuel burned in the cylinder, which leads to higher-pressure rise rate in case of premixed combustion. To avoid the higher-pressure rate, combustion phasing is delayed in order to achieve higher engine load. Figure 7.10 illustrates the RCCI combustion IMEP as function of

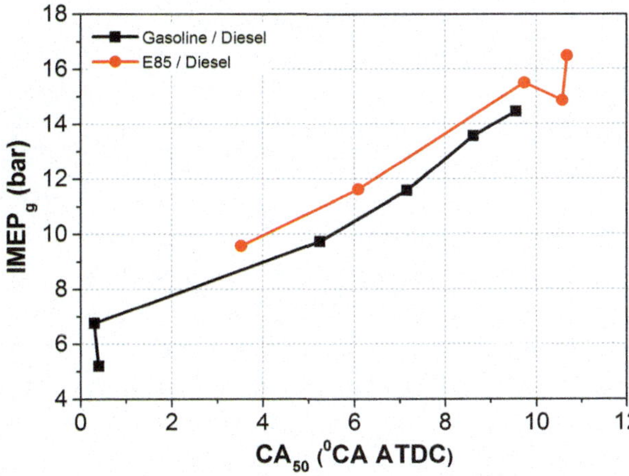

Fig. 7.10 IMEP as function of combustion phasing in RCCI engine fuelled with different low reactivity fuels (Adapted from [34])

combustion phasing for two different low reactivity fuels. In E85/diesel operation, combustion duration is approximately 50% longer due to the occurrence of longer HTHR (high temperature heat release) because of the greater RON stratification and the ethanol inhibitor effect [34]. Longer combustion duration leads to reduction in heat release rate, and thus more amount of fuel can be burned for the same peak pressure rise rate constraints. Maximum load obtained in E85 diesel RCCI is 16.5 bar at intake pressure of 2.46 bar and 47% EGR rate, whereas in gasoline/diesel RCCI, maximum IMEP is 14.5 bar at intake pressure of 2.34 bar and 57% EGR rate.

Load range extension is essential for RCCI engine operation over wide range of driving cycle map while maintaining the benefits of higher fuel conversion efficiency as well as ultralow NOx and particulate emissions. Optimization of the premixing ratio, injection events and fuel reactivity can lead to expansion of the RCCI engine operating range [35].

7.2 Engine Efficiency

The fuel conversion efficiency of an engine can be categorized into different process efficiencies in order to determine the processes, which are contributing negatively to overall engine efficiency. This also helps to compare the results between different engines or even different combustion modes. Figure 7.11 illustrates the different efficiencies defined for an engine combustion process along with associated losses. In an engine, chemical energy contained in the fuel is converted into heat energy released by combustion. Due to incomplete combustion, some of the chemical energy is lost in terms unburned or partially burned emission species. The combustion efficiency is defined as the ratio of heat released in the cylinder to total chemical energy injected from fuel. Thermodynamic efficiency is the ratio of

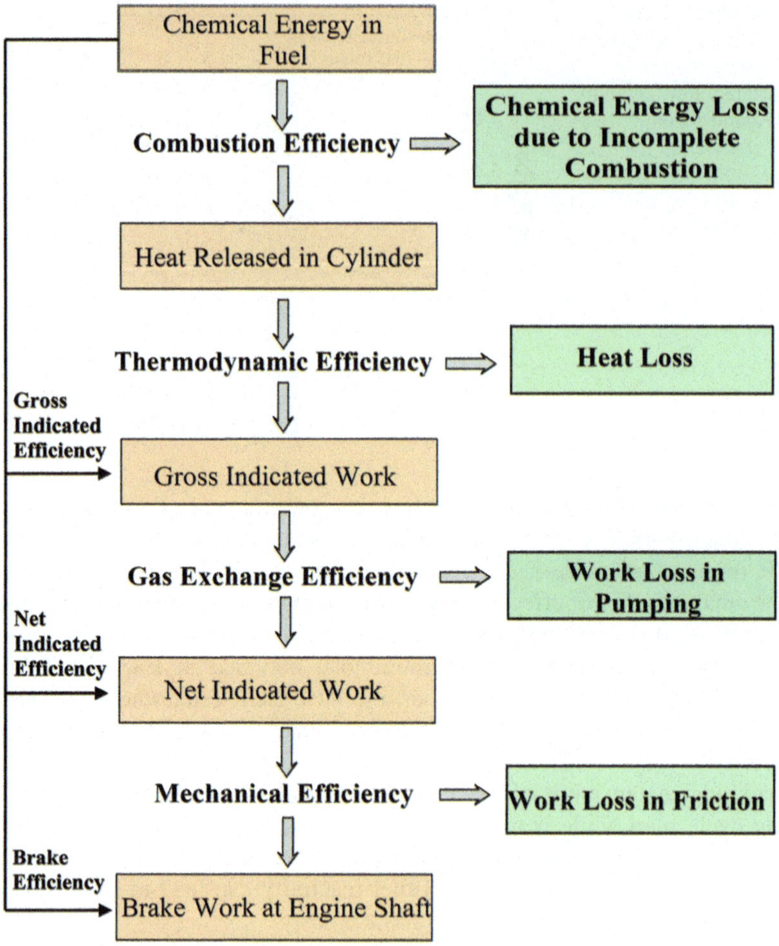

Fig. 7.11 Illustration of different efficiencies defined for the various process in a combustion engine (Adapted from [13])

indicated work to the total heat released in the engine cylinder. During the work conversion process, some of the heat released is lost in several forms. Thermodynamic efficiency is a measure of how good is the engine with respect to conversion of heat energy into work. Indicated work obtained is lost in pumping the intake and exhaust gases during intake and exhaust stroke, respectively. The efficiency of this process is measured by gas exchange efficiency. Some of the work is lost to overcome the friction in the different moving parts of the engine. The measure of this process is mechanical efficiency. The remaining work now is available as brake work at engine shaft. The brake thermal efficiency is defined as the ratio of brake work available at engine shaft to total fuel energy injected into the combustion chamber. The brake thermal efficiency is the product of combustion, thermodynamic, gas exchange and mechanical efficiencies.

In general, LTC process has higher fuel conversion efficiency potential in comparison to conventional SI engines. The LTC engine has eliminated the throttling losses, and thus pumping work loss is reduced. The LTC engine is operated at higher compression ratio than SI engines, which further improves the fuel conversion efficiency. The low combustion temperature leads to lower heat transfer loss, which also adds to efficiency improvement in LTC engines. In LTC engines, combustion duration is relatively shorter, which leads to combustion cycle close to constant volume cycles having higher efficiency. Thermodynamic efficiency of LTC engines is higher due to higher compression ratio and higher ratio of specific heat (γ) of working fluid [4].

Figure 7.12 illustrates the variations in thermodynamic, brake, combustion and gas exchange efficiencies in HCCI engine vis-à-vis SI engine. The thermodynamic efficiency presents the efficiency of the thermodynamic process of heating up the gas at high pressure, expanding the hot gas and extracting work. Figure 7.12a depicts that HCCI engines have comparatively higher thermodynamic efficiency in comparison to SI engine. Thermodynamic efficiency can reach 50% or even slightly higher in HCCI engine due to higher compression ratio and higher γ of working fluid (because of leaner mixture operation). Combustion efficiency is slightly lower in HCCI engine (detailed discussion on combustion efficiency can be found in Sect. 6.3 of Chap. 6). The overall brake thermal efficiency of HCCI engines is higher than SI engines, and it increases with increase in engine load (Fig. 7.12a).

At a particular compression ratio in HCCI engine, indicated thermal efficiency mainly depends on combustion efficiency and combustion phasing [37]. Figure 7.13 presents gross indicated thermal efficiency as function of combustion phasing in a gasoline HCCI engine at 1200 rpm in naturally aspirated conditions. Figure 7.13a depicts that engine efficiency decreases as combustion phasing is retarded (after 6 CAD) for each λ due to lower combustion temperature resulting in lower combustion efficiency. For rich mixture engine operation at $\lambda = 2.3$, combustion efficiency decreases with very advanced combustion phasing and lower than other leaner mixtures. At very advanced combustion phasing, heat release rate is very high, and higher combustion temperature is obtained in the cylinder. Thus, heat loss from cylinder wall increases resulting into relatively lower efficiency. At very advanced combustion phasing and richer mixtures, knocking may also start, which further increases the heat transfer, leading to lower engine efficiency. Therefore, optimum combustion timing depends on engine load (λ). Richer mixtures (higher load) have optimum combustion phasing later into the cycle, and for leaner mixtures (low load), optimum combustion phasing is closer to TDC. Figure 7.13b presents the indicated efficiency as function of IMEP for gasoline, ethanol, methanol and butanol for combustion timing phased for best thermal efficiency at each test conditions. Engine efficiency increases with the increase in engine load and reaches highest value and then starts declining (Fig. 7.13b). Efficiency declines at very high load due to higher combustion rate (starts knocking), which increases the heat transfer rate. Ethanol showed the highest efficiency at any given load for this engine at optimal combustion phasing (Fig. 7.13b).

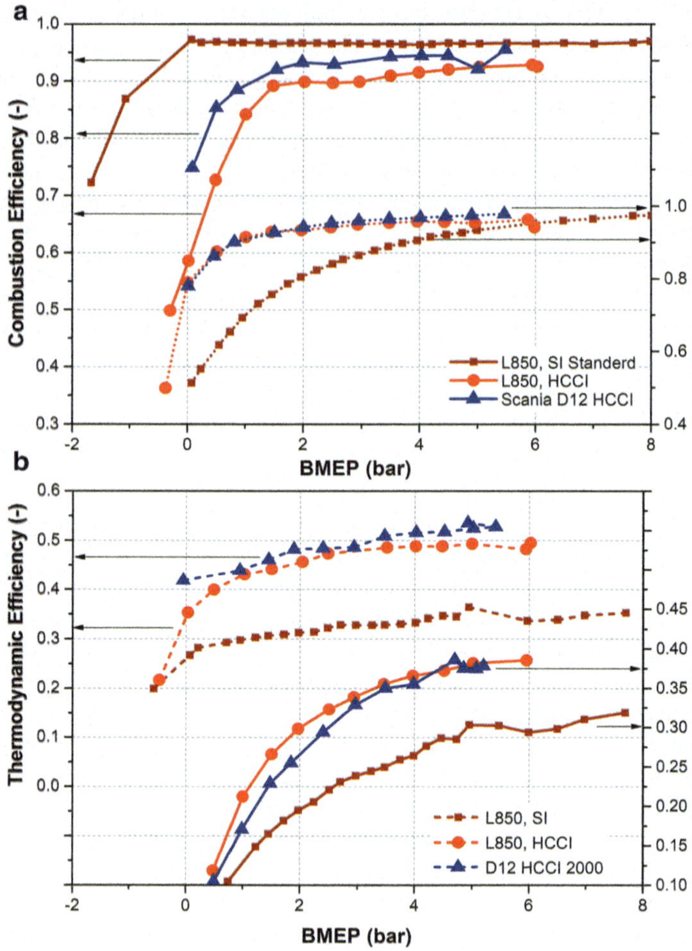

Fig. 7.12 (**a**) Thermodynamic and brake efficiency and (**b**) combustion and gas exchange efficiency as a function of BMEP in HCCI engine (Adapted from [36])

Figure 7.14 shows the variations of indicated thermal efficiency with λ and intake air temperature using gasoline and ethanol in HCCI operating range (RI < 6 MW/m^2 and COV$_{IMEP} < 3.5\%$). Figure 7.14 depicts that indicated thermal efficiency is a function of both intake air temperature and λ, because constant efficiency contour orientation is inclined. Maximum thermal efficiency in the HCCI operating range is found up to 48% for both fuels. Indicated thermal efficiency is lower near misfire boundary, and relatively higher indicated efficiency is found near the knock boundary. This trend is similar for both test fuels. Close to knocking boundary, fuel-air mixture is richer, and intake temperature is also higher, which results in higher combustion temperatures and comparatively advanced combustion phasing. Advanced combustion phasing and higher combustion efficiency due to

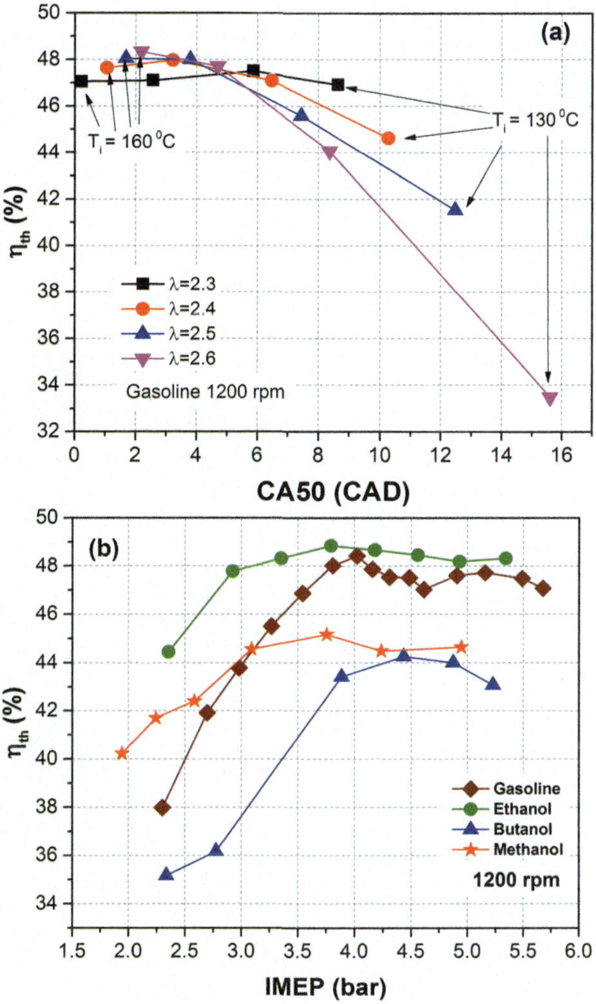

Fig. 7.13 Gross indicated thermal efficiency as a function of (**a**) combustion phasing and (**b**) IMEP at optimal combustion phasing for different test fuels, in HCCI engine [21]

higher combustion temperature lead to higher indicated thermal efficiency near the knock boundary.

To increase the load range of HCCI engine, intake pressure boost and/or fuel stratification (std PFS or E-DI) is typically used (see Sect. 7.1.2). In std-PFS strategy, majority of fuel is premixed, and thus it is required to maintain intake temperature (~60 °C) such that fuel does not condense in the manifold. However, in E-DI PFS, all the fuel is injected early in the intake stroke, and intake temperature can be reduced to ambient temperature. Many of the E-DI PFS with reduced intake temperature (T_{in}) also increases thermal efficiency along with increasing high load limit at particular intake pressure. Lower intake temperature in E-DI PFS leads to lower combustion temperature for the same amount of fuel burn (engine load) due to lower initial temperature. Lower intake temperature also increases the intake

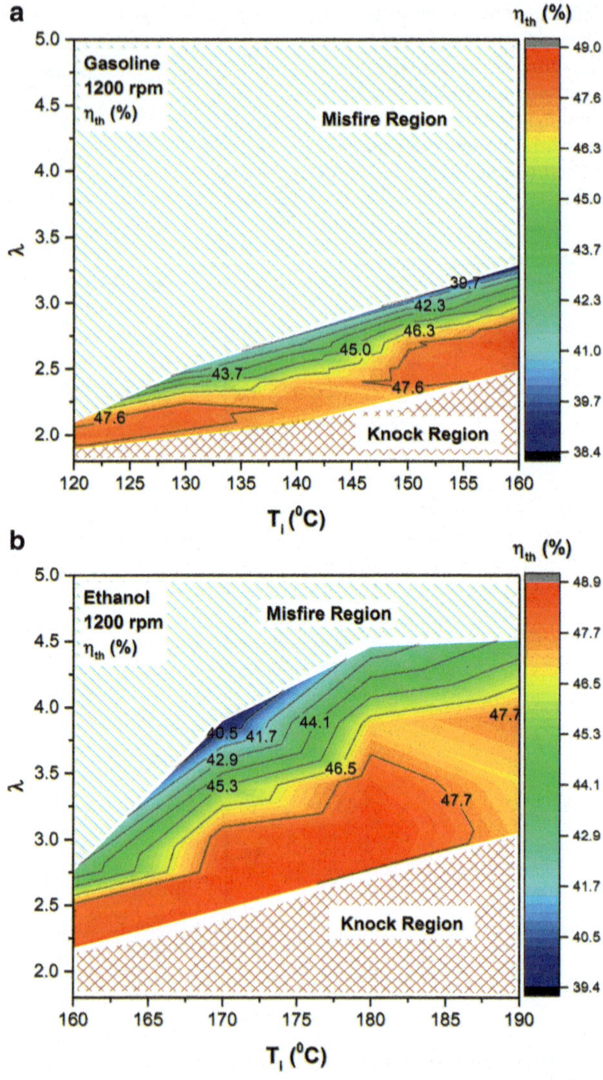

Fig. 7.14 Variations of gross indicated thermal efficiency in HCCI operating range using gasoline and ethanol [21]

mass density; thus same load can be produced with a more dilute charge which further contributes to lower combustion temperature. Lower intake temperature also significantly reduces the EGR requirement to delay the combustion phasing in order to avoid higher RI. The lower combustion temperature and lower EGR increase the specific heat ratio (γ) of the combustion products, which leads to more work extraction during the expansion stroke and increasing the thermal efficiency [18, 38]. The heat transfer losses are also reduced in E-DI PFS due to the lower combustion temperatures. Additionally, with PFS, less combustion timing retard is required to avoid the RI limit, which increases the effective expansion ratio, and therefore, higher thermal efficiency is achieved [5].

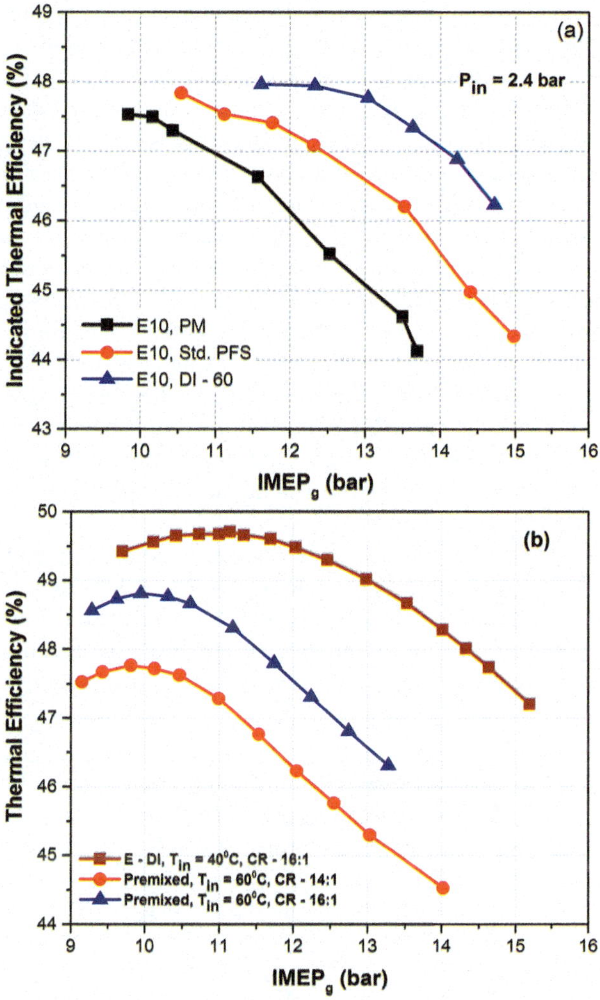

Fig. 7.15 Indicated thermal efficiency as a function of IMEP using different charge preparation strategy at (**a**) constant intake pressure (Adapted from [5, 27]) and (**b**) different compression ratios (Adapted from [5])

Figure 7.15 presents the indicated thermal efficiency as a function of IMEP using different charge preparation strategy in HCCI engine. Figure 7.15b depicts that E-DI PFS strategy has the highest thermal efficiency at constant intake pressure for a given load (particularly at high engine loads) than std-PFS and well-mixed (also known as premixed (PM)) strategy. Figure 7.15 also depicts that thermal efficiency decreases with increase in engine load due to more retarded combustion phasing at higher engine load to avoid the RI limit. Retarded combustion phasing decreases the effective expansion ratio, and thus, thermal efficiency is reduced. The well-mixed fuel strategy showed lower efficiency as more combustion phasing retard is required in the premixed (PM) case for a given load to avoid knocking. The PFS strategy increases the combustion duration, and the peak heat release rate is reduced. Therefore less combustion timing retard is required in PFS strategy.

Thermal efficiency can also be enhanced by increasing the geometric compression ratio (CR) of the engine. Higher compression ratio increases the expansion ratio, and thus higher thermal efficiency is obtained. Additionally at higher compression ratio, lower intake temperature is required for autoignition as it leads to higher temperature at the end of compression. Higher intake pressure enhances the autoignition reaction rate, and thus same combustion timing can be obtained with lower near-TDC temperatures [5]. This leads to reduction in the required intake temperature for naturally aspirated and low-boost conditions. The lower intake temperature increases the charge density, which acts to increase both the thermal efficiency and maximum load at given intake pressure (as discussed above). However, for higher boost levels, intake temperature is reduced to the minimum allowable level (60 °C for premixed and 30–40 °C for DI). To control combustion timings, EGR is added, which is required in higher quantity for higher compression ratio that can cause the maximum load to become oxygen limited [5]. Figure 7.15b depicts that higher compression ratio engine operation leads to the higher thermal efficiency. The E-DI PFS at higher compression ratio has even higher thermal efficiency.

To further increase the thermal efficiency of single E-DI PFS strategy, recently a double direct fuelling strategy (D-DI) is demonstrated [4]. In this strategy, first early direct injection during intake stroke creates a well-mixed charge with moderated stratification. The second late direction injection during compression stroke creates the greater fuel stratification in the cylinder. The effect of dual direct

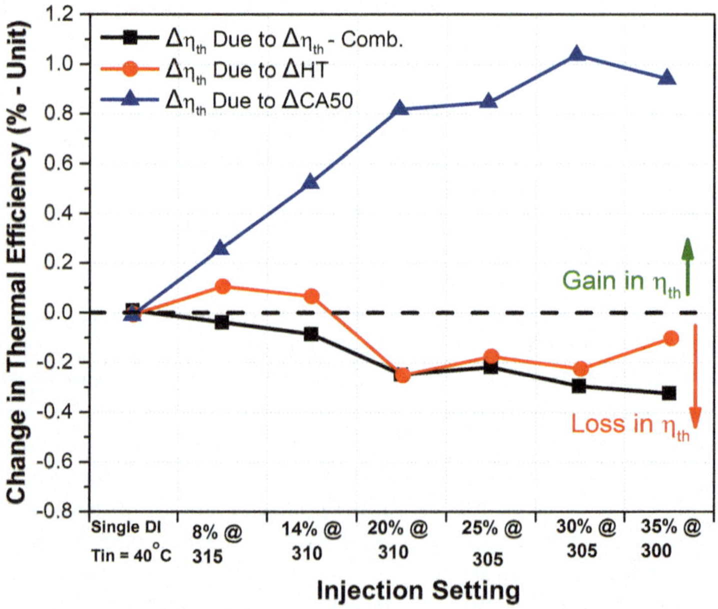

Fig. 7.16 Magnitude of factors affecting the thermal efficiency for the different D-DI fraction from S-DI (Adapted from [4])

injection on heat release rate curve is discussed in Chap. 6 (Fig. 6.16). The D-DI strategy significantly lowers the heat release rate by increasing the combustion duration. The double-DI fuelling strategy demonstrated an improvement of the thermal efficiency by up to ~0.6% units compared to the S-DI (single direct injection) over a given load range [4]. However, with this strategy, the efficiency improvement is less than expected from the allowable CA_{50} advancement. Figure 7.16 shows the magnitude of factors affecting the thermal efficiency in D-DI strategy in comparison to S-DI strategy. Loss in thermal-gained thermal efficiency is mainly by heat transfer and lower combustion efficiency. The lower combustion efficiency is obtained due to locally overly rich regions produced by fuel stratification. Heat transfer is also enhanced because of the advanced CA_{50}, and the late injection might increase the convective heat transfer coefficient [4]. To avoid local overly rich regions, multiple-DI strategy can be used, which may increase the gain in thermal efficiency (by another 0.4–0.5% units).

The partially premixed combustion (PPC) is another LTC strategy, which often uses direct injection gasoline-like fuels by creating heavy stratification in the cylinder while maintaining the benefits of HCCI combustion (detailed discussion is provided in Chap. 2). The heavy stratification is created using diesel-type direct injectors and high fuel injection pressure. Fuel injection strategy in PPC engine is discussed in Sect. 7.1.2. Figure 7.17 shows gross indicated thermal efficiency as function of IMEP in PPC engine operated at two different compression ratios. Fuels designated as G.ON x/y represents gasoline octane number, where 'x' is RON, and 'y' is MON. Figure 7.17 depicts that more than 55% indicated efficiency is achieved in PPC engine using different octane gasoline. Gasoline having RON of 70.3 showed the highest thermal efficiency at higher compression ratio operation in PPC mode. At higher compression ratio (17.1), indicated efficiency is in the range

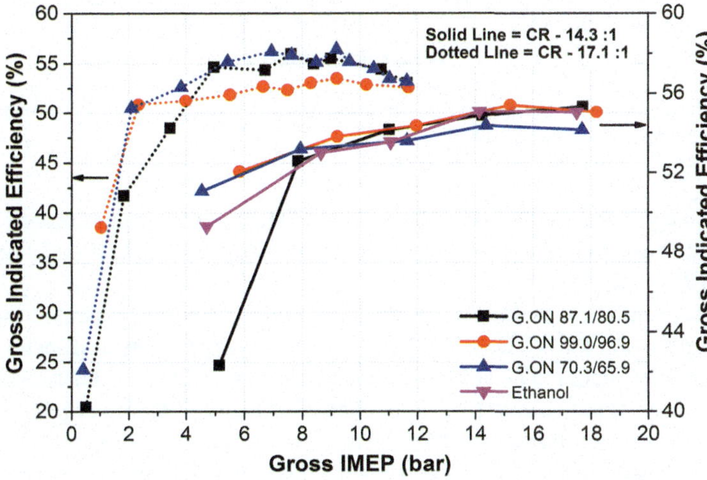

Fig. 7.17 Gross indicated thermal efficiency as a function of IMEP in PPC engine at two compression ratios using gasoline-like fuels (Adapted from [39, 40])

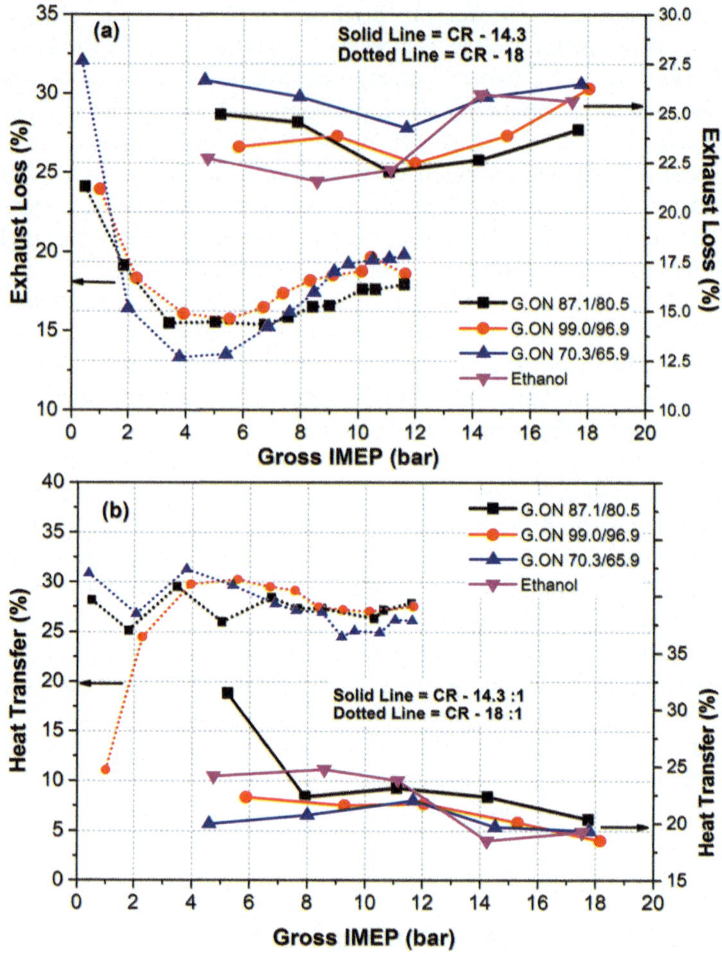

Fig. 7.18 Relative exhaust and heat transfer losses form PPC engine at two compression ratios using gasoline-like fuels (Adapted from [40, 41])

of 52–57% for engine load higher than 4 bar IMEP; however at compression ratio 14.3, indicated efficiency is in the range of 52–56% for engine load higher than 8 bar IMEP.

The three major losses during compression, combustion and expansion stroke are (1) the combustion loss (CO, HC and H2 in the exhaust), (2) heat transfer loss and (3) exhaust energy loss. The combustion loss is estimated by combustion efficiency (ratio of combustion loss and fuel input energy). The combustion efficiency of PPC engine using different gasoline-like fuels is presented in Fig. 6.41b of Chap. 6. Relative exhaust and heat transfer loss with respect to fuel energy is presented in Fig. 7.18. Figure 7.18 depicts that for engine loads higher than 4 bar gross IMEP, the combustion losses are negligible (below 5%), the heat transfer is between 25% and

30% and the exhaust losses are not higher than 20%. Low heat transfer and low exhaust loss can be simultaneously achieved by high-boost and appropriate combustion timing and duration [41]. Heat transfer is mainly a function of surface area and temperature gradient between bulk volume and wall (independent of mass inside the cylinder). Thus, heat transfer is independent of boost level, and boosting the engine increases the internal energy of the combustion volume increases (more mass). The exhaust losses are approximately constant with boost because the increase in exhaust mass balances with the reduction in exhaust temperature due to the higher mixture dilution. Heat transfer can be optimized with combustion phasing (typically 5–15 CAD ATDC depending on engine load). If combustion duration is in the range of 10–30 CAD, the expansion stroke duration is maximized, and heat transfer is minimized [41]. Combustion duration longer than 10–30 CAD reduces the effective expansion ratio, and very short combustion duration also increases the heat transfer loss. Exhaust losses increased from 20% to 25% in lower compression ratio PPC engine operation; however higher engine load can be achieved. Despite the increment in exhaust loss, the gross indicated efficiency is still higher due to reduction in heat transfer from 30% to 25% (Fig. 7.18). A study proposed two reasons for lower heat transfer loss: (1) lower mixing period for engine load higher than 7 bar IMEP; thus most of the combustion is confined in the piston bowl, and the squish area works as thermal insulator; and (2) combustion duration is in the range of 10 and 25 CAD from mid to high loads [40].

Dual-fuel RCCI engine is another LTC strategy having the potential of higher thermal efficiency along with ultralow NOx and soot emissions. In RCCI engine, lower injection pressure and EGR rate are required than PPC engine, and it is still able to achieve higher thermal efficiency. Figure 7.19 shows the RCCI engine thermal efficiency as function of engine load using different low reactivity fuels. Figure 7.19a illustrates the effect of engine load on brake thermal efficiency (BTE) in RCCI engine vis-à-vis CDC at 1500 rpm. The BTE increases with engine load for both the combustion modes, and RCCI combustion has higher BTE (Fig. 7.19a). The RCCI combustion with 20% ethanol/diesel (E20/D) obtained higher thermal efficiency as compared to gasoline/diesel (G/D) RCCI. The E20/D RCCI combustion also achieved higher peak engine load as compared to G/D RCCI operation due to comparatively lower reactivity of E20. At higher engine speed of 2600 rpm, RCCI operation has 7% higher BTE as compared to CDC [42]. Another study reported that at 8 bar BMEP and 3000 rpm condition, RCCI strategy has 40.6% brake thermal efficiency, which is 5% higher as compared to CDC engine at the same load condition. Figure 7.19b presents the indicated thermal efficiency as function of IMEP in a RCCI combustion engine. Figure 7.19b depicts that E85/diesel RCCI operation has higher indicated efficiency than gasoline/diesel RCCI. As indicated in Fig. 7.10, at higher engine loads, E85/diesel RCCI has lower requirement of EGR, and thus ratio of specific heat (γ) of working fluid production is higher leading to higher work extraction [34]. The lower EGR requirement to reduce the pressure rise rate is due to longer HTHR combustion duration (approximately 50%) with E85/diesel operation. Figure 7.19b also demonstrates that around 58% indicated thermal efficiency can be achieved in RCCI combustion engine.

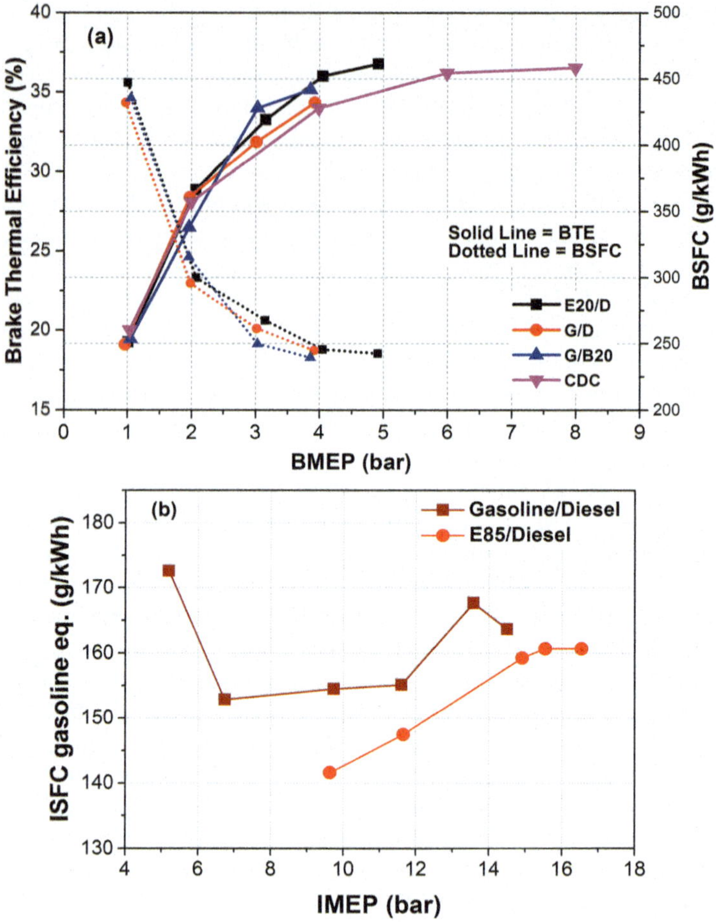

Fig. 7.19 (**a**) Brake thermal efficiency (BTE) as a function of BMEP at 1500 rpm (Adapted from [42, 43]) and (**b**) indicated thermal efficiency as a function of IMEP (Adapted from [34]) in RCCI engine

Figure 7.20 presents the gross indicated thermal efficiency as function of global equivalence ratio (Φ) and intake temperature in a RCCI engine fuelled with 3.5% EHN + gasoline/E85 and ULSD/E85 at constant load and TDC combustion phasing in RCCI engine. Figure 7.20 depicts that the region of maximum efficiency approximately occurs within the self-imposed combustion efficiency limit (>97%) and maximum pressure rise rate (MPRR) limit (<12 bar/CA), which indicates the region of maximum indicated efficiency is at least partially correlated to MPRR and combustion efficiency. However, properties of directly injected fuel might have significant role in the determination of engine efficiency. This point is confirmed in Fig. 7.20b where ULSD operation produced a lower indicated efficiency in comparison to 3.5% EHN + gasoline. Figure 7.20 also demonstrates that RCCI

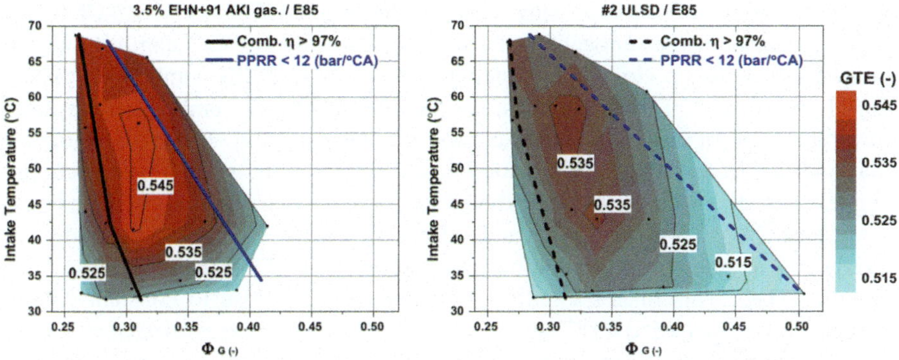

Fig. 7.20 Gross indicated thermal efficiency in global Φ and intake temperature space in RCCI engine (Adapted from [44])

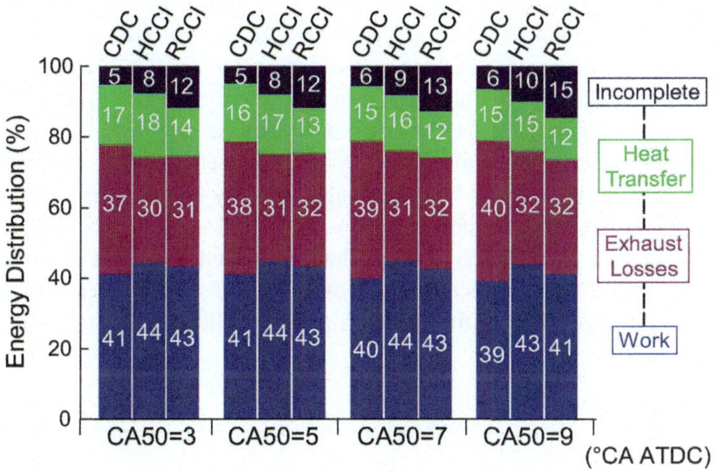

Fig. 7.21 Energy distributions in different processes of CDC, HCCI and RCCI engine at various combustion phasing [45]

combustion has the capability for engine operation at constant combustion timing and engine load over a wide range of intake conditions by adjusting the ratio of port-to-direct injected fuel [44].

Figure 7.21 demonstrates the energy distribution in different process of CDC, HCCI and RCCI engine at various combustion phasing (CA_{50}). Figure 7.21 depicts that incomplete combustion loss progressively increases with retarded CA_{50} for the three combustion modes. The CDC has the highest combustion efficiency at all the combustion phasing. The HCCI combustion has relatively more complete combustion than RCCI due to premixed fuelling. In RCCI combustion due to stratification, the fuel–air mixture close to the cylinder wall is leaner and less reactive than that in the HCCI conditions. Therefore, the lower temperature is obtained near-wall boundary in RCCI than that of HCCI, which leads to higher HC and CO emissions [45].

At constant combustion phasing, heat transfer loss in HCCI is close to CDC, and RCCI demonstrates lower heat transfer than CDC as well as HCCI combustion. The temperature gradient close to cylinder wall is lower in RCCI combustion than in HCCI combustion. The exhaust losses of HCCI and RCCI combustion are much lower than that of CDC at constant combustion phasing (Fig. 7.21). This indicates that both HCCI and RCCI combustion could extract more work from the charge, and less heat energy is rejected in exhaust gas. Combustion duration between CA_{10}-CA_{50} and CA_{50}-CA_{90} are significantly different for three combustion modes. At fixed CA_{50} engine operation, combustion duration in HCCI leads to an early CA_{90} close to TDC. This makes HCCI combustion process close to constant-volume combustion compared to CDC and RCCI, which leads to more work extraction and lower exhaust losses. In contrast, CDC has delayed CA_{90} timings due to relatively longer combustion duration. Thus, large fraction of energy is prone to increase enthalpy of exhaust gases and less conversion to work [45].

Direct injection dual-fuel stratification (DDFS) strategy combines the benefits of RCCI and PPC by injecting both gasoline and diesel directly into the engine cylinder (see Sect. 2.6.4 of Chap. 2). In this strategy, gasoline is injected during intake stroke, and it is effectively premixed. Similar to RCCI, diesel is injected 40–60° before TDC to produce reactivity gradient in the cylinder. In DDFS strategy, load expansion is obtained by late gasoline injection, and combustion stability is considerably improved [46]. The DDFS enables the control over the in-cylinder distribution of both low and high reactivity fuels. Figure 7.22 illustrates energy distributions in different processes of RCCI and DDFS RCCI strategy. Figure 7.22 depicts that DDFS strategy has improved the thermal efficiency. In this strategy, heat loss is reduced, while exhaust losses are increased, and it also reduces the turbocharging requirements. A more complete discussion can be found in the original study [46].

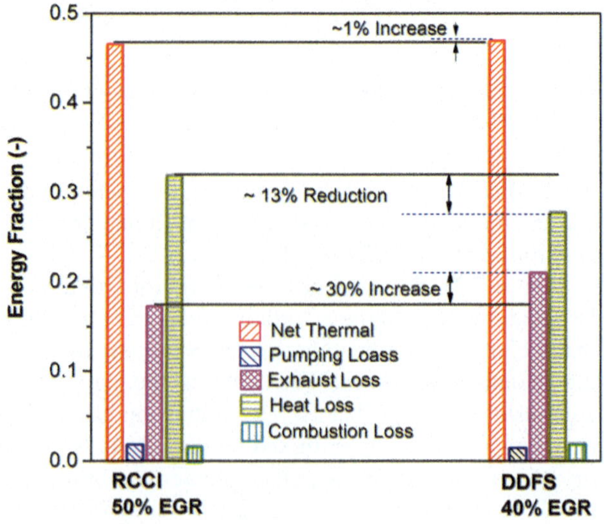

Fig. 7.22 Energy distributions in different processes of RCCI and DDFS RCCI strategy (Adapted from [46, 47])

7.3 Specific Fuel Consumption

Fuel consumption from engine is typically expressed in terms of power (brake or indicated) specific basis, which is the ratio of fuel consumption to the power produced. Brake specific fuel consumption (BSFC) and indicated specific fuel consumption (ISFC) terms are generally used to measure the fuel economy of the engine. The BSFC is typically used to compare the efficiency of engine with shaft power output. The BSFC or ISFC allows direct comparison of fuel conversion efficiency of different engines of various sizes or operating of different modes. For the user of vehicle, it indicates the required fuel carrying capacity for a particular power out.

Figure 7.23 shows the ISFC of HCCI combustion engine using gasoline at 1200 rpm in λ and intake temperature space. To calculate the ISFC, the indicated power is calculated from the pressure–volume curve. The ISFC has an inverse relationship with the indicated thermal efficiency, i.e. the maximum indicated efficiency corresponds to the minimum ISFC. At lower thermal efficiency, higher energy (higher amount of fuel) is required to satisfy the same engine load. Figure 7.23 depicts that ISFC is higher close misfire boundary due to lower indicated thermal efficiency. Near misfire boundary air fuel mixture is leaner at particular temperature, which leads to lower combustion temperature resulting in lower combustion efficiency. Figure 7.14 also confirms that indicated thermal efficiency

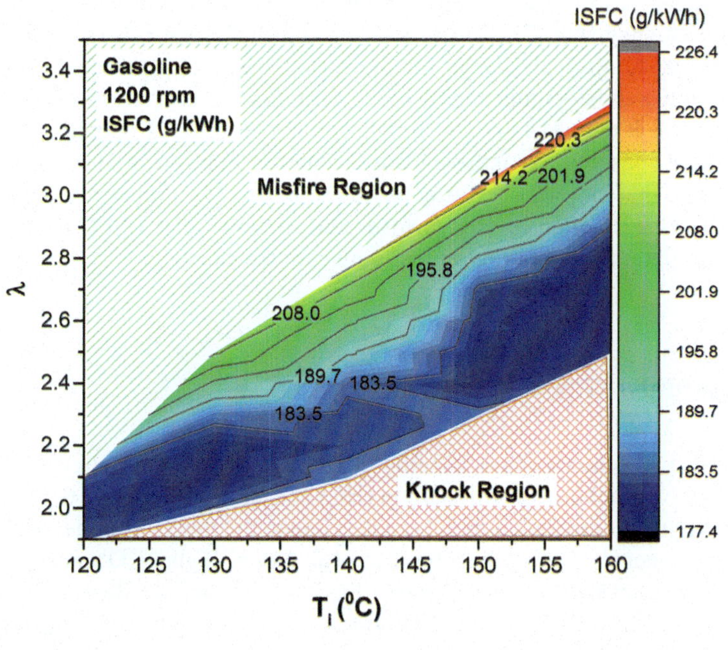

Fig. 7.23 Variations of ISFC in HCCI operating range for gasoline at 1200 rpm (Adapted from [21])

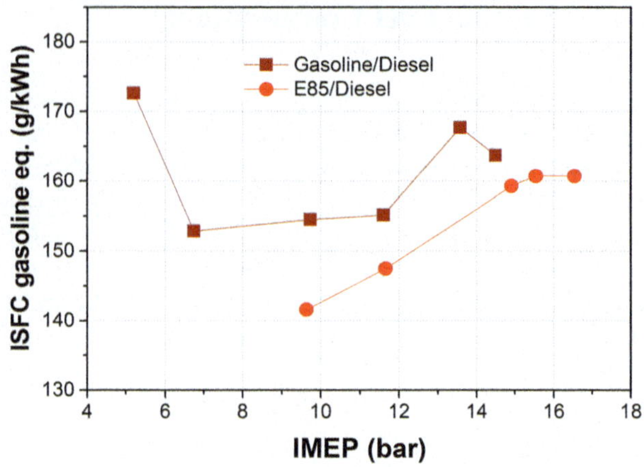

Fig. 7.24 Variation of ISFC in RCCI engine using gasoline/diesel and E85/diesel (Adapted from [33])

is lower near misfire boundary and, therefore, higher ISFC. The ISFC is lower closer to knock boundary due to higher thermal efficiency. Contour lines are similar to thermal efficiency curve, and values have an inverse relationship (Fig. 7.14).

Figure 7.24 shows the variations of ISFC in RCCI engine using gasoline/diesel and E85/diesel on gasoline equivalent basis. Ethanol has lower calorific value than gasoline, and hence higher volume is required for the same energy input. Therefore, it is normalized and gasoline equivalent basis is presented. Figure 7.24 shows that E85/diesel has lower ISFC than gasoline/diesel RCCI due to higher indicated thermal efficiency (Fig. 7.19b). Reason for higher thermal efficiency is discussed in Fig. 7.19. Additionally, with all the tested fuels, the indicated specific fuel consumption trends demonstrate a local minimum of approximately 150 g/kWh (gross based), at the 6–12 bar IMEP in gasoline/diesel RCCI operating conditions. The low fuel consumption at these loads is believed to be due to the more extensive optimization that has been performed at these loads [34]. Figure 7.19a also depicts the BSFC as a function of engine load, and it shows that BSFC decreases with increase in the load due to higher fuel conversion efficiency.

7.4 Exhaust Gas Temperature

In general, LTC engines have benefits of ultralow NOx and negligible soot emissions. However, unburned hydrocarbon (HC) and carbon monoxide (CO) emissions are very high as compared to diesel engines due to the low in-cylinder temperatures caused by leaner charge combustion. The problems of higher CO and HC emissions are severe particularly at lower engine load conditions (see Sect. 8.2 and 8.3 of Chap. 8). Reduction of HC and CO emissions from engine exhaust is relatively easier than reducing NO and soot emissions [48]. Higher HC and CO emissions can be eliminated from engine exhaust by using oxidation catalysts, which can reach up to 95% conversion efficiencies in fully warmed-up conditions [49]. It is essential to reach the operating temperature of the catalytic converter for effective conversion of HC and CO emissions. The light-off temperature (the temperature at which the catalyst becomes more than 50% effective) is about 250–300 °C for most of the catalysts [48, 50]. The exhaust gas temperature (T_e) is lower in HCCI engines as compared to conventional engines due to combustion of highly diluted mixture and the high burn rates. HCCI engine exhaust gas temperature can be as low as 120 °C; therefore elimination of HC and CO emissions by catalytic converters is a challenging problem [51]. In HCCI engine, higher amount of HC and CO is emitted at lower engine load, where exhaust temperature is also lower. This makes the problem even more difficult. Lower exhaust gas temperature in HCCI engine also leads to availability of less exhaust energy available to the turbocharger, which is often used to extend high load HCCI operating range [48]. Therefore it is essential to understand the effect of various parameters influencing HCCI exhaust temperature for a better modulation of engine charge variables to extend operating range of HCCI engine.

Figure 7.25 shows the effect of λ on exhaust gas temperature using gasoline at 1200 rpm. Figure 7.25 depicts that exhaust gas temperature shows almost linear dependence on λ at each intake air temperature. Exhaust temperature decreases by 62 °C as λ increases from 2.0 to 2.5 at $T_i = 130$ °C and CA_{50} retards by 9 CAD in the same conditions. Exhaust gas temperature is higher for richer mixture due to higher injected fuel energy content. At very advanced timing at $T_i = 150$ °C, exhaust gas temperature decreases for richer mixtures ($\lambda < 2.3$). Ignition timing location has an opposite effect on the exhaust gas temperature. Influence of energy input has stronger effect on T_e as compared to ignition timing. The trend in variations of

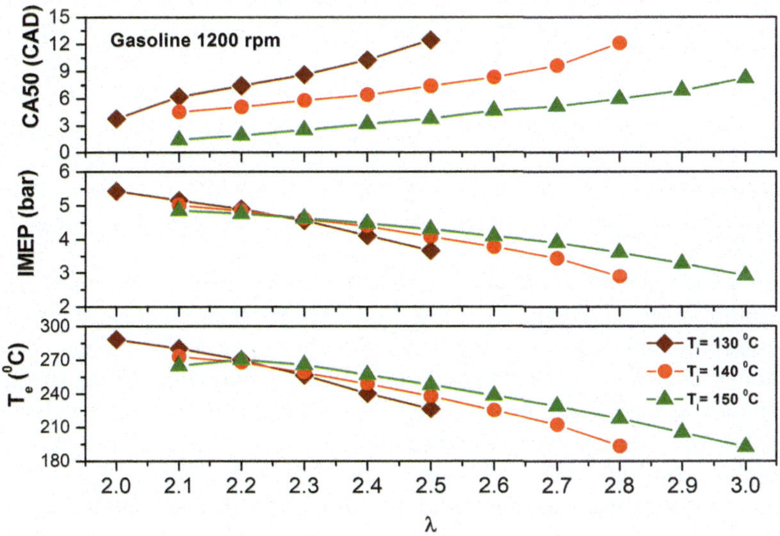

Fig. 7.25 Variation in HCCI engine exhaust gas temperature with λ at 1200 rpm for gasoline [21]

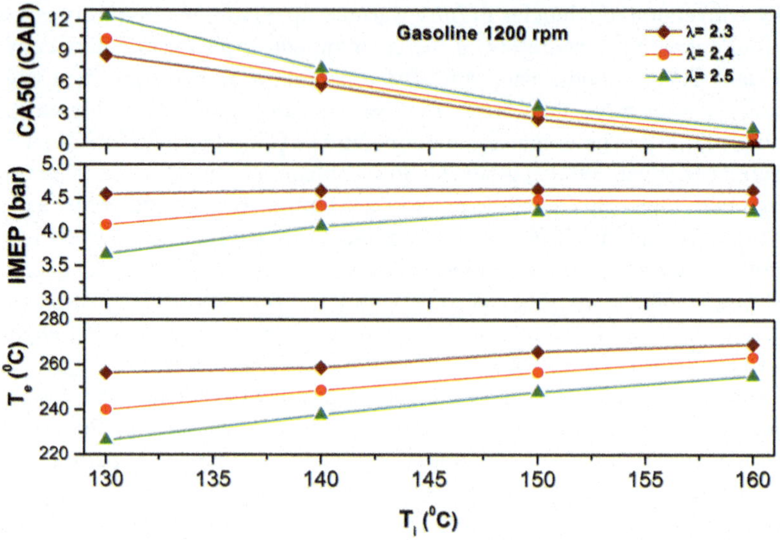

Fig. 7.26 Variations in HCCI engine exhaust gas temperature with intake temperature at 1200 rpm for gasoline [21]

exhaust gas temperature correlates well with the trends of variation in IMEP (Fig. 7.25).

Figure 7.26 shows the effect of intake manifold temperature (T_m) on HCCI exhaust temperature using gasoline at 1200 rpm. Figure 7.26 shows that exhaust gas temperature increases with the increase in intake air temperature at each λ. The variation of T_e is lower for richer mixture for the same variation of T_m. Figure 7.26 also depicts that IMEP also has a similar trend and correlates well with T_e. For $\lambda = 2.3$, exhaust gas temperature increases by 12 °C with increases in T_i from 130 °C to 160 °C, and CA_{50} advances by 8.5 CAD in the same conditions. The main factor governing increase in exhaust gas temperature is higher initial intake air temperature, which leads to higher in-cylinder gas temperature in the engine cycle. Significant variations in the combustion phasing do not affect much to exhaust gas temperature; thus variation in T_e with T_m is not as large as with λ ($\Delta T_e/\Delta T_m = 0.4$ °C/°C).

Figure 7.27 shows the effect of engine speed and energy input on exhaust gas temperature in HCCI engine. Effect of engine speed on HCCI exhaust temperature at constant fuelling is shown in Fig. 7.27a. On increasing engine speed, the exhaust temperature increases at constant fuelling in HCCI engine (Fig. 7.27a). Exhaust gas temperature increases significantly from 159 °C to 245 °C, when speed increases from 1200 to 2400 rpm. The higher exhaust gas temperature at higher engine speed is a result of delayed combustion phasing and reduced heat losses. Lower heat loss is expected at higher engine speed due to shorter time availability during combustion cycle.

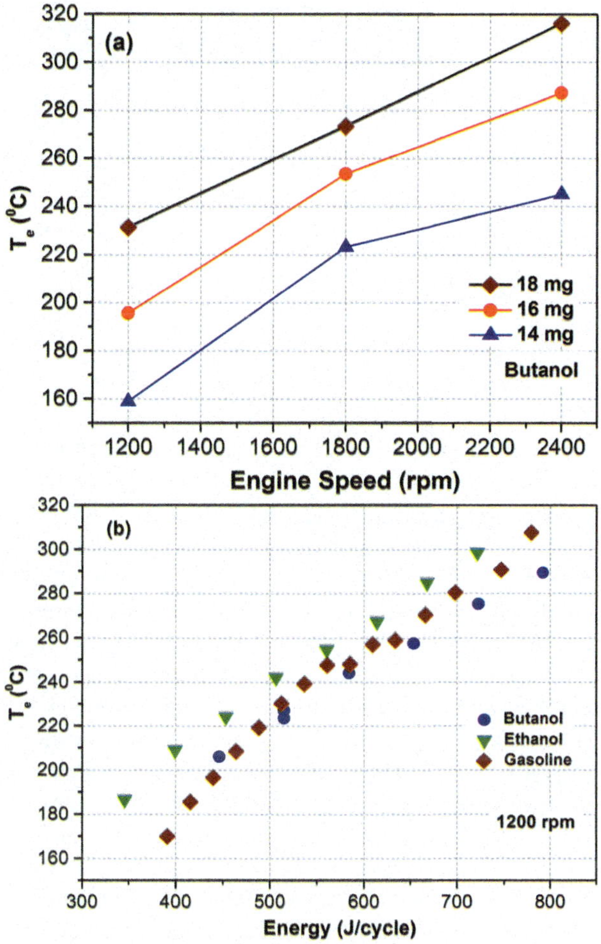

Fig. 7.27 Variations in HCCI engine exhaust gas temperature with (**a**) engine speed and (**b**) energy input for different test fuels [21]

The effect of different fuels on HCCI exhaust temperature with the different amount of injected fuel energy is shown in Fig. 7.27b. Figure 7.27b depicts that on increasing the energy input to the combustion cycle, exhaust gas temperature increases significantly for all test fuels. Ethanol has higher exhaust gas temperature as compared to gasoline and butanol at any particular energy input. The possible reason for this observation is relatively higher octane number of ethanol than the gasoline and butanol. Higher octane fuel requires high in-cylinder temperature for auto-ignition, which may lead to higher exhaust gas temperatures. For constant operating conditions, higher octane fuels have delayed combustion phasing, which leads to higher exhaust temperature in HCCI combustion.

Figure 7.28 provides the summary of various factors influencing the HCCI exhaust gas temperature. The main engine input parameters such as intake manifold pressure (P_m), fuel octane number (ON), engine speed (N) and coolant temperature (T_c), intake manifold temperature (T_m) and equivalence ratio (Φ) directly influence

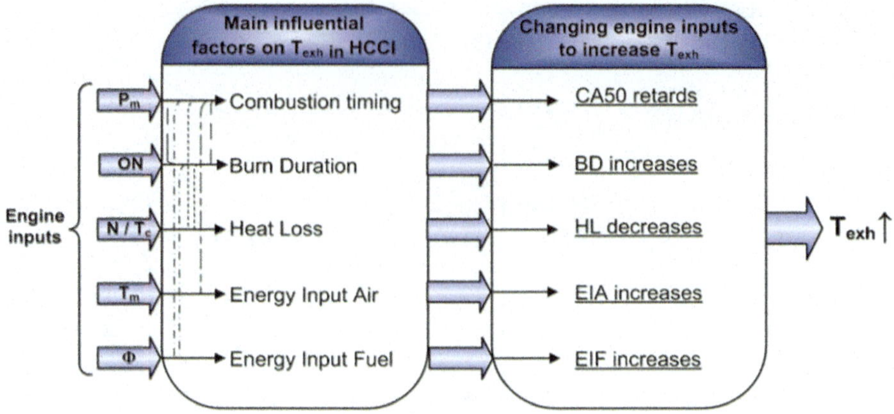

Fig. 7.28 Summary of different factors affecting the HCCI exhaust temperature [48]

combustion phasing (CA_{50}), burn duration (BD), heat loss (HL), energy input air (EIA) and energy input fuel (EIF), respectively (Fig. 7.28). Interaction of these factors sometimes makes complex behaviour of exhaust gas temperature, when conflicting nature of effect occurs. The exhaust temperature is mainly affected by change in combustion phasing when only intake manifold pressure varies, while keeping other factors constant. Similar behaviour is also observed for octane number (ON) of fuel. In the case of engine speed variation (while other factors are constant), two parallel factors (delayed combustion timings and decreased heat loss) govern the exhaust gas temperature. Increased heat loss and delayed combustion timing are the two opposing factors that occur by lowering coolant temperature while other operating parameters are constant. However, exhaust temperature can either increase or decrease with increase in equivalence ratio. The exhaust temperature depends on the competition of two opposing factors between the advanced combustion phasing (position of heat release) and increasing amount of heat released with higher input of fuel energy. First factor (advanced CA_{50}) decreases exhaust temperature, while second factor increases the exhaust temperature.

A study used steady-state experimental data of 221 test conditions using ethanol, butanol and gasoline for naturally aspirated conditions to determine empirical correlation for providing clear understanding of the variation in HCCI exhaust temperature. Parameters selected to characterize the HCCI exhaust gas temperature are IMEP (bar) (magnitude parameter), engine speed (N) (magnitude parameter), CA_{50} (location parameter) and combustion duration (CA_{90-10}) (dispersion parameter). The empirical correlation found to estimate exhaust temperature well for the experimental data is represented by the equation 7.3.

$$T_e = a + b(\mathrm{IMEP}) + c\,(\mathrm{CA}_{50}) + d\,(\mathrm{CA}_{90-10}) + e(N) \tag{7.3}$$

where CA_{90-10} is combustion duration in CAD, N is engine speed in rpm and a, b, c, d and e are the constants determined using regression of experimental data.

Resulting values of constants are found as $a = 45.0791$, $b = 37.697$, $c = -0.434$, $d = -0.939$ and $e = 0.052$, respectively ($R^2 = 0.9456$). The results indicate that this simple correlation captures the trend of T_e variations and presents good agreement with the experimental data for the majority of operating conditions. An uncertainty of ±16.2 °C is found in the predictions using this correlation.

Another study also found a correlation to characterize the exhaust gas temperature in HCCI engine at constant speed. The empirical correlation found to estimate exhaust temperature well for the experimental data is represented by the equation 7.4 [48]

$$T_e = a + b\frac{q_f}{C_{v,\mathrm{mix}}T_{\mathrm{ad}}} + c\,[\ln(\mathrm{CA}_{50})]^2 \tag{7.4}$$

where a, b and c are empirical constants, q_f is specific fuel energy input (kJ/kg), $C_{v,\mathrm{mix}}$ is constant volume specific heat of mixture and T_{ad} is mean constant volume adiabatic flame temperature. An uncertainty of ±23.7 °C is found in the predictions using this correlation.

At higher engine loads (IMEP > 8 bar), exhaust gas temperature in HCCI engine is found to be higher than 300 °C for well-mixed as well as partially stratified charge conditions [28]. The higher exhaust gas temperature can be more effective for HC reduction using oxidation catalyst as well as more effective for driving a turbocharger to achieve higher engine loads.

The exhaust gas temperature is also lower at lower engine loads in other LTC strategies such as RCCI combustion. Figure 7.29 shows the exhaust gas temperature

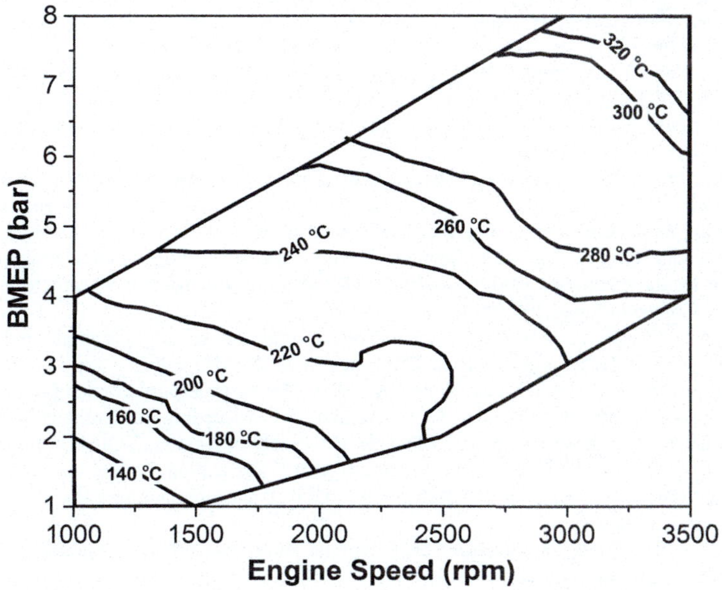

Fig. 7.29 RCCI exhaust gas temperature variations with engine load and speed (Adapted from [52])

variations with engine load and speed in RCCI combustion using gasoline/diesel. The RCCI map is determined using self-imposed constraints of peak pressure rise rate of 10 bar/CAD (high load) and a CO emission limit of 5000 ppm (lower load limit). The RCCI map is estimated without EGR by using a mix of single and split diesel injections for best brake thermal efficiency and lowest possible NO_x emissions. The low combustion temperatures from RCCI engine result in lower exhaust temperatures. Figure 7.29 depicts that there is large portions of the RCCI operating map having exhaust temperature below 200 °C. Lower exhaust gas temperature conditions combined with higher CO and HC emissions present a significant challenge for exhaust after-treatment devices such as diesel oxidation catalysts.

References

1. Johansson T, Borgqvist P, Johansson B, Tunestal P, Aulin H (2010) HCCI heat release data for combustion simulation, based on results from a turbocharged multi cylinder engine (No. 2010-01-1490). SAE technical paper
2. Eng JA (2002) Characterization of pressure waves in HCCI combustion (No. 2002-01-2859). SAE technical paper
3. Saxena S, Bedoya ID (2013) Fundamental phenomena affecting low temperature combustion and HCCI engines, high load limits and strategies for extending these limits. Prog Energy Combust Sci 39(5):457–488
4. Dernotte J, Dec J, Ji C (2017) Efficiency improvement of boosted low-temperature gasoline combustion engines (LTGC) using a double direct-injection strategy (No. 2017-01-0728). SAE technical paper
5. Dec JE, Dernotte J, Ji C (2017) Increasing the load range, load-to-boost ratio, and efficiency of low-temperature gasoline combustion (LTGC) engines. SAE Int J Engines 10(2017–01-0731):1256–1274
6. Dec JE, Yang Y, Dronniou N (2012) Improving efficiency and using E10 for higher loads in boosted HCCI engines. SAE Int J Engines 5(2012-01-1107):1009–1032
7. Andreae MM, Cheng WK, Kenney T, Yang J (2007) On HCCI engine knock (No. 2007-01-1858). SAE technical paper
8. Maurya RK, Akhil N (2016) Numerical investigation of ethanol fuelled HCCI engine using stochastic reactor model. Part 2: Parametric study of performance and emissions characteristics using new reduced ethanol oxidation mechanism. Energy Convers Manag 121:55–70
9. Maurya RK, Akhil N (2017) Development of a new reduced hydrogen combustion mechanism with NO x and parametric study of hydrogen HCCI combustion using stochastic reactor model. Energy Convers Manag 132:65–81
10. Sjöberg M, Dec JE (2007) Comparing late-cycle autoignition stability for single-and two-stage ignition fuels in HCCI engines. Proc Combust Inst 31(2):2895–2902
11. Ghazimirsaied A, Shahbakhti M, Koch CR (2009, May) Partial-burn crankangle limit criteria comparison on an experimental HCCI engine. In Proceeding of Combustion Institute-Canadian Section Spring Technical Meeting, University of Montreal, Quebec, pp 11–13
12. Chen T, Xie H, Li L, Yu W, Zhang L, Zhao H (2012) Expanding the low load limit of HCCI combustion process using EIVO strategy in a 4VVAS gasoline engine (No. 2012-01-1121). SAE technical paper
13. Johansson, T. (2010). Turbocharged HCCI engine, improving efficiency and operating range. PhD thesis, Lund University, Sweden

14. Johansson T, Johansson B, Tunestål P, Aulin H. (2010) Turbocharging to extend HCCI operating range in a multi cylinder engine-benefits and limitations. FISITA 2010 Paper F2010A037
15. Sjöberg M, Dec JE, Babajimopoulos A, Assanis DN (2004) Comparing enhanced natural thermal stratification against retarded combustion phasing for smoothing of HCCI heat-release rates (No. 2004-01-2994). SAE technical paper
16. Sjöberg M, Dec JE (2007) EGR and intake boost for managing HCCI low-temperature heat release over wide ranges of engine speed (No. 2007-01-0051). SAE technical paper
17. Sjöberg M, Dec JE, Hwang W (2007) Thermodynamic and chemical effects of EGR and its constituents on HCCI autoignition (No. 2007-01-0207). SAE technical paper
18. Dec JE, Yang Y (2010) Boosted HCCI for high power without engine knock and with ultra-low NOx emissions-using conventional gasoline. SAE Int J Engines 3(2010-01-1086):750–767
19. Saxena S (2011) Maximizing power output in homogeneous charge compression ignition (HCCI) engines and enabling effective control of combustion timing. PhD thesis, University of California, Berkeley
20. Maurya RK, Agarwal AK (2014) Experimental investigations of performance, combustion and emission characteristics of ethanol and methanol fueled HCCI engine. Fuel Process Technol 126:30–48
21. Maurya RK (2012) Performance, emissions and combustion characterization and close loop control of HCCI engine employing gasoline like fuels. PhD thesis, Indian Institute of Technology Kanpur, India
22. Urushihara T, Yamaguchi K, Yoshizawa K, Itoh T (2005) A study of a gasoline-fueled compression ignition engine~ expansion of HCCI operation range using SI combustion as a trigger of compression ignition~ (No. 2005-01-0180). SAE technical paper
23. Maurya RK, Agarwal AK (2015) Combustion and emission characterization of n-butanol fueled HCCI engine. J Energy Resour Technol 137(1):011101
24. Kasseris EP (2006) Comparative analysis of automotive powertrain choices for the near to mid-term future. Doctoral dissertation, Massachusetts Institute of Technology
25. Christensen M, Johansson B, Amnéus P, Mauss F (1998) Supercharged homogeneous charge compression ignition (No. 980787). SAE technical paper
26. Christensen M, Johansson B (2000) Supercharged homogeneous charge compression ignition (HCCI) with exhaust gas recirculation and pilot fuel (No. 2000-01-1835). SAE technical paper
27. Dec JE, Yang Y, Dernotte J, Ji C (2015) Effects of gasoline reactivity and ethanol content on boosted, premixed and partially stratified low-temperature gasoline combustion (LTGC). SAE Int J Engines 8(2015-01-0813):935–955
28. Dec JE, Yang Y, Dronniou N (2011) Boosted HCCI-controlling pressure-rise rates for performance improvements using partial fuel stratification with conventional gasoline. SAE Int J Engines 4(2011-01-0897):1169–1189
29. Bessonette PW, Schleyer CH, Duffy KP, Hardy WL, Liechty MP (2007). Effects of fuel property changes on heavy-duty HCCI combustion (No. 2007-01-0191). SAE technical paper
30. Kalghatgi GT, Risberg P, Ångström HE (2007) Partially pre-mixed auto-ignition of gasoline to attain low smoke and low NOx at high load in a compression ignition engine and comparison with a diesel fuel (No. 2007-01-0006). SAE technical paper
31. Manente V, Johansson B, Cannella W (2011) Gasoline partially premixed combustion, the future of internal combustion engines? Int J Engine Res 12(3):194–208
32. Curran S, Gao Z, Szybist J, Wagner R (2014) Fuel effects on RCCI combustion: performance and drive cycle considerations, presentation. CRC workshop on advanced fuels and engine efficiency, 24–26 February 2014, Baltimore
33. Curran S, Gao Z, Wagner R (2014) Reactivity controlled compression ignition drive cycle emissions and fuel economy estimations using vehicle systems simulations with E30 and ULSD. SAE Int J Engines 7(2014-01-1324):902–912

34. Splitter D, Hanson R, Kokjohn S, Reitz RD (2011) Reactivity controlled compression ignition (RCCI) heavy-duty engine operation at mid-and high-loads with conventional and alternative fuels (No. 2011-01-0363). SAE technical paper
35. Molina S, García A, Pastor JM, Belarte E, Balloul I (2015) Operating range extension of RCCI combustion concept from low to full load in a heavy-duty engine. Appl Energy 143:211–227
36. Hyvönen J, Wilhelmsson C, Johansson B (2006) The effect of displacement on air-diluted multi-cylinder HCCI engine performance (No. 2006-01-0205). SAE technical paper
37. Christensen M, Johansson B (1999) Homogeneous charge compression ignition with water injection (No. 1999-01-0182). SAE technical paper
38. Dernotte J, Dec JE, Ji C (2015) Energy distribution analysis in boosted HCCI-like/LTGC engines-understanding the trade-offs to maximize the thermal efficiency. SAE Int J Engines 8 (2015-01-0824):956–980
39. Manente V (2010) Gasoline partially premixed combustion-an advanced internal combustion engine concept aimed to high efficiency, low emissions and low acoustic noise in the whole load range. PhD thesis, Lund University, Sweden
40. Manente V, Tunestal P, Johansson B, Cannella WJ (2010) Effects of ethanol and different type of gasoline fuels on partially premixed combustion from low to high load (No. 2010-01-0871). SAE technical paper
41. Manente V, Johansson B, Tunestal P, Cannella W (2009) Effects of different type of gasoline fuels on heavy duty partially premixed combustion. SAE Int J Engines 2(2009-01-2668):71–88
42. Curran SJ, Hanson RM, Wagner RM (2012) Reactivity controlled compression ignition combustion on a multi-cylinder light-duty diesel engine. Int J Engine Res 13(3):216–225
43. Hanson RM, Reitz RD (2016) Effects of biofuel blends on transient reactivity-controlled compression ignition engine combustion. Int J Engine Res 17(8):857–865
44. Splitter DA, Reitz RD (2014) Fuel reactivity effects on the efficiency and operational window of dual-fuel compression ignition engines. Fuel 118:163–175
45. Li Y, Jia M, Chang Y, Kokjohn SL, Reitz RD (2016) Thermodynamic energy and exergy analysis of three different engine combustion regimes. Appl Energy 180:849–858
46. Wissink ML (2015) Direct injection for dual fuel stratification (DDFS): improving the control of heat release in advanced IC engine combustion strategies. Doctoral dissertation, The University of Wisconsin-Madison
47. Reitz RD (2016) Reactivity controlled compression ignition (RCCI) for high-efficiency clean IC engines. Aurel Stodola Lecture, ETH Zurich. November 9th, 2016
48. Shahbakhti M, Ghazimirsaied A, Koch CR (2010) Experimental study of exhaust temperature variation in a homogeneous charge compression ignition engine. Proc Inst Mech Eng D J Automob Eng 224(9):1177–1197
49. Wanker RJ, Wurzenberger JC, Schuemie HA (2008) Three-way catalyst light-off during the NEDC test cycle: fully coupled 0D/1D simulation of gasoline combustion, pollutant formation and aftertreatment systems. SAE Int J Fuels Lubr 1(2008-01-1755):1373–1386
50. Ferrari V (2007) Achieving EURO III and EURO IV with ultra-low precious metal loadings (No. 2007-01-2565). SAE technical paper
51. Williams S, Hu LR, Nakazono T, Ohtsubo H, Uchida M (2008) Oxidation catalysts for natural gas engine operating under HCCI or SI conditions. SAE Int J Fuels Lubr 1(2008-01-0807):326–337
52. Curran S, Hanson R, Wagner R, Reitz RD (2013) Efficiency and emissions mapping of RCCI in a light-duty diesel engine (No. 2013-01-0289). SAE technical paper

Chapter 8
Emission Characteristics

Abstract The main motivation for the study of low temperature combustion (LTC) mode is its potential of significant reduction in the nitrogen oxides (NO_x) and particulate matter (PM) emissions as compared to conventional engine combustion modes. This chapter presents regulated and unregulated emission characteristics of LTC engines using ethanol and methanol fuel vis-à-vis conventional fuels (gasoline and diesel). Formation of NO_x, unburned hydrocarbons, carbon monoxide (CO) and PM from LTC engine at various engine operating conditions are discussed in detail. This chapter also discusses the particle number and size distribution analysis of LTC engines using gasoline-like fuels. Total particle number emission of soot particles emitted from different LTC strategies (HCCI, PPC and RCCI) is presented. Detailed discussion on effect of various engine operating parameters on exhaust emission is presented in this chapter. Unregulated emissions (aldehydes, alkanes, alkenes, alkynes, etc.) from LTC engines are also presented and discussed in this chapter.

Keywords Combustion • Emissions • NO_x • Particulate matter • CO • PAH • Unburned hydrocarbons • LTC • PM • HCCI • N_2O • EGR • PPC • RCCI • Particle number • Formaldehyde • Ethanol • Gasoline

8.1 Nitrogen Oxide Emissions

The main nitrogen oxide species produced from automotive internal combustion engines are nitric oxide (NO) and nitrogen dioxide (NO_2), which are collectively often referred as NO_x. Ultralow NO_x emission characteristics are one of the most desirable advantages of different LTC strategies (HCCI, PPC and RCCI) [1, 2]. Thus, LTC strategies are the methods for in-cylinder NO_x emission reduction from automotive combustion engines. In-cylinder NO_x production can be understood in terms of the reaction mechanisms for the formation and removal of NO_x during engine combustion process.

R.K. Maurya, *Characteristics and Control of Low Temperature Combustion Engines*, Mechanical Engineering Series, https://doi.org/10.1007/978-3-319-68508-3_8

8.1.1 NO_x Formation Mechanism

Automotive engines mainly emit NO as the primary nitrogen oxide created by oxidation of molecular nitrogen contained in the air during combustion in the cylinder. During engine combustion process, NO can be produced from N_2 by three main reaction pathways: (i) the thermal NO mechanism, (ii) the prompt NO mechanism and (iii) the N_2O mechanism [3]. There is also growing evidence of a fourth reaction mechanism involving NNH for NO_x formation [4, 5]. The thermal NO formation mechanism consists of mainly three reversible reactions, which are often known as extended Zeldovich mechanism [6–8]. The three main reactions are R1, R2 and R3:

$$\mathrm{O} + \mathrm{N_2} \leftrightarrow \mathrm{NO} + \mathrm{N} \qquad \text{(R1)}$$

$$\mathrm{N} + \mathrm{O_2} \leftrightarrow \mathrm{NO} + \mathrm{O} \qquad \text{(R2)}$$

$$\mathrm{N} + \mathrm{OH} \leftrightarrow \mathrm{NO} + \mathrm{H} \qquad \text{(R3)}$$

The key parameters leading to thermal NO formation during engine combustion process can be estimated by using an equation for the maximum NO formation rate. Expression for the maximum NO formation rate can be calculated from kinetics of the reactions (R1–R3) by using quasi-steady-state approximation for N-atoms and assuming equilibrium values in burned gases for few species (for detailed derivation, see study [3]). The simplified maximum NO formation rate derived in the study [3] is given by Eq. (8.1):

$$\left[\frac{d(\mathrm{NO})}{dt}\right]_{\max} = 1.70 \times 10^{17} T^{-1/2} \exp\left[-\frac{69750}{T(K)}\right] \times (O_2)_{\mathrm{eq}}^{1/2} (N_2)_{\mathrm{eq}} \mathrm{mol/cm^3/s} \qquad (8.1)$$

Equation (8.1) shows the stronger dependences of gas temperature after combustion and comparatively weaker dependency on the O_2 concentration in the burned gas on thermal NO formation rate. In automotive engine combustion process, thermal NO mitigation methods are directed towards reducing the NO formation rate instead of enhancing the NO removal rate [3]. The NO removal rate in the cylinder is very slow at typical engine combustion temperatures. For temperatures above 1800 K, the NO formation rate increases rapidly. Reaction (R1) is typically accepted as being the rate-determining step because of its high activation energy. The last reaction (R3) is typically negligible except in fuel-rich flames [9]. In HCCI combustion, lower combustion temperature leads to lower NO_x emission through this mechanism (thermal) for gasoline and diesel engines [8]. Equation (8.1) also reveals that the quantity of NO produced during combustion also depends on time, particularly the time for which the burned gas temperature is high. Thus, thermal NO formation reaction in engine combustion is mainly dependent on temperature, residence time and oxygen concentration [9, 10].

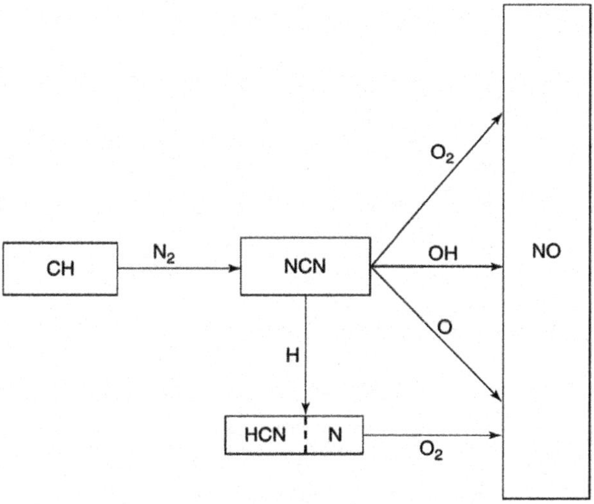

Fig. 8.1 Principal reaction pathways for prompt NO formation [3]

Measured values of NO concentrations on the air side of hydrocarbon diffusion flames or close to the flame zones of hydrocarbon–air depicted that the maximum NO formation rates are higher than the maximum NO formation rate provided by Eq. (8.1) [3]. Bowman [3] identified the three mechanisms for this higher NO formation rate: (i) super equilibrium O and OH radical concentrations, which accelerate the thermal NO reaction rates, (ii) a reaction sequence initiated by reactions of hydrocarbon radicals with molecular nitrogen [11] as shown in Fig. 8.1 and (iii) reaction between the nitrogen containing radical NNH and O atoms present in the burned gas [5, 12]. Combustion conditions in the chamber determine the relative importance of the three mechanisms of prompt NO formation [3]. Acceleration of thermal NO formation rate by super equilibrium O and OH concentrations can be important in non-premixed flames, in low-pressure premixed flames and in stirred reactors for lean fuel–air mixtures. Prompt NO formation occurs by reaction pathway of Fig. 8.1 for close to stoichiometric and rich premixed hydrocarbon–air flames and for hydrocarbon–air diffusion flames. Prompt NO formation through NNH mechanism is important in lower temperature flames [5].

Prompt NO formation mechanism can play key role during initial stages of HCCI combustion because prompt NO formation needs the presence of hydrocarbon radicals. Some amount of prompt NO can also be produced after finishing of the HCCI combustion, when hydrocarbons contained in the crevice region of the combustion chamber comes out and mixes with combustion products. During this process, temperature is not enough to completely oxidize the hydrocarbons, but NO_x formation can take place through the prompt NO_x formation mechanism [8].

In low temperature combustion strategies, theoretical chemical kinetic analysis depicts that the “prompt” and “via N_2O” formation mechanisms are significant and cannot be neglected [13–16]. The NO formation during combustion can also take place by a reaction pathway that involves the formation of N_2O by the three-body recombination reaction (R4) (M = any collision partner) [3, 17, 18]:

$$O + N_2 + M \leftrightarrow N_2O + M \qquad (R4)$$

The N_2O produced in the chamber then reacts with atomic oxygen to form NO via reaction (R5):

$$N_2O + O \leftrightarrow 2NO \qquad (R5)$$

Importance of the NO formation via the nitrous oxide pathway increases with decrease in fuel–air ratios (lean mixture), decrease in burn gas temperature or increase in pressure. The N_2O mechanism is more significant at low NO emission conditions such as HCCI engines [3]. Another study demonstrated that for moderate NO_x levels (in the range of 10 ppm), N_2O reactions play an important role in the NO_x formation, and N_2O is also expected to be present in HCCI engine exhaust [16]. Study also showed the NO_x in exhaust is highly sensitive to combustion temperature; however, N_2O formation has weak negative temperature dependence. The N_2O emissions are not regulated for automotive engines presently. However, N_2O is a very stable molecule with a lifetime of about 120 years in the troposphere and has a global warming potential of 296 times of carbon dioxide [19, 20]. The N_2O formation normally takes place at relatively low combustion temperatures. A study experimentally showed that N_2O emissions are very low (<0.5 ppm) for normal HCCI operating conditions and increases when incomplete combustion occurs in the cylinder [8]. Figure 8.2 shows the effect of air–fuel ratio and exhaust gas recirculation (EGR) on NO_x and N_2O emissions in HCCI engine. Figure 8.2a shows that comparatively high NO_x emissions (30 ppm) are found at rich mixture conditions, and NO_x emissions decrease gradually as charge becomes leaner due to reduced combustion temperature and retarded combustion phasing. Very low NO_x concentrations (<5 ppm) are observed over a wide range of air–fuel ratios, which reflects one of the desired characteristics of HCCI combustion. NO_x emission is minimum (1.4 ppm) at air–fuel ratio of 50, and it starts increasing with further leaning of the charge (increase in air-fuel ratio). As air–fuel mixture becomes further lean (air–fuel ratio >50), the mean combustion temperature is expected to further decrease and further retard combustion phasing along with lower combustion efficiency (incomplete combustion) as presented in Fig. 8.2a. Thus, increase in NO_x cannot be explained in terms of thermal NO production. NO_x formation via N_2O intermediate mechanism under fuel lean conditions is already discussed. Figure 8.2a shows that N_2O emission is almost constant for air–fuel ratios between 35 and 45, and it starts increasing with further leaner mixtures. The higher levels of N_2O can possibly enhance the formation of NO in HCCI combustion [8]. Similar phenomenon is observed with the variation of EGR at constant air–fuel ratio (Fig. 8.2b).

Figure 8.2b shows the effect of EGR on NO_x emissions at constant air–fuel ratio of 40. The NO_x emission is reduced with increase in EGR rate of up to 27%, and further increase in EGR rate increases the NO_x emissions, although mean combustion temperature is expected to decrease further along with retarded combustion phasing and lower combustion efficiency (Fig. 8.2b). Figure 8.2 also depicts an

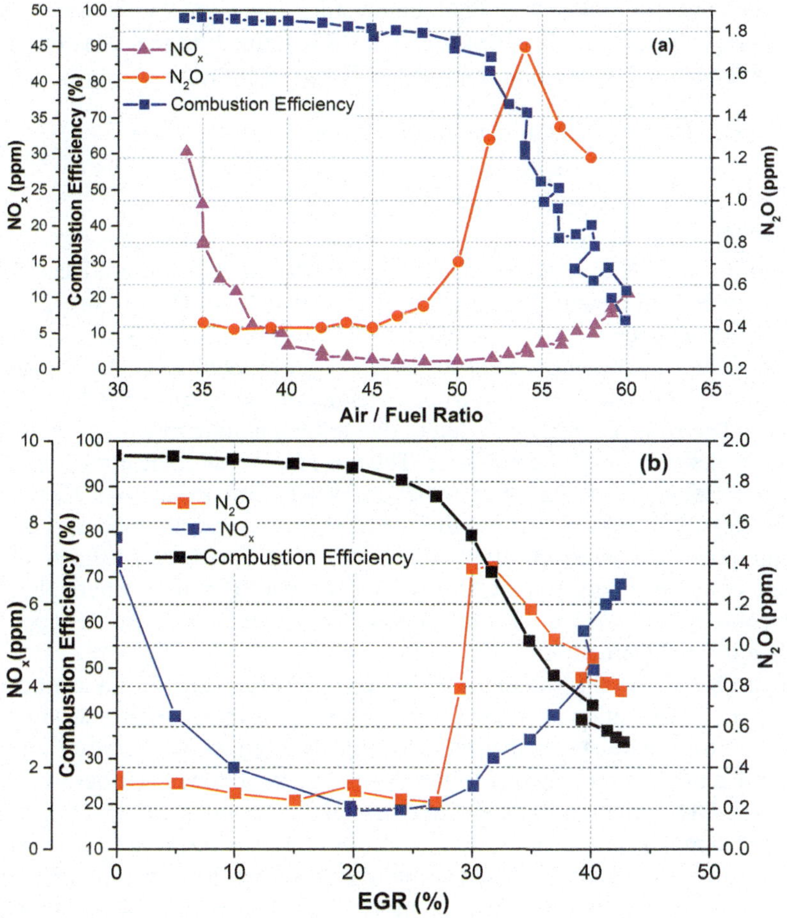

Fig. 8.2 Effect of (**a**) air–fuel ratio and (**b**) EGR (at constant air–fuel ratio) on NO_x and N_2O emissions in HCCI engine [8]

increase in N_2O emissions (for EGR rate higher than 27%), which may enhance the formation of NO_x. A peak value of N_2O emission is observed at EGR rate of 31.8%, and further increase in EGR rate decreases the N_2O emission, but NO_x emission continued to increase (Fig. 8.2b).

The NO_x emitted from combustion engines mainly consists of NO and NO_2. Measured NO_2 emissions from SI engines are comparatively lower than diesel engines (NO_2 can be 10–30% of the total NO_x in diesel engine) [21]. Main NO_2 formation reaction pathway is by reaction (R6) [3]:

$$NO + RO_2 \rightarrow NO_2 + RO \qquad (R6)$$

In this reaction, RO_2 is a peroxy radical and the R is an H atom or an alkyl group. In low temperature zones of flames, RO_2 is relatively stable. Peroxy radical (RO_2)

can react with NO produced in high temperature zones and can be transported to lower temperature zones by diffusion. The NO_2 formed by reaction (R6) can be converted back to NO through reactions (R7) and (R8) [3]:

$$NO_2 + H \rightarrow NO + OH \qquad (R7)$$

$$NO_2 + O \rightarrow NO + O_2 \qquad (R8)$$

The NO_2 removal by (R7) and (R8) is fast at higher temperatures (T > 1500 K) due to the availability of large radical pool in the combustion products. Typical lifetime of NO_2 in combustion products at 1500 K is lower than 10 ms [3]. Hence, presence of NO_2 in the engine exhaust is because of the reaction quenching by mixing with comparatively cooler fluids. The high proportions of NO_2 found from some HCCI engines are an effect from the simultaneous presence of hydrocarbons and NO. The presence of NO_2 is likely due to temperature inhomogeneities, poor mixing and slow overall combustion. Thus, it may be an indicator of poor combustion efficiency and large heat losses in HCCI combustion [16]. Flow reactor experiments also showed that certain hydrocarbons, especially propane and ethylene, promote the conversion from NO to NO_2 [22]. The NO_x emission at leaner mixture conditions may contain higher NO_2 fraction (Fig. 8.2), where NO_x emissions continue to increase after the peak of N_2O emission, which is supposed to enhance the NO formation.

Figure 8.3 shows the variations of NO and NO_2 formation in the cylinder with respect of crank position in ethanol HCCI engine at different relative air–fuel ratios ($\lambda = 4, 5, 6$). Figure shows that both NO and NO_2 emissions decrease rapidly as mixture becomes comparatively leaner due to the reduction in combustion temperature. The NO formation in the cylinder is maximum during combustion zone due to higher combustion temperature, which decreases as piston moves away from TDC and cylinder volume increases. Figure 8.3b shows the NO_2 formation reaches peak and remains constant at later crank position. However, NO formation in the cylinder achieves a maximum value and then decreases due to reburn (NO removal) in next few crank angle.

The final amount of NO produced from engine combustion is the balance of NO formation (by mechanism described above) and NO removal via reburn [3]. The important reaction pathways for reburn are the reactions of the radicals CH_3 and HCCO with NO to produce cyano-species [3, 23]. Elementary reactions that participate in the reburn (NO removal) process are presented by reactions (R9–R11):

$$CH_3 + NO \rightarrow HCN + H_2O \qquad (R9a)$$

$$CH_3 + NO \rightarrow H_2CN + OH \qquad (R9b)$$

$$HCCO + NO \rightarrow HCN + CO_2 \qquad (R10a)$$

$$HCCO + NO \rightarrow HCNO + CO \qquad (R10b)$$

At higher temperatures, the reaction of NO with CH contributes to NO removal by reaction (R11):

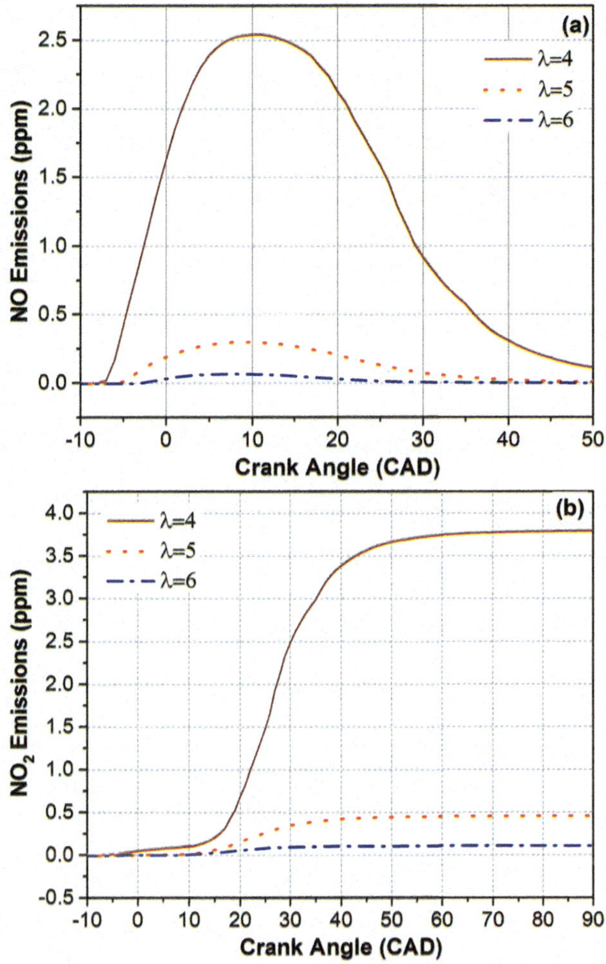

Fig. 8.3 In-cylinder history of (**a**) NO (**b**) NO_2 formation for different λ during ethanol HCCI combustion [13]

$$CH + NO \rightarrow HCN + O \qquad (R11)$$

The NO_x emission characteristics in different LTC strategies and effect of different engine operating conditions on NO_x emission are discussed in the next subsection.

8.1.2 *LTC Engines' NO_x Emission Characteristics*

One of the main advantages of LTC engines is the ultralow NO_x emission. However, NO_x emission depends on the engine operating conditions as well as fuel for a particular LTC strategy. Figure 8.4 shows the effect of different engine operating

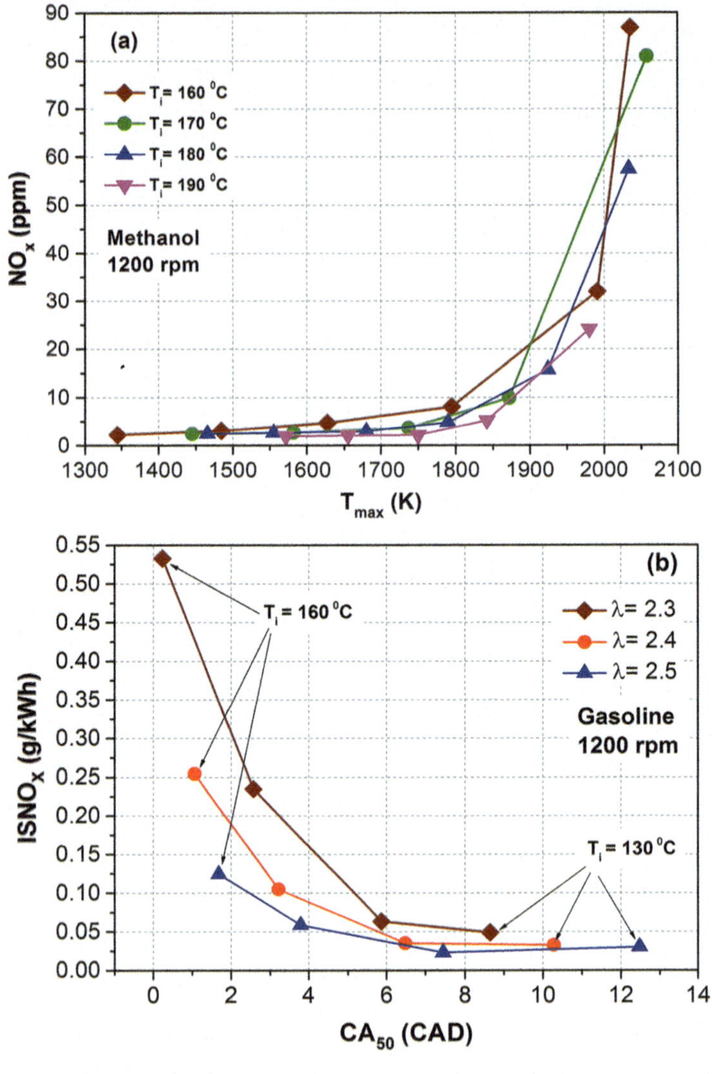

Fig. 8.4 (**a**) Measured NO_x as function of maximum mean gas temperature [1]. (**b**) Effect of combustion phasing (CA_{50}) on NO_x emissions for different λ [24]. (**c**) Effect of IMEP and inlet temperature (T_i) on NO_x emission for gasoline and ethanol at 1200 rpm (Adapted from [24]). (**d**) Measured NO_x as function of λ for different fuels and IVO timings (Adapted from [25])

parameters on NO_x emission in HCCI engine. Figure 8.4a shows the effect of maximum mean gas temperature (T_{max}) on NO_x emission in methanol HCCI engine. Different maximum mean gas temperatures are obtained using engine operation at various intake air temperatures (T_i) and λ. Figure 8.4a depicts that the NO_x formation rate increases rapidly (almost exponentially) with increased mean cylinder temperature over 1800 K. Trend of graph is similar to the maximum NO_x formation rate depicted by Eq. (8.1). Figure 8.4b shows the effect of

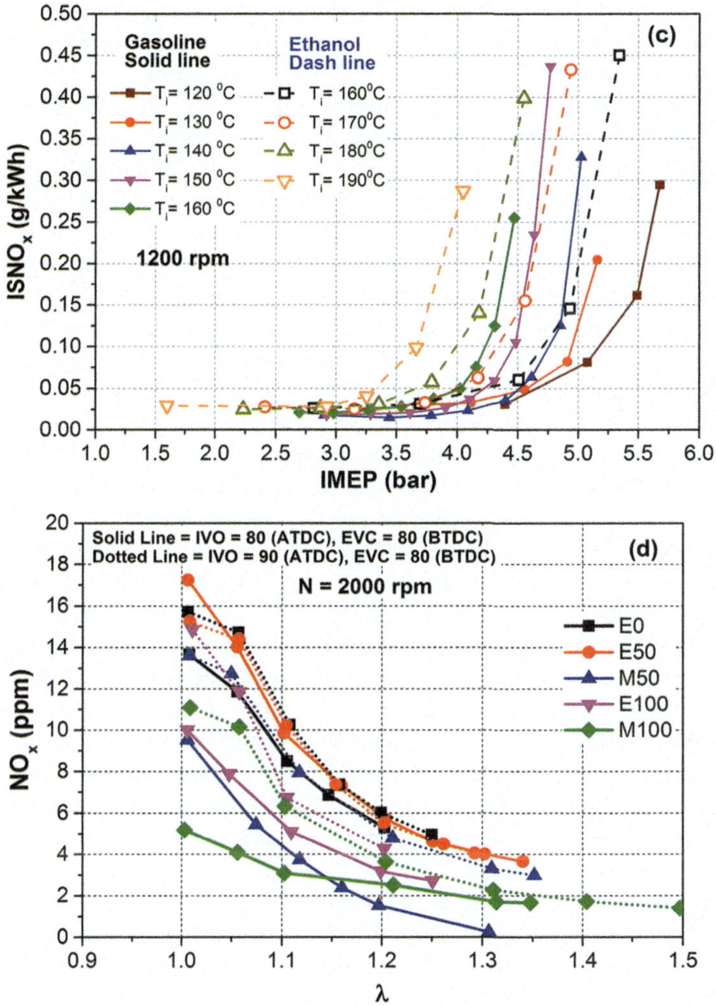

Fig. 8.4 (continued)

combustion phasing (CA_{50}) and intake air temperature on NO_x emission at different λ for gasoline HCCI combustion at 1200 rpm. Figure 8.4b depicts that NO_x increases as CA_{50} advances at each λ due to higher combustion chamber temperature at advanced combustion phasing because of early ignition at higher intake temperatures. For combustion phasing after 6 CAD aTDC, formation of NO_x is very low (<0.05 g/kWh). Figure 8.4c shows the effect of engine load on NO_x emissions at different intake temperature for gasoline and ethanol. Figure depicts that NO_x formation increases drastically with engine load at each intake air temperature for both the fuels. On increasing engine load, combustion chamber temperature increases due to combustion of higher amount of fuel burned in the cylinder in order to produce higher IMEP at each intake air temperature. NO_x emission is very low (<0.05 g/kWh) for IMEP up to 4 bar (Fig. 8.4c) for both gasoline and ethanol. Figure 8.4d shows the variation of NO_x as function of λ for different fuels (gasoline, ethanol and methanol) at two different intake valve opening (IVO) timings for residual trapped HCCI combustion. Figure 8.4d depicts that the NO_x emission decreases with the increase in λ because the mixture becomes leaner, less fuel is burnt and the combustion temperature decreases as well. Additionally, pure ethanol (E100), pure methanol (M100) and 50% methanol blend with gasoline (M50) can well reduce NO_x emission in most of the test conditions as compared to gasoline. At the both sets of valve timings, E50 cannot well reduce the NO_x emission compared with gasoline and emission level is similar to gasoline [25].

Figure 8.5 presents the variations of NO_x emission in HCCI operating range using gasoline, ethanol and methanol at 1200 rpm. Engine operating conditions having RI < 6 MW/m^2 (knock region limit) and COV_{IMEP} < 3.5% (misfire region limit) are considered in HCCI operating range. Figure 8.5 depicts that HCCI combustion mode leads to huge reduction in NO_x emissions compared to conventional SI combustion for all test fuels. For example, typical NO_x emission in convention SI combustion is approximately 2500 ppm (11 g/kWh) at $\lambda = 1.1$, 1600 rpm and $\eta_v = 50\%$ [21], while NO_x emissions in HCCI combustion are typically an order of magnitude or even two orders of magnitude (in some conditions) lower. This ultralow NO_x emission is one of the major benefits of HCCI combustion. NO_x emissions are highest for richer mixtures closer to the knock boundary and lowest for leaner mixtures closer to the misfire boundary for all test fuels (Fig. 8.5) due to lower combustion temperatures at leaner mixture. Figure 8.5 shows that the maximum raw emission of NO_x is 45 ppm (0.26 g/kWh, $ISNO_x$) in HCCI operating range for all test fuels. NO_x contour lines are inclined more towards horizontal axis (Fig. 8.5) indicating higher contribution of λ in NO_x emissions in comparison to inlet temperature in HCCI operating range. Thus, in HCCI operating range, NO_x emissions are weakly sensitive to inlet temperature in comparison to λ [1].

Figure 8.6 shows the variations of NO_x and N_2O emission at different engine speed and load conditions for most efficient combustion phasing (achieved by adjusting inlet temperature) in HCCI operating range obtained for ethanol. The HCCI operating range is limited by combustion efficiency (>85%) and maximum pressure rise rate (<5 MPa/ms). Engine operating condition satisfying this criteria is considered in the HCCI operating range. Figure 8.6a shows that NO_x emission is ultralow (<0.04 g/kWh up to 4.5 bar IMEP) at lower engine loads for all the engine speeds because of lower combustion temperature. Highest NO_x is produced at higher engine loads around 2000 rpm, which is typical engine speed corresponding to the maximum torque for a diesel engine. Figure 8.6b depicts the N_2O emission map at the most efficient combustion phasing. Maximum N_2O emission is observed up to 2.7 ppm (Fig. 8.6b). N_2O emission is higher at lower engine load (leaner mixtures) and high speed conditions. This result is supported by experimental study, which showed that N_2O emission increases significantly at highly retarded combustion phasing and incomplete combustion for heptane HCCI combustion [8].

The partially premixed combustion (PPC) is another LTC strategy, which often uses direct injection gasoline-like fuels (to create stratification in cylinder), while

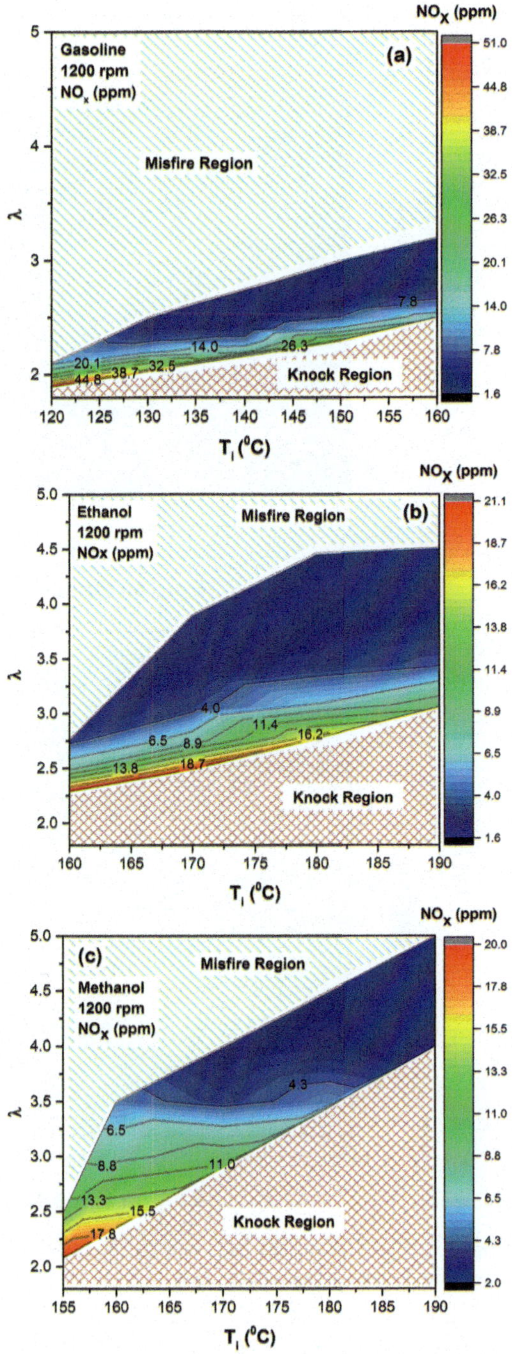

Fig. 8.5 NO_x emissions in HCCI operating range at 1200 rpm for (**a**) gasoline, (**b**) ethanol (**c**), methanol [23]

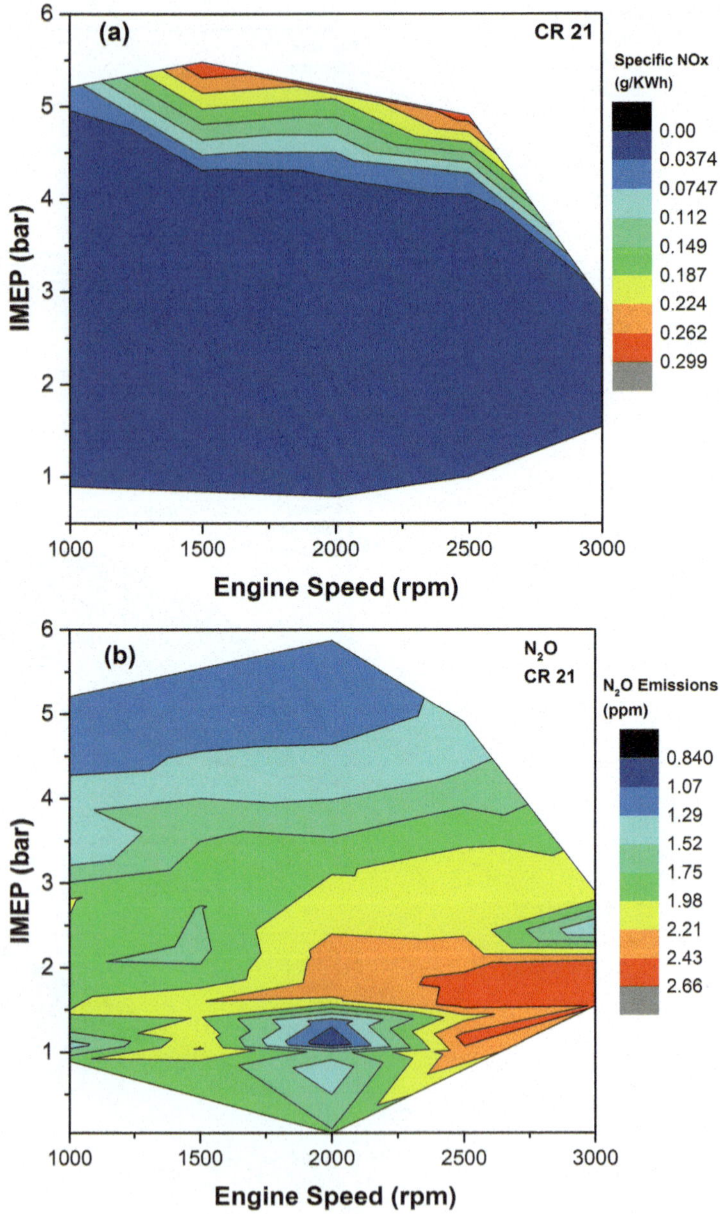

Fig. 8.6 Variation of (**a**) NO_x and (**b**) N_2O emissions with engine load and speed in HCCI operating range (Adapted from [14])

maintaining the benefits of HCCI combustion (detailed discussion is provided in Chap. 2). Figure 8.7 shows the effect of engine operating conditions on NO_x emission in PPC engine. Figure 8.7a shows the NO_x emission as function of

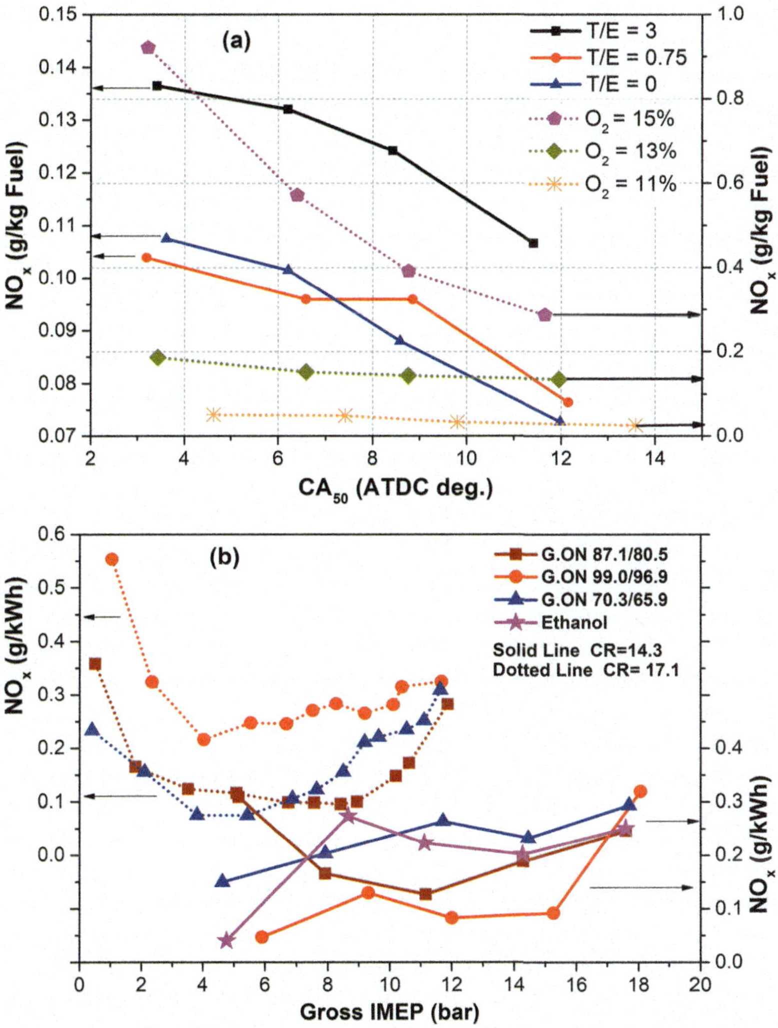

Fig. 8.7 (**a**) NO_x emissions as function of combustion phasing for different fuel composition (similar ID and RON) and different oxygen concentration in PPC combustion (Adapted from [26]), (**b**) NO_x emission as function of engine load for different gasoline-like fuels in PPC engine (Adapted from [30, 31])

combustion phasing for different fuel composition and intake oxygen concentration in PPC engine. Fuels designated in the graph show toluene to ethanol (T/E) ratio in TERF (toluene ethanol reference fuel), which is used as oxygenated gasoline surrogate. NO_x emissions are found to be closely related to the combustion phasing and decrease by retarding the combustion phasing independent on the fuels' octane number (Fig. 8.7a). Thus, NO_x emissions are a function of cylinder conditions rather than fuel properties directly. The time between the position where the thermal NO_x formation starts and position when NO_x reaction chemistry is frozen by low combustion temperature due to piston movement is restricted at late CA50s [26]. The levels of NO_x mostly depend on the cylinder conditions during the period of highest temperatures, which result in NO_x formation. Figure 8.7a shows the NO_x levels decrease with decreasing inlet O_2 concentration. The inlet O_2 concentration is controlled by the amount of EGR. Cooled EGR serves as a thermal sink and thus decreases the combustion temperature leading to reduction in NO_x emission. Additionally, lowering the inlet O_2 concentration decreases the ratio of specific heat of the cylinder charge leading to lower compression temperatures, which further decreases maximum combustion temperature [27].

NO_x emissions in PPC engine do not correlate to RON numbers of fuel [26]. Treating PRF (primary reference fuel) as a base fuel, it is evident that adding toluene increases both NO_x and smoke levels. Ethanol has the effect of decreasing both NO_x and smoke [26]. Fuel composition has an effect on PPC engine emissions. At fixed value of the RON (research octane number) and ID (ignition delay), it is easier to see the fuel composition effect on PPC engine emissions. For fixed value of RON and ID, the ratio of toluene to ethanol is varied with values of 0, 0.75 and 3 (Fig. 8.7a). Figure 8.7a depicts that higher level of toluene in oxygenated gasoline surrogate fuel leads to higher level of NO_x emissions. Detailed investigation found that increasing the ratio of toluene to ethanol (T/E) increases the emissions of NO_x, smoke and HC [28]. Another study also showed that fuels with aromatics extend the duration of NO_x formation due to local high temperature regions with combustion of decomposed hydrocarbons [29].

Figure 8.7b shows the NO_x emissions as function of engine load in PPC engine at 1300 rpm for two different compression ratios [30, 31]. Fuels designated as G.ON x/y on the figure represents gasoline octane number, where ‘x’ is RON, and ‘y’ is MON. Advanced fuel injection strategy consisting of two injections is developed in order to run high octane number fuels in a compression ignition engine (PPC) and simultaneously achieve high efficiency and low emissions [32]. The first injection is performed very early in the compression stroke, and it is used to create a homogeneous mixture, while the second is injected around TDC, and its function is to trigger the combustion event. The fuel amount in the first injection (always at −60 TDC) is independent on the engine load, and it is a function of only compression ratio, reactivity of the fuel and EGR level. The load and combustion phasing are controlled through the second injection. In this strategy, definite amount of EGR has to be added to avoid pre-ignition of fuel injected by first injection [33]. Figure 8.7b shows that for compression ratio 17.1, NO_x emission steadily decreases by decreasing the load from 12 to 4 bar IMEP due to lower combustion temperature. Below 4 bar IMEP with most of the fuels, the specific NO_x rises because a decrease in gross indicated efficiency (NO_x emission plotted with power basis). Despite the different trends, 50% of EGR was enough to keep NO_x below the Euro VI threshold value of 0.40 g/kWh with most of the fuels [33]. At lower compression ratio (CR = 14.3), higher engine load is achieved using the same fuel injection strategy. The combination of 50% of EGR and 14.3 in compression ratio allowed keeping NO_x below 0.3 g/kWh even at 18 bar gross IMEP. NO_x

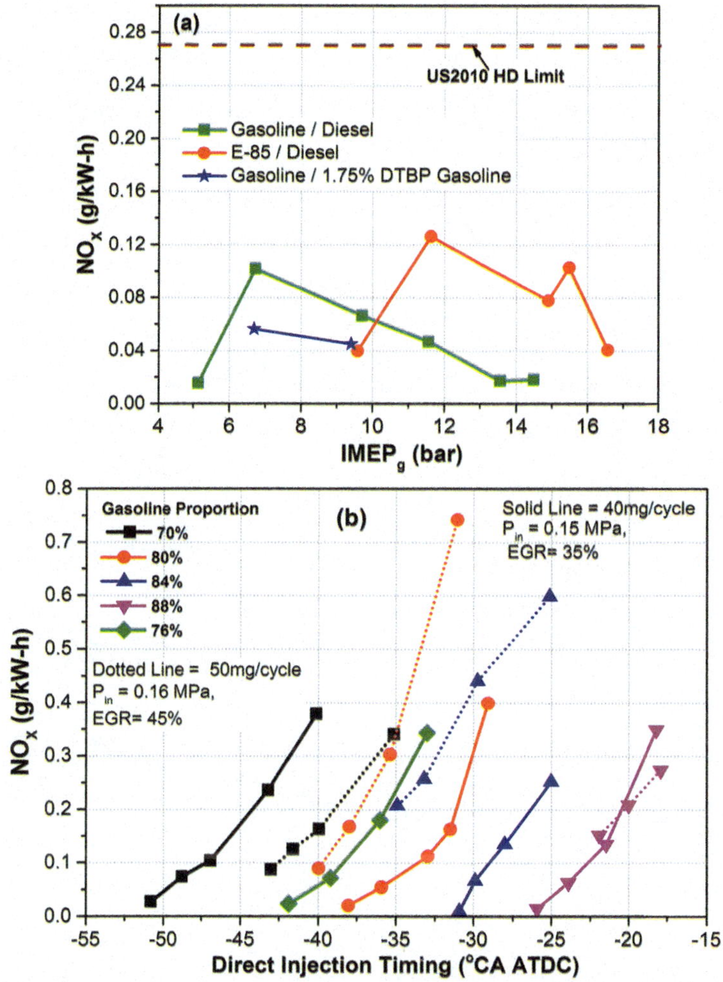

Fig. 8.8 (**a**) NO_x emissions as function of engine load in RCCI combustion with different fuels (Adapted from [34]), (**b**) NO_x emission as function of diesel injection timing for different gasoline proportions in RCCI engine (Adapted from [35]), (**c**) NO_x emission as function of EGR percentage in RCCI engine (Adapted from [36]), (**d**) effect of multi-injection on the NO_x emissions in RCCI combustion (Adapted from [36])

emission decreases with decrease in engine load for this case also because of lower combustion temperature [30].

Dual fuel reactivity controlled compression ignition (RCCI) is another LTC strategy, used to achieve higher fuel conversion efficiency along with simultaneous reduction of NO_x and PM emissions by using two fuels of different reactivity (detailed discussion is provided in Chap. 2). Figure 8.8 presents the effect of engine operating conditions on NO_x emissions in RCCI engine. Figure 8.8a shows the NO_x as function of engine load in RCCI engine using gasoline/diesel (G/D), gasoline–ethanol blend (E85)/diesel (E85/D) and gasoline/gasoline with di-tertiary butyl

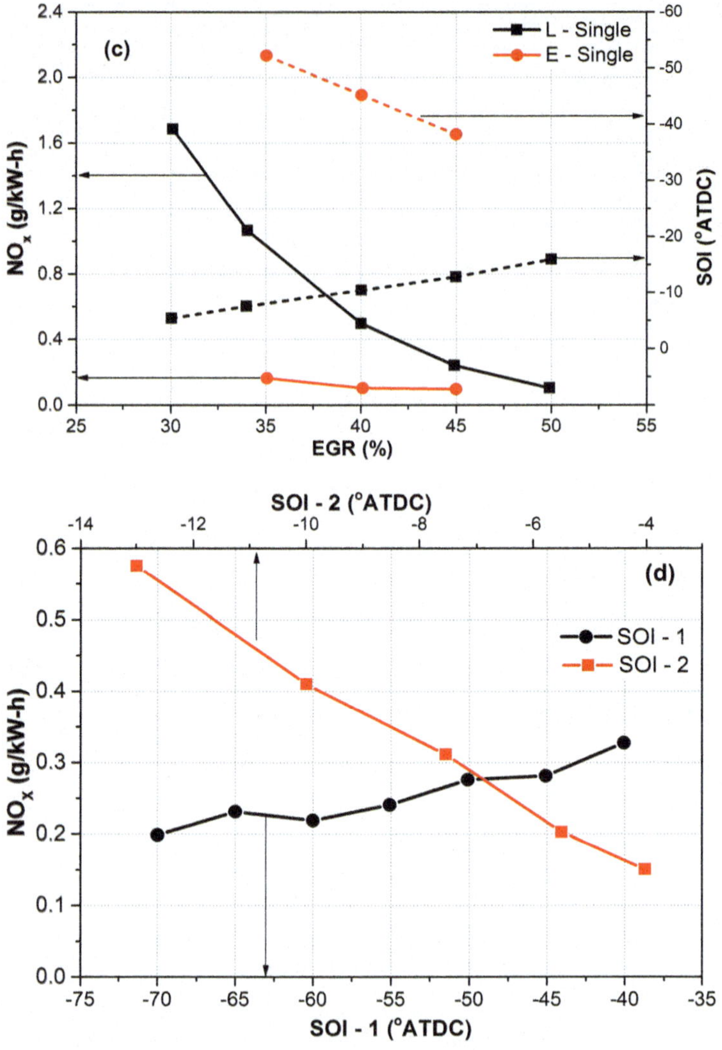

Fig. 8.8 (continued)

peroxide (DTBP) additives (G/1.75%DTBP). Figure 8.8a depicts that RCCI combustion strategy has lower NO_x emissions below the 2010 EPA HD limit without any exhaust after-treatment devices for all the test fuel combinations. Since the formation of NO_x is sensitive to the in-cylinder combustion temperature, the NO_x formation rate increases rapidly above the 1800 K combustion temperature. Low temperature combustion in RCCI combustion mode is achieved using charge dilution (mainly by EGR), which leads to the lower NO_x emissions. NO_x emission levels are slightly higher for E85/D than G/D RCCI combustion due to the different requirement of EGR to control the pressure rise rate in the cylinder. RCCI engine operation with E85 had significant differences in the EGR requirement as compared to gasoline/diesel RCCI operation. Specifically, mid-load E85/diesel RCCI operation did not require EGR, while gasoline/diesel required approximately 40% [34]. The decreased EGR requirement of E85/diesel operation as compared to gasoline/diesel offers additional benefits (such as higher thermal efficiency due to increased ratio of specific heat) beyond the different air handling requirements [34].

In RCCI combustion, fuel injection of timing of diesel affects the combustion and emission characteristics. Figure 8.8b shows the effect of diesel injection timing on NO_x emission for different gasoline proportions by using 40 and 50 mg/cycle of gasoline in RCCI engine. Figure 8.8b indicates that NO_x emission increases with retard in fuel injection timings (for before TDC injection), and this trend is similar for all the gasoline proportion. Retarded injection timing with fixed fraction of gasoline or reduced fraction of gasoline at fixed diesel injection timing leads to increase the charge stratification in the cylinder. At this condition, fuel reactivity and local equivalence ratio significantly increased. In addition to this, combustion duration and ignition delay also reduce, which results into advanced combustion phasing [35]. At the advanced combustion phasing, peak pressure rise rate and in-cylinder combustion temperature increases, which leads to higher NO_x emissions.

Figure 8.8c presents the effect of EGR and injection timings on NO_x emission in RCCI engine. Data presented in this figure is obtained at constant combustion phasing for early and late injecting timings shown in Fig. 8.8c. Early (E-single) and late (L-single) injection timings are defined with respect to injection timing of 27° bTDC, which leads to most advanced combustion phasing [36]. Advanced or retard injection timing from 27° bTDC results in late combustion phasing than the combustion phasing achieved at 27° bTDC injection timing. Very early or too late diesel injection timing results into heavily retarded combustion phasing from the TDC position, which results in unstable combustion [36]. For constant combustion phasing at 8° ATDC, the effect of EGR on the NO_x emission for early or late single injection timings is presented in Fig. 8.8c. It can be noticed that NO_x emission reduces with an increases in the EGR percentage due to lower combustion temperature with increase in dilution by EGR. Early single injection strategy prepares more homogeneous and lean charge, while late single injection strategy prepares inhomogeneous charge, and fuel may burn close to stoichiometric condition. With early single injection, combustion of lean charge results into comparatively lower combustion temperature. Therefore, comparatively lower NO_x emission is observed with early single injection strategy. Additionally, early single injection strategy is less sensitive to EGR rate as compared to late single injection strategy [36]. RCCI combustion mode achieved lower NO_x level at higher loads even with comparatively low EGR rate by the using fuel reactivity gradient inside combustion chamber.

Figure 8.8d depicts the effect of multiple direct injection timings on NO_x emission in RCCI engine. Data shown in figure is obtained by first varying pilot injection timing (SOI-1) by keeping main injection timing constant, and then by varying the main injection timing (SOI-2) by keeping pilot injection timing constant. NO_x emission reduces with advanced pilot injection timings (Fig. 8.8d) because with advanced pilot injection timings at constant main injection timing, peak in-cylinder pressure and rate of heat release are lower. In addition to this, combustion phasing is also retarded with an advanced pilot injection timing. For advanced pilot injection, the injected fuel mixed well with premixed charge due to higher availability of time for mixing. This leads to reduction in the local equivalence ratio, which leads to retarded combustion phasing, resulting in lower combustion temperature and lower peak heat release [36]. Therefore, NO_x emissions are lower when advancing the pilot fuel injection timing. Figure 8.8d also depicts that NO_x emission reduces with retarded main injection timing. At fixed pilot injection timing, the peak in-cylinder pressure and rate of heat release reduces with retarded main injection timing due to retarded combustion phasing. It leads to lower combustion temperature and, thus, NO_x emission reduces with retarded main injection timings.

Figure 8.9 shows the NO_x emission mapping with engine load and speed in RCCI operation using gasoline/diesel (G/D) and ethanol–gasoline blend (E20)/diesel–biodiesel blend (B7). Figure 8.9 demonstrates the potential of RCCI engine to achieve EURO VI NO_x levels in a heavy-duty engine. Figure 8.9a reveals that RCCI engine mapping was only possible up to 50% engine load (load referred to CDC at 1200 rpm) whatever the engine speed using gasoline/diesel, while maximum operable load to fulfil the different constraints simultaneously was 35% at all engine speeds using ethanol blend (E20) and diesel blends (Fig. 8.9b) [37, 38]. Engine operation methodology ensures to achieve the imposed limitations of maximum PRR and COV_{IMEP} without limiting the upper portion of the RCCI engine map. Detailed methodology of engine operation is provided in original study [37, 38]. Study illustrated that the main parameters to achieve stable RCCI operation are the diesel fuel injection timing and the fuel ratio of low and high reactivity fuels. The EGR rate is also another important parameter to meet the required NO_x level. The NO_x emission values are below EURO VI limit (0.4 g/kW h) for all the test conditions (Fig. 8.9). Additionally, an increase in NO_x levels is noticed as engine load is increased, and NO_x emission trends depicted a stronger dependency on engine speed (Fig. 8.9a) especially for gasoline/diesel RCCI operation.

8.2 Carbon Monoxide Emissions

Carbon monoxide (CO) is one of the main regulated emissions from internal combustion engines. The CO emission is a result of partial or incomplete oxidation of the fuel in the cylinder, and it strongly depends on the combustion temperature for homogeneous combustion of lean mixtures. Reaction mechanism involved in the oxidation of fuel in LTC engines during low and high temperature heat release is discussed in Sect. 2.2 of Chap. 2. Figure 2.1 (Chap. 2) presents the CO formation region as function of local equivalence ratio and combustion temperature. The CO formation is higher at comparatively lower combustion temperature and higher

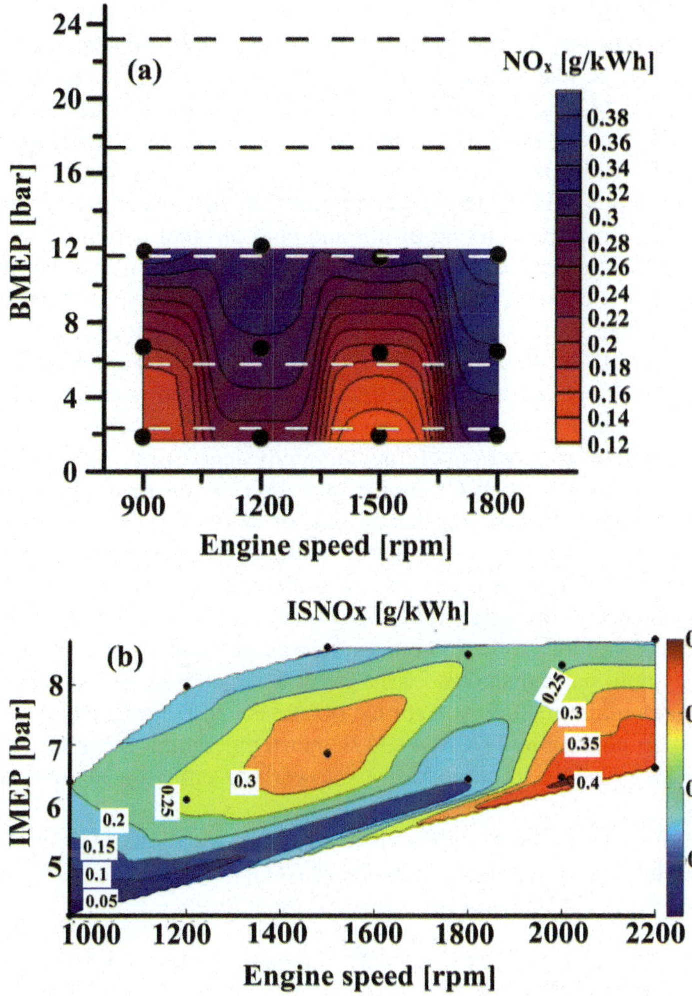

Fig. 8.9 NO_x emission mapping of RCCI operation using (**a**) gasoline/diesel (Adapted from [37]), and (**b**) ethanol–gasoline blend (E20)/diesel–biodiesel blend (B7) (Adapted from [38])

equivalence ratios. The CO-to-CO_2 conversion reactions are especially sensitive to the combustion temperature. Incomplete combustion occurs in HCCI engine because the combustion temperatures at very dilute conditions are too low to complete the CO-to CO_2 reactions before quenching due to expansion cooling. For HCCI operation with iso-octane and combustion phasing close to TDC, a critical maximum temperature of 1450 K is identified for CO formation at 1200 rpm engine speed [39, 40]. In HCCI combustion, CO emission is dependent on air–fuel ratio and combustion phasing. Richer fuel–air mixture and advanced combustion phasing lead to comparatively lower CO emission, and at leaner mixtures with retarded combustion phasing results into very high CO emission.

Close to rich limit of HCCI combustion and with early combustion phasing, very little CO is produced in the cylinder. But close to lean limit, high amount of CO can be generated due to incomplete combustion [2, 41]. The HCCI engine operation with incomplete bulk-gas reactions is not only undesirable from an emissions standpoint but also can result into a substantial reduction in fuel economy due to heavy deterioration in combustion efficiency [42]. A study investigated the sources of the CO emissions in HCCI engines and found that the CO emission is influenced by fluid–wall interactions, mixing of hot and cold air–fuel particles and the wall temperature [43]. Inhomogeneities introduced by these factors play an important role in the formation of CO emissions. The inhomogeneities occurring during compression stroke affect the CO formation rate, and inhomogeneities persisting in the expansion stroke dictate the level of HCCI exhaust CO emissions.

Figure 8.10 presents the effect of different engine operating conditions in HCCI combustion engine. Figure 8.10a shows the variation of fraction of fuel carbon converted into CO as function of fuel/air equivalence ratio (ϕ) for two fuels (iso-octane and PRF80) in HCCI engine operation at constant combustion phasing at TDC. Figure 8.10a depicts that for higher ϕ, the CO-to-CO_2 conversion is very effective for both fuels due to higher combustion temperatures. At higher ϕ, a small amount of fuel carbon is left as CO in the engine exhaust, which is mostly credited to late and incomplete oxidation of fuel contained in crevices [42, 44]. As the ϕ is gradually reduced, the CO emission increases, first progressively and then more quickly until more than 60% of the fuel carbon ends up as CO for the lowest Φ. Intake temperature requirement for iso-octane is higher than PRF80 to maintain the combustion phasing at TDC. Due to higher temperature operation, CO oxidation takes place leading to comparatively lower CO emission in iso-octane than PRF80.

A study investigated the reactions involved in the oxidation of CO into CO_2, and it suggested three most important pathways of CO oxidation (R12–R14) along with the fraction of total CO_2 produced by each of them [42]:

$$CO + OH \rightarrow CO_2 + H \text{ (78\% of total } CO_2) \quad \text{(R12)}$$

$$CO + HO_2 \rightarrow CO_2 + OH \text{ (7.2\% of total } CO_2) \quad \text{(R13)}$$

$$CO + O_2 \rightarrow CO_2 + O \text{ (1.5\% of total } CO_2) \quad \text{(R14)}$$

Reaction R12 dominates the CO oxidation, and it is dependent on OH radical concentration, which is sensitive to the charge temperature. The OH level drops off rapidly by decreasing combustion temperature in the cylinder. Therefore, the OH level becomes so low that the CO oxidation does not go to completion below a particular combustion temperature. Combustion temperature is affected by combustion phasing, which in turn affects the CO emission.

Figure 8.10b shows the variations of CO emission with combustion phasing using gasoline at 1200 rpm for different λ in HCCI combustion. Combustion phasing has almost linear correlation with CO emissions (Fig. 8.10b). On advancing the combustion phasing, CO emission decreases due to increase in combustion

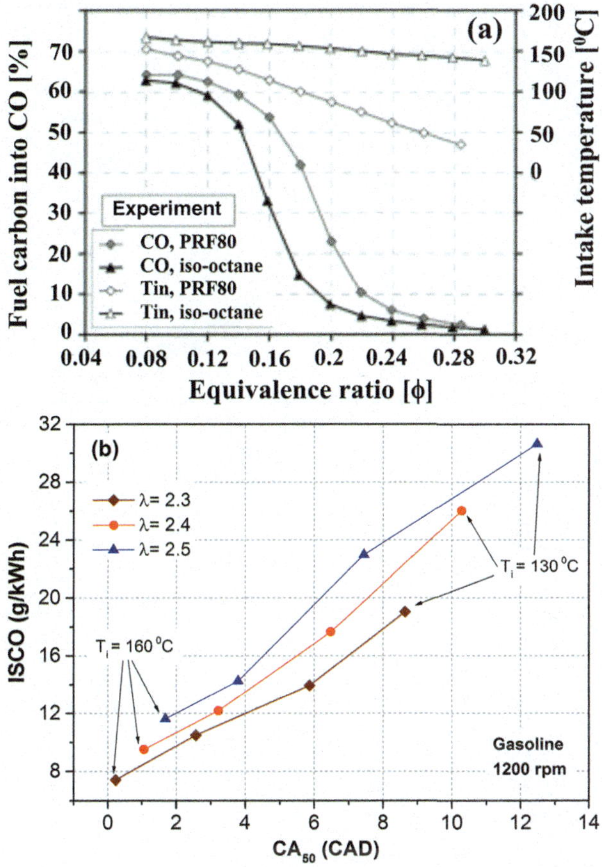

Fig. 8.10 (**a**) CO emission as function of equivalence ratio [42]. (**b**) Effect of combustion phasing on CO emission at different λ using gasoline [24], (**c**) CO emission as function of maximum mean gas temperature at 1200 rpm [24], and (**d**) effect of IMEP and T_i on CO emission using ethanol at 1200 rpm [24] in HCCI engine

chamber temperature, and there is more time available for in-cylinder oxidation. Figure 8.10b also reveals that the richer mixtures (lower λ) have lower CO emissions as compared to leaner mixtures. Late combustion phasing and leaner mixtures result in higher CO emission from HCCI engines due to lower combustion temperature. Figure 8.10c shows the variation of CO emission with maximum mean gas temperature of combustion chamber at 1200 rpm. Figure depicts that on increasing the combustion chamber temperature, CO emission drastically reduce for both fuels. The correlation of CO emission with T_{max} is stronger than linear correlation, and at combustion temperature below 1800 K, CO emission increases rapidly. Figure 8.10d shows the CO emission as function of IMEP and intake air temperature (T_i) using ethanol at 1200 rpm. The CO emission decreases radically with increasing engine load for each T_i. This is due to increase in combustion chamber temperature as higher amount of fuel burnt to produce higher IMEP.

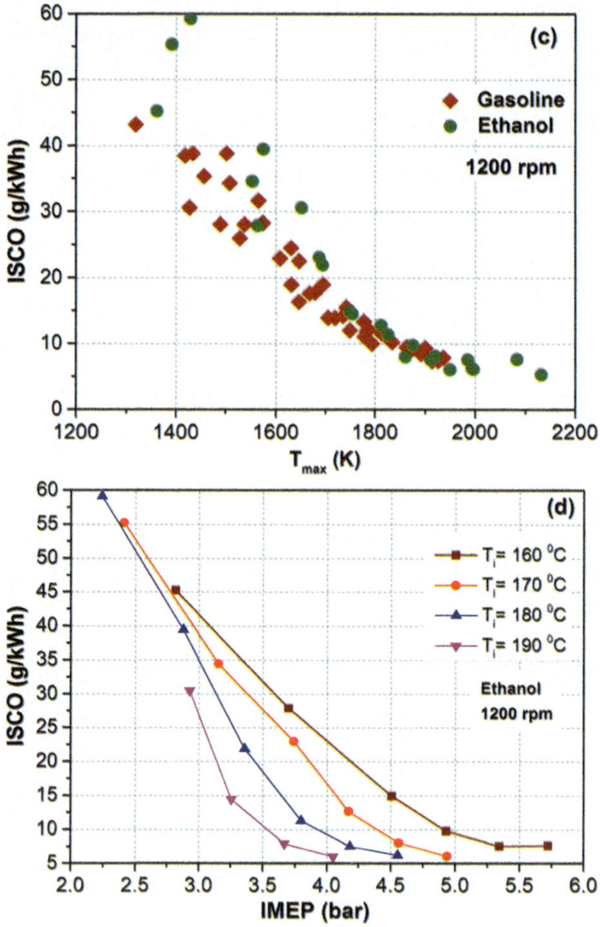

Fig. 8.10 (continued)

Therefore, CO emission increases quickly with reduction in engine load, and it is higher as compared to conventional engine combustion modes.

Figure 8.11 shows the CO emission as function of λ and T_i for gasoline, ethanol and methanol in HCCI operating range at 1200 rpm. Figure shows that the contour lines of CO emission are inclined, which indicates that CO is dependent on λ and T_i. Higher horizontal inclination represents more dependency on λ. Operating conditions close to the rich limit and knocking region lead to the advanced combustion phasing and result into lower CO emission for all test fuels. But operating conditions close to the lean limit and misfire region produce very high CO for all fuels. The maximum CO emission is as high as 55 g/kWh for methanol. The CO emission in HCCI operating range using gasoline is lower as compared to ethanol and methanol at 1200 rpm. A study summarized the CO emissions from several conventional CI engines used in various studies and found that CO emission from

Fig. 8.11 CO emission as function of λ and T_i in HCCI operating range for gasoline, ethanol and methanol (Adapted from [1])

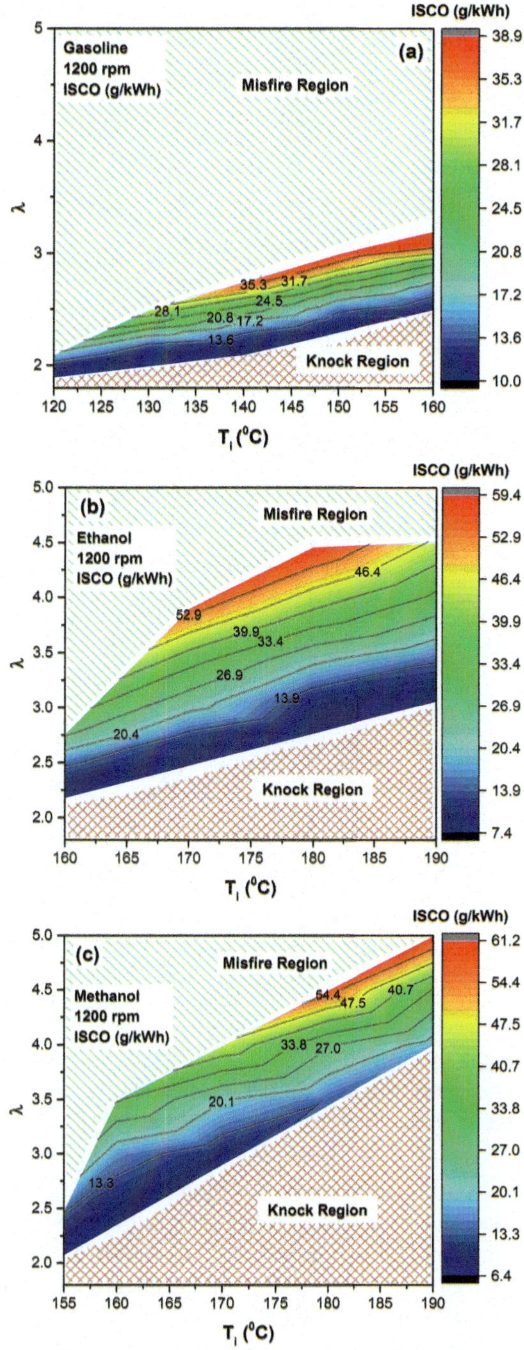

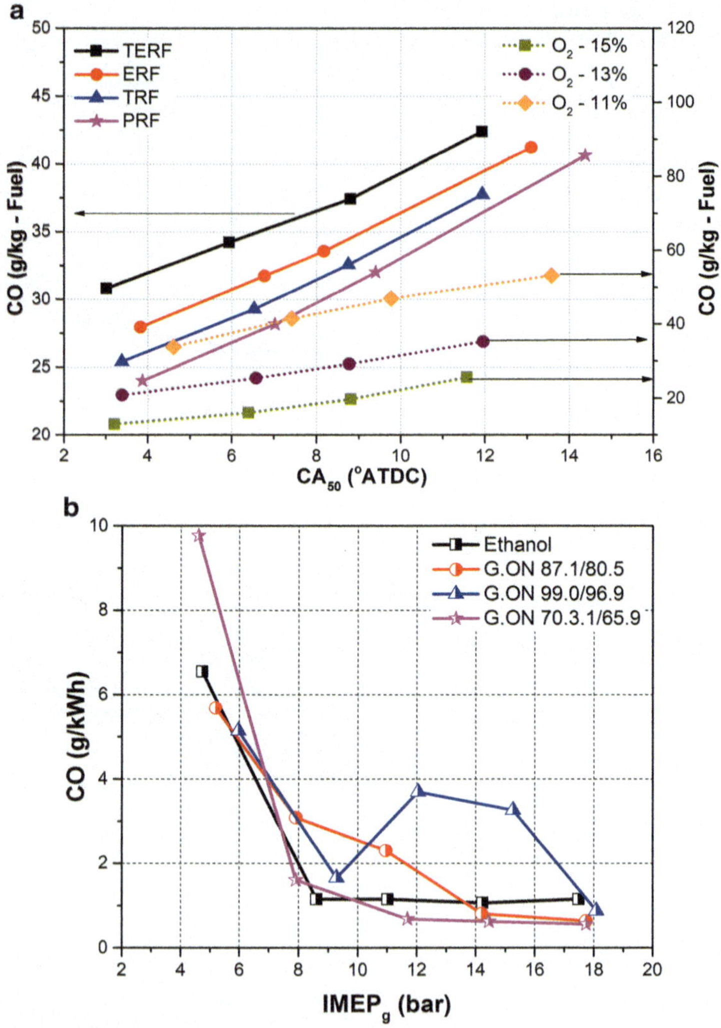

Fig. 8.12 (**a**) CO emissions as function of combustion phasing for different fuels and different oxygen concentrations in PPC combustion (Adapted from [26]), (**b**) CO emission as function of engine load for different gasoline-like fuels in PPC engine (Adapted from [30])

conventional CI engines is in the range of 0.5–15.0 g/kWh, depending on the engine load and a variety of other factors [45]. Indicated specific CO emission (ISCO) in Fig. 8.11 is in range of 5.0–55.0 g/kWh, depending on fuel and combustion timings. Therefore, CO emission in HCCI engine is generally higher in comparison to conventional CI engines.

Figure 8.12 presents the effect of combustion phasing and engine load on CO emissions in PPC engines. CO emissions increased with retarded CA_{50} in PPC engine (Fig. 8.12a). An increase in HC and CO emissions implies an incomplete

PPC combustion. There are several potential reasons for an incomplete combustion, but the main factor in relation to CA_{50} is the decrease in the in-cylinder temperature during expansion stroke. In expansion stroke, cylinder temperature reaches a point when the combustion is not able to complete while CA_{50} is retarded. The decrease in temperature leads to lower reaction rate [26]. As the inlet O_2 concentration is decreased (dilution by EGR), the local combustion temperatures become lower, leading to substantial increases in CO emissions. Fuel composition did not show an effect on CO emissions; the observed trend in Fig. 8.12a is due to research octane number (RON) variations in TERF, TRF (toluen reference fuel), ERF (ethanol reference fuel) and PRF [26]. Figure 8.12b shows that CO emission decreases with increase in engine load of PPC engines. The CO emission values in PPC engine (figure 8.12b) are comparatively lower than in HCCI engine (Fig. 8.11). For engine load greater than 8 bar IMEP, the CO emission values are lower than EURO VI legislation values (4 g/kWh). The low values of CO (Fig. 8.12b) and NO_x (Fig. 8.7b) emissions are suggesting that with high octane number fuels running in PPC mode, it is possible to burn the fuel–air mixture in the temperature range between 1500 and 2000 K. Combustion temperature higher than 1500 K is necessary to promote the reactions of CO to CO_2 conversion; on the other hand, it is important to keep combustion temperature below 2000 K to avoid thermal NO_x formation [30].

Figure 8.13 shows the effect of engine operating conditions on CO emissions of dual fuel RCCI engine. The RCCI combustion is a LTC strategy achieved by controlling the reactivity in cylinder using comparatively lower amount of EGR. Figure 8.13a shows the variation of CO and CO_2 emission with engine load in RCCI combustion using different fuels. Figure reveals that with an increase in the engine load, CO emission decreases due to the higher combustion temperature with increase in the engine load for all the fuels [46]. Similar trend is also observed in another RCCI study [47].

Since the CO formation is strongly dependent on fuel–air ratio, thus, premixing of fuel has significant effect on the CO emissions. Figure 8.13b presents the variations of CO emission with methanol and gasoline premixing ratio in RCCI combustion. The diesel injection timing has to be adjusted with increases in gasoline proportion to maintain constant CA_{50} at any EGR rate. Figure indicates that with an increase in the fuel premixing ratio, CO emission decreases for both the fuels (i.e. gasoline and methanol). With an increase in the gasoline proportion, the global reactivity of the mixture reduces, premixed combustion increases, and thus, comparatively higher fuel will burn simultaneously to keep constant combustion phasing. This leads to comparatively higher combustion temperature resulting in lower CO emission. Figure 8.13b also depicted that CO emission reduces monotonically with an increase in the methanol proportion. With an increase in the methanol proportion, mean in-cylinder temperature increases (temperature distribution is shown in original study [48]), which leads to reduction in the CO emissions. Oxygenated content of methanol may also contribute in lower CO emission.

The premixed LTC engines are very sensitive to initial thermodynamic conditions. Figure 8.13c shows the effect of intake temperature on the CO emission for

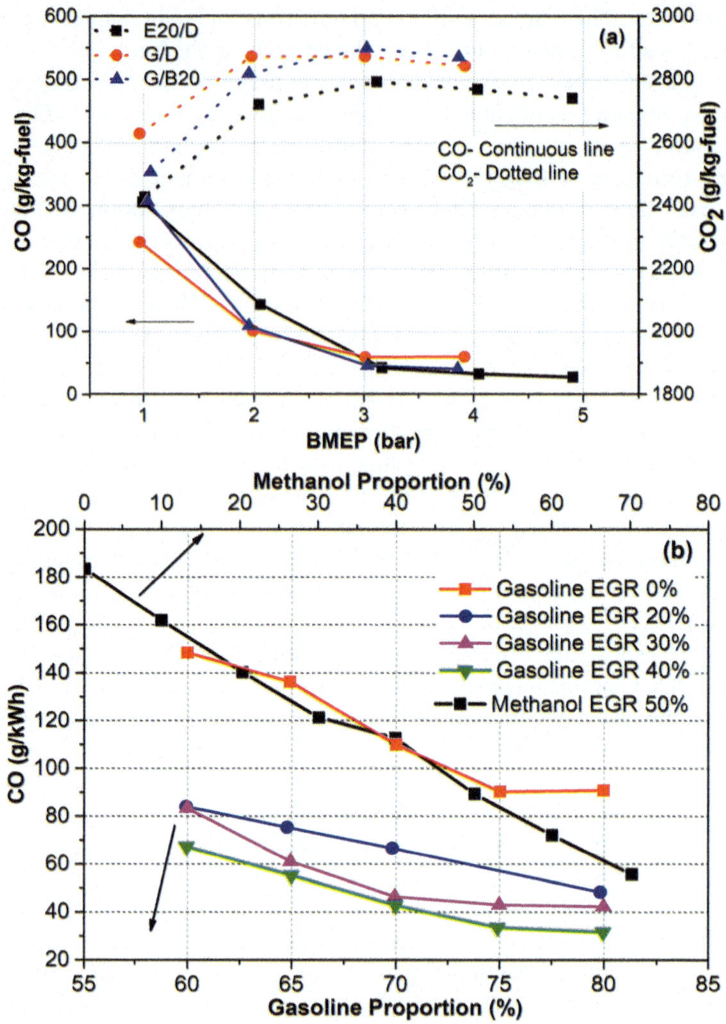

Fig. 8.13 (**a**) CO emissions as function of engine load in RCCI combustion with different fuels (Adapted from [46]), (**b**) CO emission as function of different gasoline and methanol proportions in RCCI engine (Adapted from [35, 48]), (**c**) CO emission as function of intake temperature in single and dual fuel RCCI engine (Adapted from [49]). (**d**) Effect of multi-injection on the CO emissions in RCCI combustion (Adapted from [36])

gasoline/diesel (dual fuel) and gasoline/gasoline + 2-EHN (single fuel) RCCI engine operation. With an increase in the intake temperature, combustion phasing advances and peak heat release rate increases, which results in higher in-cylinder combustion temperature. Thus, CO emission reduces with increase in intake temperature for both the fuels. Figure 8.13c also reveals that gasoline/diesel RCCI operation has a lower CO emissions as compared to gasoline + 2-EHN RCCI operation.

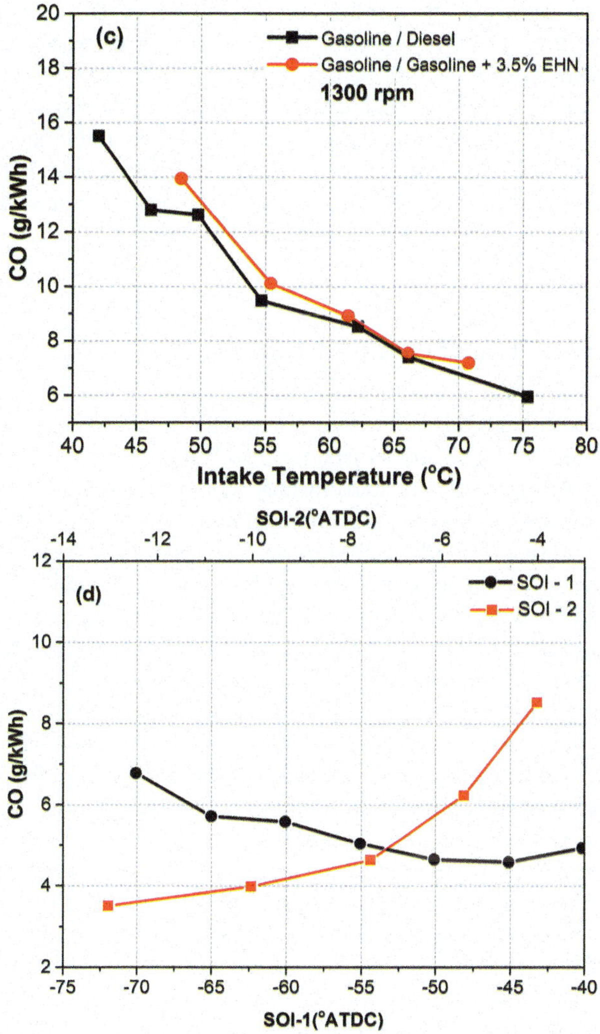

Fig. 8.13 (continued)

Figure 8.13d shows the effect multiple injection timings on the CO emission in RCCI combustion. Data shown in figure is obtained by first varying pilot injection timing (SOI-1) by keeping main injection timing constant and then by varying the main injection timing (SOI-2) by keeping pilot injection timing constant. With advanced pilot injection timing (SOI-1), diesel is injected into cylinder at comparatively lower charge temperature. Due to higher ignition delay, charge gets more timing for mixing, and it becomes more homogeneous and reduces the local fuel–air equivalence ratio. The combustion of charge with lower local equivalence ratio retards the ignition timing and lower peak of high temperature heat release rate achieved leading to lower in-cylinder combustion temperature. Thus, higher CO emission is observed at advanced pilot injection, and it decreases as pilot retards. Figure 8.13d also depicts that with retard in the main injection timing (SOI-2) while keeping pilot injection timing constant, the CO emission increases due to retarded combustion phasing. Retarded combustion phasing obtained due to late main injection timing leads to lower peak in-cylinder pressure and heat release rate [36]. It leads to reduction in the in-cylinder combustion temperature; thus, CO emission increases with retarded main injection timing.

8.3 Unburned Hydrocarbon Emissions

Although LTC strategies are praised for potentially high fuel conversion efficiency as well as ultralow NO_x and soot emissions, LTC engines produce higher unburned hydrocarbons (HC) and CO than conventional SI and CI engines, particularly at part load engine operating conditions. In homogeneous charge spark ignition engines, six significant sources of HC emission are identified: (i) combustion chamber crevice volumes, (ii) flame quenching both at the wall and in the bulk, and incomplete combustion, (iii) lube oil layers (fuel vapour absorption/desorption), (iv) deposits on the combustion chamber walls (fuel vapour/liquid absorbed in porous deposits), (v) liquid fuel in the cylinder (fuel does not evaporated and mix with sufficient air to burn) and (vi) exhaust valve leakage [50]. In conventional diesel engine, five main sources of HC emissions are identified: (i) overmixing of fuel and air beyond lean flammability limits; (ii) under-mixing to fuel–air ratios that are too rich for complete combustion; (iii) spray over penetration, i.e. impingement of fuel spray on the cylinder wall; (iv) bulk quenching of combustion reactions due to mixing with cool air or volume expansion; and (v) poorly atomized fuel from the nozzle sac volume and nozzle holes after end of injection [51]. Several sources of unburned HC emission are present in LTC engine depending on the combustion strategies. In HCCI combustion, charge preparation is similar to SI engine and combustion imitated by autoignition. Therefore, main sources of HC emissions in HCCI engines are incomplete combustion (partial burning and misfiring), crevice volumes, wall quenching and adsorption/desorption of fuel vapours into oil layers of the cylinder walls. The crevice volume between the piston and cylinder liner with narrow entrances is one of the primary sources of HC emissions by trapping fuel vapours [52]. Crevice volume can also contribute to HC emission in LTC strategies such as PCCI and PPC, where early direct injection of fuels is used for charge preparation as fuel gets more time to mix with air. Bulk quenching is another mechanism of HC emission in LTC strategies due to incomplete oxidation reaction because of lower combustion temperature and/or local equivalence ratio is too fuel lean or too fuel rich [53].

Figure 8.14 presents the effect of different engine operating conditions on HC emissions from HCCI combustion engine. Figure 8.14a shows the variation of total unburned hydrocarbon (THC) emission with combustion phasing (CA_{50}) for

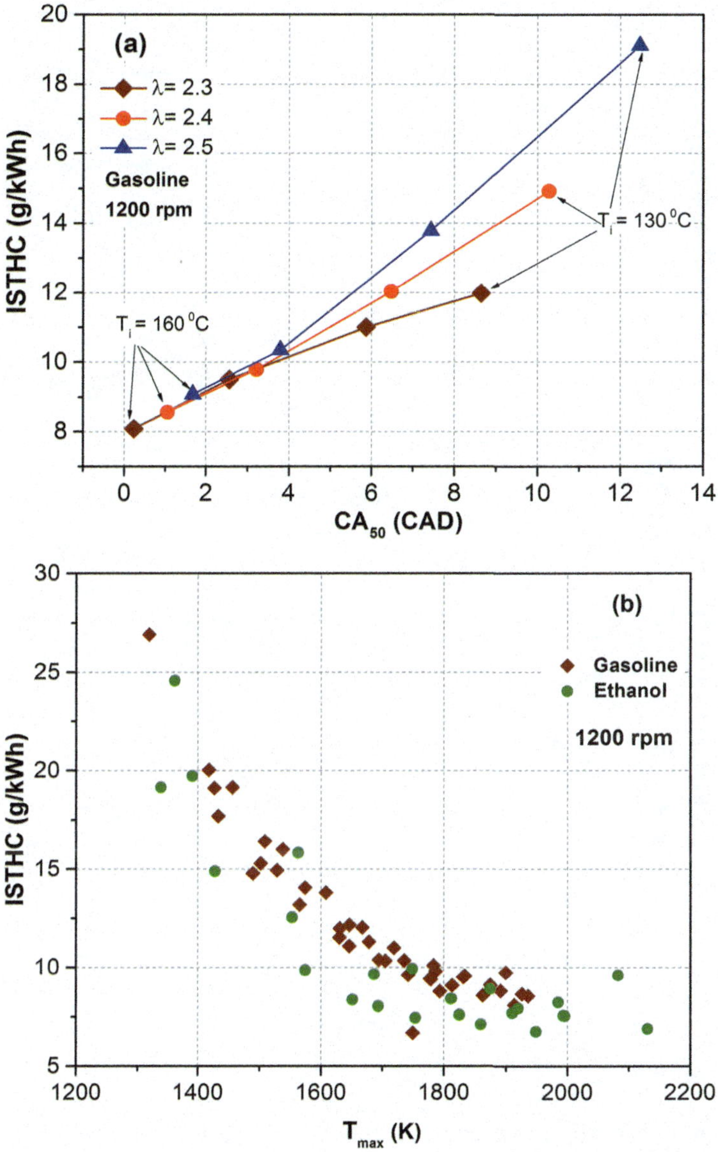

Fig. 8.14 (**a**) Effect of CA_{50} on HC emissions at different λ for gasoline [24], (**b**) HC emissions as function of maximum mean gas temperature. (**c**) Effect of IMEP and T_i on HC emissions for gasoline and ethanol [24], and (**d**) HC emissions as function of EGR (Adapted from [54]) in HCCI engine

gasoline at 1200 rpm at different λ. Combustion phasing has almost linear correlation with THC emissions (Fig. 8.14a). The trend of THC emissions is very similar to the trends of CO emissions (Fig. 8.10b). On advancing the combustion phasing or

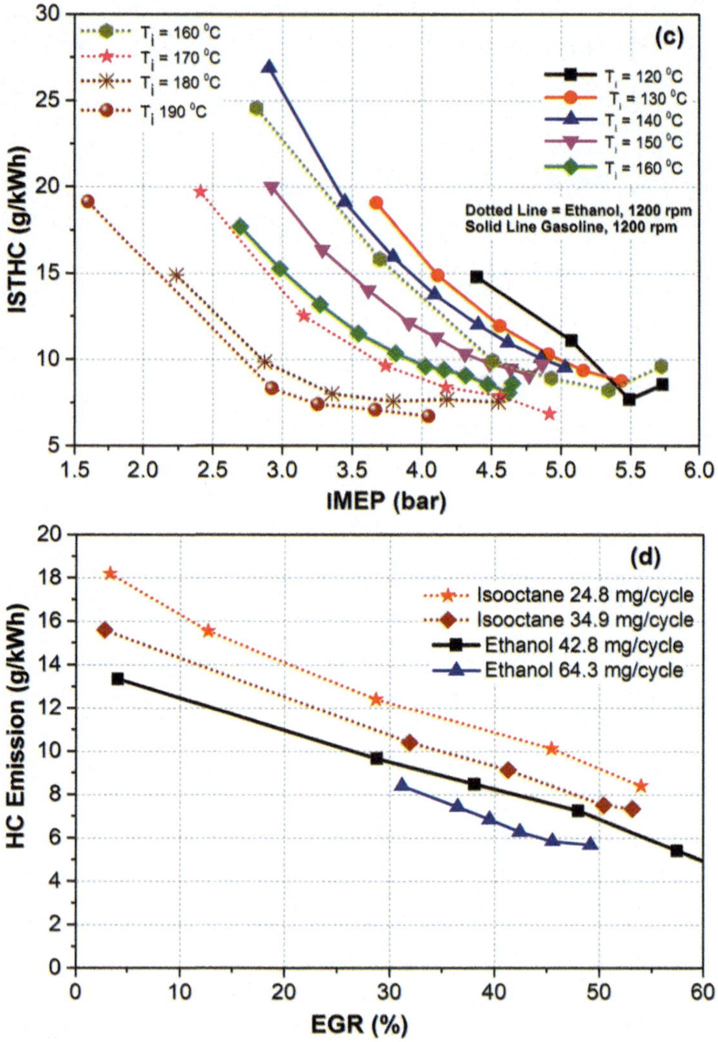

Fig. 8.14 (continued)

increasing intake air temperature, THC emission decreases due to increase in combustion chamber temperature. It can also be noticed that richer mixtures have lower THC emission as compared to leaner mixtures. Figure 8.14b shows the variations of THC emission with maximum mean gas temperature (T_{max}) of combustion chamber at 1200 rpm in HCCI engine. Figure shows that in increasing the combustion chamber temperature, THC emissions are drastically reduced for both fuels. It is also observed that correlation of THC emissions with T_{max} is stronger than linear correlation and close to exponential correlation. As temperature of combustion chamber falls below than 1700 K, THC emissions increase rapidly. Figure 8.14c shows the effect of IMEP and intake air temperature (T_i) on THC formation using gasoline and ethanol at 1200 rpm. THC formation decreases drastically with engine load at each intake air temperature for both fuels. On increasing engine load, combustion chamber temperature increases due to higher fuel quantity burnt to produce higher IMEP at each intake air temperature. It can be noted that THC emission is very high (>10 g/kWh) for IMEP values less than 4 bar for gasoline at each intake air temperature, and THC emissions for ethanol are lower as compared to gasoline at 1200 rpm (Fig. 8.14c). Figure 8.14d shows the variation of HC emissions with EGR in HCCI engine operated at constant fuel quantity using ethanol and iso-octane. The use of EGR retards the start of combustion and slows down the rate of combustion, which makes the engine to operate with less combustion noise. To initiate the combustion near TDC, higher intake temperature is used for higher EGR rate. The specific HC emissions decrease with EGR (Fig. 8.14d). This may be due to reduced flow of exhaust gases with increase in the amount of EGR as the volume fraction of unburned hydrocarbons does not show any obvious dependence on the EGR [54]. Utilization of EGR increases the exhaust gas temperature, which increases the possibility of using oxidizing catalyst, to oxidize the unburned hydrocarbons from HCCI engine.

Figure 8.15 shows the variations of THC emission using gasoline, ethanol and methanol in HCCI operating range at 1200 rpm. Inclined contour lines of THC emission indicate that THC emission depends on both λ and T_i. Typically, level of THC emissions from HCCI engine is higher over the entire operating range in comparison to conventional diesel engines. THC emissions increase with increase in λ (i.e. leaner mixture) due to comparatively lower combustion temperature achieved at lean engine operation. These conditions lead to lower oxidation rate of the fuel and incomplete combustion of some fuel/air mixtures during expansion stroke. Figure 8.15 also shows that mixture strength close to the rich limit and knocking region produces lower THC due to advanced combustion phasing and higher combustion temperatures. However, engine operating points near lean limit or misfire region generates higher HC due to late combustion phasing and lower temperature [1]. Maximum specific HC emission in this study is around 18 g/kWh for gasoline and ethanol at lowest engine loads. Methanol is found to have comparatively lower HC emissions than gasoline and ethanol in HCCI operating range (Fig. 8.15).

Piston top land crevices are one of main factors for higher HC emission in HCCI engines. The top ring-land crevice region is fairly cold, and a temperature gradient exists from the wall outwards through the boundary layer to the bulk gas, which is at the highest temperature [55]. The temperature gradient is quite steep near the wall and at the entrance to the ring-land crevice. This temperature gradient becomes progressively less steep going away from the wall and is quite shallow as the boundary layer merges into the bulk gas [55, 56]. The temperature gradient in the boundary layer is very large during the late part of the compression stroke and the early part of the expansion stroke. The temperature difference across the boundary layer can exceed 1000 K as the mean cylinder wall temperature is about 450–500 K and the maximum mean gas temperature is in the order of 1500–2000 K, depending on the operating conditions. At particular equivalence ratio, combustion raises the temperature above this initial distribution, by an amount proportional to the

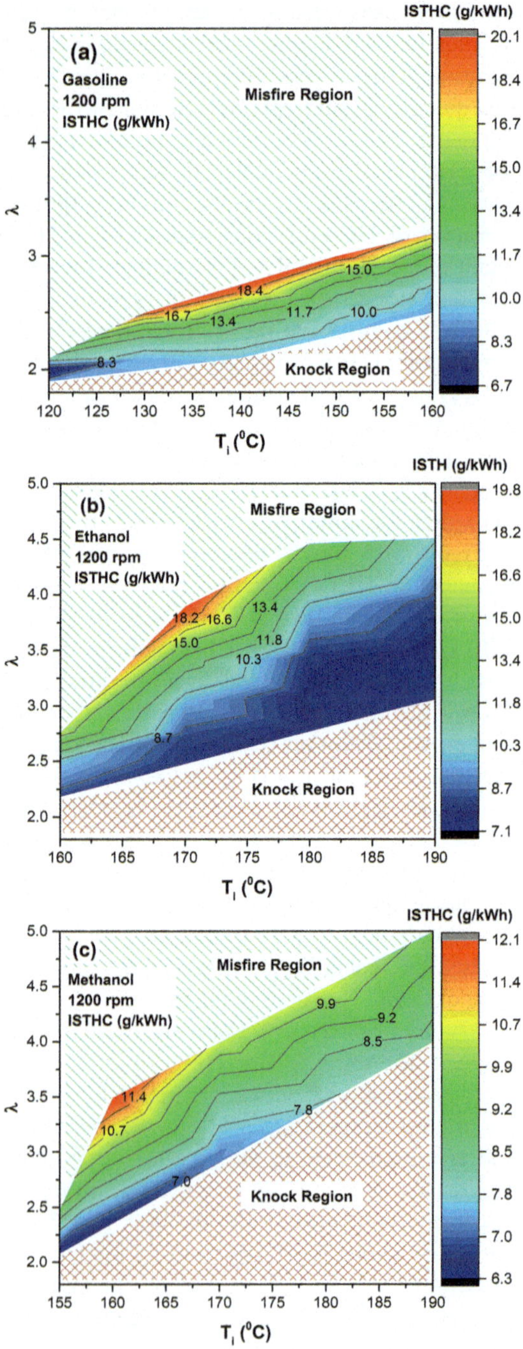

Fig. 8.15 THC emission as function of λ and T_i in HCCI operating range for gasoline, ethanol and methanol (Adapted from [1])

combustion heat release. As equivalence ratio is progressively reduced, the reduced combustion heat release causes increasingly greater regions of the charge mass to fall below the temperature of complete combustion due to the shape of the boundary-layer temperature gradient. Therefore, lowering equivalence ratio leads to progressive change in the HC emissions [56]. The importance of piston top land geometry on HCCI engine HC emission is investigated numerically and experimentally [44, 57].

Figure 8.16a shows the variations of HC emission with λ for different top land width in HCCI engine at different engine speeds. With the smallest top land width

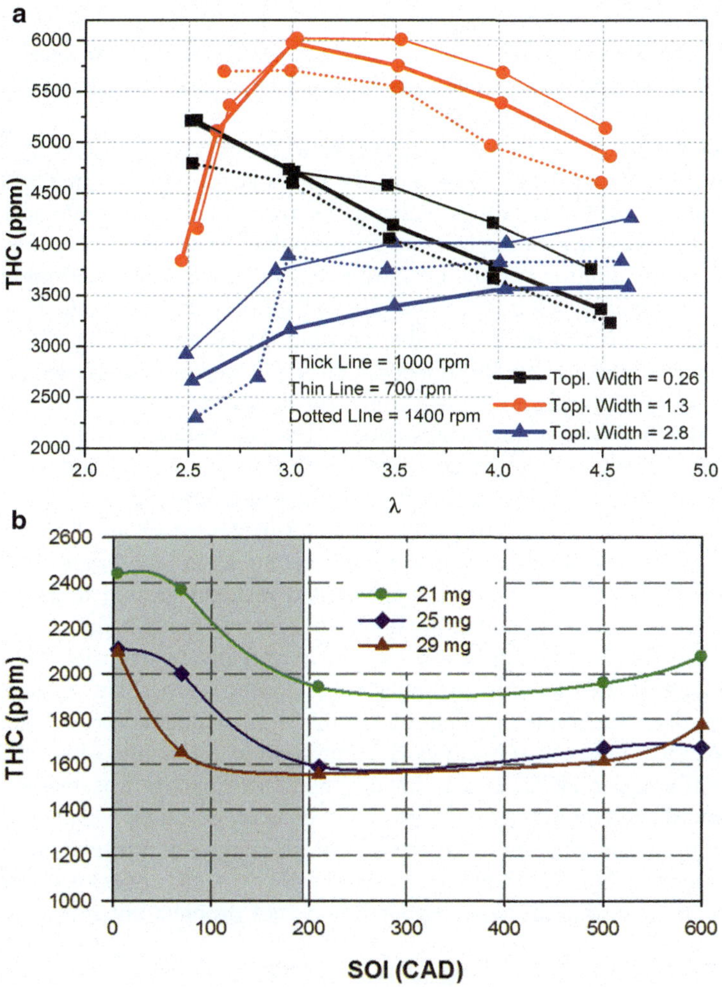

Fig. 8.16 (**a**) Dependence of λ on HC emission for different top land widths at various engine speeds in HCCI combustion (Adapted from [57]), (**b**) HC emission as function PFI injection timings at 1500 rpm in HCCI engine for different fuel quantity injected per cycle [24]

(0.26 mm), HC decreases with increased λ (leaner mixture) because less fuel will be trapped in the top land and will not participate in the normal combustion process (Fig. 8.16a). For lean mixtures (λ = 3.5–4.5), an increase in top land width to 1.3 mm results in more HC and the trend is similar to that of the smallest top land. At $\lambda = 3$, a changing trend can be observed, and a further enrichment leads to a decrease in HC, and at $\lambda = 2.5$, the amount of HC is lower compared to the 0.26 mm top land width case (Fig. 8.16a). This observation suggests that some of the charge trapped in the top land takes part in the combustion [57]. The largest top land width (2.8 mm) shows the lowest overall HC, except at $\lambda = 4.5$, where the 0.26 mm case is the lowest. At $\lambda = 2.5$, HC is reduced by 45% compared to the 0.26 mm case at 700 rpm. Increase in engine speeds leads to lower HC emission for different top land width except 2.8 mm case for $\lambda > 3.0$ (Fig. 8.16a). Study also found that the HC emissions are much more sensitive to combustion phasing with a large top land width and with a narrow top land geometry; only a modest change in HC emission is noted with combustion phasing [57].

In HCCI engine, premixed charge preparation is typically done by port fuel injection (PFI) strategy. In PFI engines, the fuel injector typically injects the liquid gasoline at the back of the intake valve and surrounding intake port surface areas. Only a small part of evaporates from fuel droplets produced during atomization, and most of fuel impinge on the back of intake valve and associated port surfaces. Thus, the behaviour of the liquid fuel after injection is strongly influenced by the PFI timings relative to the inlet valve opening (IVO) and closing (IVC) timings [50]. In closed intake valve PFI injection, the rapid backflow of burned gases from the cylinder immediately following after IVO assists the fuel vaporization process off the valve and port surfaces substantially [58], and almost all the gasoline fuel enters the cylinder as vapour in fully warmed-up engine operating conditions. However, in open valve PFI injection, higher amount of fuel can enter the cylinder as liquid, as droplets or as ligaments entrained in the intake air flow. During the intake and compression process, major fraction of entered liquid fuel evaporate and mix with the in-cylinder air, fuel vapour, and residual gas, to form the normal combustible mixture. The remaining liquid fuel entered in the cylinder (particularly the less volatile fuel components) may be stored in the deposits, oil layers or crevices, and come out into the bulk gases during the expansion and exhaust processes. The incomplete oxidation of this fuel would contribute to the HC emissions [50]. Figure 8.16b shows HC emissions as a function of the PFI start of injection timings in HCCI engine for different fuel quantity injected per cycle at 1500 rpm. Shaded region in the figure shows the IVO duration. Figure 8.16b shows that total unburned HC emissions are almost constant and independent of the fuel injection timings in IVC timings as expected. However, HC emissions increase with IVO injection timings because higher liquid fuel droplets enter the combustion chamber.

Premixed charge compression ignition (PCCI) is low temperature combustion strategy typically used for diesel-like fuels (detailed description can be found in Chap. 2). In PCCI strategy, typically diesel is injected early in the compression stroke to prepare the premixed charge. A study compared the effect of crevice

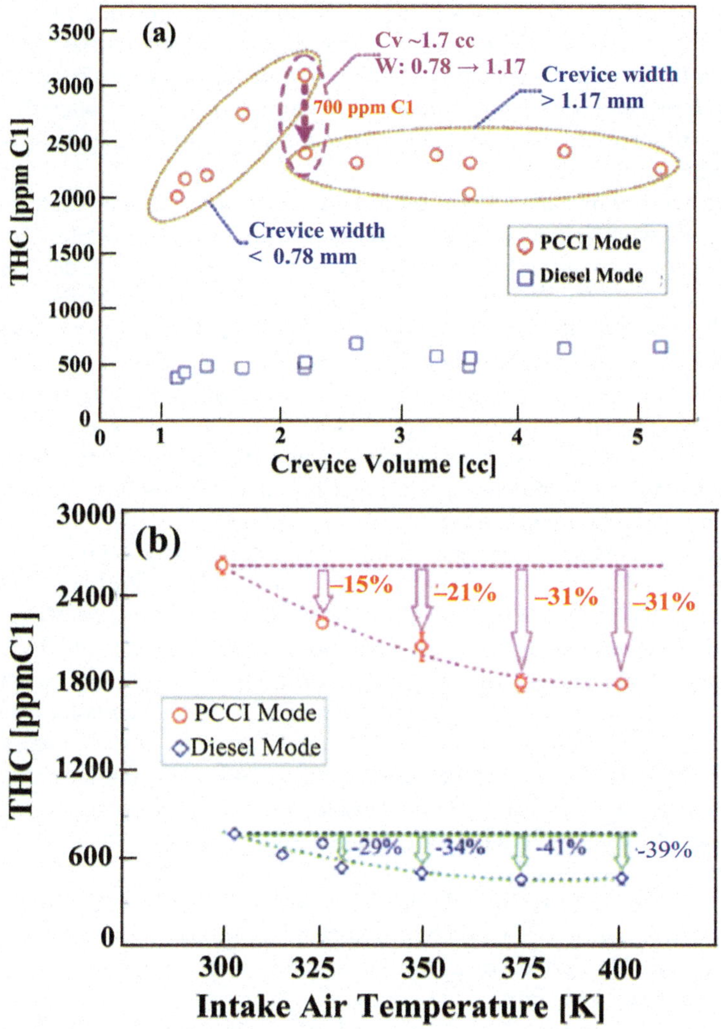

Fig. 8.17 HC emissions as function of (**a**) the piston top land crevice volume, and (**b**) intake air temperature in PCCI and conventional diesel mode [59]

volume and intake temperature (bulk quenching) in PCCI and conventional diesel mode [59]. Figure 8.17a shows the effect of crevice volume on HC emissions using different piston designs in PCCI and conventional diesel combustion (CDC) mode. In CDC mode, the HC concentration is almost constant indicating no correlation with crevice width and crevice volume. This is due to less time availability for fuel to reach ring crevices as it rapidly burns because of diesel injection close to TDC. However, in PCCI combustion mode, the HC emission increases proportionally with the crevice width (for gap 1 mm or narrower) and remains at the same level for higher width. Comparatively longer ignition delay (ID) in PCCI mode (due to early fuel injection at lower combustion temperature) provides more time for mixing of fuel with air and to reach the fuel to the circumferential crevice. Fuel trapped in the crevice volume is not able to burn if crevice width is narrower than the quench distance leading to increase in HC emissions. Study concluded that the piston top-ring crevice is one of the major sources of hydrocarbon emissions in the PCCI engine [59].

Figure 8.17b shows the HC emission as function of intake temperature in PCCI and CDC mode. Variations of HC emissions with respect to intake temperature provide a good estimate of HC emissions by bulk quenching. Figure depicts that HC emissions decreased as the intake air temperature increases in both the combustion modes. The reduction effect gradually vanishes with increasing air temperature more than 375 K. The HC emission reduction is observed to be up to 31% in PCCI mode and 40% in conventional diesel mode in comparison to 300 K intake temperature. The HC reduction effect is lower in PCCI combustion (in %, though higher in concentration), in comparison to conventional diesel mode because the bulk quenching contribution to the HC emissions in PCCI combustion is lower than in conventional diesel combustion [59].

Figure 8.18 shows the effect of injection timings and engine load on HC emission in PPC engine operated on diesel- and gasoline-like fuels. The typical behaviour of the HC emissions with respect to start of injection (SOI) is independent of the fuel, and a minimum HC emissions occur at an SOI of 21° bTDC for both diesel and biodiesel (SME) (Fig. 8.18a) [60]. The HC emissions increase as the injection timing is advanced or retarded from these timings. Increased emissions at advanced injection timings are attributed to larger amounts of fuel penetrating into the squish region [61, 62], and the increase at retarded timings to formation of lean mixtures in the squish region incapable of completing combustion [61]. Figure 8.18b shows the HC emission as function of engine load in PPC engine using different gasoline-like fuels. Figure shows that except one conditions (IMEP = 5 bar for ethanol), HC emission below 2 g/kWh is achieved. This is typically lower than in HCCI combustion (Fig. 8.14c) at similar engine loads. At high load operating conditions, it is impossible to get the same values of HC emission in diesel PPC [63].

Figure 8.19 shows the HC emissions as function of engine load, premixing ratio, pilot and main diesel injection timings in RCCI engine. Figure 8.19a shows the effect of engine load on HC emissions for conventional diesel and biodiesel combustion mode as well as RCCI combustion mode. Conventional diesel combustion mode has lower HC emissions as compared to RCCI combustion mode (Fig. 8.19a). RCCI combustion has higher HC emission due to trapping of premixed charge in the crevice volume and combustion at comparatively lower temperature leading to incomplete combustion. CNG/biodiesel RCCI emitted lower unburned HC in comparison to CNG/diesel RCCI operation (Fig. 8.19a). Since the biodiesel has a higher oxygen content and cetane number, its combustion results in higher in-cylinder combustion temperature due to advanced combustion phasing and shorter ignition delay. Higher combustion temperature in CNG/biodiesel RCCI mode leads to lower HC emissions [64]. Figure 8.19b presents the effect of fuel

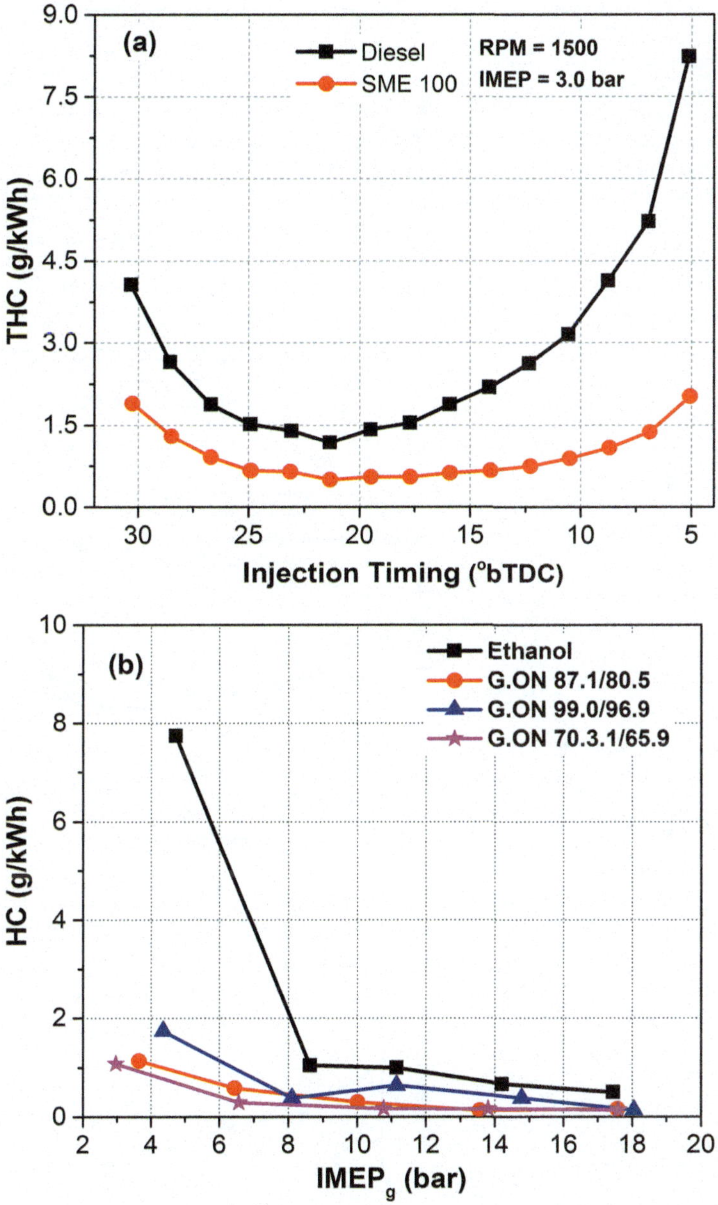

Fig. 8.18 HC emissions as function of (**a**) fuel injection timings using diesel-like fuels (Adapted from [60]), and (**b**) IMEP (engine load) using gasoline-like fuels in PPC engine (Adapted from [30])

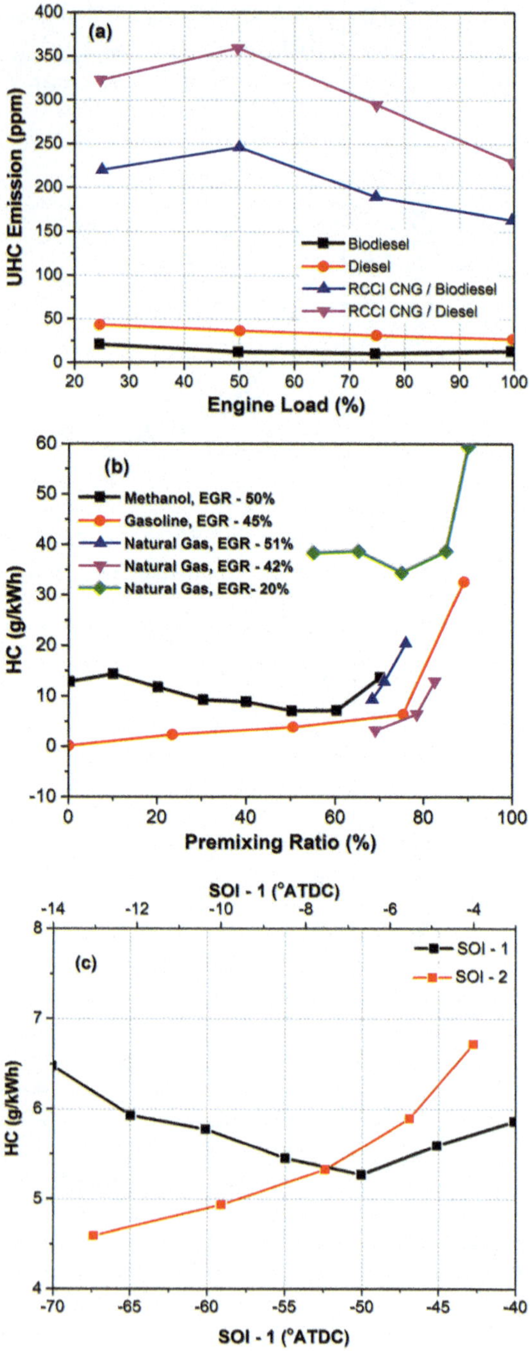

Fig. 8.19 (**a**) HC emissions as function of engine load in RCCI engine [64]. (**b**) Effect of fuel premixing ratio on HC emissions in RCCI engine (Adapted from [48, 65–67]). (**c**) Effect of pilot and main injection timings on the HC emissions in RCCI engine (Adapted from [36])

premixing ratio on the HC emission in RCCI engine using different fuels. Fuel premixing ratio plays a significant role in the formation of the HC emissions. HC emission increases with an increase in the gasoline/natural gas premixing ratio (Fig. 8.19b). With increase in the gasoline/natural gas fraction, global reactivity of charge in the cylinder reduces, which leads to reduce the oxidation rate; therefore, HC emission increases. With further increase in the fraction of gasoline/natural gas premixing ratio, HC emission increases drastically. It attributes to poor combustion process, which results in lower adiabatic flame temperature, and thus, higher HC emission is observed. Figure 8.19b also indicates that in the case of methanol/diesel RCCI operation, the HC emissions reduce up to 60% premixing ratio and further increase in the premixing ratio, HC emission starts increasing. HC emission initially reduces due to increase in the in-cylinder combustion temperature, with increase in the methanol fraction. Higher combustion temperature leads to increase of the oxidation rate. Later on HC emission increases with increase in methanol fraction due to the combined effect of shorter reaction time and longer delay period, which reduces the combustion temperature [48]. Figure 8.19c shows the effect of pilot and main injection timings on HC emission in RCCI engine. Figure demonstrates that with an advanced pilot injection timing (SOI-1), the charge becomes more homogeneous and reduces the local fuel–air equivalence ratio. The combustion of charge with lower local equivalence ratio retards the ignition timing and lower peak of HTHR obtained. It results in lower in-cylinder combustion temperature, which leads to increase in the HC emissions. Figure 8.19c also indicates that with retard of main injection timing (SOI-2), the HC emission increases due to lower combustion temperature because of late combustion phasing [36].

Figure 8.20 shows the HC mapping with engine load and speed in RCCI operation using gasoline/diesel (G/D) and ethanol–gasoline blend (E20)/diesel–biodiesel blend (B7). Figure 8.20a reveals that RCCI engine mapping was only possible up to 50% load (load referred to CDC at 1200 rpm) whatever the engine speed using gasoline/diesel, while maximum operable load to fulfil the different constraints simultaneously was 35% at all engine speeds using ethanol blend (E20) and diesel blends (Fig. 8.20b) [37, 38]. Methodology ensures to achieve the imposed limitations of maximum PRR and COV_{IMEP} without limiting the upper portion of the RCCI engine map. Detailed methodology of engine operation is provided in original study [37, 38]. Figure 8.20a shows that HC emission decreases with increase in BMEP. The lower HC emissions are present in the region of the map having high pressure rise rate and also higher combustion stability [37] due to higher combustion temperature. Inside this region, the slight differences in the values of HC emission are the result of variations in diesel injection timings, stability and gasoline fraction of fuel. Figure 8.20 shows that RCCI combustion produces higher HC emissions than conventional diesel combustion, but RCCI operation with ethanol–gasoline blend (Fig. 8.20b) emits lower HC than RCCI operation using gasoline/diesel (Fig. 5.20a).

Higher unburned hydrocarbon and CO emissions from LTC engines can be reduced using the oxidation catalyst in exhaust line of the engine. Conversion efficiency of catalyst depends on the exhaust gas temperature. In LTC strategies,

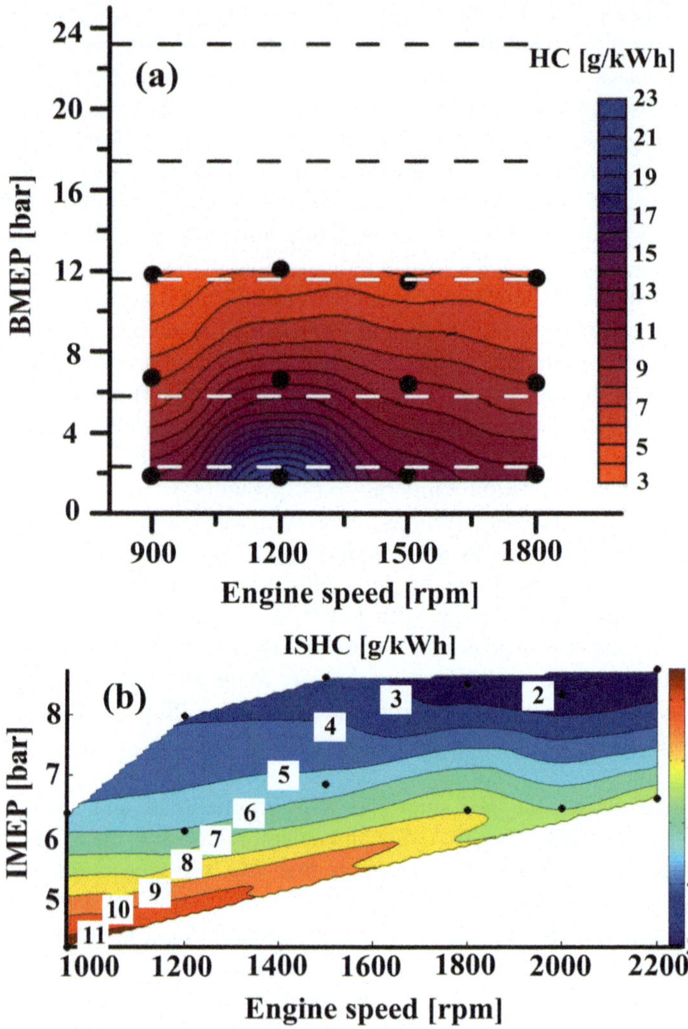

Fig. 8.20 HC emission mapping of RCCI operation using (**a**) gasoline/diesel (Adapted from [37]), and (**b**) ethanol–gasoline blend (E20)/diesel–biodiesel blend (B7) (Adapted from [38])

exhaust gas temperature is rather low due to lean engine operation and fast combustion in combination with a high compression ratio. Figure 8.21 shows the effect of oxidation catalyst on HC and CO reduction as function of pre-catalyst temperature in HCCI combustion engine. Figure 8.21a shows the λ and intake temperature used in HCCI engine operation to achieve the reduction in HC and CO shown in Fig. 8.21b. Figure 8.21b depicts the reduction in CO and HC emissions as function of exhaust gas temperature for different catalytic coatings. The HC reduction up to 90% is achieved at exhaust gas temperature of 340 °C

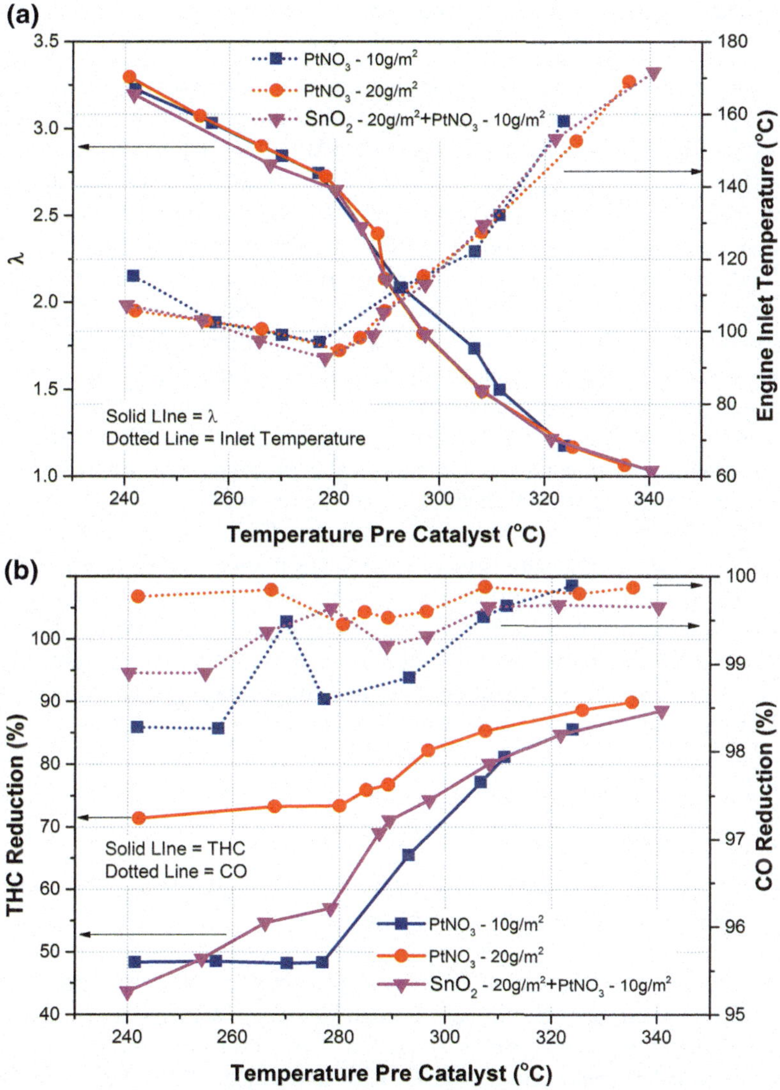

Fig. 8.21 (**a**) Engine intake temperature and λ in HCCI operation and (**b**) Reduction in total unburned HC and CO in HCCI engine using different oxidation catalyst (Adapted from [68])

and at lower exhaust gas temperature reductions is comparatively very low. CO oxidation has no problem with this catalyst above 240 °C, and reduction is more than 98% for all the test conditions. In this temperature range, CO oxidation is less dependent on the exhaust gas temperature as compared to HC (Fig. 2.21b). Catalyst with higher concentration of platinum nitrate (PtNO3) provided the best HC reduction characteristics among all catalysts evaluated at low exhaust gas temperature [68].

8.4 Particulate Matter Emissions

One of the main motivations for investigation of LTC engines is the potential of ultralow particulate matter emissions. Particulate matter emission from automotive engines is regulated by legislative agencies. Particulate matter formation in LTC combustion engine is discussed in Chap. 2. Particulate matter is typically characterized on mass basis (soot emission) as well as particle number basis. In latest emissions norms, particle number is also regulated from automotive engines. Particulate matter emitted from combustion engines has great impact on human health and urban air quality, especially small-sized particles [69]. Diesel exhaust particles mainly contain highly agglomerated solid carbonaceous materials and ash, volatile organic carbon and sulphur compounds. A small fraction of fuel and evaporated lubricating oil escape oxidation and appear as volatile or soluble organic compound in the exhaust [70]. The soluble organic fraction (SOF) contains polycyclic aromatic compounds containing oxygen, nitrogen and sulphur [71]. The exhaust particle composition depends upon where and how they are sampled or collected on the filter paper. Exhaust is typically diluted and cooled during sampling process for the measurement of particulate matter. The processes of nucleation, condensation and adsorption transforms volatile materials into solid and liquid particulate matter [2]. The composition of particulate matter depends on the engine operating parameters, conditions in the exhaust system and sampling system. In the following subsection, soot emissions (on mass basis) and particle number emissions from LTC engine are discussed in detail.

8.4.1 *Soot Emission*

8.4.1.1 Soot Formation and Composition

Carbonaceous particulate matter or soot is the product of fuel-rich mixture combustion. Higher concentration of soot is manifested as black smoke emission in the engine exhaust. Dry soot is primarily carbonaceous fraction of particulate, and its typical chemical representation is C_8H, C_9H and $C_{10}H$ [72]. About 5 to 10% of mass oxygen and 0.5% nitrogen are also present and typical empirical formula of dry soot would be $CH_{0.11}O_{0.065}N_{0.005}$ [51]. Electron micrographs show that soot formed in the engine cylinder is aggregate of primary spherical particle of size ranging from 10 to 80 nm, and most of them being in the range of 15–30 nm size range [73]. A single spherical soot particle contains 10^5 to 10^6 carbon atoms. Initially, soot particles have one hydrogen atom to about 8 carbon atom $(C_8H)_n$, and during expansion, more hydrocarbons may be adsorbed on soot particles. The primary soot particle form aggregates in the combustion chamber of about 100–200 nm in size containing over 100 primary particles. These aggregates may further agglomerate to particles as large as 1 μm [51]. Fuel composition plays important role in soot formation, and effect of fuel type on soot formation is different for premixed and diffusion combustion. For premixed flames, soot formation tendency for different fuel types in the decreasing order is as follows [74]:

Aromatics > alcohols > paraffins > olefins > acetylene

For diffusion flames, pyrolysis of fuel molecule occurs and soot forming tendency is in the following order:

Aromatics > acetylenes> olefins > paraffins > alcohols

Soot formation is kinetically controlled process and typically formed at C/O ratio of 0.5–0.8 in practical systems. The critical C/O ratio for soot formation increases with increase in temperature. Soot formation is strongly influenced by combustion chamber pressure, and higher pressure results into higher soot formation [51]. Soot formation consists of four main stages, namely, (i) inception and nucleation, (ii) surface growth, (iii) coagulation and (iv) oxidation.

Soot nucleation or inception process is a non-equilibrium process. Typically three routes of soot formation is suggested [51]: (i) pyrolysis or thermal cracking in oxygen-deficient environment, where large fuel molecule breaks into smaller molecule (most likely mechanism for soot formation), (ii) condensation reactions and polymerization that result in larger molecules and (iii) dehydrogenation process that increase C/H ratio of fuel molecule. The path of soot formation depends on combustion temperature. At lower temperatures (<1700 K), condensation reactions of aromatics and pyrolysis of other highly unsaturated hydrocarbons are likely to form soot. Fragmentation of fuel molecules to smaller unsaturated molecules and their consequent polymerization dictates the soot nucleation at intermediate temperatures (>1800 K). Acetylene, polyacetylenes and poly-unsaturated hydrocarbon radicals produced because of fragmentation reactions lead to soot formation. The aromatic condensation method for soot nucleation is more direct and fast method [51]. Acetylene and polyaromatic hydrocarbons (PAH) are known as the most likely precursors leading to soot formation. The smallest PAH and building block is benzene, which can be produced through many routes [75]. This PAH building block grows further by an entropy-driven process of acetylene addition and hydrogen abstraction reaction called the HACA mechanism [76]. The conceptual model of soot particle nucleation by three pathway (A, B,C) is presented in Fig. 8.22a. Path A is illustrated by the growth of two-dimensional PAHs into curved structures; paths B and C comprise the physical coalescence of moderate-sized PAHs into stacked clusters and the chemical coalescence of PAHs into crosslinked three-dimensional structures respectively [76].

Inception or nucleation produces a large number of small particles, which grow in size due to gas phase deposition of intermediate combustion products on the surface of nuclei. Important process involved in soot growth is shown in Fig. 8.22b. During coagulation process, collision of two spherical soot particles may form larger spherical particles. This process occurs in the beginning of particle surface growth process, when the particles are still small (<10 nm). Once the particles have become larger and solidified, surface growth decreases, and the particles resulting

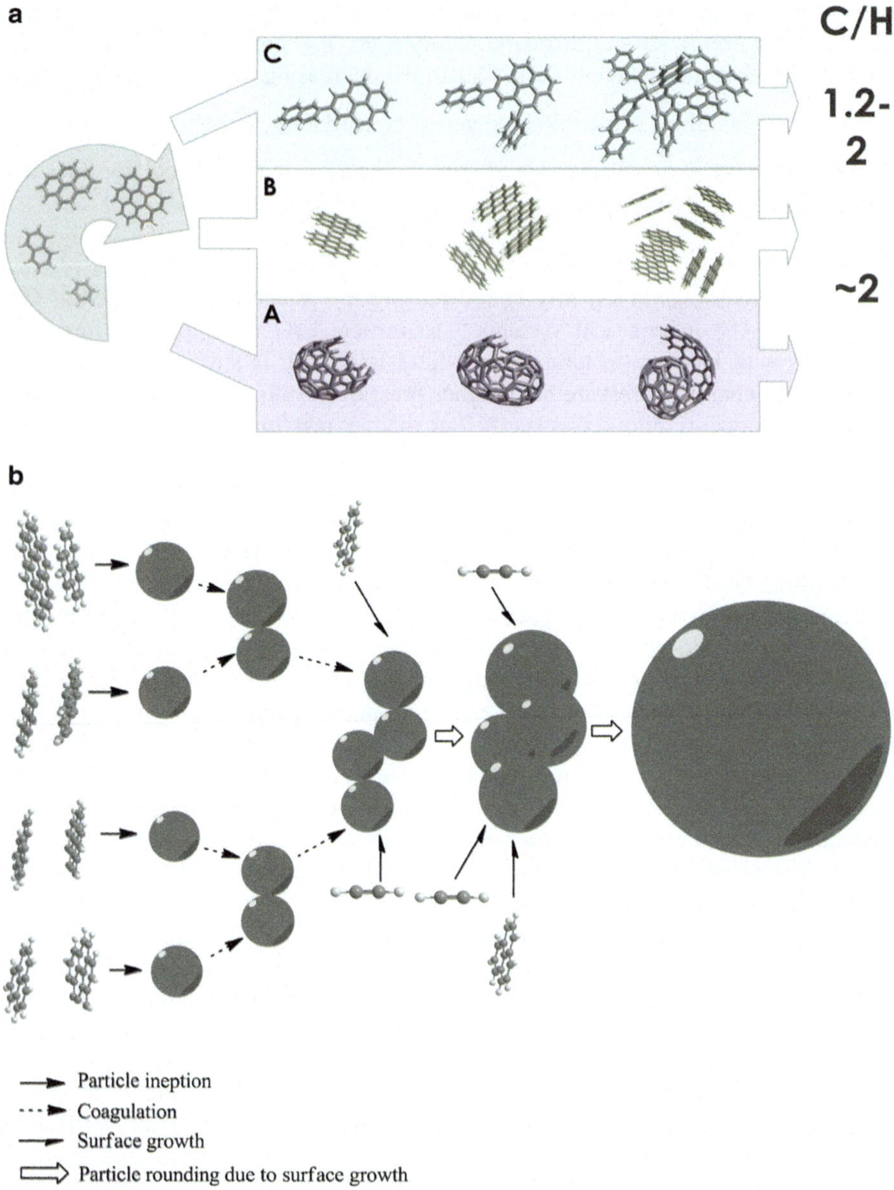

Fig. 8.22 (**a**) Conceptual mechanisms of soot particle nucleation [76]. (**b**) Important process in soot growth [77]

from collision resemble a cluster like a bunch of grapes [51]. Once the surface growth ceases, particles coalesce in the shape of chain-like structure, and this process is termed as aggregation (Fig. 8.23).

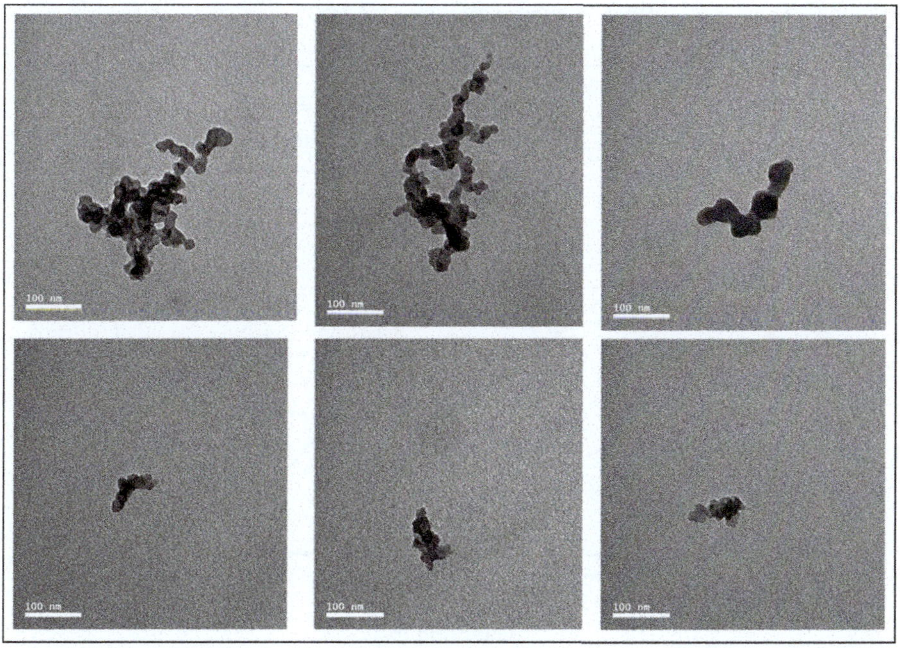

Fig. 8.23 Soot aggregates collected from RCCI engine exhaust at 4.4 bar BMEP and 1300 rpm [78]

Oxidation of soot can occur at the stages of formation of precursors, nuclei, primary soot particles and aggregates. A large fraction of soot formed early in the combustion process is oxidized within the cylinder. Soot can be oxidized on reactions with O, O_2 and OH, which are potent oxidants [51]. The OH pathway has been identified to be predominant for soot oxidation. From 10% to 30% of collisions between a soot particle and OH radicals results into oxidation. As a result of soot oxidation, the structure of soot particles changes and can lead to fragmentation, which can result in a substantial increase in particle number [79].

8.4.1.2 Soot Emission in LTC Engines

Soot emission from engine is manifested as visible black smoke. All factors that affect soot formation and oxidation also influence smoke. Smoke is generally a result of very fuel-rich combustion and mainly related with combustion in diffusion flames. With combustion of premixed stoichiometric or lean mixtures, smoke will not be generated if the fuel–air mixture is well prepared. Typically, well-mixed charge is used in HCCI combustion engines. Therefore, HCCI combustion does not produce any detectable smoke with high volatility fuels, like gasoline and alcohol, or with gaseous fuels, like natural gas [1, 45, 80]. Diesel fuel contains some very heavy fractions that do not vaporize at normal induction conditions, and diesel

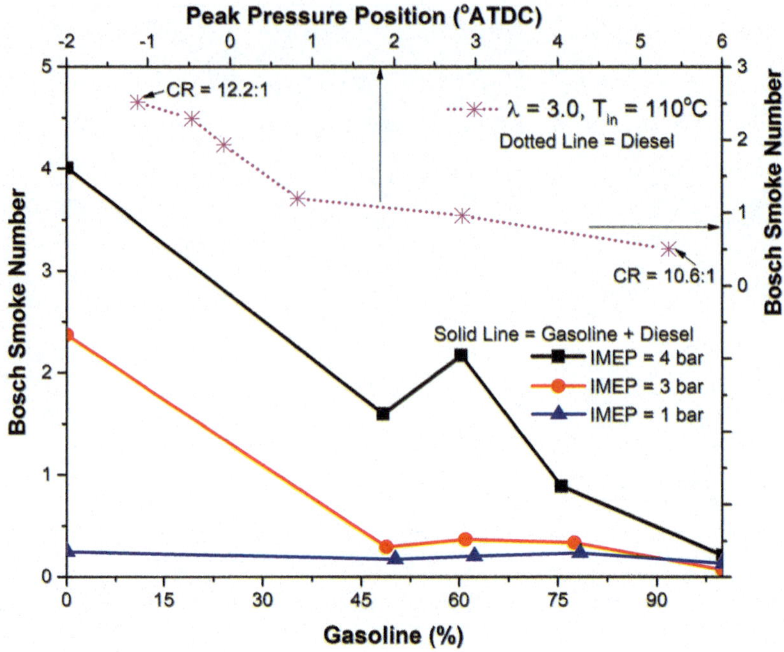

Fig. 8.24 Bosch smoke number as function of gasoline fraction of fuel and peak pressure position in diesel HCCI engine (Adapted from [80, 81])

HCCI combustion can generate smoke in some cases. Figure 8.24 shows the variation of smoke number in HCCI engine operated with gasoline-diesel blend and neat diesel fuel. Figure shows that increasing the fraction of diesel fuel leads to more smoke, when engine is operated with mixtures of gasoline and diesel fuel. Figure 8.24 also shows an interesting behaviour found with diesel fuel regarding soot formation in HCCI engine [81]. The soot emission increased strongly for engine operation with advanced combustion phasing achieved by increased compression ratio (Fig. 8.24). The HCCI engine operation with diesel produces much smoke, and thus, diesel fuel port injection is not a suitable fuel injection strategy in HCCI engine.

Figure 8.25 shows the smoke emission as function of combustion phasing and engine load in PPC engines using different gasoline-like fuels. Figure 8.25a shows soot emission as function of combustion phasing in PPC engine at 8 bar IMEP and 1500 rpm engine speed using a single injection strategy using TRF (15% toluene, 40% n-heptane and 45% iso-octane) and ERF (10% toluene, 40% n-heptane and 50% iso-octane). The fuel injection pressure strongly affected the smoke levels (Fig. 8.25a). The smoke levels reduce by approximately 0.4 FSN (filter smoke number) for TRF fuel as the fuel injection pressure increased from 800 to 1200 bar. Three main factors are responsible for lower soot emission with higher injection pressure. First, the residence time in the region with rapid soot formation in the fuel jet is decreased due to higher penetrating speed of the jet at high injection pressure.

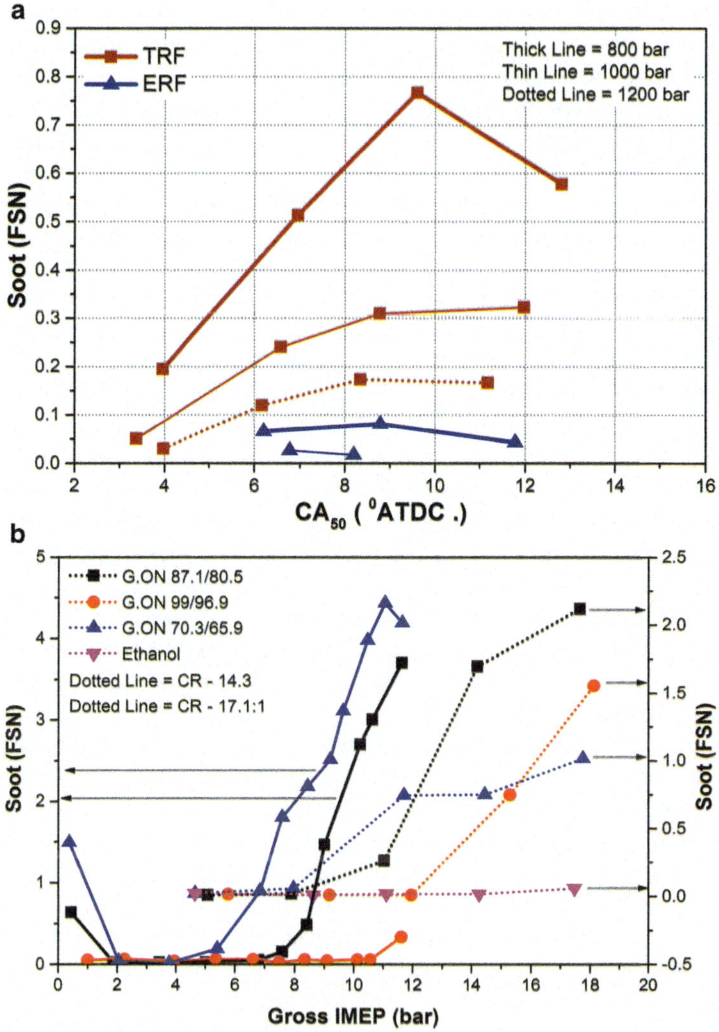

Fig. 8.25 Smoke number as function of (**a**) combustion phasing (Adapted from [26, 27]) and (**b**) IMEP in a PPC engine (Adapted from [30, 31])

Second, at higher injection pressure, the spray breakup and fuel–air mixing are more effective in preventing the formation of locally rich regions, which would produce significant amount of soot. Third, the turbulence is increased providing more effective soot oxidization [26]. The ERF fuel produced lower levels of smoke than the TRF fuels in PPC engine (Fig. 8.25a) due to oxygen content in ethanol, leading to different characteristics of the soot formation and also a more effective oxidization [82, 83]. Figure 8.25b depicts the soot emission as function of engine load at two different compression ratios in PPC engine using different gasoline-like fuels at 1300 rpm. Soot emission increases with increase in engine load (Fig. 8.25b).

Figure shows that by reducing the compression ratio from 17.1 to 14.3, soot emissions also lowered. Compression ratio is reduced to maintain a positive mixing period at highest engine load conditions and, hence, obtaining kinetically controlled combustion in the entire load range [31]. Engine load could also be increased from 12 to 18 bar gross IMEP by decreasing the compression ratio. Soot formation is strongly governed by small variations in λ rather than minor variation in EGR rate. Better tuning of λ at compression ratio 14.3 leads to roughly 60% reduction in soot emission than compression ratio 17.1 for similar mixing period at high load and the 1300 rpm engine speed. The worst case soot of 2 FSN is observed at compression ratio 14.3 at the highest engine load, which still considered relatively high and can be eliminated with an up to date injection system [31]. Ethanol produces almost zero soot (below 0.1 FSN) at 18 bar gross IMEP in PPC engine.

The RCCI combustion is another LTC strategy, which uses dual fuel of different reactivity. It is already discussed in the Chap. 2 that RCCI combustion has potential to reduce the soot and NO_x emissions simultaneously to a very low level. Figure 8.26 shows the effect of injection timing on the soot emissions for different fuel blends with various premixing ratios at low, medium and higher engine load conditions. Green dashed line in the figure represents the EURO VI soot limit (i.e. <0.01 g/kW-h) for heavy duty diesel engine. Figure 8.26 indicates that at lower engine load condition, the soot emissions are ultralow for E20-95, E10-98 and E10-95. Additionally, soot emissions are lower than the minimum detection limit of equipment (AVL 415S smoke meter) at some of the engine operating conditions. With an enough advance injection timing, there is a sufficient time for the mixing of charge before start of combustion. Thus, combustion of charge with lower local equivalence ratio leads to very low soot emissions. In the case of E85 fuel, soot emissions are higher for lower fuel premixing ratio with advanced injection timings. In general, at lower engine load, with lower fuel premixing ratio, higher soot emissions are observed due to higher fraction of diesel fuel. At medium engine load conditions, soot emission increases as compared to lower load condition but still below the EURO VI limit at most of the conditions. The soot emissions are lower with higher fuel premixing ratio, and injection timing has not much significant effect at mid-load condition except delayed injection timing with lower premixing ratio (Fig. 8.26). At higher engine load condition, the effect of fuel premixing ratio can be easily observed. Figure indicates that at higher engine load condition, soot emissions are higher for lower fuel premixing ratio, and for lowest premixing ratio (i.e. 49%), soot emissions are higher than the EURO VI limit. Study reported [84] that the difference in the concentration of soot emission for various fuel blends is due to the difference in mixing time of fuel and air.

Figure 8.27 depicts the soot emission mapping with engine load and speed in RCCI engine operation using gasoline/diesel (G/D) and ethanol–gasoline blend (E20)/diesel–biodiesel blend (B7). Figure 8.27a reveals that RCCI engine mapping was only possible up to 50% load (load referred to CDC at 1200 rpm) whatever the engine speed using gasoline/diesel, while maximum operable load to fulfil the different constraints simultaneously was 35% at all engine speeds using ethanol blend (E20) and diesel blends (Fig. 8.27b) [37, 38]. Detailed methodology of engine

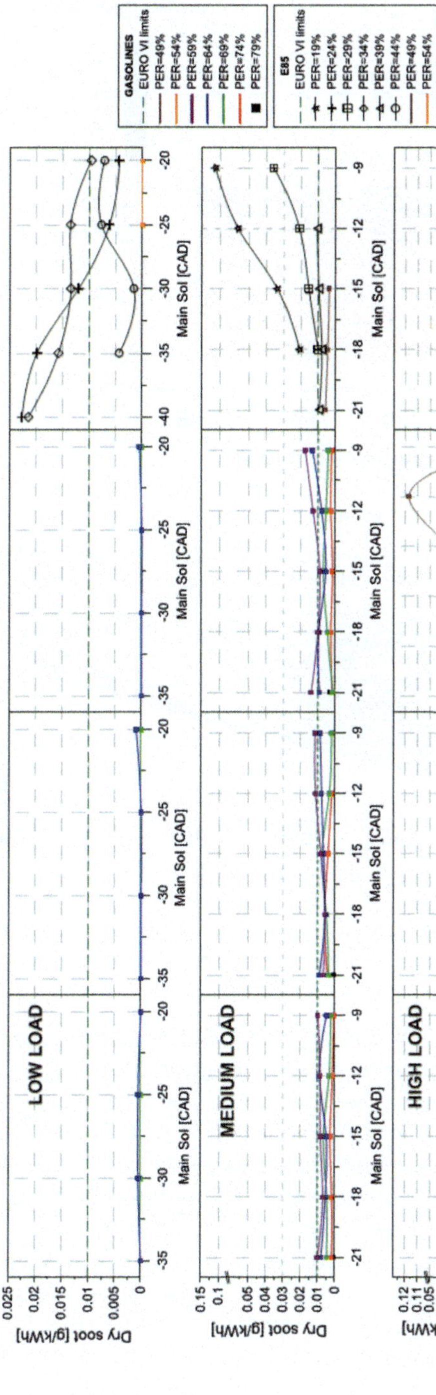

Fig. 8.26 Effect of injection timing on the soot emissions for different fuel blends with various premixing ratios at low, medium and higher load conditions [84]

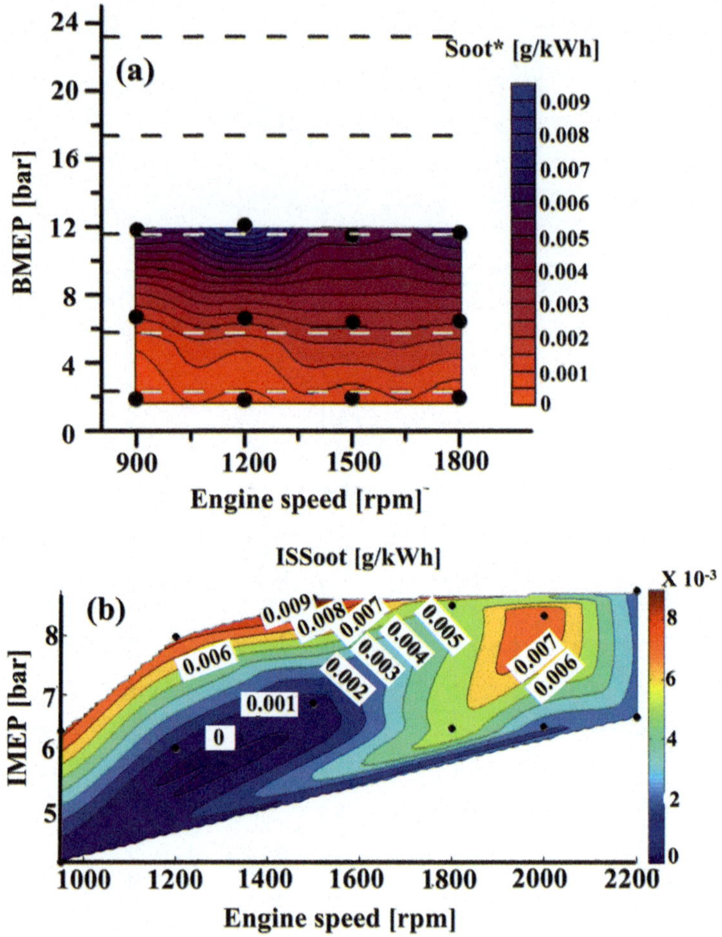

Fig. 8.27 Soot emission mapping of RCCI operation using (**a**) gasoline/diesel (Adapted from [37]) and (**b**) ethanol–gasoline blend (E20)/diesel–biodiesel blend (B7) (Adapted from [38])

operation is provided in original study [37, 38]. During experiments soot emission levels are imposed as a key constraint with limit of 0.01 g/kWh, which is achieved in the entire map. At 50% engine load, soot emission is very close to 0.01 g/kWh particularly at 1200 rpm (Fig. 8.27a). Further increase in engine load exceeds the imposed soot limit even for more than 90% premixed gasoline fraction [37]. These high soot emissions are results of required delay in diesel SOI to avoid excessive pressure rise rate at higher engine load. The mixing time for diesel injection is reduced, and the richer mixture distribution at start of combustion promotes the higher soot formation, which limits the RCCI engine map to 50% load of CDC. Using ethanol–gasoline blends, also similar levels of soot emission are obtained, and soot emissions are comparatively higher at higher engine loads (Fig. 8.27b).

Soot emissions below the 2010 EPA HD limit can be maintained without using any exhaust after-treatment devices at the engine loads achieved in RCCI combustion modes.

8.4.2 Particle Number and Size Distribution

The HCCI combustion concept is well known for its potential of having ultralow low NO_x and particulate matter emissions. However, recent studies have demonstrated that although the total particulate mass is indeed negligible, significant numbers of particles are emitted of the size below 100 nm mobility diameter [85–88]. Three different modes (nucleation, accumulation and coarse mode) of particles are observed in engine exhaust as illustrated in Fig. 8.28. The nucleation mode typically contains mainly volatile organic and sulphur compounds in particle size range of about 3–30 nm, and comprises less than 10% of the particle mass but more than 90% of the total particle number. The accumulation mode particles have typical diameter ranges of 20–500 nm and contain mainly carbonaceous agglomerates and adsorbed material. Most of the particle mass is in the accumulation mode. The size ranges and boundaries of these modes may shift and overlap with the variations in engine, fuel and engine operating conditions; however, fundamental structure remains the same [89]. The transition among the modes, the modal

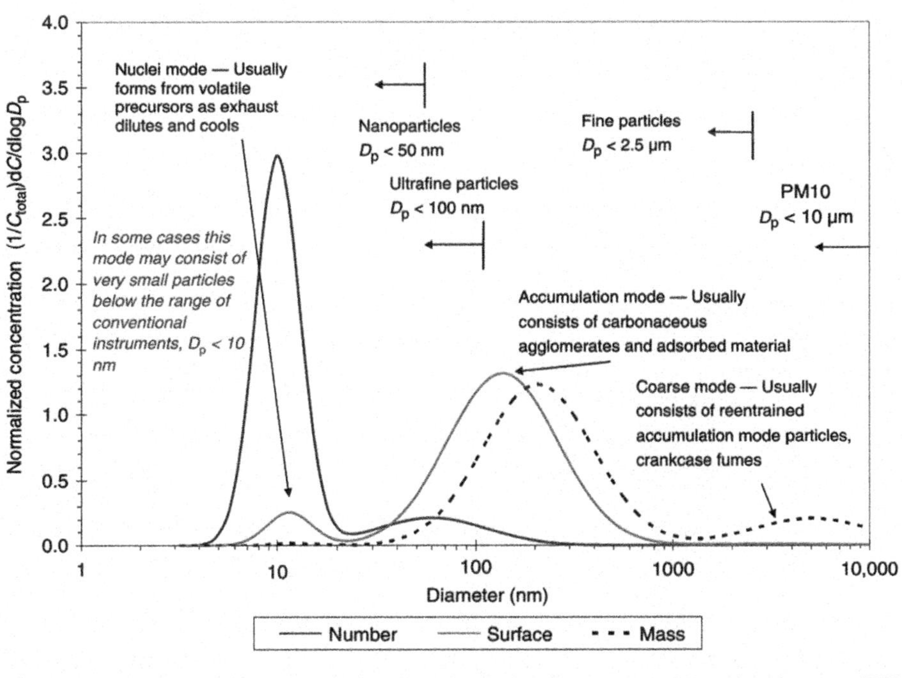

Fig. 8.28 Typical engine exhaust particle size distribution by mass, number and surface area [89]

diameter and concentration can also vary depending on dilution and sampling condition [89]. Figure 8.28 also depicts standard definitions of PM10, fine particles, ultrafine particles and nanoparticles for comparison.

In HCCI combustion engine, premixed charge is typically prepared by PFI or early direct injection of fuel. There are various factors that govern the particulate formation and suggest that particulate number is non-negligible, especially when particulate matter is defined as any dispersed solid or liquid particle-like material suspended in the exhaust gas flow [88]. These factors include charge inhomogeneity, piston and wall wetting (in direct injection case), condensation nucleation, ash and metal emissions, etc. Charge inhomogeneity (discussed in Sect. 2.2.4.6 of Chap. 2) is one of the factors that can lead to particulate formation. Charge inhomogeneity can be due to direct injection of fuel, and it is present in even PFI, which has comparatively larger time to mix with air. Charge heterogeneity is also partly a consequence of the wall wetting in the case of direct injection HCCI. In a nominally stoichiometric mixture, if there is heterogeneity then locally rich region occurs, which is the required condition for soot formation. Condensation nucleation occurs after the combustion in the exhaust stroke and in the exhaust system, when condensable species are forced into super-saturation by the cooling of the exhaust gases. This material, which consists mainly of partially burned or unburned fuel and lubricating oil, will condense homogeneously if the saturation ratio is high enough. It can also condense onto existing nucleation sites or adsorb to the surface of aggregated and agglomerated solid carbon particles, and this depends on the number density of the accumulation mode PM [88]. The condensable particulate fraction is expected to be significant in HCCI because of low temperature combustion and higher unburned hydrocarbon emissions due to lower combustion efficiency.

Figure 8.29 depicts the particle size number distribution for gasoline, ethanol and methanol for different air–fuel equivalence ratio (λ) at 1200 rpm. Particle number concentration increases with increase in rich mixture engine operation and leaner mixture have comparatively lower particle number concentration for all the fuels at constant intake temperature (Fig. 8.29). Peak concentration of particles depends on the mixture strength (i.e. engine load). Methanol HCCI combustion showed lower peak normalized number concentration in comparison to gasoline and ethanol (Fig. 8.29). Figures also reveals that the particles in the size range of 6–150 nm have significant concentration of particles. Mobility diameters correspond to peak particle concentration shift towards lower diameter for leaner air–fuel mixtures for all test fuels at fixed intake temperature (Fig. 8.29). Mobility diameters corresponding to peak particle number concentration are in the range of 22–40 nm, 25–45 nm and 34–80 nm using ethanol, methanol and gasoline, respectively, which indicated that ethanol and methanol HCCI particles are comparatively smaller than gasoline. These mobility diameters corresponding to peak concentration in gasoline HCCI is larger in comparison to gasoline SI engine (<25 nm) [90]. Figure 8.29 also depicts that gasoline particle size number distribution curves have different behaviour than ethanol and methanol for mobility diameter less than 20 nm [85]. Particle number concentration in the size range close to 10 nm is in the

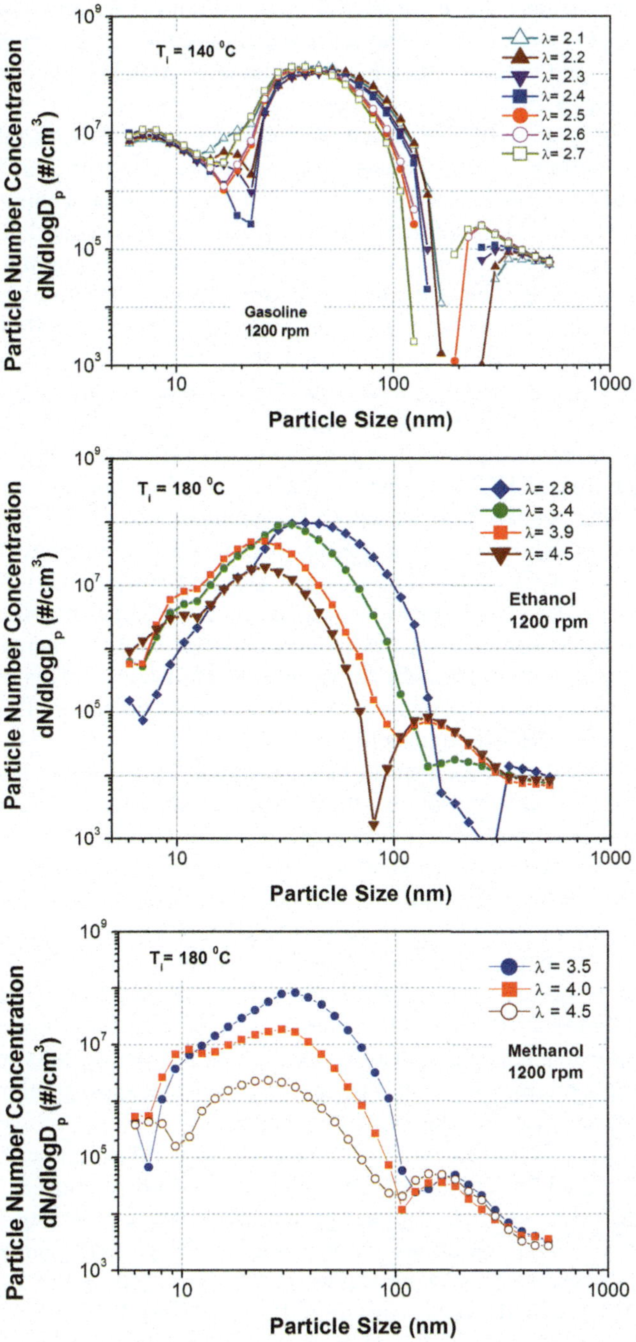

Fig. 8.29 Particle size number distributions for gasoline, ethanol and methanol HCCI combustion at 1200 rpm for different λ (Adapted from [85])

order 10^6 particles/cm^3 for ethanol and methanol and even lower at low intake temperatures. However, gasoline particle concentration in the size range of 10 nm is in the order 10^7 particles/cm^3, which is an order of magnitude higher than methanol and ethanol (Fig. 8.29).

Another study showed that nearly all particles emitted from ethanol HCCI engine using various control strategies are volatile, which means there are no solid accumulation mode particles found in measureable concentration [87]. In the absence of solid nucleation mode particles, only gas-to-particle conversion processes take place as exhaust gas is diluted and cooled, leading to homogeneous nucleation and condensation, which result in volatile nucleation mode particle formation. In conventional CI engines, lubricating oil dominates the composition of volatile nucleation mode particles and their characteristics with increasing engine loads [91]. In HCCI combustion, higher combustion temperature might lead to higher lubricating oil-related nucleation mode particle formation. Higher combustion temperature can be achieved by richer mixture or higher intake temperature operation of HCCI engine. Another study also showed that the volatile particulates are composed of at least 95% compounds created from unburned lubricating oil at light to moderate engine loads in CDC [92]. Lubricating system configuration of HCCI engine is similar to CDC, and thus, delivery of lubricating oil to the combustion chamber and associated processes in the combustion chamber may also be quite similar [87]. Moreover, particle numbers increase at higher temperature while unburned hydrocarbon emissions decrease. This might also be contributed due to relatively higher breakdown of the lubricating oil film at higher pressure rise rate in HCCI engines [85].

Figure 8.30 presents the effect of intake temperature and λ on the total particle number concentration for gasoline–ethanol and methanol HCCI combustion. Generally, total particle number concentration reduced as charge becomes leaner. This indicates that on increasing the fuelling (engine load), total particle number concentration increases. Total particle number concentration also increases with increasing intake air temperature at constant λ (Fig. 8.30), possibly due to increase in nucleation mode particles at higher intake temperature [85]. Total particle concentration is observed in the range of 9.31×10^6–8.53×10^7 particles/cm^3 (Fig. 8.30). Figure 8.30 depicts that methanol has comparatively lower total particle concentration than ethanol and gasoline except two test points. Maximum value of the total particle number concentration increases with increasing engine speeds for gasoline but decreased for ethanol and methanol [85].

Figure 8.31 shows the effect of PFI timings on the particle number concentration in methanol fuelled HCCI engine. Start of injection (SOI) timings 5, 70, 210, 500 and 600 CAD are selected in open (0 and 70 CAD) and closed intake valve (201, 500, 600 CAD) conditions for investigation of particle number emissions [93]. Various degrees of fuel unmixedness are caused by different SOI timings as different SOI timings affect fuel mixing differently [94]. SOI in intake vale open (IVO) conditions (5 and 70 CAD) have higher peak particle concentration in comparison to SOI in intake valve closed (IVC) conditions because of higher inhomogeneity is created by IVO fuel injection timings. For lower particle

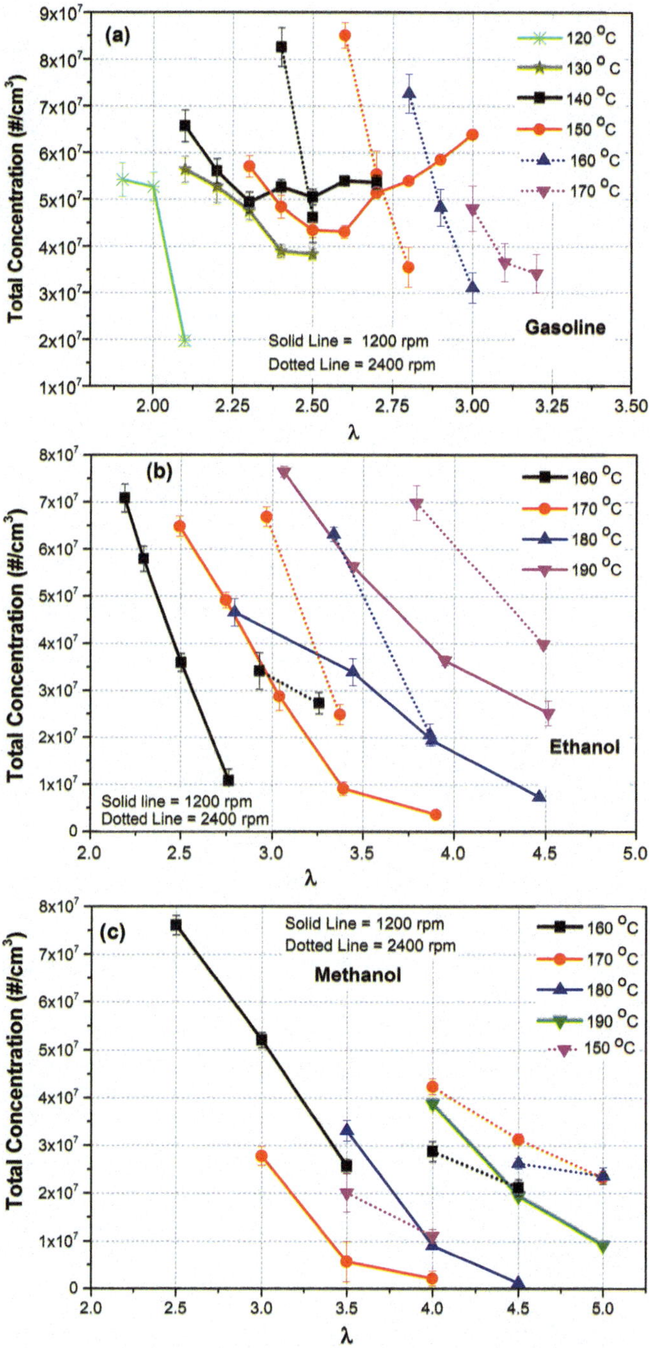

Fig. 8.30 Total particle number concentration as function of λ for gasoline, ethanol and methanol HCCI combustion (Adapted from [85])

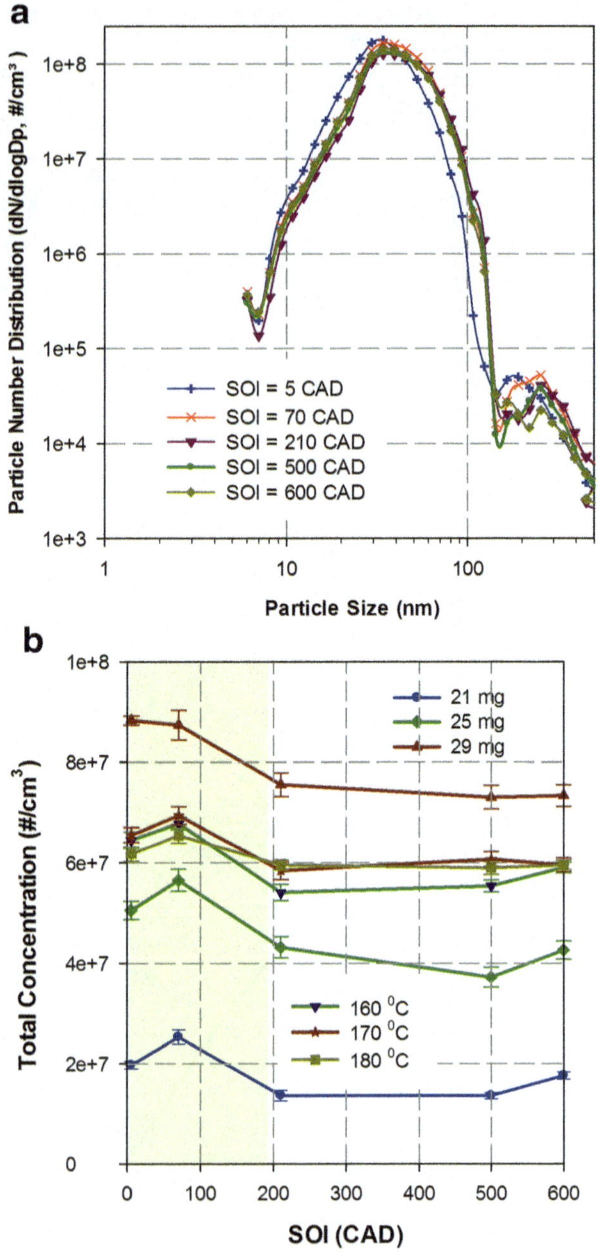

Fig. 8.31 (**a**) Particle size number distribution as function of PFI injection timing and (**b**) total particle concentration as function of PFI injection timing for different intake temperatures and fuel quantity in HCCI engine (Adapted from [93])

emission, fuel injection must be in IVC conditions in HCCI engines [93]. The peak particle number concentration occurs around the mobility diameter of 40 nm, and significant number of particles appears in the size range of 10–150 nm (Fig. 8.31a).

Figure 8.31b shows the total particle number concentration as function of PFI timings for different intake temperature and λ. Shaded yellow region in the figure represent IVO duration. Total particle concentration increases with increase in fuel quantity per cycle at constant intake temperature of 150 °C (Fig. 8.31b). Total particle number concentration increases as SOI timings get closer to IVO timings, and concentration is highest for IVO injection. Figure 8.31b also depicts the effect of intake air temperature on the total particle concentration for constant (25 mg/stroke) fuel injection quantity. Total particle number increases with increasing intake air temperature up to 170 °C, and further increasing intake temperature does not increase the total particle number significantly. Total particle number is found higher for IVO timings, and SOI timings between 210–500 CAD do not show significant variations in the total particle number for most of the test conditions (Fig. 8.31b) [93].

Figure 8.32 shows the effect of EGR on the particle number, and particle mass distribution in an ethanol fuelled HCCI engine operated at intermediate load (3.2 bar IMEP). Increasing EGR levels lead to reduction in both total number concentration and particle mobility diameter (Fig. 8.32a). The combined effect of these two characteristics is a very significant reduction in particle mass concentration (Fig. 8.32b). Higher EGR level leads to reduction in combustion temperature due to higher specific heat species present in residuals. This effect is similar to decrease in intake temperature (Fig. 8.30) and thus particle number trend is also similar.

Count mean diameter (CMD) represents the average particle size in the distribution. The CMD provides a basis for comparing overall size of particles at different engine operating conditions and fuels. The CMD is calculated as product of particle diameter and corresponding concentration divided by total concentration. Similarly mass mean diameter (MMD) can be calculated [85]. Figure 8.33 shows the CMD and MMD as function of total concentration and total mass, respectively, for gasoline, methanol and ethanol HCCI combustion at 1200 rpm. Figure shows that the CMD of particles in gasoline HCCI combustion is in the range of 32–80 nm. Ethanol and methanol have similar CMD in the range of 25–50 nm. This finding indicates that ethanol and methanol HCCI combustion has comparatively smaller mean mobility diameter than gasoline HCCI combustion. Lower CMD typically obtained at lower concentration of accumulation mode particles or higher concentration of nucleation mode. Thus, ethanol and methanol HCCI contains more number of nucleation mode particles. Lower total number of particles is also obtained for conditions having comparatively lower CMD using methanol and ethanol. Particle mass is computed using constant particle density of 1.0 g/cm^3. Figure 8.33b shows that total mass emission from gasoline is comparatively higher than ethanol and methanol. The MMD of gasoline HCCI particles is also larger than ethanol and methanol for most of the conditions. The MMD is in the range of 50–200 nm for all test fuels. The MMD is typically higher than CMD, which indicates that higher diameter particles contribute more to particulate mass.

In comparison to HCCI, the PPC engine has advantages in terms of lower unburned hydrocarbon and carbon monoxide emissions; however, soot emissions

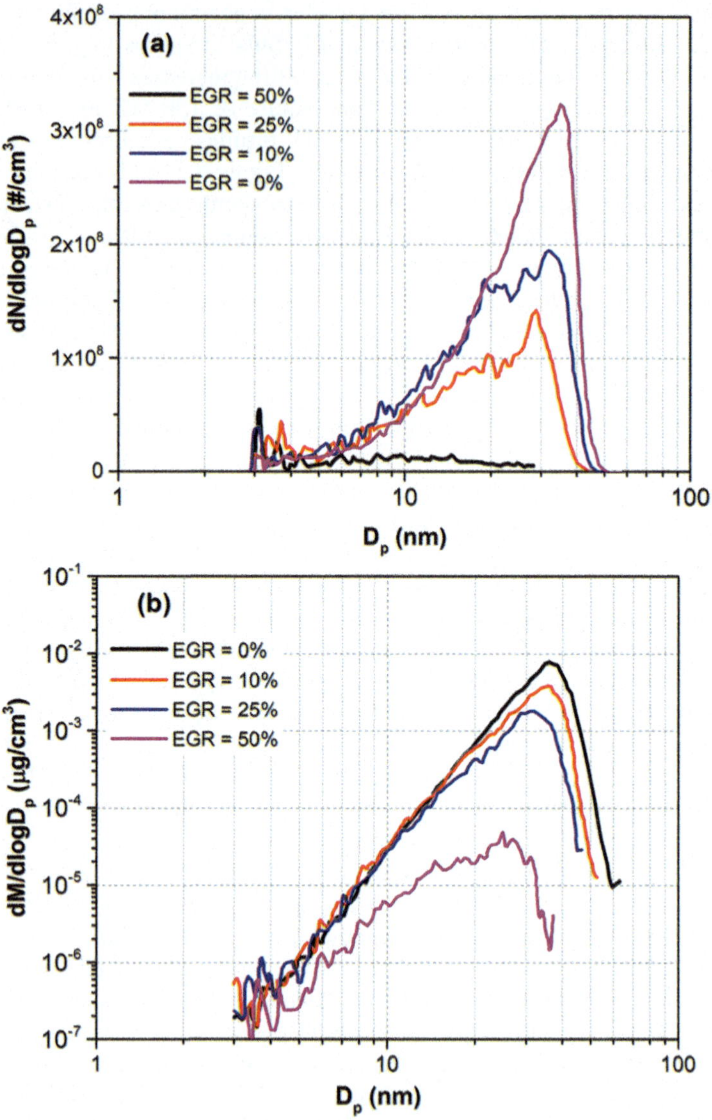

Fig. 8.32 Effect of EGR on (**a**) mobility size-number distribution, (**b**) size-mass distributions in ethanol HCCI combustion at 1500 rpm for fixed fuelling (Adapted from [87])

can be a challenge because of heavy fuel stratification in PPC and higher EGR is used. In PPC strategy, NO_x and soot can be reduced but this strategy comes with a trade-off; increasing the EGR rate increases the soot formation, while lowering the EGR rate increases the NO_x formation [95]. This trade-off can be avoided completely in PPC mode by using short carbon chain fuel such as alcohols, which are known for very little soot formation during combustion [96]. The soot

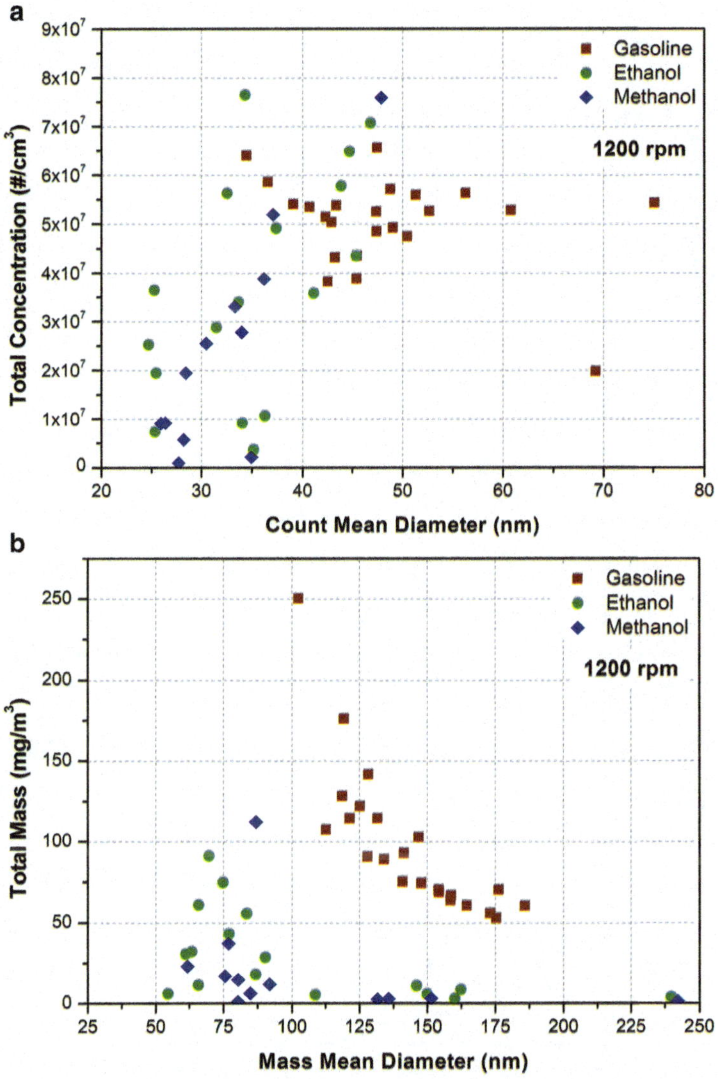

Fig. 8.33 (**a**) Total concentration as function of CMD, (**b**) total mass as function of MMD for gasoline, ethanol, and methanol HCCI combustion at 1200 rpm (Adapted from [85])

emissions in PPC mode have strong dependency on the EGR rate of the engine. Soot emissions touch a local maximum, when the inlet O_2 concentration is decreased to approximately 11–13% [97]. The intake O_2 concentration lower than this point will reduce the soot emission because of a decreased soot formation rate leading to a "soot bump" [98].

Figure 8.34 shows the particle number concentrations as function of fuel injection timings in PPC engine without and with EGR operation using gasoline. Very

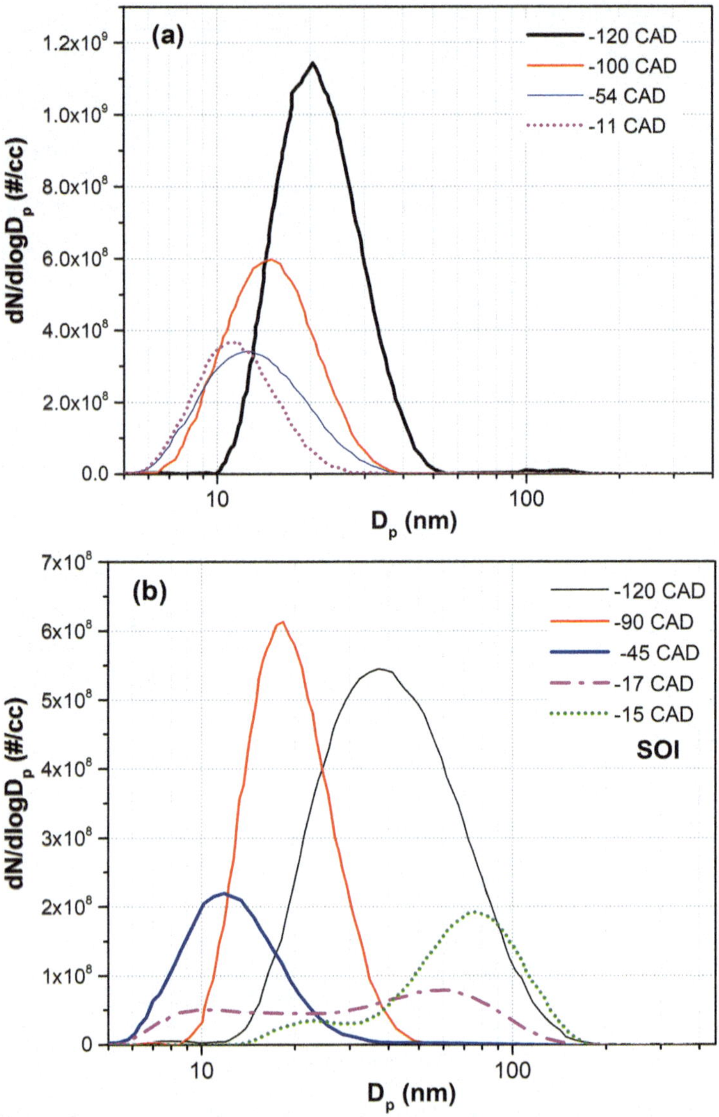

Fig. 8.34 Particle size and number distribution for different fuel injection timings using gasoline for (**a**) no EGR, and (**b**) with EGR operation of PPC engine (Adapted from [99])

early fuel injection timings have HCCI-like engine operation because of well-mixed charge due to more availability of time for mixing the fuel with air. Figure 8.34a shows the particle size distribution without EGR case. Figure 8.34a shows a unimodal particle size distribution for PPC engine for different fuel injection timings, unlike the traditional bimodal shape of the diesel exhaust particle size distribution, which consists of nucleation and accumulation mode particles

(Fig. 8.28). Nucleation mode particles highly dominate all injection timings and have the peak concentration around 15 nm except the HCCI-like conditions (SOI 120 CAD). The particle number concentrations from different injection timings are quite comparable except earliest injection timings, which have uniquely high particle number possibly due to spray impingement [99].

Figure 8.34b shows the effect of injection timings on the particle number distribution in PPC engine operated with EGR (40–44%). Figure 8.34b shows that early injection timings have similar trend of particle size distributions as the no EGR case. With late injection (45 to 15 CAD BTDC) in the PPC regime, unlike the without EGR cases, particle size distribution changes from a unimodal shape eventually into a bimodal shape when retarding injection timing (Fig. 8.34b) [99]. In the whole injection timing sweep, smallest nucleation mode is centred on 15 nm. Study also showed that retard in very early injection timings leads to a relatively slight change of decreased nucleation mode particle number concentration on the unimodal shape particle size distribution, and this slight change indicates that fuel targets at the piston top-land area [99]. At -17 CAD ATDC injection timing, the presence of large particles in accumulation mode makes the size distribution bimodal, as an indication of fuel targets into the bowl [99]. Further injection retard leads to increase in accumulation mode particle number and dominates the particle size distribution upon further injection retard. Along with increase in particle number, the size distribution shifts to bigger size for both nucleation mode and accumulation mode particles. In this stratified charge mode (delayed injection timings), the charge inhomogeneity and the diffusive combustion of fuel droplets that are not yet vaporized can be the main sources of particle formation [99].

Figure 8.35 shows the effect of fuel injection pressure and intake temperature on particle size distribution in PPC mode for gasoline and ethanol. Fuel injection pressure has different effects on particle size distributions for high and low reactivity fuels. In diesel PPC, particle size distribution shifts to smaller size with increase in injection pressure and particle number decreases [100]. In the case of gasoline (RON69), a characteristic particle mode shift occurs when increasing injection pressure (Fig. 8.35b). With increase in injection pressure first, accumulation mode particles decrease with increase in nucleation mode particles. With further increase in injection pressure, it mainly contains only nucleation mode particles, and particle number also increases. In the case of ethanol, nucleation mode particles dominate the size distributions, and higher injection pressure mainly increases the particle number (Fig. 8.35b). Figure 8.35a shows the effect of intake temperature on particle size distribution of ethanol and gasoline in PPC engine. In ethanol PPC, a unimodal particle size distribution dominated by nucleation mode is observed (Fig. 8.35a). However, in gasoline PPC, a bimodal size distribution highly dominated by accumulation mode particles is found. Typically, increase in intake temperature increases peak particle number concentration and corresponding mobility diameter in PPC [100]. In ethanol PPC, mostly particles are in nucleation mode and the size distribution is centred around 5 nm. Despite increased particle number (with increase in intake temperature) and size, no significant increase in mass concentration is observed for ethanol PPC due to the small size of particles [100].

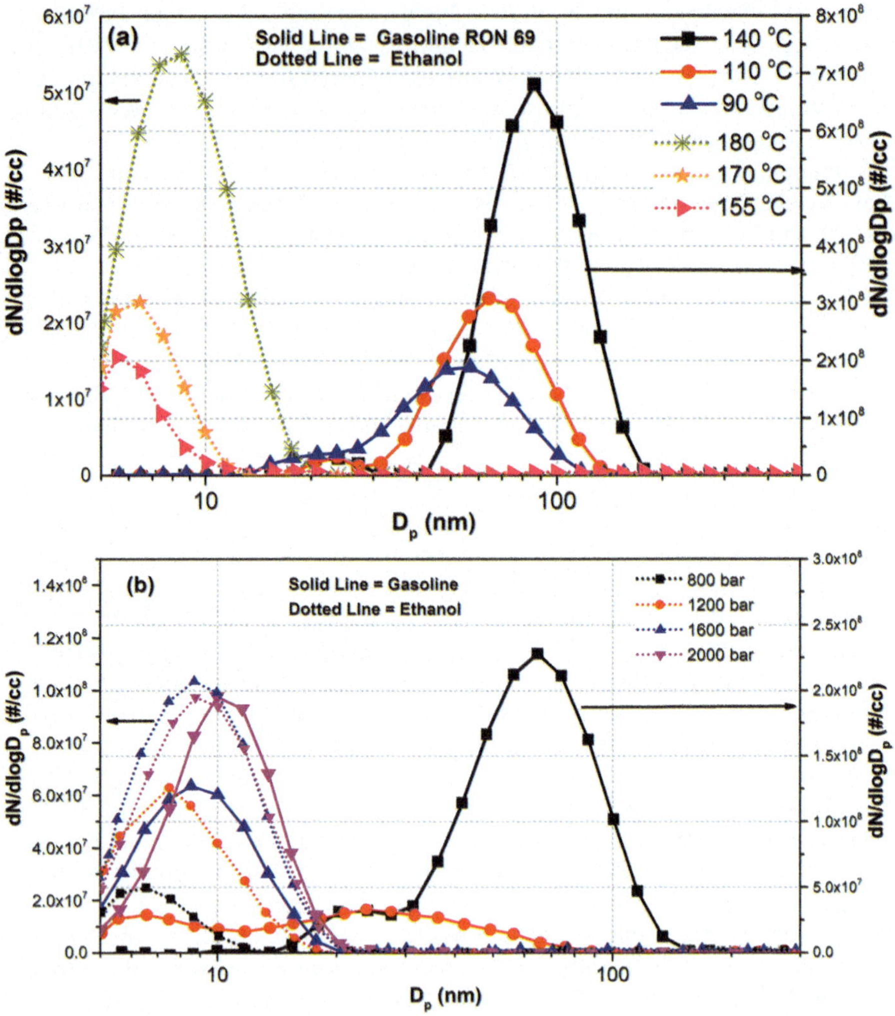

Fig. 8.35 Particles size distribution in PPC engine for (**a**) different intake temperatures, and (**b**) different fuel injection pressure for gasoline and ethanol (Adapted from [100])

Figure 8.36 shows the variation of total particle number as function of injection timings and intake oxygen concentration in PPC engine. Figure 8.36a depicts that in no EGR case, it generates mostly small particles (less than 35 nm) throughout the injection sweep as the difference in accumulation mode and total particle concentration is relatively large. Study also indicated that in EGR case a larger portion of particles are in accumulation mode for very early injection in HCCI-like combustion and late injections in PPC regime [99]. Figure 8.36b shows the total specific particle number emissions, that is, the amount of particles emitted per kWh. Figure 8.36b depicts that naphtha gasoline and methanol are mainly affected by

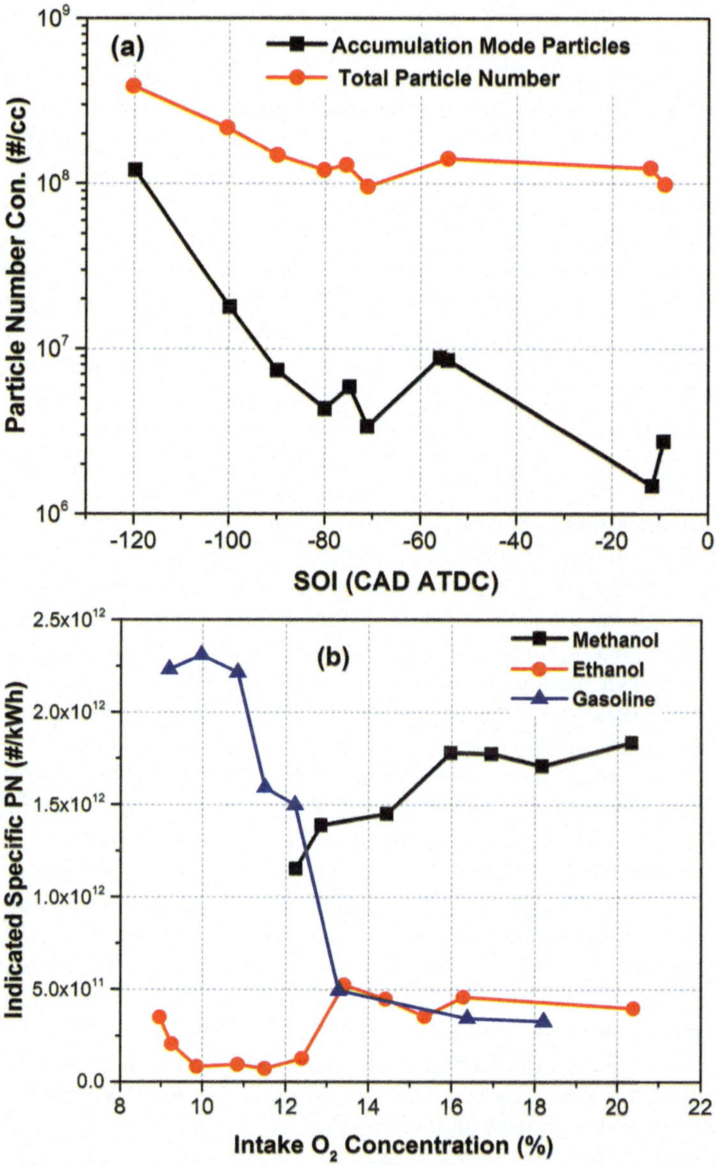

Fig. 8.36 (**a**) Total particle number and accumulation mode particle variations with SOI without EGR (Adapted from [99]), and (**b**) particle number (PN) as function of intake oxygen concentration for gasoline, ethanol and methanol PPC (Adapted from [101])

inlet O_2 concentration in terms of specific particle number emission. However, ethanol is not affect significantly by inlet O_2 concentration in terms of specific particle number emission. Figure also depicts that methanol consistently produces a significantly high amount of particle number. Particle number emissions of ethanol combustion is lowest for most of the test cases (Fig. 8.35b).

The RCCI is another LTC strategy used to achieve higher thermal efficiency along with lower NO_x and PM emissions using dual fuel of different reactivity. In RCCI mode, low reactivity fuel is injected in the intake manifold, and high reactivity fuel (typically diesel) is injected directly in the cylinder. Although smoke emissions are extremely lower, particle number emission needs to be investigated. A study compared particle number emissions from the CDC, diesel PCCI and RCCI strategies, and showed that RCCI combustion has lower particle concentration compared to CDC and diesel PCCI, and large fraction of particles are nucleation mode particles [102]. Figure 8.37a compared the different LTC strategies (HCCI, RCCI, GCI) with CDC for particle number emissions. Lowest overall particle concentration is obtained in HCCI mode particularly for particles larger than 20 nm (Fig. 8.37a). The gasoline compression ignition (GCI) strategies showed the highest concentrations for particles smaller than 10 nm, but in the larger diameter range, relatively lower concentrations were found. The RCCI combustion also has lower particle concentration than CDC.

Figure 8.37b shows the particle size number distribution from RCCI combustion with a single-stage micro dilution tunnel system along with a PMP-style two-stage dilution system. Single-stage dilution depicts a large number of nucleation mode particles (Fig. 8.37b), and with second-stage dilution, these nucleation mode particles vanished particularly at lower load (4 bar BMEP). This indicates that these particles consist of volatile compounds. Higher volatile compounds at lower engine loads are found possibly due to higher level of HC emission at lower loads.

A RCCI combustion study conducted using dual direct injection showed a bimodal particle size distribution that was sensitive to the gasoline start of injection (SOI), and the gasoline/diesel ratio [105]. Results also showed that advancing in-cylinder gasoline SOI timings caused the number of nucleation mode particles to decrease, while the number of accumulation mode particles increased, particularly for the SOI range from −340 to −360° aTDC. Another study conducted for investigating particle size distribution sensitivity to fuels with three combinations of fuel sets diesel/gasoline, diesel-E85 (85% ethanol +15% gasoline) and B20 (20% biodiesel +80% diesel)-gasoline [78]. Study showed no significant difference in particle size distribution among fuel cases, indicating that particle size distribution from RCCI combustion is somewhat insensitive to fuel properties.

In RCCI combustion with dominating nucleation mode particles and high level of unburned HC, it is expected that sampling and dilution conditions have a greater impact on particulate matter concentrations and size distributions in comparison to CDC. A study demonstrated that both the nucleation and the accumulation mode particles are reduced with increased dilution ratio [105]. Another study also compared the particulate matter emissions from RCCI, HCCI and CDC and demonstrated that nucleation mode particle size is reduced and peak concentrated move towards smaller particle diameters with increased dilution ratio [106]. Figure 8.38 shows the effect of dilution ratio and dilution temperature in RCCI combustion using diesel-gasoline (D-G) and diesel-ethanol (D-E) along with conventional

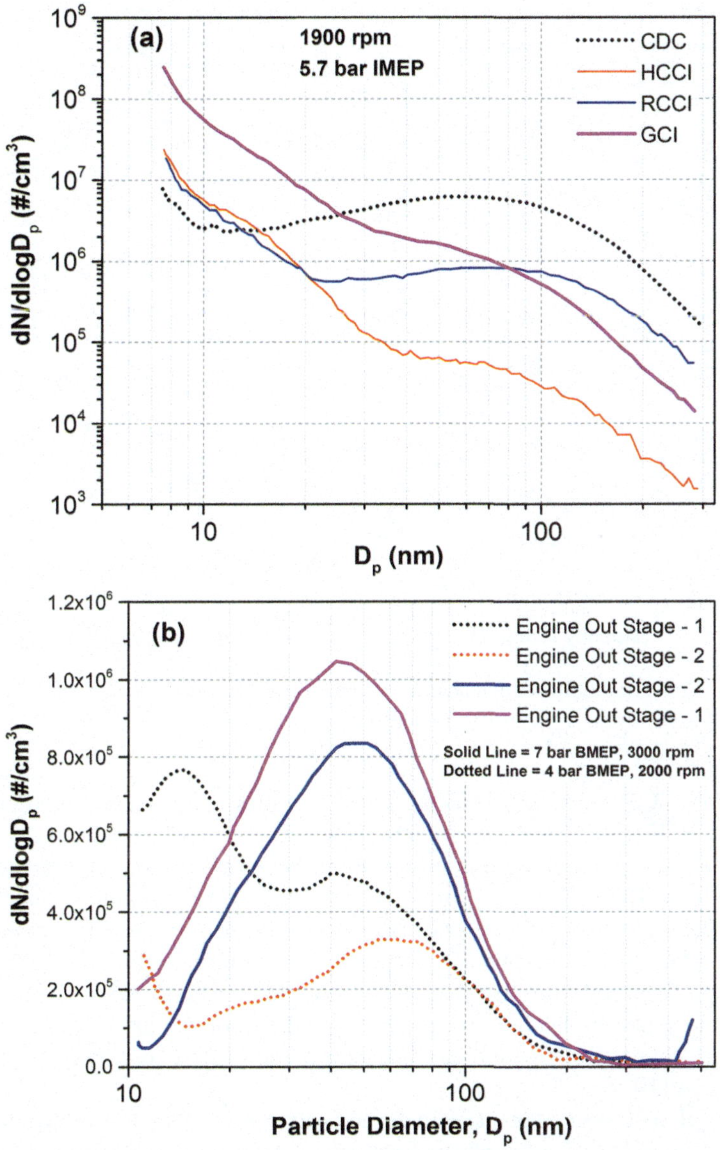

Fig. 8.37 (**a**) Comparison of different LTC strategies for particle number emission (Adapted from [103]), and (**b**) particle size number distribution from RCCI combustion at two different loads for first and second-stage dilution (Adapted from [104])

diesel combustion (CDC) mode. Study was conducted at low dilution temperature and low dilution ratio (LDT + LDR), low dilution temperature and high dilution ratio (LDT + HDR), high dilution temperature and low dilution ratio (HDT + LDR) and high dilution temperature and high dilution ratio (HDT + HDR) conditions for

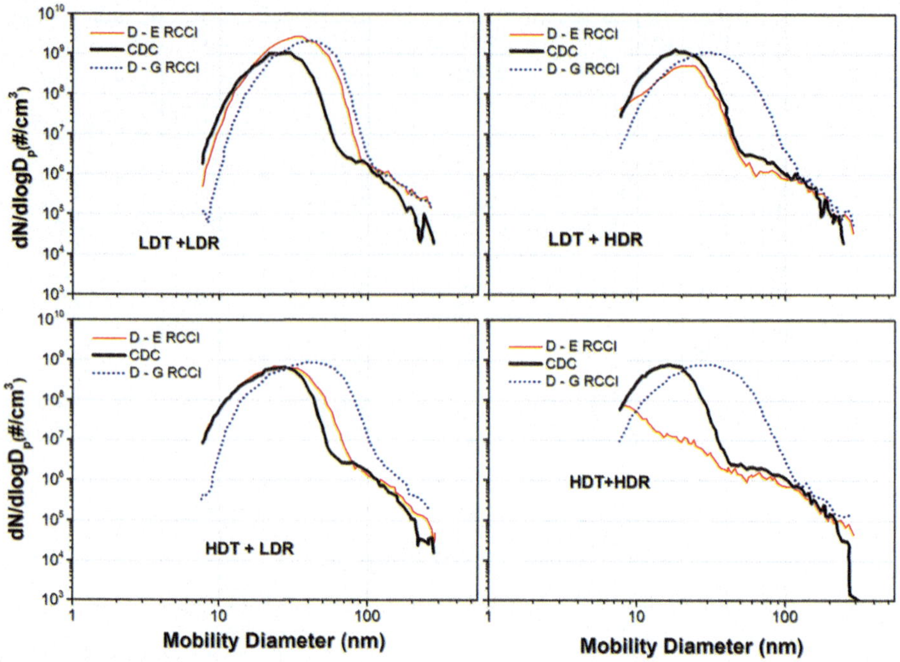

Fig. 8.38 Effect of dilution temperature and dilution ratio on particle size distribution in RCCI combustion vis-à-vis CDC (Adapted from [107])

CDC and RCCI mode. Details about dilution conditioning can be found in original study [107]. Figure 8.38 shows the ethanol/diesel (D-E) RCCI is the most sensitive to the dilution ratio and dilution temperature. Peak of nucleation mode particles in ethanol/diesel RCCI is reduced and moved towards shorter mobility diameters with either increasing dilution ratio or dilution temperature. This results into significant drop in total particle number and volume concentrations. For the high dilution ration and high dilution temperature condition, the peak of nucleation mode particles is vanished, leading to a particle size distribution curve that progressively decreases from the nucleation to the accumulation mode in ethanol/diesel RCCI combustion [107].

Study also showed that particulate emissions from RCCI combustion mainly consist of volatile and semivolatile organic compounds; however, small fraction of solid carbonaceous particles also existed because of inhomogeneity of the high reactivity fuel (diesel) distribution. The semivolatile fraction in the RCCI particulate matter emissions can be up to 99% by volume under LDT + LDR conditions [107]. In the ethanol/diesel RCCI combustion, semivolatile fraction is in the range of 60–99% by volume depending on dilution conditions, whereas it is consistently over 98% by volume for gasoline/diesel RCCI. Study suggested that the difference in dilution sensitivity of RCCI particle emissions is not directly related to the low reactivity fuel based on its high volatility. The low reactivity fuel's influence on unburned diesel, engine lubricating oil evaporation and heat release is most likely the reason of differences in particulate volatility similar to other premixed CI engine combustion modes [107]. A study investigated the volatility of particles emitted from RCCI engines showed that particles consist of a wide variety of volatile materials, having evaporation rates similar to pure C_{28} alkane at low temperatures and greater than evaporation rates for pure C_{32} at higher temperatures [108]. Volatility range of RCCI particles mostly falls within the volatility range of alkanes similar to components present in engine lubricating oil.

Figure 8.39 shows the particle number emission in full engine map from an optimized combustion strategy combining RCCI and dual fuel diesel/gasoline (dual-mode dual fuel combustion strategy). Details of engine operation strategy of these modes can be found in original study [109]. Figure 8.39a presents the smoke emissions and total particle number maps for the dual-mode dual fuel combustion strategy. High similarity is observed between smoke emission and particle numbers for the engine load more than 14 bar IMEP, where combustion is by diffusion dual fuel strategy (Fig. 8.39a). Higher FSN values in this load range indicate that particulates tend to be more like diesel combustion with increase in engine load. However, relatively less similarities are observed between smoke and particle number below 14 bar IMEP, where RCCI combustion strategy (more premixed combustion) is used for achieve the required engine load. In this region, FSN is not related to particulate number emission as RCCI particles are consists of semivolatile compounds [109]. Figure 8.39b shows particle numbers in accumulation mode and nucleation mode particles. LTC favoured the appearance of nucleation mode particles, and they mainly consist of condensed hydrocarbon species [102]. The accumulation mode particles consist of elemental carbon species

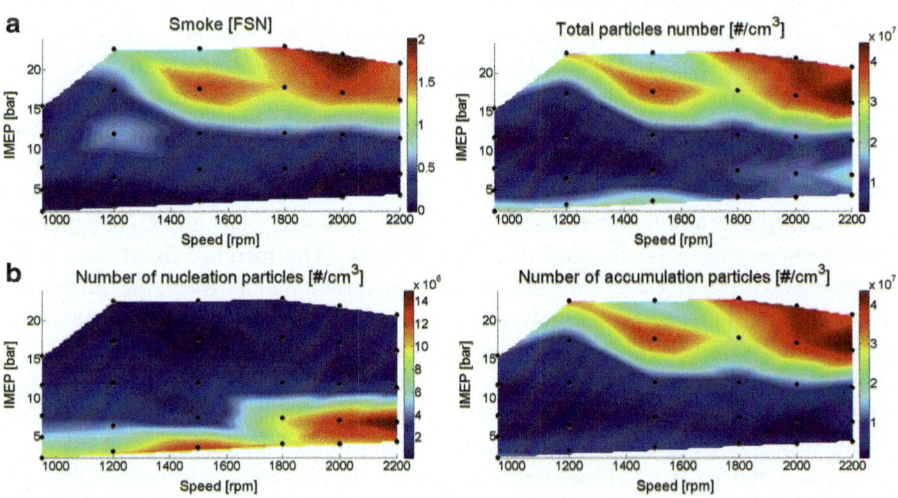

Fig. 8.39 (**a**) Smoke emissions and total particle number and (**b**) total number of nucleation (Dp <30 nm) and accumulation (30 < Dp <250 nm) particles for the dual-mode dual fuel combustion strategy [109]

(i.e. soot particles) similar to particles in CDC, and act as a sink for volatile species. Figure 8.39b shows that nucleation mode particles dominates in RCCI strategy (below 14 bar IMEP), and accumulation mode particles dominate in diffusive dual mode (above 14 bar IMEP). The nucleation mode particles are higher at higher engine speed in RCCI strategy (Fig. 8.39b). A study showed that conventional diesel oxidation catalysts (DOC) are effective to reduction of nucleation mode RCCI particles (<23 nm), while larger size PM emissions (>23 nm) are unaffected by the DOC [110].

8.5 Unregulated Emissions

8.5.1 Hydrocarbon Species

The LTC engines emit higher unburned hydrocarbon (HC) emissions in comparison to conventional engines. The sources of HC emissions and their characteristics in LTC engine are discussed in Sect. 8.3. In emission legislation, total amount of unburned HCs are regulated on mass basis and individual hydrocarbon species are not regulated by legislation. Some of the hydrocarbon species present in significant quantity in engine exhaust affect human health adversely and have higher greenhouse gas (GHG) potential [111]. Effect of principal pollutants emitted from IC engines on human health is summarized in Fig. 1.2 (Chap. 1). The hydrocarbon species emitted from engines can be divided into two main categories: oxygenated hydrocarbon (OHC) species (such as aldehyde, ketones, alcohols, etc.) and pure hydrocarbon (HC) species (alkanes, alkenes, alkynes, etc.). Figure 8.40a shows the variations of HC and OHC emissions with equivalence ratio in a HCCI engine fuelled with iso-octane. The HC curve presents the total percentage of fuel carbon in all the HC species, but not any fuel carbon from the OHC species. Similarly, the OHC curve presents all the fuel carbon in the OHC species, both the carbon atoms that are attached to an oxygen and those that are not. Total HC and OHC emissions provide a basis for determining the contribution of each emission species to the total unburned hydrocarbon emissions. Figure 8.40a shows that HC and OHC emission decreases as engine is operated on richer mixture. The increase in HC and OHC emissions with reduced fuelling (lower Φ) is governed mainly by an increase in the contributions of species with a smaller molecular size than iso-octane (fuel) [56]. The HC and OHC species emitted can be broadly separated into three categories: (i) the parent fuel, iso-octane and species that are closely related in molecular structure to the parent fuel; (ii) species that have one or more carbon atoms less than iso-octane, but still have related molecular structure; and (ii) small molecules that have no direct resemblance to the fuel molecule and could have resulted from the oxidation reactions of almost any hydrocarbon fuel [56]. Studies found that third category small molecules contribute more towards higher unburned hydrocarbon emission in leaner HCCI conditions than rich conditions.

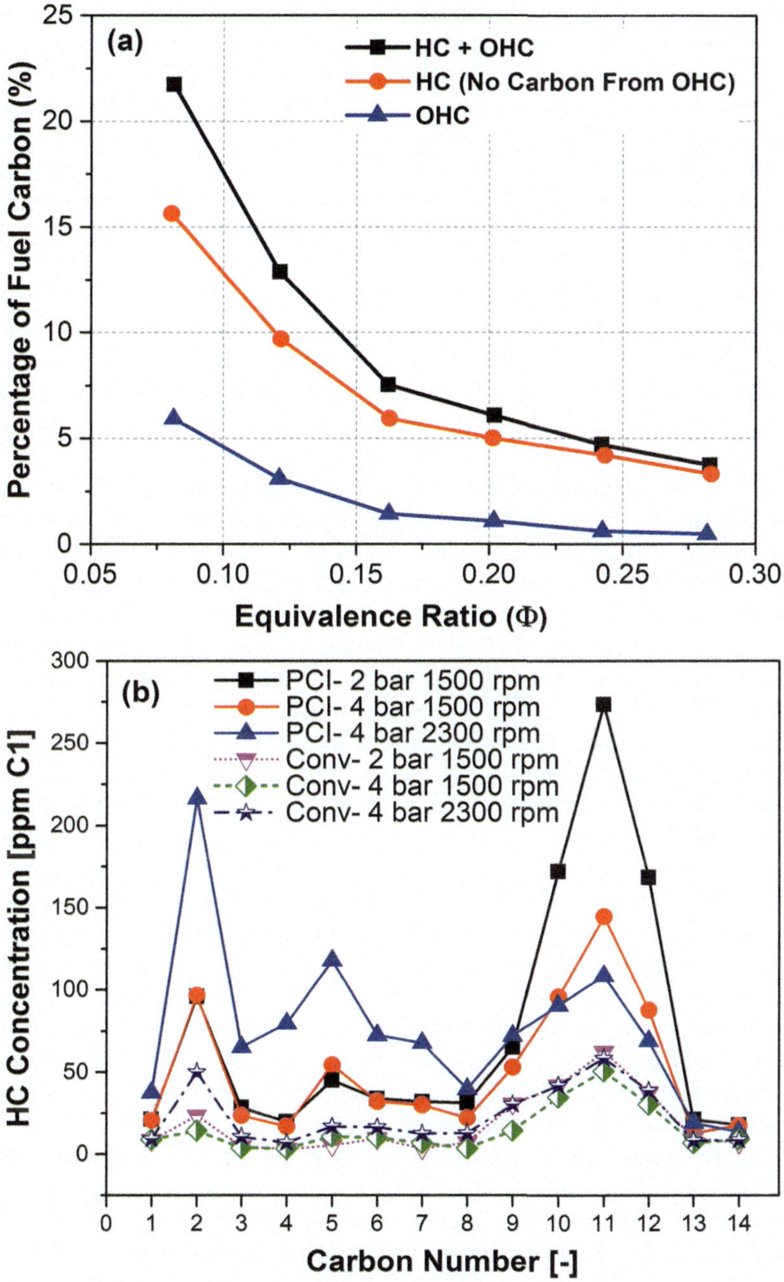

Fig. 8.40 (**a**) Unburned hydrocarbon emission as function of Φ in HCCI engine (Adapted from [56]), and (**b**) PCI and CDC engine-out HC distribution at different engine speed and load by carbon number bin [112]

Figure 8.40b shows the HC emission distribution by carbon number bin for premixed charge compression ignition (PCI) and conventional diesel combustion (CDC). In the case of PCI combustion at engine speed 1500 rpm, the partially burned HC (lower carbon number bins) distribution of 2 bar BMEP is comparable to that of 4 bar BMEP; however, larger carbon number (unburned HC's) concentration at 2 bar BMEP is very higher than at 4 bar BMEP. At 4 bar BMEP in PCI combustion, the lower carbon number bin (partially burned HCs) concentration at 2300 rpm is higher than that at 1500 rpm [112]. The HC emission for each carbon number bin is the lowest for 1500 rpm at 4 bar in CDC.

Figure 8.41 shows the variation of two paraffins' (CH_4 and C_2H_6) and two olefins' (C_2H_4 and C_3H_6) emissions with CA_{50} for different λ in HCCI combustion. Species shown in Fig. 8.41 are obtained computationally. Increasing intake air temperature (advanced combustion phasing) emission of each species decreases for each test condition (Fig. 8.41). This trend is justified due to higher combustion temperature at advanced combustion phasing. At constant intake temperature, unburned hydrocarbon emission also decreases for richer mixture (lower λ) due to higher combustion chamber temperature as higher amount of fuel is burnt in rich operating conditions. Figure 8.41a shows that CH_4 emissions are higher for retarded combustion phasing and lean mixture operating conditions. Range of methane emissions is similar to experimental CH_4 emission reported in reference [114]. Trends in variation of C_2H_6 is similar to variation in CH_4. Amount of C_2H_6 emission is lower in comparison to CH_4. Possible reason for this observation could be that higher strength of C-H bonds in CH_4 molecule leads to difficulty in oxidizing it. As a result, significant amount of unburned CH_4 is emitted as compared to C_2H_6 during exhaust.

Figure 8.41b shows the two alkenes (C_2H_4 and C_3H_6) emitted in significant concentrations. Alkenes are formed during combustion due to the incomplete combustion of the alkanes present in the fuel [114]. During combustion, pyrolysis reactions (long chain hydrocarbon thermally decomposes to form smaller hydrocarbon) and synthesis reactions (smaller hydrocarbon molecules combine and form larger hydrocarbon) occur in some extent. C_2H_4 and C_3H_6 are typical intermediates in diesel fuel pyrolysis process. Production of these intermediate compounds is governed by the two competing (opposite) processes (fuel thermal pyrolysis and oxidation rate) in addition to composition of fuel–air mixture and combustion temperature [115]. Low temperature combustion and leaner mixture in the cylinder decreases both fuel pyrolysis and oxidation rate, which leads to higher amount of intermediate compounds (Fig. 8.41b). At richer mixture engine operation (higher loads) and advanced combustion phasing, concentrations of intermediate species decrease due to greater oxidation rate at higher combustion temperature.

Figure 8.42 shows the concentration of HC and OHC (measured by FTIR analyser) along with their mole fractions in THC at different operating conditions in SI and HCCI engine. The HCCI combustion is achieved using two negative valve overlap (NVO) positions at two different engine loads (presented as four cases in Fig. 8.42). Two cases of SI combustion present the data for two different spark timings. Switching between SI and HCCI combustion changes a trade-off between

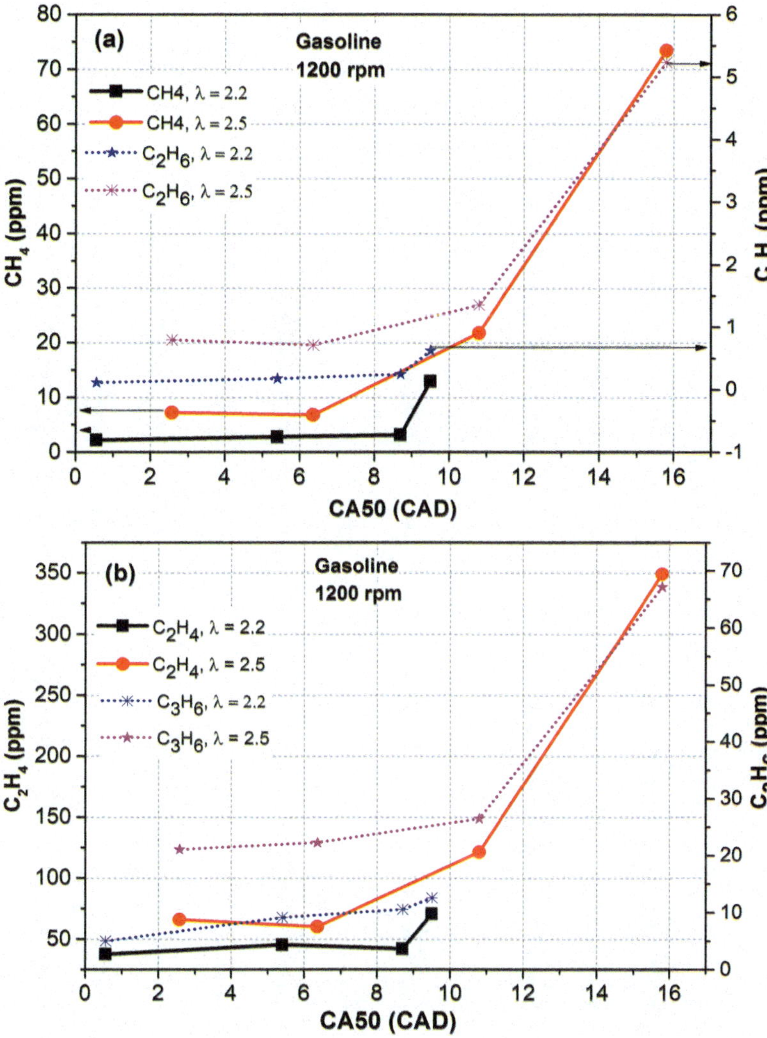

Fig. 8.41 Variations of (**a**) CH_4 and C_2H_6. (**b**) C_2H_4 and C_3H_6 emission with combustion phasing (CA_{50}) for different relative air–fuel ratios in HCCI engine (Adapted from [113])

hydrocarbon species [116]. The HCCI combustion emits comparatively lower methane and light unsaturated hydrocarbons than SI combustion. However, for early NVO injection, HCCI combustion emitted the highest concentrations of aforementioned species. Concentrations of ethane and pentanes reveal opposite trend. In HCCI combustion mode, trade-off between hydrocarbon concentrations is found for early and late NVO injections. This is possibly due to exhaust-fuel reactions that occur during the NVO for early injection, which break long-chained hydrocarbons and produce methane, light unsaturated hydrocarbons and

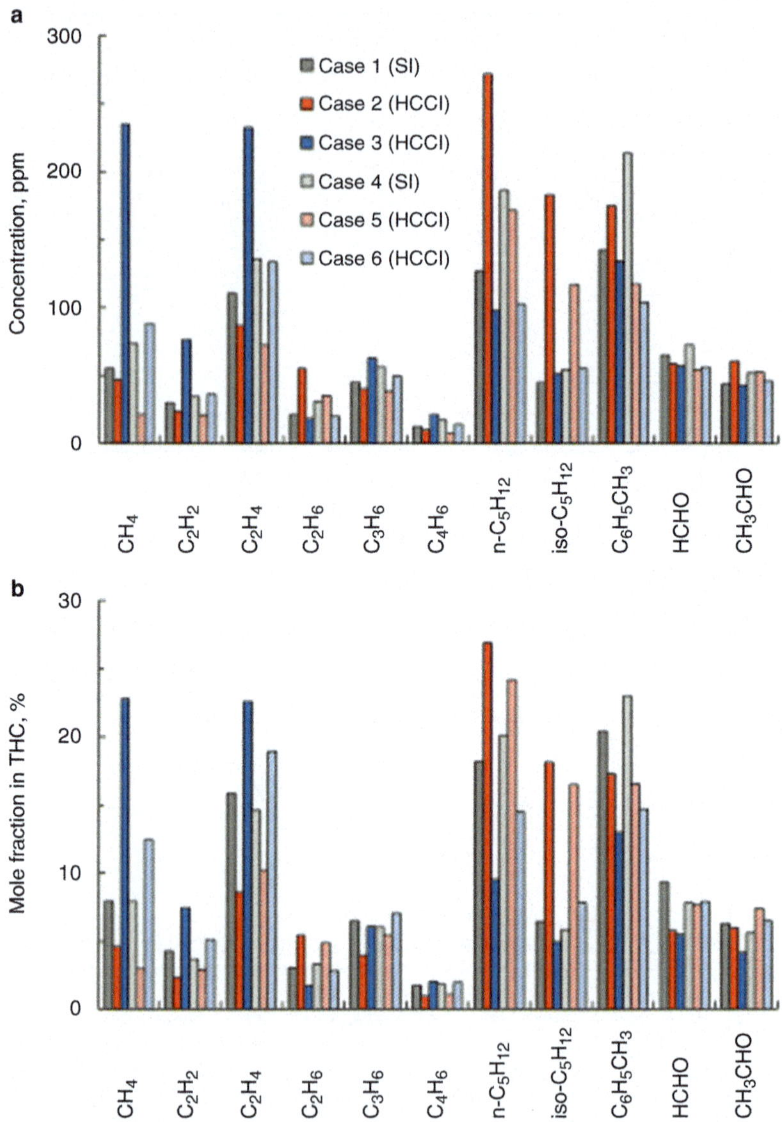

Fig. 8.42 (**a**) Concentrations of hydrocarbons and oxygenated hydrocarbons and (**b**) mole fraction of the species in THC [116]

aldehydes [116]. Therefore, the fuel composition changes can modify exhaust hydrocarbon structure.

Fig. 8.43 shows the variation of concentrations of non-oxygenated hydrocarbon species for different charge dilution levels and the diesel SOI timings in combustion of in-cylinder blending of gasoline/diesel. The oxygen concentration is affected by EGR dilution (Fig. 8.43a). Concentrations of exhaust hydrocarbon species

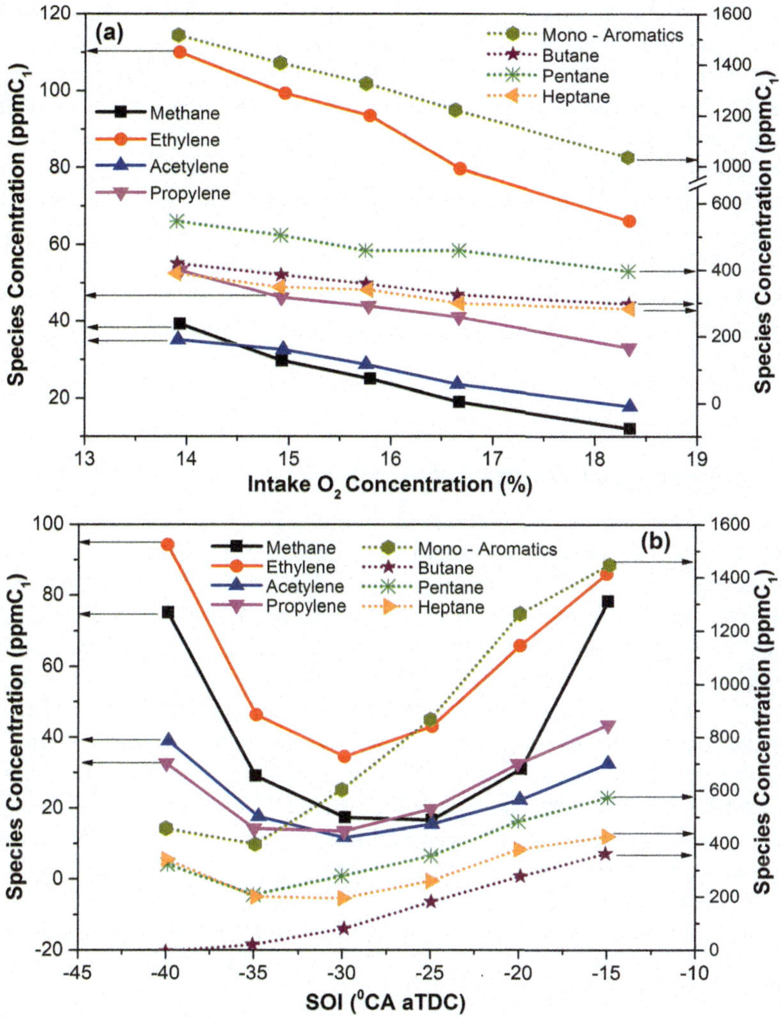

Fig. 8.43 Concentrations of non-oxygenated hydrocarbon species at (**a**) different charge dilution levels, and (**b**) the diesel SOI timings in combustion of in-cylinder blending of gasoline/diesel (Adapted from [117])

generally increased with charge dilution using EGR because of lower availability of oxygen and the lower combustion temperatures associated with the increased EGR rate. Diesel/gasoline combustion showed comparable amounts of light hydrocarbons as in neat diesel combustion [117]. Figure 8.43a shows significantly higher concentration of C4–C7 alkanes and mono-aromatics for combustion using in-cylinder blending of diesel-gasoline, which is approximately 10 times higher C4–C7 alkanes and 30 times higher mono-aromatics in comparison to pure diesel combustion. Mono-aromatics are mostly single-ring aromatics with short alkyl chains in gasoline fuels (e.g. toluene), which are considerably harder to oxidize than the mono-aromatics contained in diesel [117]. Figure 8.43b shows the HC speciation data for different SOI timings of diesel. The concentrations of mono-aromatics and C4–C7 alkanes decreased sharply with diesel SOI advanced until the further advance of SOI made the fuel–air mixture overmixed and began to retard combustion phasing. The advance of diesel SOI mitigated the over-stratification and led to a more reactive in-cylinder charge through better mixing between gasoline and diesel across the combustion chamber as noted by the drop in UHCs [117]. Results depict that there is an optimum timing such that the cold, low reactivity regions are effectively minimized and corresponding emissions are minimum.

8.5.2 Oxygenated Hydrocarbon Species

In LTC engines, oxygenated unburned hydrocarbon (OHC) species are emitted in significant quantity. In OHC species, mainly aldehydes and few ketones (such as acetone) are emitted from LTC engines. Figure 8.44a shows the typical variation in different OHCs in well-mixed HCCI combustion engine. Figure shows that different OHC species can vary up to 1.4% of fuel carbon, depending on engine operating conditions. The OHC species emissions decrease with increase in equivalence ratio due to increase in combustion temperature, similar to HC species. Figure 8.44b shows the variation of HC and OHC with direct injection timings in partially stratified HCCI engine using iso-octane fuel. Both HC and OHC emissions decrease with increase in stratification (delayed SOI timings) of fuel (Fig. 8.44b). It can also be noticed from the figure that HC emission reduction is comparatively higher than OHC emission reduction by stratification of charge.

To quantify the carbonyl species emission in HCCI emission, a comparative study investigated 14 different carbonyl species in HCCI and SI engines. Figure 8.45 shows the concentration of different carbonyl species in exhaust of SI and HCCI engine using commercial gasoline. The main carbonyl compounds detected in reasonable quantity are acetaldehyde followed by acrolein, benzaldehyde, o-tolualdehyde, m-tolualdehyde and 2,5-Dimethyl-benzaldehyde (Fig. 8.45). Although emissions of carbonyl species are very low in comparison to THC emissions, carbonyl species can be found in the range of 2.7% and 5.2% of total HC based on engine operating conditions and engine operating modes [118].

Figure 8.45 shows that acetaldehyde is the highest emitted carbonyl species in HCCI engine at lower engine load. Acrolein is the second highest carbonyl species in the engine exhaust, which is produced by the oxidation of propyl radicals. Nearly all normal alkanes having three or more carbon atoms can produce propyl radicals, which acts as a precursor for formation of acrolein, which is present in engine exhaust [118]. Figure 8.45 also depicts that the aliphatic aldehydes reduces with increase in engine operating load or combustion mode shift from HCCI to SI;

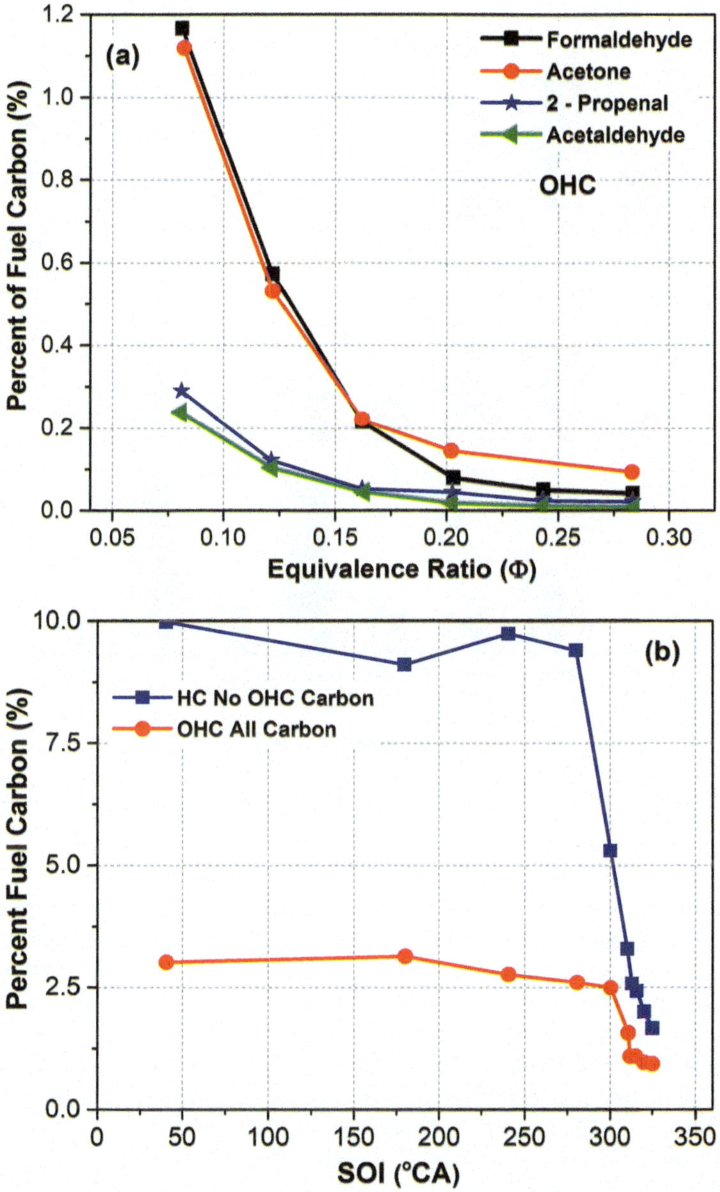

Fig. 8.44 (**a**) OHC species as function of Φ in well-mixed HCCI engine, and (**b**) variation of OHC with SOI timing in partially stratified HCCI engine (Adapted from [56])

however, the aromatic aldehydes showed reverse trend. The formation of aromatic aldehydes increases with the increasing engine speed in addition to the change in engine operation mode from low load HCCI to part load HCCI to low load SI. Detailed description and justification can be found in original study [118].

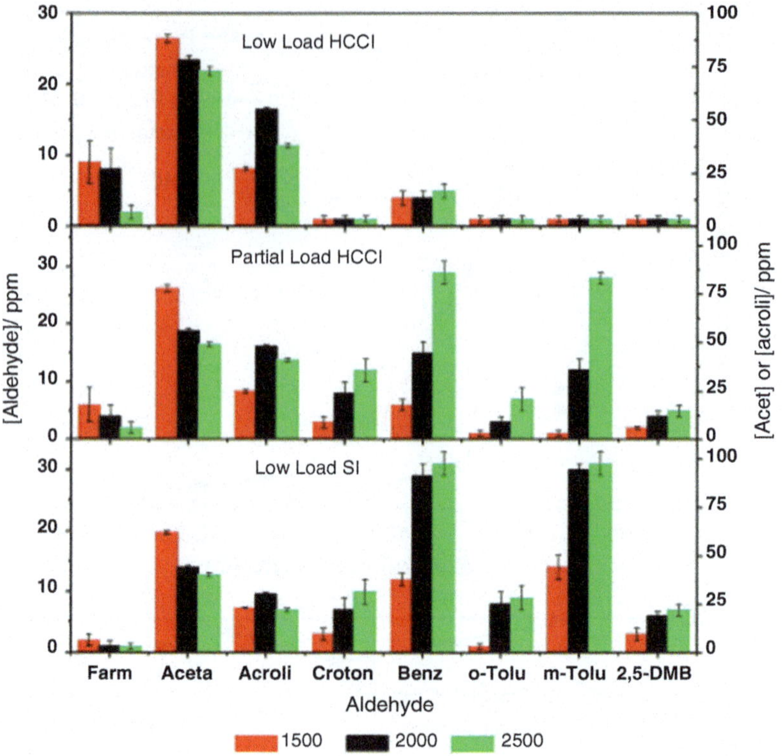

Fig. 8.45 Carbonyl species emissions in different SI and HCCI engines at different engine speeds [118]

Figure 8.46 shows the variation of acetaldehyde (CH_3CHO) and formaldehyde (HCHO) with different speeds and load conditions in ethanol HCCI operating range achieved at most efficient combustion phasing. HCHO and CH_3CHO are two of the main aldehydes produced from HCCI combustion particularly engine operation with alcohols, and these aldehydes are known to be highly toxic and harmful to human health [115]. Acetaldehyde emission is higher in comparison to formaldehydes (Fig. 8.46), and both the species are produced in significant quantity in ethanol HCCI combustion. Figure 8.46 shows that higher engine loads (where relatively lower intake temperature required for optimal combustion phasing) coupled with higher engine speeds lead to incomplete combustion, which results into higher acetaldehyde and formaldehyde emissions. At lower engine loads, aldehyde emission is comparatively lower, and there is lesser variation with engine speed. However, at higher engine load, acetaldehyde and formaldehyde emission significantly depends on engine speed.

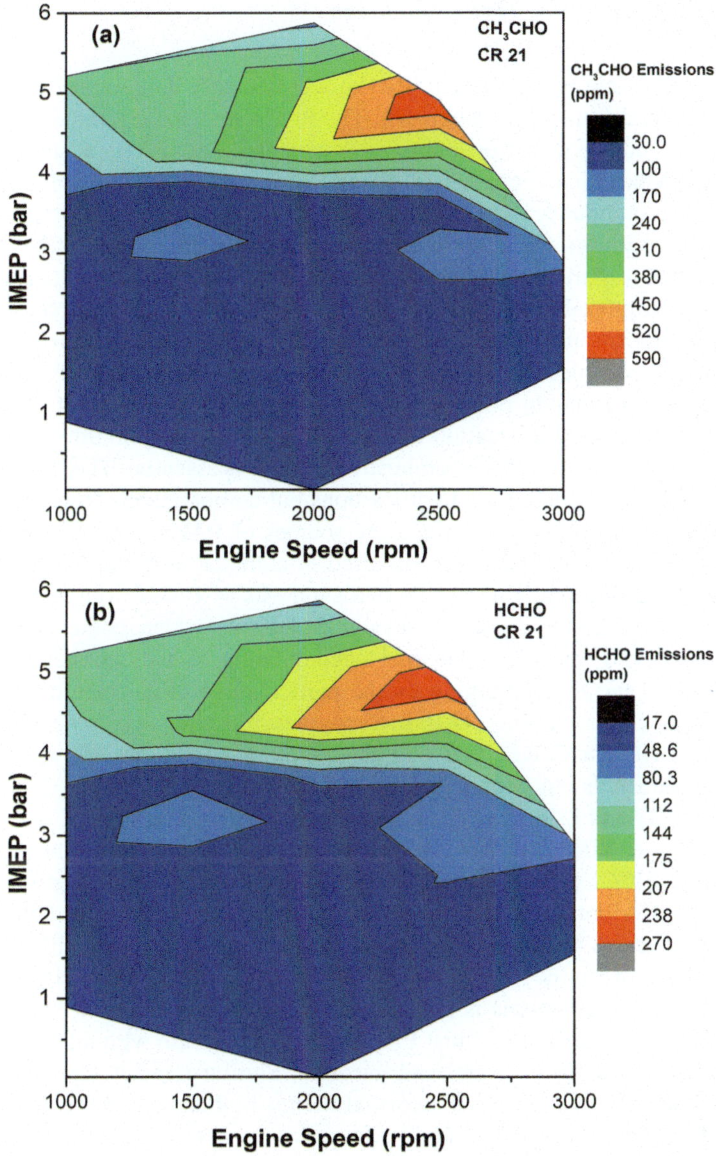

Fig. 8.46 Variation in emission of (**a**) CH_3CHO and (**b**) HCHO with engine and load in ethanol HCCI operating region for compression ratio 21 [14]

8.5.3 Polyaromatic Hydrocarbons

Polycyclic aromatic hydrocarbons (PAHs) are presently unregulated pollutants in the internal combustion engine exhaust. The PAHs are released in the environment either in the vapour phase or along with soot particles (PAHs adsorbed on fine particles) produced in the engine. Although PAHs are unregulated pollutants, these hydrocarbon compounds (PAHs) can lead to several environmental issues (such as photochemical smog) and adverse effect on human health. Some of the PAHs are now identified as carcinogenic species and also resulted into an increased human morbidity and mortality rates [119]. In diesel exhaust, major sources of mono-aromatic hydrocarbons (MAHs) and PAHs are pyrolysis and structural amendments in unburned fuel molecules (which also contains aromatics in significant amount) during diesel combustion process [120]. Incomplete combustion of hydrocarbon fuels can result into the formation of PAHs and particulate, and therefore, investigation and understanding of the combustion process is essential. The LTC strategies typically emit higher unburned hydrocarbons but comparatively lower particulate. Lea-Langton et al. [121] categorized the sources of PAHs in conventional diesel engines in five possible paths: (i) fuel used in combustion event, (ii) pyrolysis of fuel to produce different PAHs, (iii) pyrosynthesis of PAHs from alkanes during combustion, (iv) PAHs absorbed in engine deposits which come later into the exhaust and (v) PAHs from the fuel accumulating in the lubricating oil with blow-by gases and then emerging on the particulate with the lube oil fraction of the soluble oil fraction (SOF) [122]. In SI engines, fuel chemical composition, types of lubricant, fuel additives and engine operating conditions are major factors governing the PAHs emission [123].

Figures 8.47 and 8.48 show the variations of particle bound PAH w.r.t. relative air–fuel ratio at different intake air temperature using ethanol and gasoline at 1200 rpm in HCCI engine. Total particle mass per unit volume (estimated from particle size distribution) is also presented for same test for the sake of comparison. Figures 8.47 and 8.48 show that PAH emissions increase drastically as mixture becomes richer or engine load increases for both fuels. Elghawi et al. also found that reduction in PAH emissions is proportional to reduction in engine load in HCCI mode [122]. Figures also show that PAH emissions increase with increase in intake air temperature for both ethanol and gasoline. Similar trend is also observed for total particle mass variations. Factors affecting higher particulate emission are discussed above in Sect. 8.4.2. The PAH emissions are dependent on combustion phasing and engine load. Figures show that in most of the conditions, total PAHs are in the range of 50–600 ng/m^3. Only in one case of very rich condition, ethanol emitted PAH mass of 920 ng/m^3. It can also be noticed that at lower load conditions (leaner mixtures), ethanol has lower PAH mass emission as compared to gasoline. The total particulate mass is comparatively lower in the case of ethanol as compared to gasoline. Ethanol is an oxygenated fuel, which leads to lower total particulate emission in the exhaust due to superior combustion, being an oxygenated fuel. Detailed PAH characterization of HCCI as well as other LTC strategies is needed for more fundamental understanding.

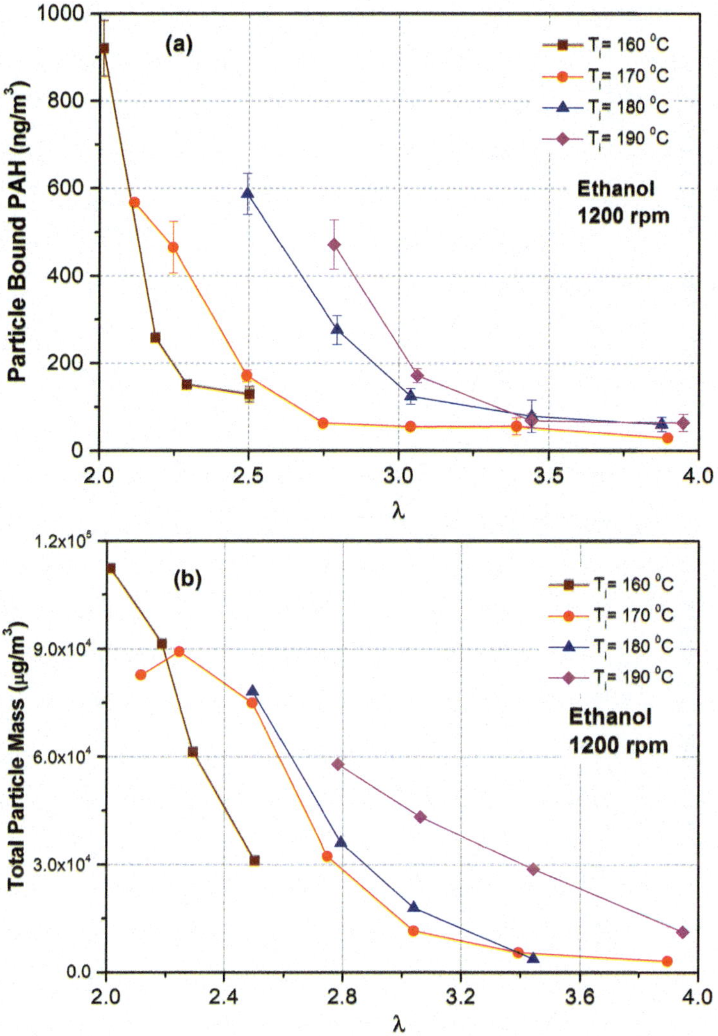

Fig. 8.47 Effect of λ and T_i on variation of **(a)** particle bound PAH emissions and **(b)** total particle mass for ethanol at 1200 rpm in HCCI engine [24]

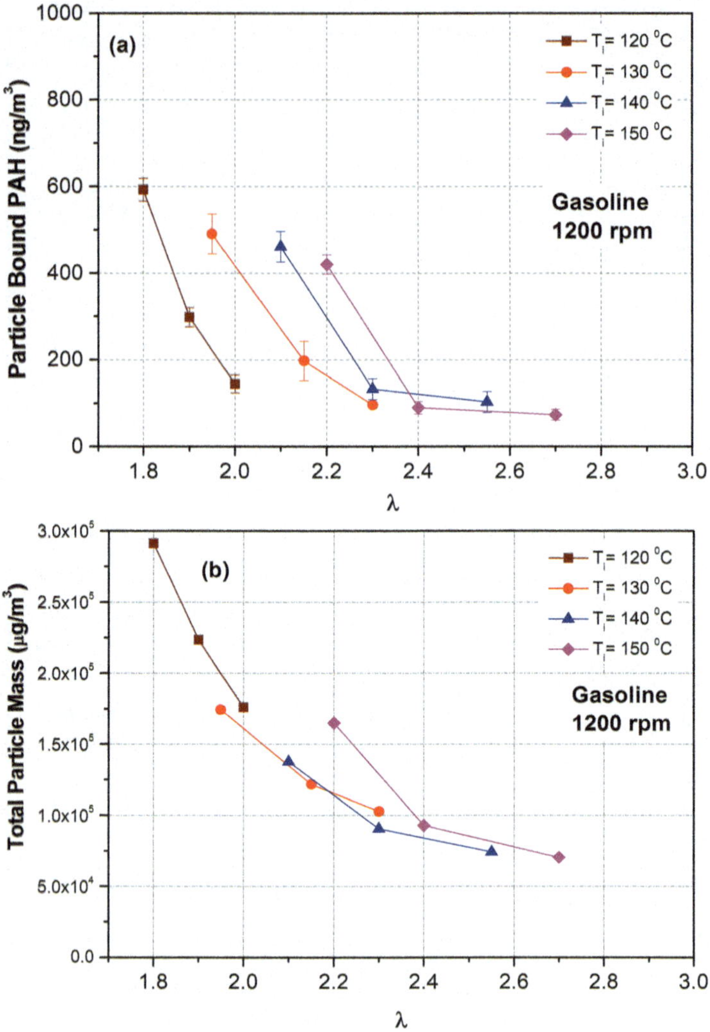

Fig. 8.48 Effect of λ and T_i on variation of (**a**) particle bound PAH emissions and (**b**) total particle mass for gasoline at 1200 rpm [24]

References

1. Maurya RK, Agarwal AK (2014) Experimental investigations of performance, combustion and emission characteristics of ethanol and methanol fueled HCCI engine. Fuel Process Technol 126:30–48
2. Agarwal AK, Singh AP, Maurya RK (2017) Evolution, challenges and path forward for low temperature combustion engines. Prog Energy Combust Sci 61:1–56
3. Bowman CT (2015) NOx formation and models. In: Crolla D (ed) Encyclopaedia of automotive engineering. Wiley, Chichester, UK. https://doi.org/10.1002/9781118354179.auto117

4. Tomeczek J, Gradoń B (2003) The role of N2O and NNH in the formation of NO via HCN in hydrocarbon flames. Combust Flame 133(3):311–322
5. Bozzelli JW, Dean AM (1995) O+ NNH: a possible new route for NOx formation in flames. In J Chem Kinet 27(11):1097–1109
6. Zeldovich YB (1946) The oxidation of nitrogen in combustion and explosions. Acta Physicohim URRS 21:577–628
7. Gardiner WC (ed) (1999) Gas-phase combustion chemistry. Springer, New York
8. Li H, Neill WS, Guo H, Chippior W (2012) The NOx and N2O emission characteristics of an HCCI engine operated with n-heptane. J Energy Resour Technol 134(1):011101
9. Hill SC, Smoot LD (2000) Modeling of nitrogen oxides formation and destruction in combustion systems. Prog Energy Combust Sci 26(4):417–458
10. Hayhurst AN, Vince IM (1980) Nitric oxide formation from N2 in flames: the importance of "prompt" NO. Prog Energy Combust Sci 6(1):35–51
11. Fenimore CP (1971) Formation of nitric oxide in premixed hydrocarbon flames. Symp Combust 13(1):373–380. Elsevier
12. Harrington JE, Smith GP, Berg PA, Noble AR, Jeffries JB, Crosley DR (1996) Evidence for a new NO production mechanism in flames. Symp Combust Vol. 26(2):2133–2138. Elsevier
13. Maurya RK, Akhil N (2016) Numerical investigation of ethanol fuelled HCCI engine using stochastic reactor model. Part 1: Development of a new reduced ethanol oxidation mechanism. Energy Convers Manag 118:44–54
14. Maurya RK, Akhil N (2016) Numerical investigation of ethanol fuelled HCCI engine using stochastic reactor model. Part 2: Parametric study of performance and emissions characteristics using new reduced ethanol oxidation mechanism. Energy Convers Manag 121:55–70
15. Yoshikawa T, Reitz RD (2008) Development of an improved NOx reaction mechanism for low temperature diesel combustion modeling. SAE Int J Engines 1:1105–1117. (2008-01-2413)
16. Amnéus P, Mauss F, Kraft M, Vressner A, Johansson B (2005) NOx and N2O formation in HCCI engines (No. 2005–01-0126). SAE technical paper
17. Malte PC, Pratt DT (1974) The role of energy-releasing kinetics in NOx formation: fuel-lean, jet-stirred CO-air combustion. Combust Sci Technol 9(5–6):221–231
18. Wolfrum J (1972) Formation of nitric oxide in combustion. Chem Ing Tec 44:656–659
19. Wuebbles DJ (2009) Nitrous oxide: no laughing matter. Science 326(5949):56–57
20. Robertson GP, Grace PR (2004) Greenhouse gas fluxes in tropical and temperate agriculture: the need for a full-cost accounting of global warming potentials. In: Tropical agriculture in transition—opportunities for mitigating greenhouse gas emissions? Springer Netherlands, Dordrecht, pp 51–63
21. Heywood JB (1988) Internal combustion engine fundamentals. McGraw-Hill, New York
22. Hori M, Matsunaga N, Marinov N, William P, Charles W (1998) An experimental and kinetic calculation of the promotion effect of hydrocarbons on the NO-NO2 conversion in a flow reactor. Symp Combust 27(1):389–396. Elsevier
23. Miller JA, Durant JL, Glarborg P (1998) Some chemical kinetics issues in reburning: the branching fraction of the HCCO+ NO reaction. Symp Combust 27(1):235–243. Elsevier
24. Maurya RK (2012) Performance, emissions and combustion characterization and close loop control of HCCI engine employing gasoline like fuels. PhD thesis, Indian Institute of Technology Kanpur, India
25. Xie H, Wei Z, He B, Zhao H (2006) Comparison of HCCI combustion respectively fueled with gasoline, ethanol and methanol through the trapped residual gas strategy (No. 2006-01-0635). SAE technical paper
26. Solaka H (2014) Impact of fuel properties on partially premixed combustion. PhD thesis, Lund University, Sweden
27. Solaka H, Tuner M, Johansson B, Cannella W (2013) Gasoline surrogate fuels for partially premixed combustion, of toluene ethanol reference fuels (No. 2013-01-2540). SAE technical paper

28. Aronsson HS, Tuner M, Johansson B (2014) Using oxygenated gasoline surrogate compositions to map RON and MON (No. 2014-01-1303). SAE technical paper
29. Kee SS, Mohammadi A, Kidoguchi Y, Miwa K (2005) Effects of aromatic hydrocarbons on fuel decomposition and oxidation processes in diesel combustion (No. 2005-01-2086). SAE technical paper
30. Manente V, Tunestal P, Johansson B, Cannella WJ (2010) Effects of ethanol and different type of gasoline fuels on partially premixed combustion from low to high load (No. 2010-01-0871). SAE technical paper
31. Manente V (2010) Gasoline partially premixed combustion-an advanced internal combustion engine concept aimed to high efficiency, low emissions and low acoustic noise in the whole load range. PhD thesis, Lund University, Sweden
32. Manente V, Johansson B, Tunestal P (2009) Partially premixed combustion at high load using gasoline and ethanol, a comparison with diesel (No. 2009-01-0944). SAE technical paper
33. Manente V, Johansson B, Tunestal P, Cannella W (2009) Effects of different type of gasoline fuels on heavy duty partially premixed combustion. SAE Int J Engines 2:71–88. (2009-01-2668)
34. Splitter D, Hanson R, Kokjohn S, Reitz RD (2011) Reactivity controlled compression ignition (RCCI) heavy-duty engine operation at mid-and high-loads with conventional and alternative fuels (No. 2011-01-0363). SAE technical paper
35. Wang Y, Yao M, Li T, Zhang W, Zheng Z (2016) A parametric study for enabling reactivity controlled compression ignition (RCCI) operation in diesel engines at various engine loads. Appl Energy 175:389–402
36. Ma S, Zheng Z, Liu H, Zhang Q, Yao M (2013) Experimental investigation of the effects of diesel injection strategy on gasoline/diesel dual-fuel combustion. Appl Energy 109:202–212
37. Benajes J, Pastor JV, García A, Monsalve-Serrano J (2015) The potential of RCCI concept to meet EURO VI NOx limitation and ultra-low soot emissions in a heavy-duty engine over the whole engine map. Fuel 159:952–961
38. Benajes J, García A, Monsalve-Serrano J, Balloul I, Pradel G (2016) An assessment of the dual-mode reactivity controlled compression ignition/conventional diesel combustion capabilities in a EURO VI medium-duty diesel engine fueled with an intermediate ethanol-gasoline blend and biodiesel. Energy Convers Manag 123:381–391
39. Sjöberg M, Dec JE (2003) Combined effects of fuel-type and engine speed on intake temperature requirements and completeness of bulk-gas reactions for HCCI combustion (No. 2003-01-3173). SAE technical paper
40. Dec JE, Sjöberg M (2003) A parametric study of HCCI combustion-the sources of emissions at low loads and the effects of GDI fuel injection (No. 2003-01-0752). SAE technical paper
41. Christensen M, Johansson B, Einewall P (1997) Homogeneous charge compression ignition (HCCI) using isooctane, ethanol and natural gas-a comparison with spark ignition operation (No. 972874). SAE technical paper
42. Sjöberg M, Dec JE (2005) An investigation into lowest acceptable combustion temperatures for hydrocarbon fuels in HCCI engines. Proc Combust Inst 30(2):2719–2726
43. Bhave A, Kraft M, Montorsi L, Mauss F (2006) Sources of CO emissions in an HCCI engine: a numerical analysis. Combust Flame 144(3):634–637
44. Aceves SM, Flowers DL, Espinosa-Loza F, Martinez-Frias J, Dibble RW, Christensen M et al (2002) Piston-liner crevice geometry effect on HCCI combustion by multi-zone analysis (No. 2002-01-2869). SAE technical paper
45. Maurya RK, Agarwal AK (2015) Combustion and emission characterization of n-butanol fueled HCCI engine. J Energy Resour Technol 137(1):011101
46. Hanson RM, Reitz RD (2016) Effects of biofuel blends on transient reactivity-controlled compression ignition engine combustion. Int J Engine Res 17(8):857–865
47. Hanson R, Curran S, Wagner R, Reitz RD (2013) Effects of biofuel blends on RCCI combustion in a light-duty, multi-cylinder diesel engine. SAE Int J Engines 6:488–503. (2013-01-1653)

48. Li Y, Jia M, Liu Y, Xie M (2013) Numerical study on the combustion and emission characteristics of a methanol/diesel reactivity controlled compression ignition (RCCI) engine. Appl Energy 106:184–197
49. Hanson R, Kokjohn S, Splitter D, Reitz RD (2011) Fuel effects on reactivity controlled compression ignition (RCCI) combustion at low load. SAE Int J Engines 4:394–411. (2011-01-0361)
50. Cheng WK, Hamrin D, Heywood JB, Hochgreb S, Min K, Norris M (1993) An overview of hydrocarbon emissions mechanisms in spark-ignition engines (No. 932708). SAE technical paper
51. Pundir BP (2010) IC engines: combustion and emissions, 1st edn. Alpha Science International, Oxford
52. Zeng W, Xie M (2008) A novel approach to reduce hydrocarbon emissions from the HCCI engine. Chem Eng J 139(2):380–389
53. Kashdan JT, Mendez S, Bruneaux G (2007) On the origin of unburned hydrocarbon emissions in a wall guided, low NOx diesel combustion system (No. 2007-01-1836). SAE technical paper
54. Christensen M, Johansson B (1998) Influence of mixture quality on homogeneous charge compression ignition (No. 982454). SAE technical paper
55. Aceves SM, Flowers DL, Espinosa-Loza F, Martinez-Frias J, Dec JE, Sjöberg M et al (2004) Spatial analysis of emissions sources for HCCI combustion at low loads using a multi-zone model (No. 2004-01-1910). SAE technical paper
56. Dec JE, Sjöberg M, Hwang W, Davisson ML, Leif RN (2008) Detailed HCCI exhaust speciation and the sources of hydrocarbon and oxygenated hydrocarbon emissions. SAE Int J Fuels Lubr 1:50–67. (2008-01-0053)
57. Christensen M, Johansson B, Hultqvist A (2001) The effect of piston topland geometry on emissions of unburned hydrocarbons from a homogeneous charge compression ignition (HCCI) engine (No. 2001-01-1893). SAE technical paper
58. Cheng CO, Cheng WK, Heywood JB, Maroteaux D, Collings N (1991) Intake port phenomena in a spark-ignition engine at part load (No. 912401). SAE technical paper
59. Jung GS, Sung YH, Choi BC, Lee CW, Lim MT (2012) Major sources of hydrocarbon emissions in a premixed charge compression ignition engine. Int J Automot Technol 13 (3):347–353
60. Petersen BR, Ekoto IW, Miles PC (2010) Optical investigation of UHC and CO sources from biodiesel blends in a light-duty diesel engine operating in a partially premixed combustion regime. SAE Int J Fuels Lubr 3:414–434. (2010-01-0862)
61. Kim D, Ekoto I, Colban WF, Miles PC (2008) In-cylinder CO and UHC imaging in a light-duty diesel engine during PPCI low-temperature combustion. SAE Int J Fuels Lubr 1:933–956. (2008-01-1602)
62. Kook S, Bae C, Miles P., Choi D, Pickett LM (2005) The influence of charge dilution and injection timing on low-temperature diesel combustion and emissions (No. 2005-01-3837). SAE technical paper
63. Noehre C, Andersson M, Johansson B, Hultqvist A (2006) Characterization of partially premixed combustion (No. 2006-01-3412). SAE technical paper
64. Gharehghani A, Hosseini R, Mirsalim M, Jazayeri SA, Yusaf T (2015) An experimental study on reactivity controlled compression ignition engine fueled with biodiesel/natural gas. Energy 89:558–567
65. Benajes J, Molina S, García A, Belarte E, Vanvolsem M (2014) An investigation on RCCI combustion in a heavy duty diesel engine using in-cylinder blending of diesel and gasoline fuels. Appl Therm Eng 63(1):66–76
66. Jia Z, Denbratt I (2015) Experimental investigation of natural gas-diesel dual-fuel RCCI in a heavy-duty engine. SAE Int J Engines 8:797–807. (2015-01-0838)

67. Poorghasemi K, Saray RK, Ansari E, Irdmousa BK, Shahbakhti M, Naber JD (2017) Effect of diesel injection strategies on natural gas/diesel RCCI combustion characteristics in a light duty diesel engine. Appl Energy 199:430–446
68. Erlandsson, O., Johansson, B., & Silversand, F. A. (2000). Hydrocarbon (HC) reduction of exhaust gases from a homogeneous charge compression ignition (HCCI) engine using different catalytic mesh-coatings (No. 2000-01-1847). SAE technical paper
69. Hall DE, King DJ, Morgan TBD, Baverstock SJ, Heinze P, Simpson BJ (1998) A review of recent literature investigating the measurement of automotive particulate; the relationship with environmental aerosol, air quality and health effects (No. 982602). SAE technical paper
70. Kittelson DB (1998) Engines and nanoparticles: a review. J Aerosol Sci 29(5):575–588
71. Farrar-Khan JR, Andrews GE, Williams PT, Bartle KD (1992) The influence of nozzle sac volume on the composition of diesel particulate fuel derived SOF (No. 921649). SAE technical paper
72. Eastwood P (2008) Particulate emissions from vehicles, vol 20. Wiley, Chichester
73. Smith OI (1981) Fundamentals of soot formation in flames with application to diesel engine particulate emissions. Prog Energy Combust Sci 7(4):275–291
74. Borman GL, Ragland KW (1998) Combustion engineering. McGraw-Hill Science/Engineering/Math, Boston
75. Frenklach M, Wang H (1991) Detailed modeling of soot particle nucleation and growth. Symp Combust 23(1):1559–1566. Elsevier
76. Wang H (2011) Formation of nascent soot and other condensed-phase materials in flames. Proc Combust Inst 33(1):41–67
77. Sander M, Patterson RI, Braumann A, Raj A, Kraft M (2011) Developing the PAH-PP soot particle model using process informatics and uncertainty propagation. Proc Combust Inst 33 (1):675–683
78. Storey JM, Curran SJ, Lewis SA, Barone TL, Dempsey AB, Moses-DeBusk M et al (2016) Evolution and current understanding of physicochemical characterization of particulate matter from reactivity controlled compression ignition combustion on a multicylinder light-duty engine. Int J Engine Res. https://doi.org/10.1177/1468087416661637
79. Lighty JS, Romano V, Sarofim AF (2009) Soot oxidation. In: Bockhorn H, D'Anna A, Sarofim A, Wang H (eds) Combustion generated fine carbonaceous particles. KIT Scientific Publishing, Karlsruhe, pp 487–500
80. Christensen M (2002) HCCI combustion-engine operation and emission characteristics. PhD thesis, Department of Heat and Power Engineering, Lund University
81. Christensen M, Hultqvist A, Johansson B (1999) Demonstrating the multi fuel capability of a homogeneous charge compression ignition engine with variable compression ratio (No. 1999-01-3679). SAE technical paper
82. He BQ, Wang JX, Yan XG, Tian X, Chen H (2003) Study on combustion and emission characteristics of diesel engines using ethanol blended diesel fuels (No. 2003-01-0762). SAE technical paper
83. Shen M, Tuner M, Johansson B (2013) Close to stoichiometric partially premixed combustion-the benefit of ethanol in comparison to conventional fuels (No. 2013-01-0277). SAE technical paper
84. Benajes J, Molina S, García A, Monsalve-Serrano J (2015) Effects of direct injection timing and blending ratio on RCCI combustion with different low reactivity fuels. Energy Convers Manag 99:193–209
85. Maurya RK, Agarwal AK (2015) Experimental investigations of particulate size and number distribution in an ethanol and methanol fueled HCCI engine. J Energy Resour Technol 137 (1):012201
86. Maurya RK, Agarwal AK (2014) Effect of intake air temperature and air–fuel ratio on particulates in gasoline and n-butanol fueled homogeneous charge compression ignition engine. Int J Engine Res 15(7):789–804

87. Franklin L (2010) Effects of homogeneous charge compression ignition (HCCI) control strategies on particulate emissions of ethanol fuel. Doctoral dissertation, University of Minnesota
88. Price P, Stone R, Misztal J, Xu H, Wyszynski, M., Wilson, T., & Qiao, J. (2007). Particulate emissions from a gasoline homogeneous charge compression ignition engine (No. 2007-01-0209). SAE technical paper
89. Kittelson D, Kraft M (2015) Particle formation and models. In: Encyclopedia of automotive engineering. Wiley, Chichester
90. Gupta T, Kothari A, Srivastava DK, Agarwal AK (2010) Measurement of number and size distribution of particles emitted from a mid-sized transportation multipoint port fuel injection gasoline engine. Fuel 89(9):2230–2233
91. Tobias HJ, Beving DE, Ziemann PJ, Sakurai H, Zuk M, McMurry PH et al (2001) Chemical analysis of diesel engine nanoparticles using a nano-DMA/thermal desorption particle beam mass spectrometer. Environ Sci Technol 35(11):2233–2243
92. Sakurai H, Tobias HJ, Park K, Zarling D, Docherty KS, Kittelson DB et al (2003) On-line measurements of diesel nanoparticle composition and volatility. Atmos Environ 37 (9):1199–1210
93. Maurya RK, Agarwal AK (2011) Effect of start of injection on the particulate emission from methanol fuelled HCCI engine. SAE Int J Fuels Lubr 4:204–222. (2011-01-2408)
94. Herold RE, Foster DE, Ghandhi JB, Iverson RJ, Eng JA, Najt PM (2007) Fuel unmixedness effects in a gasoline homogeneous charge compression ignition engine. Int J Engine Res 8 (3):241–257
95. Lewander M, Ekholm K, Johansson B, Tunestål P, Milovanovic N, Keeler N et al (2008) Investigation of the combustion characteristics with focus on partially premixed combustion in a heavy duty engine. SAE Int J Fuels Lubr 1:1063–1074. (2008-01-1658)
96. Roberts G, Johnson B, Edwards C (2014) Prospects for high-temperature combustion, neat alcohol-fueled diesel engines. SAE Int J Engines 7:448–457. (2014-01-1194)
97. Gao T, Jeftic M, Bryden G, Reader G, Tjong J, Zheng M (2016) Heat release analysis of clean combustion with ethanol ignited by diesel in a high compression ratio engine (No. 2016-01-0766). SAE technical paper
98. Shen M, Malmborg V, Gallo Y, Waldheim BB, Nilsson P, Eriksson A et al (2015) Analysis of soot particles in the cylinder of a heavy duty diesel engine with high EGR (No. 2015-24-2448). SAE technical paper
99. Shen M, Tuner M, Johansson B, Tunestal P, Pagels J (2016) Influence of injection timing on exhaust particulate matter emissions of gasoline in HCCI and PPC (No. 2016-01-2300). SAE technical paper
100. Shen M (2016) Particulate matter emissions from partially premixed combustion with diesel, gasoline and ethanol. Doctoral dissertation, Lund University, LUND Sweden
101. Shamun S, Shen M, Johansson B, Tuner M, Pagels J, Gudmundsson A, Tunestal P (2016) Exhaust PM emissions analysis of alcohol fueled heavy-duty engine utilizing PPC. SAE Int J Engines 9:2142–2152. (2016-01-2288)
102. Prikhodko VY, Curran SJ, Barone TL, Lewis SA, Storey JM, Cho K et al (2010) Emission characteristics of a diesel engine operating with in-cylinder gasoline and diesel fuel blending. SAE Int J Fuels Lubr 3:946–955. (2010-01-2266)
103. Zhang Y (2017) Comparisons of particulate size distributions from multiple combustion strategies. Doctoral dissertation, The University of Wisconsin-Madison
104. Dempsey A, Curran S, Storey J, Eibl M, Pihl J, Prikhodko V et al (2014) Particulate matter characterization of reactivity controlled compression ignition (RCCI) on a light duty engine (No. 2014-01-1596). SAE technical paper
105. Kolodziej C, Wissink M, Splitter D, Hanson R, Reitz RD, Benajes J (2013) Particle size and number emissions from RCCI with direct injections of two fuels (No. 2013-01-1661). SAE technical paper

106. Zhang Y, Ghandhi J, Rothamer D (2014) Comparison of particulate size distributions from advanced and conventional combustion-part I: CDC, HCCI, and RCCI. SAE Int J Engines 7:820–834. (2014-01-1296)
107. Fang W, Kittelson DB, Northrop WF (2017) Dilution sensitivity of particulate matter emissions from reactivity controlled compression ignition combustion. J Energy Resour Technol 139:032204–032201
108. Lucachick G, Curran S, Storey J, Prikhodko V, Northrop WF (2016) Volatility characterization of nanoparticles from single and dual-fuel low temperature combustion in compression ignition engines. Aerosol Sci Technol 50(5):436–447
109. Benajes J, García A, Monsalve-Serrano J, Boronat V (2017) An investigation on the particulate number and size distributions over the whole engine map from an optimized combustion strategy combining RCCI and dual-fuel diesel-gasoline. Energy Convers Manag 140:98–108
110. Prikhodko VY, Curran SJ, Barone TL, Lewis SA, Storey JM, Cho K et al (2011) Diesel oxidation catalyst control of hydrocarbon aerosols from reactivity controlled compression ignition combustion. In: ASME 2011 international mechanical engineering congress and exposition, January, pp 273–278
111. Yang B, Yao M, Cheng WK, Zheng Z, Yue L (2014) Regulated and unregulated emissions from a compression ignition engine under low temperature combustion fuelled with gasoline and n-butanol/gasoline blends. Fuel 120:163–170
112. Han M, Assanis DN, Bohac SV (2008) Comparison of HC species from diesel combustion modes and characterization of a heat-up DOC formulation. Int J Automot Technol 9 (4):405–413
113. Maurya RK, Mishra P (2016) Computational analysis of emissions from gasoline HCCI engine using stochastic reactor model. In: Proceedings of international conference on computing in mechanical engineering, Research Publishing Services, Singapore, ISBN:978-981-09-6278-4
114. Hasan AO, Leung P, Tsolakis A, Golunski SE, Xu HM, Wyszynski ML, Richardson S (2011) Effect of composite after treatment catalyst on alkane, alkene and monocyclic aromatic emissions from an HCCI/SI gasoline engine. Fuel 90(4):1457–1464
115. Zhou JH, Cheung CS, Leung CW (2014) Combustion, performance, regulated and unregulated emissions of a diesel engine with hydrogen addition. Appl Energy 126:1–12
116. Hunicz J, Medina A (2016) Experimental study on detailed emissions speciation of an HCCI engine equipped with a three-way catalytic converter. Energy 117:388–397
117. De Ojeda W, Zhang Y, Xie K, Han X, Wang M, Zheng M (2012) Exhaust hydrocarbon speciation from a single-cylinder compression ignition engine operating with in-cylinder blending of gasoline and diesel fuels (No. 2012-01-0683). SAE technical paper
118. Elghawi UM, Mayouf AM (2014) Carbonyl emissions generated by a (SI/HCCI) engine from winter grade commercial gasoline. Fuel 116:109–115
119. World Health Organization (1998) Selected non-heterocyclic polycyclic aromatic hydrocarbons, Environmental health criteria, vol 202. World Health Organization, Stuttgart
120. Correa SM, Arbilla G (2006) Aromatic hydrocarbons emissions in diesel and biodiesel exhaust. Atmos Environ 40(35):6821–6826
121. Lea-Langton A, Li H, Andrews GE (2008) Comparison of particulate PAH emissions for diesel, biodiesel and cooking oil using a heavy duty DI diesel engine (No. 2008-01-1811). SAE technical paper
122. Elghawi UM, Mayouf A, Tsolakis A, Wyszynski ML (2010) Vapour-phase and particulate-bound PAHs profile generated by a (SI/HCCI) engine from a winter grade commercial gasoline fuel. Fuel 89(8):2019–2025
123. Mi HH, Lee WJ, Tsai PJ, Chen CB (2001) A comparison on the emission of polycyclic aromatic hydrocarbons and their corresponding carcinogenic potencies from a vehicle engine using leaded and lead-free gasoline. Environ Health Perspect 109(12):1285

Chapter 9
Closed-Loop Combustion Control

Abstract Low temperature combustion (LTC) is an engine combustion mode that yields ultralow NO_x and soot emission levels along with high fuel conversion efficiency. Typically, LTC engines use premixed fuel–air mixture, and combustion is mainly governed by chemical kinetics. The LTC strategies such as partially premixed combustion (PPC) and reactivity-controlled compression ignition (RCCI) have some level of direct control on combustion phasing due to direct injection of fuel in the engine cylinder. However, homogeneous charge compression ignition (HCCI) combustion strategy lacks the direct control on combustion phasing. In HCCI combustion, ignition timings are kinetically controlled and affected by pressure and temperature history of the charge in the engine cylinder (indirect control). Therefore, HCCI combustion requires the combustion feedback control for its very operation. The present chapter describes the closed-loop combustion control in LTC engines. First, the need of closed-loop combustion control and control variables are discussed. Then, combustion feedback sensors and combustion control actuators are described in detail. Typically, cylinder pressure sensor and ion current sensors are used for sensing of combustion phasing in HCCI combustion. The last section presents the combustion control methods and various controllers used in different LTC strategies such as HCCI, PPC and RCCI, for closed-loop combustion control.

Keywords Closed-loop control • MPC • PID • Control • Actuators • Sensors • Combustion • LTC • HCCI • PPC • RCCI

9.1 Need of Combustion Control

In general, control means governing the system to behave in a desired manner. In combustion engines, control is about achieving reliable and durable performance while fulfilling the constraints on engine efficiency, emissions and combustion noise levels. The LTC engines are more sensitive to temperature and mixture conditions in comparison to conventional spark ignition (SI) and compression ignition (CI) engines. This sensitivity leads to a relatively higher demand upon

R.K. Maurya, *Characteristics and Control of Low Temperature Combustion Engines*, Mechanical Engineering Series, https://doi.org/10.1007/978-3-319-68508-3_9

engine controller design, where appropriate engine operating conditions have to be set with more accuracy for acceptable engine performance.

The HCCI combustion does not have the direct actuators for control of combustion timings such as spark ignition timing or fuel injection timings similar to conventional SI and CI engines. In SI engine, combustion timings are controlled by spark ignition timings, whereas in CI engine, injection timings govern the ignition timings. The HCCI engine lacks these combustion actuators; therefore combustion timing is very sensitive to initial charge conditions such as temperature, pressure and equivalence ratio of fuel–air mixture [1]. Other LTC strategies such as PPC and RCCI combustion have relatively higher degree of combustion control because fuel stratification is controlled by direct injection timings which essentially governed the combustion timings. However, in HCCI combustion using a well-mixed charge, there is no way to affect the ignition timings after initial conditions are set. In the absence of accurate control on cylinder pressure, temperature and equivalence ratio of charge, occurrence of misfire, excessive peak pressure rise rate or very high peak pressure can happen, which can result into even damage of engine or high NO_x emissions [2]. Therefore, in order to obtain higher engine load range, it is essential to be able to control the combustion timing in HCCI engine.

Very advanced HCCI combustion timings lead to excessive pressure rise rate (knocking) along with very high peak cylinder pressure and temperature. This engine operating condition leads to very high NO_x emission and increases the fuel consumption. In severely prolonged engine operation in knocking, conditions may lead to the engine damage. In case of highly retarded combustion phasing, combustion is quenched prior to completion during compression stroke, which leads to higher fuel consumption along with higher unburned hydrocarbon (HC) and carbon monoxide (CO) emissions. In the worst case of excessively delayed combustion, a complete misfire can occur and repeated operation can stall the engine. In this operating condition, engine can also enter into a vicious circle of misfire and strong combustion (misfire followed by strong combustion that followed by misfire) [3].

Some of the HCCI engine operating conditions are unstable, which is another complexity in HCCI combustion process. Therefore, HCCI engine cannot be operated reliably based on engine map. Small variations in engine operating conditions may lead to a large effect on the combustion phasing. Therefore, closed-loop combustion control is essential to ensure the correct combustion phasing [2, 3]. Figure 9.1 illustrates the variation of combustion phasing after closed-loop combustion is turned off in stable and unstable conditions. Open loop control of HCCI combustion engine becomes difficult at higher engine operating loads and retarded combustion phasing operation [4, 5]. Figure 9.1 depicts that engine can enter into unstable condition leading to combustion phasing run-away either towards misfire conditions (more retarded combustion timings) or towards advanced combustion timings. In case of advanced combustion phasing, fuel injection has to be turned off to avoid engine damage. For advanced combustion timings, cylinder walls heat up due to increase in heat transfer at earlier combustion phasing. Higher wall temperature leads to the start of combustion even earlier in next cycles, which causes higher heat transfer and so on. Kinetically controlled event such as HCCI combustion needs to be stabilized at this point [4].

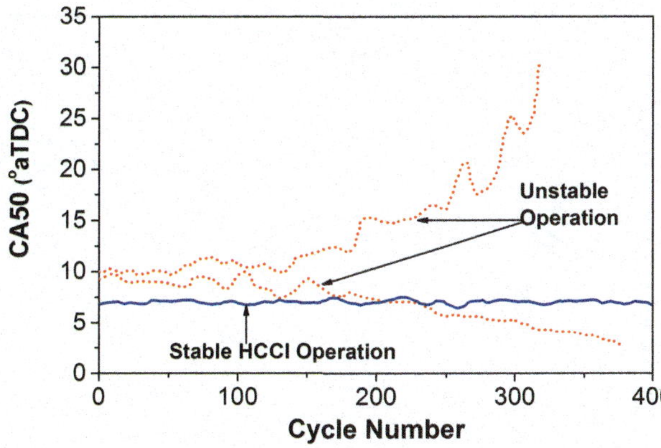

Fig. 9.1 Variation of combustion phasing with time after closed-loop control is turned off in HCCI engine (Adapted from [4])

A study proposed the instability criteria for the HCCI combustion given by Eq. (9.1) [4]:

$$\frac{dT_w}{d\mathrm{CA}_{50}} \cdot \frac{d(\mathrm{CA}_{50})}{dT_w} \geq 1 \tag{9.1}$$

where T_w is wall temperature and CA_{50} is combustion phasing. Two derivatives have different origins and thus, equation is not equal to 1. The first term reflects the effect of combustion timings on the wall temperature and heat transfer, while the second term reflects the effect of wall temperature on gas temperature which affects the combustion phasing and chemical kinetics. The first derivative is governed by heat transfer during combustion, and the second derivative is governed by heat transfer during compression stroke [4].

In HCCI combustion, there are other mechanisms of knock runaway, instead of wall temperature. Figure 2.24a (Chap. 2) depicts the summary of different routes of runaway (combustion instability) in HCCI combustion using different EGR operating conditions. Figure 9.2 illustrates the wall heating-induced runaway and EGR-induced runaway in HCCI combustion operation using iso-octane at 4.85 bar IMEP and PRF80 at 6.11 bar IMEP, respectively. Data plotted is curve fit of experimental data points, and the pressure rise rate (PRR) presented in the Fig. 9.2 is the maximum value of PRR in a combustion cycle. Figure 9.2 illustrates that wall heating-induced runaway is relatively a slow process, and slope of maximum pressure rise rate curve is smaller. However, EGR-induced runaway is very fast as shown by very steep curves of combustion phasing and PRR. In this case, enhancement of autoignition is caused by trace species such as (NO_x) present in EGR (for more details see Sect. 6.2.3) [6].

The combustion instability conditions in HCCI combustion make a compulsory closed-loop combustion control for the very operation of HCCI engine. This requirement differentiates HCCI engine from conventional SI and CI engines, which runs on engine map quite well without closed-loop control [3]. Therefore, the design and development of appropriate closed-loop controller are required to achieve the desired performance in HCCI engines.

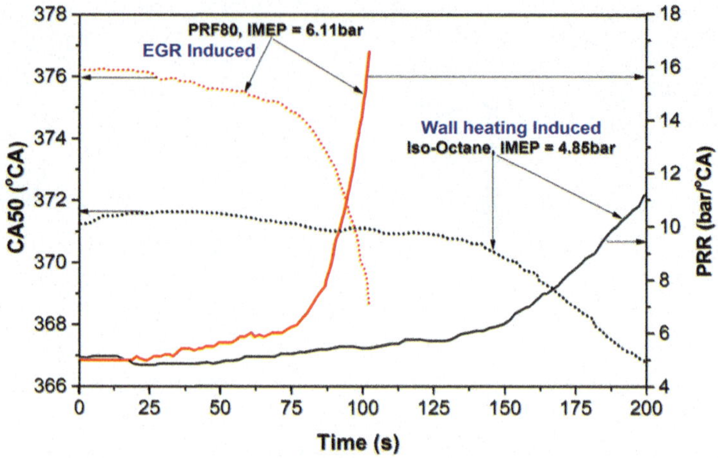

Fig. 9.2 Illustration of wall heating and EGR-induced runaway mechanisms in HCCI combustion (Adapted from [6, 7])

9.2 Combustion Control Variables

The measured and controlled condition or quantity is termed control variable or parameter [8]. The main objective of combustion control approaches is to expand the stable and high-efficiency operation range of LTC engines, particularly HCCI combustion engines. The unstable HCCI operating conditions can be avoided by controlling some of the combustion parameters. Advanced model-based combustion control strategies are typically used to maintain optimal combustion timings, regulate cycle-to-cycle variation (CA_{50} and IMEP) and adjust combustion rate to control peak pressure and PRR [8]. The HCCI combustion control variables can be divided into four major groups, namely, (1) control of combustion phasing, (2) control of engine load, (3) control of exhaust gas after-treatment efficiency (exhaust gas temperature) and (4) control of HCCI dynamics for combustion mode switching between HCCI and conventional SI or CI modes [9]. In PPC strategy, ignition delay is also used as control variable as it can be independently controlled [10]. However, ignition delay and combustion phasing are strongly coupled. Combustion control variables in LTC engines are discussed in the following subsections.

9.2.1 *Combustion Phasing*

Optimum combustion timing can decrease the emission levels (HC and CO emissions), improve the cyclic variability and extend the HCCI operating range. The absence of direct actuator to initiate the combustion and complexity of HCCI combustion control require a sophisticated control strategy that can adjust charge

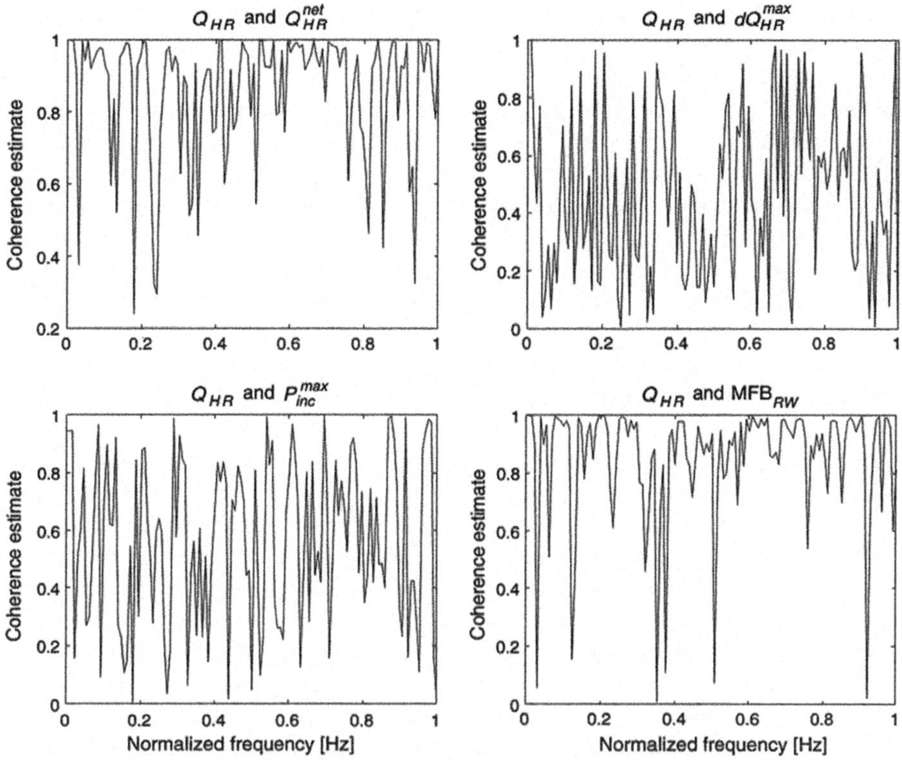

Fig. 9.3 Coherence estimates between CA_{50} determined from full heat release and CA_{50} calculated from other methods in HCCI combustion [11]

properties on cycle to cycle basis for achieving the desired optimal combustion phasing [9]. Crank angle position corresponding to 50% heat release (CA_{50}) is the most commonly used combustion phasing parameter in HCCI combustion control [11]. However, crank angle position of maximum cylinder pressure (θ_{pmax}) is also used for HCCI control as combustion phasing parameter in some of the studies [12, 13]. A robust parameter of combustion phasing is required for the closed-loop combustion control. Start of combustion (SOC) estimation is necessary for combustion control, but it is not adequate for model-based HCCI combustion control because for the same start of combustion, different combustion durations can occur that significantly affect the engine performance [8]. The CA_{50} is a robust feedback parameter of combustion phasing, in terms of both SOC and combustion duration. Different methods of estimation of CA_{50} from measured cylinder pressure are compared, and coherence with different calculation methods is shown in Fig. 9.3.

Different possible methods for calculation of CA_{50} are full heat release (Q_{HR}), net heat release with constant γ (Q_{HR}^{net}), maximum heat release rate (dQ_{HR}^{max}), maximum pressure increase by combustion (P_{inc}^{max}), maximum pressure (P_{max}) and Rassweiler and Withrow method for 50% mass fraction burned (MFB_{RW}) [11]. The accuracy of CA_{50} calculation from full heat release analysis depends on well tuning of measurement parameters and removal of signal noise from measured pressure data [14, 15]. A more complete detail on computation methods of CA_{50} can be found in [11]. These parameters are evaluated over a wide range of engine operations including very early and late combustion phasing HCCI combustion. The CA_{50} based on pressure parameters has significant deviation from actual combustion timings at some of the operating conditions. The best indicator of CA_{50} is found from full heat release (Q_{HR}) analysis of pressure signal. The coherence estimate between Q_{HR} and other candidates shows that Q_{HR}^{net} and MFB_{RW} have highest coherence spectrum, which is close to 1 (Fig. 9.3). Calculation of CA_{50} by using MFB_{RW} for combustion phasing feedback is found to work, but it has relatively lower performance for the same computational cost [11]. The CA_{50} determination from Q_{HR}^{net} is found to be best suited for real-time combustion control because of low computation complexity and robustness to measurement noise, and gives satisfactory result even with low sampling rate and resolution of cylinder pressure signal.

Besides calculation of CA_{50} from cylinder pressure analysis, it can also be calculated by using signals from ion current sensor, knock sensor, microphone sensor and on-board torque sensor (see the next section) or even by modelling [8]. The effects of combustion phasing on combustion, performance and emission characteristics in LTC engines are discussed in Chaps. 6, 7 and 8, respectively. Fast combustion phasing control is essential for reliable engine operation as combustion timing sets the performance limitations [5]. In other LTC strategies such as PPC and RCCI combustion, combustion phasing is also a control variable for engine combustion control [10, 16, 17].

9.2.2 Ignition Delay

Ignition delay is typically defined as the duration between the start of fuel injection and start of combustion. In LTC strategies such as PPC, ignition delay should be sufficiently large to ensure the adequate air and fuel mixing before start of combustion. Ignition delay can be increased by influencing the state of cylinder gas at the time of fuel injection by varying the direct fuel injection timings. Ignition delay can also be extended by diluting the intake air by cooled EGR, which lowers the peak cylinder temperatures and slows down the early chemical reactions. The EGR reduces the oxygen concentration and increases the heat capacity of inducted charge, which leads to lower temperature rise that increases the ignition delay of air-fuel mixture.

To obtain higher high thermal efficiency, combustion needs to be phased such that the highest possible pressure is achieved during the expansion stroke without excessive heat transfer losses. Unless fuel injection timing is very advanced, combustion phasing depends both on the ignition delay and the injection timing in PPC engine [10]. Combustion phasing and ignition delay are strongly coupled, and decoupling of these parameters provides an interesting multiple input, multiple output control problem where fuel injection and air control systems need to combine. In order to maintain efficient PPC operation, the ignition delay and the combustion phasing have to be controlled simultaneously. A study used a centralized controller design that controls ignition delay and combustion phasing by executing coordinated control action in the fuel injection system and the gas flow system [10].

9.2.3 Engine Load

Desired engine operating load is another important control variable that should be met while satisfying the other operating limitations such as ringing limit, combustion stability limit, peak pressure limit and emission limit (see Sect. 7.1.1 of Chap. 7). Typically, combustion phasing and engine load are simultaneously controlled to realize the advantages of higher fuel conversion efficiency and emissions. To control the engine load in LTC combustion engine, indicated mean effective pressure (IMEP) is used as a common control parameter [17–21]. Peak cylinder pressure is also used as control parameter in some of the studies [13, 22]. Engine load is mainly governed by the amount of fuel burned in the cylinder. Higher amount of fuel leads to higher peak pressure rise rate and peak cylinder pressure in premixed combustion. To avoid these limits, combustion phasing is typically retarded, which may lead to reduction in thermal efficiency and higher combustion instability. Thus, combustion phasing and engine load control are interconnected, and engine controller needs to address both of them.

9.2.4 Exhaust Gas Temperature

One of the major challenges in LTC is the reduction of HC and CO emissions. The formation of HC and CO in LTC engines and their characteristics are discussed in Chap. 8. All the strategies in LTC regime produce higher HC and CO emissions in comparison to conventional SI and CI engines. High levels of HC and CO emissions can be reduced by employing oxidation catalyst in the exhaust [23]. However, lower exhaust temperature of HCCI engines limits the conversion (oxidation) of HC and CO effectively as oxidation catalyst requires temperature of the exhaust gas above catalyst light-off temperature. Figure 8.21 (Chap. 8) illustrates the conversion of HC and CO emissions with respect to exhaust gas temperature for different catalysts. Reduction of HC and CO emissions in HCCI engines is more challenging due to very low exhaust gas temperature at some of the engine operating conditions. Therefore, exhaust gas temperature control in HCCI engine is essential to meet catalyst light-off temperature requirement of the oxidation catalyst. Thus, exhaust gas temperature is another control variable for HCCI combustion engine

[18, 24, 25]. A simultaneous control of exhaust gas temperature and maximum cylinder pressure is investigated in [22], and a constraint-based control technique is used for exhaust gas temperature control [25]. In a recent study, a three-input three-output controller of combustion phasing, load and exhaust gas temperature has been designed with sufficient accuracy by modulating octane rating, fuel mass flow rate and intake manifold pressure [26]. An integrated control of all the three major engine parameters (i.e. CA_{50}, engine load and exhaust temperature) is essential for realizing HCCI engine as a viable solution for automotive applications.

9.2.5 Combustion Mode Switching

Engine load range achieved in LTC engines is typically lower than conventional engines. To extend the load range of LTC engines, one possible solution is to enable the switching on the fly between the LTC and conventional combustion strategies. Technical challenges of enabling and ensuring the stable engine operation in LTC and conventional combustion modes are significantly different, which makes combustion mode switching further complex [27]. Additionally, the LTC engines are intrinsically more sensitive to small variations in the engine operating conditions, and hence, the control system should be able to satisfactorily respond to such disturbances on a cycle-to-cycle basis. Figure 9.4 illustrates the complex dynamics of heat release during transition from SI to HCCI combustion mode. The figure

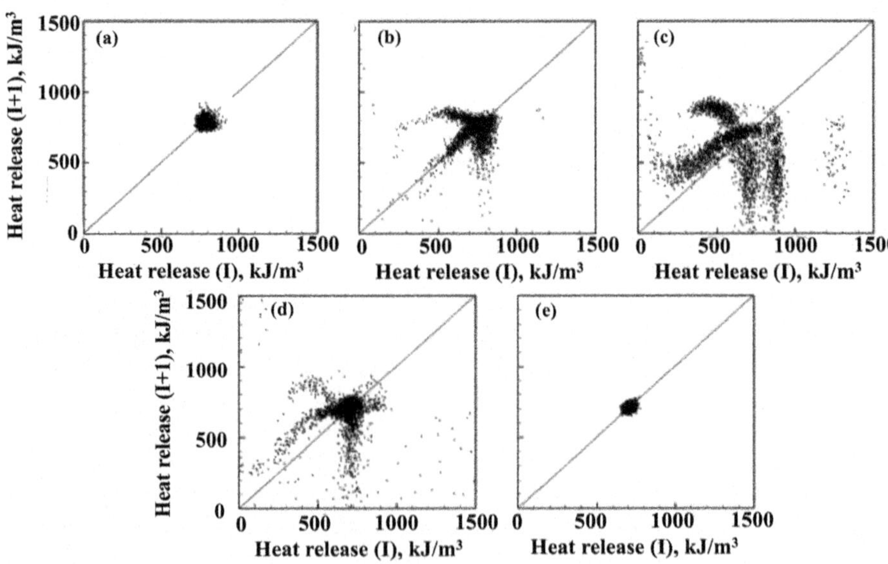

Fig. 9.4 Heat release return maps illustrating the complex dynamics encountered during transitioning from (**a**) stable spark ignition to (**e**) HCCI operation using higher levels of internal EGR [28]

depicts the major differences in combustion modes as shown in return maps, which are created from time-series experimental data by plotting pairs of consecutive values as points in two dimensions. In SI combustion (Fig. 9.4a), the consecutive data points are concentrated in a small unstructured cluster of around a fixed point that represents the nominal flame propagation heat release. Dispersed data points in return map indicates the stochastic or high-dimensional component [28]. There are significant cyclic combustion variations (Fig. 9.4b) when EGR is increased. The original fixed point starts to destabilize in certain specific ways (i.e. points scatter mainly in particular directions on the return map), which indicates unstable manifolds in a low-dimensional phase space [29]. The level of destabilization is maximum in Fig. 9.4c (further increased EGR), and combustion again starts becoming more stable (Fig. 9.4d). At last combustion mode transition to a stable HCCI combustion is complete (Fig. 9.4e).

The multimode engine operation needs the switching between the combustion modes at appropriate set points in the engine operating regime while fulfilling the load demand requested by the driver, without departing from the low emission characteristics [27]. The challenges of implementing the combustion mode switching are twofold: (1) the stability and efficiency of each combustion mode must be guaranteed, and (2) the control system should be able to ensure seamless transition between the modes. Figure 9.5 illustrates the HCCI to CI engine coordination control. First, desired combustion mode is determined by a state machine based on engine state information and combustion mode map. After determination of the desired combustion mode, the mode switching commands are delivered to subcontrol systems, such as EGR controller, variable nozzle turbine (VNT) controller, injection controller as well as rail pressure controller, step by step with predefined sequences. The VNT and EGR are controlled by a boost pressure closed-loop PID controller and an EGR rate closed-loop controller, respectively [30].

Control of combustion mode switching is investigated for mode switching between HCCI and either conventional SI [25] or CI [27, 30, 31] engines. Mode

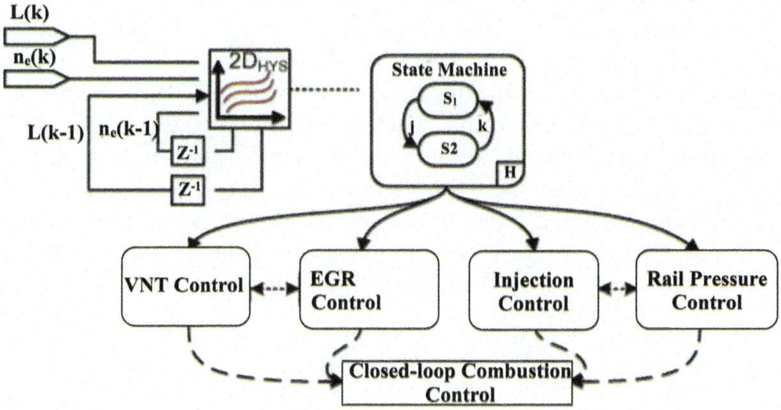

Fig. 9.5 Schematic diagram of HCCI/CI coordination control [30]

switching between HCCI engine and electric machine in hybrid electric vehicle (HEV) powertrain is also investigated [9]. Simulation studies are conducted for energy management between the dual mode SI-HCCI and electric machine at different level of hybridization and different HEV configurations [32–34].

9.3 Combustion Feedback Sensors

In order to implement closed-loop combustion control, a robust combustion feedback parameter is required. For combustion control variables of engine load and combustion phasing, the measurement of IMEP and CA_{50} is required to provide feedback for the next successive combustion cycle. Typically, sensing of combustion phasing can be performed by cylinder pressure sensor, ion current sensor, microphone or knock sensor and even by speed and torque fluctuations. The various types of sensors used for estimation of combustion phasing are based on their complexity, measurement accuracy and cost.

9.3.1 In-Cylinder Pressure

In-cylinder pressure signal has been used in internal combustion engine research from the beginning of engine development for combustion diagnostics. A summary of performance and combustion parameters that can be estimated by processing of measured in-cylinder pressure data is presented in Table A2 (Appendix 1). Typical measurement system required for the in-cylinder pressure of engine is also given in Appendix 1. The in-cylinder pressure signal can be also exploited for a multitude of further applications because it contains a lot of information about the engine combustion process. The evolution of the engine cycle with all its phases (compression, combustion, expansion, gas exchange) is directly affected by the engine boundary conditions such as engine temperature, intake and exhaust conditions, fuel quality, etc. The evolution of the in-cylinder pressure is the result of these effects over the engine cycle, and this information is captured by the pressure sensor and therefore available inside the acquired cylinder pressure signal [35]. The cylinder pressure contains information of thermodynamic state of the charge. Combustion rate in the engine cylinder can be estimated by employing the first law of thermodynamics and some simplifying assumptions (see details in Appendix 2). The net heat release is computed from measured cylinder pressure by Eq. (9.2) (Appendix 2), by neglecting the heat transfer and crevice losses:

$$\frac{\partial Q}{\partial \theta} = \frac{\gamma}{\gamma - 1} p \frac{dV}{d\theta} + \frac{1}{\gamma - 1} V \frac{\partial p}{\partial \theta} \tag{9.2}$$

The combustion phasing (CA_{50}) can be calculated by integrating this equation, and crank angle position of 50% heat release can be estimated.

In real-time control application, it is inconvenient to include the pressure derivative in the calculation because the derivative process enhances the signal noise particularly in combination with a very high sampling rate [36]. Instead of using Eq. (9.2) for net heat release calculation, another Eq. (9.3) is proposed for net heat release calculation [36, 37]. This heat release calculation equation is also derived from the same energy conservation equation, but it does not have pressure derivative term. The optional method of heat release calculation is given by Eq. (9.3) [37]:

$$Q = \frac{1}{\gamma - 1}\int_{\theta_{start}}^{\theta} d(pV) + \int_{\theta_{start}}^{\theta} pdV =$$

$$= \underbrace{\frac{1}{\gamma - 1}p(\theta)V(\theta) + \int_{\theta_{start}}^{\theta} p(\theta)\frac{dV}{d\theta}d\theta}_{\textit{Calculated} \text{ on } \textit{Line}} - \underbrace{\frac{1}{\gamma - 1}\{p(\theta_{start})V(\theta_{start})\}}_{\textit{Constant Added offline}} \tag{9.3}$$

Combustion phasing can be easily calculated from heat release by assuming constant value of ratio of specific heat (γ). The method of combustion phasing calculation is implemented for automatic engine control [36].

Until recent times, the cylinder pressure transducers are not employed on production engine. Today, there exist several low-cost piezo transducers, and the accuracy of these low-cost pressure sensors is probably sufficient for feedback control of HCCI combustion phasing [38, 39].

9.3.2 Ion Current

The present piezoelectric pressure transducer used for engine combustion detection is expensive and sometimes fragile, which cannot be used for production engines. Ion current is one possible solution for combustion detection in engine cylinder by a low-cost and more robust sensor. Amounts of ions formed during combustion are precisely related with the combustion event in a compression ignition engine. The ion current sensor can be conveniently installed on commercial engines similar to the ubiquitous, low-cost, rugged, spark plug [40]. Figure 9.6 illustrates the ion current measurement system on the engine. A spark plug is used as an ion current sensor, which employed central electrode of spark plug as the positive bias to measure the in-cylinder ion signal [40, 41]. To obtain stronger ion signals, a bias voltage of 237 V is used across the sparkplug electrode along with a resistance of 241 kΩ. The ion current signal is measured by dividing the 241 kΩ into the voltage drop across the resistor [41].

Ion sensors have been used in SI engine for detection of knock and misfire conditions [42, 43]. Near-stoichiometric fuel concentration and higher combustion

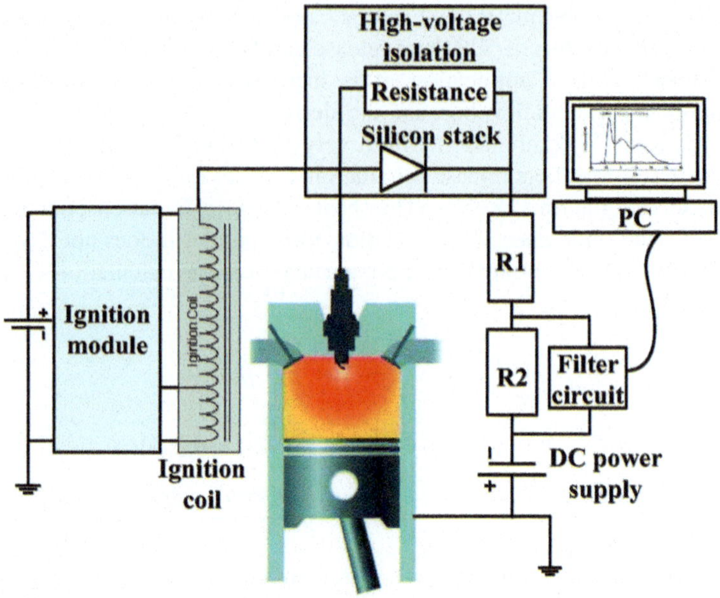

Fig. 9.6 Schematic diagram of ion detection setup on an engine [41]

temperature form strong ion current signals. The HCCI combustion is a lean and low temperature process leading to relatively lower quantities of ions during combustion events [40, 44]. The ions during HCCI combustion process are formed by a chemi-ionization process governed by three principal reactions (R1–R3) [40, 44]:

$$CH + O \leftrightarrow CHO^{+} + e^{-} \qquad (R1)$$

$$CHO^{+} + H_2O \leftrightarrow H_3O^{+} + CO \qquad (R2)$$

$$H_3O^{+} + e^{-} \leftrightarrow H_2O + H \qquad (R3)$$

Studies demonstrated that H_3O^+ is the most dominant ion, and the associated electron is responsible for the ion current that can be used for combustion detection [45].

The ion current signal depends on the fuel properties and engine operating conditions. Ion current sensing provides the local information, but if charge is homogeneous, a local measurement can be sufficient [2]. Figure 9.7 shows the distribution of H_3O^+ ion in cylinder of HCCI combustion. Fuel is injected as double injection, and injection ratio (Inj_{ratio}) is the ratio of fuel injected in the first injection event to the second injection event. The figure depicts the ion distribution for two injection ratios, and in the injection ratio 0.2, more fuel is injected during the compression stroke. Figure 9.7 shows that the ion concentration is more heterogeneous at CA10 (10% heat release crank position) in comparison to injection ratio

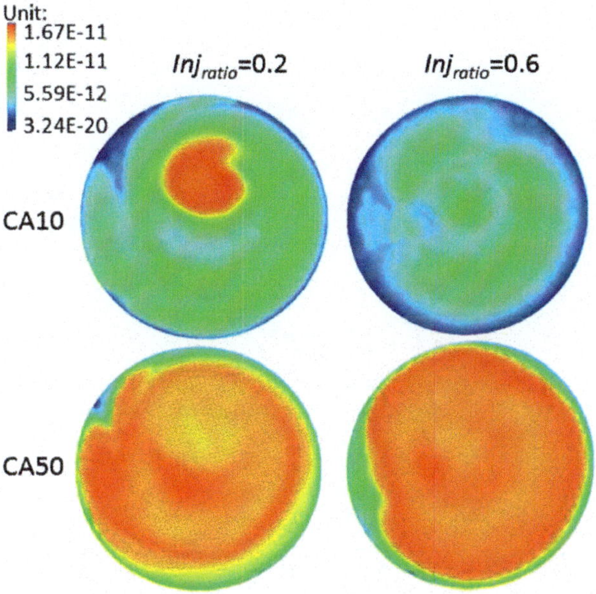

Fig. 9.7 In-cylinder H_3O^+ ion distribution under different injection ratio conditions in HCCI combustion [46]

0.6. The ion concentration is distributed significantly at CA_{50} and injection ratio 0.2 has relatively lower concentration [46]. These ion predictions are in good agreement with the signal amplitude variations in experimental engines, and hence, these tools can be used for optimizing the ion sensing system.

Figure 9.8 shows the comparison between the combustion phase (CA_{50}), the ion current phase (Ion50) and the position with maximum H_3O^+ ion concentration (Pion_max). Figure 9.8a depicts the linear relation between combustion phase and ion current phase at different equivalence ratios. Position of maximum concentration is also correlated well with the Ion50. Combustion phasing is advanced with increase in injection ratio from 0.2 to 0.6. The linear correlation between Ion50 and CA_{50} still exists in this condition (Fig. 9.8b). There exist an offset in combustion phasing predicted by ion current sensor depending on engine operating conditions. A more complete discussion can be found in the study [47].

9.3.3 *Microphone and Knock Sensor*

Theoretically, combustion timings can be possibly measured by sensing from both inside and outside the engine cylinder. Ion current and cylinder pressure sensor provides the combustion phasing by direct in-cylinder sensing of combustion process. However, microphone, knock sensors and torque sensors estimate the combustion phasing indirectly by sensing the effect of combustion event on external engine components [3]. Sound sensors (microphone) are typically used in SI

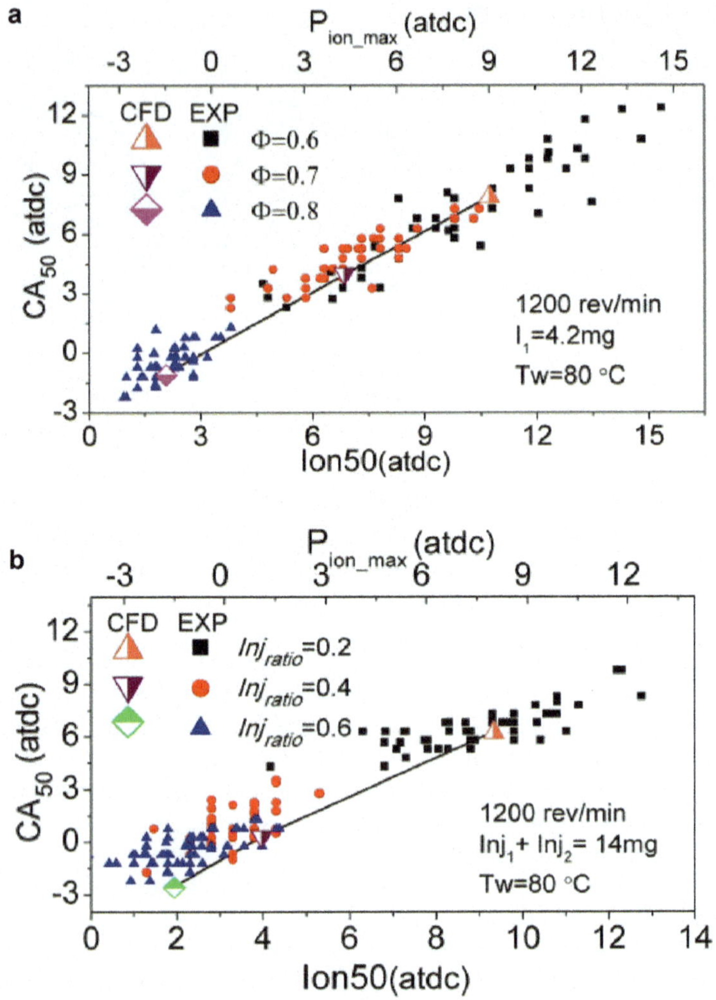

Fig. 9.8 Phase comparison between the CA_{50}, Ion50 and position of maximum ion concentration (P_{ion_max}) [46]

engine for knock detection purpose. Some of the techniques used for knock detection in SI engine can also be used for estimation of combustion phasing in HCCI combustion. A combination of cylinder pressure, engine block vibration and sound pressure has been used to detect knock in SI engines [48].

A study demonstrated the use of microphones and knock sensors for feedback control of HCCI combustion [49]. The study showed that combustion phasing information from sensor output deteriorates with decrease in engine load, and these sensors can be used for closed-loop control of the combustion phasing in a HCCI engine at medium to high engine loads [49]. Engine block-mounted accelerometers can also be used to extract information about the combustion by analysing the acoustic properties in the cylinders [50]. The accelerometer signals require a lot of signal processing, and the quality of the measurements is heavily dependent on the actual location of the sensor on the engine [51].

9.3.4 Engine Torque and Speed Fluctuations

During combustion, cylinder pressure acting on the piston top is transferred to engine crankshaft through the connecting rod, and hence, information of combustion can be extracted from torque fluctuations on the crankshaft [3]. Extraction of combustion phasing information from crank shaft torque measurements is a challenging task due to complexity of torsional dynamics of the crankshaft and its dependency on engine load and crank shaft position. Crankshaft torque measurements have been used to estimate the mass fraction burned profile for a SI engine [51]. Crankshaft-mounted torque sensors give similar information as the instanteneous angular velocity measurements, and since the torque is proportional to the derivative of the angular velocity, it can give a more accurate measurement [51, 52].

The engine speed is essentially the time integral of the net crank shaft torque (excess torque than that is required to maintain a constant engine speed). Measurements of angular velocity give easily accessible information on the fluctuations in rotational speed, which can be used to estimate torque and/or pressure [53, 54]. The signal needs to be differentiated (is a noise-sensitive procedure) in order to be transformed into torque, however even more challenging to extract the cylinder pressure information from this integrated signal, since the low-frequency information tends to drown the more interesting high-frequency information [3]. Reconstruction of cylinder pressure and crank shaft torque based on measurement of engine speed fluctuation has been performed for engine control [55]. These approaches for combustion phasing estimation can also be used for HCCI combustion control.

9.4 Combustion Control Actuators

Enabling LTC with acceptable combustion phasing is more challenging than conventional SI or CI engines. Cycle-to-cycle control of LTC engines enables accurate tracking even during transient engine operation and also reduces the cyclic variations, thereby extending the operating range. Different control strategies used in LTC engine are discussed in Chap. 5. Several actuators are proposed to control the LTC engines and discussed in the next subsections.

9.4.1 Fuel Injection System

Fuel injection system is one of the key actuators in all the LTC strategies. The main benefit of fuel injection system as control actuator comes from the fact that today most production engines are equipped with direct fuel injection system, which enables cycle-to-cycle and cylinder-to-cylinder control of fuel injection timing and quantity [8]. This control actuation is typically used in LTC strategies such as PPC and RCCI combustion, where fuel is typically injected directly in the cylinder [10, 16, 17, 56].

Double fuel injection strategy is used to control PPC engine along with other actuators [16, 56]. Figure 9.9 demonstrates the effect of start of injection timings (θ_{SOI}) on combustion phasing (θ_{50}) and effect of pilot injection on IMEP and gross indicated efficiency (η_{GIE}). Figure 9.9a depicts the combustion phasing as a function of injection timings for three fuel injection durations (θ_{DOI}) at 800 bar fuel injection pressure. With advance in fuel injection timings, gain in combustion phasing decreases, and it becomes negative for very advanced fuel injection timings (Fig. 9.9a). This result provides a lower bound of fuel injection timing for effective control of combustion phasing. The study kept fuel injection timing above -20 CAD after TDC to control the combustion phasing using PI controller with manual tuning.

Pilot fuel injection is used to increase the combustion stability and ignition delay of main fuel injection event. Figure 9.9b depicts the variation in gross indicated efficiency and IMEP as a function of pilot and main injection duration for combustion phasing close to TDC and pilot injection 20 CAD before the main injection timings [16]. The figure shows that with lower values of IMEP, gross thermal efficiency increases with increase in pilot duration, and effect becomes less significant with increased IMEP. The study concluded to keep the pilot injection duration at 0.4, in order to increase the thermal efficiency. Fuel injection timing along with premixing ratio of fuels is also used for control of RCCI combustion [17].

9.4.2 Variable Valve Actuation

Variable valve actuation (VVA) strategy is used to vary the effective compression ratio of the engine and amount of trapped internal residuals. Different variable valve actuation strategies used to enable and control the HCCI combustion are discussed in Sect. 5.1.3 of Chap. 5. Intake valve closing (IVC) timings decide the effective compression ratio by fixing the start of compression. Compression ratio significantly affects the combustion phasing. At particular IVC, modulating intake valve opening (IVO) and exhaust valve closing (EVC) determine the quantity of inducted reactants and re-inducted product. These three valve timings collectively can be used for independent control of both effective compression ratio and the ratio of re-inducted products [12]. Therefore, desired combustion phasing and engine load can be controlled concurrently by adjusting IVO, EVC and IVC.

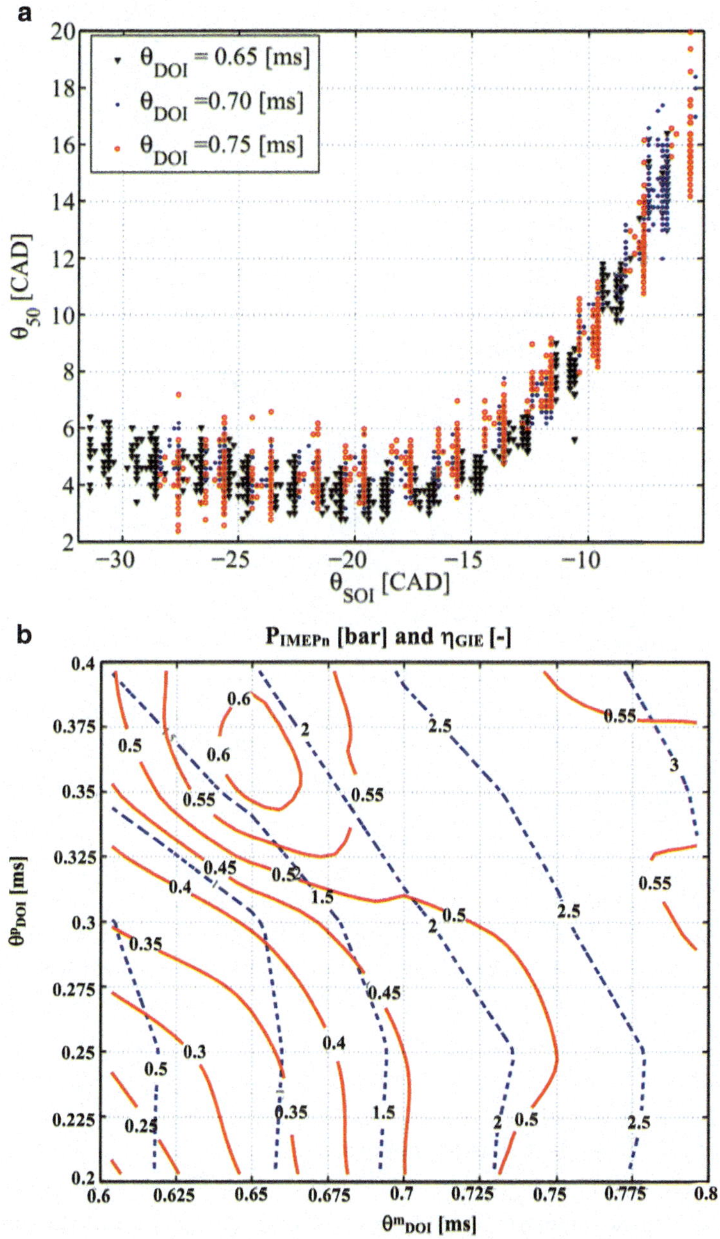

Fig. 9.9 Illustration of the effect of start of fuel injection timings (θ_{SOI}) on combustion phasing (θ_{50}) and effect of pilot injection on IMEP and gross indicated efficiency (η_{GIE}) in PPC engine [16]

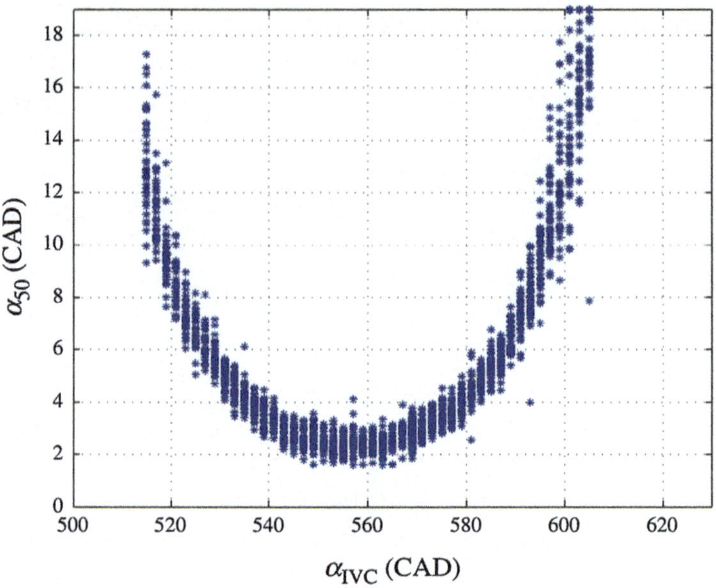

Fig. 9.10 The effect of IVC position on variation in combustion phasing (CA_{50}) at steady-state HCCI combustion [57]

Figure 9.10 illustrates the effect of early and late IVC on combustion phasing in HCCI combustion using VVA at steady-state operating condition. A change of IVC timing has a nonlinear effect on combustion phasing due to the nonlinear piston movement and higher sensitivity of delayed combustion phasing to initial conditions (Fig. 9.10) [57]. The figure demonstrates that there are two options to achieve the same combustion phasing, either during the compression after BDC or by closing the inlet valve during the expansion before BDC.

To improve low load combustion stability and efficiency of PPC engine, variable valve actuation is used, and strategies using negative valve overlap (NVO), rebreathing and split main fuel injection are able to increase the low load performance [58, 59].

9.4.3 Fast Thermal Management

Fast thermal management is used to control the temperature of air–fuel mixture at the beginning of the compression stroke. Intake thermal management strategies are described in Sec. 5.1.1 of Chap. 5. A method to achieve fast thermal management is by controlling the intake temperature using a source of cold ambient air and a source of hot recovery heated air (Fig. 5.2 of Chap. 5). Intake temperature strongly affects the combustion phasing of LTC engines. Fast thermal management is used to control the HCCI combustion as well as PPC engines [16].

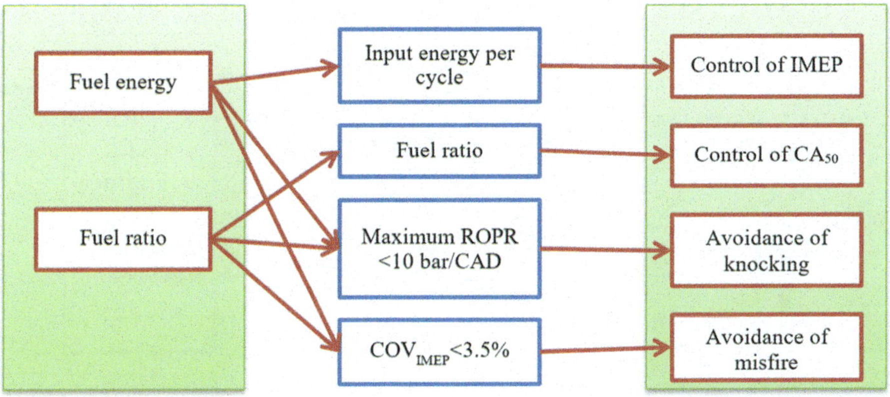

Fig. 9.11 Illustration of a dual fuel HCCI combustion control concept [60]

9.4.4 Dual Fuel (Fuel Octane/Reactivity)

In dual fuel-type controls, two fuels with different auto-ignition reactivity are used. Typically, the dual fuel system has a main fuel with a high octane number and a secondary fuel with low octane number. Dual fuel strategy of LTC control is discussed in Sect. 5.2.3 of Chap. 5. Different autoignition reactivity of dual fuel system is used to control the HCCI combustion phasing by blending the two fuels in different fuel ratios. Different fuel ratios vary the overall reactivity (octane number) of the fuel used for charge preparation. Typical examples of the fuels considered in LTC engines include mixtures of n-heptane and diesel, mixtures of ethanol and n-heptane, mixtures of methanol and n-heptane and mixtures of gasoline and diesel [60]. Figure 9.11 illustrates the HCCI combustion control concept using two fuels of different reactivity. The engine load (IMEP) is controlled by the amount of fuel energy supplied to the engine as IMEP is strongly dependent on input energy [1, 61, 62]]. The combustion phasing (CA_{50}) is controlled by varying the fuel ratio at constant energy input. Increase in high reactivity fuel advances the combustion phasing, and hence, CA_{50} can be used a process output, which is controlled by fuel ratio. Additionally, two constraints are imposed to avoid the excessive pressure rise rate and misfire conditions during combustion.

Dual fuel HCCI combustion control is typically implemented using block diagram shown in Fig. 9.12. In this particular diagram, engine speed and intake air temperature controller (shaded region) work independently to maintain the desired engine speed and inlet temperature using separate controller. The combustion phasing and IMEP controller take the input of CA_{50} and IMEP respectively that are computed from measured in-cylinder pressure data online. Depending on the reference set value of the CA_{50}, combustion phasing controller determines the fuel ratio (R_f) and sends it to the pulse width (PW) calculator. The PW calculator

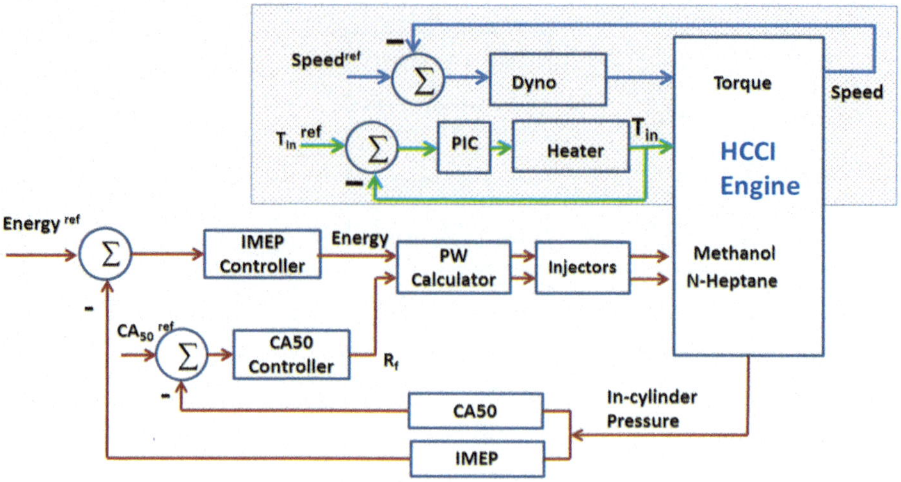

Fig. 9.12 Block diagram of the control structure of dual fuel HCCI combustion control [60]

determines the PW of fuel injection for both fuel injectors based on the given fuel ratio and fuel energy at particular engine load (IMEP). Based on the input received, injection driver injects the required quantity of methanol and n-heptane.

Dual fuel combustion control is also used in RCCI combustion engine. In RCCI combustion, high reactivity fuel is directly injected and low reactivity fuel is injected typically in the intake port of the engine. Directly injected high reactivity fuel creates the reactivity stratification in the combustion chamber. The combustion phasing depends on the reactivity stratification distribution of fuel in the cylinder. The premixing fuel ratio and direct fuel injection timings are used to control the combustion phasing and load in RCCI combustion engine [17].

9.5 Control Methods and Controllers

Control means measuring the value of the control parameter of the system and implementing the control signal on the system to correct or limit deviation of the measured value from the desired value [8]. In closed-loop control of the system, controller tends to reduce the difference between system output and reference set value, and the control operation is based on this difference. The LTC engines are sensitive to initial conditions and, thus, adequate control is required. In a well-premixed HCCI combustion, closed-loop control is essential to operate the engine at medium to high loads. Open loop control can sufficiently work at low engine load conditions below 5 bar IMEP [63]. Various control approaches are used to control HCCI combustion engines. Figure 9.13 presents the summary of different control approaches used for LTC engine control.

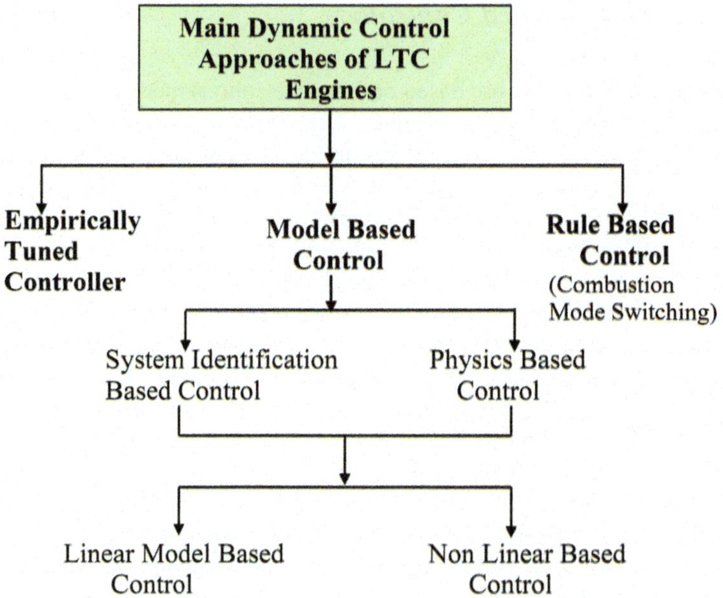

Fig. 9.13 Summary of LTC control strategies (Adapted from [9])

The main dynamic controllers for LTC engine combustion phasing are grouped into two main categories: (1) empirical controllers and (2) model-based controllers (MBC). The empirical controllers include empirically tuned proportional integral (PI) and proportional integral derivative (PID) controllers. For model-based controllers, an accurate model is essential to dynamically predict engine operation, and this type of controller uses integral state feedback, linear quadratic regulator (LQR) and linear quadratic Gaussian (LQG), model predictive control (MPC), sliding mode, H2 and constraint-based controllers to control the engine [9]. Detailed description of design and implementation of these controllers can be found in standard books [64–66]. To control the mode switching (HCCI to HEV), rule-based controllers (RBCs) are used [32, 33]. An accurate model is necessary for prediction of combustion phasing in model-based controllers. The model-based controllers can be further split into two categories, system identification-based controllers and physics-based controllers, depending on the type of model used in the synthesis of the controller. The first category of controllers is based on empirical models such as black-box modelling or system identification [67, 68]. The second category of controllers relates to MBC designed from physical models such as thermodynamic models of combustion cycle [9]. Model-based controller can also be categorized into linear and nonlinear model-based controllers (Fig. 9.13). A comprehensive review of modelling and design of controller architecture for HCCI combustion control can be found in [8].

9.5.1 Manually Tuned Controllers

Manually tuned controllers are based on intuitive control methodologies and do not require any modelling of the engine. This type of controller can be easily implemented, but their performance in nonlinear system is variable and has lag in response to large disturbances [8]. This type of controller has the combination of proportional control action, integral control action and derivative control action which is termed as PID control. These controllers are intuitively tuned manually.

Closed-loop control of the combustion phasing and engine load is implemented by using a gain-scheduled experimentally tuned PID controller in a HCCI engine [19]. Variation in fuel ratio of n-heptane and iso-octane (octane number variation) is used to control the combustion phasing, and fuel input energy is used to control the engine load (similar to Fig. 9.12) [19]. The combustion phasing and IMEP, calculated from cylinder pressure, are used as feedback to the engine controller. A study implemented PI controller for HCCI combustion by modulating the fraction of external residuals, as well as the temperature of the intake air/residual mixture [69]. This study used a combination of two PI controllers, a slow one for IMEP and a faster one for CA_{50}. It is also demonstrated that microphones can be used to provide feedback signals for combustion timing control [49]. An exhaust back pressure valve is used to regulate the amount of residuals, which were used to control the combustion timing. A PID controller is designed to change the ratio of the hot to cold intake air for combustion timing control in HCCI engine, while fuel input energy is used to control the engine load [70]. It is found that fuel type and its low temperature heat release characteristics have a large influence on closed-loop response of the controller. A PID control strategy for combustion timing control is demonstrated by using valve timings. NVO duration and IVC timing are used as valve timing strategies. NVO is used to vary the intake valve closing temperature by trapping the residual gas, and IVC is used to regulate effective compression ratio for engine load control [71]. In a variable compression ratio HCCI engine, compression ratio is used to control combustion phasing with a PID controller [72]. The controller demonstrated acceptable performance with disturbances in intake charge temperature, fuelling rate and engine speed.

9.5.2 Model-Based Controllers

Model-based controller uses a mathematical model of the system for control action, and it adapts automatically to the process change. However, model-based controller cannot account for unmodelled disturbances and has poor performance with inaccurate models [8]. There exist several LTC engine models in published literature with different level of complexity and accuracy. These engine models range from simple linearized transfer function models to very complex and detailed cycle simulation models. Different strategies used to model the HCCI combustion

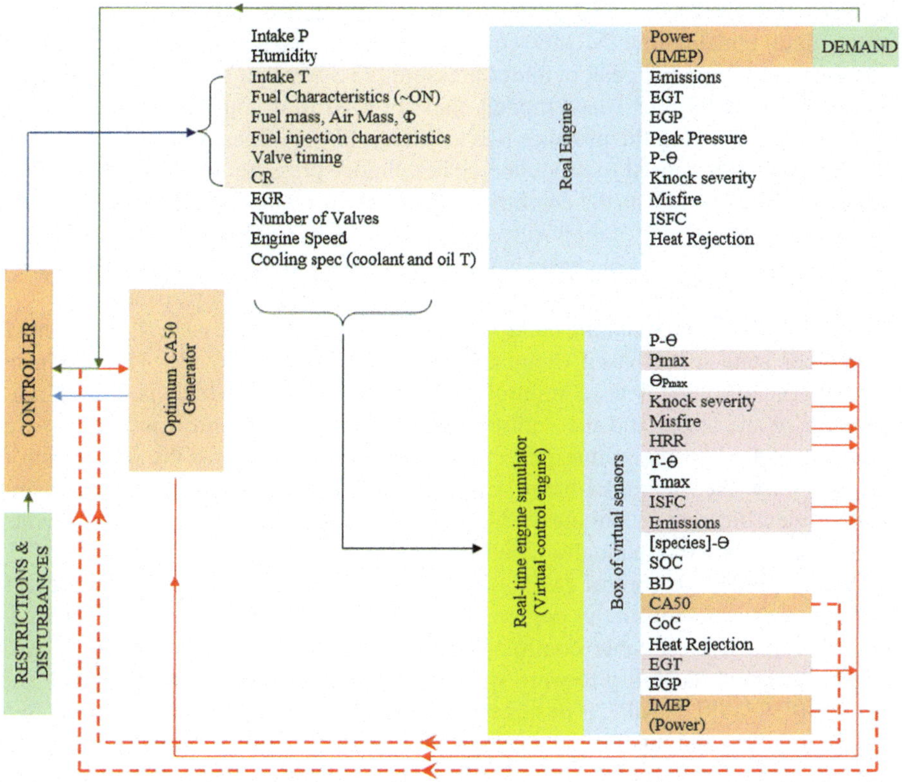

Fig. 9.14 Schematic of a LTC engine control structure [8]

behaviour include empirical modelling, black-box modelling, mean value modelling, chemical kinetic-based modelling and grey box modelling [8]. Each of these strategies has different levels of information about process/system, computation time and complexity in real-time implementation.

An accurate HCCI combustion model is of great importance and helps in model-based controllers, and this type of controllers can be ported across different test beds. The HCCI combustion needs the controller that can achieve desired optimum state within minimum number of cycles while having good robustness to physical disturbances. Figure 9.14 depicts a typical control structure for this type of engine for optimal performance. This engine controller can adjust the performance to the desired set point. Desired performance can be achieved if proper combustion timing suggested by optimum CA_{50} generator can occur. In the meantime, a rapid control-oriented engine model plays the role of an engine simulator (real-time engine simulator, virtual control engine or box of virtual sensors) to provide a real-time prediction of engine performance-/emission-related parameters to be used in the controller and optimum CA_{50} generator [8].

Several studies have used model-based controllers for LTC engine control. A multi-input–multi-output (MIMO) controller is developed to decouple the control of combustion timing and peak cylinder pressure [73]. Another study developed both data-driven and physics-based models that are used to design controllers to track load, while regulating the pressure rise and CA_{50} [74]. A linear quadratic regulator (LQR) controller is used to modulate peak cylinder pressure, while a proportional integral controller adjusts the combustion phasing in an engine with VVA [75]. The VVA system adjusted the composition of intake gas for controlling the peak cylinder pressure on a cycle-to-cycle basis while using inlet valve closing at relatively slow timescale to change the effective compression ratio and control phasing.

To design optimal dynamic controllers, linear quadratic Gaussian (LQG) control is another state-space-based optimal control methodology [5, 8]. The synthesis problem is formulated as a minimization of a criterion, which is a quadratic function of the states and the control signal. The optimal control problem is the problem of finding the admissible control function minimizing the loss function [8]. A study used a model-based LQG controller, which was developed using system identification techniques [76]. The LQG controller showed slightly better performance than the manually tuned PID controller.

Model predictive control (MPC) is an optimal model-based control approach that requires a model of the engine combustion for control action. The MPC is demonstrated as a suitable control strategy for HCCI engine due to its MIMO capabilities, and its ability to work with explicit constraints on control signals and outputs [8, 77, 78]. A MPC is designed and implemented on HCCI engine with four inputs and three outputs [79]. To design MPC control system, identification-based modelling is used with four inputs fuelling rate, ethanol fraction, engine speed and inlet temperature, and three outputs combustion timing, engine load and pressure rise rate. Constraints are imposed on the actuators and pressure rise rate. The MPC showed robust performance in tracking the combustion phasing and engine load. Another study developed MPC based on system identification with IVC timing, intake manifold temperature, injected fuel energy and engine speed as the inputs. Combustion phasing and IMEP are controlled using IVC timing and fuelling rate as main actuators considering constraints on pressure rise rate [80]. Three control structures (MPC, LQG and PID controllers) are implemented on six-cylinder heavy-duty engine for cycle-to-cycle control of combustion timing [57]. Dynamic models for HCCI combustion are obtained by system identification for LQG and MPC controllers. The MPC shows better performance in controlling HCCI combustion phasing and engine load in comparison to the other two controllers.

In PPC engine, the MPC design is used to control the ignition delay while keeping the combustion phasing within an acceptable range using the fuel injection timings [81]. Another study used MPC for closed-loop control of the ignition delay and combustion phasing on cycle-to-cycle basis in a multi-cylinder PPC engine operated on a gasoline [10]. The control of the output parameters is implemented by modulating the injection timings, the valve positions of a dual path EGR system and the valve positions of an inlet manifold fast thermal management system. Reliable and sufficiently good performance is shown by MPC control in PPC engine, which can be further improved by extending the model.

References

1. Maurya RK, Agarwal AK (2013) Experimental investigation of close-loop control of HCCI engine using dual fuel approach (No. 2013-01-1675). SAE technical paper
2. Bengtsson J (2004) Closed-loop control of HCCI engine dynamics. PhD Theses. Lund University, Sweden
3. Tunestål P, Johansson B (2007) HCCI control. In: CAI and HCCI engines for the automotive industry. Woodhead Publishing Limited, Cambridge, IL, USA, pp 164–184
4. Olsson JO, Tunestål P, Johansson B, Fiveland S, Agama R, Willi M, Assanis DN (2002) Compression ratio influence on maximum load of a natural gas fueled HCCI engine (No. 2002-01-0111). SAE technical paper
5. Strandh P, Bengtsson J, Johansson R, Tunestål P, Johansson B (2005) Variable valve actuation for timing control of a homogeneous charge compression ignition engine (No. 2005-01-0147). SAE technical paper
6. Sjöberg M, Dec JE (2009) Influence of EGR quality and unmixedness on the high-load limits of HCCI engines. SAE Int J Engines 2. (2009-01-0666):492–510
7. Sjöberg M, Dec JE (2008) Influence of fuel autoignition reactivity on the high-load limits of HCCI engines. SAE Int J Engines 1(2008-01-0054):39–58
8. Fathi M, Jahanian O, Shahbakhti M (2017) Modeling and controller design architecture for cycle-by-cycle combustion control of homogeneous charge compression ignition (HCCI) engines – a comprehensive review. Energy Convers Manag 139:1–19
9. Bidarvatan M (2015) Physics-based modeling and control of powertrain systems integrated with low temperature combustion engines. PhD Thesis, Michigan Technological University, Houghton
10. Ingesson G, Yin L, Johansson R, Tunestal P (2015) Simultaneous control of combustion timing and ignition delay in multi-cylinder partially premixed combustion. SAE Int J Engines 8(2015-24-2424):2089–2098
11. Bengtsson J, Strandh P, Johansson R, Tunestål P, Johansson B (2004) Closed-loop combustion control of homogeneous charge compression ignition (HCCI) engine dynamics. Int J Adapt Control Signal Process 18(2):167–179
12. Shaver GM, Gerdes JC, Roelle M (2004) Physics-based closed-loop control of phasing, peak pressure and work output in HCCI engines utilizing variable valve actuation. In: American control conference, 2004. Proceedings of the 2004, vol 1. IEEE, Massachusetts, USA, pp 150–155, June
13. Shaver GM, Gerdes JC, Roelle MJ (2009) Physics-based modeling and control of residual-affected HCCI engines. J Dyn Syst Meas Control 131(2):021002
14. Maurya RK, Agarwal AK (2013) Investigations on the effect of measurement errors on estimated combustion and performance parameters in HCCI combustion engine. Measurement 46(1):80–88
15. Maurya RK, Pal DD, Agarwal AK (2013) Digital signal processing of cylinder pressure data for combustion diagnostics of HCCI engine. Mech Syst Signal Process 36(1):95–109
16. Ingesson G, Yin L, Johansson R, Tunestal P (2016) Control of the low-load region in partially premixed combustion. In: Journal of physics: conference series, vol 744, no. 1. IOP Publishing, p 012106, September
17. Arora JK, Shahbakhti M (2017). Real-time closed-loop control of a light-duty RCCI engine during transient operations (No. 2017-01-0767). SAE technical paper
18. Bidarvatan M, Thakkar V, Shahbakhti M, Bahri B, Aziz AA (2014) Grey-box modeling of HCCI engines. Appl Therm Eng 70(1):397–409
19. Olsson JO, Tunestål P, Johansson B (2001) Closed-loop control of an HCCI engine (No. 2001-01-1031). SAE technical paper
20. Bidarvatan M, Shahbakhti M (2013) Two-input two-output control of blended fuel HCCI engines (No. 2013-01-1663). SAE technical paper

21. Ravi N, Liao HH, Jungkunz AF, Widd A, Gerdes JC (2012) Model predictive control of HCCI using variable valve actuation and fuel injection. Control Eng Pract 20(4):421–430
22. Tandra V, Srivastava N (2009) Optimal peak pressure and exhaust temperature tracking control for a two-zone HCCI engine model with mean burn duration (No. 2009-01-1130). SAE technical paper
23. Williams S, Hu LR, Nakazono T, Ohtsubo H, Uchida M (2008) Oxidation catalysts for natural gas engine operating under HCCI or SI conditions. SAE Int J Fuels Lubr 1(2008-01-0807):326–337
24. Bidarvatan M, Shahbakhti M (2014) Integrated HCCI engine control based on a performance index. J Eng Gas Turbines Power 136(10):101601
25. Gorzelic P, Hellström E, Stefanopoulou A, Li J (2012) Model-based feedback control for an automated transfer out of SI operation during SI to HCCI transitions in gasoline engines. In: ASME 2012 5th annual dynamic systems and control conference joint with the JSME 2012 11th motion and vibration conference, Florida, USA, pp 359–367, October
26. Bidarvatan M, Kothari D, Shahbakhti M (2015) Integrated cycle-to-cycle control of exhaust gas temperature, load, and combustion phasing in an HCCI engine. In: American control conference (ACC). IEEE, Chicago, IL, USA, pp 7–12, July
27. Asad U, Divekar P, Chen X, Zheng M, Tjong J (2012) Mode switching control for diesel low temperature combustion with fast feedback algorithms. SAE Int J Engines 5(2012-01-0900):850–863
28. Daw CS, Wagner RM, Edwards KD, Green JB (2007) Understanding the transition between conventional spark-ignited combustion and HCCI in a gasoline engine. Proc Combust Inst 31 (2):2887–2894
29. Abarbanel H (1996) Analysis of observed chaotic data. Springer, New York
30. Fang C, Yang F, Ouyang M, Gao G, Chen L (2013) Combustion mode switching control in a HCCI diesel engine. Appl Energy 110:190–200
31. Wang J (2008) Hybrid robust air-path control for diesel engines operating conventional and low temperature combustion modes. IEEE Trans Control Syst Technol 16(6):1138–1151
32. Delorme A, Rouseau A, Wallner T, Ortiz-Soto E, Babajimopolous A, Assanis D (2010) Evaluation of homogeneous charge compression ingition (HCCI) engine fuel savings for various electric drive powertrains. In: Proceeding of the 25th world battery, hybrid and fuel cell electric vehicle symposium and exhibition, EVS-25, Shenzhen, China
33. Lawler B, Ortiz-Soto E, Gupta R, Peng H, Filipi Z (2011) Hybrid electric vehicle powertrain and control strategy optimization to maximize the synergy with a gasoline HCCI engine. SAE Int J Engines 4(2011-01-0888):1115–1126
34. Ahn K, Whitefoot J, Babajimopoulos A, Ortiz-Soto E, Papalambros PY (2013) Homogeneous charge compression ignition technology implemented in a hybrid electric vehicle: system optimal design and benefit analysis for a power-split architecture. Proc Inst Mech Eng D J Automob Eng 227(1):87–98
35. Saracino R, Gaballo MR, Mannal S, Motz S, Carlucci A, Benegiamo M (2015) Cylinder pressure-based closed loop combustion control: A valid support to fulfill current and future requirements of diesel powertrain systems (No. 2015-24-2423). SAE technical paper
36. Wilhelmsson C (2007) Field programmable gate arrays and reconfigurable computing in automatic control. Lund University, Lund
37. Tunestål P (2001) Estimation of the in-cylinder air/fuel ratio of an internal combustion engine by the use of pressure sensors. PhD Thesis, Lund Institute of Technology, 1025
38. Sellnau MC, Matekunas FA, Battiston PA, Chang CF, Lancaster DR (2000) Cylinder-pressure-based engine control using pressure-ratio-management and low-cost non-intrusive cylinder pressure sensors (No. 2000-01-0932). SAE technical paper
39. Shimasaki Y, Kobayashi M, Sakamoto H, Ueno M, Hasegawa M, Yamaguchi S, Suzuki T (2004) Study on engine management system using in-cylinder pressure sensor integrated with spark plug (No. 2004-01-0519). SAE technical paper

40. Bogin G, Chen JY, Dibble RW (2009) The effects of intake pressure, fuel concentration, and bias voltage on the detection of ions in a homogeneous charge compression ignition (HCCI) engine. Proc Combust Inst 32(2):2877–2884
41. Chen Y, Dong G, Mack JH, Butt RH, Chen JY, Dibble RW (2016) Cyclic variations and prior-cycle effects of ion current sensing in an HCCI engine: a time-series analysis. Appl Energy 168:628–635
42. Saitzkoff A, Reinmann R, Mauss F, Glavmo M (1997) In-cylinder pressure measurements using the spark plug as an ionization sensor (No. 970857). SAE technical paper
43. Wang Y, Zhou L (2003) Investigation of the detection of knock and misfire of a spark ignition engine with the ionic current method. Proc Inst Mech Eng D J Automob Eng 217(7):617–621
44. Strandh P, Christensen M, Bengtsson J, Johansson R, Vressner A, Tunestål P, Johansson B (2003) Ion current sensing for HCCI combustion feedback (No. 2003-01-3216). SAE technical paper
45. Prager J, Riedel U, Warnatz J (2007) Modeling ion chemistry and charged species diffusion in lean methane–oxygen flames. Proc Combust Inst 31(1):1129–1137
46. Dong G, Chen Y, Li L, Wu Z, Dibble R (2017) A skeletal gasoline flame ionization mechanism for combustion timing prediction on HCCI engines. Proc Combust Inst 36(3):3669–3676
47. Dong G, Li L, Wu Z, Zhang Z, Zhao D (2013) Study of the phase-varying mechanisms of ion current signals for combustion phasing in a gasoline HCCI engine. Fuel 113:209–215
48. Lee JH, Hwang SH, Lim JS, Jeon DC, Cho YS (1998) A new knock-detection method using cylinder pressure, block vibration and sound pressure signals from a SI engine (No. 981436). SAE technical paper
49. Souder JS, Mack JH, Hedrick JK, Dibble RW (2004) Microphones and knock sensors for feedback control of HCCI engines. In: ASME 2004 internal combustion engine division fall technical conference. American Society of Mechanical Engineers, California, USA, pp 77–84, January
50. Urlaub M, Böhme JF (2004) Reconstruction of pressure signals on structure-borne sound for knock investigation (No. 2004-01-0521). SAE technical paper
51. Larsson S, Andersson I (2005) An experimental evaluation of torque sensor based feedback control of combustion phasing in an SI-engine (No. 2005-01-0060). SAE technical paper
52. Schagerberg S, McKelvey T (2003) Instantaneous crankshaft torque measurements-modeling and validation (No. 2003-01-0713). SAE technical paper
53. Guezennec YG, Gyan P (1999) A novel approach to real-time estimation of the individual cylinder combustion pressure for SI engine control (No. 1999-01-0209). SAE technical paper
54. Taraza D, Henein NA, Bryzik W (1998) Determination of the gas-pressure torque of a multicylinder engine from measurements of the crankshaft's speed variation (No. 980164). SAE technical paper
55. Lee B, Rizzoni G, Guezennec Y, Soliman A, Cavalletti M, Waters J (2001) Engine control using torque estimation (No. 2001-01-0995). SAE technical paper
56. Ingesson G, Yin L, Johansson R, Tunestål P (2016) A double-injection control strategy for partially premixed combustion. IFAC-PapersOnLine 49(11):353–360
57. Bengtsson J, Strandh P, Johansson R, Tunestål P, Johansson B (2006) Hybrid control of homogeneous charge compression ignition (HCCI) engine dynamics. Int J Control 79 (05):422–448
58. Borgqvist P, Tunestal P, Johansson B (2012) Gasoline partially premixed combustion in a light duty engine at low load and idle operating conditions (No. 2012-01-0687). SAE technical paper
59. Borgqvist P, Tunestal P, Johansson B (2013) Comparison of negative valve overlap (NVO) and rebreathing valve strategies on a gasoline PPC engine at low load and idle operating conditions. SAE Int J Engines 6(2013-01-0902):366–378
60. Maurya RK (2012) Performance, emissions and combustion characterization and close loop control of HCCI engine employing gasoline like fuels. PhD Thesis, Indian Institute of Technology Kanpur, India

61. Maurya RK, Agarwal AK (2014) Experimental investigations of performance, combustion and emission characteristics of ethanol and methanol fueled HCCI engine. Fuel Process Technol 126:30–48
62. Maurya RK, Agarwal AK (2015) Combustion and emission characterization of n-butanol fueled HCCI engine. J Energy Resour Technol 137(1):011101
63. Haraldsson G, Tunestål P, Johansson B, Hyvönen J (2005) Transient control of a multi cylinder HCCI engine during a drive cycle (No. 2005-01-0153). SAE technical paper
64. Maciejowski JM (2002) Predictive control: with constraints. Pearson Education, London
65. Bubnicki Z (2005) Modern control theory. Springer, New York
66. Wang L (2009) Model predictive control system design and implementation using MATLAB®. Springer-Verlag, London
67. Audet A, Koch CR (2009) Actuator comparison for closed loop control of HCCI combustion timing (No. 2009-01-1135). SAE technical paper
68. Bengtsson J, Strandh P, Johansson R, Tunestål P, Johansson B (2007) Hybrid modelling of homogeneous charge compression ignition (HCCI) engine dynamics – a survey. Int J Control 80(11):1814–1847
69. Ohmura T, Ikemoto M, Iida N (2006) A study on combustion control by using internal and external EGR for HCCI engines fuelled with DME (No. 2006-32-0045). SAE technical paper
70. Haraldsson G, Tunestål P, Johansson B, Hyvönen J (2004) HCCI closed-loop combustion control using fast thermal management (No. 2004-01-0943). SAE technical paper
71. Agrell F, Ångström HE, Eriksson B, Wikander J, Linderyd J (2003) Integrated simulation and engine test of closed loop HCCI control by aid of variable valve timings (No. 2003-01-0748). SAE technical paper
72. Haraldsson G, Tunestål P, Johansson B, Hyvönen J (2003) HCCI combustion phasing with closed-loop combustion control using variable compression ratio in a multi cylinder engine (No. 2003-01-1830). SAE technical paper
73. Shaver GM, Roelle M, Gerdes JC (2005) Decoupled control of combustion timing and work output in residual-affected HCCI engines. In: American control conference, 2005. Proceedings of the 2005. IEEE, pp 3871–3876, June
74. Kulzer A, Hathout JP, Sauer C, Karrelmeyer R, Fischer W, Christ A (2007) Multi-mode combustion strategies with CAI for a GDI engine (No. 2007-01-0214). SAE technical paper
75. Shaver GM, Roelle MJ, Caton PA, Kaahaaina NB, Ravi N, Hathout JP et al (2005) A physics-based approach to the control of homogeneous charge compression ignition engines with variable valve actuation. Int J Engine Res 6(4):361–375
76. Strandh P, Bengtsson J, Johansson R, Tunestål P, Johansson B (2004) Cycle-to-cycle control of a dual-fuel HCCI engine (No. 2004-01-0941). SAE technical paper
77. Widd A, Ekholm K, Tunestal P, Johansson R (2012) Physics-based model predictive control of HCCI combustion phasing using fast thermal management and VVA. IEEE Trans Control Syst Technol 20(3):688–699
78. Janakiraman VM, Nguyen X, Assanis D (2016) An ELM based predictive control method for HCCI engines. Eng Appl Artif Intell 48:106–118
79. Bengtsson J, Strandh P, Johansson R, Tunestal P, Johansson B (2006) Model predictive control of homogeneous charge compression ignition (HCCI) engine dynamics. In: Computer aided control system design, 2006 I.E. international conference on control applications, 2006 I.E. international symposium on intelligent control, 2006 IEEE. IEEE, pp 1675–1680, October
80. Bengtsson J, Strandh P, Johansson R, Tunestål P, Johansson B (2006) Multi-output control of a heavy duty HCCI engine using variable valve actuation and model predictive control (No. 2006-01-0873). SAE technical paper
81. Lewander M, Johansson B, Tunestål P, Keeler N, Milovanovic N, Bergstrand P (2008) Closed loop control of a partially premixed combustion engine using model predictive control strategies. In: Proceedings of AVEC, vol 8. Japan

Chapter 10
Closure

Abstract Research towards the development of internal combustion engine having high fuel conversion efficiency and ultralow emissions is driven by stringent emission legislations, degradation of ambient environmental conditions, depletion of fossil resources, energy security and global warming. The low temperature combustion (LTC) engines are one of the potential options to fulfil the objective of high fuel conversion efficiency along with ultralow emissions of NOx and particulate matter. The LTC engines are radically different from conventional spark ignition and compression ignition engines. Research has been conducted on various LTC concepts using conventional and alternative fuels on both light-duty (LD) and heavy-duty (HD) engines. Performance, combustion and emissions characteristics along with different control strategies of LTC engines are discussed in previous chapters of the present book. Summary of main findings regarding performance, combustion and emissions characteristics of various LTC strategies is presented in this chapter, and recommendations for further work are also outlined.

Keywords LTC • HCCI • PPC • RCCI • Dual fuel • Combustion control • Engine • Gasoline • Diesel • Stratification • Emissions • IMEP • Ignition • SI • CI

10.1 Summary

The internal combustion engines, fuelled mostly by petroleum-derived liquid fuels, have been the main source of transport power over the past century and are likely to remain so in the foreseeable future, though alternatives such as electrification of transport play a role. However, combustion engines are continuously changing, primarily driven by the need to be more efficient and meet the stringent emission legislations along with satisfying the customer demands such as driveability at an affordable cost. These requirements stimulate new developments in both conventional and alternative engines as well as fuels. To address the challenges in fuels and combustion engines, there are four possible approaches, i.e. (1) improvement of conventional engines, (2) improvement of conventional fuels, (3) development of alternative engine and (4) utilization of alternative fuels, and their combinations. Conventional and alternative fuels are continuously evolving to minimize the local

R.K. Maurya, *Characteristics and Control of Low Temperature Combustion Engines*, Mechanical Engineering Series, https://doi.org/10.1007/978-3-319-68508-3_10

and global environmental impact in their manufacturing process and in their use in combustion engines. Premixed LTC is an alternative combustion concept for reciprocating internal combustion engines, which offers prominent benefits in terms of simultaneous reduction of both NO_x and particulate emissions to ultralow levels along with reduction in fuel consumption.

Among the LTC strategies, homogeneous charge compression ignition (HCCI) concept is evolved as third major engine combustion modes in addition to the conventional spark ignition (SI) and compression ignition (CI) modes of engine combustion. The HCCI combustion mode used well-mixed fuel–air mixture, and hence, simultaneous autoignition of entire mixture in the cylinder leads to very high-pressure rise rate and combustion noise. To control the combustion rate, HCCI engines are operated on diluted lean fuel–air mixture, which results into lower engine operating range. Fuel and thermal stratification in the combustion chamber is found beneficial in reduction of heat release rate by sequential autoignition starting from most favourable conditions (locally richer mixture/hotter region). Fuel stratification in the cylinder can be easily controlled by direct fuel injection system. Therefore, different levels of fuel stratification are used to increase the load range of HCCI combustion while keeping its benefit of ultralow NO_x and soot emissions. All the LTC strategies using gasoline direct injection are grouped in the three main categories based on stratification of charge: (1) partial fuel stratification (PFS), (2) moderate fuel stratification (MFS) and (3) heavy fuel stratification (HFS). Partially premixed combustion (PPC) in HFS regime is able to achieve the engine load similar to CI engines using gasoline-like fuels. To control the high combustion rate, a very high amount (~50% or more) of exhaust gas reticulation (EGR) is used. To reduce the requirement of EGR in PPC strategy, dual fuel reactivity controlled compression ignition (RCCI) strategy is developed, where fuel stratification as well as reactivity stratification is used to control the combustion rate. Presently, PPC and RCCI are the most widely investigated LTC strategies for automotive engine application. To further improve the RCCI engine efficiency and combustion control, direct injection dual fuel stratification (DDFS) strategy is demonstrated, where both low and high reactivity fuels are directly injected into the engine cylinder.

Fuel properties and composition play an important role in all the physical and chemical processes involved in LTC process. Autoignition chemistry depends on the fuel composition and mixture quality (equivalence ratio and homogeneity). Creation of premixed charge in the combustion chamber is the key feature of LTC engines. Quality of fuel–air mixture governs the combustion process and its rate. Formation of premixed charge is required prior to the start of combustion in LTC engine. Depending on LTC strategy, different qualities of premixed charge (degree of homogeneity) are required for higher thermal efficiency and to control the combustion rate. The premixed charge can be created by injecting fuel, outside the engine cylinder (external charge preparation) or inside the cylinder (internal charge preparation), depending on fuel properties and combustion strategy. Typically, in well-mixed HCCI combustion, external charge preparation is used, and in other LTC strategies, internal charge preparation is also used. To enable and control the LTC engines, two main approaches are used: (1) altering pressure–temperature history of the charge in the cylinder and (2) altering fuel reactivity of charge. Temperature history of the charge in the cylinder can be altered by several parameters such as intake conditions (temperature and pressure), EGR, variable valve timings (VVT), variable compression ratio (VCR), water injection, supercharging and fuel injection strategies. Fuel reactivity of charge in the cylinder can be altered by various parameters such as equivalence ratio (Φ), fuel stratification, fuel additives, ozone additions and dual fuel.

The LTC operating range is typically constrained by several limiting factors such as combustion noise, combustion instability, maximum cylinder pressure, oxygen availability, excessive reactivity and emission limits. Typically, ringing intensity (RI) is used to define the high load limit. Acceptable RI value mostly used is below 5 MW/m^2 in LTC engines. The coefficient of variation in indicated mean effective pressure (IMEP) is typically used to define the lower engine operating range with limiting value of 3.5%. Using these operating limits, maximum IMEP achieved is typically around 5 bar in naturally aspirated HCCI engine using well-mixed fuel–air mixture. In HCCI operating range, the IMEP mainly depends on the amount of energy input in the cylinder or maximum equivalence ratio of engine operation. It is demonstrated that high octane fuels (gasoline-like fuels) can be best utilized in compression ignition engines. The maximum engine load up to 20 bar IMEP in well-mixed HCCI combustion and 25 bar IMEP with PPC combustion is achieved while maintaining the emission level of NOx and soot below EURO VI limits. This engine loads achieved in HCCI and PPC engines are comparable to the conventional diesel engines. However, higher boost pressure up to 3.6 bar is used to achieve this engine load, and this level of boost requirement cannot be achieved by present turbocharger used in vehicles. It is demonstrated that partial fuel stratification can reduce the requirement of boost pressure for the same maximum engine load operation in well-mixed HCCI combustion.

The HCCI engine has higher thermodynamic efficiency in comparison to the conventional SI engine. Higher thermal efficiency is one of the main benefits of LTC combustion engines. Gross indicated thermal efficiency up to 57% in PPC mode using gasoline-like fuels and up to 58% in dual fuel RCCI engine is demonstrated (Chap. 7). Higher thermal efficiency leads to reduction in specific fuel consumption of the engine and saves the fuel. Thus, fuel economy of vehicles with LTC engine powertrain is better than conventional engines.

Depending on the fuel used in LTC engines, single-stage and two-stage heat release characteristics are observed during combustion. In single-stage heat release fuel, only high temperature heat release (HTHR) is observed, while in two-stage heat release fuels, low temperature heat release (LTHR) is also observed along with HTHR. Fuels exhibiting LTHR typically require less intake temperature for enabling the autoignition in the cylinder. However, LTHR is affected by intake pressure and temperature, EGR, and engine speed in addition to fuel type. Some fuels like gasoline exhibit LTHR or intermediate temperature heat release (ITHR) at higher intake (boost) pressure, while in naturally aspirated conditions, only single-stage HTHR is observed. Typically, LTC engine has very high heat release rate (HRR) due to premixed combustion, and it can be controlled by using high amount of EGR, dilution and stratification. Typically, EGR has higher specific heat that lowers the combustion temperature. Lower combustion temperature due to EGR retards the combustion phasing which lowers the peak heat release rate. However, trace species present in EGR such as NO can also advance the start of combustion due to participation of trace species of EGR in autoignition reactions. In LTC engines, combustion instability (cycle-to-cycle variations) can be very high in some of the operating conditions particularly at retarded combustion phasing and leaner mixture operation at lower engine loads. Cyclic variations in LTC combustion can be random normal variations as well as deterministic or periodic patterns.

Due to its nature, premixed LTC strategy has the potential to generate very low levels of NO_x and soot emissions. Typically, all the LTC strategies are able to achieve NO_x and PM emissions below EURO VI norms in LTC operating range without any exhaust after-treatment system. The ultralow NO_x and PM emissions are the most beneficial part in LTC engines. However, LTC engines generally have very high amount of unburned hydrocarbon (HC) and carbon mono-oxide (CO) emissions, which can be easily handled by oxidation catalyst. For effective conversion efficiency, oxidation catalyst requires higher operating temperature than its light-off temperature. However, LTC engines have very low exhaust gas temperatures particularly at lower engine loads, where HC and CO emissions are also higher. This is one of the challenges to be tackled in LTC engines. Particulate matter emitted from LTC engines is generally lower on mass basis; however particle number emissions are significantly emitted. Most of the particles emitted from the LTC engines are of volatile nature depending upon the LTC strategy and fuel used.

The LTC engines, particularly HCCI combustion, need closed-loop combustion control, and map-based engine operation is not possible as some of the engine operating conditions are unstable (Chap. 9). Typically, cylinder pressure sensor and ion current sensors are used for sensing of combustion phasing of LTC engine for closed-loop control. The HCCI combustion control variables are divided into four major groups, namely, (1) control of combustion phasing, (2) control of engine load, (3) control of exhaust treatment efficiency (exhaust gas temperature) and (4) control of HCCI dynamics for combustion mode switching between HCCI and conventional SI or CI modes. Closed-loop control of combustion phasing and engine load is demonstrated using manually tuned controller and several model-based controllers in HCCI, PPC and RCCI engines. Advanced controllers such as adaptive controller and model-based predictive controller are more effective in controlling combustion phasing in a wide range of operating conditions of LTC engines.

10.2 Future Directions

Research activities from the last few decades lead to the development of different LTC strategies such as HCCI, PPC and RCCI combustion modes that demonstrated higher fuel conversion efficiency as well as very low NOx and PM emissions. Several technical challenges are still to be resolved before mass commercialization of high-efficiency, emissions-compliant LTC engine-based powertrains in heavy- and light-duty vehicles. Research and development effort is required in the following direction for improving the present LTC mode engines.

1. *Load range extension*: Engine operating range of LTC combustion strategies is typically lower than conventional engines. However, PPC combustion has demonstrated IMEP up to 25 bar at fixed engine speed with a very high boost pressure (~3.6 bar). This engine load is similar to the engine load possibly achieved by CI engines; however required boost pressure level is very high and it cannot be achieved by turbocharger used in present CI engines. Research is required to reduce the boost to engine load ratio in LTC engines by optimizing fuel injection strategies, engine operating conditions (EGR, intake temperature, compression ratio) and fuel. Several renewable fuels are being developed that have potential for load range extension of LTC engines in combination with optimized engine operating strategies. Fundamental combustion studies (modelling and optical engine experiments) need to be conducted for different LTC strategies (RCCI, DDFS, PPC) with alternative and conventional fuels to better understand the combustion process, which helps in operating range extension. Comprehensive study of combustion noise level and ringing is also required, which limits the load range of LTC engines.
2. *Optimization of air handling system*: The LTC strategies such as HCCI and RCCI are highly sensitive to initial intake conditions, which are strongly affected by the gas exchange process. Air handling systems including EGR, turbocharger/supercharger performance and valve timing affect the gas exchange process of engine. Therefore, optimization of air handling system is required for different LTC strategies, especially determination of system parameters and air handling configurations for high engine load operations. Redesign of manifold and variable valve actuation (VVA) strategies can be investigated for reducing pumping and friction work during engine operation. To improve the turbocharger efficiency, a newly proposed strategy called active control turbocharging [1] could be explored for LTC engine applications. Utilization of mixed flow turbine is another strategy to maximize the work extraction and increase the turbocharger efficiency [2], and it may be used for LTC engine turbocharging. Optimization and closed-loop control of fuel and air handling system can lead to optimal combustion characteristics of LTC engines.
3. *Premixed charge optimization*: Premixed charge preparation plays an important role in LTC strategies. Fuel and reactivity distribution in the cylinder determines the combustion characteristics. The optimization of injection strategies (number of injection, fuel injection timings, fuel injection pressure, injector cone angle

and nozzle geometry), swirl and combustion chamber geometry is required for desired combustion characteristics. In more complex strategies like DDFS, optimization of the injection strategy of both high and low reactivity fuels along with other parameters involved in charge preparation could lead to improved engine efficiency and reduced emissions. Flow characteristics such as swirl affect the mixing process of air and fuel. Therefore, manifold design and manifold wave dynamics can also affect the charge preparation, and hence, it needs to be optimally designed for particular LTC strategy.

4. *Cold start and transient operation control*: The LTC engines, particularly HCCI combustion, are very sensitive to initial charge conditions, and minor variations can significantly affect the combustion phasing. Additionally, intake temperature required for autoignition varies with engine operating conditions and fuel properties. Under cold start conditions, it is difficult to autoignite the charge due to lower compressed gas temperature achieved in the engine cylinder during cold start condition. One possible suggestion to address the cold start issue is to start the engine in conventional mode and then shift to LTC mode. In this case, combustion mode transition needs to be controlled efficiently. Most of the LTC studies are conducted on steady-state operation. Automotive engine needs to operate on frequent transition in a driving cycle. Therefore, transient characteristics need to be investigated, and the required control strategies for transient engine operation needs to be developed for LTC engines.
5. *Particle emission characterization*: Although LTC engines have very low particulate matter on mass basis, the particle number emissions are significant. Particle number distribution in various LTC strategies including HCCI, PPC and RCCI combustion is investigated with certain range of operating conditions. Particles emitted from LTC modes have large fraction of volatile particles. Volatile and non-volatile particle characterization from different LTC strategies with different fuels over a wide range of engine operating conditions needs to be conducted. Fuel composition and physical properties have great impact on particulate formation, and hence, investigation is required for particulate emission with oxygenated fuels and fuels with short carbon chains (low C/H ratio), which are potential candidate for low particulate emissions. Air–fuel mixing affects the soot formation, and hence, soot reduction can be improved further by enhancing the fuel-wall interactions in the cylinder. To find a good combination of turbulence and injection process (for a specific fuel) for particle emission in LTC engine, several parameters such as swirl ratio, injection spray angle, piston geometry and injector parameters should be investigated. Particles emission from LTC engines needs further investigation including particle chemical composition, optical properties and micro- and nanostructure using different fuels. Study of other unregulated species (such as PAH, carbonyl compounds, etc.) and their toxicity can also be investigated on LTC engines using conventional and alternative fuels.
6. *Exhaust after-treatment development*: In LTC engines, NO_x and soot emissions are controlled in cylinder, and no exhaust after-treatment system is required to meet the emission legislation for NO_x and PM. However, HC and CO emissions

are very high in LTC engines. To maintain acceptable level of emissions, development of suitable exhaust after-treatment along with appropriate control strategy is required. To mitigate the HC and CO emissions, typically oxidation catalyst is used, which works efficiently above its light-off temperature. Exhaust gas temperature in LTC engines is very low at some of the operating conditions. This makes the emission control problem further difficult. Two possible approaches can be used to handle this issue. First, exhaust gas temperature control system can be developed to maintain the required exhaust gas temperature in the exhaust line. Second option is to develop the catalyst that can work efficiently at lower exhaust temperatures achieved in LTC engines.

7. *Closed-loop control development*: Closed-loop control is required in HCCI engines particularly at high engine load operation. Closed-loop control of combustion phasing and engine load is demonstrated in HCCI, PPC and RCCI combustion strategies using different actuators (Chap. 9). Different aspects of control structure needs to be thoroughly addressed, and precisely validated with experimental data to achieve the benefits of HCCI combustion with all its advantages. In current scenario, exhaust gas temperature is also a control variable along with combustion phasing and engine load in HCCI engines. Model-based predictive control (MPC) is found to be a better option for closed-loop control of LTC engines. To effectively utilize and implement, accurate predictive model of LTC engine is required including all the parameters over a wide range of engine operating conditions. For closed-loop control, sensing of combustion phasing is typically based on cylinder pressure measurement. Low-cost cylinder pressure sensor development is required to use it on production engine. Different control structure issues need to be further explored so that a final viable control structure can be developed and implemented in LTC engines.
8. *Mode switching and integration of LTC engines with other powertrains*: Dual-mode engine operation (SI-HCCI or CI-HCCI) seems to be a potential solution for full load operation in production engines by adapting HCCI mode. Research and development efforts are required for development of mode transition control and robust operation of LTC mode. In dual-mode engine operation, exhaust after-treatment for NO_x and PM needs to be installed in exhaust line. During conventional mode engine operation, to mitigate NO_x and PM emission, after-treatment system is required. Integration of HCCI engine into hybrid electric vehicles (HEVs) may become a viable solution to address some of the HCCI engine control challenges. Depending on the HEV architecture, most of the vehicle transient operations can be handled by the electric motor, which facilitates the steady-state operation of HCCI engine at fixed load. HCCI engine operation can be better controlled during steady-state conditions. In HEV application, option of supercharger instead of turbocharger can also be explored to increase the engine load range of HCCI engine.

Review papers in reference [3–10] may be helpful for additional details about the present developments of LTC engines and future directions for path to be followed for further improvement in current LTC strategies.

References

1. Rajoo S, Pesiridis A, Martinez-Botas R (2014) Novel method to improve engine exhaust energy extraction with active control turbocharger. Int J Engine Res 15(2):236–249
2. Karamanis N, Martinez-Botas RF (2002) Mixed-flow turbines for automotive turbochargers: steady and unsteady performance. Int J Engine Res 3(3):127–138
3. Agarwal AK, Singh AP, Maurya RK (2017) Evolution, challenges and path forward for low temperature combustion engines. Prog Energy Combust Sci 61:1–56
4. Boot MD, Tian M, Hensen EJ, Sarathy SM (2017) Impact of fuel molecular structure on auto-ignition behavior–design rules for future high performance gasolines. Prog Energy Combust Sci 60:1–25
5. Fathi M, Jahanian O, Shahbakhti M (2017) Modeling and controller design architecture for cycle-by-cycle combustion control of homogeneous charge compression ignition (HCCI) engines–a comprehensive review. Energy Convers Manag 139:1–19
6. Reitz RD, Duraisamy G (2015) Review of high efficiency and clean reactivity controlled compression ignition (RCCI) combustion in internal combustion engines. Prog Energy Combust Sci 46:12–71
7. Saxena S, Bedoya ID (2013) Fundamental phenomena affecting low temperature combustion and HCCI engines, high load limits and strategies for extending these limits. Prog Energy Combust Sci 39(5):457–488
8. Musculus MP, Miles PC, Pickett LM (2013) Conceptual models for partially premixed low-temperature diesel combustion. Prog Energy Combust Sci 39(2):246–283
9. Lu X, Han D, Huang Z (2011) Fuel design and management for the control of advanced compression-ignition combustion modes. Prog Energy Combust Sci 37(6):741–783
10. Yao M, Zheng Z, Liu H (2009) Progress and recent trends in homogeneous charge compression ignition (HCCI) engines. Prog Energy Combust Sci 35(5):398–437

Appendix 1
Important Ethanol Reactions Rates & Cylinder Pressure Measurement

R.K. Maurya, *Characteristics and Control of Low Temperature Combustion Engines*,
Mechanical Engineering Series, https://doi.org/10.1007/978-3-319-68508-3

Table A1 Reaction rate coefficients (k) of important ethanol oxidation reactions in HCCI engines [1]. ($k = AT^b \exp(-E_a/RT)$)

Important reactions	A	b	E_a
$H + O_2 (+ M) [=] HO_2 (+ M)$ Third body efficiencies $CH_4 = 10.0$ $CO_2 = 3.8$ $CO = 1.9$	1.05E + 0007	1.257	0
$H + HO_2 [=] O + H_2O$	3.0401E + 0007	0	7200.664
$OH [=] O + H_2O$	3.6057E − 0002	2.4	8836.608
$OH (+ M) [=] H_2O_2 (+ M)$	1.2524E + 0008	0.37	0
$CH_3 + HO_2 [=] OH + CH_3O$	7.07E + 0006	0	0
$CH_3 + HO_2 [=] CH_4 + O_2$	3.03E + 0006	0	0
$CH_3 + O_2 [=] O + CH_3O$	1.4645E + 0007	0	122210.456
$CH_3 + O_2 [=] CH_2O + OH$	253,510	0	61253.76
$CH_2O + OH [=] HCO + H_2O$	3464.3	1.18	−1870.248
$CO + OH [=] H + CO_2$	9.5142E − 0003	2.25	−9836.584
$CO + O + M [=] CO_2 + M$	623.17	0	12,552
$CO + HO_2 [=] CO_2 + OH$	5.858E + 0007	0	95955.856
$C_2H_5OH (+ M) [=] CH_3 + CH_2OH (+ M)$ Third body efficiencies $H_2 = 2.0$ $CO_2 = 3.0$ $CO = 2.0$ $H_2O = 5.0$	6.059394E + 0023	−1.68	381425.992
$OH + C_2H_5OH [=] H_2O + C_2H_4OH$	175,740	0.27	2510.4
$OH + C_2H_5OH [=] H_2O + CH_3CHOH$	468,640	0.15	0
$OH + C_2H_5OH [=] H_2O + CH_3CH_2O$	753,460	0.3	6836.656
$CH_3 + C_2H_5OH [=] CH_4 + C_2H_4OH$	2.2119E − 0004	3.18	40258.448
$CH_3 + C_2H_5OH [=] CH_4 + CH_3CHOH$	7.3528E − 0004	2.99	33254.432
$HO_2 + C_2H_5OH [=] H_2O_2 + CH_3CHOH$	8.2E − 0003	2.55	44,978
$HO_2 + C_2H_5OH [=] H_2O_2 + C_2H_4OH$	1.2423E − 0002	2.55	65,898
$CH_3CH_2O + M [=] H + CH_3HCO + M$	1.1716E + 0029	−5.89	105746.416
$CH_3CH_2O + M [=] CH_3 + CH_2O + M$	1.3635E + 0032	−6.96	99579.2
$OH + CH_3HCO [=] H_2O + CH_3CO$	9.3324	1.5	−4025.008
$OH + CH_3HCO [=] H_2O + CH_2HCO$	0.17372	2.4	3409.96
$OH + CH_3HCO [=] CH_3 + HCOOH$	3.03E + 0009	1.076	0
$CH_3 + CH_3HCO [=] CH_4 + CH_3CO$	3.939E − 0013	5.8	9204.8
$HO_2 + CH_3HCO [=] H_2O_2 + CH_3CO$	2.424E + 0013	−2.2	58701.52
$HO_2 + CH_3HCO [=] H_2O_2 + CH_2HCO$	234,320	0.4	62190.976
$O_2 + CH_3HCO [=] HO_2 + CH_3CO$	1.01E + 0008	0	176564.8
$OH + C_2H_4 [=] H_2O + C_2H_3$	2.0402E + 0007	0	24836.224
$CH_3 + C_2H_4 [=] CH_4 + C_2H_3$	6.6862E − 0006	3.7	39,748
$O_2 + C_2H_3 [=] CH_2O + HCO$	1.717E + 0023	−5.312	27,196
$O_2 + C_2H_3 [=] O + CH_2HCO$	5.61055E + 0008	0.611	22007.84
$O_2 + CH_2HCO [=] CH_2O + CO + OH$	30,300	0	0
$HO_2 + CH_2HCO [=] O_2 + CH_3HCO$	3.03E + 0006	0	0

A units, $m^3\ mol^{-1}\ s^{-1}$; E_a units, J/mol

Table A2 Summary of performance and combustion parameters that can be estimated by processing of measured in-cylinder pressure data [2]

Data derived from in-cylinder pressure	Information regarding engine	Signal processing methods/ equations
Cycle averaged in-cylinder pressure	Firing cycle crank angle (CA)-based combustion event and quality of combustion	Averaging of cylinder pressure of different consecutive combustion cycle on crank angle (CA) basis
Motoring cylinder pressure, peak compression pressure and its CA position	Used for tuning heat release parameter, TDC correction, engine cylinder condition and blow-by estimation	Measuring signal in non-firing engine cycle and averaging to number of cycle. Calculation of TDC position using thermodynamic method using measured pressure signal
Peak pressure	Mechanical load on piston-cylinder	Maximum value computation in in-cylinder pressure signal
Rate of pressure rise	Knock limit of engine, combustion noise	Derivative of pressure signal
Crank angle at peak pressure, rate of heat release curve and energy conversion positions	Overall efficiency, combustion efficiency, qualitative exhaust values, quality of ignition system	Using first law of thermodynamics, ROHR calculation using equation $\frac{dQ}{d\theta}=\frac{\gamma}{\gamma-1}P\frac{dV}{d\theta}+\frac{1}{\gamma-1}V\frac{dP}{d\theta}+\frac{\partial Q_w}{d\theta}$ Energy conversion points are calculated by integrating ROHR curve
IMEP, FMEP, PMEP	Cylinder work output, combustion stability (cyclical fluctuations), friction losses, gas exchange losses	Computation of work using the cylinder pressure signal and volume curve generated from engine geometry $W_{\text{net}}=\frac{2\pi}{360}\int_{-360}^{360}\left(P(\theta)\frac{dV}{d\theta}\right)d\theta$ Calculation of work in high pressure component and low pressure component to calculate IMEP, PMEP and FMEP
PV and log(PV) diagram	Work output and pumping losses, determination of TDC position, polytropic coefficient of mixture in compression and expansion	Computation of area under curve, calculation of logarithmic value of pressure and volume curve, slope of curve in compression and expansion stroke to find polytropic coefficient
Gas temperature and wall heat transfer	Qualitative exhaust values, heat transfer	Using empirical equations developed using in-cylinder pressure and mean gas temperature
High-frequency component of vibration	Knocking, combustion noise, ringing intensity	Filtering using bandpass filter, wavelet analysis, power spectrum
Ignition delay	Formation of air–fuel mixture, premixed charge quality	Calculated from ignition (SI) or injection point (CI) and start of combustion calculated from heat release curve

(continued)

Table A2 (continued)

Data derived from in-cylinder pressure	**Information regarding engine**	**Signal processing methods/ equations**
Mass flow rate	Air mass flow estimation, residual exhaust gas in cylinder, backflow	Application of the Δp-method for estimating the air mass
Compression ratio estimation	Actual compression ratio	Computation using polytropic compression model and optimization algorithm
Air–fuel ratio calculation	Cylinder mixture strength	Cylinder pressure moment-based approach, net heat release-based estimation
Torque estimation	Engine torque	Indicated torque is estimated from the peak pressure and its location, and the load torque is observed on the basis of the estimated indicated torque
COV_{IMEP}, CA position for different heat release and mass burn fraction	Cycle-to-cycle dispersion, misfire and partial burn cycle	Statistical analysis, symbol sequence analysis, return maps
Emissions estimation	Engine emissions, quality of combustion	Neural network-based algorithm, regression-based approach
Control parameter calculations	EGR control, noise control, emission control, online combustion failure detection, start of injection control	Online signal processing (filtering) and computation of control parameter

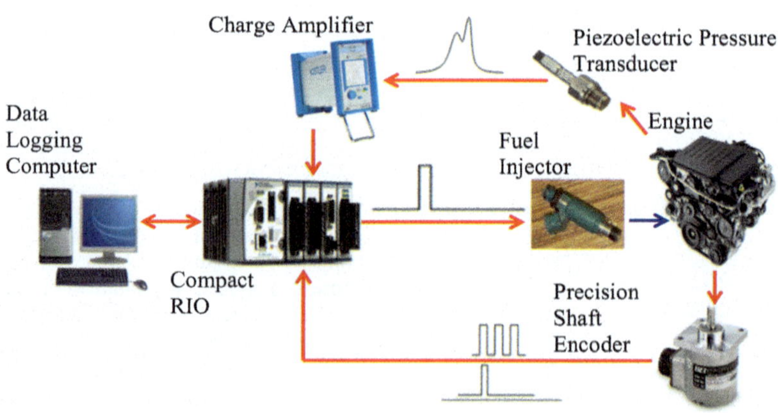

Fig. A1 Schematic diagram of in-cylinder pressure measurement system [3]

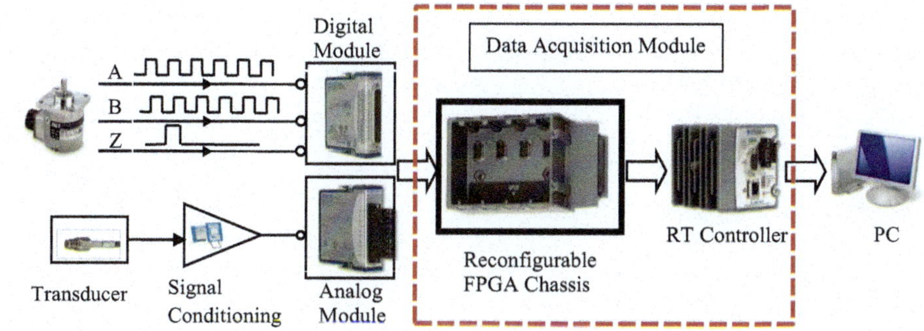

Fig. A2 Schematic diagram of typical high speed data acquisition system cylinder pressure measurement [3]

References

1. Maurya RK, Akhil N (2016) Numerical investigation of ethanol fuelled HCCI engine using stochastic reactor model. Part 1: Development of a new reduced ethanol oxidation mechanism. Energ Convers Manag 118:44–54
2. Maurya RK, Pal DD, Agarwal AK (2013) Digital signal processing of cylinder pressure data for combustion diagnostics of HCCI engine. Mech Syst Signal Process 36(1):95–109
3. Maurya RK (2012) Performance, emissions and combustion characterization and close loop control of HCCI engine employing gasoline like fuels. PhD Thesis, Indian Institute of Technology Kanpur, India

Appendix 2
Measured Cylinder Pressure Data Analysis

In-Cylinder Pressure Data Analysis

In-cylinder pressure is typically measured by using piezoelectric pressure sensor that has charge output depending on the pressure in cylinder. The charge output from the piezoelectric pressure transducer is supplied to the charge amplifier, which converts the charge output to a proportional voltage signal. The voltage signal is acquired using high speed data acquisition system by converting analog voltage signal into digital signal. Piezoelectric pressure sensor typically measures the relative pressure in the cylinder. Therefore, to convert the pressure data into absolute pressure value, a known pressure value at certain crank angle (CA) position in the cycle is provided (called "pegging" or referencing) [1]. Presently two types of methods are used for estimating the cylinder pressure pegging: (1) the methods which require additional absolute pressure reference and (2) the methods which utilize the polytropic compression curve [2]. Absolute pressure values are calculated by assigning a known absolute pressure at pegging position in the cycle, and calculated using the following equation [3]:

$$P(\theta) = P_{\text{peg}} + \text{cal}\left[v_t(\theta) - v_t\left(\theta_{\text{peg}}\right)\right]$$

where P_{peg} is the known pressure value at the pegging point, "Cal" is calibration factor of the sensor (bar/Volt), v_t is the measured voltage and θ_{peg} is the CA position of pegging point. The voltage difference between the point of interest and the pegging point is multiplied by the appropriate calibration factor and added to the pegging pressure.

To obtain an average value in a noisy pressure signal, it is necessary to average multiple points (10–15) to determine $v_t(\theta_{\text{peg}})$, as shown below:

R.K. Maurya, *Characteristics and Control of Low Temperature Combustion Engines*, Mechanical Engineering Series, https://doi.org/10.1007/978-3-319-68508-3

$$v_t(\theta_{\text{peg}}) = \frac{1}{w} \sum_{z=-k}^{k} v_t(\theta_z) \quad \text{where } k = \frac{w-1}{2}$$

Typically, pressure pegging point is chosen as bottom dead centre (BDC) of piston during intake stroke, which is considered to be equal to intake manifold pressure. Incorrect pegging affects calculated parameters such as the heat release, mass-fraction burned, charge temperature, etc. [1]. The signal from the pressure transducer is required to be processed in order to obtain vital information about engine combustion. Often in-cylinder pressure data processing consists of pegging or referencing, phasing with crank angle, cycle averaging and filtering of experimental pressure signals.

Cylinder Volume Calculations

Combustion chamber volume at any CA position is calculated purely from engine geometry. When piston is at TDC, the volume of combustion chamber is equal to the clearance volume. For any other CA position of piston, the volume of combustion chamber is the sum of "clearance volume" and "cylindrical volume" with diameter equal to engine bore and height equal to piston displacement (Fig. B1):

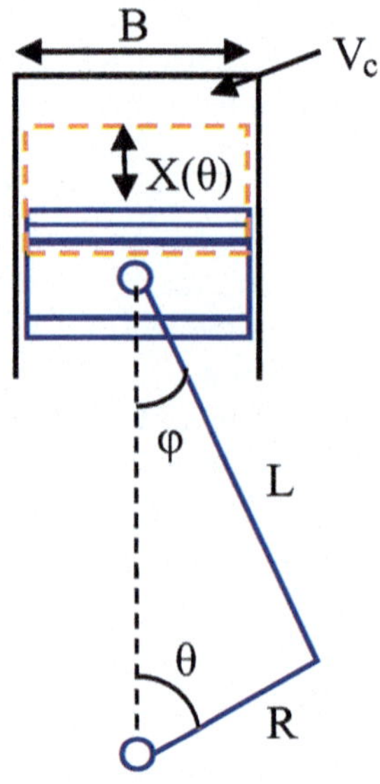

Fig. B1 Geometric dimension nomenclature of the engine

$$V(\theta) = V_c + X(\theta)\frac{\pi B^2}{4}$$

Here V_c is clearance volume and B is bore diameter of the engine cylinder. $X(\theta)$ is the instantaneous length of piston movement from the TDC position.

Displacement volume V_d and clearance volume V_c is calculated from bore and stroke of the engine by the following equation:

$$V_d = \frac{\pi B^2 S}{4} \text{ and } V_c = \frac{\pi B^2 R}{2(r_c - 1)}$$

Here S is the stroke of engine and $S = 2R$, where R is the crank radius. r_c is the compression ratio of the engine.

Instantaneous length of movement of piston is calculated by the following equation:

$$X(\theta) = (R + L) - (R\cos(\theta) + L\cos(\varphi))$$

Here L is the connecting rod length.

Hence, combustion chamber volume can be expressed as a function of CA rotation by the following equation [3]:

$$V(\theta) = v_c + \frac{\pi B^2}{4}\left(L + R - R\cos\theta - \sqrt{L^2 - R^2\sin^2\theta}\right)$$

Mean Gas Temperature Analysis

In the HCCI engine, combustion is occurring homogeneously in the entire combustion chamber except in the proximity of walls. Therefore, assumption of uniform temperature inside the combustion chamber is close to actual temperature in the HCCI combustion mode. Assuming a uniform temperature within the cylinder, ideal gas equation is used to estimate the bulk average cylinder temperature of the charge. The results of temperature calculation by this method are valid between IVC (intake valve closing) and EVO (exhaust valve opening) position of engine cycle. The mean gas temperature as a function of CA is given by the following equation [3]:

$$T(\theta) = \frac{P(\theta)V(\theta)n(\theta)}{P_{\mathrm{IVC}}V_{\mathrm{IVC}}n_{\mathrm{IVC}}}T_{\mathrm{IVC}}$$

Here P is in-cylinder pressure, V is volume of combustion chamber and T is temperature of gases inside combustion chamber.

Temperature of charge at IVC is an important parameter required for calculation of temperature at any CA position during engine cycle.

Temperature at IVC (T_{IVC}) is calculated as follows [4]:

$$T_{\mathrm{IVC}} = \frac{1}{(n_a + n_f + n_R)} \times \frac{P_{\mathrm{IVC}} V_{\mathrm{IVC}}}{R}$$

Here n_a is moles of air, n_f is moles of fuel and n_R is mole of residual gases in the cylinder. R is the universal gas constant.

Pressure and volume at IVC can be easily calculated from measured pressure and calculated volume by engine geometry. Moles of air and fuel are calculated as ratio of measured quantity with their respective molecular weights. Moles of residual gas are calculated by the following equation:

$$n_R = \frac{P_{\mathrm{EVC}} V_{\mathrm{EVC}}}{T_{\mathrm{EVC}} R}$$

While the pressure and volume are known at EVC, the temperature at EVC is not known. The temperature is measured in the exhaust port; however, this measured temperature represents a multicycle averaged exhaust temperature, and because of heat transfer, it is lower than the in-cylinder temperature at EVC. However, it was found that if the exhaust port temperature and pressure were used with the ideal gas law instead of T_{EVC} and P_{EVC}, the estimated trapped gas fraction was very close to that measured using the in-cylinder CO_2 approach [4].

Instantaneous Work and Mean Effective Pressure Analysis

Work done during an engine cycle is calculated by using measured in-cylinder pressure and calculated volume of combustion chamber for complete engine cycle. Net work done during an engine cycle is calculated by the following equation [3]:

$$W_{\mathrm{net}} = \frac{2\pi}{360} \int_{-360}^{360} \left(P(\theta) \frac{dV}{d\theta} \right) d\theta$$

The mechanical work transferred from the cylinder gases to the piston during compression and expansion stroke is called indicated work. Indicated work is calculated using the following equation:

$$W_{\text{ind}} = \frac{2\pi}{360} \int_{-180}^{180} \left(P(\theta) \frac{dV}{d\theta} \right) d\theta$$

By definition, mean effective pressure is the ratio of work done per cycle and volumetric displacement of the engine, and this has units of pressure. Mean effective pressure (MEP) can be represented in terms of average pressure over a cycle in the engine combustion chamber producing the same work. This parameter is useful for comparing the performance of very different size engines. Net mean effective pressure (NMEP) is calculated as follows:

$$\text{NMEP} = \frac{W_{\text{net}}}{V_d}$$

Indicated mean effective pressure (IMEP) is the ratio of indicated work and displacement volume (V_d) as

$$\text{IMEP} = \frac{W_{\text{ind}}}{V_d}$$

$$W_{\text{net}} = W_{\text{ind}} + W_{\text{pump}}$$

Fuel MEP is defined as fuel energy supplied during engine cycle equivalent mean effective pressure:

$$\text{FuelMEP} = \frac{m_f Q_{\text{LHV}}}{V_d}$$

Heat release (HR) MEP is the amount of fuel energy converted into heat and defined as

$$Q_{\text{HR}}\text{MEP} = \frac{Q_{\text{HR}}}{V_d} = \frac{\int dQ_{\text{HR}}}{V_d}$$

Mass Burn Fraction Analysis

Mass burn fraction is calculated using Rassweiler and Withrow method, which is a well-established method for estimating MFB [5]. It was calculate as

$$(\Delta p_c) = p_i - p_{i-1}\left(\frac{V_{i-1}}{V_i}\right)^{\gamma}$$

Here Δp_c is the increase in-cylinder pressure due to combustion and γ is the polytropic exponent; MFB can now be computed as

$$\frac{m_{b(i)}}{m_{b(\text{total})}} = \frac{\sum_{j=0}^{i} \Delta(p_c)_j}{\sum_{j=0}^{N} \Delta(p_c)_j}$$

Here it is assumed that sample 0 is between inlet valve closing and the start of combustion and that sample N is after combustion has completed.

Ringing Intensity Analysis

Severe cylinder pressure oscillations can be detected in HCCI combustion engine in some regimes of operation. These pressure oscillations are similar to knock in a spark ignition engine (oscillation frequency is different), but due to the differences between HCCI and SI combustion, these are termed as "ringing". These pressure oscillations generate a very large amount of noise in the engine structure. For assessment of the combustion noise from HCCI engines, "ringing intensity" is calculated using the following equation [6]:

$$RI = \frac{\sqrt{\gamma R T_{\max}}}{2\gamma P_{\max}}\left[\beta\left(\frac{dP}{dt}\right)_{\max}\right]^2$$

where $(dP/dt)_{\max}$ is the maximum pressure rise rate, $P_{\max}$ is the peak in-cylinder pressure and $T_{\max}$ is the maximum of mass averaged in-cylinder temperature (calculated using ideal gas law). γ is the ratio of specific heats (C_p/C_v) and R is the gas constant. β is a tuning parameter, which relates the amplitude of pressure pulsations to the maximum pressure rise rate, and it has typical value of 0.05.

Other method used to define knocking boundary for HCCI combustion is by using the pressure rise rate (PRR). Excessive PRR leads to not only noisy operation but also structural damage to the engine. The pressure rise rate is typically calculated using the fourth-order central difference method from measured digital cylinder pressure data by the following equation:

$$\frac{dP}{d\theta} = \frac{-P(\theta + 2h) + 8P(\theta + h) - 8P(\theta - h) + P(\theta - 2h)}{12h}$$

Engine Efficiency Analysis

Gross indicated thermal efficiency ($\eta_{i,g}$) is defined as the ratio between the work on the piston during the compression and expansion stroke ($W_{\text{ind},g}$) to the input fuel energy [7]:

$$\eta_{i,g} = \frac{W_{\text{ind},g}}{m_f q_{\text{LHV}}}$$

Here m_f is fuel mass injected per cycle and q_{LHV} is the lower heating value of the fuel.

Gas exchange efficiency η_{ge}: Gas exchange efficiency is defined as the ratio of the indicated work during the complete cycle and the closed part of the cycle. Gas exchange efficiency can be calculated as

$$\eta_{\text{ge}} = \frac{\text{IMEP}_n}{\text{IMEP}_g}$$

Combustion efficiency (η_{com}) is typically calculated from the following equation [8]:

$$\eta_c = 1 - \frac{\sum_i x_i Q_{HVi}}{\left[\dot{m}_f / \left(\dot{m}_a + \dot{m}_f\right)\right] Q_{HV_f}}$$

where x_i are the mass fractions of CO, H_2, HC and particulates, respectively. Q_{HV} are the lower heating values of these species, and the subscripts "*f*" and "*a*" denote fuel and air.

The combustion efficiency can also be estimated from the calculated rate of heat release data by following equations [9]:

$$\eta_{\text{com}} = \frac{\sum \text{ROHR}}{Q_{\text{in}}} \times 100$$

Here $\sum$ROHR is the integrated value of heat release rate; Q_{in} is total heat content of the introduced fuel.

Heat Release Rate Analysis

Heat release analysis is a very useful method to characterize combustion in internal combustion engines. The first law of thermodynamics is implemented to combustion chamber gases, and cylinder is considered as a closed system for the combustion event as shown in Fig. B2. In order to implement the equation, engine

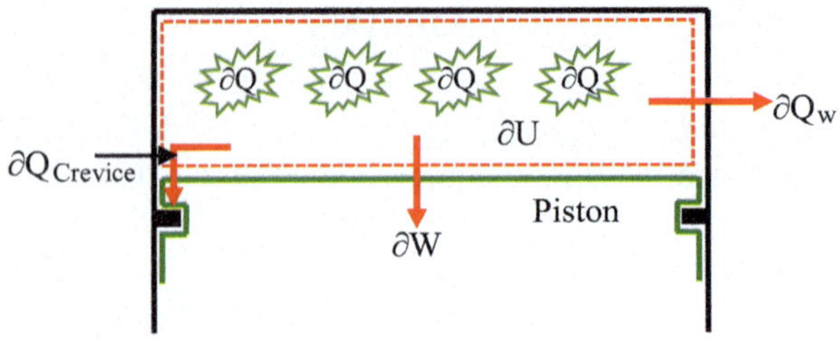

Fig. B2 Energy balance for engine combustion chamber with chemical heat release

combustion chamber is considered as a single zone, where no temperature gradient exists and the reactants and products are completely mixed. The reactants and products are assumed to have the same properties.

The chemical heat released from combustion in the cylinder can be written by the following equation [10]:

$$\partial Q = \partial U + \partial W + \partial Q_w + \partial Q_{\text{Crevice}}$$

Here ∂U is the change in internal energy, ∂Q_w is the instantaneous heat transfer to the cylinder walls, ∂W is the work on the piston and $\partial Q_{\text{Crevice}}$ is mass flow across the system boundary.

The work on the piston can also be written in terms of pressure and volume change by equation:

$$\partial W = PdV$$

The internal energy depends mainly on temperature and is defined in terms of mass, specific heat and temperature as $U = mC_vT$. Changes in internal energy are given by

$$\partial U = dmC_vdT + mC_vdT$$

If loss of mass to the crankcase (blow-by) is considered to be small, then $dm = 0$ and change in internal energy is given as

$$\partial U = mC_vdT$$

By assuming that combustion chamber gases follow ideal gas law ($PV = mRT$), changes in temperature can be calculated by following equations:

$$\partial T = \frac{dPV + PdV}{mR}$$

By substituting the values of change in work and internal energy in heat release equation, the equation can be rewritten as

$$\partial Q = \left(1 + \frac{C_v}{R}\right) PdV + \frac{C_v}{R} dPV + \partial Q_w + \partial Q_{\text{Crevice}}$$

Since gases are assumed as ideal gas, then $C_p + C_v = R$, and $\gamma = C_p/C_v$. Substituting the values of γ, the final heat release equation is written as

$$\frac{\partial Q}{\partial \theta} = \frac{\gamma}{\gamma - 1} p \frac{dV}{d\theta} + \frac{1}{\gamma - 1} V \frac{\partial p}{\partial \theta} + \frac{\partial Q_w}{\partial \theta} + \frac{\partial Q_{\text{Crevice}}}{\partial \theta}$$

The ratio of specific heat can be calculated by the following equation [11]:

$$\gamma = \gamma_0 - \frac{k}{100} \frac{T}{1000}$$

γ_0 is the γ value at some reference temperature, usually 300 Kelvin. γ_0 is dependent on the gas composition. For atmospheric air, γ_0 is 1.4 and for lean air–fuel mixtures 1.38 is a usable value. The constant k is usually set at 8.

Rate of heat loss through the combustion chamber walls is calculated using the following equation:

$$\frac{\partial Q_w}{\partial \theta} = \frac{h(T - T_w)A}{360N} \quad [\text{J/CAD}]$$

Here h is heat transfer coefficient, T_w is the average cylinder wall temperature, A is the actual cylinder wall area and N is the engine speed. It is normally assumed that T_w is constant over the entire cycle. h, T and A are functions of crank angle position.

Heat transfer coefficient can be calculated by Hohenberg model given by the following equation [12]:

$$h = \alpha_s V^{-0.06} p^{0.8} T^{-0.4} \left(S_p + 1.4\right)^{0.8}$$

Here S_p is the mean piston speed; α_s is scaling factor; and P, V and T are pressure, volume and temperature of combustion chamber, respectively.

Another correlation is typically used for calculation of heat transfer coefficient by Woschini analysis [13] and given by the following equation:

$$h = 131 \times (\text{Bore})^{-0.2} \times P^{0.8} \times T^{-0.55} W^{0.8}$$

where

$$w = C_1 \times 2.28 \times S_p + C_2 \times 3.24 \times 10^{-3} \times \frac{V_d}{V_{\text{IVC}}} \times \left(\frac{P_{\text{fir}} - P_{\text{mot}}}{P_{\text{IVC}}}\right) \times T_{\text{IVC}}$$

where S_p is the mean piston speed and C_1 and C_2 are constants.

Typically, heat loss through the crevices is assumed to be small and neglected.

The combustion phasing corresponding to 10%, 50% and 90% heat release is calculated from cumulative heat release curve calculated by integrating heat release rate curve. Typically crank angle position corresponding to 10% heat release (CA_{10}) is considered as start of combustion (SOC). The difference in crank angle position between 90% and 10% heat release ($CA_{90\text{-}10}$) is defined as combustion duration. Ignition position is important in HCCI combustion as there is no direct control on ignition timings. Ignition timings are estimated by different methods in various studies. A study used one crank angle before the 1% heat release position as ignition timing in HCCI engine [10]. Sometimes crank angle position corresponding to 1% heat release is also used as ignition position in HCCI combustion. Another study used the threshold value of third derivative of pressure to calculate ignition timings [14] as shown by equation

$$\frac{d^3P}{d\theta^3} \rangle \frac{d^3P}{d\theta^3}\bigg|_{\lim}$$

where $\frac{d^3P}{d\theta^3}\big|_{\lim} = 25\left(\frac{kPa}{CAD^3}\right)$ is selected for the study.

A threshold value of 0.2 J/CAD of heat release rate is also used for calculation ignition timing and corresponding ignition pressure and temperature in HCCI engine [15]. This study also used 1% burnt fraction to calculate ignition temperature, and both strategies have similar trend with different data values. Another study used the crank angle position where heat release rate becomes positive as a start of LTHR position in partially premixed combustion strategy [16].

Combustion Variability Quantification Analysis

The most common method to quantify cycle-to-cycle variability includes statistical methods such as standard deviation and coefficient of variation (COV) of IMEP. Standard deviation of IMEP quantifies how widely IMEP values are dispersed from the mean and calculated by the following equations:

$$\text{STD of IMEP } (\sigma_{\text{IMEP}}) = \sqrt{\frac{1}{N-1}\sum_{i=1}^{N}\left(\text{IMEP}_i - \overline{\text{IMEP}}\right)^2}$$

Here $\overline{\text{IMEP}} = \sum_{i=1}^{N}\frac{\text{IMEP}_i}{N}$; i is the sample of interest and N is the number of the samples.

COV of IMEP can be calculated using the following equations:

$$\text{COV of IMEP} = \frac{\sigma_{\text{IMEP}}}{\text{IMEP}} \times 100$$

In general, COV of any combustion parameter can be calculated using following equation:

$$\text{COV}(x) = \frac{\sigma}{\bar{x}} \times 100\%$$

Here $\bar{x} = \sum_{i=1}^{n} x_i/n$ and standard deviation (σ) $\sigma = \sqrt{\sum_{i=1}^{n} (x_i - \bar{x})^2/(n-1)}$

References

1. Maurya RK, Agarwal AK (2013) Investigations on the effect of measurement errors on estimated combustion and performance parameters in HCCI combustion engine. Meas 46 (1):80–88
2. Lee K, Yoon M, Sunwoo M (2008) A study on pegging methods for noisy cylinder pressure signal. Control Eng Pract 16(8):922–929
3. REVelation operator reference manual. Madison (USA): Hi-Techniques Inc. 2004
4. Scaringe RJ (2009) Extension of the high load limit in the homogenous charge compression ignition engine. PhD Thesis, Massachusetts Institute of Technology, Cambridge
5. Rassweiler GM, Withrow L (1938) Motion pictures of engine flames correlated with pressure cards (No. 380139). SAE technical paper
6. Eng JA (2002) Characterization of pressure waves in HCCI combustion (No. 2002-01-2859). SAE technical paper
7. Christensen M, Hultqvist A, Johansson B (1999) Demonstrating the multi fuel capability of a homogeneous charge compression ignition engine with variable compression ratio (No. 1999-01-3679). SAE technical paper
8. Heywood J (1988) Internal combustion engine fundamentals. McGraw-Hill Education
9. Iida N (2007) Natural gas HCCI engines. In: Zhao H (ed) HCCI and CAI engines for the automotive industry. Woodhead Publishing, England
10. Christensen M (2002) HCCI combustion-engine operation and emission characteristics. Department of Heat and Power Engineering, Lund university
11. Gatowski JA, Balles EN, Chun KM, Nelson FE, Ekchian JA, Heywood JB (1984) Heat release analysis of engine pressure data (No. 841359). SAE technical paper
12. Hohenberg GF (1979) Advanced approaches for heat transfer calculations (No. 790825). SAE technical paper
13. Woschni G (1967) A universally applicable equation for the instantaneous heat transfer coefficient in the internal combustion engine (No. 670931). SAE technical paper
14. Shahbakhti M, Koch CR (2009) Dynamic modeling of HCCI combustion timing in transient fueling operation. SAE Int J Engines 2:1098–1113. (2009-01-1136)
15. Truedsson I, Tuner M, Johansson B, Cannella W (2013) Pressure sensitivity of HCCI auto-ignition temperature for oxygenated reference fuels. J Eng Gas Turbines Power 135 (7):072801
16. Aronsson HS, Truedsson I, Tuner M, Johansson B, Cannella W (2014) Comparison of fuel effects on low temperature reactions in PPC and HCCI combustion (No. 2014-01-2679). SAE technical paper

Appendix 3
Fuel Properties

R.K. Maurya, *Characteristics and Control of Low Temperature Combustion Engines*,
Mechanical Engineering Series, https://doi.org/10.1007/978-3-319-68508-3

Fuel	Molecular weight	O_2 (wt. %)	LHV (MJ/l)	Stoichiometric air–fuel ratio	Boiling point (°C)	RON	MON	ΔH (kJ/kg from 25 °C)	Solubility in water at 25 °C, wt%	Specific gravity at 20 °C
Methanol	32.04	0.50	15.8	6.46	64.7	109	89	1168	Miscible	0.792
Ethanol	46.06	0.35	21.4	8.98	78	109	90	919.6	Miscible	0.794
n-Butanol	74.11	0.22	26.9	11.17	118	98	85	707.9	7.4	0.81
Iso-butanol	74.11	0.22	26.6	11.17	108	105	90	686.4	8.1	0.802
n-Pentanol	88.14	0.18	28.5	11.74	138	80	74	647.1	2.2	0.816
Gasoline	111.19	0.00	30-33	14.58	27–225	88–98	80–88	~351	Negligible	0.72–0.78
Diesel	198.4	0.00	35.66	14.95	125–400	<0	<0	~232	Negligible	0.81–0.89

Source: Sarathy SM, Oßwald P, Hansen N, Kohse-Höinghaus K (2014) Alcohol combustion chemistry. Prog Energy Combust Sci 44:40–102

Index

R.K. Maurya, *Characteristics and Control of Low Temperature Combustion Engines*, Mechanical Engineering Series, https://doi.org/10.1007/978-3-319-68508-3

H

I

The manufacturer's authorised representative in the EU is Springer Nature Customer Service Centre GmbH, Europaplatz 3, 69115 Heidelberg, Germany. If you have any concerns regarding our products, please contact ProductSafety@springernature.com

Printed and bound by CPI Group (UK) Ltd, Croydon, CR0 4YY
30/09/2026
02225501-0018